Structural Mechanics –
A Unified Approach

Structural Mechanics – A Unified Approach

Alberto Carpinteri

Department of Structural Engineering, Politecnico di Torino, Italy

CRC Press
Taylor & Francis Group
Boca Raton London New York

CRC Press is an imprint of the
Taylor & Francis Group, an **informa** business

A TAYLOR & FRANCIS BOOK

CRC Press
Taylor & Francis Group
6000 Broken Sound Parkway NW, Suite 300
Boca Raton, FL 33487-2742

First issued in paperback 2019

ISBN-13: 978-0-419-19160-5 (hbk)
ISBN-13: 978-0-367-44834-9 (pbk)

Visit the Taylor & Francis Web site at
http://www.taylorandfrancis.com

and the CRC Press Web site at
http://www.crcpress.com

Typeset by EXPO Holdings, Malaysia.

A catalogue record for this book is available from the British Library

Library of Congress Catalog Card Number: 94-66939

To my daughters
Margherita and Sofia

Contents

CONTENTS

ix

CONTENTS

Preface

This text intends to provide a complete and uniform treatment of the fundamental themes of Structural Mechanics, ranging from the more traditional to the most advanced.

The **mechanics of linear elastic solids** (beams, plates, shells, three-dimensional bodies) is studied adopting a matrix approach, which is particularly useful for numerical applications. The kinematic, static and constitutive equations, once composed, provide an operator equation which has as its unknown the generalized displacement vector. Moreover, constant reference is made to duality, i.e. to the strict correspondence between statics and kinematics that emerges as soon as the corresponding operators are rendered explicit, and it is at once seen how each of these is the adjoint of the other. In this context the Finite Element Method is illustrated as a method of discretization and interpolation for the approximate solution of elastic problems.

The **theory of beam systems** (statically determinate, statically indeterminate and hypostatic) is then presented, with the solution of numerous examples and the plotting of the corresponding diagrams of axial force, shearing force and bending moment, obtained both analytically and graphically. For the examination of framed structures, approached on the basis of the method of displacements, automatic computation procedures, normally involving the use of computers, are introduced in both the static and the dynamic regime. In addition, the energy aspects and their usefulness in reaching solutions are emphasized.

Finally, the more frequently occurring **phenomena of structural failure** are studied: instability of elastic equilibrium, plastic collapse and brittle fracture. The unifying aspects, such as those regarding post-critical states and the discontinuous phenomena of snap-back and snap-through are underlined. Numerous examples regarding frames previously examined in the elastic regime are once more taken up and analysed incrementally in the plastic regime. Furthermore, comparison of the results based on the two theorems of plastic limit analysis (the static and kinematic theorems) is made. As regards fracture mechanics, the conceptual distinction between 'concentration' and 'intensification' of stresses is highlighted, and the stress treatment and energy treatment are set in direct correlation. Finally, size scale effects, as well as the associated ductile–brittle transition, are discussed.

To this it may be added that all the topics regarding structural symmetry (frames and plates) are gathered together in a single chapter, whereas subjects regarding dynamics are recalled in various chapters, for the purpose of emphasizing how dynamics is, in any case, a generalization of statics. Topics of *considerable current interest from the applicational standpoint*, such as those regarding anisotropic and/or heterogeneous materials, are dealt with in appendices.

Whereas Chapters 11 and 12 present continuity with Chapters 7, 8, 9 and 10, based on the common operator formulation of the elastic problem, Chapters 13, 14, 15 and 16 represent the completion of the theory of beam systems introduced in Chapters 3, 4, 5 and 6. Elementary mechanisms of structural collapse are dealt with in Chapters 17, 18 and 20, where the mutual interactions of these mechanisms are also touched on. Chapter 19 deals with plane problems of thin plates and cylindrical or prismatic solids of large thickness, and constitutes an indispensable basis for Linear Elastic Fracture Mechanics (Chapter 20).

The book has been written to be used as a text for graduate or undergraduate students of either Architecture or Engineering, as well as to serve as a useful reference for research workers and practising engineers. A suitable selection of various chapters may constitute a convenient support for different types of courses, from the more elementary to the more advanced, and from short monographic seminars to courses covering an academic year.

This text is the fruit of many years of teaching in Italian universities, formerly at the University of Bologna and currently at the Politecnico di Torino, where I have been Professor of Structural Mechanics since 1986. A constant reference and source of inspiration for me in writing this book has been the tradition of the Italian School, to which I am sincerely indebted. At the same time, it has been my endeavour to update and modernize a basic, and in some respects dated, discipline by merging classical topics with ones that have taken shape in recent times. The logical sequence of the subjects dealt with makes it possible in fact to introduce, with a minimum of effort, even topics that are by no means elementary and that are of differing nature, such as the shell theory, the finite element method, the automatic computation of frames, the dynamics of structures, the theory of plasticity and the mechanics of fracture.

Finally, I wish to express my gratitude to all those colleagues, collaborators and students, who, attending my lessons or reading the original manuscript, have, with their suggestions and comments, contributed to the text as it appears in its definitive form. I further wish to thank in advance all those who in future will have the courtesy to point out to me any mistakes or omissions that may have been overlooked in this first edition.

Alberto Carpinteri

About the Author

Alberto Carpinteri was awarded a PhD in Nuclear Engineering *cum laude* (1976), and a PhD in Mathematics *cum laude* (1981), by the University of Bologna. After two years at the Consiglio Nazionale delle Ricerche he was appointed Assistant Professor at the University of Bologna in 1980.

After moving to the Politecnico di Torino in 1986 as Professor, he was appointed Director and Head of Department of Structural Engineering in 1989. He was a Founding Member and Director of the Graduate School of Structural Engineering in 1990.

He served as Secretary of RILEM Technical Committee 89-FMT on Fracture Mechanics of Concrete (1986-91) and as Chairman of ESIS Technical Committee 9 on Concrete Structures since 1991. He is a member of the Editorial Board of *Theoretical and Applied Fracture Mechanics* and has been a reviewer for various associations and international publications.

He has given invited courses and lectures throughout Europe, and in the USA, Mexico, South Africa, India, Japan and Australia, and was a visiting Professor at Lehigh University, Bethlehem, Pennsylvania, 1982-83.

Professor Carpinteri has written and edited seven books on fracture mechanics of concrete, localized damage, and composite materials, and seven books in Italian. He has written over 150 papers on fracture mechanics, material fatigue, thermoelasticity, seismic structures and reinforced concrete.

Among the awards he has received are: the Robert L'Hermite International Prize, RILEM, 1982; Japan Society of Mechanical Engineers Medal, 1993; Doctor of Physics Honoris Causa, Constantinian University, Cranston-Rhode Island, USA, 1994; International Cultural Diploma of Honor, American Biographical Institute, 1995.

1 Introduction

1.1 Preliminary remarks

Structural Mechanics is the science that studies the **structural response** of solid bodies subjected to external loading. The structural response takes the form of **strains and internal stresses**.

The variation of shape generally involves relative and absolute displacements of the points of the body. The simplest case that can be envisaged is that of a string, one end of which is held firm while a tensile load is applied to the opposite end. The percentage lengthening or stretching of the string naturally implies a displacement, albeit small, of the end where the force is exerted. Likewise, a membrane, stretched by a system of balanced forces, will dilate in two dimensions and its points will undergo relative and absolute displacements. Also three-dimensional bodies, when subjected to stress by a system of balanced forces, undergo, point by point and direction by direction, a dilation or a contraction, as well as an angular distortion. Similarly, beams and horizontal plates bend, imposing a certain curvature, respectively to their axes and to their middle planes, and differentiated deflections to their points.

As regards internal stresses, these can be considered as exchanged between the single (even infinitesimal) parts which make up the body. In the case of the string, the tension is transmitted continuously from the end on which the force is applied right up to the point of constraint. Each elementary segment is thus subject to two equal and opposite forces exerted by the contiguous segments. Likewise, each elementary part of a membrane will be subject to four mutually perpendicular forces, two equal and opposite pairs. In three-dimensional bodies, each elementary part is subject to normal and tangential forces. The former generate dilations and contractions, whilst the latter produce angular distortions. Finally, each element of beam or plate that is bent is subject to self-balanced pairs of moments.

In addition to the shape and properties of the body, it is the external loading applied and the constraints imposed that determine the structural response. The constraints react to the external loads, exerting on the body additional loads called **constraint reactions**. These reactions are *a priori* unknown. In the case where the constraints are not redundant from the kinematic point of view, the calculation of the constraint reactions can be made considering the body as being perfectly rigid and applying only the cardinal equations of statics. In the alternative case where the constraints are redundant, the calculation of the constraint reactions requires, in addition to **equations of equilibrium**, the so-called **equations of congruence**. These equations are obtained by eliminating the redundant constraints, replacing them with the constraint reactions exerted by them and imposing the abeyance of the constraints that have been eliminated. *The procedure presupposes that the strains and displacements, produced both by the external loading and by the reactions of the constraints that have been eliminated, are known. A simple example may suffice to illustrate these concepts.*

1

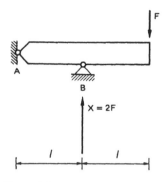

Figure 1.1

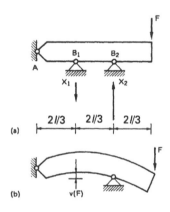

(a)

(b)

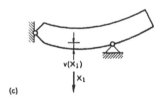

(c)

Figure 1.2

Let us consider a bar hinged at point A and supported at point B, subjected to the end force F (Figure 1.1). The reaction X produced by the support B is obtained by imposing equilibrium with regard to rotation about hinge A:

$$F(2l) = Xl \Rightarrow X = 2F \tag{1.1}$$

The equation of equilibrium with regard to vertical translation provides, on the other hand, the reaction of hinge A. The problem is thus **statically determinate** or **isostatic**.

Let us now consider the same bar hinged, not only at A but also at two points B_1 and B_2, distant $\frac{2}{3}l$ and $\frac{4}{3}l$ respectively from point A (Figure 1.2 (a)). The condition of equilibrium with regard to rotation gives us an equation in two unknowns:

$$F(2l) + X_1 \frac{2}{3}l = X_2 \frac{4}{3}l \tag{1.2}$$

Thus the pairs of reactions X_1 and X_2 which ensure rotational equilibrium are infinite, but only one of these also ensures congruence, i.e. abeyance of the conditions of constraint. The vertical displacement both in B_1 and B_2 must in fact be zero.

To determine the constraint reactions, we thus proceed to eliminate one of the two hinges B_1 or B_2, for example B_1, and we find out how much point B_1 rises owing to the external force F (Figure 1.2(b)) and how much it drops owing to the unknown reaction X_1 (Figure 1.2(c)). The condition of congruence consists of putting the total displacement of B_1 equal to zero:

$$v(F) = v(X_1) \tag{1.3}$$

The equation of equilibrium (1.2) and the equation of congruence (1.3) together solve the problem, which is said to be **statically indeterminate** or **hyperstatic**.

1.2 Classification of structural elements

As has already been mentioned in the preliminary remarks, the structural elements which combine to make up the load-bearing structures of civil and industrial buildings, as well as any naturally occurring structure such as rock masses, plants or skeletons, can fit into one of three distinct categories:

1. one-dimensional elements (e.g. ropes, struts, beams, arches);
2. two-dimensional elements (e.g. membranes, plates, slabs, shells, vaults);
3. three-dimensional elements (stubby solids).

In the case of one-dimensional elements, for example beams (Figure 1.3), one of the three dimensions, the length, is much larger than the other two, which compose the cross section. Hence, it is possible to neglect the latter two dimensions and consider the entire element as concentrated along the line forming its centroidal axis. In our calculations, features which represent the geometry of the cross section and, consequently, the three-dimensionality of the element, will thus be used. Ropes are elements devoid of flexural and compressive stiffness, and are able only to bear states of tensile stress. Bars, however, present a high axial stiffness, both in compression (struts) and in tension (tie rods), whilst their flexural stiffness is poor. Beams and, more generally, arches (or curvilinear beams), also present a high degree of flexural stiffness,

2

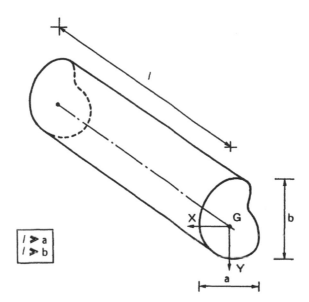

Figure 1.3

provided that materials having particularly high tensile strength are used. In the case of stone materials and concrete, which present very low tensile strength, straight beams are reinforced to stand up to bending stresses, whilst arches are traditionally shaped so that only internal compressive stresses are produced.

When, in the cross section of a beam, one dimension is clearly smaller than the others (Figure 1.4), the beam is said to be **thin-walled**. Beams of this sort can be easily produced by rolling or welding metal plate, and prove to be extremely efficient from the point of view of the ratio of flexural strength to the amount of material employed.

In the case of two-dimensional elements, for example flat plates (Figure 1.5 (a)) or plates with double curvature (Figure 1.5 (b)), one of the three dimensions, the thickness, is much smaller than the other two, which compose the middle surface. It is thus possible to neglect the thickness and to consider the entire element as being concentrated in its middle surface. Membranes are elements devoid of flexural and compressive stiffness, and are able to withstand only states of biaxial traction. Also plates that are of a small thickness present a low flexural stiffness and are able to bear loads only in their middle plane. Thick plates (also referred to as slabs), instead, also withstand bending stresses, provided that materials having particularly high tensile strength are used. In the case of stone materials and concrete, flat plates are, on the other hand, ribbed and reinforced, while vaults and domes are traditionally shaped so that only internal compressive stresses are produced (for instance, in arched dams).

Finally, in the case of so-called stubby solids, the three dimensions are all comparable to one another and hence the analysis of the state of strain and internal stress must be three-dimensional, without any particular simplifications or approximations.

3

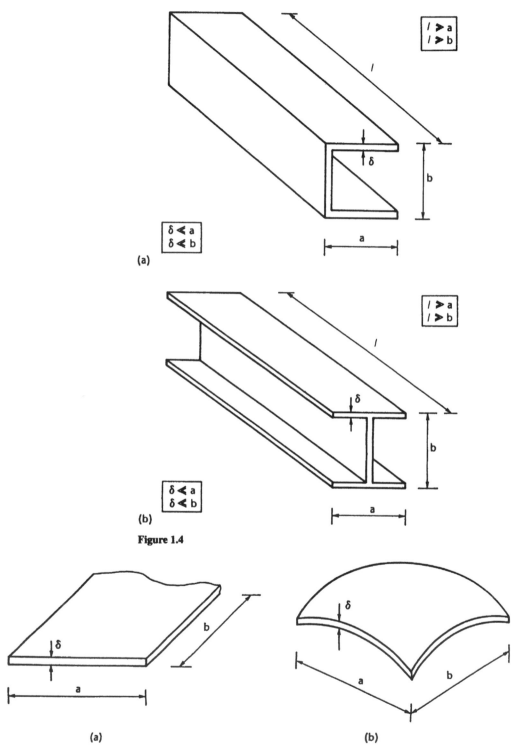

Figure 1.4

Figure 1.5

4

1.3 Structural types

The single structural elements, introduced in the previous section, are combined to form load-bearing structures. Usually, for buildings of a civil type, one-dimensional and two-dimensional elements are connected together. The characteristics of the individual elements and the way in which they are connected one to another and to the ground, together define the structural type, which can be extremely varied, according to the purposes for which the building is designed.

In many cases, the two-dimensional elements do not have a load-bearing function (for example, the walls of buildings in reinforced concrete), and hence it is necessary to highlight graphically and calculate only the so-called **framework**, made up exclusively of one-dimensional elements. This framework, according to the type of constraint which links together the various beams, will then be said to be **trussed** or **framed**. In the former case, the calculation is made by inserting hinges which connect the beams together, whereas in the latter case the beams are considered as built into one another. In real situations, however, beams are never connected by frictionless hinges or with perfectly rigid joints. Figures 1.6–1.11 show some examples of load-bearing frameworks: a timber-beam bridge, a truss in reinforced concrete, an arch centre, a plane steel frame, a grid and a three-dimensional frame.

Also in the case of **bridges** it is usually possible to identify a load-bearing structure consisting of one-dimensional elements. The road surface of an **arch bridge** is supported by a parabolic beam which is subject to compression, and, if well-designed, is devoid of dangerous internal flexural stresses. The road surface can be built to rest above the arch by means of struts (Figure 1.12), or

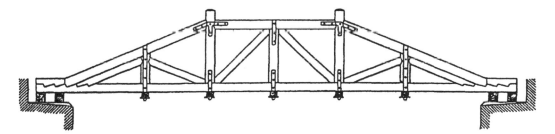

Figure 1.6

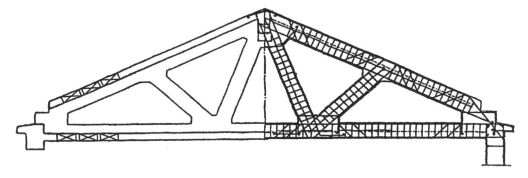

Figure 1.7

5

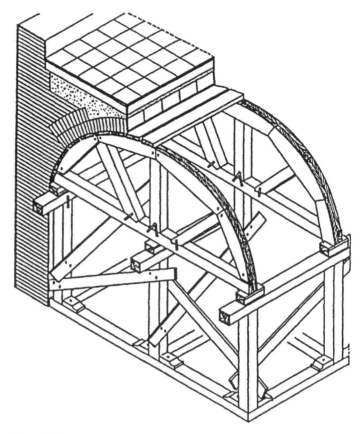

Figure 1.8

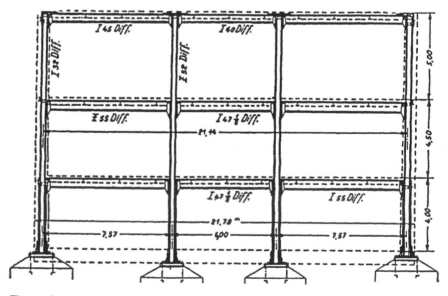

Figure 1.9

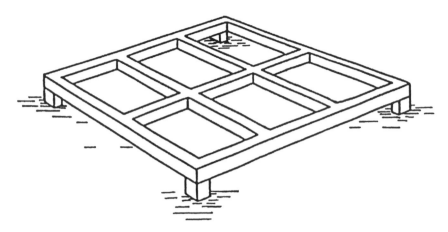

Figure 1.10

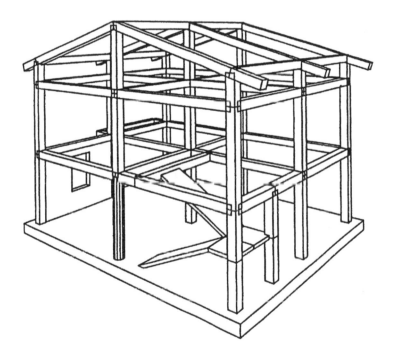

Figure 1.11

can be suspended beneath the arch by means of tie rods (Figure 1.13). Inverting the static scheme and using a primary load-bearing element subject exclusively to tensile stress, we arrive at the structure of **suspension bridges** (Figure 1.14). In these, the road surface hangs from a parabolic cable by means of tie rods. The cable is, of course, able to withstand only tensile stresses, which are, however, transmitted onto two compressed piers.

As regards two-dimensional structural elements, it is advantageous to exploit the same static principles already met with in the case of bridges. To avoid, for example, dangerous internal stresses of a flexural nature, the usual

7

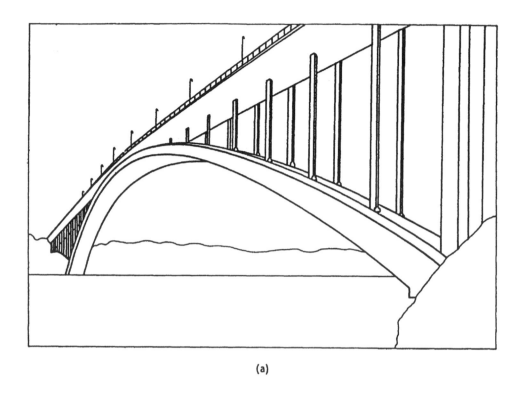

(a)

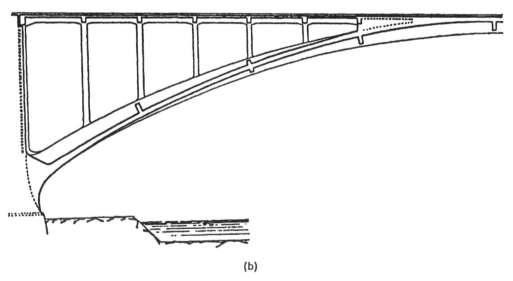

(b)

Figure 1.12

approach is to use **vaults** or **domes** having double curvature, which present parabolic sections in both of the principal directions (Figure 1.15 (a)). A variant is provided by the so-called **cross vault** (Figure 1.15 (b)), consisting of two mutually intersecting cylindrical vaults. Membranes, on the other hand,

8

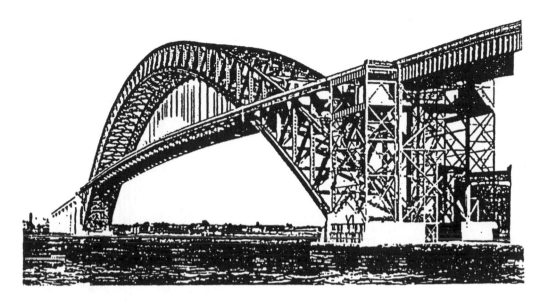

(a)

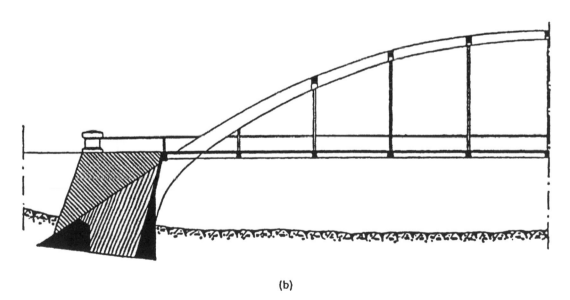

(b)

Figure 1.13

can assume the form of hyperbolic paraboloids, with saddle points and curvatures of opposite sign (Figure 1.15 (c)). In the so-called **prestressed membranes**, both those cables with the concavity facing upwards and those with the concavity facing downwards are subject to tensile stress.

9

(a)

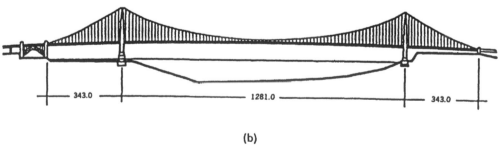

(b)

Figure 1.14

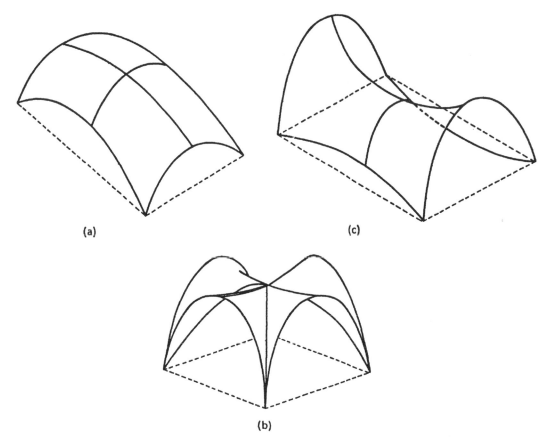

(a)

(c)

(b)

Figure 1.15

1.4 External loading and constraint reactions

The strains and internal stresses of a structure obviously depend on the external loads applied to it. These can be of varying nature according to the structure under consideration. In the civil engineering field, the loads are usually represented by the **weight load**, both of the structural elements themselves (**permanent loads**) and of persons, vehicles or objects (**live loads**).

Figure 1.16 represents two load diagrams, used in the early decades of this century, of horse-drawn carts and carriages. The forces are considered as concentrated and, of course, proceeding over the road surface. Figure 1.17 illustrates the load diagram of a locomotive engine, and Figure 1.18 that of a hoisting device. Figure 1.19 compares the permanent load diagrams of two beams, one with constant cross section, and the other with linearly variable cross section.

Other loads of a mechanical nature are **hydraulic loads** and **pneumatic loads**. Figure 1.20 shows how the thrust of water against a dam can be represented with a triangular distributed load. Then there are **inertial forces**, which act on rotating mechanical components, such as the blades of a turbine, or on

11

the floors of a storeyed building, following ground vibration caused by an earthquake (Figure 1.21). A similar system of horizontal forces can represent the action of the wind on the same building.

In addition to external loading, the structural elements undergo the action of the other structural elements connected to them, including the action of the foundation. These kinds of action are more correctly termed **constraint reactions**, those exchanged between elements being **internal**, and those exchanged with the foundation being **external**. The nature of the constraint

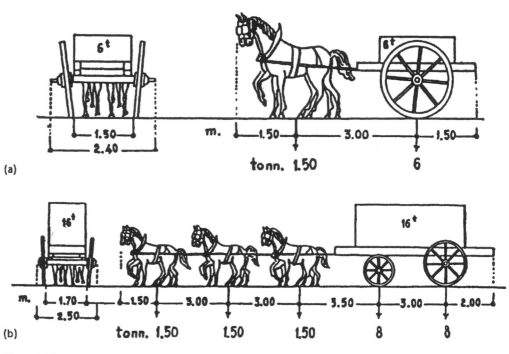

(a)

(b)

Figure 1.16

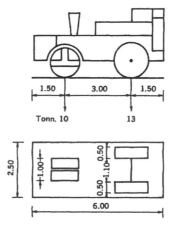

Figure 1.17

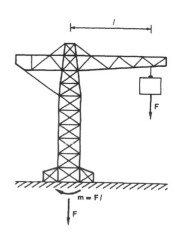

Figure 1.18

12

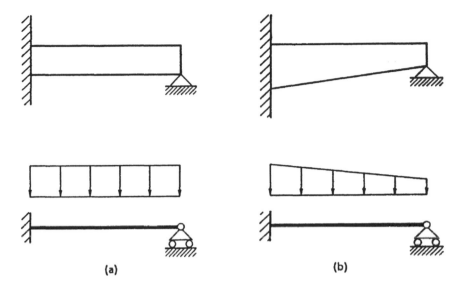

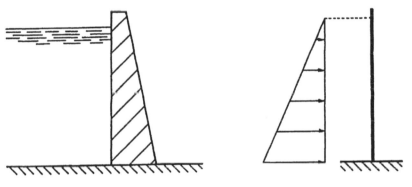

Figure 1.19

Figure 1.20

reaction depends on the conformation and mode of operation of the constraint which connects the two parts.

Figure 1.22 gives examples of some types of **beam support** to the foundation. In the case of Figure 1.22 (a), we have a pillar in reinforced concrete; in Figure 1.22 (b), we have joints that are used in bridges, and in Figure 1.22 (c), a roller support. In all cases, the constraint reaction exchanged between the foundation and the structural part is constituted by a vertical force, no constraint being exerted horizontally, except for friction.

Figure 1.23 shows the detailed scheme of a hinge connecting a part in reinforced concrete to the foundation. The hinge allows only relative rotations between the two connected parts and hence reacts with a force that passes through its centre. In the case illustrated, there will thus be the possibility of a horizontal reaction, as well as a vertical one.

13

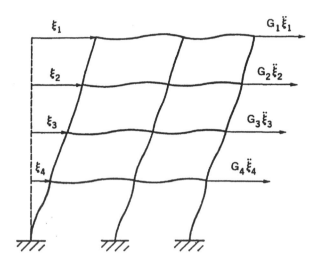

Figure 1.21

Figure 1.24 illustrates the joint between two timber beams, built with joining plates and riveting. Similar joints are made for steel girders by means of bolting or welding. This constraint is naturally more severe than a simple hinge, and yet in practice it proves to be much less rigid than a perfectly fixed joint. In the designing of trusses, it is customary to model the joint with a hinge, neglecting the exchange of moment between the two parts. The effect of making such an assumption is, in fact, that of guaranteeing a greater margin of safety.

1.5   Structural collapse

If the loading exerted on a structure exceeds a certain limit, the consequence is the complete collapse or, at any rate, the failure of the structure itself. The loss of stability can occur in different ways depending on the shape and dimensions of the structural elements, as well as on the material of which these are made. In some cases the constraints and joints can fail, with the result that rigid mechanisms are created, with consequent large displacements, toppling over, etc. In other cases, the structural elements themselves can give way; the mechanisms of structural collapse can be divided schematically into three distinct categories:

1. buckling
2. yielding
3. brittle fracturing.

In real situations, however, many cases of structural collapse occur in such a way as to involve two of these mechanisms, if not all three.

Buckling, or instability of elastic equilibrium, is the type of structural collapse which involves slender structural elements, subject prevalently to compression, such as struts of trusses, columns of frameworks, piers and arches of bridges, valve stems, crankshafts, ceiling shells, submarine hulls, etc. This kind of collapse often occurs even before the material of which the element is made has broken or yielded.

Unlike buckling, **yielding**, or plastic deformation, involves also the material itself and occurs in a localized manner in one or more points of the struc-

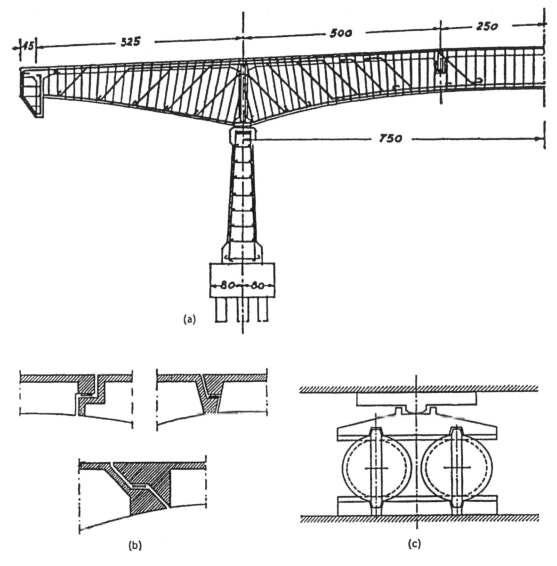

(a)

(b)   (c)

Figure 1.22

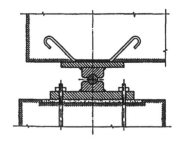

Figure 1.23

ture. When, with the increase in load, plastic deformation has taken place in a sufficient number of points, the structure can give way altogether since it has become hypostatic, i.e. it has become a mechanism. This type of generalized structural collapse usually involves structures built of rather ductile material, such as metal frames and plates, which are mainly prone to bending.

Finally, **brittle fracturing** is of a localized origin, as is plastic deformation, but spreads throughout the structure and hence constitutes a structural collapse of a generalized nature. This type of collapse affects prevalently one- and two-dimensional structural elements of considerable thickness (bridges, dams, ships, large ceilings and vessels, etc.), large three-dimensional elements (rock masses, the Earth's crust, etc.), brittle materials (high-strength steel and concrete, rocks, ceramics, glass, etc.) and tensile conditions.

15

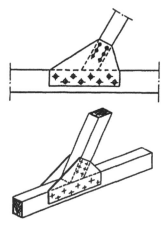

Figure 1.24

As, with the decrease in their degree of slenderness, certain structures, subject prevalently to compression and bending, very gradually pass from a collapse due to buckling to one due to plastic deformation, likewise, as we move down the size scale, other structures, prone to tension and bending, gradually pass from a collapse due to brittle fracturing to one due to plastic deformation.

1.6 Numerical models

With the development of electronics technology and the production of computers of ever-increasing power and capacity, structural analysis has undergone in the last two or three decades a remarkable metamorphosis. Calculations which were carried out manually by individual engineers, with at most the help of the traditional graphical methods can now be performed using computer software.

Up to a few years ago, since the calculation of strains and internal stresses of complex structures could not be handled in such a way as to obtain an exact result, such calculations were carried out using a procedure of approximation. These approximations, at times, were somewhat crude and, in certain cases, far from being altogether realistic. Today numerical models

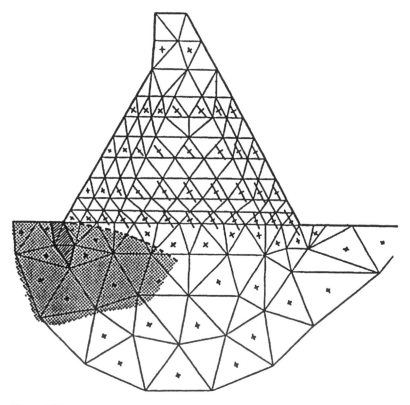

Figure 1.25

16

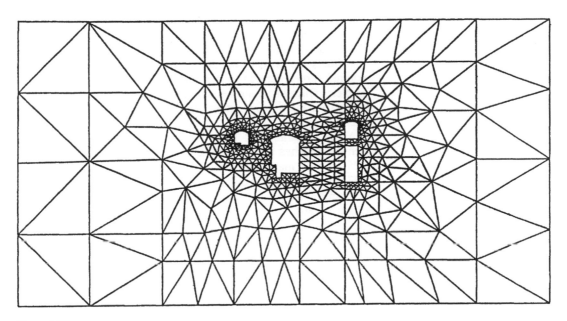

Figure 1.26

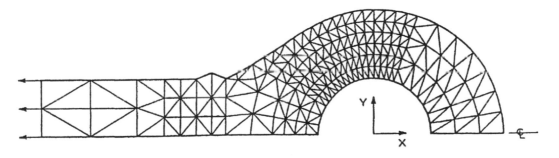

Figure 1.27

allow us to consider enormous numbers of points, or nodes, with their corresponding displacements and corresponding strains and internal stresses. The so-called **finite-element method** is both a discretization method, since it considers a finite number, albeit a very large one, of structural nodes, and an interpolation method, since it allows us to estimate the static and kinematic quantities even outside the nodes.

The enormous amount of information to be handled is organized and ordered in a matrix form by the computer. In this way, the language itself of structural analysis has taken on a different appearance, undoubtedly more synthetic and homogeneous. This means that, for every type of structural element, it is possible to write static, kinematic and constitutive equations having the

17

same form. Once discretized, these provide a matrix of global stiffness which presents a dimension equal to the number of degrees of freedom considered. This matrix, multiplied by the vector of the nodal displacements, which constitutes the primary unknown of the problem, provides the vector of the external forces applied to the nodes; this represents the known term of the problem. Once this matrix equation has been resolved, taking into account any boundary conditions, it is then possible to arrive at the nodal strains and nodal internal stresses.

As an illustration of these mathematical techniques, a number of **finite-element meshes** are presented. They correspond to: a buttress dam (Figure 1.25), a rock mass with a tunnel system (Figure 1.26), an eye hook (Figure 1.27), two mechanical components having supporting functions (Figure 1.28), a concrete vessel for a nuclear reactor (Figure 1.29) and an arch dam (Figure 1.30).

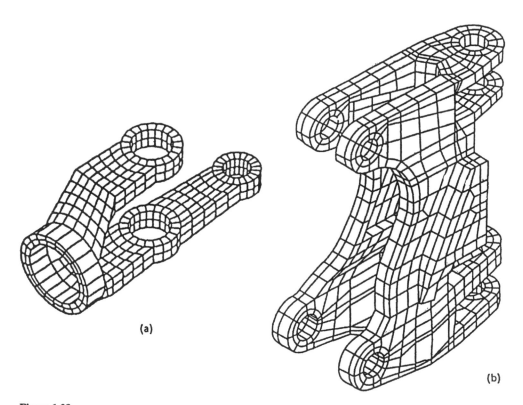

(a)

(b)

Figure 1.28

18

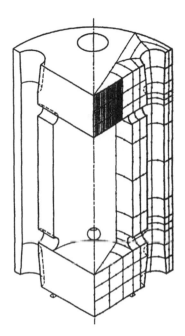

Figure 1.29

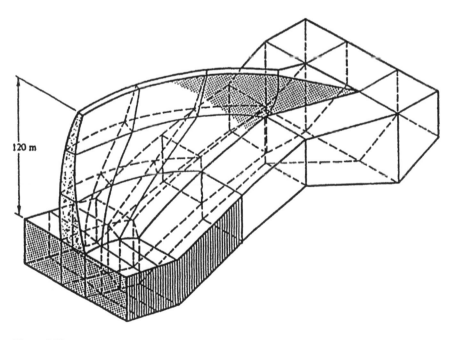

120 m

Figure 1.30

19

2 Geometry of areas

2.1 Introduction

When analysing beam resistance, it is necessary to consider the geometrical features of the corresponding right sections. These features, as will emerge more clearly hereafter, amount to a scalar quantity, the **area**, a vector quantity, the position of the **centroid**, and a tensor quantity, consisting of the **central directions** and the **central moments of inertia**.

The laws of transformation, by translation and rotation of the reference system, both of the vector of static moments and of the tensor of moments of inertia, will be considered. It will thus become possible also to calculate composite sections, consisting of the combination of a number of elementary parts, and the graphical interpretation (due to Mohr) of this calculation will be given.

Particular attention will be paid to the cases of sections presenting symmetry, whether axial or polar, and of thin-walled beam sections, which have already been mentioned in the introductory chapter and for which a simplified calculation is possible. A number of examples will close the chapter.

2.2 Laws of transformation of the position vector

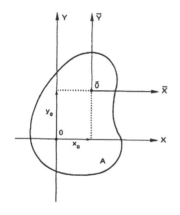

Figure 2.1

The coordinates x, y of a point of the plane in the XY reference system are linked to the coordinates $\bar{x}, \bar{y}$ of the same point in the **translated reference system** $\overline{XY}$ (Figure 2.1) by the following relations:

$$\bar{x} = x - x_0 \tag{2.1a}$$

$$\bar{y} = y - y_0 \tag{2.1b}$$

where x_0, y_0 are the coordinates of the origin $\overline{O}$ of the translated system, with respect to the original XY axes.

The laws of transformation (2.1) can be reproposed in a vector form as follows:

$$\{\bar{r}\} = \{r\} - \{r_0\} \tag{2.2}$$

where $\{r\}$ indicates the position vector $[x, y]^T$ of the generic point in the original reference system with $\{\bar{r}\}$ being the position vector $[\bar{x}, \bar{y}]^T$ of the same point in the translated reference system and with $\{r_0\}$ being the position vector $[x_0, y_0]^T$ of the origin $\overline{O}$ of the translated system in the original reference system.

The coordinates $\bar{x}, \bar{y}$ of a point of the plane $\overline{XY}$ are linked to the coordinates $\bar{x}^*, \bar{y}^*$ of the same point in the rotated reference system $\overline{X}^*\overline{Y}^*$ (Figure 2.2) via the following relations:

$$\bar{x}^* = \bar{x} \cos \vartheta + \bar{y} \sin \vartheta \tag{2.3a}$$

$$\bar{y}^* = -\bar{x} \sin \vartheta + \bar{y} \cos \vartheta \tag{2.3b}$$

where ϑ indicates the angle of rotation of the second reference system with respect to the first (positive if the rotation is counterclockwise).

20

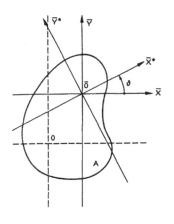

Figure 2.2

These transformation laws can be reproposed in a matrix form, as follows:

$$\{\bar{r}^*\} = [N]\{\bar{r}\} \tag{2.4}$$

where

$$[N] = \begin{bmatrix} \cos\vartheta & \sin\vartheta \\ -\sin\vartheta & \cos\vartheta \end{bmatrix} \tag{2.5}$$

is the orthogonal matrix of rotation.

2.3 Laws of transformation of the static moment vector

Consider the area A in the XY reference system (Figure 2.1). The definition of **static moment vector**, relative to the area A and calculated in the XY reference system, is given by the following two-component vector:

$$\{S\} = \begin{bmatrix} S_y \\ S_x \end{bmatrix} = \begin{bmatrix} \int_A x \, dA \\ \int_A y \, dA \end{bmatrix} = \int_A \{r\} \, dA \tag{2.6}$$

The static moment vector, again referred to the area A, calculated in the translated $\overline{XY}$ system, can be expressed in the following way:

$$\{\bar{S}\} = \begin{bmatrix} S_{\bar{y}} \\ S_{\bar{x}} \end{bmatrix} = \begin{bmatrix} \int_A \bar{x} \, dA \\ \int_A \bar{y} \, dA \end{bmatrix} = \int_A \{\bar{r}\} \, dA \tag{2.7}$$

Applying the transformation law (2.2), equation (2.7) becomes

$$\{\bar{S}\} = \int_A \{r\} \, dA - \{r_0\} \int_A dA \tag{2.8}$$

since $\{r_0\}$ is a constant vector. Recalling the definition (2.6), we obtain finally the static moment vector transformation law for translations of the reference system:

$$\{\bar{S}\} = \{S\} - A\{r_0\} \tag{2.9}$$

The vector relation (2.9) is equivalent to the following two scalar relations:

$$S_{\bar{y}} = S_y - Ax_0 \tag{2.10a}$$

$$S_{\bar{x}} = S_x - Ay_0 \tag{2.10b}$$

The reference system, translated with respect to the original one, for which both static moments vanish, is determined by the following position vector:

$$x_G = \frac{S_y}{A} \tag{2.11a}$$

$$y_G = \frac{S_x}{A} \tag{2.11b}$$

The origin G of this particular reference system is termed the **centroid** of area A, and is a characteristic point of the area itself, in the sense that it is altogether independent of the choice of the original XY system.

21

5 {4}}

Now consider the reference system $\overline{X}^*\overline{Y}^*$, rotated with respect to the $\overline{XY}$ system (Figure 2.2). The static moment vector, relative to area A and calculated in the rotated system $\overline{X}^*\overline{Y}^*$, may be expressed using the law (2.4):

$$\{\overline{S}^*\} = \int_A \{\overline{r}^*\}\, dA = [N]\int_A \{\overline{r}\}\, dA \tag{2.12}$$

where $[N]$ is the constant matrix (2.5). Finally, recalling the definition (2.7), the static moment vector transformation law for rotations of the reference system is obtained:

$$\{\overline{S}^*\} = [N]\{\overline{S}\} \tag{2.13}$$

The matrix relation (2.13) is equivalent to the following two scalar relations:

$$S_{\overline{y}*} = S_{\overline{y}}\cos\vartheta + S_{\overline{x}}\sin\vartheta \tag{2.14a}$$

$$S_{\overline{x}*} = -S_{\overline{y}}\sin\vartheta + S_{\overline{x}}\cos\vartheta \tag{2.14b}$$

From equations (2.14) two important conclusions may be drawn.

1. The static moments are zero with respect to any pair of centroidal orthogonal axes.
2. If the origin $\overline{O}$ of the reference system does not coincide with the centroid G of area A, there exists no angle of rotation ϑ of the reference system for which the static moments both vanish. In fact, from equations (2.14) we obtain

$$S_{\overline{y}*} = 0 \text{ for } \vartheta = \arctan\left(-\frac{S_{\overline{y}}}{S_{\overline{x}}}\right) \tag{2.15a}$$

$$S_{\overline{x}*} = 0 \text{ for } \vartheta = \arctan\left(\frac{S_{\overline{x}}}{S_{\overline{y}}}\right) \tag{2.15b}$$

The conditions (2.15) are not, however, compatible.

If we consider a reference system $\overline{X}^*\overline{Y}^*$, obtained by translating and then rotating the original XY system (Figures 2.1 and 2.2), it is possible to formulate the general static moment vector transformation law for rototranslations of the reference system, combining the foregoing partial laws (2.9) and (2.13):

$$\{\overline{S}^*\} = [N]\left(\{S\} - A\{r_0\}\right) \tag{2.16}$$

The inverse rototranslation formula may be obtained from the previous one by premultiplying both members by $[N]^T = [N]^{-1}$

$$\{S\} = [N]^T\{\overline{S}^*\} + A\{r_0\} \tag{2.17}$$

2.4 Laws of transformation of the moment of inertia tensor

Consider the following matrix product (referred to as the **dyadic product**):

$$\{r\}\{r\}^T = \begin{bmatrix} x \\ y \end{bmatrix}[x\ y] = \begin{bmatrix} x^2 & xy \\ yx & y^2 \end{bmatrix} \tag{2.18}$$

The definition of the **moment of inertia tensor**, relative to area A and calculated in the XY reference system, is given by the following symmetric (2×2) tensor:

$$[I] = \begin{bmatrix} I_{yy} & I_{xy} \\ I_{yx} & I_{xx} \end{bmatrix} = \begin{bmatrix} \int_A x^2 \, dA & \int_A xy \, dA \\ \int_A yx \, dA & \int_A y^2 \, dA \end{bmatrix} \quad (2.19)$$

Taking into account relation (2.18), definition (2.19) can be expressed in the following compact form:

$$[I] = \int_A \{r\}\{r\}^T \, dA \quad (2.20)$$

The moment of inertia tensor, relative again to area A and calculated in the translated reference system $\overline{XY}$ (Figure 2.1), can be expressed as follows:

$$[\bar{I}] = \int_A \{\bar{r}\}\{\bar{r}\}^T \, dA \quad (2.21)$$

and thus, applying the position vector transformation law for the translations of the reference system (equation (2.2)), we obtain

$$[\bar{I}] = \int_A (\{r\} - \{r_0\})(\{r\} - \{r_0\})^T \, dA \quad (2.22)$$

Since the transpose of the sum of two matrices is equal to the sum of the transposes, we have

$$[\bar{I}] = \int_A (\{r\} - \{r_0\})(\{r\}^T - \{r_0\}^T) dA$$

$$= \int_A \{r\}\{r\}^T \, dA - \int_A \{r\}dA\{r_0\}^T - \{r_0\}\int_A \{r\}^T \, dA + \quad (2.23)$$

$$\{r_0\}\{r_0\}^T \int_A dA$$

Finally, recalling definitions (2.6) and (2.20), we obtain the law of transformation of the moment of inertia tensor for translations of the reference system

$$[\bar{I}] = [I] + A\{r_0\}\{r_0\}^T - \{r_0\}\{S\}^T - \{S\}\{r_0\}^T \quad (2.24)$$

The matrix relation (2.24) can be rendered explicit, as follows:

$$I_{\overline{xx}} = I_{xx} + Ay_0^2 - 2y_0 S_x \quad (2.25a)$$

$$I_{\overline{yy}} = I_{yy} + Ax_0^2 - 2x_0 S_y \quad (2.25b)$$

$$I_{\overline{xy}} = I_{\overline{yx}} = I_{xy} + Ax_0 y_0 - x_0 S_x - y_0 S_y \quad (2.25c)$$

The above relations simplify in the case where the origin of the primitive XY reference system coincides with the centroid G of area A. In this case, we have

$$S_x = S_y = 0 \quad (2.26)$$

and equations (2.25) assume the form of the well-known **Huygens' laws**:

$$I_{\overline{xx}} = I_{x_G x_G} + Ay_0^2 \quad (2.27a)$$

$$I_{\overline{yy}} = I_{y_G y_G} + Ax_0^2 \quad (2.27b)$$

$$I_{\overline{xy}} = I_{x_G y_G} + Ax_0 y_0 \quad (2.27c)$$

As regards relations (2.27a) and (2.27b), it may be noted how the centroidal moment of inertia is the minimum of all those corresponding to an infinite number of parallel straight lines.

Now consider the moment of inertia tensor, relative to area A and calculated in the rotated reference system $\bar{X}^*\bar{Y}^*$ (Figure 2.2):

$$[\bar{I}^*] = \int_A \{\bar{r}^*\}\{\bar{r}^*\}^T \, dA \qquad (2.28)$$

Using the law (2.4) of transformation of the position vector for rotations of the reference system, we have

$$[\bar{I}^*] = \int_A ([N]\{\bar{r}\})\,([N]\{\bar{r}\})^T \, dA \qquad (2.29)$$

Now applying the law by which the transpose of the product of two matrices is equal to the inverse product of the transposes, we have

$$[\bar{I}^*] = \int_A ([N]\{\bar{r}\})\,(\{\bar{r}\}^T[N]^T) \, dA \qquad (2.30)$$

Exploiting the associative law and carrying the constant matrices $[N]$ and $[N]^T$ outside the integral sign, equation (2.30) becomes

$$[\bar{I}^*] = [N]\int_A \{\bar{r}\}\{\bar{r}\}^T \, dA \, [N]^T \qquad (2.31)$$

Finally, recalling definition (2.21), we obtain the law of transformation of the moment of inertia tensor for rotations of the reference system

$$[\bar{I}^*] = [N][\bar{I}][N]^T \qquad (2.32)$$

The matrix relation (2.32) can be rendered explicit as follows:

$$I_{\bar{x}^*\bar{x}^*} = I_{\bar{x}\bar{x}} \cos^2 \vartheta + I_{\bar{y}\bar{y}} \sin^2 \vartheta - 2I_{\bar{x}\bar{y}} \sin \vartheta \cos \vartheta \qquad (2.33a)$$

$$I_{\bar{y}^*\bar{y}^*} = I_{\bar{x}\bar{x}} \sin^2 \vartheta + I_{\bar{y}\bar{y}} \cos^2 \vartheta + 2I_{\bar{x}\bar{y}} \sin \vartheta \cos \vartheta \qquad (2.33b)$$

$$I_{\bar{x}^*\bar{y}^*} = I_{\bar{y}^*\bar{x}^*} = I_{\bar{x}\bar{y}} \cos 2\vartheta + \frac{1}{2}(I_{\bar{x}\bar{x}} - I_{\bar{y}\bar{y}})\sin 2\vartheta \qquad (2.33c)$$

Two important conclusions can be derived from equations (2.33).

1. The sum of the two moments of inertia I_{xx} and I_{yy} remains constant as the angle of rotation ϑ varies. We have in fact

$$I_{\bar{x}^*\bar{x}^*} + I_{\bar{y}^*\bar{y}^*} = I_{\bar{x}\bar{x}} + I_{\bar{y}\bar{y}} \qquad (2.34)$$

This sum is the first scalar invariant of the moment of inertia tensor and can be interpreted as the **polar moment of inertia** of area A with respect to the origin of the reference system:

$$I_p = \int_A r^2 \, dA \qquad (2.35)$$

2. Equating to zero the expression of the product of inertia $I_{\bar{x}*\bar{y}*}$, it is pos—sible to obtain the angle of rotation ϑ_0 which renders the moment of inertia tensor diagonal:

$$I_{\bar{x}*\bar{y}*} = I_{\bar{y}*\bar{x}*} = 0 \text{ for} \tag{2.36}$$

$$\vartheta_0 = \frac{1}{2}\arctan\left(\frac{2I_{\overline{xy}}}{I_{\overline{yy}} - I_{\overline{xx}}}\right), \quad -\frac{\pi}{4} < \vartheta_0 < \frac{\pi}{4}$$

Substituting equation (2.36) in (2.33a, b), the so-called **principal moments of inertia** are determined. The two orthogonal directions defined by the angle ϑ_0 are referred to as the **principal directions of inertia**. It can be demonstrated how the principal moments of inertia are, in one case, the minimum, and the other, the maximum, of all the moments of inertia $I_{\bar{x}*\bar{x}*}$ and $I_{\bar{y}*\bar{y}*}$, which we have as the angle of rotation ϑ varies. When the axes, in addition to being principal are also centroidal, they are referred to as **central**, as are the corresponding moments of inertia.

The general law of transformation of the moment of inertia tensor for rototranslations of the reference system (Figures 2.1 and 2.2) is obtained by combining the partial laws (2.24) and (2.32):

$$[\bar{I}^*] = [N]([I] + A\{r_0\}\{r_0\}^T - \{r_0\}\{S\}^T - \{S\}\{r_0\}^T)[N]^T \tag{2.37}$$

The inverse rototranslation formula may be obtained from (2.37) by premulti-plying both sides of the equation by $[N]^T$ and postmultiplying them by $[N]$ and inserting equation (2.17):

$$[I] = [N]^T[\bar{I}^*][N] + [N]^T\{\bar{S}^*\}\{r_0\}^T + \tag{2.38}$$
$$\{r_0\}\{\bar{S}^*\}^T[N] + A\{r_0\}\{r_0\}^T$$

2.5 Principal axes and moments of inertia

Using well-known trigonometric formulas, relation (2.33a) becomes

$$I_{\bar{x}*\bar{x}*} = I_{\overline{xx}}\frac{1 + \cos 2\vartheta}{2} + I_{\overline{yy}}\frac{1 - \cos 2\vartheta}{2} - I_{\overline{xy}}\sin 2\vartheta \tag{2.39}$$

Via equation (2.36) we obtain

$$I_{\bar{x}*\bar{x}*}(\vartheta_0) = \frac{I_{\overline{xx}} + I_{\overline{yy}}}{2} + \frac{I_{\overline{xx}} - I_{\overline{yy}}}{2}\cos 2\vartheta_0 + \tag{2.40}$$
$$\frac{I_{\overline{xx}} - I_{\overline{yy}}}{2}\tan 2\vartheta_0 \sin 2\vartheta_0$$

and hence

$$I_{\bar{x}*\bar{x}*}(\vartheta_0) = \frac{I_{\overline{xx}} + I_{\overline{yy}}}{2} + \frac{I_{\overline{xx}} - I_{\overline{yy}}}{2}\frac{1}{\cos 2\vartheta_0} \tag{2.41}$$

Since we know from trigonometry that

$$\frac{1}{\cos 2\vartheta_0} = \left(1 + \tan^2 2\vartheta_0\right)^{\frac{1}{2}} \tag{2.42}$$

25

it is possible to apply once more equation (2.36)

$$\frac{1}{\cos 2\vartheta_0} = \left(1 + \frac{4I^2_{\bar{x}\bar{y}}}{(I_{\bar{y}\bar{y}} - I_{\bar{x}\bar{x}})^2}\right)^{\frac{1}{2}} \tag{2.43}$$

$$= \begin{cases} \dfrac{1}{I_{\bar{x}\bar{x}} - I_{\bar{y}\bar{y}}}\left((I_{\bar{x}\bar{x}} - I_{\bar{y}\bar{y}})^2 + 4I^2_{\bar{x}\bar{y}}\right)^{\frac{1}{2}} \text{ when } I_{\bar{x}\bar{x}} > I_{\bar{y}\bar{y}} \\ \dfrac{1}{I_{\bar{y}\bar{y}} - I_{\bar{x}\bar{x}}}\left((I_{\bar{x}\bar{x}} - I_{\bar{y}\bar{y}})^2 + 4I^2_{\bar{x}\bar{y}}\right)^{\frac{1}{2}} \text{ when } I_{\bar{x}\bar{x}} < I_{\bar{y}\bar{y}} \end{cases}$$

Then, indicating $I_{\bar{x}^*\bar{x}^*}(\vartheta_0)$ with the simpler notation I_ξ, we have

$$I_\xi = \begin{cases} \dfrac{I_{\bar{x}\bar{x}} + I_{\bar{y}\bar{y}}}{2} + \dfrac{1}{2}\left((I_{\bar{x}\bar{x}} - I_{\bar{y}\bar{y}})^2 + 4I^2_{\bar{x}\bar{y}}\right)^{\frac{1}{2}} \text{ when } I_{\bar{x}\bar{x}} > I_{\bar{y}\bar{y}} \\ \dfrac{I_{\bar{x}\bar{x}} + I_{\bar{y}\bar{y}}}{2} - \dfrac{1}{2}\left((I_{\bar{x}\bar{x}} - I_{\bar{y}\bar{y}})^2 + 4I^2_{\bar{x}\bar{y}}\right)^{\frac{1}{2}} \text{ when } I_{\bar{x}\bar{x}} < I_{\bar{y}\bar{y}} \end{cases} \tag{2.44}$$

Likewise, indicating $I_{\bar{y}^*\bar{y}^*}(\vartheta_0)$ with I_η, we have

$$I_\eta = \begin{cases} \dfrac{I_{\bar{x}\bar{x}} + I_{\bar{y}\bar{y}}}{2} - \dfrac{1}{2}\left((I_{\bar{x}\bar{x}} - I_{\bar{y}\bar{y}})^2 + 4I^2_{\bar{x}\bar{y}}\right)^{\frac{1}{2}} \text{ when } I_{\bar{x}\bar{x}} > I_{\bar{y}\bar{y}} \\ \dfrac{I_{\bar{x}\bar{x}} + I_{\bar{y}\bar{y}}}{2} + \dfrac{1}{2}\left((I_{\bar{x}\bar{x}} - I_{\bar{y}\bar{y}})^2 + 4I^2_{\bar{x}\bar{y}}\right)^{\frac{1}{2}} \text{ when } I_{\bar{x}\bar{x}} < I_{\bar{y}\bar{y}} \end{cases} \tag{2.45}$$

We can thus conclude that, when the $\overline{XY}$ axes, by rotation, become the principal axes, the order relation is conserved:

$$I_{\bar{x}\bar{x}} > I_{\bar{y}\bar{y}} \Rightarrow I_\xi > I_\eta \tag{2.46a}$$

$$I_{\bar{x}\bar{x}} < I_{\bar{y}\bar{y}} \Rightarrow I_\xi < I_\eta \tag{2.46b}$$

When

$$I_{\bar{x}\bar{x}} = I_{\bar{y}\bar{y}}, \quad I_{\bar{x}\bar{y}} \neq 0 \tag{2.47}$$

relation (2.36) is not defined and thus it makes no difference whether the $\overline{XY}$ reference system is rotated by $\pi/4$ clockwise or counterclockwise ($\vartheta_0 = \pm\pi/4$) in order to obtain the principal directions.

Moreover, when

$$I_{\bar{x}\bar{x}} = I_{\bar{y}\bar{y}}, \quad I_{\bar{x}\bar{y}} = 0 \tag{2.48}$$

all the rotated reference systems $\overline{X}^*\overline{Y}^*$ are principal systems, for any angle of rotation ϑ_0. The areas that satisfy the conditions of equation (2.48) are said to be **gyroscopic**. As will be seen in the next section, it is possible to give a synthetic graphical interpretation of cases (2.47) and (2.48).

2.6 Mohr's circle

With the aim of introducing the graphical method of **Mohr's circle**, let us consider the inverse problem of the one previously solved: given an area A, and its principal axes of inertia $\xi\eta$ and the corresponding principal moments I_ξ, I_η known, with respect to a point O of the plane (Figure 2.3), we intend to express the moments of inertia with respect to a reference system rotated by

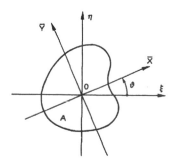

Figure 2.3

26

an angle ϑ, counterclockwise with respect to the principal $\xi\eta$ reference system.

Applying equations (2.33), and since $I_{\xi\eta} = 0$, we have

$$I_{\overline{xx}} = I_\xi \cos^2\vartheta + I_\eta \sin^2\vartheta \tag{2.49a}$$

$$I_{\overline{yy}} = I_\xi \sin^2\vartheta + I_\eta \cos^2\vartheta \tag{2.49b}$$

$$I_{\overline{xy}} = \frac{I_\xi - I_\eta}{2}\sin 2\vartheta \tag{2.49c}$$

The trigonometry formulas used previously give

$$I_{\overline{xx}} = \frac{I_\xi + I_\eta}{2} + \frac{I_\xi - I_\eta}{2}\cos 2\vartheta \tag{2.50a}$$

$$I_{\overline{yy}} = \frac{I_\xi + I_\eta}{2} - \frac{I_\xi - I_\eta}{2}\cos 2\vartheta \tag{2.50b}$$

$$I_{\overline{xy}} = \frac{I_\xi - I_\eta}{2}\sin 2\vartheta \tag{2.50c}$$

Relations (2.50a,c) constitute the parametric equations of a circumference having as its centre

$$C\left(\frac{I_\xi + I_\eta}{2},0\right) \tag{2.51a}$$

and as its radius

$$R = \frac{I_\xi - I_\eta}{2} \tag{2.51b}$$

in Mohr's plane $I_{\overline{xx}}I_{\overline{xy}}$ (Figure 2.4). The above circumference represents all the pairs $(I_{\overline{xx}},I_{\overline{xy}})$ which succeed one another as the angle ϑ (Figure 2.3) varies. Note that, since $I_{\overline{xx}}$ is in any case positive, we have in fact a Mohr's half-plane.

Let us now reconsider the direct problem: given the moments of inertia with respect to the two generic orthogonal axes $\overline{XY}$ (Figure 2.3), determine the principal axes and moments of inertia. This determination has already been made analytically in section 2.5. We shall now proceed to repropose it graphically using Mohr's circle (Figure 2.5).

1. The first operation to be carried out is to identify the two notable points P and P' on Mohr's plane:

$$P(I_{\overline{xx}},I_{\overline{xy}}), \quad P'(I_{\overline{yy}},-I_{\overline{xy}}) \tag{2.52}$$

2. The intersection C of the segment PP' with the axis $I_{\overline{xx}}$ identifies the centre of Mohr's circle, while the segments CP and CP' represent two radii of that circle.
3. Draw through the point P the line parallel to the axis $I_{\overline{xx}}$ and through P' the line parallel to the axis $I_{\overline{xy}}$. These two lines meet in point P^*, called the **pole**, again belonging to Mohr's circle.

27

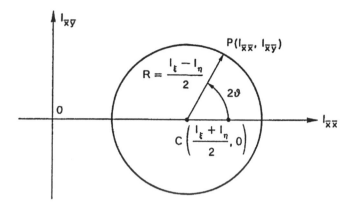

Figure 2.4

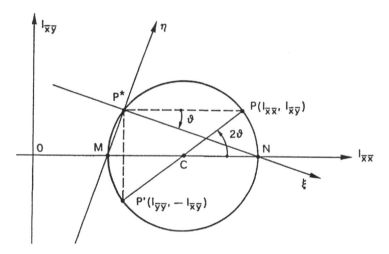

Figure 2.5

4. The lines joining pole P^* with points M and N of the $I_{\overline{xx}}$ axis, which are intersections of the circumference with the axis, give the directions of the two principal axes of inertia. Naturally, points M and N each have as abscissa the value of one of the two principal moments of inertia. In particular, in Figure 2.5, the abscissa of M is I_η, while the abscissa of N is I_ξ, since we have assumed $I_{\overline{xx}} > I_{\overline{yy}}$. Pole P^* can obviously also fall in one of the three remaining quadrants corresponding to Mohr's circle.

The graphical construction described above and shown in Figure 2.5 is justified by noting that the circumferential angle PP^*N is half of the corresponding central angle $PCN = 2\vartheta$, and that thus its amplitude is equal to the angle ϑ.

2.7 Areas presenting symmetry

An area is said to present **oblique axial symmetry** (Figure 2.6.(a)) when there exists a straight line s which cuts the area into two parts, and a direction s', conjugate with this straight line, such that, if we consider a generic point P,

28

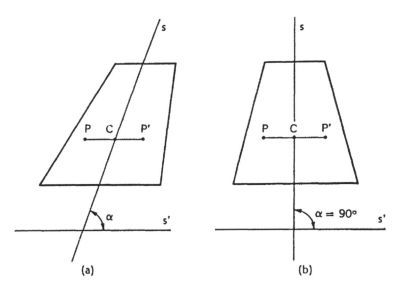

Figure 2.6

belonging to the area and the line $\overline{PC}$, parallel to the direction s', and we draw on that line the segment $\overline{CP'} = \overline{PC}$ on the opposite side of P with respect to s, the point P' still belongs to the area. When the angle α between the directions of the lines s and s' is equal to 90°, then we have **right axial symmetry** (Figure 2.6(b)).

It is easy to verify that the centroid of a section having axial symmetry lies on the corresponding axis of symmetry. The centroid relative to the pair of symmetrical elementary areas located in P and P' coincides in fact with point C (Figure 2.6). Applying the so-called distributive law of the centroid, it is possible to think of concentrating the whole area on the axis of symmetry s, and thus the global centroid is sure to lie on the same line s.

In the case of an area presenting right symmetry (Figure 2.6 (b)), the axis of symmetry is also a central axis of inertia. In fact, it is centroidal and, with

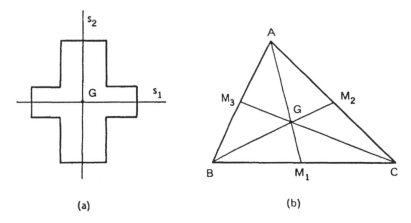

Figure 2.7

29

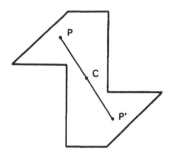

Figure 2.8

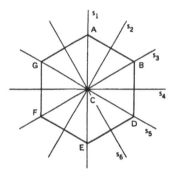

Figure 2.9

respect to it and to any orthogonal axis, the product of inertia $I_{ss'}$ vanishes by symmetry.

When there are two or more axes of symmetry (oblique or right), since the centroid must belong to each axis, it coincides with their intersection (Figure 2.7). In the case of double right symmetry (Figure 2.7(a)), the axes of symmetry are also central axes of inertia.

An area is said to present **polar symmetry** (Figure 2.8) when there exists a point C such that, if we consider a generic point P belonging to the area and the line PC joining the two points, and we draw on this line the segment $\overline{CP'}$ = $\overline{PC}$ on the side opposite to P with respect to C, the point P' still belongs to the area.

It is immediately verifiable that the centroid of a section having polar symmetry coincides with its geometrical centre C. The centroid corresponding to the pair of symmetrical elementary areas in P and P' coincides, in fact, with point C (Figure 2.8). Applying the distributive law of the centroid, it is possible to think of concentrating the whole area in point C, and thus the global centroid must certainly coincide with the same point C.

It is interesting to note how an n-tuple right symmetry area, with n being an even number ($2 \leqslant n < \infty$), is also a polar symmetry area (Figure 2.9), whereas a polar symmetry area is not necessarily also an n-tuple right symmetry area (Figure 2.8).

Areas having n-tuple right symmetry, with n being an odd number ($3 \leqslant n < \infty$), do not, however, present polar symmetry, even though they are gyroscopic areas, as also are those with n even.

2.8 Elementary areas

If, on an XY plane, we assign n areas, $A_1, A_2, \ldots, A_n$, the distributive law of static moments, and, respectively, that of the moments of inertia, are defined as follows (Figure 2.10):

$$S\left(\bigcup_{i=1}^{n} A_i\right) = \sum_{i=1}^{n} S(A_i) \qquad (2.53a)$$

$$I\left(\bigcup_{i=1}^{n} A_i\right) = \sum_{i=1}^{n} I(A_i) \qquad (2.53b)$$

where S and I indicate generically a static moment and a moment of inertia, calculated with respect to the coordinate axes ($S_x, S_y; I_{xx}, I_{yy}, I_{xy}$).

In determining the static and inertial characteristics of composite areas, it is necessary to exploit the above laws. These derive from the integral nature of the definitions which have previously been given of first and second order moments. The first law expresses the fact that the static moment of a composite area (i.e. of the union of more than one elementary area) is equal to the sum of the static moments of the single areas. The second law refers to the moments of inertia and is altogether analogous.

Since it is therefore possible to reduce the calculation of composite areas to the calculation of simpler areas, the importance of calculating once and for all the static and inertial features of elementary areas emerges clearly. In the sequel, we shall examine the rectangle, the right triangle and the annulus sector.

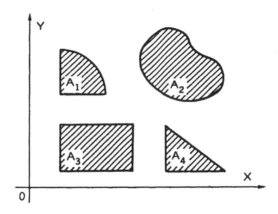

Figure 2.10

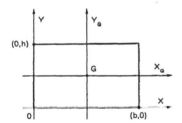

Figure 2.11

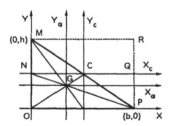

Figure 2.12

Consider the **rectangle** having base b and height h (Figure 2.11). From the definition of centroid, we obtain immediately the static moments in the XY reference system:

$$S_x = Ay_G = \frac{1}{2}bh^2 \tag{2.54a}$$

$$S_y = Ax_G = \frac{1}{2}hb^2 \tag{2.54b}$$

For the central moments of inertia, we have

$$I_{x_G x_G} = \int_{-b/2}^{+b/2}\int_{-h/2}^{+h/2} y^2 dx\, dy = \frac{bh^3}{12} \tag{2.55a}$$

and likewise

$$I_{y_G y_G} = \frac{hb^3}{12} \tag{2.55b}$$

It is then possible to obtain the inertia tensor in the XY reference system by applying Huygens' laws (2.27):

$$I_{xx} = I_{x_G x_G} + Ay_0^2 = \frac{bh^3}{12} + \frac{bh^3}{4} = \frac{bh^3}{3} \tag{2.56a}$$

$$I_{yy} = I_{y_G y_G} + Ax_0^2 = \frac{hb^3}{12} + \frac{hb^3}{4} = \frac{hb^3}{3} \tag{2.56b}$$

$$I_{xy} = I_{x_G y_G} + Ax_0 y_0 = 0 + bh\left(-\frac{b}{2}\right)\left(-\frac{h}{2}\right) = \frac{b^2h^2}{4} \tag{2.56c}$$

Consider the **right triangle** having base b and height h (Figure 2.12). As is well-known, its centroid coincides with the point of intersection of the three medians, which are at the same time axes of oblique symmetry. The moment of inertia with respect to the axis X_C of the triangle MOP is equal to the moment of inertia with respect to the axis X_C of rectangle $NOPQ$. The latter, in fact, is obtained from the former by suppressing triangle MNC and adding

31

triangle *CQP*. These two triangles are equal and arranged symmetrically with respect to the axis X_C. Hence

$$I_{x_C x_C} = \frac{1}{3} b \left(\frac{h}{2} \right)^3 = \frac{bh^3}{24} \tag{2.57}$$

Applying the inverse of Huygens' law, we obtain

$$I_{x_G x_G} = \frac{bh^3}{24} - \left(\frac{bh}{2} \right) \left(\frac{h}{6} \right)^2 = \frac{bh^3}{36} \tag{2.58}$$

Finally, applying Huygens' law, we have

$$I_{xx} = \frac{bh^3}{36} + \left(\frac{bh}{2} \right) \left(\frac{h}{3} \right)^2 = \frac{bh^3}{12} \tag{2.59}$$

Likewise

$$I_{yy} = \frac{hb^3}{12} \tag{2.60}$$

As regards the product of inertia I_{xy}, its integral definition can be applied:

$$I_{xy} = \int_0^b \int_0^{h(b-x)/b} xy \, dx \, dy = \frac{b^2 h^2}{24} \tag{2.61}$$

The product of inertia $I_{x_G y_G}$ may be obtained from I_{xy}, via the inverse application of Huygens' law:

$$I_{x_G y_G} = \frac{b^2 h^2}{24} - \left(\frac{bh}{2} \right) \left(-\frac{b}{3} \right) \left(-\frac{h}{3} \right) = -\frac{b^2 h^2}{72} \tag{2.62}$$

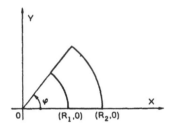

Figure 2.13

Consider the **annulus sector** of internal radius R_1, external radius R_2 and angular amplitude φ (Figure 2.13). The static moment of the sector with respect to the X axis is

$$S_x = \int_0^\varphi \int_{R_1}^{R_2} (r \sin \varphi) \, r \, d\varphi = \frac{1}{3} (1 - \cos \varphi)(R_2^3 - R_1^3) \tag{2.63a}$$

Likewise, the static moment with respect to the Y axis is

$$S_y = \int_0^\varphi \int_{R_1}^{R_2} (r \cos \varphi) r \, dr \, d\varphi = \frac{1}{3} \sin \varphi (R_2^3 - R_1^3) \tag{2.63b}$$

Also in the calculation of the moments of inertia it is possible to apply the definition:

$$I_{xx} = \int_0^\varphi \int_{R_1}^{R_2} (r \sin \varphi)^2 r \, dr \, d\varphi = \frac{1}{8} (\varphi - \sin \varphi \cos \varphi)(R_2^4 - R_1^4) \tag{2.64a}$$

$$I_{yy} = \int_0^\varphi \int_{R_1}^{R_2} (r \cos \varphi)^2 r \, dr \, d\varphi = \frac{1}{8} (\varphi + \sin \varphi \cos \varphi)(R_2^4 - R_1^4) \tag{2.64b}$$

$$I_{xy} = \int_0^\varphi \int_{R_1}^{R_2} (r \sin \varphi)(r \cos \varphi) r \, dr \, d\varphi = \frac{1}{16} (1 - \cos 2\varphi)(R_2^4 - R_1^4) \tag{2.64c}$$

In the particular case of a circle of radius R, we have

$$R_1 = 0, R_2 = R, \varphi = 2\pi$$

and thus

$$S_x = S_y = 0 \qquad (2.65a)$$

$$I_{xx} = I_{yy} = \frac{\pi R^4}{4}, \quad I_{xy} = 0 \qquad (2.65b)$$

The static moments and the product of inertia are zero by symmetry. Another way to obtain the moments of inertia I_{xx} and I_{yy} of the circle is that of calculating the polar moment as given by equation (2.35):

$$I_p = I_{xx} + I_{yy} = \int_0^{2\pi} \int_0^R (r^2)\, r\, dr\, d\varphi = \frac{\pi R^4}{2} \qquad (2.66)$$

Since $I_{xx} = I_{yy}$, once more we obtain equation (2.65b).

2.9 Thin-walled sections

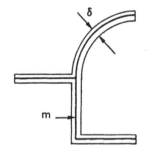

A section is said to be **thin-walled** when one of its dimensions (the thickness δ) is clearly smaller than the others (Figure 2.14). In these cases, the section is represented and calculated as if its whole area were concentrated in its midline m. This approximate calculation approaches the exact result, the smaller the thickness is, compared to the other dimensions of the section.

Consider a **rectilinear segment** of length l and thickness δ (Figure 2.15). The moment of inertia $I_{x_G x_G} = I_{xx} = l\delta^3/12$ can be neglected with regard to all the other quantities. This moment is, in fact, an infinitesimal of a higher order, as it is proportional to the infinitesimal quantity δ raised to the third power, whilst the other quantities are proportional to the quantity δ raised to the first power.

Figure 2.14

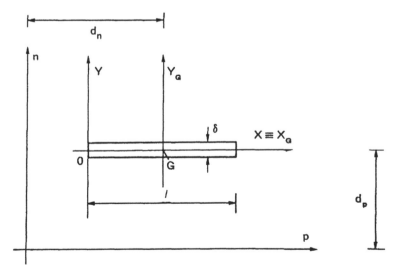

Figure 2.15

33

When Huygens' law is applied, for the calculation of the moments of inertia of a rectilinear segment with respect to the translated axes p and n (Figure 2.15), it is important to distinguish the two cases. In fact, for the calculation of I_{pp}, it is possible to consider the entire area concentrated in the centroid and to neglect the local moment:

$$I_{pp} = \delta l d_p^2 \qquad (2.67a)$$

while for the calculation of I_{nn}, in addition to the contribution of translation, it is necessary to include also the local contribution, which is not a negligible quantity:

$$I_{nn} = \frac{\delta l^3}{12} + \delta l d_n^2 \qquad (2.67b)$$

If we imagine inclining the rectilinear segment by an angle α with respect to the X axis (Figure 2.16), the moment of inertia with respect to the X axis then becomes

$$I_{xx} = \int_A y^2 \, dA = \int_{-l/2}^{+l/2} (z \sin \alpha)^2 \delta dz \qquad (2.68)$$

where Z is the longitudinal axis of the segment. Evaluating the integral, we obtain

$$I_{xx} = \frac{\delta l^3}{12} \sin^2 \alpha \qquad (2.69a)$$

and, likewise

$$I_{yy} = \frac{\delta l^3}{12} \cos^2 \alpha \qquad (2.69b)$$

$$I_{xy} = \frac{\delta l^3}{12} \sin \alpha \cos \alpha \qquad (2.69c)$$

It may be noted how, for $\alpha = 0, \pi/2$, we obtain once more the results already found.

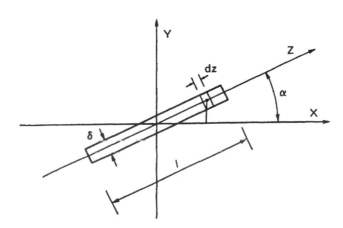

Figure 2.16

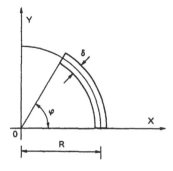

Figure 2.17

Consider an **arc of circumference** of radius R, angular amplitude φ and thickness δ (Figure 2.17). To define the static and inertial characteristics, it is possible to reconsider the formulas of section 2.8 and to particularize them for

$$R_1 \simeq R_2 \simeq R; \quad R_2 - R_1 = \delta \ll R \tag{2.70}$$

For example, equation (2.63a) is transformed as follows:

$$S_x = \frac{1}{3}(1 - \cos\varphi)(R_2 - R_1)(R_2^2 + R_2 R_1 + R_1^2) \simeq (1 - \cos\varphi)\delta R^2 \tag{2.71a}$$

and likewise, equation (2.63b) becomes

$$S_y = \sin\varphi \, \delta R^2 \tag{2.71b}$$

Then, as regards the moments of inertia, from equations (2.64) and (2.70) we obtain

$$I_{xx} = \frac{1}{2}(\varphi - \sin\varphi \cos\varphi)\delta R^3 \tag{2.72a}$$

$$I_{yy} = \frac{1}{2}(\varphi + \sin\varphi \cos\varphi)\delta R^3 \tag{2.72b}$$

$$I_{xy} = \frac{1}{4}(1 - \cos 2\varphi)\delta R^3 \tag{2.72c}$$

2.10 Examples of calculation

Five examples of calculation are given below for five sections, without any particular comments. In each case, the areas, the static moments, the coordinates of the centroid, the moments of inertia in the centroidal system, the central directions and moments of inertia are calculated and listed, in order and for each elementary part of the section, as well as for the entire section.

Example 1 (Figure 2.18(a)) concerns an L-section made up of two rectangular-section plates and of a triangular-section angle iron (Figure 2.18(b)). This section does not present particular symmetries and thus, for the determination of the global centroid, requires the calculation of the areas and the static moments. The latter are obtained by multiplying the partial areas by the coordinates of the partial centroids. For the calculation of the moments of inertia in the centroidal system $X_G Y_G$, use has, instead, been made of Huygens' formulas (2.27), which add the local moment of inertia to the moment of translation. Angle ϑ_0 of counterclockwise rotation, which provides the central directions, may be obtained analytically from equation (2.36), just as the central moments may be deduced from equations (2.44) and (2.45). Figure 2.18(c) gives the graphical construction of Mohr's circle, with the definition of the aforesaid quantities.

Example 2 (Figure 2.19(a)) concerns an H-section which may be obtained ideally by removing a square and a semicircle from the rectangle circumscribing the section (Figure 2.19(b)). The axis of symmetry Y is also centroidal and principal, and thus central. The ordinate y_G of the global centroid remains unknown, and hence only the static moments with respect to the X axis are to be calculated. Whereas the moments $I_{x_G}^{(1)}$ and $I_{x_G}^{(2)}$, relative to the rectangle and to the square are obtained by applying equation (2.27a), for the moment $I_{x_G}^{(3)}$ of the semicircle, equation (2.25a) has been resorted to. The partial prod-

35

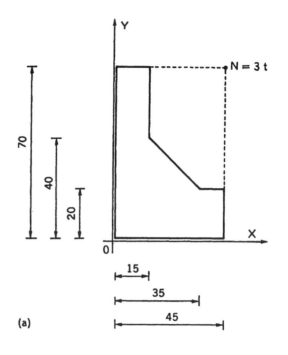

(a)

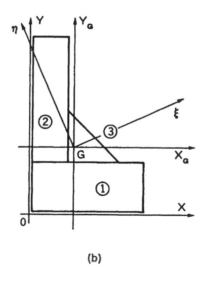

(b)

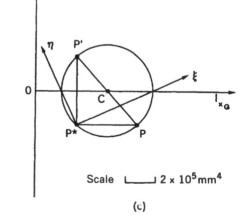

(c)

Figure 2.18

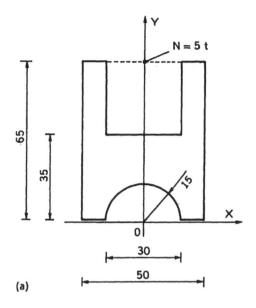

(a)

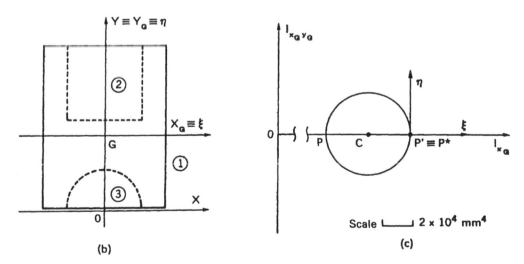

(b)

(c)

Scale ⌞___⌟ 2 × 10⁴ mm⁴

Figure 2.19

37

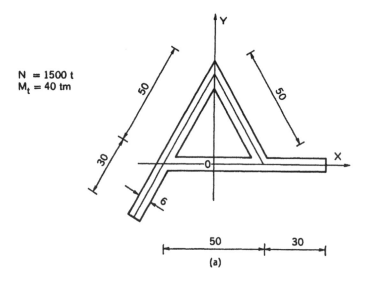

N = 1500 t
M$_t$ = 40 tm

(a)

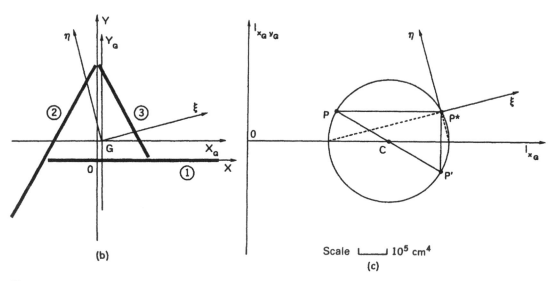

(b)

Scale ⌞____⌟ 10^5 cm^4

(c)

Figure 2.20

38

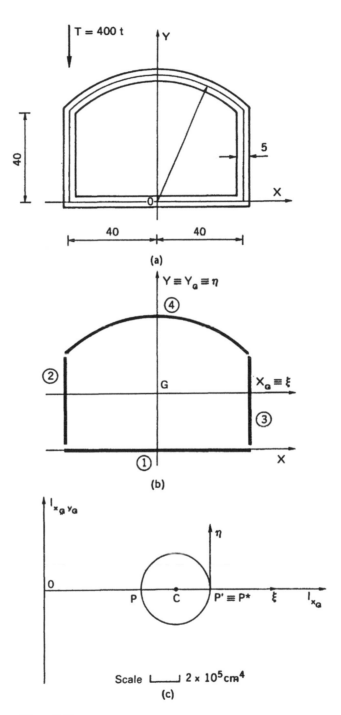

Figure 2.21

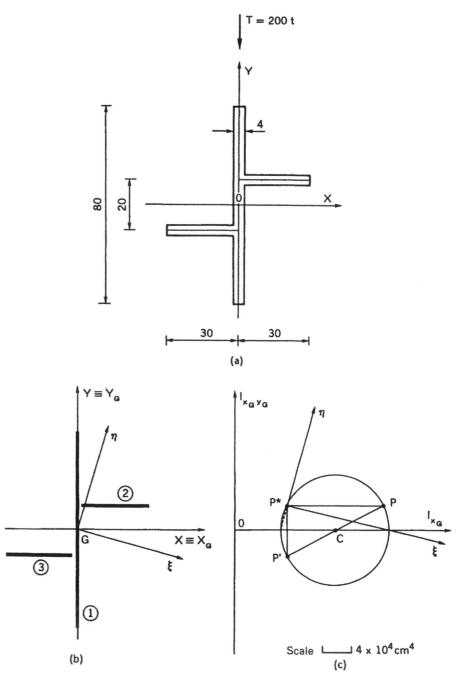

(a)

(b)

(c)

Scale ⊢———⊣ $4 \times 10^4 \, cm^4$

Figure 2.22

ucts of inertia are all zero by symmetry. In this case Mohr's graphical construction is of little significance, on account of the axial symmetry (Figure 2.19 (c)).

Example 3 (Figure 2.20 (a)) concerns a closed thin-walled section made up of three plates inclined by 60° with respect to one another (Figure 2.20 (b)). To calculate the moments of inertia, Huygens' laws (2.27) have been used, whilst the local moments of the inclined segments have been evaluated using equations (2.69). Figure 2.20 (c) presents Mohr's graphical construction.

Example 4 (Figure 2.21 (a)) concerns a closed thin-walled section made up of three plane plates and one cylindrical plate (Figure 2.21 (b)). The static moment $S_x^{(4)}$, relative to the circular segment, has been calculated according to equation (2.71 b), while the moment of inertia $I_{x_G}^{(4)}$ has been evaluated, applying the law of transformation by translation (2.25 a) and, locally, equation (2.72 b). The moment of inertia $I_{y_G}^{(4)}$ has, instead, been evaluated by simply applying equation (2.72 a). Mohr's circle for this case is represented in Figure 2.21 (c).

Finally, Example 5 (Figure 2.22 (a)) regards a thin-walled section having polar symmetry, made up of three plane plates (Figure 2.22 (b)). In this case, calculation of the position of the centroid serves no purpose, as the polar symmetry causes it to coincide with the geometrical centre of the area. Mohr's circle (Figure 2.22 (c)) provides confirmation of the central directions and moments determined analytically.

EXAMPLE 1

$$\begin{cases} A^{(1)} = 45 \times 20 \text{ mm}^2 = 900 \text{ mm}^2 \\ A^{(2)} = 50 \times 15 \text{ mm}^2 = 750 \text{ mm}^2 \\ A^{(3)} = \frac{1}{2} \times 20 \times 20 \text{ mm}^2 = 200 \text{ mm}^2 \end{cases}$$

$$A = \sum_{i=1}^{3} A^{(i)} = 1\,850 \text{ mm}^2$$

$$\begin{cases} S_x^{(1)} = A^{(1)}y_G^{(1)} = 900 \times 10 = 9\,000 \text{ mm}^3 \\ S_x^{(2)} = A^{(2)}y_G^{(2)} = 750 \times 45 = 33\,750 \text{ mm}^3 \\ S_x^{(3)} = A^{(3)}y_G^{(3)} = 200 \times 26.67 = 5\,333 \text{ mm}^3 \end{cases}$$

$$S_x = \sum_{i=1}^{3} S_x^{(i)} = 48\,083 \text{ mm}^3$$

$$\begin{cases} S_y^{(1)} = A^{(1)}x_G^{(1)} = 900 \times 22.5 = 20\,250 \text{ mm}^3 \\ S_y^{(2)} = A^{(2)}x_G^{(2)} = 750 \times 7.5 = 5\,625 \text{ mm}^3 \\ S_y^{(3)} = A^{(3)}x_G^{(3)} = 200 \times 21.67 = 4\,333 \text{ mm}^3 \end{cases}$$

$$S_y = \sum_{i=1}^{3} S_y^{(i)} = 30\,208 \text{ mm}^3$$

$$\left\{ \begin{aligned} x_G &= \frac{S_y}{A} = 16.33 \text{ mm} \\ y_G &= \frac{S_x}{A} = 25.99 \text{ mm} \end{aligned} \right.$$

$$\left\{ \begin{aligned} I_{x_G}^{(1)} &= \frac{45 \times 20^3}{12} + 900 \times 15.99^2 = 260\,112 \text{ mm}^4 \\ I_{x_G}^{(2)} &= \frac{15 \times 50^3}{12} + 750 \times (45 - 25.99)^2 = 427\,285 \text{ mm}^4 \\ I_{x_G}^{(3)} &= \frac{20 \times 20^3}{36} + 200 \times (26.67 - 25.99)^2 = 4\,536 \text{ mm}^4 \end{aligned} \right.$$

$$I_{x_G} = \sum_{i=1}^{3} I_{x_G}^{(i)} = 691\,933 \text{ mm}^4$$

$$\left\{ \begin{aligned} I_{y_G}^{(1)} &= \frac{20 \times 45^3}{12} + 900 \times (22.5 - 16.33)^2 = 186\,100 \text{ mm}^4 \\ I_{y_G}^{(2)} &= \frac{50 \times 15^3}{12} + 750 \times (16.33 - 7.5)^2 = 72\,583 \text{ mm}^4 \\ I_{y_G}^{(3)} &= \frac{20 \times 20^3}{36} + 200 \times (21.67 - 16.33)^2 = 10\,133 \text{ mm}^4 \end{aligned} \right.$$

$$I_{y_G} = \sum_{i=1}^{3} I_{y_G}^{(i)} = 268\,816 \text{ mm}^4$$

$$\left\{ \begin{aligned} I_{x_G y_G}^{(1)} &= -900 \times (22.5 - 16.33)(25.99 - 10) = -88\,792 \text{ mm}^4 \\ I_{x_G y_G}^{(2)} &= -750 \times (16.33 - 7.5)(45 - 25.99) = -125\,941 \text{ mm}^4 \\ I_{x_G y_G}^{(3)} &= -\frac{20^2 \times 20^2}{72} + 200 \times (21.67 - 16.33)(26.67 - 25.99) \\ &= -1\,500 \text{ mm}^4 \end{aligned} \right.$$

$$I_{x_G y_G} = \sum_{i=1}^{3} I_{x_G y_G}^{(i)} = -216\,186 \text{ mm}^4$$

$$\vartheta_0 = \frac{1}{2} \arctan \frac{2 \times (-216\,186)}{268\,816 - 691\,933} = 22.81°$$

$$I_\xi = \frac{691\,933 + 268\,816}{2} + \frac{1}{2} \left((691\,933 - 268\,816)^2 + 4 \times (216\,186)^2 \right)^{\frac{1}{2}}$$
$$= 480\,374 + 302\,478 = 782\,852 \text{ mm}^4$$

$$I_\eta = 480\,374 - 302\,478 = 177\,896 \text{ mm}^4$$

EXAMPLE 2

$$\begin{cases} A^{(1)} = 50 \times 65 = 3\ 250 \text{ mm}^2 \\[4pt] A^{(2)} = -30 \times 30 = -900 \text{ mm}^2 \\[4pt] A^{(3)} = -\dfrac{1}{2} \times 3.1415 \times 15^2 = -353.42 \text{ mm}^2 \end{cases}$$

$$A \;=\; \sum_{i=1}^{3} A^{(i)} = 1\ 996.58 \text{ mm}^2$$

$$\begin{cases} S_x^{(1)} = A^{(1)} y_G^{(1)} = 3\ 250 \times 32.5 = 105\ 625 \text{ mm}^3 \\[4pt] S_x^{(2)} = A^{(2)} y_G^{(2)} = -900 \times 50 = -45\ 000 \text{ mm}^3 \\[4pt] S_x^{(3)} = -\dfrac{2}{3}(15)^3 = -2\ 250 \text{ mm}^3 \end{cases}$$

$$\begin{cases} S_x \;=\; \displaystyle\sum_{i=1}^{3} S_x^{(i)} = 58\ 375 \text{ mm}^3 \\[10pt] S_y \;=\; 0 \end{cases}$$

$$\begin{cases} x_G \;=\; 0 \\[6pt] y_G \;=\; \dfrac{S_x}{A} = 29.24 \text{ mm} \end{cases}$$

$$\begin{cases} I_{x_G}^{(1)} \;=\; \dfrac{50 \times 65^3}{12} + 3\ 250 \times (32.5 - 29.24)^2 = 1\ 178\ 810.5 \text{ mm}^4 \\[10pt] I_{x_G}^{(2)} \;=\; -\dfrac{30 \times 30^3}{12} - 900 \times (50 - 29.24)^2 = -455\ 379.84 \text{ mm}^4 \\[10pt] I_{x_G}^{(3)} \;=\; -\dfrac{\pi}{8} \times 15^4 - 353.42 \times (29.24)^2 + \\[6pt] \qquad\qquad 2 \times 29.24 \times \dfrac{2}{3} \times 15^3 = -190\ 465.99 \text{ mm}^4 \end{cases}$$

$$I_{x_G} = \sum_{i=1}^{3} I_{x_G}^{(i)} = 532\ 956 \text{ mm}^4$$

$$\begin{cases} I_{y_G}^{(1)} = \dfrac{65 \times 50^3}{12} = 677\ 083.33 \text{ mm}^4 \\[10pt] I_{y_G}^{(2)} = -\dfrac{30 \times 30^3}{12} = -67\ 500 \text{ mm}^4 \\[10pt] I_{y_G}^{(3)} = -\dfrac{\pi}{8} \times 15^4 = -19\ 879.80 \text{ mm}^4 \end{cases}$$

$$I_{y_G} = \sum_{i=1}^{3} I_{y_G}^{(i)} = 589\ 702\ \text{mm}^4$$

$$\begin{cases} I_\xi = I_{x_G} = 532\ 956\ \text{mm}^4 \\ I_\eta = I_{y_G} = 589\ 702\ \text{mm}^4 \end{cases}$$

EXAMPLE 3

$$\begin{cases} A^{(1)} = 80 \times 6 = 480\ \text{cm}^2 \\ A^{(2)} = 80 \times 6 = 480\ \text{cm}^2 \\ A^{(3)} = 50 \times 6 = 300\ \text{cm}^2 \end{cases}$$

$$A = \sum_{i=1}^{3} A^{(i)} = 1\ 260\ \text{cm}^2$$

$$\begin{cases} S_x^{(1)} = 0 \\ S_x^{(2)} = A^{(2)}y_G^{(2)} = 480 \times 8.66 = 4\ 156.8\ \text{cm}^3 \\ S_x^{(3)} = A^{(3)}y_G^{(3)} = 300 \times 21.65 = 6\ 495\ \text{cm}^3 \end{cases}$$

$$S_x = \sum_{i=1}^{3} S_x^{(i)} = 10\ 652\ \text{cm}^3$$

$$\begin{cases} S_y^{(1)} = A^{(1)}x_G^{(1)} = 480 \times 15 = 7\ 200\ \text{cm}^3 \\ S_y^{(2)} = A^{(2)}x_G^{(2)} = -480 \times 20 = -9\ 600\ \text{cm}^3 \\ S_y^{(3)} = A^{(3)}x_G^{(3)} = 300 \times 12.5 = 3\ 750\ \text{cm}^3 \end{cases}$$

$$S_y = \sum_{i=1}^{3} S_y^{(i)} = 1\ 350\ \text{cm}^3$$

$$\begin{cases} x_G = \dfrac{S_y}{A} = 1.07\ \text{cm} \\ y_G = \dfrac{S_x}{A} = 8.45\ \text{cm} \end{cases}$$

$$\begin{cases} I_{x_G}^{(1)} = 480 \times (8.45)^2 = 34\ 273.2\ \text{cm}^4 \\ I_{x_G}^{(2)} = \dfrac{6 \times 80^3}{12}(0.866)^2 + 480 \times (8.66 - 8.45)^2 = 192\ 009.9\ \text{cm}^4 \\ I_{x_G}^{(3)} = \dfrac{6 \times 50^3}{12}(0.866)^2 + 300 \times (21.65 - 8.45)^2 = 99\ 144.25\ \text{cm}^4 \end{cases}$$

$$I_{x_G} = \sum_{i=1}^{3} I_{x_G}^{(i)} = 325\ 446\ \text{cm}^4$$

$$\begin{cases} I_{y_G}^{(1)} = \dfrac{6 \times 80^3}{12} + 480 \times (15 - 1.07)^2 = 349\ 000\ \text{cm}^4 \\[3mm] I_{y_G}^{(2)} = \dfrac{6 \times 80^3}{12} \times (0.5)^2 + 480 \times (20 + 1.07)^2 = 277\ 000\ \text{cm}^4 \\[3mm] I_{y_G}^{(3)} = \dfrac{6 \times 50^3}{12} \times (0.5)^2 + 300 \times (12.5 - 1.07)^2 = 55\ 000\ \text{cm}^4 \end{cases}$$

$$I_{y_G} = \sum_{i=1}^{3} I_{y_G}^{(i)} = 681\ 000\ \text{cm}^4$$

$$\begin{cases} I_{x_G y_G}^{(1)} = -480 \times (15 - 1.07) \times 8.45 = -56\ 500\ \text{cm}^4 \\[3mm] I_{x_G y_G}^{(2)} = \dfrac{6 \times 80^3}{12} \times (0.866) \times (0.5) - \\[3mm] \qquad 480 \times (8.66 - 8.45) \times (20 + 1.07) = 109\ 000\ \text{cm}^4 \\[3mm] I_{x_G y_G}^{(3)} = \dfrac{6 \times 50^3}{12} \times (0.866) \times (-0.5) + \\[3mm] \qquad 300 \times (12.5 - 1.07) \times (21.65 - 8.45) = 18\ 000\ \text{cm}^4 \end{cases}$$

$$I_{x_G y_G} = \sum_{i=1}^{3} I_{x_G y_G}^{(i)} = 70\ 500\ \text{cm}^4$$

$$\vartheta_0 = \frac{1}{2} \arctan \frac{2 \times 70\ 500}{681\ 000 - 325\ 000} = 10.8°$$

$$I_\xi = \frac{325\ 000 + 681\ 000}{2} - \frac{1}{2}\left((681\ 000 - 325\ 000)^2 + 4 \times (70\ 500)^2\right)^{\frac{1}{2}}$$
$$= 503\ 000 - 191\ 000 = 312\ 000\ \text{cm}^4$$

$$I_\eta = 503\ 000 + 191\ 000 = 694\ 000\ \text{cm}^4$$

EXAMPLE 4

$$\begin{cases} A^{(1)} = 80 \times 5 = 400\ \text{cm}^2 \\[2mm] A^{(2)} = A^{(3)} = 40 \times 5 = 200\ \text{cm}^2 \\[2mm] A^{(4)} = \dfrac{\pi}{2} \times (56.57) \times 5 = 444\ \text{cm}^2 \end{cases}$$

$$A = \sum_{i=1}^{4} A^{(i)} = 1\,244 \text{ cm}^2$$

$$\begin{cases} S_x^{(1)} = 0 \\ S_x^{(2)} = S_x^{(3)} = 200 \times 20 = 4\,000 \text{ cm}^3 \\ S_x^{(4)} = 2 \times 5 \times (56.57)^2 \times (0.707) = 22\,600 \text{ cm}^3 \end{cases}$$

$$S_x = \sum_{i=1}^{4} S_x^{(i)} = 30\,600 \text{ cm}^3$$

$$S_y = 0$$

$$\begin{cases} x_G = 0 \\ y_G = \dfrac{S_x}{A} = 24.61 \text{ cm} \end{cases}$$

$$\begin{cases} I_{x_G}^{(1)} = 400 \times (24.61)^2 = 242\,000 \text{ cm}^4 \\ I_{x_G}^{(2)} = I_{x_G}^{(3)} = \dfrac{5 \times 40^3}{12} + 200 \times (24.61 - 20)^2 = 31\,000 \text{ cm}^4 \\ I_{x_G}^{(4)} = \left(\dfrac{\pi}{4} + \dfrac{1}{2}\right) \times 5 \times (56.57)^3 + 444 \times (24.61)^2 - \\ \qquad 2 \times (24.61) \times (22\,600) = 320\,000 \text{ cm}^4 \end{cases}$$

$$I_{x_G} = \sum_{i=1}^{4} I_{x_G}^{(i)} = 624\,000 \text{ cm}^4$$

$$\begin{cases} I_{y_G}^{(1)} = \dfrac{5 \times 80^3}{12} = 213\,000 \text{ cm}^4 \\ I_{y_G}^{(2)} = I_{y_G}^{(3)} = 200 \times 40^2 = 320\,000 \text{ cm}^4 \\ I_{y_G}^{(4)} = \left(\dfrac{\pi}{4} - \dfrac{1}{2}\right) \times 5 \times (56.57)^3 = 258\,000 \text{ cm}^4 \end{cases}$$

$$I_{y_G} = \sum_{i=1}^{4} I_{y_G}^{(i)} = 1\,111\,000 \text{ cm}^4$$

$$\begin{cases} I_\xi = I_{x_G} = 624\,000 \text{ cm}^4 \\ I_\eta = I_{y_G} = 1\,111\,000 \text{ cm}^4 \end{cases}$$

EXAMPLE 5

$$
\left\{
\begin{array}{l}
I_{x_G}^{(1)} = \dfrac{4 \times 80^3}{12} = 170\,667 \text{ cm}^4 \\[2mm]
I_{x_G}^{(2)} = I_{x_G}^{(3)} = 120 \times 10^2 = 12\,000 \text{ cm}^4
\end{array}
\right.
$$

$$
I_{x_G} = \sum_{i=1}^{3} I_{x_G}^{(i)} = 194\,667 \text{ cm}^4
$$

$$
\left\{
\begin{array}{l}
I_{y_G}^{(1)} = 0 \\[2mm]
I_{y_G}^{(2)} = I_{y_G}^{(3)} = \dfrac{4 \times 30^3}{3} = 36\,000 \text{ cm}^4
\end{array}
\right.
$$

$$
I_{y_G} = \sum_{i=1}^{3} I_{y_G}^{(i)} = 72\,000 \text{ cm}^4
$$

$$
\left\{
\begin{array}{l}
I_{x_G y_G}^{(1)} = 0 \\[2mm]
I_{x_G y_G}^{(2)} = I_{x_G y_G}^{(3)} = 120 \times 15 \times 10 = 18\,000 \text{ cm}^4
\end{array}
\right.
$$

$$
I_{x_G y_G} = \sum_{i=1}^{3} I_{x_G y_G}^{(i)} = 36\,000 \text{ cm}^4
$$

$$
\begin{aligned}
I_\xi &= \frac{194\,667 + 72\,000}{2} + \frac{1}{2}\left((194\,667 - 72\,000)^2 + 4 \times (36\,000)^2\right)^{\frac{1}{2}} \\
&= 133\,000 + 71\,000 = 204\,000 \text{ cm}^4
\end{aligned}
$$

$$
I_\eta = 133\,000 - 71\,000 = 62\,000 \text{ cm}^4
$$

$$
\vartheta_0 = \arctan \frac{2 \times 36\,000}{72\,000 - 194\,667} = -15.2°
$$

3 Kinematics and statics of rigid systems

3.1 Introduction

The kinematics and statics of rigid systems are intimately connected. The movements prevented by mutual and external constraints are in fact strictly related to the reactive forces exerted by the constraints themselves. More particularly, it will be noted, and subsequently rigorously demonstrated, how the static matrix is the transpose of the kinematic one and *vice versa*, a property that will re-present itself also in other chapters devoted to elastic body systems. In every case this property will show itself to be a consequence of the Principle of Virtual Work, each of the two theorems implying the other.

After defining plane constraints from the twin viewpoints of kinematics and statics and investigating in depth the concept of duality from the algebraic standpoint, the same concept will be reproposed from the graphical point of view, with the presentation of various examples of statically indeterminate, statically determinate and hypostatic constraint. Particular attention will be paid to the condition of ill-disposed constraint (for which the system has a rigid deformed configuration) in the framework of the hypothesis of linearized constraints. In this case the solution of the equilibrium equations proves to be impossible.

3.2 Degrees of freedom of a mechanical system

The **degrees of freedom** of a mechanical system represent the number of generalized coordinates that are necessary and sufficient to describe its configuration. A system with g degrees of freedom can thus be arranged according to ∞^g different configurations.

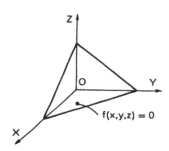

Figure 3.1

Consider, for instance, the case of a material point forced to move, in three-dimensional space, on a surface of equation $f(x,y,z) = 0$ (Figure 3.1). The degrees of freedom, which originally are three, are reduced to two by the constraint f which binds the coordinates of the point. In the same way, it may be stated that, in the case where the point is forced to follow a skew curve of equations $f_1(x,y,z) = 0, f_2(x,y,z) = 0$, the degrees of freedom are further reduced to one, the curve being a geometric variety of one dimension only.

Imagine then connecting a material point A with a fixed system of reference (e.g. the foundation) by means of a rigid rod OA (Figure 3.2) and connecting, by means of another rod AB, the point A to a second point B. If we assume that both connecting rods are not extensible, the **rigidity constraints** which these impose on the two points may be represented by the equations of two circumferences: the first, with radius l_1 centred in the origin O, and the second, with radius l_2, with the centre travelling along the first circumference

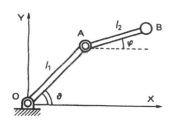

Figure 3.2

$$x_A^2 + y_A^2 = l_1^2 \tag{3.1a}$$

$$(x_A - x_B)^2 + (y_A - y_B)^2 = l_2^2 \tag{3.1b}$$

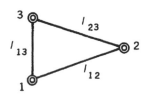

Figure 3.3

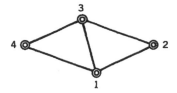

Figure 3.4

The number of residual degrees of freedom (two) will be given by the difference between the degrees of freedom of the two unconstrained material points (four) and the degrees of constraint (two).

In the case of three material points connected together by three connecting rods (rigidity constraints) as in Figure 3.3, we have originally six degrees of freedom. In the plane, in fact, the position of each point can be identified by two coordinates. On the other hand, the three rigidity constraints

$$(x_1 - x_2)^2 + (y_1 - y_2)^2 = l_{12}^2 \qquad (3.2\text{a})$$

$$(x_2 - x_3)^2 + (y_2 - y_3)^2 = l_{23}^2 \qquad (3.2\text{b})$$

$$(x_3 - x_1)^2 + (y_3 - y_1)^2 = l_{13}^2 \qquad (3.2\text{c})$$

reduce the degrees of freedom of the system to three ($g = 6 - 3 = 3$). Note that the system made up of three points and three connecting rods is rigid and its position in the plane can be defined, once three relevant items are known (e.g. the coordinates of the centroid and the angle of orientation).

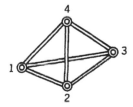

By adding a fourth point to the above system and connecting it to two of the previous points by means of a pair of connecting rods, we again obtain a rigid system with three degrees of freedom in the plane ($g = 8 - 5 = 3$). In fact, two new degrees of freedom are introduced, but at the same time these are eliminated with the two connecting rods (Figure 3.4). The same happens if a fifth point is added, and so on.

Whereas in the plane, five is the minimum number of connecting rods required to connect four material points rigidly (Figure 3.4), in space this number rises to six (Figure 3.5), so as to form a tetrahedron. In fact, the original $4 \times 3 = 12$ degrees of freedom are reduced to six, which is the number of degrees of freedom of a rigid body in three-dimensional space. The generalized coordinates of a body can therefore be considered the cartesian coordinates of the centroid plus the three Euler angles.

Figure 3.5

Often, in the pages that follow, the rigidity constraint will be, to use the term generally adopted, **linearized**. This is to say that only infinitesimal displacements about the initial configuration will be considered, and it will thus be possible to equate the circular trajectories with the rectilinear tangential ones. The simplest case is that of a material point connected to the foundation by a connecting rod. Obviously, such an elementary system has one degree of freedom and can be defined as **hypostatic**. The trajectory imposed on the point is the circumference of centre O and radius r (Figure 3.6), although, circumscribing the

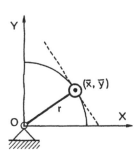

Figure 3.6

49

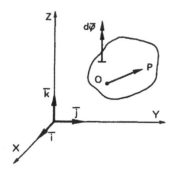

Figure 3.7

kinematic analysis about any initial position of coordinates $(\bar{x}, \bar{y})$, it is possible to assume as local trajectory an infinitesimal segment of the tangent

$$x\bar{x} + y\bar{y} = r^2 \tag{3.3}$$

The elementary displacements are hence considered to be vectors perpendicular to the radius vector. This statement derives from the kinematic theory of the rigid body in three-dimensional space.

As is well-known from rational mechanics, the relation which links the elementary displacements of two generic points P and O of a rigid body undergoing a rototranslational motion is the following (Figure 3.7):

$$\{ds_P\} = \{ds_O\} + \{d\varphi\} \wedge \{P - O\} \tag{3.4}$$

where $\{d\varphi\}$, termed **rotation vector**, is that vector which has as its axis of application, that of instantaneous rotation, as its sense, the feet–head sense of an observer who sees the body rotate counterclockwise, and as its magnitude, the value of the infinitesimal angle of rotation.

With $\bar{i}, \bar{j}, \bar{k}$ as the unit vectors of the reference axes X, Y and Z, relation (3.4) can take on the following form:

$$(u_P - u_O)\,\bar{i} + (v_P - v_O)\,\bar{j} + (w_P - w_O)\,\bar{k} \tag{3.5}$$

$$= \det \begin{bmatrix} \bar{i} & \bar{j} & \bar{k} \\ \varphi_x & \varphi_y & \varphi_z \\ (x_P - x_O) & (y_P - y_O) & (z_P - z_O) \end{bmatrix}$$

where the determinant of the formal matrix on the right-hand side provides the components of the vector product which appears in relation (3.4), while u, v, w indicate the components along the axes X, Y, Z of the elementary displacements, and φ_x, φ_y, φ_z indicate the components of the rotation vector.

Relation (3.5) can alternatively be presented in the form of a product of an antisymmetric matrix, called a **rotation matrix**, for the position vector of the point P with respect to the point O,

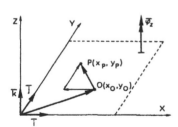

Figure 3.8

$$\{ds_P\} - \{ds_O\} = \begin{bmatrix} 0 & -\varphi_z & \varphi_y \\ \varphi_z & 0 & -\varphi_x \\ -\varphi_y & \varphi_x & 0 \end{bmatrix} \begin{bmatrix} x_P - x_O \\ y_P - y_O \\ z_P - z_O \end{bmatrix} \tag{3.6}$$

In the particular case of an elementary rotation of a two-dimensional rigid body in its XY plane (Figure 3.8), equation (3.5) is particularized as follows:

$$(u_P - u_O)\,\bar{i} + (v_P - v_O)\,\bar{j} = \begin{bmatrix} \bar{i} & \bar{j} & \bar{k} \\ 0 & 0 & \varphi_z \\ (x_P - x_O) & (y_P - y_O) & 0 \end{bmatrix} \tag{3.7}$$

50

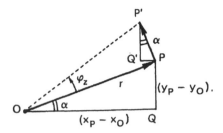

Figure 3.9

Evaluating the determinant of the symbolic matrix on the right-hand side of the equation, we obtain

$$u_P - u_O = -(y_P - y_O)\,\varphi_z \qquad (3.8a)$$

$$v_P - v_O = (x_P - x_O)\,\varphi_z \qquad (3.8b)$$

Equations (3.8) are linear relations which describe algebraically the concept of linearization of the constraint, already introduced previously on a more intuitive basis. A geometrical interpretation (Figure 3.9) of equation (3.8) may then be given, considering the elementary rotation φ_z of P about O. The point P will move to P', at a distance equal to $r\varphi_z$ (except for infinitesimals of a higher order). The horizontal relative displacement, represented in Figure 3.9 by segment PQ', thus equals

$$u_P - u_O = PQ' = -r\varphi_z \sin\alpha \qquad (3.9)$$

Since triangles OQP and $P'Q'P$ are similar, it follows that $r\sin\alpha = (y_P - y_O)$, so that from equation (3.9) we obtain again equation (3.8a). In the same way it is possible to verify also the meaning of the linear relation (3.8b).

3.3 Kinematic definition of plane constraints

The constraints that we shall hereafter assume to be connecting the plane rigid body to the fixed reference system are referred to as **external constraints**. These can be classified on the basis of the elementary movements of the constrained point P which can be prevented, these movements consisting of the two translations u_P and v_P and of the elementary rotation φ_P. The subscript P has been applied to the latter quantity, even though this is a characteristic of the act of rigid motion and thus of each point of the body.

There is thus created a hierarchy of constraints, from those which restrain the body more weakly (single constraints) to those which more effectively inhibit its movements (triple constraints or fixed joints).

The simplest kind of constraint, frequently used in technical applications, is the **roller support** (Figure 3.10(a)) or the **connecting rod** (Figure 3.10(b)). This constraint imposes on the point P a movement along the straight line p. It should be noted that in the case of the connecting rod of Figure 3.10(b), the constraint is

51

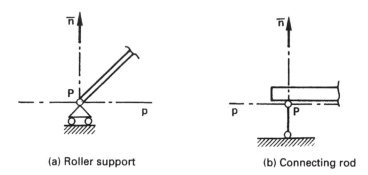

(a) Roller support   (b) Connecting rod

Figure 3.10

(a) Hinge

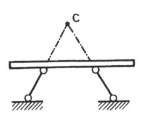

(b) Ideal hinge

Figure 3.11

linearized, as was clarified in the previous section. In mathematical terms, the symbols of Figure 3.10 impose the following condition:

$$\{ds_P\}^T \{n\} = 0 \qquad (3.10)$$

i.e. that the elementary displacement of the point P cannot present components different from zero on the straight line n, perpendicular to the straight line p. The elementary rotation φ_z can, on the other hand, be different from zero. Since equation (3.10) is a scalar relation (it represents, in fact, the scalar product of the displacement and the unit vector of the straight line n), it can be stated that the roller support or the connecting rod are single constraints. Single constraints require that any centre of instantaneous rotation must lie on a straight line. In the case of the roller support or the connecting rod, the centre of instantaneous rotation must lie on the straight line n, since this is perpendicular to the instantaneous trajectory p.

Another constraint which has a wide application is the **hinge** (Figure 3.11(a)). This imposes on the constrained point P to remain fixed in the plane, while the elementary rotation φ_z can be about the same point P, which thus coincides with the centre of instantaneous rotation. In mathematical terms,

$$\{ds_P\} = \{0\} \qquad (3.11)$$

The foregoing condition is of a vector nature and thus the hinge can be classified as a double constraint.

We may then consider that, by suitably combining two single constraints, the result for the body can be a double constraint. The most typical case is that of two non-parallel connecting rods (Figure 3.11(b)), the axes of which come together in the centre of instantaneous rotation C. Each connecting rod in fact conditions the centre to belong to its axis, and the two conditions are compatible, except when the connecting rods are parallel, in which case the centre would be at infinity. The centre C is said to constitute an **ideal hinge.**

When the two connecting rods are parallel, the centre of instantaneous rotation coincides with the point at infinity of the axes of the connecting rods (Figure 3.12(a)). This means that the rigid motion becomes one of pure translation in the p direction perpendicular to the axis n. In mathematical terms, we have the following two scalar conditions:

$$\{ds_P\}^T\{n\} = 0 \qquad (3.12a)$$

$$\varphi_z = 0 \qquad (3.12b)$$

The **double connecting rod** is thus a double constraint and can be represented, in an altogether equivalent way, by a **sliding joint** (Figure 3.12(b)).

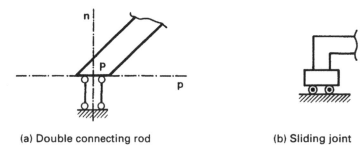

(a) Double connecting rod   (b) Sliding joint

Figure 3.12

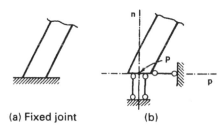

(a) Fixed joint   (b)

Figure 3.13

The only triple constraint of a local type is the **fixed joint** or **built-in support** (Figure 3.13(a)), which, by definition, prevents all three movements (hence, no centre of rotation exists):

$$\{ds_P\} = \{0\} \tag{3.13a}$$

$$\varphi_z = 0 \tag{3.13b}$$

Obviously, the triple constraint may be obtained by suitably combining a double constraint with a single one (Figure (3.13(b)).

In the hierarchy of plane constraints outlined above just one case is missing, one which, as a rule, is rarely applied in building practice, but which may be defined mathematically and may prove useful in applications of the Principle of Virtual Work. This constraint is that which allows motion of translation in all directions, while it inhibits rotation:

$$\varphi_z = 0 \tag{3.14}$$

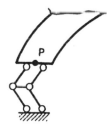

Double articulated
parallelogram

Figure 3.14

This can be represented by a **double articulated parallelogram** (Figure 3.14) and is a single constraint which obliges the centre of instantaneous rotation to remain on the straight line at infinity.

So far only point constraints have been introduced, i.e. constraints concentrated in one point or, at the most, acting within an infinitesimal area of the body. The only exception is that of the ideal hinge (Figure 3.11(b)) in which two connecting rods are applied, at a finite distance from one another. On the other hand, in building practice, the constraints are arranged in different points of the structural element, in such a way as to prevent its movement. When the constraints are insufficient to fix the position of the rigid body in the plane, the constraint condition is said to be **hypostatic**. In the case of Figure 3.15(a), for instance, a single hinge allows rotation of the plate

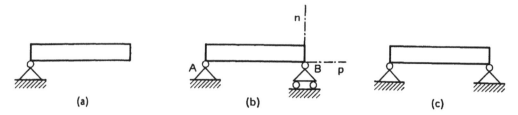

Figure 3.15

constrained by it about its axis. In the case of Figure 3.15(b), a hinge and a roller support prove to constitute a strictly sufficient condition to hold the body in position; the body is then said to be constrained **isostatically**. While, in fact, the hinge A forces any rotation to be centred in A, the roller support B forces the centre to lie along the normal n. These two conditions are evidently mutually incompatible, as the point A does not lie along the line n. Likewise, it may be observed how the hinge A imposes the trajectory n on the point B, while the roller support B imposes the trajectory p on the same point B. These two trajectories are again incompatible, in that they have in common only the point B. In the case of Figure 3.15(c), finally, the two hinges constitute a redundant condition of constraint. The body is then said to be **hyper-statically constrained**.

On the other hand, it should at once be pointed out that, if the constraints are not suitably disposed, they can lose their effectiveness. Thus it is that bodies which are apparently isostatic or even hyperstatic, may then in fact prove to be hypostatic. The hinge and the roller support of Figure 3.16(a), for example, are **ill-disposed** constraints, because the centre of rotation allowed by the roller support B can come to find itself on the straight line n, which passes through the hinge A. Hence, it is possible for the centre of rotation to be in A.

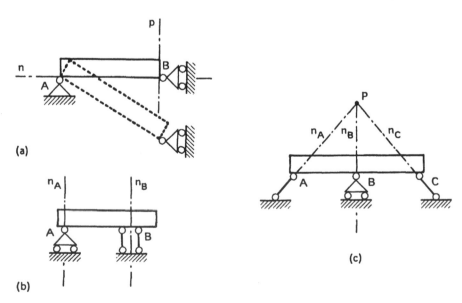

Figure 3.16

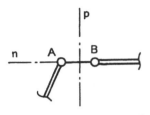

Internal connecting rod

Figure 3.17

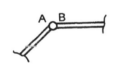

Internal hinge

Figure 3.18

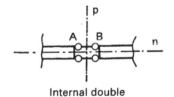

Internal double
connecting rod

Figure 3.19

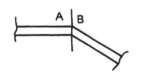

Internal fixed joint

Figure 3.20

In other words, the trajectories imposed on the point B by both constraints coincide in the straight line p. Likewise, the roller support and the double connecting rod of Figure 3.16(b) are ill-disposed, because both of them permit horizontal translations of the body. The centre of rotation will thus be at infinity, as the meeting point of the normals n_A and n_B. Finally, the case illustrated in Figure 3.16(c) is yet another example of ill-disposition of the constraints in that the three normals, n_A, n_B and n_C, all meet at the point P, i.e. there is a centre of instantaneous rotation at the point where all three lines meet, since this centre must belong to each one of them.

Just as external constraints impose particular conditions on absolute elementary displacements of points held to the foundation, so **internal constraints** impose similar conditions on the relative elementary displacements of points belonging to different rigid bodies. Consider the case of the **internal connecting rod** (Figure 3.17). This prevents discontinuity of displacement in the direction of its axis n:

$$\{ds_A - ds_B\}^T\{n\} = 0 \qquad (3.15a)$$

while it allows discontinuity of displacement in the perpendicular direction p:

$$\{ds_A - ds_B\}^T\{p\} \neq 0 \qquad (3.15b)$$

just as it allows discontinuity of rotation, i.e. relative rotation:

$$\varphi_A - \varphi_B \neq 0 \qquad (3.15c)$$

Any centre of relative rotation must lie on the line n. Internal constraints may also be considered, in a complementary way, as **disconnections** of degree $(3 - g)$, where g is the degree of constraint; i.e. it is possible to start from a single rigid body that has two parts which are firmly joined and to reduce it, via elementary disconnections, to the case under examination. In this context, the connecting rod is a single constraint and, at the same time, a double disconnection.

The **internal hinge** eliminates any relative displacement of the two points which it connects, while it allows relative rotation (Figure 3.18):

$$\{ds_A - ds_B\} = \{0\} \qquad (3.16a)$$

$$\varphi_A - \varphi_B \neq 0 \qquad (3.16b)$$

Any centre of relative rotation will coincide with the hinge, which is in fact a double constraint and, at the same time, a single disconnection.

The **double internal rod** (Figure 3.19) eliminates relative displacement in the direction n as well as relative rotation:

$$\{ds_A - ds_B\}^T\{n\} = 0 \qquad (3.17a)$$

$$\varphi_A - \varphi_B = 0 \qquad (3.17b)$$

while it allows the relative displacement in the perpendicular direction p:

$$\{ds_A - ds_B\}^T\{p\} \neq 0 \qquad (3.17c)$$

Any centre of relative rotation will be the point at infinity of the straight line n, and thus the only relative motion allowed will be the translational one in the p direction. The double rod is a double constraint or can be considered as a single disconnection. The **internal fixed joint** (Figure 3.20) firmly joins one

Internal double articulated
parallelogram

Figure 3.21

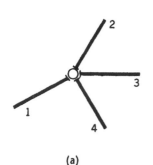

(a)

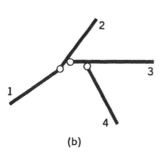

(b)

Figure 3.22

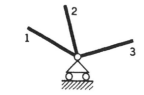

Figure 3.23

portion of the body with the other and is a triple constraint. In other words, there is no internal disconnection.

To bring our discussion of internal constraints to a close, let us consider the case of the **internal double articulated parallelogram** (Figure 3.21), which allows any relative translation but not relative rotation. It is thus a single constraint or a double disconnection. The relative centre lies on the straight line at infinity and can be any point on that line.

The internal hinge often connects more than two elements (Figure 3.22 (a)), in such a way that the relative centres all come to coincide with it. In these cases, it is logical to consider $(n-1)$ mutual connections, if n is the number of the connected elements (Figure 3.22 (b)). In the case, therefore, of four bars mutually connected as in Figure 3.22 (a), the residual degrees of freedom are $g = (4 \times 3) - (3 \times 2) = 6$. We have, in fact, 12 original degrees of freedom and six degrees of constraint. More synthetically, it is possible to arrive at the same result again by considering as generalized coordinates of the system the cartesian coordinates of the hinge plus the four angles of orientation of the bars. In the same way, the roller support of Figure 3.23, with three hinged bars, forms a mechanical system having four degrees of freedom. A less immediate calculation again gives $g = (3 \times 3) - (2 \times 2) - 1 = 4$, as there are nine original degrees of freedom, and five degrees suppressed by two double (internal) constraints and one single (external) constraint.

In the case where the bars are built into one another so as to form one or more closed configurations, the system is said to be internally hyperstatic and there will be three degrees of hyperstaticity for each closed configuration. For example, for the rectangle of Figure 3.24(a), formed by four bars built into one another, the calculation $g = (4 \times 3) - (4 \times 3) = 0$ does not hold good, as it would suggest, quite falsely, that the system is isostatic. In actual fact, the doubly-connected body presents three external degrees of freedom (it is not externally constrained), and three degrees of internal indeterminacy, since the various parts would remain firmly joined even if a triple disconnection (i.e. a cut) were made in any point of the axis (Figure 3.24(b)). The triple disconnection could, on the other hand, be distributed in two or three points (Figures 3.24(c), (d)). It is necessary, however, to take care in arranging such disconnections. The hinge on the axis of the connecting rod or the three aligned hinges, for example, would be ill-disposed, with the consequent development of an internal mechanism (Figures 3.24(e), (f)).

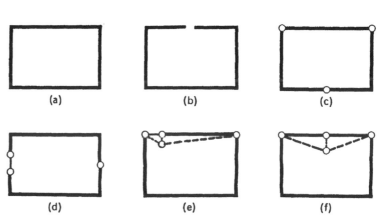

(a) (b) (c)

(d) (e) (f)

Figure 3.24

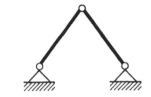

Figure 3.25

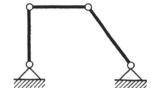

Figure 3.26

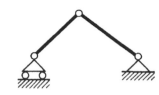

Figure 3.27

Two bars connected both to the foundation and to one another by three hinges constitute a fundamental isostatic scheme called a **three-hinged arch** (Figure 3.25). Naturally, for the same reasons already seen previously, the three hinges must not be collinear; otherwise, a mechanism is created. By inserting a fourth hinge (single disconnection), a mechanism with one degree of freedom is obtained (Figure 3.26); this is known as an **articulated parallelogram** ($g = 9 - 8 = 1$). Alternatively, yet another simple mechanism is obtained by disconnecting, with respect to the horizontal translation, one of the two hinges connected to the foundation ($g = 6 - 5 = 1$). In this case the **crank mechanism** of Figure 3.27 is obtained.

3.4 Algebraic study of kinematics of rigid systems

We shall now approach, from the algebraic point of view, the problem of the kinematics of rigid systems, thus giving a rigorous interpretation to the degenerate case of the ill-disposition of constraints. Progressively more complex examples will be introduced, with an increasing number of bodies.

As a first case consider the L-shaped beam of Figure 3.28(a). This is constrained by a roller support hinged in A and by a hinge in B. The unknowns of the problem will be the displacements of a representative point of the rigid body, chosen arbitrarily, referred to as the **pole** or **centre of reduction**. This point is chosen to fall at the convergence O of the two bars. A cartesian system is chosen with centre in O and axes parallel to the two bars. The unknowns are then u_O, v_O, φ_O, i.e. the two elementary translations of the point O with

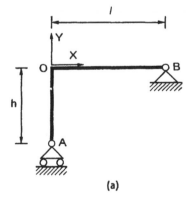 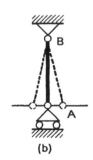

(a)   (b)

Figure 3.28

reference to the axes X and Y and the elementary rotation of the entire rigid body. On the other hand, there are also three constraint equations:

$$v_A = 0 \tag{3.18a}$$

$$u_B = 0 \tag{3.18b}$$

$$v_B = 0 \tag{3.18c}$$

The task, therefore, will be to render explicit equations (3.18) as a function of the elementary displacements of the pole. Applying equations (3.8), we have

$$v_A = v_O + (x_A - x_O)\varphi_O = v_O \tag{3.19a}$$

$$u_B = u_O - (y_B - y_O)\varphi_O = u_O \tag{3.19b}$$

$$v_B = v_O + (x_B - x_O)\varphi_O = v_O + l\varphi_O \tag{3.19c}$$

and thus the constraint equations become

$$v_O = 0 \tag{3.20a}$$

$$u_O = 0 \tag{3.20b}$$

$$v_O + l\varphi_O = 0 \tag{3.20c}$$

The system of linear algebraic equations (3.20) may be resolved very quickly by substitution, thus affording the obvious or trivial solution. The displacements of the pole O are all zero, which means that the constraints are well-disposed. Since, however, it is our purpose to reason in terms that are more general and that can be more readily extrapolated to cases of greater complexity, it is necessary to introduce the matrix notation

$$\begin{bmatrix} 0 & 1 & 0 \\ 1 & 0 & 0 \\ 0 & 1 & l \end{bmatrix} \begin{bmatrix} u_O \\ v_O \\ \varphi_O \end{bmatrix} = \begin{bmatrix} 0 \\ 0 \\ 0 \end{bmatrix} \tag{3.21}$$

It is well-known from linear algebra that a necessary and sufficient condition for a homogeneous system of linear algebraic equations to admit of the obvious solution is that the determinant D of the matrix of the coefficients should be different from zero. In the case of the system of equations (3.21), we have $D = -l$. When $l > 0$, as appears in Figure 3.28(a), the solution is thus the trivial one. On the other hand, carrying out a parametric study on the variable l, we find that, for $l = 0$, the kinematic solutions become ∞^1 and the system becomes a mechanism, since an ill-disposition of constraints is produced. The normal n to the plane in which the roller support moves, in fact, in this case contains the hinge B (Figure 3.28(b)), so that a centre of instantaneous rotation arises in the same point B.

As a second example, consider again the L-shaped beam met with earlier, this time, however, restrained in A with a roller support having its plane of movement inclined at the angle α with respect to the vertical direction (Figure 3.29(a)). The first of equations (3.18) will then be substituted by the following:

$$\{ds_A\}^T \{n\} = 0 \tag{3.22a}$$

or more explicitly by

$$u_A \cos\alpha + v_A \sin\alpha = 0 \tag{3.22b}$$

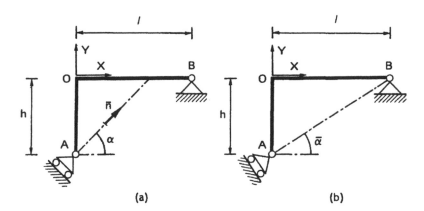

Figure 3.29

Writing the left-hand side expression as a function of the displacements of the pole O, we have

$$(u_O + h\varphi_O)\cos\alpha + v_O \sin\alpha = 0 \qquad (3.22c)$$

In the matrix of the coefficients which appears in equation (3.21), only the first line will vary with respect to the previous case:

$$\begin{bmatrix} \cos\alpha & \sin\alpha & h\cos\alpha \\ 1 & 0 & 0 \\ 0 & 1 & l \end{bmatrix} \begin{bmatrix} u_O \\ v_O \\ \varphi_O \end{bmatrix} = \begin{bmatrix} 0 \\ 0 \\ 0 \end{bmatrix} \qquad (3.23)$$

The determinant of the matrix of the coefficients has the value: $D = -(l\sin\alpha - h\cos\alpha)$, and is generally different from zero. Assuming as parameter the angle α, we find that D vanishes for $\alpha = \bar{\alpha} = \arctan(h/l)$. In this case, in fact, the plane in which the roller support moves is set perpendicular to the line joining A and B (Figure 3.29(b)).

We now examine the case of a portal frame made up of two L-shaped beams, connected together by a hinge and to the foundation by a hinge and by a horizontal double rod, respectively (Figure 3.30(a)). The points A and C of external constraint are assumed as poles. The six kinematic unknowns will

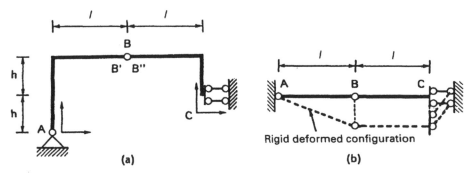

Figure 3.30

thus consist of the elementary displacements of the two poles: u_A, v_A, φ_A, u_C, v_C, φ_C. On the other hand, the six constraint equations will impose

$$u_A = 0 \qquad (3.24a)$$

$$v_A = 0 \qquad (3.24b)$$

$$u_{B'} - u_{B''} = 0 \qquad (3.24c)$$

$$v_{B'} - v_{B''} = 0 \qquad (3.24d)$$

$$u_C = 0 \qquad (3.24e)$$

$$\varphi_C = 0 \qquad (3.24f)$$

While in the first and last pairs of equations the unknowns of the problem appear directly, the intermediate ones, which exclude relative displacements in B, must be expressed in terms of the generalized displacements of the poles A and C. Applying equations (3.8), the displacements of the central ends B' and B'' of the two beams are given by

$$u_{B'} = u_A - 2h\varphi_A \qquad (3.25a)$$

$$v_{B'} = v_A + l\varphi_A \qquad (3.25b)$$

$$u_{B''} = u_C - h\varphi_C \qquad (3.25c)$$

$$v_{B''} = v_C - l\varphi_C \qquad (3.25d)$$

Substituting equations (3.25) into equations (3.24(c),(d)) we obtain

$$u_A = 0 \qquad (3.26a)$$

$$v_A = 0 \qquad (3.26b)$$

$$u_A - 2h\varphi_A - u_C + h\varphi_C = 0 \qquad (3.26c)$$

$$v_A + l\varphi_A - v_C + l\varphi_C = 0 \qquad (3.26d)$$

$$u_C = 0 \qquad (3.26e)$$

$$\varphi_C = 0 \qquad (3.26f)$$

and, in matrix form

$$
\begin{bmatrix}
1 & 0 & 0 & 0 & 0 & 0 \\
0 & 1 & 0 & 0 & 0 & 0 \\
1 & 0 & -2h & -1 & 0 & h \\
0 & 1 & l & 0 & -1 & l \\
0 & 0 & 0 & 1 & 0 & 0 \\
0 & 0 & 0 & 0 & 0 & 1
\end{bmatrix}
\begin{bmatrix}
u_A \\
v_A \\
\varphi_A \\
u_C \\
v_C \\
\varphi_C
\end{bmatrix}
=
\begin{bmatrix}
0 \\
0 \\
0 \\
0 \\
0 \\
0
\end{bmatrix}
\qquad (3.27)
$$

The determinant of the matrix of the coefficients has the value $D = -2h$. For $h = 0$, this vanishes and we end up with an ill-disposition of the constraints, as emerges from the rigid deformed configuration of Figure 3.30(b). In this

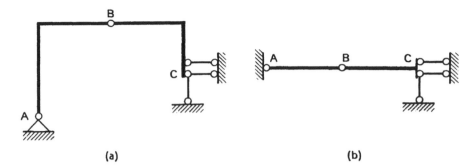

(a)   (b)

Figure 3.31

case, it may be noted how the two absolute centres and the relative one are aligned.

If the portal frame just considered is further constrained in C, by adding a vertical connecting rod, the constraint equations become seven, whilst the kinematic unknowns remain the six previously introduced. In the case of $h > 0$ (Figure 3.31(a)), the solution will still be the obvious one. On the other hand, even when $h = 0$ (Figure 3.31(b)), the solution is the obvious one, because, from the matrix of the coefficients (7×6), it will be possible to extract a non-zero minor of order 6, since the added line is not linearly dependent on the others.

If, instead, one of the two rods in C of the portal frame of Figure 3.30(a) is suppressed, the constraint equations are reduced to five. In the case of $h > 0$ (Figure 3.32(a)), it will be possible to extract a nonzero minor of order 5, and thus the solutions will be ∞^1 and the system will have one degree of freedom. On the other hand, when $h = 0$ (Figure 3.32(b)), it will be possible to extract a nonzero minor of order 4, and thus the solutions will be ∞^2, which means that the mechanical system is hypostatic to the second degree. In fact, two angles of rotation will be necessary, for example, the absolute angle ϑ and the relative angle φ (Figure 3.32(b)), to describe its deformed configuration.

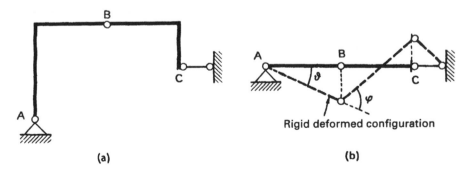

Rigid deformed configuration

(a)   (b)

Figure 3.32

61

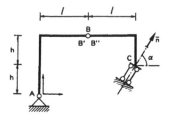

Figure 3.33

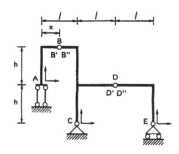

Figure 3.34

Now imagine parameterizing, in the portal frame of Figure 3.33, the angle of orientation α of the double connecting rod in C. Thus only equation (3.26e) is altered:

$$u_C \cos \alpha + v_C \sin \alpha = 0 \qquad (3.28)$$

and consequently in the matrix of the coefficients (3.27) only the fifth row varies:

$$
\begin{bmatrix}
1 & 0 & 0 & 0 & 0 & 0 \\
0 & 1 & 0 & 0 & 0 & 0 \\
1 & 0 & -2h & -1 & 0 & h \\
0 & 1 & l & 0 & -1 & l \\
0 & 0 & 0 & \cos \alpha & \sin \alpha & 0 \\
0 & 0 & 0 & 0 & 0 & 1
\end{bmatrix}
\begin{bmatrix}
u_A \\
v_A \\
\varphi_A \\
u_C \\
v_C \\
\varphi_C
\end{bmatrix}
=
\begin{bmatrix}
0 \\
0 \\
0 \\
0 \\
0 \\
0
\end{bmatrix}
\qquad (3.29)
$$

The determinant of the matrix of the coefficients has the value $D = -2h \cos \alpha + l \sin \alpha$, and vanishes for $\alpha = \bar{\alpha} = \arctan(2h/l)$, i.e. when the double connecting rod is parallel to the line joining A and B. In this case the constraints are ill-disposed, since, for example, the hinge A on the one hand, and the double rod C on the other, allow the point B the same trajectory about the original configuration, viz. the normal for B to the line joining A and B. Also in this example, when a condition of ill-disposition of the constraints and consequently of freedom is reached, the absolute centres and the relative centre are aligned.

Finally, consider a structure made up of three rigid parts, connected together with hinges and attached to the foundation by a vertical double rod, a hinge and a roller support (Figure 3.34). There are, in all, nine original degrees of freedom and nine degrees of constraint (four double constraints and one single constraint). We shall study the system, assuming as variable parameter the distance of x of the internal hinge B from the left upper vertex of the structure. It is convenient to choose as pole of each rigid part the point constrained to the foundation, so that the nine kinematic unknowns will be

$$u_A, \ v_A, \ \varphi_A, \ u_C, \ v_C, \ \varphi_C, \ u_E, \ v_E, \ \varphi_E$$

The constraint conditions are the following:

$$v_A = 0 \qquad (3.30\text{a})$$
$$\varphi_A = 0 \qquad (3.30\text{b})$$
$$u_{B'} - u_{B''} = 0 \qquad (3.30\text{c})$$
$$v_{B'} - v_{B''} = 0 \qquad (3.30\text{d})$$
$$u_C = 0 \qquad (3.30\text{e})$$
$$v_C = 0 \qquad (3.30\text{f})$$
$$u_{D'} - u_{D''} = 0 \qquad (3.30\text{g})$$
$$v_{D'} - v_{D''} = 0 \qquad (3.30\text{h})$$
$$v_E = 0 \qquad (3.30\text{i})$$

In the third, fourth, seventh and eighth, the kinematic parameters of the poles do not appear directly, and thus it will be necessary to have recourse to equations (3.8)

$$u_{B'} = u_A - h\varphi_A \tag{3.31a}$$
$$v_{B'} = v_A + x\varphi_A \tag{3.31b}$$
$$u_{B''} = u_C - 2h\varphi_C \tag{3.31c}$$
$$v_{B''} = v_C + (x-l)\varphi_C \tag{3.31d}$$
$$u_{D'} = u_C - h\varphi_C \tag{3.31e}$$
$$v_{D'} = v_C + l\varphi_C \tag{3.31f}$$
$$u_{D''} = u_E - h\varphi_E \tag{3.31g}$$
$$v_{D''} = v_E - l\varphi_E \tag{3.31h}$$

The equations of the problem appear, then, as follows:

$$v_A = 0 \tag{3.32a}$$
$$\varphi_A = 0 \tag{3.32b}$$
$$u_A - h\varphi_A - u_C + 2h\varphi_C = 0 \tag{3.32c}$$
$$v_A + x\varphi_A - v_C + (l-x)\varphi_C = 0 \tag{3.32d}$$
$$u_C = 0 \tag{3.32e}$$
$$v_C = 0 \tag{3.32f}$$
$$u_C - h\varphi_C - u_E + h\varphi_E = 0 \tag{3.32g}$$
$$v_C + l\varphi_C - v_E + l\varphi_E = 0 \tag{3.32h}$$
$$v_E = 0 \tag{3.32i}$$

and in a matrix form, we have

$$
\begin{bmatrix}
0 & 1 & 0 & 0 & 0 & 0 & 0 & 0 & 0 \\
0 & 0 & 1 & 0 & 0 & 0 & 0 & 0 & 0 \\
1 & 0 & -h & -1 & 0 & 2h & 0 & 0 & 0 \\
0 & 1 & x & 0 & -1 & l-x & 0 & 0 & 0 \\
0 & 0 & 0 & 1 & 0 & 0 & 0 & 0 & 0 \\
0 & 0 & 0 & 0 & 1 & 0 & 0 & 0 & 0 \\
0 & 0 & 0 & 1 & 0 & -h & -1 & 0 & h \\
0 & 0 & 0 & 0 & 1 & l & 0 & -1 & l \\
0 & 0 & 0 & 0 & 0 & 0 & 0 & 1 & 0
\end{bmatrix}
\begin{bmatrix}
u_A \\ v_A \\ \varphi_A \\ u_C \\ v_C \\ \varphi_C \\ u_E \\ v_E \\ \varphi_E
\end{bmatrix}
=
\begin{bmatrix}
0 \\ 0 \\ 0 \\ 0 \\ 0 \\ 0 \\ 0 \\ 0 \\ 0
\end{bmatrix}
\tag{3.33}
$$

The determinant of the matrix of the coefficients has the value $D = (l - x)\, l$, and vanishes for $x = l$, when the hinge B is at the right end of the upper horizontal beam. In the next section we shall discuss the type of mechanism that is created in this last particular case (Figure 3.39).

3.5 Graphical study of kinematics of systems having one degree of freedom

The kinematic study of systems having one degree of freedom, herein also referred to as **kinematic chains**, will be proposed in this section from the graphical point of view. Also certain cases of the foregoing section will be taken up again, with the aim of defining the displacement vector field and the deformed configuration of the rigid bodies (here all assumed as being one-dimensional) which make up the chain.

The graphical study of kinematic chains is founded on two theorems, which here for reasons of brevity will not be demonstrated, but only stated.

First Theorem of Kinematic Chains (applicable when the chain is made up of at least two rigid bodies). A necessary and sufficient condition for the mechanical system to be hypostatic is that, for each pair of bodies i and j, the absolute centres of rotation C_i and C_j and the relative centre C_{ij} should be aligned.

Second Theorem of Kinematic Chains (applicable when the chain is made up of at least three rigid bodies). A necessary and sufficient condition for the mechanical system to be hypostatic is that, for each group of three bodies i, j and k, the three relative centres C_{ij}, C_{jk} and C_{ki} should be aligned.

Let us consider the crank mechanism of Figure 3.35(a) in its initial configuration, and let us proceed to study its elementary movements about that configuration. The absolute centre of rotation C_1 of the connecting rod I coincides with the point hinged to the foundation, just as the relative centre C_{12} will coincide with the internal hinge. The absolute centre of the bar II is not known *a priori*. Since one end of this bar is hinged to a roller support, the straight line n on which that centre must lie is, however, known. By the First Theorem of Kinematic Chains, the centre C_2 must then be collinear to centres C_1 and C_{12}. We are thus provided with the information indicating that centre C_2 belongs to two different straight lines:

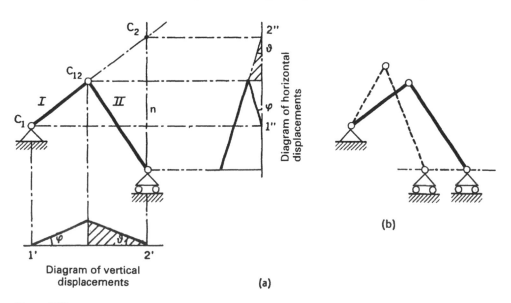

Diagram of horizontal displacements

Diagram of vertical displacements

(a)

(b)

Figure 3.35

$$C_2 \in n, \; C_2 \in (C_1, C_{12})$$

so that it must consequently lie at the intersection of these two lines (Figure 3.35(a)).

Once the centres of instantaneous rotation have been identified, these and the other notable points of the system will be projected onto two orthogonal lines (Figure 3.35(a)). It will thus be possible to draw the diagrams of the horizontal and vertical displacements, respectively, on the vertical and horizontal fundamental lines. To be able to do this, we shall, then, have to take into consideration the linear equations (3.8), the points where the displacement functions vanish (coinciding with the projections of the absolute centres) and the conditions of continuity imposed by the internal constraints on the displacements and the rotations.

The actual procedure will be to draw a segment of a line, inclined at an arbitrary angle φ with respect to the horizontal fundamental line (Figure 3.35(a)) which will represent the vertical displacements of the corresponding points projected from the connecting rod I. As regards the bar II, the projection of the absolute centre is known, as is also the vertical displacement of the end hinged to the connecting rod I. The internal hinge, in fact, prevents relative displacements in that point, i.e. it imposes the continuity of the vertical (and horizontal) displacement function. It may be stated, therefore, that the vertical displacements of the points of the bar II will be represented by the segment of straight line which joins the projection of the centre C_2 to the right end of the previous linear diagram. This segment is thus rotated clockwise by an angle ϑ, while the rod I is rotated counterclockwise by an angle φ.

The horizontal displacements can then be read in reference to the vertical fundamental line (Figure 3.35(a)). Draw a segment rotated counterclockwise by an angle φ about the projection of the centre C_1. Next, consider the line that joins the projection of centre C_2 to the point representative of the horizontal displacement of the internal hinge. The horizontal projection of bar II on this line represents the horizontal displacements of the points of the bar itself. Note that the rotations on the two displacement diagrams must be the same for each rigid body. It can be demonstrated in fact that, by exploiting the similitude of the hatched triangles in Figure 3.35(a), that also on the diagram of the horizontal displacements the bar II rotates clockwise by the angle ϑ.

To conclude, it may be of interest for a verification of the results to draw the deformed configuration of the system on the basis of the diagrams obtained. The procedure is to compose horizontal and vertical displacements of the notable points of the chain and to reconnect these points, transformed by the movement, with straight line segments. In the case of the crank mechanism of Figure 3.35(a), the internal hinge is displaced leftwards and upwards (by amounts which can be deduced from the diagrams described above), while the roller support moves towards the left, dragged by the rotation of the connecting rod I. Of course, when we come to draw the deformed configuration, we must bear in mind that the linearization of the constraints, which is justified in a context of infinitesimal displacements, instead deforms rigid bodies when these displacements are amplified to meet the requirements of graphical clarity. In the crank mechanism, for example, the connecting rod I appears dilated owing to the movement (Figure 3.35(b)).

As a second example, consider the hypostatic arch of Figure 3.36(a). This time the roller support moves in a vertical plane. The absolute centre C_2 is

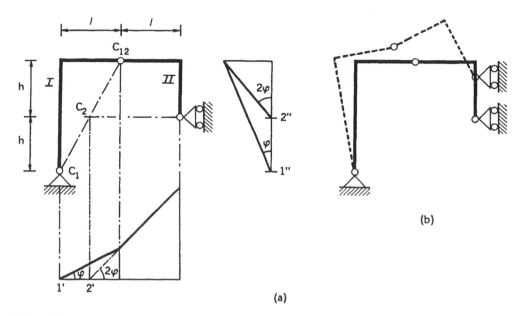

Figure 3.36

identifiable as the intersection of the normal to the plane in which the roller support moves with the line joining the centres C_1 and C_{12}. With the absolute and relative centres projected on the two fundamental lines, and taking into account the conditions of continuity imposed on the displacements by the internal constraint, we can proceed to draw the diagrams of the horizontal and vertical displacements. It may be noted (Figure 3.36(a)) how the absolute centres of rotation undergo null displacements (this is true by definition), just as the extreme points of the two rigid bodies connected by the internal hinge are displaced by the same amounts, both vertically and horizontally. Both bodies rotate counterclockwise, the first by an arbitrary (and infinitesimal) angle φ, the second by an angle 2φ, twice as much on account of the particular geometrical proportions of the system (the upright on the left is twice as long as the one on the right). These angles emerge from both diagrams of horizontal and vertical displacements. While, therefore, the body I rotates about the external hinge, the body II is dragged in such a way as to rotate twice as much, pulling the roller support upwards with a displacement equal to $3\varphi l$ (Figure 3.36(b)).

We are now confronted with the case of a hypostatic arch (Figure 3.37) consisting of two rigid L-shaped bodies, connected together by a horizontal double rod and to the foundation with a hinge and a roller support, respectively. Also in this case, the centre C_2 is obtained as the intersection of the normal to the plane of movement of the roller support with the line joining C_1 and $C_{12}(\infty)$. Since $C_{12}(\infty)$ is the point at infinity of the horizontal lines, the line joining C_1 and C_{12} is the horizontal line passing through C_1. With the absolute centres projected on the two fundamentals, the two rigid bodies must rotate by the same angle (since the double rod does not allow relative rotations), while the horizontal displacement cannot present discontinuity in correspondence

66

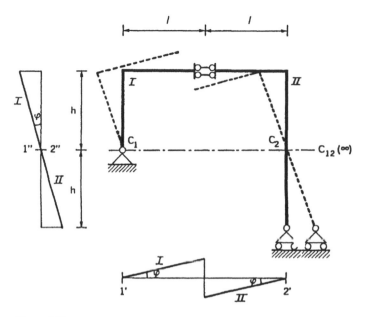

Figure 3.37

with the double rod. There is, instead, a discontinuity in the diagram of the vertical displacement equal to $2\varphi l$, so that the end to the left of the double rod, belonging to body I, will be displaced upwards (as well as leftwards), while the end to the right, belonging to body II, will be displaced downwards (as well as leftwards). The roller support will, meanwhile, be displaced an amount φh towards the right.

So far we have considered systems that are hypostatic for manifest insufficiency of constraints. We shall now consider two cases (already met with in the previous section) which concern structures having ill-disposed constraints. The difference, therefore, in this case lies in the fact that the initial configuration is not arbitrary, as instead occurs in the case of crank mechanisms and, in general, for all mechanisms having one degree of freedom.

Consider the arch with ill-disposed constraints of Figure 3.38, already studied algebraically earlier. The three centres, two of which absolute, C_1 and $C_2(\infty)$, and one relative, C_{12}, as stated above, are aligned and thus, according to the First Theorem of Kinematic Chains, the structure is hypostatic. The projections of the centre $C_2(\infty)$ fall at the points at infinity of the two fundamentals, so that the diagrams of the horizontal and vertical displacements of the points of body II are two segments parallel to their respective fundamentals. These points translate downwards by the amount φl and rightwards by $2\varphi h$, and thus, globally, translate in a direction orthogonal to that of the axes of the rods. At the same time, body I rotates clockwise by an angle φ, so that the internal hinge finds itself translated rightwards by $2\varphi h$ and downwards by φl.

To conclude, reconsider the structure made up of three rigid bodies, in the case of ill-disposition of the constraints (Figure 3.39). The absolute centre $C_1(\infty)$ is the point at infinity of the vertical straight lines. The centre C_2 is

67

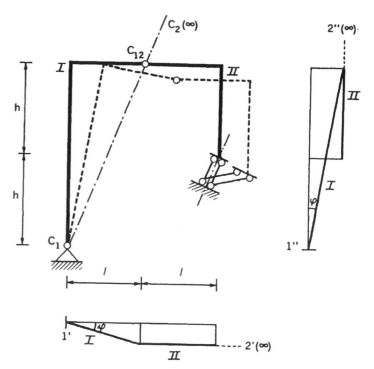

Figure 3.38

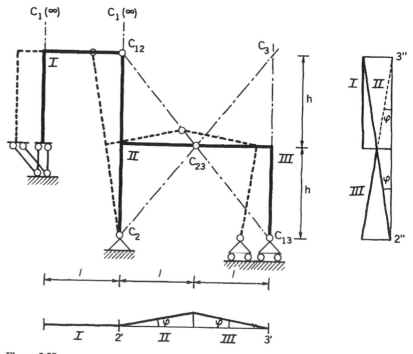

Figure 3.39

found on the external hinge, while the centre C_3 is unknown *a priori*. The relative centres C_{12} and C_{23} coincide with the internal hinges.

The centre C_3 is found at the intersection of the normal to the plane of movement of the roller support with the line joining C_2 and C_{23}. On the other hand, the third relative centre C_{13} is found at the intersection of the line joining C_{12} and C_{23} (Second Theorem of Kinematic Chains) with the line joining $C_1(\infty)$ and C_3, which is the vertical line through C_3 (First Theorem of Kinematic Chains). In short, C_{13} is found to coincide with the hinge of the roller support.

Once all the centres are known, both absolute and relative, it is possible to project them on the two fundamentals and to draw the linearized functions of displacement graphically, turning, so to speak, on these projections. As is natural, the body *I* translates only horizontally, while the bodies *II* and *III* both rotate by the same angle, the former, however, counterclockwise, the latter clockwise. The roller support moves towards the left by the amount $2\varphi h$, as does the upper internal hinge. The lower internal hinge translates leftwards by φh and upwards by φl.

3.6 Cardinal equations of statics

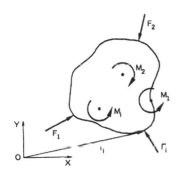

Figure 3.40

Consider a plane rigid body subjected to the action of n concentrated forces F_i and of m concentrated moments (or couples) M_i (Figure 3.40). A necessary and sufficient condition for this body to be in equilibrium is that the system of loads should satisfy the cardinal equations of statics:

$$\{R\} = \sum_{i=1}^{n} \{F_i\} = \{0\} \tag{3.34a}$$

$$M(O) = \sum_{i=1}^{m} M_i + \sum_{i=1}^{n} (\{r_i\} \wedge \{F_i\})^{\mathrm{T}} \{k\} = 0 \tag{3.34b}$$

where $\{R\}$ is the resultant force and $M(O)$ is the resultant moment (scalar because the system is plane) with respect to an arbitrary pole O of the plane. The arbitrariness of the pole is permitted by the condition whereby the resultant force is zero. In fact, the resultant moment with respect to a different pole O' is linked to the foregoing one by the following relation:

$$M(O') = M(O) + ((O - O') \wedge \{R\})^{\mathrm{T}} \{k\} \tag{3.35}$$

A system of loads which satisfies the conditions (3.34) is said to be **balanced** or **equivalent to zero** (two systems being equivalent when they possess equal resultant forces and equal resultant moments). On the other hand, two systems of loads are said to be one the **equilibrant** of the other when their sum is a balanced system. It follows from this that an equilibrant of a system of loads is the opposite of an equivalent system. It will be seen later how the system of external loads and the system of constraint reactions balance one another, their sum necessarily constituting a system equivalent to zero.

3.7 Static definition of plane constraints

Plane constraints, both internal and external, have already been introduced and defined from the kinematic point of view in Section 3.3. We shall now give a static definition, which means that we shall specify the reactions that

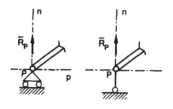

Figure 3.41

the constraints themselves are able to exert. There is a perfect correspondence, known as **duality**, between the kinematic definition and the static one. To each elementary movement (or generalized displacement) prevented by the constraint, there corresponds a generalized force exerted by the constraint on the body (and *vice versa*) in the case of external constraint, and mutually between two connected bodies in the case of internal constraint.

Recalling that for the **roller support** or **connecting rod** the kinematic conditions are (Figure 3.41)

$$\{ds_P\}^T\{p\} \neq 0 \tag{3.36a}$$

$$\{ds_P\}^T\{n\} = 0 \tag{3.36b}$$

$$\varphi_z \neq 0 \tag{3.36c}$$

to these there will correspond in a complementary (or dual) fashion the following elementary reactions:

$$\{R_P\}^T\{p\} = 0 \tag{3.37a}$$

$$\{R_P\}^T\{n\} \neq 0 \tag{3.37b}$$

$$M_P = 0 \tag{3.37c}$$

Hence, only the reaction orthogonal to the plane of movement of the roller support is different from zero, while the reaction parallel to this plane and the reaction moment are zero. Absence of friction (smooth constraint) is therefore assumed. Note that, as a consequence of the duality, the total work of the constraint reactions is zero:

$$\{R_P\}^T\{ds_P\} + M_P\varphi_z = 0 \tag{3.38}$$

This relation is valid, however, only when the constraint is ideal; that is, when the constraint is **rigid** and **smooth**, the conditions whereby displacements and reactions cancel, rigorously holding good.

The **hinge**, which is a double constraint, reacts with a force whose line of action passes through the hinge itself (Figure 3.42). The kinematic and static conditions are

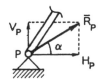

Figure 3.42

$$\{ds_P\} = \{0\} \qquad \Rightarrow \{R_P\} \neq \{0\} \tag{3.39a}$$

$$\varphi_z \neq 0 \qquad \Rightarrow M_P = 0 \tag{3.39b}$$

The parameters which identify the reaction force are in any case two: the two orthogonal components, or magnitude and angle of orientation.

Also the **double rod** is a double constraint, which reacts with a force parallel to the axes of the rods. The kinematic and static conditions are (Figure 3.43)

$$\{ds_P\}^T\{p\} \neq 0 \quad \Rightarrow \{R_P\}^T\{p\} = 0 \tag{3.40a}$$

$$\{ds_P\}^T\{n\} = 0 \quad \Rightarrow \{R_P\}^T\{n\} \neq 0 \tag{3.40b}$$

$$\varphi_z = 0 \qquad \Rightarrow M_P \neq 0 \tag{3.40c}$$

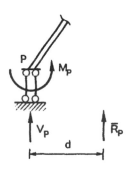

Figure 3.43

There are two parameters that identify the reaction: force and reaction moment, or translated force (after its composition with the reaction moment) and distance of the line of action from the constraint (Figure 3.43). Note that, considering the double rod as an ideal hinge at infinity, it may be stated that the reaction is a force whose line of action passes through this ideal hinge.

70

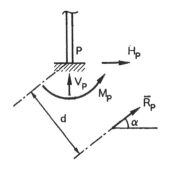

Figure 3.44

The triple constraint, or **fixed joint**, inhibits all elementary movements, and thus all three elementary reactions can be different from zero (Figure 3.44):

$$\{ds_P\} = \{0\} \quad \Rightarrow \{R_P\} \neq \{0\} \tag{3.41a}$$

$$\varphi_z = 0 \quad \Rightarrow M_P \neq 0 \tag{3.41b}$$

There are three parameters that identify the reaction: two orthogonal components of the reaction force and reaction moment, or magnitude and angle of orientation of the translated force and distance of the line of action from the constraint (Figure 3.44).

Finally, the last external constraint that remains to be considered is the **double articulated parallelogram** (Figure 3.45), which is a single constraint that inhibits rotation and that will thus react only with the moment

$$\{ds_P\} \neq \{0\} \quad \Rightarrow \{R_P\} = \{0\} \tag{3.42a}$$

$$\varphi_z = 0 \quad \Rightarrow M_P \neq 0 \tag{3.42b}$$

For **internal constraints**, the considerations are altogether similar to those just set forth for external constraints. On the other hand, just as for external constraints the reactions are understood as mutual actions (i.e. equal and opposite) exerted by the foundation on the body and *vice versa*, so for internal constraints the mutual action is to be understood as being exerted between the two bodies connected by the constraint.

The **connecting rod** exerts an equal and opposite force on the two bodies (Figure 3.46). This force has the line of action coinciding with the axis of the connecting rod, and thus the only static parameter involved is its magnitude (in addition to the sense). The force perpendicular to its axis and the reaction moment are not transmitted by the connecting rod.

The **hinge** exerts a force which passes through its own centre (Figure 3.47). Since it is a double constraint, there are two static parameters involved: the two orthogonal components, or the magnitude and the angle of orientation. The reaction moment is zero since relative rotations are allowed. The hinge is usually said not to react to moment.

The **double rod** transmits a force parallel to the axes of the rods, of which the magnitude and the distance from the constraint are to be defined (Figure 3.48). In an equivalent manner it may be stated that the double rod transmits a force, with line of action coincident with the axes of the rods (which are at an infinitesimal distance), and a reaction moment. Considering the double rod as an ideal hinge at infinity, the reaction is a force passing through this ideal hinge.

The **double articulated parallelogram** transmits the reaction moment only (Figure 3.49).

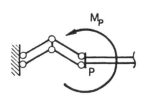

Figure 3.45

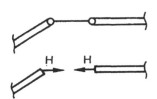

Figure 3.46

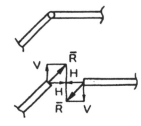

Figure 3.47

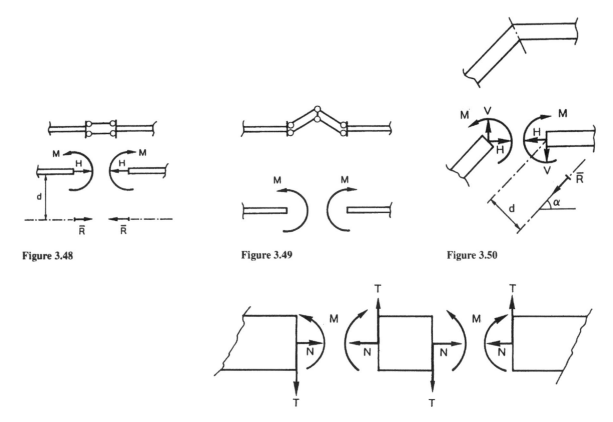

Figure 3.48   Figure 3.49   Figure 3.50

Figure 3.51

The **internal joint**, since it is a triple constraint, transmits all three elementary reactions: horizontal and vertical components of the force and the reaction moment (Figure 3.50). It is equivalent, on the other hand, to consider the global reaction obtained by summing up the single components: this can be any force, however oriented in the plane and however distant from the constraint.

It is important to note at this point that a rigid beam can be considered as a succession of infinite triple constraints which connect the infinite elementary segments that make it up (Figure 3.51). In this case, the elementary reactions assume peculiar structural meanings, as will be seen in the sequel, and particular denominations. Axial reaction is also referred to as **normal reaction**; reaction perpendicular to the axis is known as **transverse** or **shear reaction**; reaction moment is called **bending moment**. We shall hereafter refer to these elementary reactions in general as **characteristics of internal reaction**.

3.8 Algebraic study of statics of rigid systems

In this section we shall show how hypostaticity, isostaticity and hyperstaticity of rigid systems also find a highly meaningful algebraic interpretation in the field of statics. Moreover, we shall see that whereas in the kinematic

72

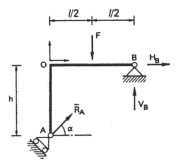

Figure 3.52

field, the condition of the system is an intrinsic property of the system itself, in the static field it is also a function of the external load.

As an introduction to the subject, consider again the L-shaped beam of Figure 3.52, loaded by a vertical force concentrated in the centre of the horizontal part. The external constraint reactions consist of the force R_A, perpendicular to the plane of movement of the roller support A, passing through the hinge of the roller support, and assumed to be acting upwards, and of the components H_B and V_B of the reaction exerted by the hinge B, assumed to be acting rightwards and upwards, respectively. These three parameters, R_A, H_B, V_B, represent the unknowns of the static problem, while the equations for resolving them are the cardinal equations of statics. If, once the solution has been obtained, a parameter is found to have a negative value, this means that the corresponding elementary reaction has a sense contrary to the one initially assumed.

To bring out the correspondence, or duality, between statics and kinematics, it is necessary to choose the static pole about which equilibrium to rotation is to be expressed, corresponding to the same kinematic pole that characterized the displacements of the individual rigid body (Figure 3.29(a)). This point O is thus chosen, also in this case, at the point of convergence of the two sections, horizontal and vertical. We must then express, in order, equilibrium to horizontal translation, equilibrium to vertical translation and equilibrium to rotation about the pole (or centre of reduction) O:

$$R_A \cos \alpha + H_B = 0 \tag{3.43a}$$

$$R_A \sin \alpha + V_B - F = 0 \tag{3.43b}$$

$$R_A \, h\cos \alpha + V_B l - F\frac{l}{2} = 0 \tag{3.43c}$$

Note that the vertical component $R_A \sin \alpha$ of the reaction of the roller support and the horizontal component H_B of the reaction of the hinge do not appear in the third equation, as these elementary reactions have a zero arm with respect to the pole O. The term $R_A \, h \cos \alpha$ may alternatively be read as the product of the magnitude R_A of the roller support reaction and the corresponding moment arm (Figure 3.52).

A matrix version of equations (3.43) may be given as

$$\begin{bmatrix} \cos \alpha & 1 & 0 \\ \sin \alpha & 0 & 1 \\ h\cos \alpha & 0 & l \end{bmatrix} \begin{bmatrix} R_A \\ H_B \\ V_B \end{bmatrix} = - \begin{bmatrix} 0 \\ -F \\ -F\dfrac{l}{2} \end{bmatrix} \tag{3.44}$$

The vector of the known terms is the opposite of the so-called **vector of reduced external forces**. The latter represents a system of loads equivalent to the system of external forces and acting precisely at the pole. In the case in point, it is the vertical force F translated at O, plus the moment of translation $-Fl/2$ (negative as it is clockwise).

Note that the matrix of the coefficients of relation (3.44) is exactly the transpose of the matrix in equation (3.23). This, as shall be seen more clearly later, is a property altogether general, known as **static–kinematic duality of rigid body systems**. The matrix of the static coefficients is therefore the transpose of the matrix of the kinematic coefficients.

73

To discuss the system (3.44), it is necessary to refer to the well-known Rouché–Capelli Theorem, valid for systems of non-homogeneous linear algebraic equations. It is useful here to recall the statement of this theorem.

Rouché–Capelli Theorem

A necessary and sufficient condition for a system of m linear equations in n unknowns to possess a solution is that the matrix of the coefficients and the matrix made up of the coefficients and the known terms, the so-called augmented matrix, should have the same rank.

We recall that the rank of a matrix is the integer expressing the maximum order of its non-zero minors.

The determinant of the matrix which appears in the system (3.44) vanishes, as has already been seen, for $\alpha = \bar{\alpha} = \arctan{(h/l)}$. The determinant of a square matrix is in fact equal to that of its transpose. When, therefore, the line joining A and B is perpendicular to the plane of movement of the roller support, the matrix of the coefficients presents rank 2 whereas the augmented matrix admits of rank 3. On the basis of the Rouché–Capelli Theorem, the system, therefore, does not possess a solution. This means that, in a condition of ill-disposition of the constraints and hence of hypostaticity, calculation of the constraint reactions is **impossible**.

As a second example, consider the arch made up of two rigid bodies, already analysed from the kinematic point of view in Section 3.3 (Figure 3.53(a)). Let this be loaded by an oblique force F. In this case the unknowns consist of the constraint reactions of the external hinge, H_A, V_A, of the internal hinge, H_B, V_B, and of the double rod, R_C, M_C. The centres of reduction are chosen coincident with the external constraints, at A and C respectively, while, as is customary, the horizontal forces directed towards the right, the vertical forces directed upwards and the counterclockwise moments are considered positive. These choices are, in actual fact, altogether arbitrary and conventional.

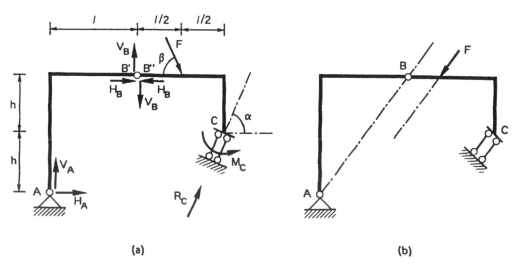

(a) (b)

Figure 3.53

The six equilibrium equations, the first three relative to the left-hand body and the following three relative to the right-hand body, are presented as follows:

$$H_A + H_B = 0 \tag{3.45a}$$

$$V_A + V_B = 0 \tag{3.45b}$$

$$-2hH_B + V_B l = 0 \tag{3.45c}$$

$$R_C \cos\alpha - H_B + F\cos\beta = 0 \tag{3.45d}$$

$$R_C \sin\alpha - V_B - F\sin\beta = 0 \tag{3.45e}$$

$$H_B h + V_B l - h\,F\cos\beta + \frac{l}{2}\,F\sin\beta + M_C = 0 \tag{3.45f}$$

In matrix form we obtain

$$
\begin{bmatrix}
1 & 0 & 1 & 0 & 0 & 0 \\
0 & 1 & 0 & 1 & 0 & 0 \\
0 & 0 & -2h & l & 0 & 0 \\
0 & 0 & -1 & 0 & \cos\alpha & 0 \\
0 & 0 & 0 & -1 & \sin\alpha & 0 \\
0 & 0 & h & l & 0 & 1
\end{bmatrix}
\begin{bmatrix}
H_A \\ V_A \\ H_B \\ V_B \\ R_C \\ M_C
\end{bmatrix}
= -
\begin{bmatrix}
0 \\ 0 \\ 0 \\ F\cos\beta \\ -F\sin\beta \\ -hF\cos\beta + \dfrac{l}{2}F\sin\beta
\end{bmatrix}
\tag{3.46}
$$

The opposite of the vector of the known terms represents the vector of the external forces reduced at the poles A and C. The first three terms are zero because the left-hand body is not subjected to external loads.

The static matrix of system (3.46) is the transpose of the kinematic matrix of system (3.29). When the determinant is different from zero, both the matrix of the coefficients and its augmented matrix evidently present the same rank 6. By the Rouché–Capelli Theorem, the algebraic system then possesses only one solution, i.e. an orderly set of six values. In this case the mechanical system is said to be **statically determinate** or **isostatic**.

The determinant, on the other hand, vanishes for $\alpha = \overline{\alpha} = \arctan(2h/l)$. In this case the matrix of the coefficients is of rank 5 while its augmented matrix in general is of rank 6. The algebraic system is thus impossible and does not possess any solution, while the mechanical system is **hypostatic** and loaded by external forces that cannot be balanced in any way.

An even more particular case is where $\alpha = \overline{\alpha} = \arctan(2h/l)$, $\beta = \pi - \overline{\alpha} = \arctan(-2h/l)$. Also the external force becomes parallel to the line joining A and B, as well as the double rod at C (Figure 3.53(b)). In this case the augmented matrix will be as follows:

$$
\left[
\begin{array}{ccccccc}
1 & 0 & 1 & 0 & 0 & 0 & 0 \\
0 & 1 & 0 & 1 & 0 & 0 & 0 \\
0 & 0 & -2h & l & 0 & 0 & 0 \\
0 & 0 & -1 & 0 & \cos\overline{\alpha} & 0 & F\cos\overline{\alpha} \\
0 & 0 & 0 & -1 & \sin\overline{\alpha} & 0 & F\sin\overline{\alpha} \\
0 & 0 & h & l & 0 & 1 & -hF\cos\overline{\alpha} - (l/2)F\sin\overline{\alpha}
\end{array}
\right]
\tag{3.47}
$$

The seventh column, that of the known terms, is a linear combination of the fifth column (multiplied by F) and the sixth (multiplied by $-hF \cos \bar{\alpha} - (l/2) F \sin \bar{\alpha}$). Consequently, both the matrix of the coefficients and the augmented matrix present rank 5. By the Rouché–Capelli Theorem, there exist then ∞^1 solutions. The mechanical system is thus intrinsically hypostatic, but is loaded by an external force that can be balanced in ∞^1 different ways (Figure 3.53(b)). The mechanical system is thus in equilibrium on account of the particular load condition. But this equilibrium presents itself as **statically indeterminate**, or **hyperstatic**.

To conclude, let us re-examine the double portal frame of Figure 3.54(a), loaded by a horizontal force F_1 and by a vertical force F_2. In this case, we have nine equilibrium equations in nine unknowns, which, in a matrix form, are presented as follows:

$$
\begin{bmatrix}
0 & 0 & 1 & 0 & 0 & 0 & 0 & 0 & 0 \\
1 & 0 & 0 & 1 & 0 & 0 & 0 & 0 & 0 \\
0 & 1 & -h & x & 0 & 0 & 0 & 0 & 0 \\
0 & 0 & -1 & 0 & 1 & 0 & 1 & 0 & 0 \\
0 & 0 & 0 & -1 & 0 & 1 & 0 & 1 & 0 \\
0 & 0 & 2h & l-x & 0 & 0 & -h & l & 0 \\
0 & 0 & 0 & 0 & 0 & 0 & -1 & 0 & 0 \\
0 & 0 & 0 & 0 & 0 & 0 & 0 & -1 & 1 \\
0 & 0 & 0 & 0 & 0 & 0 & h & l & 0
\end{bmatrix}
\begin{bmatrix}
V_A \\ M_A \\ H_B \\ V_B \\ H_C \\ V_C \\ H_D \\ V_D \\ V_E
\end{bmatrix}
= -
\begin{bmatrix}
F_1 \\ 0 \\ -F_1 h \\ 0 \\ -F_2 \\ 0 \\ 0 \\ 0 \\ 0
\end{bmatrix}
\qquad (3.48)
$$

The matrix of the coefficients is the transpose of that of equation (3.33).

When $0 \le x < l$, the matrix of the coefficients is of rank 9, as is the augmented matrix. The algebraic system thus possesses a solution and the mechanical system is said to be **statically determinate**.

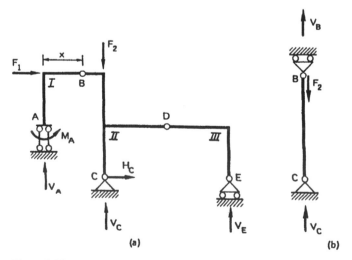

Figure 3.54

When $x = l$ and $F_2 = 0$, the matrix of the coefficients is of rank 8, while the augmented matrix is again of rank 9. The algebraic system is thus impossible, just as the mechanical system is hypostatic and loaded by an unbalanceable force.

Finally, when $x = l$ and $F_1 = 0$, the matrix of the coefficients is again of rank 8, as is the augmented matrix. The column of the known terms is in fact proportional to the sixth column of the static matrix. The algebraic system thus possesses ∞^1 solutions, the rigid system being hypostatic but in equilibrium on account of the particular load condition. This equilibrium can be obtained in ∞^1 different ways. This can be immediately justified if the body *I* is replaced by a roller support which moves horizontally in *B*, and the body *III* is eliminated since it does not react (Figure 3.54(b)). In this way, having reduced the system to its essential elements, we may verify how the force F_2 can be balanced by an infinite number of pairs of vertical reactions V_C and V_B, such that $V_B + V_C - F_2 = 0$.

3.9 Static–kinematic duality

We close this chapter with a summary of its contents, underlining the aspect of duality between kinematics and statics. A rigorous demonstration will therefore be given of the fact that the **static matrix** is the transpose of the **kinematic matrix**, and *vice versa*.

The **kinematic equations** of a rigid system, with g original degrees of freedom and v degrees of constraint, can be represented in a compact form as follows:

$$\underset{v \times g}{[C]} \underset{g \times 1}{\{s_O\}} = \underset{v \times 1}{\{s_X\}} = \underset{v \times 1}{\{0\}} \tag{3.49}$$

where $[C]$ = kinematic matrix
$\{s_O\}$ = vector of displacements of poles
$\{s_X\}$ = vector of displacements (absolute or relative) of constraint points

When $v < g$, the algebraic system possesses at least ∞^{g-v} solutions, and the mechanical system is at least $(g - v)$ times **hypostatic**.
When $v = g$, the algebraic system possesses at least the obvious solution, and the mechanical system is **isostatic**, if there is no ill-disposition of the constraints.
When $v > g$, the algebraic system possesses at least the obvious solution, and the mechanical system is $(v - g)$ times **hyperstatic**, if there is no ill-disposition of the constraints.

On the other hand, the **static equations** of the same rigid system can be represented as follows:

$$\underset{g \times v}{[A]} \underset{v \times 1}{\{X\}} = -\underset{g \times 1}{\{F\}} \tag{3.50}$$

where $[A]$ = static matrix
$\{X\}$ = vector of constraint reactions
$\{F\}$ = vector of reduced external forces

When $v < g$, the algebraic system is generally impossible and the system of external forces cannot be balanced in any way.
When $v = g$, the algebraic system generally possesses one solution, and the mechanical system is said to be **statically determinate** or **isostatic**.

When $v > g$, the algebraic system generally possesses ∞^{v-g} solutions and the mechanical system is said to be **statically indeterminate** or **hyperstatic**.

The Principle of Virtual Work can be applied to a rigid system in equilibrium, subjected to external loads and constraint reactions. If the external forces are replaced, as is permissible, with the forces reduced to the poles, we have

$$\{s_O\}^T\{F\} + \{s_X\}^T\{X\} = 0 \tag{3.51}$$

On the other hand, using the matrix relations (3.49) and (3.50), equation (3.51) is transformed as follows:

$$-\{s_O\}^T[A]\{X\} + \{s_O\}^T[C]^T\{X\} = 0 \tag{3.52}$$

from which we obtain

$$\{s_O\}^T\left([A] - [C]^T\right)\{X\} = 0 \tag{3.53}$$

This equation, on account of the arbitrariness of $\{s_O\}$ and $\{X\}$, is satisfied if and only if

$$[A] = [C]^T \tag{3.54}$$

The fundamental relation (3.54), which arose inductively in this chapter, can thus be deemed to be rationally demonstrated.

4 Determination of constraint reactions

4.1 Introduction

This chapter presents some methods for determining constraint reactions in the case of statically determinate structures, i.e. structures constrained in a non-redundant manner. In addition to the algebraic method of auxiliary equations, which helps us to split the general solution system into two or more systems of smaller dimensions, the method based on the Principle of Virtual Work, as well as the classical graphical method using pressure lines, is also proposed.

In this chapter particular attention is again drawn to the problem of ill-disposition of constraints for which the potential centres of rotation, both absolute and relative, fall on a straight line, just as, from the static viewpoint, the force polygons do not close except at infinity.

In the case of continuous distributions of forces acting in the same direction, the differential equation of the pressure line is obtained, so revealing how this line represents, save for one factor, the diagram of the bending moment. The cases of arch bridges and suspension bridges are considered as typical examples in which, since the geometrical axis of the load-bearing member coincides with the pressure line, the member itself is either only compressed (arch) or else only stretched (cable).

4.2 Auxiliary equations

As we have already seen in the previous chapter, if the number of rigid bodies making up an isostatic, or statically determinate, structure is n, it will be possible to write $3n$ equations of partial equilibrium in $3n$ unknown elementary reactions. On the other hand, this leads to systems of equations that are unwieldy to resolve even with a low n.

An alternative way that can be followed is that of the **method of auxiliary equations**. In this case, we consider the three equations of global equilibrium, i.e. of the entire structure, with the addition of s auxiliary equations of partial equilibrium, if s is the degree of internal disconnection of the structure. We thus have $(3 + s)$ equilibrium equations in $(3 + s)$ unknown external reactions. The s auxiliary equations are chosen in such a way that the internal reactions are not involved in the system of resolution.

To clarify the process outlined above, let us examine once more the arch of Figure 4.1, consisting of two rigid parts hinged at B and connected to the foundation with a hinge and a double rod, respectively. In this case the degree of internal disconnection s is equal to one (the relative rotation allowed by the hinge B). We can, therefore, write three equations of global equilibrium plus an auxiliary equation of partial equilibrium with regard to rotation of the body AB about point B, in the four unknown external reactions, H_A, V_A, R_C, M_C. The auxiliary equation of partial equilibrium with regard to rotation of the body BC about point B is equivalent to the foregoing auxiliary equation, it being a

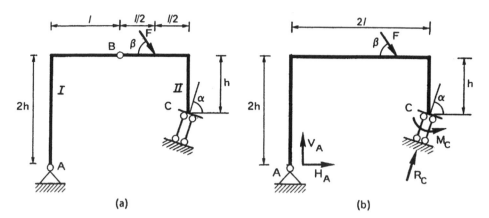

Figure 4.1

linear combination of the four equations so far considered. It is, however, more complex. It is therefore more convenient to write

$$H_A + R_C \cos\alpha + F\cos\beta = 0 \tag{4.1a}$$

$$V_A + R_C \sin\alpha - F\sin\beta = 0 \tag{4.1b}$$

$$H_A h - 2V_A l - h\,F\cos\beta + \frac{l}{2}F\sin\beta + M_C = 0 \tag{4.1c}$$

$$2H_A h - V_A l = 0 \tag{4.1d}$$

Whereas, then, the first three are the equations of equilibrium of the **braced** structure, i.e. of the hyperstatic structure obtained by replacing the internal hinge with an internal built-in constraint (Figure 4.1(b)), the fourth is the auxiliary equation which expresses equilibrium with regard to rotation of the left-hand body about the hinge. In this way, the internal unknowns H_B and V_B do not for the moment enter into the balance, and the system (4.1) consists of four equations in four unknowns, as against the six equations in six unknowns of the system (3.45).

It is, however, possible subsequently and once the external reactions have been determined, also to determine the internal reactions. These may be deduced using the equations of partial equilibrium with regard to translation, of the body AB or, alternatively, of the body BC. Since the body AB is not subjected to external loads, the first way is the more convenient:

$$H_A + H_B = 0 \tag{4.1e}$$

$$V_A + V_B = 0 \tag{4.1f}$$

Now consider the case, already introduced in the last chapter, of three rigid bodies (Figure 4.2(a)). For this structure we have $s = 2$, since two rotational single disconnections in B and D are present. There will then be five equations (three global equilibrium equations and two auxiliary equations) in the five unknown external reactions: V_A, M_A, H_C, V_C, V_E. The first three equations refer to the braced structure (twice hyperstatic) of Figure 4.2(b):

80

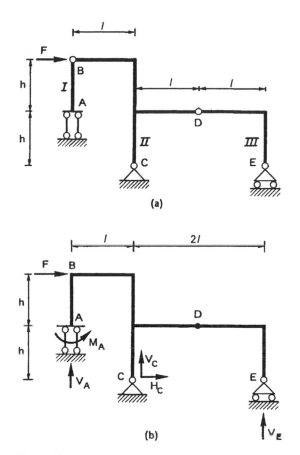

Figure 4.2

$$H_C + F = 0 \tag{4.2a}$$

$$V_A + V_C + V_E = 0 \tag{4.2b}$$

$$-V_A l + M_A + 2V_E l - 2Fh = 0 \tag{4.2c}$$

Of these, equation (4.2c) expresses equilibrium with regard to rotation about point C. It is then necessary to provide the information that there exists a hinge at B, by writing the equation of partial equilibrium with regard to rotation of the body AB about B (in this way, we avoid introducing additional unknowns),

$$M_A = 0 \tag{4.2d}$$

and, finally, that there also exists a hinge at point D, expressing the partial equilibrium with regard to rotation of the body DE about D,

$$V_E l = 0 \tag{4.2e}$$

81

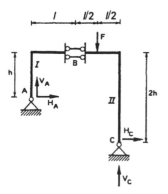

Figure 4.3

Once we have obtained the five external unknowns, it will also be possible to deduce the internal ones, by resolving the equations of equilibrium with regard to translation of the two end bodies:

$$F + H_B = 0 \tag{4.2f}$$

$$V_A + V_B = 0 \tag{4.2g}$$

$$-V_D + V_E = 0 \tag{4.2h}$$

$$-H_D = 0 \tag{4.2i}$$

More generally, it is then possible to consider structures having a generic number s of internal hinges and thus $(3 + s)$ external elementary reactions. The procedure will be altogether analogous to that presented previously. There are, however, auxiliary equations to be considered relative to partial sections, having one end internally constrained and the other externally constrained. These sections can contain possible internal disconnections.

On the other hand, the forms of internal disconnection that may be considered are not limited to the hinge. For instance, the portal frame of Figure 4.3 contains an internal disconnection to the vertical translation, i.e. a horizontal double rod. The three global equations are those of equilibrium with regard to horizontal and vertical translation and the equation of equilibrium with regard to rotation about point C:

$$H_A + H_C = 0 \tag{4.3a}$$

$$V_A + V_C - F = 0 \tag{4.3b}$$

$$-2V_A l - H_A h + F\frac{l}{2} = 0 \tag{4.3c}$$

whilst the auxiliary equation to be considered is that of partial equilibrium with regard to vertical translation of the left-hand section:

$$V_A = 0 \tag{4.3d}$$

In the case of the double rod being vertically oriented (Figure 4.4(a)), the auxiliary equation would then be that of equilibrium with regard to horizontal translation of one of the two rigid sections.

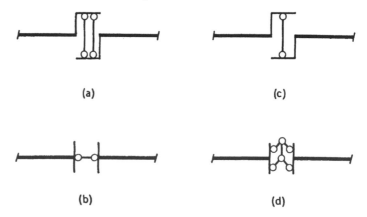

(a)   (c)

(b)   (d)

Figure 4.4

When, instead, there is a double internal disconnection, i.e. $s = 2$, which is concentrated in the same point, there are two auxiliary equations corresponding to it, and they must be formulated so that the only elementary reaction transmitted does not appear. In the case of a horizontal connecting rod, for example, it will be necessary to consider the partial equilibrium with regard to rotation about the constraint and with regard to vertical translation of one of the two sections into which the structure is separated by the connecting rod itself (Figure 4.4(b)). When, instead, the connecting rod is vertical (Figure 4.4(c)), the partial equilibrium with regard to rotation and with regard to horizontal translation will be considered. Finally, in the case of the double articulated parallelogram (Figure 4.4(d)), the two auxiliary equations will both be equations of equilibrium with regard to translation.

4.3 Principle of Virtual Work

We have previously described two algebraic methods for determining constraint reactions: (1) the **general method**, according to which each single rigid body is set in equilibrium, writing $3n$ equations in $3n$ unknowns (n = number of rigid bodies of the system); (2) the **method of auxiliary equations**, according to which the global equilibrium is considered and at the same time the information is provided that there exist s internal disconnections, by writing $3 + s$ equations in $(3 + s)$ unknowns.

A semi-graphical method will now be introduced that is based on the Principle of Virtual Work and on the theory of mechanisms, which is, on the other hand, able to provide a single elementary reaction each time. This is thus a method which can be used to advantage when we wish to determine a specific reaction, necessary, for example, for dimensioning the constraint supporting it.

As an introduction to this method, consider the bar system of Figure 4.5(a), subjected to the horizontal force F. We intend to define the value of the horizontal reaction H_C exerted by the hinge C. It will then be necessary to effect a disconnection in such a way that, apart from the external force, only the reaction sought will be able to perform work. The hinge C will then be degraded by being transformed into a horizontally moving roller support (from a double constraint to a single constraint) and the horizontal force H_C, exerted by the hinge C, will be applied, assumed to be acting towards the right. It is evident that the assigned structure has now been transformed into the mechanism already studied in Section 3.5 (Figure 4.5(b)). The reactions H_A and V_A of the external hinge do not perform work because their point of application is not displaced. The internal reactions H_B and V_B perform equal and opposite work on the two bars that make up the bar system. Finally, the reaction V_C of the roller support does not perform work, since it is displaced in a direction perpendicular to that of its line of action. There thus remain to be taken into account the external force F, which is displaced by the quantity ϑh_3, and the reaction H_C (which in this scheme has the role of an external force), which is displaced by the amount $\vartheta(h_1 + h_2 + h_3)$. Since these two forces undergo displacements opposite to their direction, the two contributions will both be negative (Figure 4.5(b)):

$$\text{Work} = -F\vartheta h_3 - H_C\vartheta(h_1 + h_2 + h_3) = 0 \tag{4.4}$$

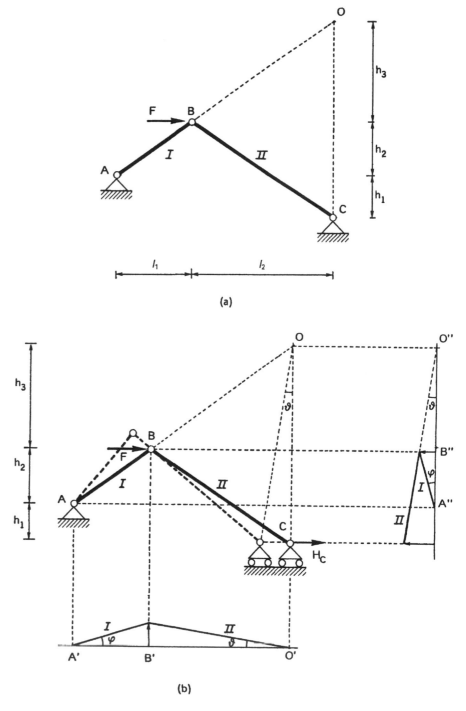

(a)

(b)

Figure 4.5

from which

$$H_C = -F \frac{h_3}{h_1 + h_2 + h_3} \qquad (4.5)$$

It may be noted that the angle of virtual rotation ϑ of the bar BC is cancelled by both the contributions and therefore does not enter into the expression (4.5). Nor do the abscissae l_1 and l_2 enter into this expression (Figure 4.5). The mechanism thus remains in equilibrium on account of the particular load condition, and H_C is negative because it is, in actual fact, acting in the opposite direction to the one assumed.

As a second example consider the asymmetrical portal frame of Figure 4.6(a), subjected to a constant distributed load q. We intend to determine the vertical reaction of the hinge C. The procedure, then, is to reduce the constraint C in a dual manner with respect to the reaction that we are seeking. The hinge C will thus be transformed into a vertically moving roller support. In this case, only the diagram of the vertical displacements defined already in Figure 3.36(a) will be used. With the aim of applying the Principle of Virtual Work, we shall consider the two partial resultants of the external load, each acting on one rigid section, plus the reaction V_C:

$$-(ql)\left(\varphi \frac{l}{2}\right) - (ql)(2\varphi l) + V_C(3\varphi l) = 0 \qquad (4.6)$$

from which

$$V_C = \frac{5}{6}ql \qquad (4.7)$$

In this case, the direction assumed proves to be the actual one.

If we intend then to determine the vertical reaction T (shearing force) transmitted by the internal hinge B (Figure 4.6(b)), we have to reduce the hinge itself and to transform it into a disconnection of a higher order which allows the vertical relative translations. If, then, a horizontal connecting rod is introduced in place of the hinge, the mechanism will undergo the horizontal and vertical displacements shown in Figure 4.6(b). The two portions both turn in the same direction, the one on the right through an angle twice as large. Note how, in correspondence with the relative centre C_{12}, relative displacements of the two bodies do not occur. While the force T, acting on the left-hand portion, performs the work $T\varphi l$, that on the right-hand portion performs twice as much work. The same applies to the external partial resultants, even though in this case the amounts of work performed are of opposite algebraic sign. Altogether we have

$$T\varphi l + 2T\varphi l - ql\left(\varphi \frac{l}{2}\right) + ql\left(2\varphi \frac{l}{2}\right) = 0 \qquad (4.8)$$

from which

$$T = -\frac{ql}{6} \qquad (4.9)$$

85

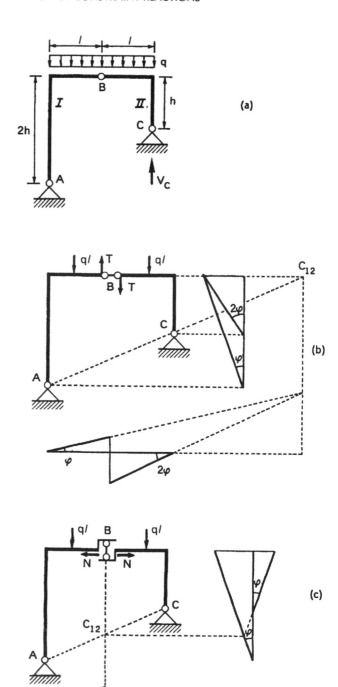

Figure 4.6

86

If, instead, we wish to obtain the horizontal reaction N (axial force) transmitted by the internal hinge B (Figure 4.6(a)), we must replace the hinge with a vertical connecting rod (Figure 4.6(c)), which allows horizontal relative translations. This time the two sections turn through the same angle but in opposite directions. Also in this case, in correspondence with the relative centre C_{12}, the two bodies are displaced by the same amounts. The total balance works out as follows:

$$-ql\left(\varphi\frac{l}{2}\right) - ql\left(\varphi\frac{l}{2}\right) + N(2\varphi h) + N(\varphi h) = 0 \tag{4.10}$$

whence

$$N = \frac{ql^2}{3h} \tag{4.11}$$

It is possible to check the results obtained using the Principle of Virtual Work, by applying the method of auxiliary equations:

$$H_A + H_C = 0 \tag{4.12a}$$
$$V_A + V_C - 2ql = 0 \tag{4.12b}$$
$$-2V_A l + H_A h + 2ql^2 = 0 \tag{4.12c}$$
$$-V_A l + 2H_A h + q\frac{l^2}{2} = 0 \tag{4.12d}$$

where the last equation is the auxiliary one (equilibrium to rotation of the section AB about B).

Resolving the system (4.12), we obtain

$$H_A = \frac{ql^2}{3h} \tag{4.13a}$$
$$V_A = \frac{7}{6}ql \tag{4.13b}$$
$$H_C = -\frac{ql^2}{3h} \tag{4.13c}$$
$$V_C = \frac{5}{6}ql \tag{4.13d}$$

Considering equilibrium with regard to translation of the left-hand section, we have then

$$H_A - N = 0 \tag{4.14a}$$
$$V_A + T - ql = 0 \tag{4.14b}$$

from which we obtain the axial and shearing forces transmitted by the hinge B

$$N = \frac{ql^2}{3h}, \quad T = -\frac{ql}{6} \tag{4.15}$$

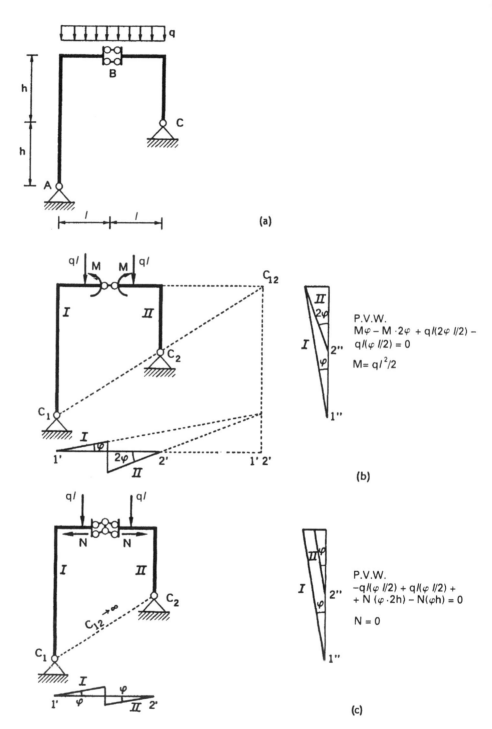

Figure 4.7

As regards the asymmetrical portal frame of Figure 4.7(a), obtained from the foregoing one by replacing the internal hinge with a horizontal double rod, the determination of the two elementary reactions transmitted by the internal constraint *B* can be arrived at, applying the Principle of Virtual Work and using the two mechanisms shown in Figures 4.7(b) and 4.7(c).

4.4 Graphical method

The graphical method for determining the constraint reactions is based on the cardinal equations of statics. In the case where we have three forces in equilibrium in the plane, these must form a triangle if laid out one after the other. This derives from the well-known **parallelogram law** and hence from the first cardinal equation of equilibrium with regard to translation. At the same time, the lines of action of the three forces must all pass through the same point of the plane. The moment of the three forces must in fact be zero with respect to any point in the plane, and thus also with respect to the intersection of each pair of lines of action: the third line must then pass through that point. This latter requirement follows directly from the second cardinal equation of equilibrium with regard to rotation.

For greater clarity, let us take an example of application of the graphical method. Examine the L-shaped beam of Figure 4.8(a), already considered

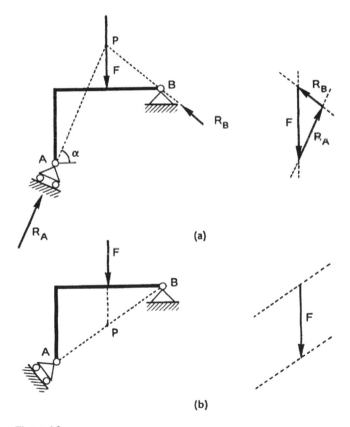

Figure 4.8

from both the static and the kinematic viewpoints. We know *a priori* two of the three lines of action for the forces involved: the line of action of the external force F and the line of action of the reaction R_A of the roller support. These two lines are concurrent in point P, and hence, for equilibrium with regard to rotation, also the reaction R_B of the hinge must pass through P. Thus we have defined the direction of the reaction R_B, which is that of the line joining B and P. The magnitudes, which were previously unknown, of the two constraint reactions can then be determined graphically, laying off to scale the given force F and drawing through its ends two lines oriented as the lines of action of R_A and R_B (Figure 4.8(a)). The triangle that is thus obtained represents half of the parallelogram of the forces R_A and R_B, which will assume a direction so that they follow the given force F. In this way the force F is equilibrant of R_A and R_B (while the opposite force is their resultant).

When the constraints are ill-disposed (Figure 4.8(b)), the lines of action of the reactions R_A and R_B both coincide with the line joining A and B and the triangle of forces does not close, other than at infinity. The reactions thus tend to infinity and equilibrium is impossible.

However, there are not always only three forces involved in ensuring equilibrium of a rigid body. On the other hand, if we compose the forces suitably, it is always possible to reduce them to three partial resultants. For example, Figure 4.9(a) shows the case of a beam constrained by three well-disposed connecting rods, i.e. ones not passing through the same point, and loaded by an external force F. Composing the connecting rods B and C, we can assume that we have an ideal hinge in D, and thus the graphical resolution will closely resemble the previous one (Figure 4.9(b)). We identify, in fact, point P, the intersection of the line of action of the external force with the axis of the rod A, and we join P with the ideal hinge D. The line PD is the line of action of the composite reaction R_D. Once we have found the magnitude of R_D using the triangle of forces (Figure 4.9(b)), we can resolve this reaction into its two elementary components R_B and R_C by drawing a new triangle of forces, which in Figure 4.9(b) appears adjacent to the foregoing one. Note, however, that while the first triangle resolves a problem of equilibrium (the arrows follow one another round), the second one resolves a problem of equivalence (parallelogram law).

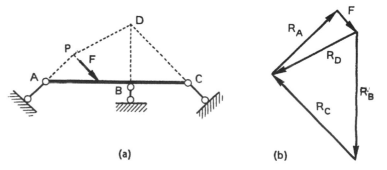

(a)   (b)

Figure 4.9

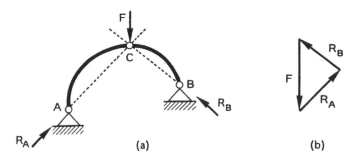

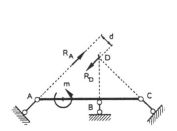

Figure 4.10

Figure 4.11

In the case where the external load consists of a concentrated couple or moment m (Figure 4.10), we must consider the straight line passing through the ideal hinge D and parallel to the axis of the rod A. The moment m will, in fact, be balanced by a couple of equal and opposite forces R_A and R_D, the magnitude of which is $R_A = R_D = m/d$, where d is the distance between the two lines of action. The reaction R_D can then be resolved into its two components R_B and R_C.

The graphical method can be used also when the structure consists of more than one rigid body. As regards three-hinged arches, a rapid and convenient application of the method requires, however, that the external load should act only on one of the two rigid sections, or else on the internal hinge.

In the case of the arch of Figure 4.11(a), the vertical external force is applied to the hinge C, which, being considered in this case as a material point, is found in equilibrium with regard to translation under the action of F, R_A and R_B (Figure 4.11(b)). The triangle of forces will therefore have one side vertical and the other two in the directions of the lines joining A and C, and B and C, respectively.

If the external force acts on the section BC (Figure 4.12(a)), the section AC performs the function of a connecting rod and we find ourselves once again in a case similar to those which we have already seen previously and which correspond to a single rigid body (Figure 4.12(b)).

In the event then that the three hinges are aligned and hence ill-disposed, also the external reactions R_A and R_C become collinear and the triangle of forces does not close, thus ruling out the possibility of any static solution.

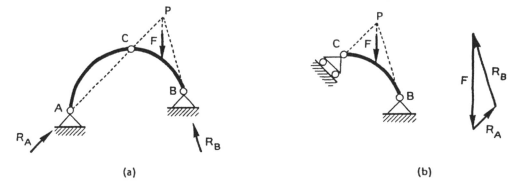

(a)

(b)

Figure 4.12

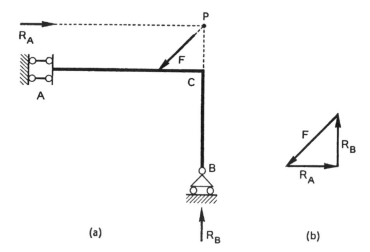

Figure 4.13

So far only hinges and roller supports have been considered, but also the double rod can be involved in a solution of a graphical type. This may, in fact, be considered as an ideal hinge at infinity, and thus be treated as we have by now already seen more than once. In the case, for instance, of an L-shaped beam constrained by a double rod and a roller support (Figure 4.13(a)) and loaded by an oblique force F, the pole P is given by the intersection of the line of action of F with the normal to the plane of movement of the roller support. The double rod will react with a horizontal force passing through P. The triangle of forces will have the hypotenuse parallel to the external force and the two catheti parallel, respectively, to the two rectilinear sections of the beam (Figure 4.13(b)).

In the case of generalized three-hinged arches, i.e. arches where double rods are also involved (Figure 4.14), our analysis will once more be based on the interpretation of the double rod as an ideal hinge and on the partial equilib-

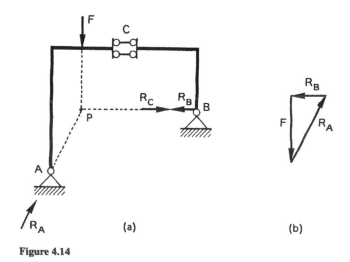

Figure 4.14

92

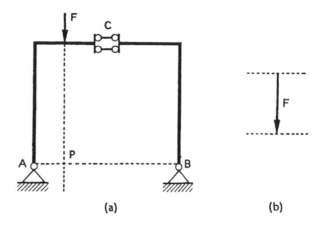

(a)   (b)

Figure 4.15

rium of the portion that is not externally loaded. The portion *BC*, in fact, is in equilibrium under the action of two equal and opposite collinear forces. These two forces must be horizontal and must pass through the hinge *B*. In other words, their line of action will be the line joining the two hinges, the real one *B* and the ideal one consisting of the point at infinity of the horizontal lines. Given this, the pole *P* is furnished by the intersection of the line of action of *F* with the horizontal line through *B*. The third force acting on the structure is the reaction R_A, whose line of action is given by the line joining *A* and *P*.

In the case where the portal frame is symmetrical, i.e. where it has uprights of equal height (Figure 4.15), the pole *P* falls on the line joining *A* and *B*, so that the triangle of forces does not close. The ill-disposition of the constraints thus renders the equilibrium of the system impossible, once this is subjected to the vertical force *F*.

Now reconsider the arch of Figure 4.16 (a), already studied in the previous chapter. The equilibrium of the portion *AB* is achieved with two equal and opposite forces on the line joining *A* and *B*. The intersection of the line *AB* with the line of action of the external force gives the pole *P*, through which also the horizontal reaction R_C of the double rod must pass. When the parameter *h* vanishes, the system becomes hypostatic and equilibrium is impossible (Figure 4.16 (b)).

Another example, already amply studied, is the portal frame of Figure 3.53(a). When the double rod is disposed parallel to the line joining *A* and *B* (Figure 4.17(a)), the system is hypostatic and the generic force *F* cannot be balanced in any way, since the triangle of forces does not close (Figure 4.17(b)). When, moreover, also the external force is directed as the straight line *AB* and the double rod, the pole *P* comes to coincide with the point at infinity of this direction, so that the line of action of the reaction R_C of the double rod remains indeterminate. On the other hand, in this case the triangle of forces is degenerate, with all three sides collinear (Figure 4.17(c)), and there exist ∞^1 pairs of vectors R_A and R_C that satisfy equilibrium. As was already seen algebraically in Chapter 3, the mechanical system is hypostatic, but in equilibrium for the particular condition of load. Since, however, there exist ∞^1 solutions, both on account of the indetermination of the line of action of the reaction R_C of the double rod and on account of the possibility of balancing

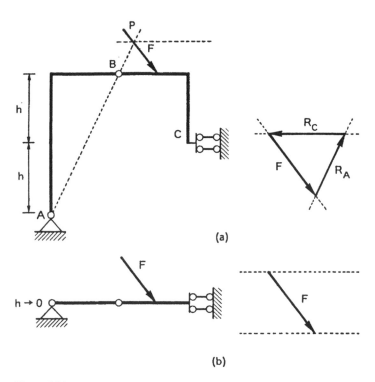

(a)

(b)

Figure 4.16

the force F with two reactions parallel to it (of which one is arbitrary), the system is said to be statically indeterminate (or once hyperstatic).

Another case that has already been amply discussed is that of the double portal frame of Figure 4.18(a). Imagine that the hinge B is in the centre of the upper horizontal beam ($x = l/2$) and that the system is loaded by the horizontal force F. To start with, it should be noted that the vertical reaction V_E of the roller support must be zero. If we were to assume *ab absurdo* that this were not so, it ought to be balanced by the vertical reaction V_D of the hinge D, and thus a couple would be formed that cannot be balanced by the only reaction H_D remaining to be considered as acting on the section *III*. This section is thus completely inactive, or, to use the term normally applied, unloaded. The remaining sections, *I* and *II*, constitute a generalized three-hinged arch of the same type as those already introduced. The section *II* is not loaded externally and hence its equilibrium develops on the line joining B and C. The hinge B, therefore, coincides with the pole P and the vertical reaction of the double rod A must pass through this point. Since the triangle of forces (Figure 4.18(b)) is geometrically similar to the triangle $BB'C$ (Figure 4.18(a)), it is immediately evident that

$$R_A = 4\,F\,h/l, \quad R_C = F\big(1 + (16\,h^2/l^2)\big)^{\frac{1}{2}} \qquad (4.16)$$

and the structure is completely resolved.

When the hinge B is at the right end of the upper horizontal beam, the beam system is transformed into a mechanism (Figure 3.39). The vertical reaction R_A of the double rod is reduced to being collinear to the reaction R_C of the hinge so that, if the external force is horizontal, the triangle of forces cannot

94

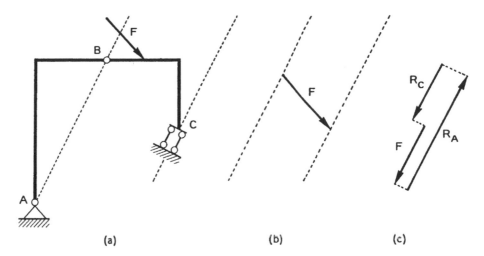

Figure 4.17

close and equilibrium is impossible (Figure 4.18(c)). If the external force is, instead, vertical (Figure 4.18(d)), the beam system reveals itself to be (once) hyperstatic owing to the particular load condition.

Consequently, with extreme economy, we have been able to arrive back at the results that are already known to us, which in Chapter 3 were brought to light algebraically, using the Rouché–Capelli Theorem.

As a final case, consider that of an internal constraint made up of two non-parallel connecting rods (Figure 4.19(a)). The system of bars can be considered as a three-hinged arch, where the internal hinge consists of the ideal hinge H, the point of concurrence of the axes of the two connecting rods, and the two rigid bodies are the bars AE and BG. The pole P is the intersection of the line of action of the external force F with the line joining B and H. The reaction R_A of the external hinge A must pass through the pole P, as well as through A. This is determined by the triangle of forces, together with the reaction R_B. The latter is equal and opposite to the resultant reaction transmitted by the two connecting rods to the section BG, while it is equivalent to the resultant reaction transmitted by the two connecting rods to the section AE. Figure 4.19(b) shows the triangles of forces for global equilibrium, for equilibrium of the section BG and for the equivalence of the reaction R_B with the sum of the two reactions R_E and R_C transmitted by the two connecting rods to the section AE.

We could equally have resolved the exercise with the method of auxiliary equations. The primary unknowns would have been the external ones, H_A, V_A, H_B, V_B, while the four resolving equations would have been those of global equilibrium plus the auxiliary equation of equilibrium with regard to the partial rotation of the section BG about the ideal hinge H.

4.5 Line of pressure

The set of lines of action of the successive resultant forces acting on a structure, or rather, that act as internal constraint reactions, proceeding from one end to the other of the structure itself, we call a **line of pressure**.

To illustrate this definition more clearly, consider the isostatic arch of Figure 4.20(a), subjected to four external forces, of which the resultant R is

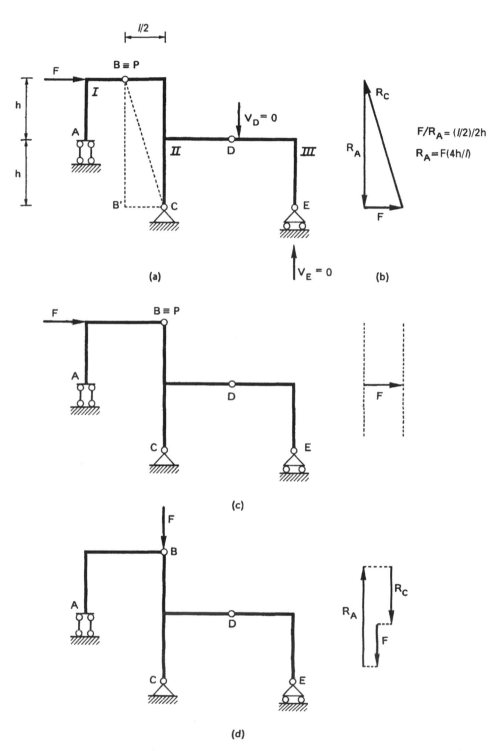

(a)

(b)

(c)

(d)

Figure 4.18

96

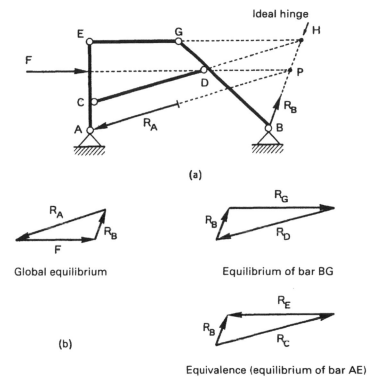

(a)

Global equilibrium

Equilibrium of bar BG

(b)

Equivalence (equilibrium of bar AE)

Figure 4.19

known, and to the two constraint reactions R_A and R_B, identified via the triangle of forces of Figure 4.20(b). Imagine then composing the constraint reaction R_A with the first external force F_1: the first partial resultant R_1 will then be given by the corresponding triangle of forces of Figure 4.20(b) and its line of action will pass through point P_1 of intersection of the lines of action of R_A and F_1. Compose then the first partial resultant R_1 with the second force F_2. The second partial resultant R_2 passes through the pole P_2 and is obtained from the triangle $R_1 F_2 R_2$ of Figure 4.20(b). We then proceed by adding each partial resultant vectorially with the following external force, thus obtaining the next partial resultant. Finally, by adding R_3 and F_4, we obtain the last resultant R_4, which must be a vector equal and opposite to the reaction R_B. In this way, the **polygon** (in this case, a hexagon) **of forces** closes (Figure 4.20(b)). It is made up of four triangular segments, each of which represents a vector sum (i.e. a problem of equivalence). Note once again that the vector that has as its foot that of F_1 and as its head that of F_4 represents the resultant R. The partial resultants R_i, $i = 1, ..., 4$, are thus the internal fixed joint reactions which each portion to the left of a generic section contained between F_i and $F_i + 1$ transmits to the complementary portion to the right of the same section. If in the triangular segments of Figure 4.20 (b) we invert the sense of the vectors R_i, $i = 1, ..., 4$, this means to consider a problem of equilibrium (instead of equivalence) and these vectors would represent, in this case, the internal fixed joint reactions which each right-hand portion transmits to the complementary left-hand portion.

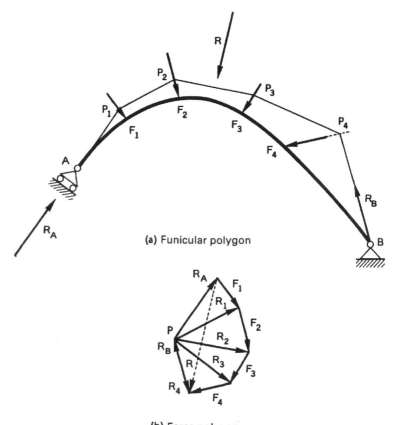

(a) Funicular polygon

(b) Force polygon

Figure 4.20

It may now be understood how the **funicular polygon** or **pressure line**, being the set of the lines of action of the successive resultants, is represented in Figure 4.20(a) by the broken line $AP_1P_2P_3P_4B$. The sides of the funicular polygon are parallel to the rays of the **polygon of forces**, represented in Figure 4.20(b) by the polygonal line $R_AF_1F_2F_3F_4R_B$.

Now examine the case of a continuous system of forces having equal direction $q(z)$ acting on the arch AB (Figure 4.21(a)). Integrating, we can obtain the resultant

$$F = \int_0^l q(z)\,dz \tag{4.17}$$

and its arm d with respect to the straight line $z = 0$

$$Fd = \int_0^l q(z)\,z\,dz \tag{4.18}$$

It will then be possible to obtain the constraint reactions R_A and R_B via the global triangle of forces (Figure 4.21(b)). In this case the polygon of forces reduces to a triangle, because the distributed forces are all acting in the same direction. This is made up of infinite infinitesimal triangular segments, each of

98

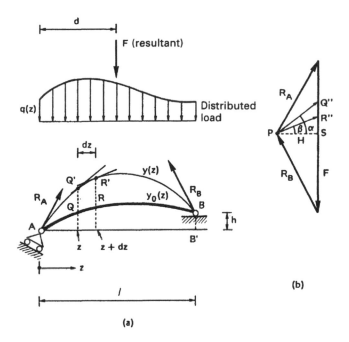

Figure 4.21

which represents the vector sum of the partial resultant with the subsequent increment of load $q(z)\,dz$.

Let us then take the pressure line to be known *a priori* and assume that it may be described analytically via the function $y(z)$. If Q and R represent two points of the arch infinitely close to one another of abscissae, respectively, z and $z + dz$, the corresponding partial resultants will be directed along the respective tangents to the pressure line (this being the envelope of the infinite lines of action of the subsequent resultants). These partial resultants will then be definable in their magnitudes via the rays PQ'' and PR'' of the triangle of forces of Figure 4.21(b). Their angles of orientation α and β will differ by an infinitesimal amount and will be given by the respective tangents to the pressure line.

The segment $Q''R''$ in Figure 4.21(b) represents the increment of distributed load

$$Q''R'' = q(z)\,dz \qquad (4.19)$$

On the other hand, some elementary geometrical considerations give

$$Q''R'' = Q''S - R''S = H(\tan\alpha - \tan\beta) \qquad (4.20)$$
$$= -H[y'(R) - y'(Q)]$$

where H indicates the magnitude of the horizontal component of R_A and R_B, whilst the prime indicates the first derivation with respect to the coordinate z. From equations (4.19) and (4.20), by the transitive law, we have

$$q(z)\,dz = -H\,dy' \qquad (4.21)$$

99

where the difference has been replaced by the differential, as the quantities are extremely close and as a pressure line is considered devoid of cusps (in fact there are no concentrated forces).

Finally, from relation (4.21) we obtain the differential equation of the pressure line for distributed loads acting in the same direction:

$$\frac{d^2 y}{dz^2} = -\frac{q(z)}{H} \qquad (4.22)$$

This is a second order differential equation and hence is to be combined with two boundary conditions, which are, in this case $y(0) = 0$, $y(l) = h$, the pressure line passing through the two end hinges. The first partial resultants which we meet going around the structure from the left or from the right are the reaction R_A or the reaction R_B, respectively. Hence the pressure line, besides passing through A and B, must be tangential at these points to the relative external reactions.

When the distributed load q is constant, the funicular curve is parabolic. We shall see various examples of this hereafter.

It is interesting to note that the pressure line, of equation $y(z)$, represents, but for one factor, the diagram of the bending moment. The partial resultant, PQ'', in fact, is made up of the horizontal force PS and the vertical force SQ'' (Figure 4.21(b)), and is applied in Q' (Figure 4.21(a)). Thus it is clear that the moment of internal reaction PQ'' is equal to the sum of the moments of PS and SQ''. While the former is equal to $H \times \overline{QQ'}$, the latter is zero because the corresponding arm vanishes.

Finally

$$M = H(y - y_0) \qquad (4.23)$$

where M is the so-called bending moment, H is the projection of the external reactions R_A and R_B on the normal to the direction of the external forces, y is the distance of the pressure line from the fundamental straight line AB' while y_0 indicates the distance of the axis of the arch from that fundamental. The segment intercepted between the pressure line and the axis of the beam thus represents, but for the factor H, the corresponding bending moment.

From relation (4.22) we then obtain

$$\frac{d^2 M}{dz^2} = -q(z) - H \frac{d^2 y_0}{dz^2} \qquad (4.24)$$

which is the differential equation of the bending moment in the case of a beam with a curvilinear axis. When the beam is rectilinear, we have $y_0 = hz/l$, and thus

$$\frac{d^2 M}{dz^2} = -q(z) \qquad (4.25)$$

If we imagine constructing an arch which presents exactly the form of the pressure line, we have $y = y_0$ and thus the bending moment vanishes at each point of the arch. Between one section and another only a compressive force would then be transmitted, as the internal reaction is always tangential to the axis of the curved beam. This is the situation that tends to occur when incoherent materials are used, i.e. those without tensile strength.

If all the forces acting on the arch, and thus also the reactions R_A and R_B, were inverted, only tensile internal reactions would be obtained. A string of

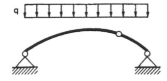

Figure 4.22

length equal to that of the pressure line would, in fact, be disposed according to the configuration of this line, it not being able to support other than tensile stresses; hence, the name **funicular curve**, from the Latin *funis* meaning cable or rope.

Consider a three-hinged parabolic arch, subjected to a uniform vertically distributed load q (Figure 4.22). The pressure line is parabolic and passes through the three hinges, where the bending moment vanishes. Recalling that only one parabola may pass through three given points, the pressure line must necessarily coincide with the axis of the arch, which will be found to be entirely in compression and devoid of bending moment.

Wide use has traditionally been, and still is, made of **arches** for buildings having wide spans, such as bridges. Usually it is the deck which transmits the vertical loads, which consist of its own weight and any live loads, to the supporting arch by means of connecting bars that can all be in compression (Figure 4.23(a)), all in tension (Figure 4.23(b)) or partly in compression and partly in tension (Figure 4.23(c)) according to the level at which the deck is disposed. Then there are **suspension bridges**, where the static scheme is inverted and the supporting element is represented by a cable in tension in a parabolic configuration (Figure 4.23(d)). In this case the elements of transmission are all tie rods.

We recall that, in the case where the vertical distributed load is not constant per unit of span, but constant per unit length of the arch, i.e. in practice it represents the weight of the arch itself assumed to be of uniform section, the pressure line is no longer exactly parabolic but assumes the form of a **catenary**.

Let us now examine another notable case in which the pressure line coincides with the axis of the arch. Let the three-hinged semicircular arch of Figure 4.24(a) be subjected to a constant radial distributed load q. For reasons of symmetry it is possible to consider only one half of the arch, if it is noted that the vertical reaction transmitted by the hinge A must vanish (Figure 4.24(b)). On the basis of the triangle of forces, the reactions R_A and R_B are thus of magnitude equal to qR. The moment transmitted by a generic section S of the arch, identified by the polar angle φ, may be calculated as the algebraic

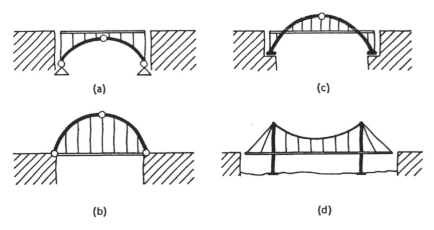

(a)

(c)

(b)

(d)

Figure 4.23

101

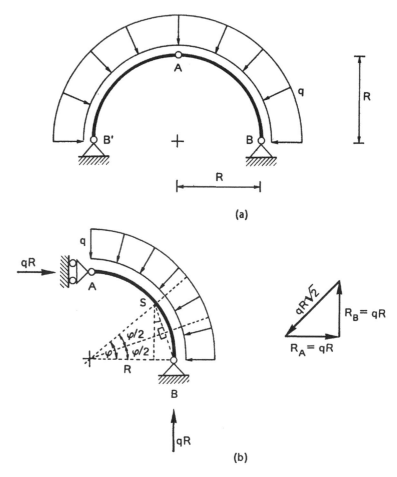

Figure 4.24

sum of the moment of the reaction R_B and the moment of forces distributed between 0 and φ. The first is counterclockwise and equal to

$$M_S(R_B) = qR^2(1 - \cos\varphi) \tag{4.26}$$

whilst the second is clockwise and equal to

$$M_S(q) = (2qR\sin(\varphi/2))\,(R\sin(\varphi/2)) \tag{4.27}$$

The resultant of a constant radial distribution q of forces is given by the product of q and the chord subtended by the arc on which the forces act. Since then from trigonometry we have

$$\sin(\varphi/2) = \left(\frac{1 - \cos\varphi}{2}\right)^{\frac{1}{2}} \tag{4.28}$$

it follows that the expressions (4.26) and (4.27) are equal and that, therefore, the moment M_S in the generic sections is zero.

To demonstrate rigorously that the pressure line coincides with the circular axis of the beam, it is necessary to demonstrate that, in addition to M_S, in each section S the radial internal reaction T_S also vanishes.

The radial reaction transmitted by the reaction R_B is equal to (Figure 4.24(b)).

$$T_S(R_B) = qR\sin\varphi \qquad (4.29)$$

whilst the radial reaction transmitted by the forces distributed between 0 and φ is equal to

$$T_S(q) = -\int_0^\varphi qR\cos(\varphi - \omega)\,\mathrm{d}\omega \qquad (4.30)$$

$qR\mathrm{d}\omega$ being the elementary contribution, acting at an inclined polar angle ω on the horizontal (Figure 4.25). Setting $x = \varphi - \omega$ and integrating equation (4.30), we obtain

$$T_S(q) = -qR\int_0^\varphi \cos x\,\mathrm{d}x \qquad (4.31)$$

$$= -qR\,[\sin x]_0^\varphi = -qR\sin\varphi$$

The contributions (4.29) and (4.31) cancel each other out and thus the pressure line coincides with the semicircumference of radius R.

The only reaction transmitted by the internal built-in constraints is the tangential one N_S. The tangential reaction transmitted by R_B is equal to (Figure 4.24(b))

$$N_S(R_B) = qR\cos\varphi \qquad (4.32)$$

while the tangential reaction transmitted by the radial forces between 0 and φ is equal to (Figure 4.25)

$$N_S(q) = \int_0^\varphi qR\sin(\varphi - \omega)\,\mathrm{d}\omega \qquad (4.33)$$

Setting again $x = \varphi - \omega$, we obtain

$$N_S(q) = qR\int_0^\varphi \sin x\,\mathrm{d}x \qquad (4.34)$$

$$= qR\,[-\cos x]_0^\varphi = qR(1 - \cos\varphi)$$

Summing up the axial forces (4.32) and (4.34) we have finally

$$N_S = qR \qquad (4.35)$$

The axial force is thus constant and compressive over the whole arch. Its absolute value coincides with that of reactions R_A and R_B.

In the simple cases considered in the previous chapter, consisting of one, two or, at the most, three rigid bodies, loaded by a concentrated force, the pressure line reduces to the set of a finite number of straight lines. In the case of the L-shaped beam of Figure 4.8(a), the pressure line consists of the line of action of the reaction R_A for all the points contained between A and the point of application of the force, and by the line of action of the reaction R_B for all the points contained between the point of application of the force and the

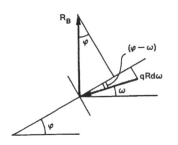

Figure 4.25

103

hinge *B*. In other words, it is a pair of straight lines and we pass discontinu-
ously from one to the other, passing over the point of application of the concentrated force.

For the bar system of Figure 4.5(a), the pressure line consists of the axes of
the two bars, whilst for the three-hinged arch of Figure 4.11, the pressure line
consists of the line joining *A* and *C* for all the points of the section *AC*, and of
the line joining *B* and *C* for all the points of the section *BC*. In the case of the
arch of Figure 4.12, the pressure line is the straight line *BP* for the points to
the right of the external force, while it is the straight line *AP* for the points
contained between *A* and the force.

For the beam constrained by the three connecting rods of Figure 4.9(a), the
pressure line consists of the line joining *A* and *P* for the points between *A* and
the point of application of the force, of the straight line *DP* for the points contained between the point of application of the force and the rod *B*, and finally
of the axis of the rod *C* for the points contained between the last two right-
hand connecting rods, *B* and *C*.

When the same beam is subjected to a concentrated moment *m* (Figure
4.10), the pressure line is represented by the axis of the rod *A* for the points
contained between *A* and the loaded point, by a parallel straight line passing
through point *D* for the points contained between *m* and *B*, and finally by the
axis of the rod *C* for the remaining points.

Now consider the three-hinged portal frame of Figure 4.26(a), subjected to
a vertical distributed load *q* on the right-hand section. The external reaction R_A
has as its line of action the line joining *A* and *B*, which intersects the line of
action of the resultant *ql* in the pole *P*. The second external reaction R_C passes
through points *C* and *P* and can be determined graphically, together with R_A,
by means of the triangle of forces. The pressure line for the left-hand section
will again be the line joining *A* and *B*, while for the section *BC'* this will be
composed of an infinite number of straight lines which have a parabolic envelope with a vertical axis. This parabola passes, of course, through the hinges *B*
and *C* and admits, in those points, of the lines of action of the reactions R_A and
R_C as its tangents. At this point the arc of parabola between *B* and *C* is defined,
since three data are sufficient to identify a second order parabola. Even though
two points of the parabola are already known with their corresponding tangents, to make the graphical construction easier, a third point will be identified
along with its tangent. Indicate by *P"* the intersection of the vertical through *P*
with the line joining the extreme points *B* and *C*, and by *P'* the midpoint of the
segment *PP"*. Then draw through *P'* a straight line *p* parallel to the line joining
B and *C*. It is possible to demonstrate that *P'* and the latter line *p* constitute the
third point and the third tangent which we had set out to obtain. It is now
extremely easy to draw the arc of parabola, since this must pass through *B*, *P'*
and *C*, and it is inscribed in the polygonal line made up of the straight lines
AP, *p* and *PC* (Figure 4.26(a)). In the section *CC'* the pressure line is represented by the straight line *CP*.

Consider the case of a horizontal distributed load, acting on the left-hand
section of the portal frame previously studied (Figure 4.26(b)). While for the
section *CA'*, which is not externally loaded, the pressure line is represented by
the line joining the two hinges *B* and *C*, for the section *AA'* the pressure line is
represented by an arc of parabola with horizontal axis, which has as its
extreme points *A* and *B* and as its tangents at those points the lines of action of
the reactions R_A and R_B. It will then be possible, as shown previously, to iden-

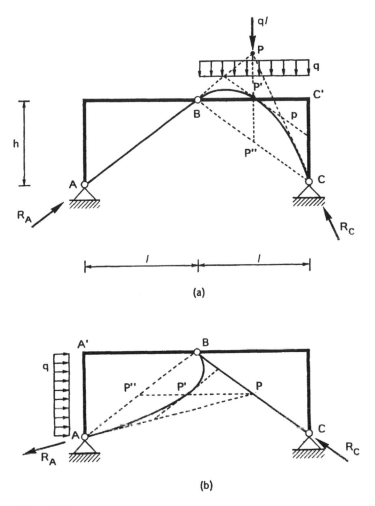

Figure 4.26

tify a third point P' and a third tangent parallel to the line joining the extreme points. Note that, if the axis of the structure were now conceived with the same form as the funicular curve found, the section AB would be found to be completely in tension and the section BC would act as a strut (connecting bar in compression).

As our last example, consider the asymmetrical portal frame of Figure 4.27, already studied more than once, loaded by a constant distribution of vertical forces on the right-hand section. The horizontal reaction of the double rod passes through the pole P and constitutes, with its line of action, one of the two extreme tangents to the arc of parabola with a vertical axis, which represents the pressure line for the section BB'. This arc must be contained in the vertical strip of the plane containing also the vertical distribution q. The second tangent to the arc of parabola is, of course, given by the straight line AB. We therefore find, finally, that the pressure line for the section CB' is represented by the line of action of the reaction R_C, for the section AB it is

105

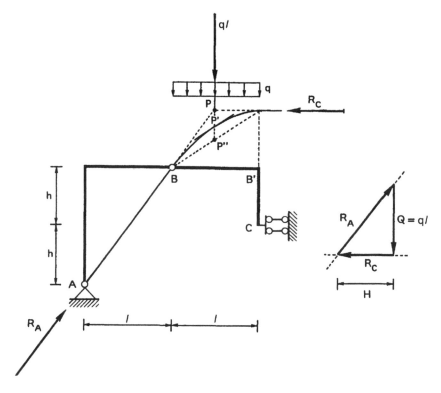

Figure 4.27

represented by the line joining A and B, while for the loaded section BB' it is represented by the arc of parabola connecting the previous two straight lines.

Note that, in the case of a concentrated load equal to the resultant of the distributed load under examination, the pressure line would be made up of the two above-mentioned extreme straight lines only, passing from one to the other with discontinuity in crossing the point of application of the force. Hence there derives from this a sort of rounding off brought about by the continuous, rather than discrete, distribution of the external forces.

106

5 Internal beam reactions

5.1 Introduction

The **characteristics of internal reaction** for a beam section are, as stated in the earlier chapters, the elementary internal reactions transmitted by the section itself. In the case of a plane beam (line of axis contained in the plane), there are three characteristics of internal reaction (Figure 5.1):

1. **axial force**, which is the component of the force tangential to the axis of the beam;
2. **shearing force** (or, more simply, **shear**), which is the component of the force perpendicular to the axis of the beam;
3. **bending moment**, which is the moment of the force that the two portions of beam transmit to one another, with respect to the section being considered.

The usual conventions regarding the signs of the characteristics in the plane are the following:

1. axial force is taken to be positive when it is tensile;
2. shearing force is taken to be positive when it tends to rotate the segment of the beam on which it acts in a clockwise direction;
3. bending moment is taken to be positive when it stretches the lower fibres and compresses the upper fibres of the beam.

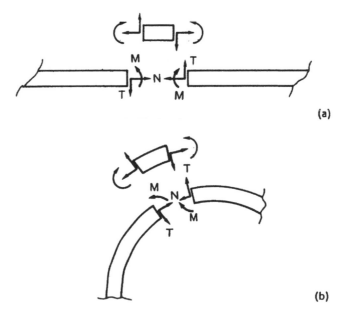

(a)

(b)

Figure 5.1

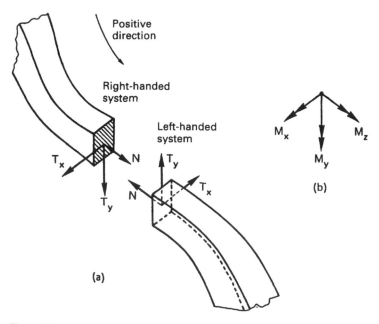

Figure 5.2

It is at once evident that these conventions are purely arbitrary and, except in the case of axial force, relative to the observer's orientation (bending moment) or half-space of observation (shear).

In the case of a beam with skewed axis, it becomes necessary to establish a direction in which we consider the axis of the beam, and to establish, for each section, an intrinsic reference system consisting of the tangent, the normal and the binormal to the curve (Figure 5.2). More precisely, the reference system will be right-handed, with the Z axis oriented according to the tangent and in the direction chosen for considering the axis, the Y axis oriented according to the normal, and the X axis according to the binormal. The force transmitted by the lower section to the upper section must be projected on these axes and the respective components are the **axial force** N, the **shearing force** T_y and the **shearing force** T_x (Figure 5.2(a)). On the other hand, the opposite force transmitted by the upper section to the lower section must be projected on to the left-handed reference system, opposite to the system previously considered, in order to obtain characteristics of the same sign as the previous one. As regards the moment vector mutually exchanged between the two beam portions (Figure 5.2(b)), conventions and considerations altogether analogous to those set out above hold good also here. The moment vector comprises three components. The one along the Z axis is called the **twisting moment** M_z, whilst the remaining two are the bending moments M_y and M_x. In the case, therefore, of the beam in three-dimensional space, there are six characteristics of the internal reaction: axial force, two shearing forces, twisting moment and two bending moments. Also in the three-dimensional case the conventions are arbitrary and the signs of the characteristics come to depend on the direction in which we consider the axis of the beam.

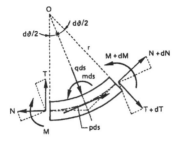

Figure 5.3

5.2 Indefinite equations of equilibrium for plane beams

We shall now deduce the differential equations that govern the equilibrium of a plane beam. For this purpose consider an infinitesimal element of the beam, bounded by two right sections, i.e. perpendicular to the axis of the beam (Figure 5.3). Indicate by O the centre of curvature of the axis of the beam in relation to the element considered. If r is the radius of curvature, ds the increment of the curvilinear coordinate in the direction of the axis of the beam and $d\vartheta$ the angle formed by the two end sections of the element (positive if the upper section is superposed on the lower section rotating in a counterclockwise direction), we have by definition.

$$ds = r\, d\vartheta \qquad (5.1)$$

In this way the radius of curvature r, as well as the increments, assumes an algebraic sign (in Figure 5.3, r is positive). We point out that, if instead the element were observed from the opposite half-space, the radius r would be negative.

The infinitesimal element of the beam is in general subjected to an axial distributed load $p(s)ds$, a transverse distributed load $q(s)ds$ and a distributed moment $m(s)ds$, as well as to the characteristics N, T, M, at the upward end, and to the incremented characteristics $N + dN, T + dT, M + dM$, at the downward end (Figure 5.3). Noting that $d\vartheta/2$ represents both the angle between the median radial line and the end sections, and the angle between the median tangent and the extreme tangents, it is possible to impose equilibrium with regard to translation of the element in the directions of the median tangent and the median radial line, respectively:

$$p\,ds - N\cos\frac{d\vartheta}{2} + (N+dN)\cos\frac{d\vartheta}{2}$$
$$+T\sin\frac{d\vartheta}{2} + (T+dT)\sin\frac{d\vartheta}{2} = 0 \qquad (5.2a)$$

$$q\,ds - N\sin\frac{d\vartheta}{2} - (N+dN)\sin\frac{d\vartheta}{2}$$
$$-T\cos\frac{d\vartheta}{2} + (T+dT)\cos\frac{d\vartheta}{2} = 0 \qquad (5.2b)$$

As regards rotational equilibrium, it is expedient to choose as centre of reduction the point of intersection of the extreme tangents, i.e. of the lines of action of the axial forces on the end sections:

$$m\,ds + dM - p\,ds\left(\frac{r}{\cos\left(\dfrac{d\vartheta}{2}\right)} - r\right) -$$
$$T r\tan\frac{d\vartheta}{2} - (T+dT)r\tan\frac{d\vartheta}{2} = 0 \qquad (5.2c)$$

Since the angle $d\vartheta/2$ is assumed to be infinitesimal, it is legitimate to take this angle as equal to its sine or its tangent, and put $\cos(d\vartheta/2) \approx 1$. By so doing, equations (5.2) are transformed as follows:

$$p\,ds + dN + T\,d\vartheta = 0 \qquad (5.3\text{a})$$

$$q\,ds - N\,d\vartheta + dT = 0 \qquad (5.3\text{b})$$

$$m\,ds + dM - Tr\,d\vartheta = 0 \qquad (5.3\text{c})$$

Dividing the foregoing equations by ds and applying the relation (5.1), we obtain the **indefinite equations of equilibrium** for the beam:

$$\frac{dN}{ds} + \frac{T}{r} + p = 0 \qquad (5.4\text{a})$$

$$\frac{dT}{ds} - \frac{N}{r} + q = 0 \qquad (5.4\text{b})$$

$$\frac{dM}{ds} - T + m = 0 \qquad (5.4\text{c})$$

In the case where the radius of curvature r is not a function of the curvilinear coordinate s, namely that of **circular arches** and **rings**, equations (5.4) can be presented in the following form:

$$\frac{dN}{d\vartheta} + T + pR = 0 \qquad (5.5\text{a})$$

$$\frac{dT}{d\vartheta} - N + qR = 0 \qquad (5.5\text{b})$$

$$\frac{dM}{d\vartheta} - TR + mR = 0 \qquad (5.5\text{c})$$

where the independent variable is represented by the angular coordinate ϑ and R is the radius of the circular axis of the beam.

Equations (5.5), like equations (5.4), form a system of three linear differential equations of the first order in the three unknown functions N, T, M. It is possible to decouple the function M from the other two and obtain a third order differential equation, where only the unknown $M(\vartheta)$ appears.

From equation (5.5c) we have

$$T = m + \frac{1}{R}\frac{dM}{d\vartheta} \qquad (5.6)$$

so that equations (5.5a, b) are transformed as follows:

$$\frac{dN}{d\vartheta} + m + \frac{1}{R}\frac{dM}{d\vartheta} + pR = 0 \qquad (5.7\text{a})$$

$$\frac{dm}{d\vartheta} + \frac{1}{R}\frac{d^2M}{d\vartheta^2} - N + qR = 0 \qquad (5.7\text{b})$$

From equation (5.7b) then we obtain

$$N = qR + \frac{dm}{d\vartheta} + \frac{1}{R}\frac{d^2M}{d\vartheta^2} \qquad (5.8)$$

110

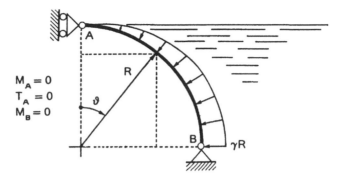

$M_A = 0$
$T_A = 0$
$M_B = 0$

Figure 5.4

and hence from equation (5.7a) there follows

$$\frac{d^3 M}{d\vartheta^3} + \frac{dM}{d\vartheta} = -R^2 \left(p + \frac{dq}{d\vartheta} \right) - R \left(m + \frac{d^2 m}{d\vartheta^2} \right) \qquad (5.9)$$

Equation (5.9) is a non-homogeneous third order differential equation, which has the following complete integral:

$$M(\vartheta) = M_0(\vartheta) + C_1 \sin \vartheta + C_2 \cos \vartheta + C_3 \qquad (5.10)$$

where $M_0(\vartheta)$ indicates the particular solution. In the case, for instance, of the circular arch of Figure 5.4, subjected to a hydrostatic load, we have

$$p(\vartheta) = m(\vartheta) = 0 \qquad (5.11a)$$

$$q(\vartheta) = -\gamma R (1 - \cos \vartheta) \qquad (5.11b)$$

where γ indicates the specific weight of the fluid exerting pressure. The three boundary conditions are those which cause the corresponding characteristics to vanish at the two ends: $M(A) = M(B) = T(A) = 0$. In order to resolve the above analytical problem, it is possible to apply the method of variation of the arbitrary constants (Appendix A).

In the case where the radius of curvature r tends to infinity, namely that of **rectilinear beams** (Figure 5.5), and in the absence of distributed moments ($m = 0$), the indefinite equations of equilibrium (5.4) reduce to the following:

$$\frac{dN}{dz} = -p(z) \qquad (5.12a)$$

$$\frac{dT}{dz} = -q(z) \qquad (5.12b)$$

$$\frac{dM}{dz} = T \qquad (5.12c)$$

The first of these tells us that the derivative of the axial force with respect to the axial coordinate is equal to the opposite of the axial distributed load; the second that the derivative of the shearing force is equal to the opposite of the distributed load perpendicular to the axis; and finally, the third that the derivative of the bending moment is equal to the shearing force. In addition, deriving both sides of equation (5.12c) and taking into account equation (5.12b), we

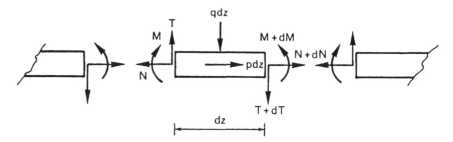

Figure 5.5

again obtain equation (4.25), which in the previous chapter was obtained on the basis of considerations involving the pressure line.

In the absence of distributed loads, then, both the axial force and the shearing force are constant, while the bending moment is a linear function of the z coordinate. The diagram of the characteristics is to be studied for the portions of beam contained between one concentrated load and another. At the points of application of the concentrated loads there emerges, instead, the discontinuity of the corresponding characteristics. When the distributed loads are constant, on the other hand, both the axial force and the shearing force are linear functions of the z coordinate, while the bending moment is a second order parabolic function. More generally, when the distributed loads are polynomial functions of order n, the axial force and the shearing force are polynomial functions of order $(n + 1)$, while the bending moment is a polynomial function of order $(n + 2)$.

We shall now apply the indefinite equations of equilibrium (5.12) to determine analytically the functions M, T, N, for the inclined rectilinear beam of Figure 5.6(a), which is subjected to a uniform vertical distributed load $q_0 = F/l$, where F is the resultant force and l the projection of the beam on the horizontal. The triangle of equilibrium gives the constraint reactions $R_A = q_0 l/\tan \beta$, $R_B = q_0 l/\sin \beta$, β being the angle that the line joining B and P forms with the horizontal (Figure 5.6(b)). The pressure line is a parabola with vertical axis which passes through the extreme points of the beam and presents as extreme tangents the lines of action of the two external reactions. Later we shall verify how this represents, but for a factor of proportionality, the diagram of the bending moment.

The vertical distributed load per unit length of the beam is

$$q^* = \frac{q_0 l}{l/\cos \alpha} = q_0 \cos \alpha \qquad (5.13)$$

where α is the angle of inclination of the beam on the horizontal and $l/\cos \alpha$ is the length of the beam. The axial component and the component perpendicular to the axis of the distributed load q^* are then equal to (Figure 5.6(c)).

$$p(z) = q_0 \cos \alpha \sin \alpha \qquad (5.14a)$$

$$q(z) = q_0 \cos^2 \alpha \qquad (5.14b)$$

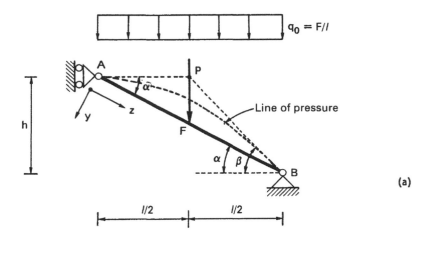

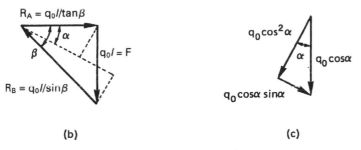

(a)

(b)   (c)

Figure 5.6

The differential equation which governs the bending moment will thus be the following:

$$\frac{d^2 M}{dz^2} = -q_0 \cos^2 \alpha \qquad (5.15)$$

with the boundary conditions

$$M(0) = M(l/\cos\alpha) = 0 \qquad (5.16)$$

The complete integral of the function $M(z)$ contains two constants which depend on the foregoing boundary conditions:

$$M(z) = -q_0 \cos^2 \alpha \frac{z^2}{2} + C_1 z + C_2 \qquad (5.17)$$

Applying the two conditions (5.16), we obtain two algebraic equations in the two unknowns C_1 and C_2:

$$M(0) = C_2 = 0 \qquad (5.18a)$$

$$M(l/\cos\alpha) = -\frac{1}{2}q_0 l^2 + C_1 \frac{l}{\cos\alpha} = 0 \qquad (5.18b)$$

113

whence

$$C_1 = \frac{1}{2}q_0 l \cos\alpha, \quad C_2 = 0 \qquad (5.19)$$

The moment function is then given by

$$M(z) = \frac{1}{2}q_0 l z \cos\alpha \left(1 - \frac{z}{l}\cos\alpha\right) \qquad (5.20)$$

for $0 \le z \le l/\cos\alpha$.

The diagram of the moment function, given in Figure 5.7(a) on the side of the fibres in tension, represents a parabola having the axis perpendicular to the beam. Note that the distribution is symmetrical with respect to the centre of the beam and that the maximum that is reached in the centre is independent of the inclination α of the beam:

$$M_{max} = M(l/2\cos\alpha) = \frac{1}{8}q_0 l^2 \qquad (5.21)$$

The shearing force can be found by derivation of the moment function (5.20):

$$T(z) = \frac{\mathrm{d}M}{\mathrm{d}z} = \frac{1}{2}q_0 l \cos\alpha \left(1 - \frac{2z}{l}\cos\alpha\right) \qquad (5.22)$$

The diagram of the shear function (Figure 5.7(b)) is linear and skew-symmetrical with respect to the centre of the beam, where it vanishes (stationary point of the bending moment).

The axial force, finally, may be obtained from the first of equations (5.12):

$$\frac{\mathrm{d}N}{\mathrm{d}z} = -q_0 \cos\alpha \sin\alpha \qquad (5.23)$$

which gives

$$N(z) = -q_0 z \cos\alpha \sin\alpha + C \qquad (5.24)$$

where the constant C is determined by imposing a suitable boundary condition. It is possible, for example, to consider that, at the end A, the axial compressive force coincides with the component of the reaction R_A along the axis of the beam (Figure 5.6(b)).

$$N(0) = C = -q_0 l \frac{\cos\alpha}{\tan\beta} \qquad (5.25)$$

hence we have

$$N(z) = -q_0 \cos\alpha \left(z \sin\alpha + \frac{l}{\tan\beta}\right) \qquad (5.26)$$

It is possible, on the other hand and in an equivalent manner, to assume that the axial force at the end B is compressive and coincides with the component of the reaction R_B along the axis of the beam (Figure 5.6(b)).

$$N(l/\cos\alpha) = -q_0 l \sin\alpha + C = -q_0 l \frac{\cos(\beta-\alpha)}{\sin\beta} \qquad (5.27)$$

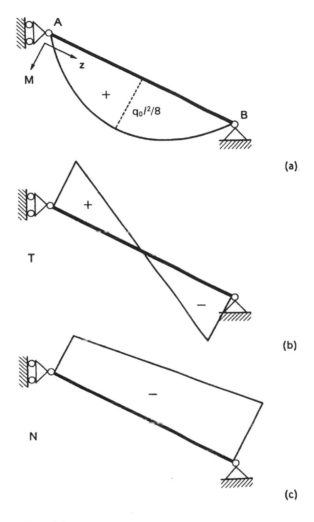

Figure 5.7

From equation (5.27) we obtain

$$C = -q_0 l \left(\frac{\sin \alpha \sin \beta + \cos \alpha \cos \beta}{\sin \beta} - \sin \alpha \right) = -q_0 l \frac{\cos \alpha}{\tan \beta} \qquad (5.28)$$

which confirms equation (5.25).

The trapezoidal diagram of the axial force appearing in Figure 5.7(c) shows how the beam is entirely in compression and how the maximum of this force is reached at the end B.

The distinction between distributed load per unit of horizontal projection and distributed load per unit length of the beam is necessary, for example, also in the case of the circular arch of Figure 5.8, subjected to a vertical distributed load q_0, which is uniform if considered per unit length of horizontal span. Just as was seen in the case of the inclined rectilinear beam (cf. equation (5.14)), the elementary components of the distributed load are

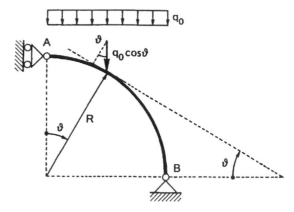

Figure 5.8

$$p(\vartheta) = q_0 \cos \vartheta \sin \vartheta \qquad (5.29a)$$

$$q(\vartheta) = -q_0 \cos^2 \vartheta \qquad (5.29b)$$

The full solution for the above example is given in Appendix B.

5.3 Diagrams of characteristics of internal reaction: direct method and graphical method

In the last section we introduced the characteristics of internal reaction and the differential equations governing them. It has thus been possible to note how the problem of the determination of these functions can be set and resolved in a purely analytical manner. In this section, we shall see how it is possible to approach the same problem from a direct point of view, that is, imposing equilibrium on finite portions of beam, subjected to external loads, to known constraint reactions and to unknown characteristics. On these bases, it is possible in many cases to draw the diagrams of the characteristics M, T, N, using purely graphical procedures.

To start with, let us consider the so-called **elementary schemes**, which, on account of their simplicity, recur very frequently, also being inserted within more complex structural schemes. A built-in rectilinear beam, known as a **cantilever beam**, is subjected to a force F perpendicular to its axis and with the point of application in the end B (Figure 5.9(a)). The built-in support A reacts with a force equal and opposite to the external one, so that the pressure line coincides with the line of action of F for all the points of the beam. On the other hand, the reaction R_A can be thought of as acting in point A together with the counterclockwise moment of transport $M_A = R_A l = Fl$. The built-in support A will react, therefore, transmitting to the beam a positive shear F and a negative bending moment $-Fl$. The diagram of the bending moment is linear, owing to the absence of distributed loads $q(z)$, and vanishes at point B, with respect to which the external force has a zero arm (Figure 5.9(b)). To draw it, it will be sufficient to perform a simple graphical operation, joining point B with the upper end of the segment that represents the moment in the built-in support $-Fl$. This operation, however commonplace it may be, involves a series of logical steps which we shall endeavour to illustrate.

116

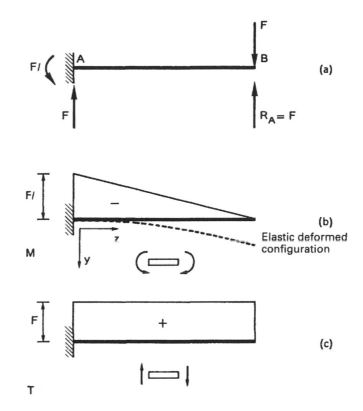

Figure 5.9

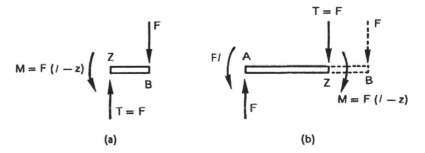

Figure 5.10

The fact that, in a generic section of the cantilever beam, the bending moment is $M(z) = -F(l - z)$ and the shear is $T(z) = F$ (Figure 5.9(c)) means that the beam portion ZB is found to be in equilibrium under the action of the external force F, the positive shear F and the counterclockwise bending moment $F(l-z)$ (Figure 5.10(a)). The last two loads are the internal reactions transmitted by the left-hand portion AZ to the one being considered, ZB. On the other hand, it also means that the portion AZ (Figure 5.10(b)) is in equilibrium under the action of the counterclockwise fixed-end moment Fl, the vertical fixed-end reaction F, the positive shear F and the clockwise bending moment $F(l - z)$. To determine,

117

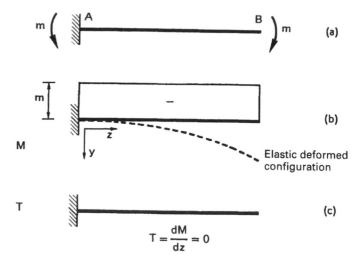

Figure 5.11

section by section, the internal characteristics M and T, it will suffice, therefore, to consider a problem of equilibrium for either of the two portions into which the beam is divided by the section under examination. Similarly, it is possible, instead, to consider a problem of equivalence and to transfer into the section under examination all the forces acting upward of the section itself, applying this equivalent system to the complementary portion of the beam.

If the cantilever beam is loaded by a concentrated moment at the end B (Figure 5.11(a)), the built-in support reacts with an equal and opposite moment, so that each partial portion of the beam, whether finite or infinitesimal, is in equilibrium under the action of two opposite moments. The line of pressure is the straight line at infinity, since the resultant force of a couple is zero and tends to act at infinity. The moment diagram is constant and negative (Figure 5.11(b)), while the shear is identically zero (Figure 5.11(c)), being equal to the derivative of a constant function and there being no vertical forces involved.

The last elementary scheme for the cantilever beam is that of the uniform distributed load q (Figure 5.12(a)). The reaction of the built-in support A is equal and opposite to the resultant ql of the distributed load. The pressure line is thus degenerate and consists of the sheaf of vertical straight lines contained between the end B and the midpoint C (Figure 5.12(a)). The bending moment in one generic section of abscissa z is thus (Figure 5.12(b)).

$$M(z) = -q\frac{(l-z)^2}{2} \tag{5.30}$$

while the shear is (Figure 5.12(c))

$$T(z) = q(l-z) \tag{5.31}$$

These two functions have been obtained by reducing, in the section under consideration, the system of forces acting to the right of the section. This reduction of the forces which precede a generic section is referred to as the **direct method**.

118

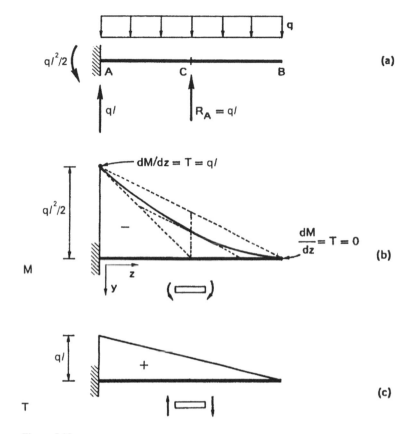

Figure 5.12

It is, on the other hand, possible to proceed in a completely graphical manner, using the moment diagram for a concentrated force equal to the resultant of the system and acting in the middle of the beam. This diagram is linear and presents the zero value in the centre C and the maximum absolute value $ql^2/2$ at the built-in support A (Figure 5.12(b)). In actual fact, the moment diagram for the distribution of forces q is parabolic, and presents a zero value at the end B and a maximum absolute value equal to the previous one of $ql^2/2$ at the built-in support A, while its tangent in A coincides with the linear diagram described formerly, the shear being transmitted by the built-in support in a like manner in both cases. The tangent at B is then horizontal, the shear vanishing at that point. We thus have the two extreme points with the corresponding tangents to the arc of parabola that is sought. It is then simple, by applying the graphical construction already illustrated in the previous chapter, to identify a third point with its corresponding tangent and to draw the diagram of $M(z)$ precisely.

As regards the shear (Figure 5.12(c)), the graphical construction of the diagram is immediate, if we join the end of the cantilever beam with the upper end of the segment that represents the vertical reaction of the built-in support ql. In this case it is also possible to verify that the function $T(z)$ is the derivative of the function $M(z)$.

119

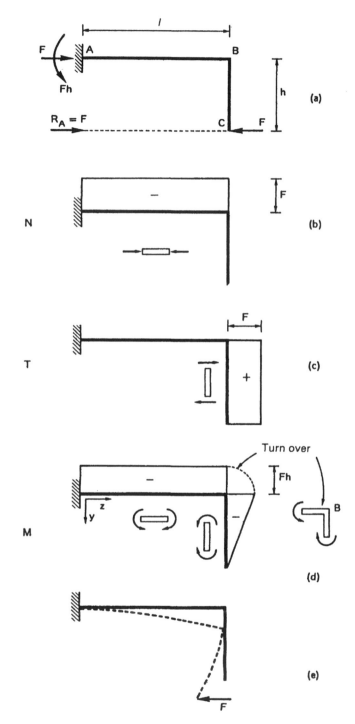

Figure 5.13

120

Now consider the L-shaped cantilever beam of Figure 5.13(a), subjected to a horizontal concentrated load at the end C. The reaction of the built-in support R_A is a force equal and opposite to the external one, so that, reduced in point A, it is equivalent to a horizontal reaction F and to a fixed-end moment equal to the moment of translation Fh. The axial force (Figure 5.13(b)) is zero on the portion CB, while it is compressive and of an absolute value equal to F on the portion BA. The shearing force (Figure 5.13(c)) is, *vice versa*, zero on the portion BA and equal to F on the portion CB. Finally, the bending moment (Figure 5.13(d)) increases linearly in absolute value proceeding from the end C to the knee B. From B to A the absolute value remains constant and equal to the product of the force and the arm h. The algebraic sign of the bending moment depends on the conventions (taking those of Figure 5.13(d) it is negative). However, what is independent of the reference system and physically important is the part (or edge) of the beam in which the longitudinal fibres are in tension. As has already been said, it is customary to draw the moment diagram (whatever sign it may have) on the side of the fibres in tension. This is illustrated by the elastic deformed configuration of Figure 5.13(e). From the graphical point of view, we use the term **overturning** of the value of the moment at B, implying by this the equilibrium to rotation of the built-in node B (Figure 5.13(d)).

If the cantilever beam has a skewed axis, the characteristics that could be present total six. An example is shown in Figure 5.14, where the intrinsic reference system has been highlighted for each rectilinear portion of the beam. In the portion AB, only two characteristics are different from zero: $T_y = F$, $M_x = Fz$. In the portion BC, we have $N = -F$, $M_x = Fb$. In the last portion CD, there also emerges the internal reaction of twisting moment: $T_x = -F$, $M_y = Fz$, $M_z = -Fb$.

Also in plane cantilever beams, if these are loaded with forces not contained in the plane, there is present the internal characteristic M_z. In the case, for example, of the semicircular cantilever beam of Figure 5.15(a), loaded

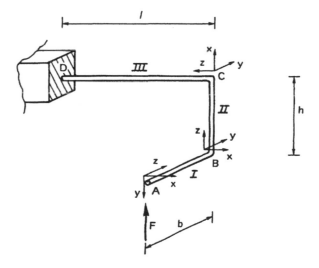

Figure 5.14

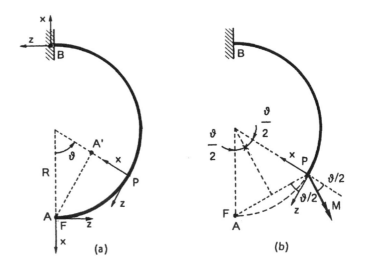

Figure 5.15

with a force F perpendicular to the plane and applied at the end A (the direction assumed looking into the plane of the diagram) we have

$$N = 0 \tag{5.32a}$$

$$T_x = 0 \tag{5.32b}$$

$$T_y = -F \tag{5.32c}$$

$$M_z = FR(1 - \cos \vartheta) \tag{5.32d}$$

$$M_x = -FR \sin \vartheta \tag{5.32e}$$

$$M_y = 0 \tag{5.32f}$$

The moments M_z and M_x can be determined in two different ways. The first consists of transferring the force F at A' (Figure 5.15(a)), adding the first moment of translation M_x, and subsequently in P, adding the second moment of translation M_z. The second way consists of considering the moment vector of the force F with respect to the generic point P (double-headed arrow in Figure 5.15(b)) and projecting this vector onto the left-handed XYZ reference system. The total moment has the magnitude

$$M = 2FR \sin(\vartheta/2) \tag{5.33}$$

and thus

$$M_z = M \sin(\vartheta/2) = FR(1 - \cos \vartheta) \tag{5.34a}$$

$$M_x = -M \cos(\vartheta/2) = -FR \sin \vartheta \tag{5.34b}$$

Now consider a rectilinear beam, hinged at one end and constrained with a horizontally moving roller support at the other (Figure 5.16(a)). This elementary scheme is referred to as the **supported beam** and will be studied in the various cases of external loading. Let the supported beam be subjected to the concentrated force F acting on its centre. The constraint reactions consist of two vertical forces, each equal to $F/2$. The horizontal reaction of the hinge is,

122

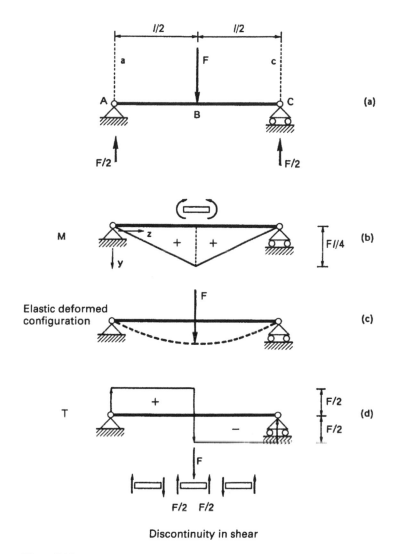

Figure 5.16

in fact, zero by equilibrium to horizontal translation. The pressure line will thus consist of the line of action of the corresponding constraint reaction for each of the two portions AB and BC. The moment diagram will consist of two linear functions, symmetrical with respect to the centre (Figure 5.16(b)):

$$M(z) = \frac{F}{2} z, \qquad \text{for} \quad 0 \leqslant z \leqslant \frac{l}{2} \qquad (5.35a)$$

$$M(z) = \frac{F}{2} z - F\left(z - \frac{l}{2}\right), \qquad \text{for} \quad \frac{l}{2} \leqslant z \leqslant l \qquad (5.35b)$$

The function (5.35b), which emphasizes the sum of the two contributions, may be rewritten as follows:

123

$$M(z) = -\frac{F}{2}z + F\frac{l}{2} \qquad (5.36)$$

The maximum of the function is obtained in the centre and equals $Fl/4$. This maximum is not a stationary point but a cusp of the function, i.e. the left-hand derivative appears different from the right-hand one. The bitriangular diagram $M(z)$ can be drawn graphically, referring to the symmetry of the problem and joining the end points A and C of the beam with the lower end of the vertical segment which represents the moment in the centre. The diagram is drawn from the side of the lower longitudinal fibres, which are the ones physically in tension (Figure 5.16(c)).

The shear diagram, on the other hand, is birectangular and skew-symmetrical with respect to the centre (Figure 5.16(d)). It represents exactly the derivative of the moment function. Where the function $M(z)$ presents a cusp, its derivative $T(z)$ presents a discontinuity of the first kind, i.e. a negative jump. The infinitesimal element of the beam straddling the centre is, in fact, in equilibrium under the action of the external force F and the two shearing forces $F/2$ both directed upwards (Figure 5.16(d)).

Let the supported beam be subjected to the concentrated moment m acting in the centre (Figure 5.17(a)). The two constraint reactions in this case will be opposite and equal to m/l so as to form a couple equal and opposite to the one applied externally. The pressure line, as before, consists of two vertical

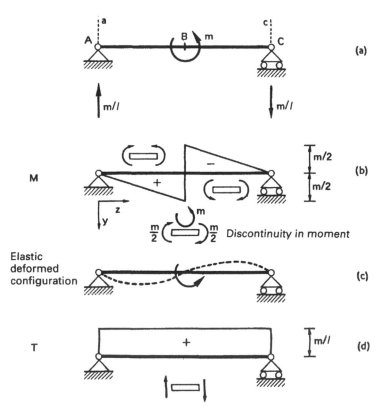

Figure 5.17

124

straight lines passing through A and through C. The moment diagram consists of two linear functions, which are skew-symmetrical with respect to the centre (Figure 5.17(b)).

$$M(z) = \frac{m}{l} z, \qquad \text{for} \quad 0 \le z \le \frac{l}{2} \qquad (5.37a)$$

$$M(z) = \frac{m}{l} z - m, \qquad \text{for} \quad \frac{l}{2} \le z \le l \qquad (5.37b)$$

In the centre a discontinuity of the first kind is thus created, i.e. a positive jump, equal to the concentrated moment applied there. The infinitesimal element straddling the centre will thus be in equilibrium with regard to rotation under the action of the counterclockwise external moment m and the two bending moments $m/2$, both clockwise. The elastic deformed configuration presents an inflection in the centre, so that the fibres in tension appear below in the portion AB and above in the portion BC (Figure 5.17(c)).

The shear diagram is constant, positive and equal to the magnitude of the constraint reactions (Figure 5.17(d)). In fact, going along the axis of the beam and encountering the moment m, no contribution is added to the vertical force. On the other hand, the derivative of the function $M(z)$ of Figure 5.17(b) is defined and equal to m/l in each section. It would not be analytically defined only in the centre, where, instead, physically it is defined and equal to the left-hand and right-hand derivatives.

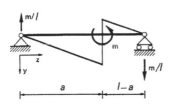

Figure 5.18

Let us imagine applying the moment m in a generic section, other than that of the centre (Figure 5.18). The constraint reactions are the same as in the previous case, so that the moment diagram appears still made up of two linear segments of equal inclination (Figure 5.18):

$$M(z) = \frac{m}{l} z, \qquad \text{for} \quad 0 \le z \le a \qquad (5.38a)$$

$$M(z) = \frac{m}{l} z - m, \qquad \text{for} \quad a \le z \le l \qquad (5.38b)$$

The shear diagram is obviously identical to the previous one (Figure 5.17(d)).

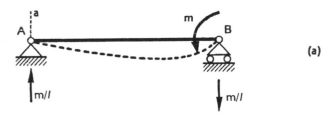

(a)

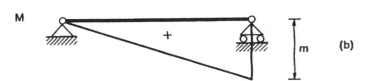

(b)

Figure 5.19

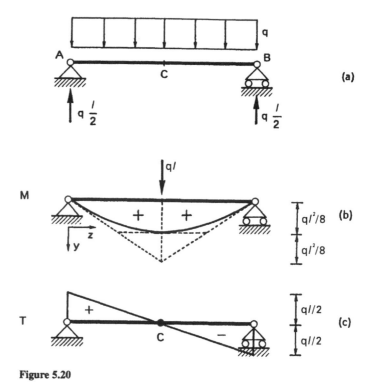

Figure 5.20

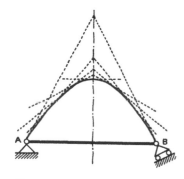

Figure 5.21

When, finally, the concentrated moment is applied to one of the two ends of the beam (Figure 5.19(a)), the moment diagram reduces to a single linear function which has as its maximum the value of the applied moment itself (Figure 5.19(b)). In this case the pressure line consists of the single straight line a, and the inflection of the elastic deformed configuration disappears, leaving the fibres in tension always underneath. The shear diagram, of course, coincides with that of Figure 5.17(d).

As a last elementary scheme, consider the supported beam subjected to the distributed load q (Figure 5.20(a)). The constraint reactions are two vertical forces having the same direction equal to $ql/2$, so that the pressure line is made up of the sheaf of vertical straight lines external to the beam. Imagine going along the axis of the beam, starting from the end A. In the end section A, the pressure line is the line of action of the constraint reaction. In a section immediately to the right of A, the reaction R_A will have to be composed with the partial resultant of the distributed forces $q\mathrm{d}z$. Since the two forces have opposite direction, their resultant will be a vertical force $(R_A - q\mathrm{d}z)$ passing to the left of A. If we increase the coordinate z, the line of action of the subsequent resultants remains vertical but departs more and more from point A, until it reaches infinity for $z = l/2$. The system of the forces acting to the left of the centre is, in fact, equivalent to a couple of moment $ql^2/8$.

Note that, if the roller support B were not moving horizontally (Figure 5.21), the pressure line would be made up of all the tangents enveloping the arc of parabola which has as its extreme points A and B and as its extreme tan-

126

gents the lines of action of the constraint reactions. Considering the foregoing case as a limiting case, we can understand then how the parabola becomes degenerate and the infinite tangents are transformed into the sheaf of vertical straight lines described above.

The moment diagram is parabolic and may be determined using the direct method, taking into account all the contributions upwards (or downwards) of a generic section of coordinate z (Figure 5.20(b)).

$$M(z) = q\frac{l}{2}z - (qz)\frac{z}{2} = q\frac{z}{2}(l - z) \qquad (5.39)$$

This function vanishes for $z = 0$ and for $z = l$ and is symmetrical with respect to the centre, where it presents a maximum equal to $ql^2/8$.

The same diagram may be determined also using a purely graphical procedure, considering the auxiliary diagram of the moment relative to a concentrated force acting in the centre and equal to the total resultant ql. The auxiliary diagram is bitriangular with the maximum equal to $ql^2/4$. The diagram sought presents the same extreme points and the same extreme tangents, the constraint reactions, and hence the shear value at the ends, as unchanging. On the other hand, according to the by now familiar graphical construction, the third point and the third tangent correspond to the stationary point in the centre.

The shear diagram is linear and skew-symmetrical with respect to the centre (Figure 5.20(c)). This vanishes at the central point, where the moment diagram is stationary. It is also possible to proceed on the basis of merely graphical considerations, joining the ends of the vertical segments which represent the values that the shear assumes at the ends A and B of the beam. On the other hand, the direct method applied to a generic section of coordinate z gives

$$T(z) = q\frac{l}{2} - qz \qquad (5.40)$$

where the first term is the contribution of the reaction R_A, while the second is the contribution of the partial distribution of external forces that extends from A to the section under examination. It may thus be verified that the function (5.40) is the derivative of the function (5.39).

In the case of the distribution of constant concentrated forces of Figure 5.22(a), the moment diagram is represented by a polygonal line with the sides contained between each pair of consecutive forces (Figure 5.22(b)).

$$M(z) = 2Fz, \qquad\qquad\qquad\qquad \text{for } 0 \leqslant z \leqslant l \qquad (5.41a)$$

$$M(z) = 2Fz - F(z - l) = Fz + Fl, \qquad \text{for } l \leqslant z \leqslant 2l \qquad (5.41b)$$

$$M(z) = Fz + Fl - F(z - 2l) = 3Fl, \qquad \text{for } 2l \leqslant z \leqslant 3l \qquad (5.41c)$$

$$M(z) = 3Fl - F(z - 3l) = 6Fl - Fz, \qquad \text{for } 3l \leqslant z \leqslant 4l \qquad (5.41d)$$

$$M(z) = 6Fl - Fz - F(z - 4l) = 10Fl - 2Fz, \quad \text{for } 4l \leqslant z \leqslant 5l \qquad (5.41e)$$

The shear diagram is represented by a step function, with discontinuity of the first kind in each point in which a force is applied (Figure 5.22(c)). This may be derived analytically from the moment diagram, or it may be obtained using the direct method and summing up algebraically all the contributions that precede a section.

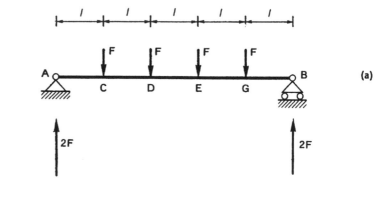

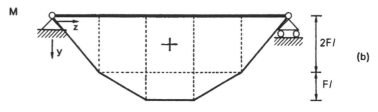

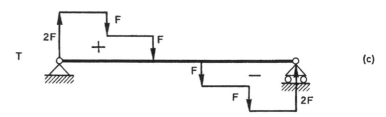

Figure 5.22

In the case of the supported beam of Figure 5.23(a), with a constant distributed load only on the intermediate portion, the constraint reactions are symmetrical, vertical and equal to $ql/2$. The pressure line is then represented by the vertical straight line a for the portion AB, by the vertical straight line d for the portion CD, and by the sheaf of vertical straight lines external to the beam for the portion BC.

The moment diagram is obtained using the direct method (Figure 5.23(b)).

$$M(z) = q\frac{l}{2}z, \qquad\qquad \text{for } 0 \leq z \leq l \qquad (5.42\text{a})$$

$$M(z) = q\frac{l}{2}z - q\frac{(z-l)^2}{2} \qquad\qquad\qquad (5.42\text{b})$$

$$= -\frac{1}{2}qz^2 + \frac{3}{2}qlz - \frac{1}{2}ql^2, \quad \text{for } l \leq z \leq 2l$$

128

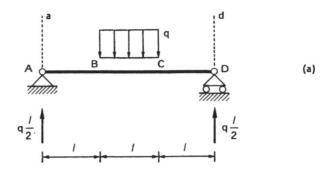

(a)

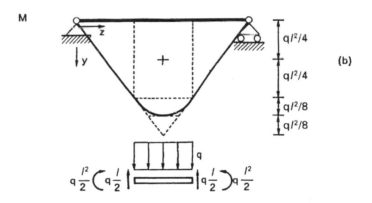

(b)

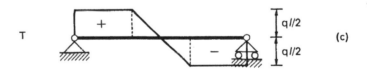

(c)

Figure 5.23

$$M(z) = q\frac{l}{2}z - ql\left(z - \frac{3}{2}l\right) \qquad (5.42c)$$

$$= -\frac{1}{2}qlz + \frac{3}{2}ql^2, \qquad \text{for} \quad 2l \leqslant z \leqslant 3l$$

The stationary point of the function $M(z)$ is obtained by equating to zero its derivative in the portion $l \leqslant z \leqslant 2l$

$$\frac{dM}{dz} = T(z) = -qz + \frac{3}{2}ql = 0 \qquad (5.43)$$

for $z = \frac{3}{2}l.$

129

INTERNAL BEAM REACTIONS

The moment diagram will then be symmetrical, with the maximum at the centre, which is equal to

$$M_{max} = M\left(\frac{3}{2}l\right) = \frac{5}{8}ql^2$$

This will be made up of two linear external segments and the parabolic intermediate segment (Figure 5.23(b)).

It is also possible, however, to draw the moment diagram graphically, using the auxiliary diagram corresponding to the resultant. This diagram is bitriangular and symmetrical and its maximum in the cusp is equal to $\frac{3}{4}ql^2$ (Figure 5.23(b)). In the outermost portions AB and CD the auxiliary diagram coincides with the one sought. In the intermediate portion BC, the diagram for the distribution q will present the same extreme values and the same extreme tangents as the auxiliary diagram. It is therefore easy to draw an arc of parabola that corresponds to these conditions and thus to refind the solution described above.

The shear diagram, as usual, may be obtained by derivation of the moment diagram, or rather, directly, considering the successive contributions of the forces acting perpendicularly to the beam (Figure 5.23(c)). In the outermost portions, the function $T(z)$ is constant, whilst it varies linearly where the distributed load is applied; in fact, the differential equation (5.12b) must hold good at all points. The cusps in the shear diagram reflect the discontinuity of the distributed load, which passes sharply from zero to q, and *vice versa*. The point of zero shear corresponds, of course, to the stationary point of $M(z)$ (Figure 5.23(b)).

Now consider a supported beam not symmetrically loaded (Figure 5.24(a)). The load weighs only on the left-hand half, so that the reaction V_A will be greater than the other reaction V_B. The respective values are: $V_A = \frac{3}{4}ql$, $V_B = \frac{1}{4}ql$. The pressure line is made up of the sheaf of vertical straight lines external to the beam, for the section AC, and by the line of action of the reaction V_B, for the section CB.

The moment diagram, as has already been seen, may be obtained using the direct method or using the graphical method. The direct method gives two functions, one parabolic and the other linear (Figure 5.24(b)).

$$M(z) = \frac{3}{4}qlz - q\frac{z^2}{2}, \qquad \text{for} \quad 0 \leqslant z \leqslant l \qquad (5.44a)$$

$$M(z) = \frac{3}{4}qlz - ql\left(z - \frac{l}{2}\right) \qquad (5.44b)$$

$$= -\frac{1}{4}qlz + \frac{1}{2}ql^2, \qquad \text{for} \quad l \leqslant z \leqslant 2l$$

We have the stationary point in the left-hand half when

$$\frac{dM}{dz} = T(z) = \frac{3}{4}ql - qz = 0 \qquad (5.45)$$

i.e. for $z = \frac{3}{4}l$. The value of the moment in the centre may be obtained more simply by going along the beam from B leftwards:

130

$$M(l) = V_B l = \frac{1}{4} q l^2 \qquad (5.46)$$

The graphical method is applied using the auxiliary diagram for the resultant. This diagram is bitriangular and presents a maximum in the cusp equal to $\frac{3}{8} q l^2$. The linear segment between B and C coincides with the diagram for the distributed load, since between B and C no external loads act. On the other hand, the two linear segments of the auxiliary diagram constitute the extreme tangents of the arc of parabola which represents the diagram $M(z)$ between A and C. Also in this case the real diagram follows the course of the auxiliary one, the cusp rounding off considerably.

The shear diagram is linear between A and C and constant between C and B (Figure 5.24(c)). There are no discontinuities in $T(z)$, as there are no concentrated forces apart from the constraint reactions, but there is a discontinuity in the derivative of $T(z)$ which reflects the discontinuity that the distributed load undergoes in the centre. The shear vanishes where the moment shows a stationary point (Figure 5.24(b)).

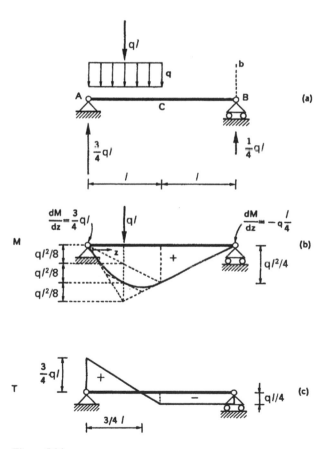

Figure 5.24

131

Let the beam with overhanging ends of Figure 5.25(a) be subjected to a constant distributed load q between the two supports and to two concentrated forces F at the ends. Each of the two constraint reactions is therefore equal to $F +(ql/2)$. The moment diagram (Figure 5.25(b)) can be drawn by summing up graphically the trapezoidal contribution due to the forces F with the parabolic one of the load q. The procedure will then be to consider as reference axis for the arc of parabola the constant central diagram equal to $Fl/2$. The sagitta of the arc of parabola is equal to $ql^2/8$, a notable value, being one already met with more than once. In the case where $F < ql/4$, a part of the parabola falls beneath the axis of the beam (Figure 5.25(b)), and at the points where the bending moment vanishes, two inflections are produced in the elastic deformed configuration; these separate the central portion, concave upwards, from the outermost portions, concave downwards (Figure 5.25(c)). For $F \geqslant ql/4$ the inflections disappear and the longitudinal fibres in tension are found only in the upper edge of the beam.

The shear diagram (Figure 5.25(d)) is constant at the overhanging ends and linear between the two supports. It undergoes two positive jumps at the points corresponding to the supports, equal to the reactions of the supports themselves. The infinitesimal element of beam straddling the support A, for example, is in equilibrium with regard to vertical translation under the action of the left-hand shear F and the right-hand shear $ql/2$, both directed downwards, and of the reaction $F + (ql/2)$, directed upwards (Figure 5.25(d)). The moment diagram shows cusps just where the shear diagram is discontinuous, and a stationary point where the shear vanishes.

Using the graphical method, we have so far examined only rectilinear beams. Now consider the beam with broken axis of Figure 5.26(a), made up of three rectilinear portions and loaded by a concentrated moment m at the centre of the horizontal beam. The constraint reactions are vertical and form a couple equal and opposite to the one applied. In the portion AB the bending moment is zero, since the reaction at A has no arm with respect to its points. In the portion BC the moment grows in linear manner up to the value $m/3$. It then undergoes a discontinuity equal to the moment applied and, in the portion CD, it decreases linearly in absolute value until it vanishes virtually at point E'. At D the moment is $-m/3$ and the representative segment on CD can be turned through $45°$ in a clockwise direction, so that it becomes the representative segment on DE. The diagram is, of course, the same but of opposite sign, going along the beam from E to A. In all cases the usual procedure is to draw the moment diagram on the side of the fibres in tension. As regards shearing force and axial force, these are represented by constant diagrams, as no distributed loads are present. The diagram of shearing force is shown in Figure 5.26(b), whilst that of axial force is shown in Figure 5.26(c).

Finally, consider the three-hinged arch of Figure 5.27(a), which presents the same polygonal line of axis as the previous beam and is loaded by a concentrated force F on the left-hand portion. The triangle of forces gives the internal and external constraint reactions (Figure 5.27(a)). The pressure line consists of the lines of action of the reactions R_A and R_C. It is important to define where the pressure line intersects the axis of the structure, since at these points the

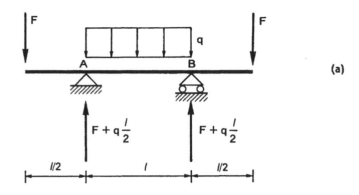

(a)

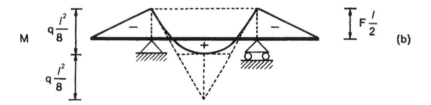

(b)

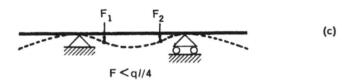

(c)

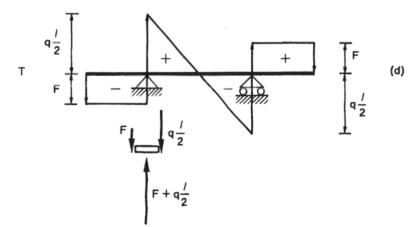

(d)

Figure 5.25

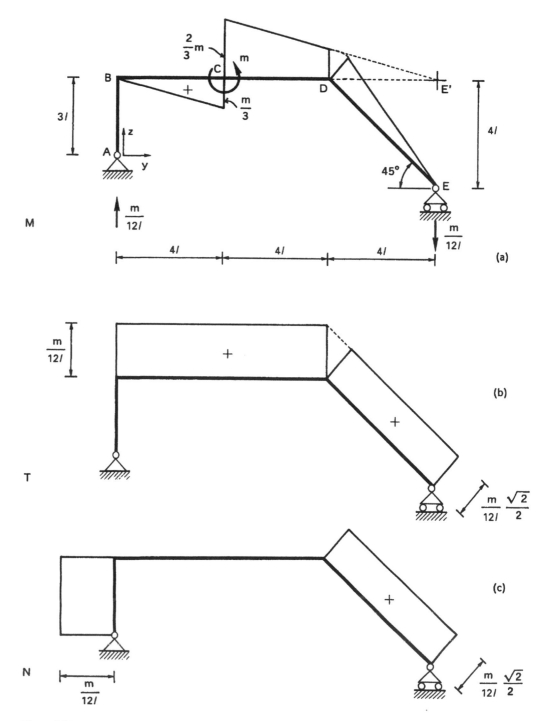

Figure 5.26

134

bending moment vanishes; in fact, the arm of the partial resultant vanishes with respect to the section. In the structure under consideration, the pressure line encounters the axis at point E, as well as at the hinges A, B, C. At these points the moment diagram will vanish (Figure 5.27(b)). In the portion AD the absolute value of moment grows linearly up to the value of $3H_A l$. At D the diagram is turned over, remaining linear between D and F and vanishes at E. At F there will be a cusp since a concentrated force is applied there. The linear diagram between F and G is obtained simply by joining the end of the segment representative of the moment at F with the hinge at B. The moment at F is thus $H_A l$, while at G it is $2H_A l$. The diagram is then turned through $45°$ and finally it is joined with the hinge C. The same diagram would, of course, have been

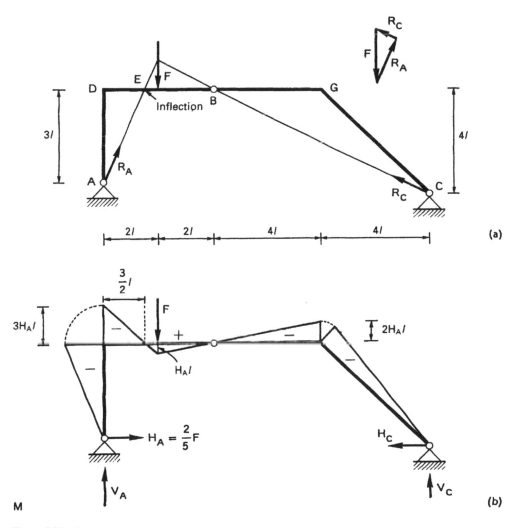

M

Figure 5.27 a, b

135

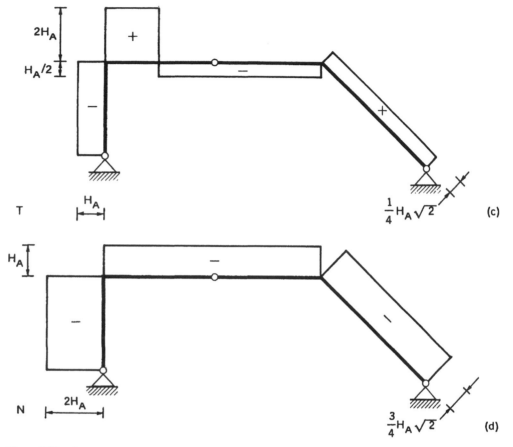

Figure 5.27 c, d

obtained by going around the structure from C to A. The absolute value of the moment at G, calculated by means of the reaction R_C, is equal to

$$M_G = (H_C - V_C)4l \tag{5.47}$$

The triangle of forces (Figure 5.27(a)) shows, on the other hand, that $H_C = H_A$ and $V_C = H_C/2 = H_A/2$, from which we obtain $M_G = 2H_Al$, the same value as before.

The shear diagram and the axial force diagram are constant in all sections and are shown in Figures 5.27(c) and 5.27(d).

5.4   Determination of characteristics of internal reaction via the Principle of Virtual Work

Just as external reactions can be calculated using the Principle of Virtual Work, by suitably reducing the corresponding external constraints, so can the characteristics of internal reaction be obtained by transforming the internal built-in constraint into a double constraint: the hinge to obtain the bending moment, and the double rod to obtain the shearing force or the axial force.

136

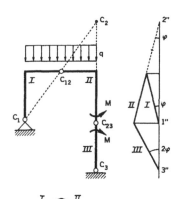

Figure 5.28

In the case, for instance, of the three-hinged portal frame of Figure 4.6(a), if we wish to determine the bending moment in the midpoint of the taller upright, it is necessary to introduce a hinge at that point and to apply the corresponding unknown moments (Figure 5.28). Using the diagrams of horizontal and vertical displacements, we have

$$-ql\left(\frac{l}{2}\varphi\right) - ql\left(\frac{l}{2}\varphi\right) - M\varphi - M(2\varphi) = 0 \qquad (5.48)$$

whence we obtain

$$M = -\frac{1}{3}ql^2 \qquad (5.49)$$

In the case where we wish to determine the shear at the same point, we must introduce a double rod parallel to the axis of the upright and apply the two unknown forces T (Figure 5.29)

$$-ql\left(\frac{l}{2}\varphi\right) - ql\left(\frac{l}{2}\varphi\right) + T(2h\varphi) + T(h\varphi) = 0 \qquad (5.50)$$

whence we obtain

$$T = q\frac{l^2}{3h} \qquad (5.51)$$

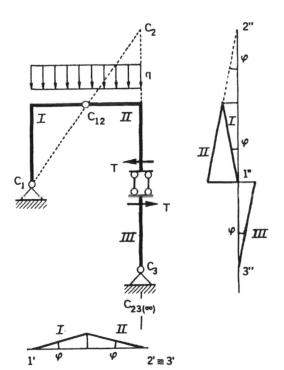

Figure 5.29

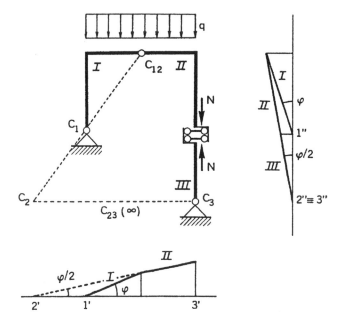

Figure 5.30

Finally, in the case where we wish to determine the axial force in the upright, we must introduce a double rod transversely to the axis and apply the two unknown forces N (Figure 5.30)

$$-ql\left(\frac{l}{2}\varphi\right) - ql\left(\frac{5}{4}l\varphi\right) - N\left(\frac{3}{2}l\varphi\right) = 0 \qquad (5.52)$$

whence we obtain

$$N = -\frac{7}{6}ql \qquad (5.53)$$

138

6 Statically determinate beam systems

6.1 Introduction

Isostatic, or statically determinate, structures made up of beams are widely used in civil and industrial buildings. As will emerge more clearly later, the internal stresses induced by mechanical loads are greater in the isostatically constrained structure than in the structure that is redundantly constrained and consequently rendered hyperstatic by further constraints. On the other hand, the internal stresses induced by the so-called thermal loads (variations of temperature, uniform or otherwise, through the depth of the beam) are zero in isostatic structures, whereas in certain cases they are considerable in hyperstatic, or statically indeterminate, structures. From this it is possible to deduce the importance, also from a practical point of view, of considering isostatic beam systems. In fact, often the mechanical loads that the structure will have to support are known to the designer, at least with a certain degree of approximation, whilst the thermal variations and the constraint settlements that the structure will undergo are not reasonably foreseeable, even as regards the order of magnitude. In these cases an isostatic structure shows ample possibilities of settling with the intervention of rigid movements only (translations and global rotations), whereas a hyperstatic structure, in view of its redundant degree of constraint, will undergo deformations also of a mechanical nature, and thus internal stresses different from zero.

These subjects will be taken up again and analysed in greater depth in the chapters on statically indeterminate structures, where the consequences induced by thermal and mechanical distortions will be studied. On the other hand, the above discussion serves to understand why, for structures having large spans, and thus subjected to notable dilations and rotations of a thermal origin, an isostatic scheme is preferred to a hyperstatic one. The two schemes most widely used for realizing structures that have large spans and are devoid of vertical encumbrance are:

1. **Gerber beams**. These consist of a rectilinear beam with a number of supports and an adequate number of disconnections, and are used, for example, in the construction of motorway bridges.
2. **Trusses**. These are made up of elements whose finer structure consists of mutually hinged connecting rods, and are traditionally used in railway bridges.

For constructions with smaller spans, or ones that present also vertical encumbrance, arched structures are traditionally used, where, as we have already seen in Chapter 4, compressive stress prevails, whilst bending stress tends to be reduced. For industrial sheds and, more in general, for all sorts of roofing (stations, gymnasia, football grounds, etc.), the following structural schemes are mainly used:

3. **Three-hinged arches**. These have been widely examined in the foregoing chapters in order to introduce the fundamental statical concepts.

139

4. **Closed-frame structures**. These are made up of chains of structural elements which close in on themselves and may in some cases also present internal statical indeterminacy.

Very often the four structural types mentioned above are found combined in the global structural scheme. A typical example is represented by arched bridges, where a gerber beam can rest on a three-hinged arch, or by suspension bridges, where a gerber beam can be hung on a truss system of tie rods and cables. In other cases, more complex schemes, also closed ones, can be reduced to simpler three-hinged arch schemes. In this chapter we shall look at some examples of these.

In many technically important cases, the mechanical loads and the structural geometry are such as to induce the designer to choose hyperstatic schemes. These cases, on the other hand, can be reduced to similar isostatic schemes, where, in addition to external loads, there act also hyperstatic loads exerted by redundant constraints. The calculation of these structures, which have few degrees of statical indeterminacy, is made by eliminating ideally the redundant constraints and replacing them with their respective constraint reactions. These reactions are obtained from considerations of congruence that regard the respect of the kinematic conditions imposed by the suppressed constraints. We shall return, however, to these aspects in Chapter 13, which is devoted to hyperstatic structures and their solution using the method of forces.

6.2 Gerber beams

Gerber beams are rectilinear beams with $(2 + s)$ supports, in which the line of axis presents s single disconnections, so as to render the structure isostatic (Figure 6.1(a)). To obtain statical determinacy also with respect to horizontal forces, the supports must be all roller supports, except for one hinged to the foundation. The s simple disconnections (which may be hinges or double rods) must be well-arranged, so as not to create hypostatic and/or hyperstatic portions (Figure 6.1(b)). Generally speaking, that is, three hinges must never be arranged consecutively (hypostatic portion) or three supports consecutively (hyperstatic portion).

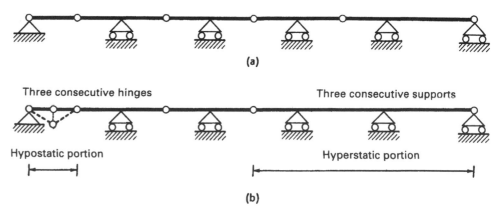

Figure 6.1

This having been said, gerber beams can be resolved using analytical and graphical methods, valid for all isostatic structures, which have been presented in the foregoing chapters. Indeed, for gerber beams it will not usually be necessary to consider equilibrium with regard to horizontal translation, since, with only vertical loads, it is identically satisfied, once the only horizontal reaction potentially present is removed. To determine the $(2 + s)$ external reactions, it is expedient to resort to the method of auxiliary equations, already introduced in Chapter 4. The global equilibrium equations are those of equilibrium with regard to vertical translation and to rotation about a suitable point of the plane. The auxiliary equations are the s equations of partial equilibrium corresponding to the portions into which the disconnections subdivide the gerber beam. The diagrams of the characteristics of internal reaction are then drawn by isolating the individual portions ideally and by applying to them the external forces, the external constraint reactions and the internal constraint reactions, transmitted by the adjacent portions.

As an example, let us consider the gerber beam of Figure 6.2(a), subjected to the distributed load q on all three spans of length l. In this case there is only one disconnection, and the two equations of global equilibrium are accompanied by the auxiliary equation of partial equilibrium with regard to rotation of the portion AB about the hinge B:

$$V_A + V_C + V_D = 3ql \qquad (6.1a)$$

$$V_A \frac{3}{2}l = V_C \frac{l}{2} + V_D \frac{3}{2}l \qquad (6.1b)$$

$$V_A l = q\frac{l^2}{2} \qquad (6.1c)$$

To express equilibrium with regard to global rotation, the midpoint of the gerber beam has been used as pole, with respect to which the moment of external load vanishes. The system (6.1) of three equations in three unknowns possesses the following solution:

$$V_A = \frac{1}{2}ql, \quad V_C = 3ql, \quad V_D = -\frac{1}{2}ql \qquad (6.2)$$

The reaction V_D of the support D turns out to be negative and thus acts in the opposite direction to the one assumed. On the other hand, the internal reaction transmitted by the hinge B is given by the equation of equilibrium with regard to vertical translation of the portion AB:

$$V_A + V_B = ql \qquad (6.3)$$

from which we have: $V_B = (1/2) \, ql$.

Applying the direct method, it is possible to identify the analytical functions $M(z)$ and $T(z)$. As regards bending moment, we have

$$M(z) = \frac{1}{2}qlz - \frac{1}{2}qz^2, \quad \text{for} \quad 0 \leqslant z \leqslant 2l \qquad (6.4a)$$

$$M(z) = \frac{1}{2}qlz - \frac{1}{2}qz^2 + 3ql(z - 2l) \qquad (6.4b)$$

$$= -\frac{1}{2}qz^2 + \frac{7}{2}qlz - 6ql^2, \quad \text{for} \quad 2l \leqslant z \leqslant 3l$$

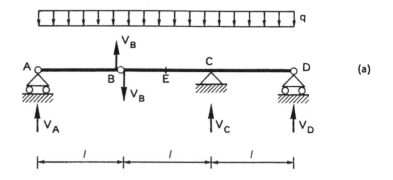

(a)

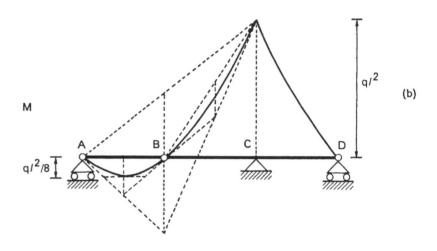

(b)

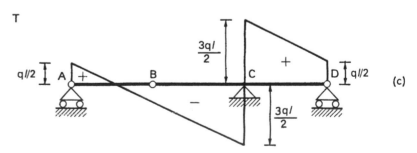

(c)

Figure 6.2

142

The shear function can then be obtained from the foregoing by derivation, or, as was done for the moment, by summing up one at a time the contributions that are encountered going along the beam:

$$T(z) = \frac{1}{2}ql - qz, \quad \text{for} \quad 0 \leq z \leq 2l \tag{6.5a}$$

$$T(z) = \frac{7}{2}ql - qz, \quad \text{for} \quad 2l \leq z \leq 3l \tag{6.5b}$$

The moment diagram is drawn in Figure 6.2(b). It is formed by two arcs of parabola with a vertical axis. The portion corresponding to the span AB is the diagram, already calculated in the last chapter, corresponding to a simply supported and uniformly loaded beam (Figure 5.20(b)). Since no concentrated forces act on the hinge B, the shear at that point will not undergo discontinuity and thus the moment will not present cusps. The third point for identifying univocally the arc of parabola between A and C is the one representing the moment on the support C: $M(2l) = -ql^2$. Note that this moment in absolute value is as much as eight times that which loads the centre of the first span. In order to construct this first arc of parabola graphically, it is possible to proceed by considering the spans AB and BC separately and then the corresponding portions of the arc; or it is possible to proceed by considering the entire portion AC at once. The above graphical constructions are given in detail in Figure 6.2(b). The parabolic diagram for the span CD may then be drawn immediately, if we note that the forces acting on the portion BD are symmetrical with respect to the vertical straight line passing through C. This arc of parabola will thus be specularly symmetrical with respect to the one for the span BC. It thus emerges that there is no point in studying the functions (6.4a,b) analytically. As we have already suggested, it is far more advantageous to proceed graphically and synthetically.

The shear diagram is given in Figure 6.2(c). It is formed by two rectilinear segments of equal slope, since the distributed load has a constant value over the entire beam. The function $T(z)$ vanishes where the moment $M(z)$ presents a stationary point, while it undergoes a positive jump equal to the reaction $V_C = 3ql$, at the support C. Also the extreme values of $T(z)$, for $z = 0$ and $z = 3l$, can be interpreted as jumps, positive and negative respectively, of the function.

As a second example, let us now examine the gerber beam of Figure 6.3(a), consisting of three spans, two of which uniformly loaded. The vertical portion AA' is constrained in the hinge A, whilst the single disconnection in C is, in this case, a double rod. Since only vertical loads are present, the upright is subjected only to the axial force V_A, while bending moment and shearing force are zero.

The equations of global equilibrium to vertical translation and to rotation about point P are

$$V_A + V_B + V_D = 3ql \tag{6.6a}$$

$$V_A \frac{5}{2}l + V_B \frac{3}{2}l = V_D \frac{3}{2}l \tag{6.6b}$$

143

whilst the auxiliary equation is that of partial equilibrium to the vertical translation of the portion CD,

$$V_D = 2ql \qquad (6.6c)$$

The external constraint reactions are then the following:

$$V_A = \frac{3}{2}ql, \quad V_B = -\frac{1}{2}ql, \quad V_D = 2ql \qquad (6.7)$$

The reactive moment transmitted by the double rod is then obtained by considering the equilibrium with regard to rotation of the portion CD: $M_C = 2ql^2$.

The shear diagram is constant in the portion $A'B$ and linear in the portion BD (Figure 6.3(b)). In B the shear undergoes a discontinuity equal to the value of the reaction, but does not change its algebraic sign. At the right of B the function $T(z)$ equals ql, whereas it vanishes in C, because the double rod does not transmit the shear. Hence, two points are known of the linear function contained between B and D, which is thus defined. It is possible to verify the diagram by going along the structure from right to left. The shear at the end D, $T(4l) = -2ql$, is in absolute value equal to the reaction V_D.

The moment diagram (Figure 6.3(c)) is linear in the portion $A'B$ and parabolic between B and D:

$$M(z) = \frac{3}{2}qlz, \quad \text{for} \quad 0 \leqslant z \leqslant l \qquad (6.8a)$$

$$M(z) = \frac{3}{2}qlz - \frac{1}{2}ql\,(z-l) - \frac{1}{2}q(z-l)^2 \qquad (6.8b)$$

$$= -\frac{1}{2}qz^2 + 2qzl, \quad \text{for} \quad l \leqslant z \leqslant 4l$$

Having set the scale and drawn the linear part, we can identify at once three values of the parabolic part: $M(l) = (3/2)ql^2$; $M(2l) = 2ql^2$; $M(4l) = 0$. The pattern of the moment diagram between B and D is thus clear, all the more so, because in correspondence with the double rod C there is a stationary point (zero shear) with a horizontal tangent. The tangent at the end D may be identified by joining D with point Q, intersection of the horizontal tangent with the vertical line through R. Finally, the graph of Figure 6.3(c) clearly indicates four points of the arc of parabola together with their respective tangents.

An alternative graphical construction can be made by joining the extreme points of the arc and drawing a vertical segment that starts from the midpoint of this line and drops by $\frac{1}{4}q\,(3l)^2 = \frac{9}{4}ql^2$. In this way again we find the intersection of the extreme tangents at the height $\frac{3}{4}ql^2 + \frac{9}{4}ql^2 = 3ql^2$, and the weak angular point produced by the discontinuity of the shear in B emerges.

The elastic deformed configuration must present the lower longitudinal fibres in tension. On the other hand, the deformed configuration of Figure 6.4(a) would create an axial tensile force on the upright AA'. The axial force is, instead, compressive, and a shortening of the upright is compatible only with a deformed configuration which presents a discontinuity of the vertical displacement in correspondence with the double rod (Figure 6.4(b)). In Chapter 16 we shall re-examine this example and calculate the elastic displacements rigorously.

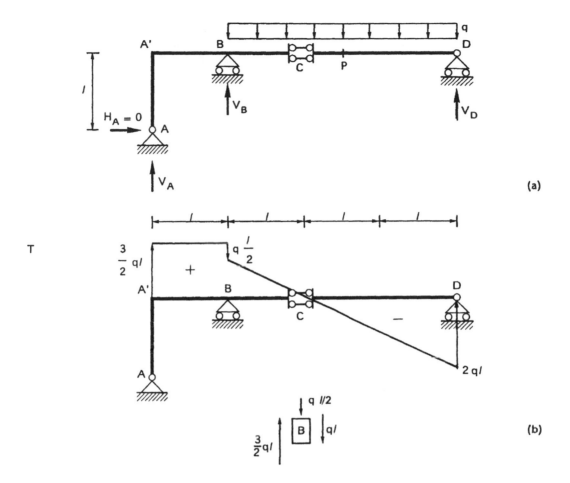

(a)

(b)

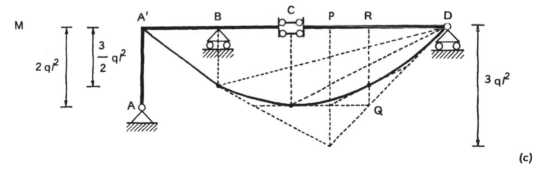

(c)

Figure 6.3

145

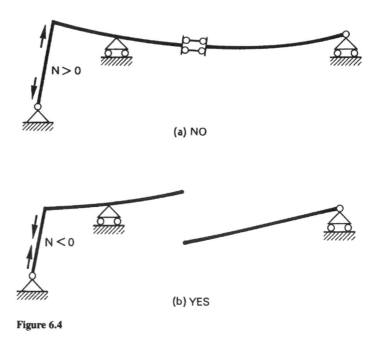

Figure 6.4

6.3 Trusses

Trusses are systems of bars connected by hinges. These hinges are referred to as **nodes** and are considered loaded by external forces and by the reactions of the bars (Figure 6.5(a)). This is to say that the hinges, which are normally considered only as constraints and hence as boundary conditions, in the case of trusses are considered as material points in equilibrium under the action of the forces involved. On the other hand, the bars, if they are not loaded directly from outside, are considered as connecting rods and thus as constraints.

In truss schemes only axial force will therefore be present as a characteristic of internal reaction. In actual situations, however, bending moment and shearing force are also present, albeit frequently not in significant measure. There are substantially two reasons for this. The first is that the external loads do not always concentrate their action on the nodes, but rather often appear as distributed along the bars or concentrated at points that are other than the nodes. The second reason is that the real connection between the bars does present a certain rotational stiffness. It would therefore be closer to the true situation to represent the connections with **semi-fixed joints**, i.e with elastic hinges. The latter will be introduced further on, when we come to deal with elastic constraints. Hence when the loads are not all concentrated on the nodes and, at the same time, the bars are fixed into one another (welded or bolted joints), trusses will work, from the static standpoint, in a way similar to that in which the so-called framed structures work. These structures will be dealt with in the sequel, since they have many degrees of redundancy and reveal a notable presence of bending moment.

From a static viewpoint, a truss presents $(a + 3)$ unknowns, if a is the number of bars and if the external constraint condition is statically determinate.

146

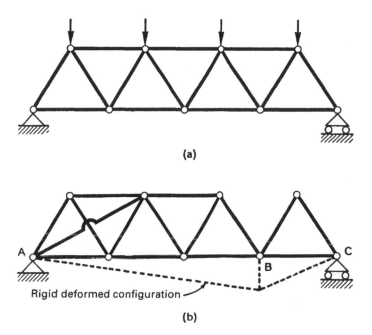

(a)

(b)

Figure 6.5

These unknowns are the internal reactions, and thus the axial forces of the bars, and the external reactions. On the other hand, the number of resolving equations is $2n$, if n is the number of hinge-nodes. They are in fact the pairs of equations of equilibrium with regard to translation corresponding to each node. A necessary, but not sufficient condition for the truss to be isostatic, is that the following relation should hold:

$$a + 3 = 2n \qquad (6.9)$$

That this relation is not of itself sufficient is shown by the truss of Figure 6.5(b), which presents, as that of Figure 6.5(a), 15 bars and nine nodes. In this structure, however, the bars are ill-arranged, so that globally the system is kinematically free or hypostatic. In fact, two portions are created which can turn about the two supports, and of which one is internally statically indeterminate and the other internally statically determinate.

A simpler and unequivocal way of judging the internal statical determinacy of a truss is that of checking whether it is made up of triangles of bars with adjacent sides, without intersections or joints through a single vertex (Figure 6.5(b)). It is therefore easy to verify the statical determinacy of the metal trusses traditionally most widely used for bridges and roofings (Figure 6.6): (a) Polonceau truss; (b) English truss; (c) Mohnié truss; (d) Howe truss; (e) Pratt truss; (f) Neville truss; (g) Nielsen parabolic truss; (h) Inverted parabolic truss; (i) Fink truss; (j) K truss.

The single elements making up a structure then often consist of substructures of a truss type. Take for example the supported arch of Figure 6.7(a), or the three-hinged arch of Figure 6.7(b), which can be realized by eliminating

147

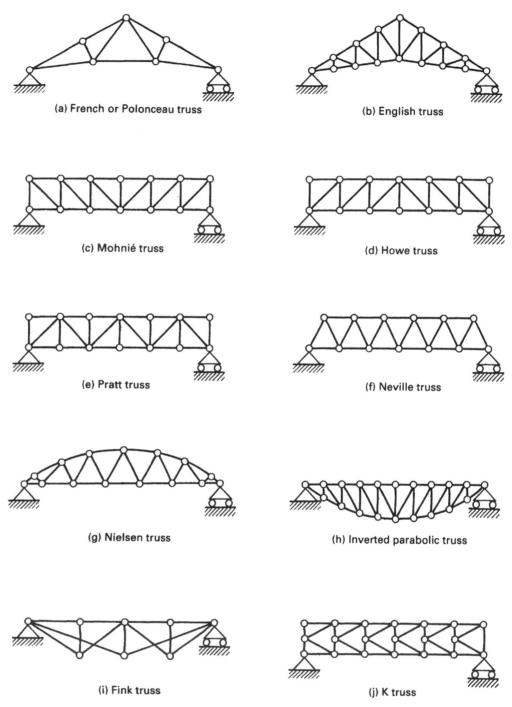

(a) French or Polonceau truss

(b) English truss

(c) Mohnié truss

(d) Howe truss

(e) Pratt truss

(f) Neville truss

(g) Nielsen truss

(h) Inverted parabolic truss

(i) Fink truss

(j) K truss

Figure 6.6

TRUSSES

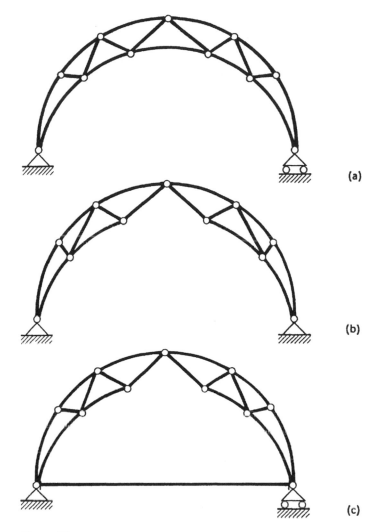

Figure 6.7

the central bar of the lower chord and adding the horizontal chain of Figure 6.7(c). This latter *structural scheme* is known as the **tied arch**, where the horizontal action of the arch on the two supports is eliminated. It presents problems of encumbrance due to the tie bar linking the two supports.

Another example of truss elements within a primary structural scheme is provided by the supported beam of Figure 6.8(a), and also by the gerber beam of Figure 6.8(b), obtained ideally from the previous one by eliminating two members of the upper chord and by adding two intermediate supports.

An example of methods of solution of trusses is given hereafter in relation to the simple structure of Figure 6.9(a). This truss is statically determinate both internally and externally. The external reactions may be determined using the triangle of forces, where the horizontal external force F, the vertical

149

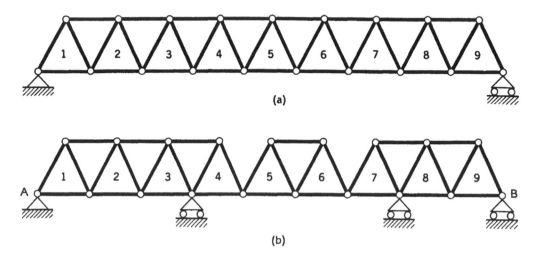

Figure 6.8

reaction V_A and the reaction R_B follow one another around the triangle (Figure 6.9(b)), their magnitudes being

$$V_A = \frac{F}{2}, \quad R_B = \left(F^2 + \frac{1}{4} F^2 \right)^{\frac{1}{2}} = \frac{\sqrt{5}}{2} F$$

To determine the axial forces of the individual bars, it is generally possible to write $2n$ equations in $(a + 3)$ unknowns, and so to check the values of the external reactions, already obtained by imposing global equilibrium. This mode of proceeding is known as the **method of equilibrium of nodes** and it is possible to give a highly significant graphical version of it, by taking each node as being in equilibrium and considering its polygon of forces.

The node A is in equilibrium under the action of the external force, the reaction of the roller support V_A, the action N_{DA} of the bar DA and the action N_{EA} of the bar EA. The trapezium of forces of Figure 6.9(c) gives the axial forces involved. Noting that the actions of the bars on the hinges are equal and opposite to the actions of the hinges on the bars, it is possible to find that $N_{DA} = -F\sqrt{2}/2$ and $N_{EA} = -F/2$, i.e. that the bars DA and EA are both struts. On the other hand, the forces N_{CA}, N_{CD}, N_{GE}, N_{GB} are all zero, by virtue of the equilibrium of the nodes C and G.

The node D is in equilibrium under the action of the forces N_{AD}, N_{ED}, N_{BD}. The first is known from the previously considered force polygon, so that also the other two are determined with the triangle of forces of Figure 6.9(d): $N_{AD} = N_{DA} = -F\sqrt{2}/2$; $N_{ED} = F/2$; $N_{BD} = -F/2$.

The node E is in equilibrium under the action of the forces $N_{AE} = N_{EA} = -F/2$; $N_{DE} = N_{ED} = F/2$; $N_{BE} = -F\sqrt{2}/2$ (Figure 6.9(e)). Finally, the node B proves to be in equilibrium under the action of the forces $N_{EB} = N_{BE} = -F\sqrt{2}/2$; $N_{DB} = N_{BD} = -F/2$ and of the constraint reaction R_B, already obtained from considerations of global equilibrium (Figure 6.9(f)). The axial forces in the individual bars may be summarized as follows:

AC unloaded
CD unloaded

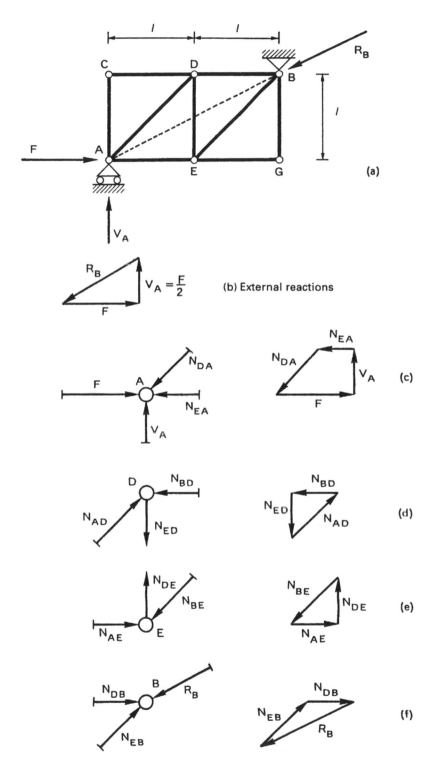

(a)

(b) External reactions

(c)

(d)

(e)

(f)

Figure 6.9

151

BG unloaded
EG unloaded
AE $-F/2$
AD $-F\sqrt{2}/2$
DE $+F/2$
DB $-F/2$
EB $-F\sqrt{2}/2$

It may thus be noted that the sides of the parallelogram *ADBE* are all struts, whilst the diagonal *DE* is the only tie bar present in the structure. Taking into account also the external forces and reactions, there is a polar-symmetrical distribution of forces with respect to the centre of the tie bar *DE*.

A verification of the solution just described may be carried out using the **method of sections** introduced by Ritter. A section of the truss is said to be a Ritter section in relation to a bar, if this section cuts, in addition to the bar under examination, other bars that are concurrent at a real point or at a point at infinity. The additional sectioned bars must therefore intersect in a single pole or be parallel. In the former case it will suffice to consider the equation of partial equilibrium with regard to rotation about the pole, and in the latter case, the equation of partial equilibrium with regard to translation orthogonally to the parallel direction, to find at once the force in the bar under consideration.

The section of Figure 6.10 is a Ritter section in relation to the bar *DE*, as the two remaining sectioned bars *DB* and *AE* are parallel and horizontal. The equation of equilibrium with regard to vertical translation of the portion that remains to the left of the section in fact gives

$$N_{DE} = +\frac{F}{2} \tag{6.10}$$

The same section is then a Ritter section also in relation to the bar *AE*, as the remaining two sectioned bars *DE* and *DB* are concurrent in *D*. The equation of equilibrium with regard to rotation about point *D* of the same portion of truss considered previously, is written

$$(F + N_{AE})l = \frac{F}{2}l \tag{6.11}$$

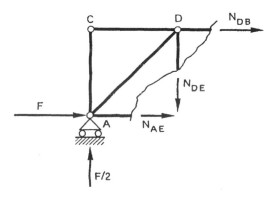

Figure 6.10

152

from which we obtain $N_{AE} = -F/2$.

Finally, the section is a Ritter section in relation also to the bar DB, as the two remaining sectioned bars AE and DE are concurrent in E. The equation of equilibrium with regard to rotation about E assumes the following form:

$$N_{DB}l + \frac{F}{2}l = 0 \tag{6.12}$$

from which we find $N_{DB} = -F/2$.

There follow three examples of trusses, resolved using the method of equilibrium of nodes. The first regards a cantilever truss with variable cross section (Figure 6.11), while the remaining two refer to trusses made up of diagonal struts (Figure 6.12) and of diagonal tie rods (Figure 6.13).

6.4 Three-hinged arches and closed-frame structures

The schemes of three-hinged arches (where the hinges are real, ideal or at infinity) have already been extensively discussed in the previous chapters. In the present section we shall highlight the existence of these schemes within more complex structures, thus bringing out more clearly how the entire structure works from a static viewpoint and, at the same time, providing an interesting graphical and synthetic approach to resolving such schemes.

Statically determinate closed structures are made up of internally isostatic closed frames, externally constrained in a non-redundant way. Each closed frame must thus present three single disconnections and be constrained to the foundation by three single constraints. In the cases where there are external forces concentrated on the internal hinges or external hinges coinciding with internal ones, and thus external reactions acting on the internal hinges, it will be convenient also to consider these hinges as bodies in equilibrium. We shall proceed by looking at three examples: in the first closed system the external and internal constraints are all separate (Figure 6.14); in the second, an external constraint coincides with an internal one (Figure 6.16), i.e. two beams converge at one external hinge, so that the hinge is, simultaneously, both external and internal; in the third system, finally, the two external constraints both coincide with internal constraints (Figure 6.18).

The closed structure of Figure 6.14(a) consists of the L-shaped beam, CAE, on which the three-hinged arch CDE rests. It is possible, in the first place, to determine the external reactions, imposing equilibrium with regard to vertical translation and rotation about point P of the entire structure:

$$H_A = 0 \tag{6.13a}$$

$$V_A + V_B = ql \tag{6.13b}$$

$$V_A \frac{3}{2}l + V_B \frac{l}{2} = 0 \tag{6.13c}$$

from which we obtain

$$V_A = -\tfrac{1}{2}\, ql, \qquad\qquad V_B = \tfrac{3}{2}\, ql$$

It is then expedient to resolve the three-hinged arch CDE (Figure 6.14(b)) determining the internal reactions H_C and R_E, and to verify the equilibrium of the beam CAE, once this is subjected to the internal reactions opposite to the previous ones and to the external reactions (Figure 6.14(c)). Note how the

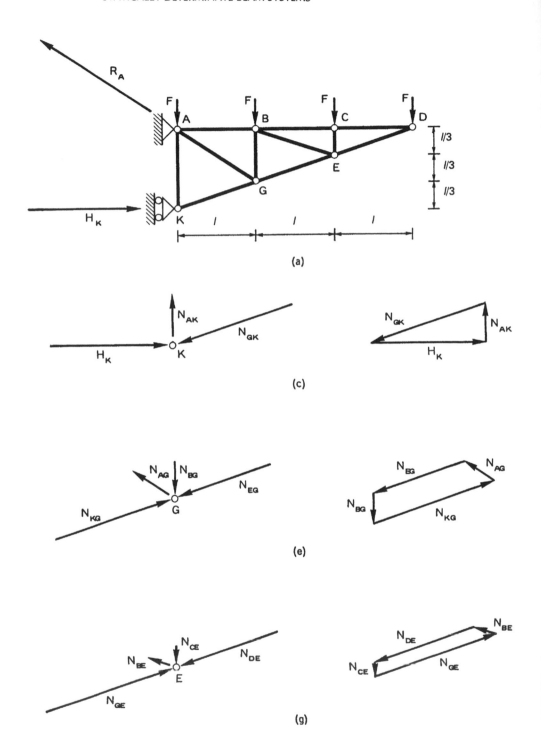

Figure 6.11

154

External reactions

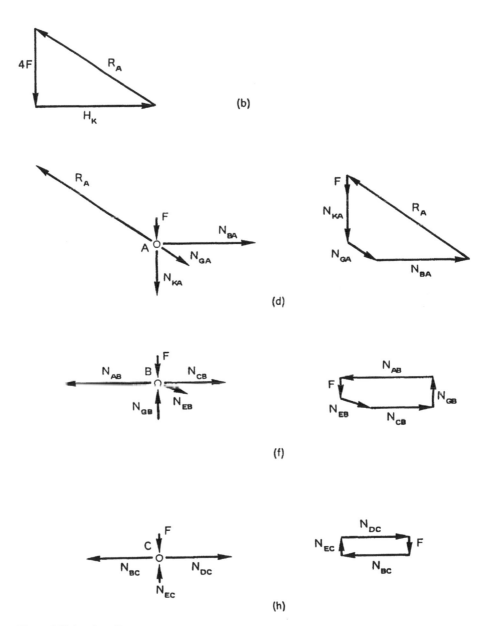

Figure 6.11 (continued)

forces transmitted by the three-hinged arch to the beam constitute a system equivalent to the distribution of external forces and, at the same time, a system that balances the external constraint reactions.

The diagrams of the characteristics of internal reaction are drawn considering each portion as isolated and subjected to all the forces involved, both active and

155

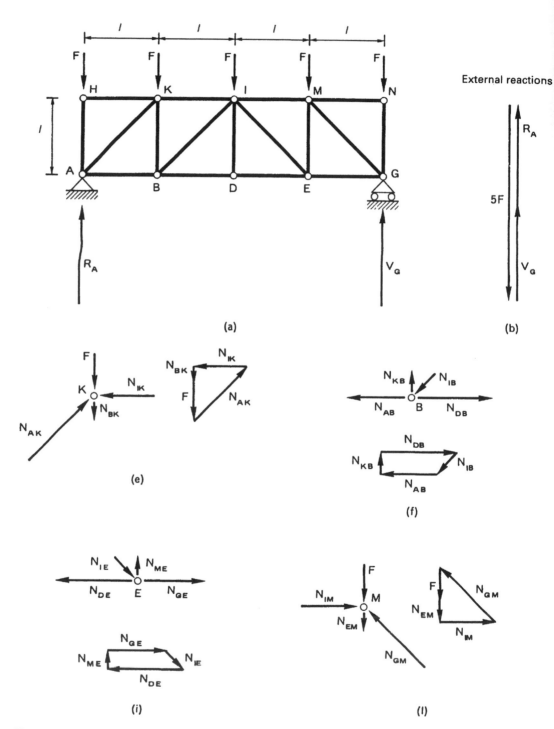

(a)

(b)

External reactions

(e)

(f)

(i)

(l)

Figure 6.12

156

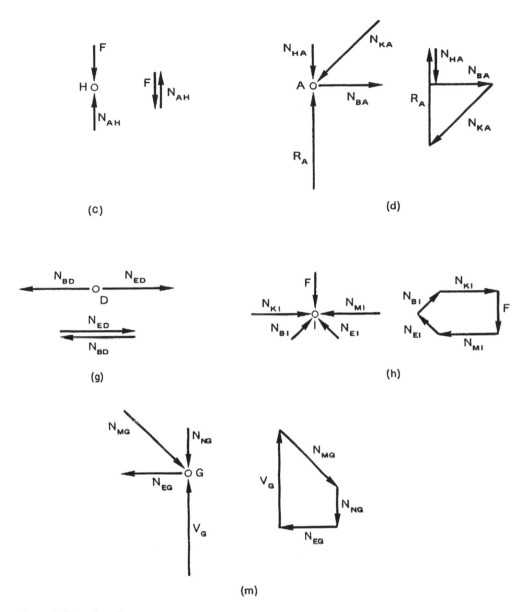

(c)

(d)

(g)

(h)

(m)

Figure 6.12 (continued)

reactive. As regards axial force (Figure 6.15(a)), in the portion CA it is zero, since the total force does not have an axial component. In the portion AE the axial force is tensile and equals $\frac{1}{2}ql$, this being the component of the resultant of all the forces which precede and follow any one of its cross sections. In the portion ED' the axial force is given by the vertical component of the reaction R_E, and hence is compressive and equal to $-ql$ (Figure 6.14(b)). Finally, in the portion $D'C$, the axial force is that of the strut CD and equals $-\frac{1}{2}ql$.

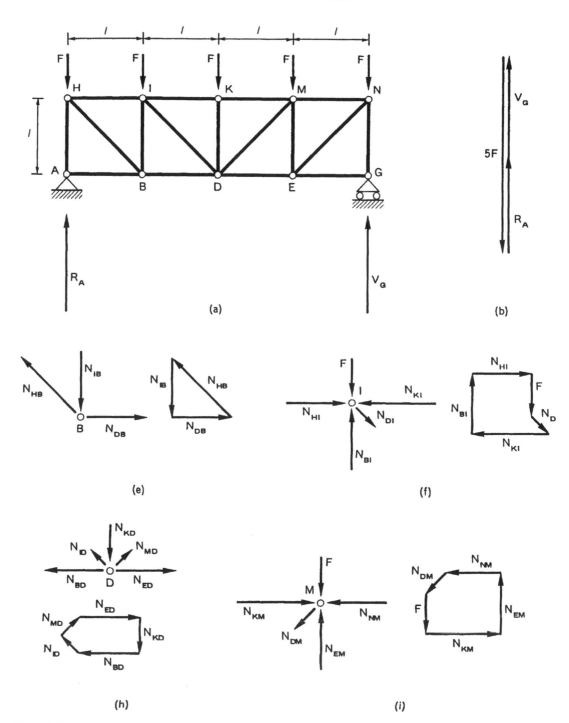

(a)

(b)

(e)

(f)

(h)

(i)

Figure 6.13

158

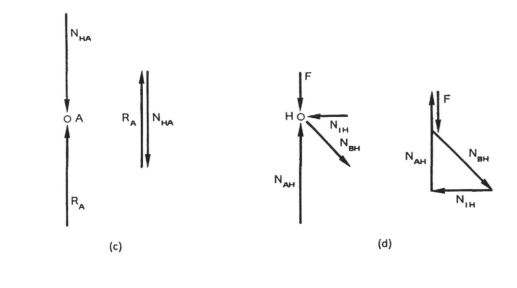

(c)   (d)

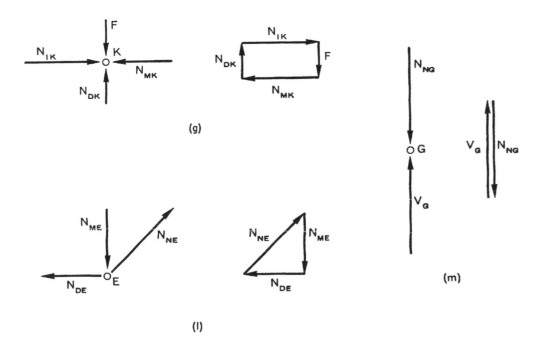

(g)

(l)   (m)

Figure 6.13 (continued)

The shearing force (Figure 6.15(b)) is equal to the reaction of the strut *CD* in the portion *CA* and to the vertical reaction V_A in the portion *AB*. The diagram then undergoes a positive jump in *B*, where the vertical reaction V_B is applied. The shear is equal to the horizontal component of the reaction R_E on the upright *ED'*, and varies linearly in the portion *D'D*, where the distributed load

159

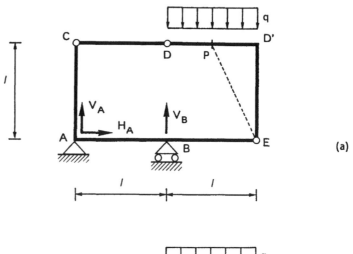

(a)

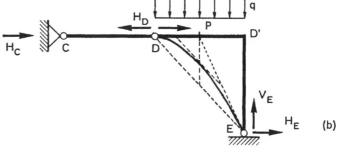

(b)

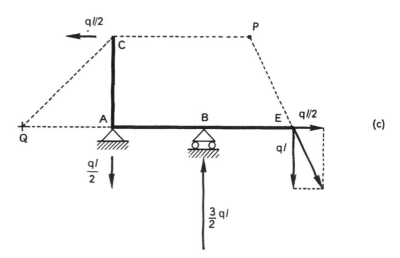

(c)

Figure 6.14

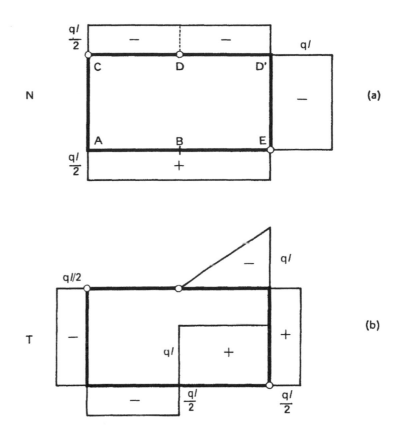

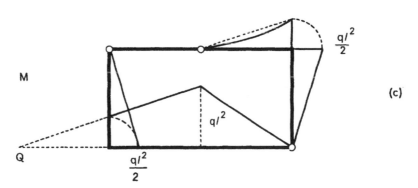

Figure 6.15

is applied. Its value at the end D' is given by the vertical component of the reaction R_E, $T(D') = -ql$, while it is zero at the end D, and also in the strut CD.

The diagram of bending moment (Figure 6.15(c)) rises linearly in absolute value between C and A, as also between A and B. The moment $M(A)$ is obtained as the product of the reaction of the strut CD and the arm l, and its value is the same to

161

the left as to the right of A, as opposed to what occurs in the case of the other two characteristics. The representative segment is therefore turned by 90° in a counter-clockwise direction and we thus obtain a point of the following linear diagram. A second point can then be represented by the pole Q, where the horizontal axis intersects the line of action of the resultant of the forces acting in C and A (Figure 6.14(c)). In the portion BE the moment diagram decreases in absolute value until it vanishes in the hinge E. The cusp that corresponds to point B reflects the discontinuity in the shear diagram (Figure 6.15(b)). The diagram then is again linear in the portion ED' and parabolic in the portion $D'D$, where the distributed load is applied. The arc of parabola is the same as that determined in the case of the cantilever beam (Figure 5.12(b)), because the strut CD transmits only an axial force.

The pressure line consists of a set of three straight lines plus the infinite number of straight lines which envelop the arc of parabola of Figure 6.14(b)). In particular, for the portions DC and CA the pressure line is the straight line DC, for the portion AB it is the straight line CQ (Figure 6.14(c)), for the portions BE and ED' it is the straight line EP and, finally, for the portion $D'D$ it is the arc of parabola already indicated.

The closed structure of Figure 6.16(a) consists of a closed rectangular frame with three internal hinges which ensure its internal isostaticity. The roller support A constrains the portion CAD to the foundation, whilst the hinge B, in addition to connecting the portions CB and DB, further constrains the structure to the foundation. To resolve the structure algebraically, we proceed by isolating each single portion of it, including the hinge B, and we replace the constraints with the actions exerted by them (Figure 6.16(b)). The connecting rod CB is subjected to two equal and opposite axial forces R_C. The portion CAD is subjected to the horizontal reaction R_C, to the external reaction of the roller support $m/2l$, and to the axial force H_D and the shearing force V_D, transmitted by the internal hinge D. The hinge B is subjected to the reaction of the connecting rod R_C, to the vertical external reaction $m/2l$, to the horizontal internal reaction H_B and to the vertical internal reaction V_B. For equilibrium with regard to translation of the hinge B, we must have $H_B = R_C$ and $V_B = m/2l$. It follows that the portion DB is subjected to the horizontal force R_C and the vertical force $m/2l$ at the end B, whilst at the end D it is subjected to the axial force H_D and the shearing force V_D, transmitted by the hinge D, as well as to the external concentrated moment m, assumed as acting on that point. There are three unknowns, R_C, H_D, V_D. Imposing equilibrium on one of the two portions, CAD or DB, we obtain the internal unknowns. Taking the portion CAD, we have for example

$$H_D = R_C \tag{6.14a}$$

$$V_D = \frac{m}{2l} \tag{6.14b}$$

$$R_C l = \frac{m}{2l} l \tag{6.14c}$$

whence it follows that $R_C = H_D = V_D = m/2l$. As a verification of this, we can note that, according to this solution, the portion DB is in equilibrium with regard to rotation, and is loaded by the external moment m and two equal and mutually concordant couples, each of moment $m/2$.

A shorter way to resolve the same closed structure of Figure 6.16(a) is to recognize in it a three-hinged arch scheme. Note that the static and kinematic function of the connecting rod CB does not vary according to the variation of

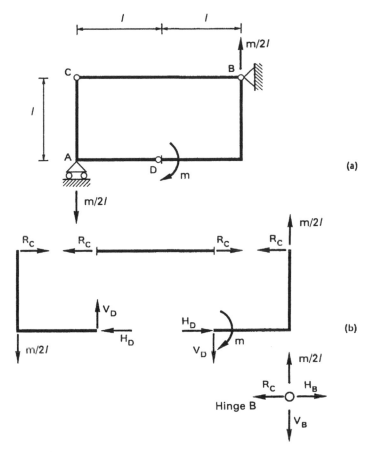

Figure 6.16

its length and thus of the position B' of the hinge to the foundation (Figure 6.17(a)). We thus come to identify an ideal hinge at point C and hence the simplified scheme of Figure 6.17(b). In this way, the static working of the structure emerges clearly, whereas in the original scheme it was not evident.

We now find graphically the reactions previously obtained proceeding algebraically. The pressure line of the fictitious structure consists of two parallel straight lines inclined at 45° to the horizontal. The moment diagram presents a discontinuity where the concentrated moment is applied (Figure 6.17(b)). The shear diagram (Figure 6.17(c)) and the axial force diagram (Figure 6.17(d)), on the other hand, do not present discontinuity in D, as no concentrated forces are applied to the structure in that point. The pressure line of the original structure (Figure 6.16(a)) is hence defined as follows:

portion CB: straight line CB
portion CA: straight line CB
portion AD: straight line CD
portion DB: straight line BQ

The closed structure of Figure 6.18(a) consists of a square frame with three internal hinges, and is constrained to the foundation by two of these hinges. The only external reaction that can oppose the horizontal load ql is the horizontal

163

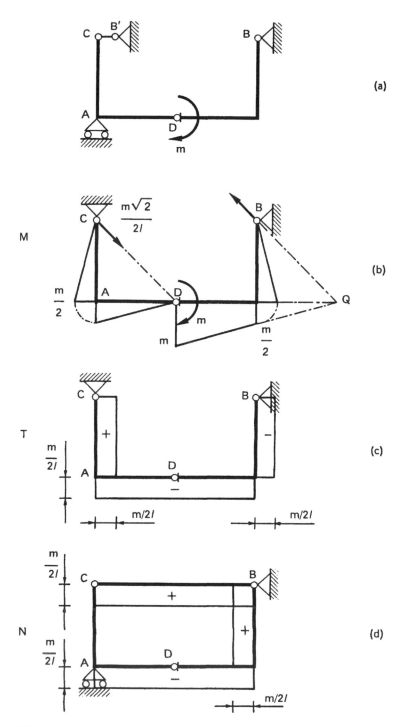

Figure 6.17

164

THREE-HINGED ARCHES AND CLOSED-FRAME STRUCTURES

one in C. There is formed, on the other hand, a clockwise couple of moment $\frac{1}{2}ql^2$, which will be balanced by an equal and opposite couple created by the vertical external reactions. Another way to identify these reactions is that of considering the pole D, where the line of action of the external resultant and the straight line perpendicular to the plane of moving of the roller support intersect. The three elements CA, AB, BC and the two hinges A and C having been isolated, the procedure will be to impose equilibrium on each of these (Figure 6.18(b)). The connecting rod AC is in equilibrium under the action of two equal and opposite forces R_C. The hinge A is in equilibrium under the action of the reaction R_C of the connecting rod, the vertical external reaction $\frac{1}{2}ql$ and the reactions H_A and V_A transmitted by the portion AB. The portion AB will thus be subjected to the reactions $H_A = R_C$ and $V_A = \frac{1}{2}ql$, the external force ql and the internal reactions H_B and V_B. The three unknowns R_C, H_B, V_B may immediately be determined, considering the equilibrium of the portion AB,

$$H_B + R_C = ql \tag{6.15a}$$

$$V_B = \frac{1}{2}ql \tag{6.15b}$$

$$\frac{1}{2}ql^2 = V_B\frac{l}{2} + H_Bl \tag{6.15c}$$

from which we obtain

$$R_C = \frac{3}{4}ql, \quad H_B = \frac{1}{4}ql, \quad V_B = \frac{1}{2}ql \tag{6.16}$$

At this point the equilibrium of the remaining elements is verified. The hinge C (Figure 6.18(b)) is subject to the horizontal external reaction ql, the vertical external one $\frac{1}{2}ql$, the reaction of the connecting rod R_C, and the reactions H_C and V_C, transmitted by the portion BC. We thus have $H_C = \frac{1}{4}ql$ and $V_C = \frac{1}{2}ql$. The portion BC is found, finally, to be subject to two equal and opposite forces acting along the line joining B and C.

The graphical approach can be adopted, if we note that the static solution is not a function of the length of the connecting rod AC'' (Figure 6.18(c)). The fundamental scheme can, that is, be reduced to the three-hinged arch ABC. We thus identify the pole P as the intersection of the line of action of the external resultant with the line joining C and B. The force triangle thus furnishes the reactions R_A and R_B. The values of equation (6.16) are again obtained if it is noted that the force triangle is geometrically similar to the triangle ACP. On the basis of the geometrical ratios of the latter, the horizontal component of R_A is three times the horizontal component of R_B, so as to give the values $H_A = \frac{3}{4}ql$ and $H_B = \frac{1}{4}ql$. At the same time the vertical components of R_A and R_B prove to be equal in magnitude: $V_A = V_C = \frac{1}{2}ql$.

The pressure line for the portion CA' is the straight line CB, whilst for the portion AA', on which the distributed load acts, it is the arc of parabola shown in Figure 6.18(c). The pressure line coincides, but for a scale factor, with the diagram of bending moment (Figure 6.18(d)). This diagram is in fact linear in the portions CC', $C'B$, BA', vanishing at the hinges C and B, and parabolic in the loaded portion AA'. Its maximum value is reached at three-quarters of the height of the upright AA',

$$M\left(\frac{3}{4}l\right) = \left(\frac{3}{4}ql\right)\left(\frac{3}{4}l\right) - \left(\frac{3}{4}ql\right)\left(\frac{3}{8}l\right) = \frac{9}{32}ql^2$$

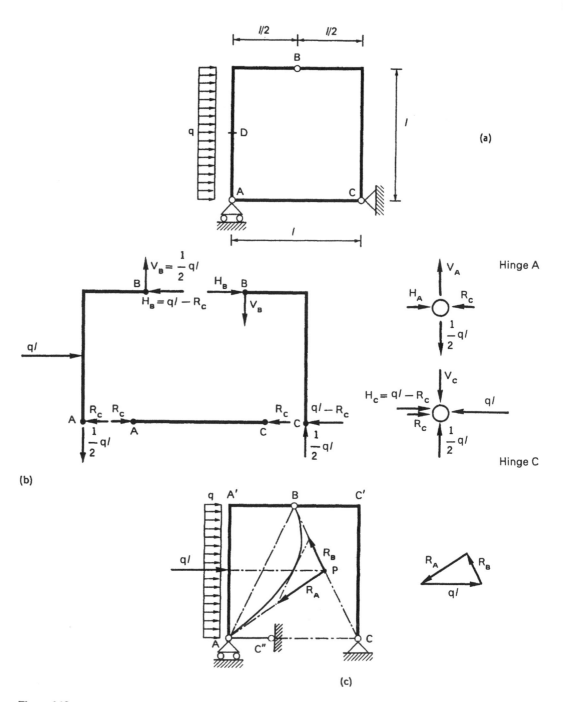

Figure 6.18

166

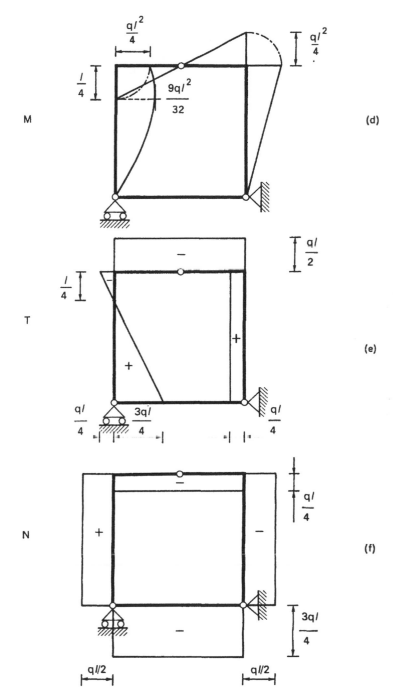

Figure 6.18 (continued)

167

where the shear vanishes (Figure 6.18(e)). The axial force diagram is given, without further comments, in Figure 6.18(f).

There follow three examples regarding three isostatic structures, along with the resolving diagrams of bending moment, shearing force, axial force and the pressure line.

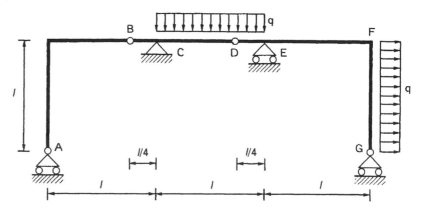

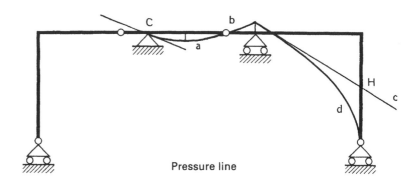

Pressure line

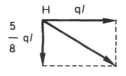

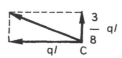

Pressure line	
Portion	Corresponding line
CD	Parabola a
DE	Parabola b
EF	Straight line c
FG	Parabola d

Example 6.1

168

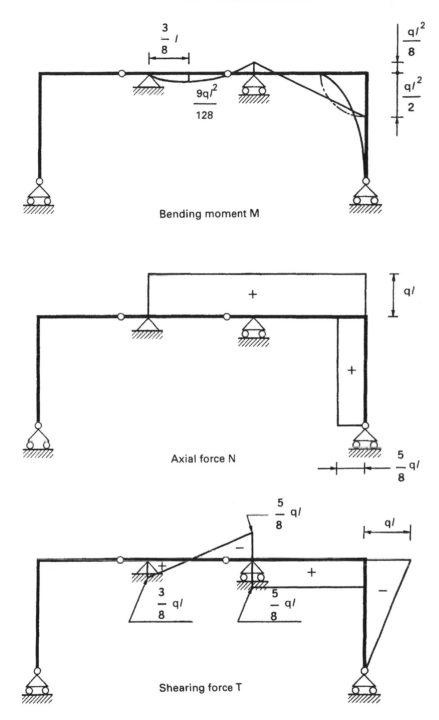

Bending moment M

Axial force N

Shearing force T

Example 6.1 (continued)

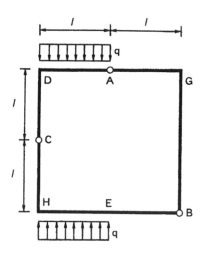

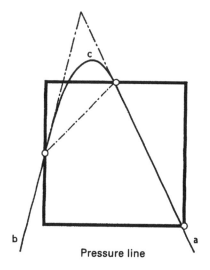

Pressure line	
Portion	Corresponding line
AB	Straight line a
DCH	Straight line b
AD	Parabola c
BE	Straight line a
HE	Parabola c

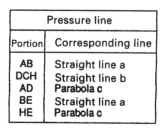

Pressure line

Example 6.2

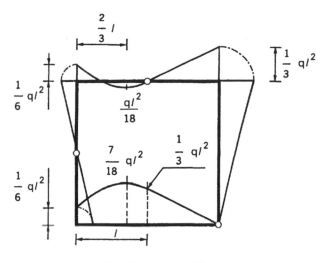

Bending moment M

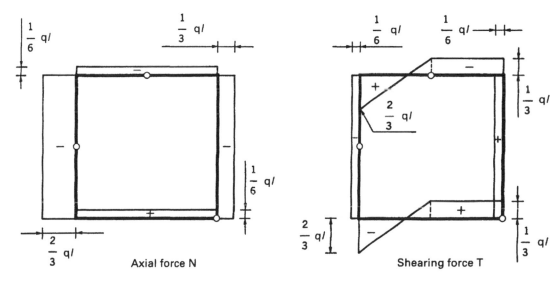

Axial force N

Shearing force T

Example 6.2 (continued)

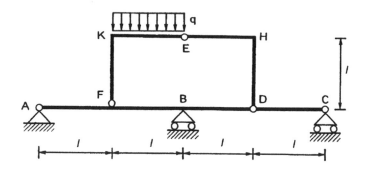

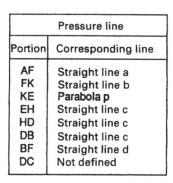

Pressure line	
Portion	Corresponding line
AF	Straight line a
FK	Straight line b
KE	**Parabola p**
EH	Straight line c
HD	Straight line c
DB	Straight line c
BF	Straight line d
DC	Not defined

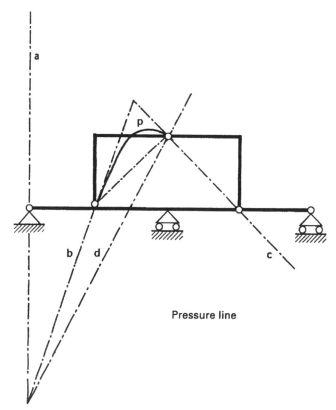

Pressure line

Example 6.3

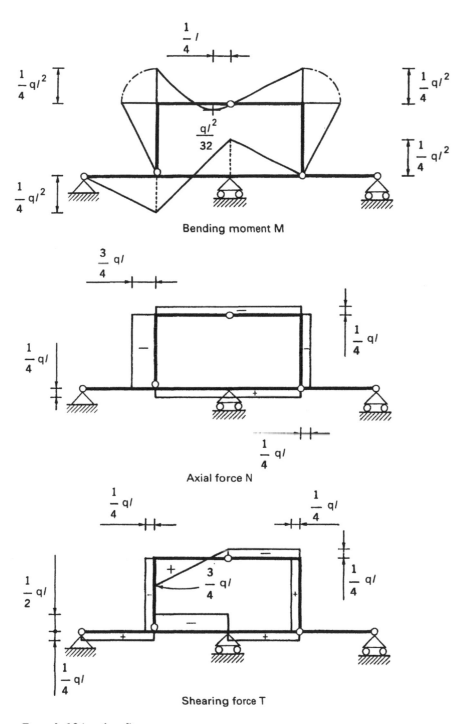

Bending moment M

Axial force N

Shearing force T

Example 6.3 (continued)

173

7 Analysis of strain and stress

7.1 Introduction

The analysis of strain and that of stress may be dealt with in a similar way, in that both these quantities, in the one case kinematic, in the other static, present a tensorial nature. In this chapter the corresponding tensors are defined: the strain tensor, consisting of dilations and shearing strains, and the stress tensor, correspondingly made up of normal stresses and shearing stresses.

On the basis of the laws of projection of the displacement vector and of the stress vector, we arrive at the laws of transformation of the strain and stress tensors for rotations of the reference system. Once the principal directions are defined as those that diagonalize the corresponding tensor, the determination of these directions and the corresponding principal values is reduced to an eigenvalue problem. There are three principal directions, at right angles to one another. An elementary cube of the solid with the sides set parallel to the principal directions is subject to dilations only (zero shearing strains) and to normal stresses only (zero shearing stresses).

As regards the principal stress values, the graphical interpretation due to Mohr, already presented in Chapter 2 in the case of the inertia of plane sections, is reproposed.

7.2 Strain tensor

In the foregoing chapters only rigid bodies have been considered, i.e. undeformable bodies in which the distance between each pair of points does not vary, even when these bodies are loaded by external forces. We have defined the rigidity constraint and the linearization of this constraint in the assumption of small displacements. On these bases we have recalled the kinematics of the rigid body, as a study of the relationships existing between the displacements of different points belonging to one rigid body undergoing rototranslational motion. In this chapter, this study will be extrapolated to the more complex case of a deformable body, i.e. we shall analyse the relationships existing between the displacements of different points (but ones sufficiently close together) belonging to one deformable body which in general undergoes rototranslational motion.

Let us thus define the **displacement function** f, as that correspondence which associates each position vector $\{r\}$ of the points of the body, in the initial position and in the undeformed configuration, to the vector $\{\eta\}$ of the displacement which these points undergo, bringing the body into the final position and into the deformed configuration (Figure 7.1). Expressed in symbols, it is

$$f : \mathcal{D} \to \mathcal{C} \tag{7.1a}$$

$$f : P \mapsto P' \tag{7.1b}$$

$$f : \{r\} \mapsto \{\eta\} \tag{7.1c}$$

where the domain $\mathcal{D}$ of the function consists of the total set of geometrical points occupied by the material points of the body in the initial state, and the

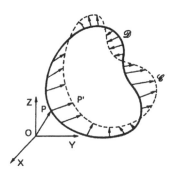

Figure 7.1

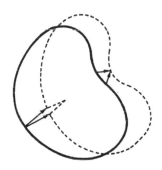

Figure 7.2

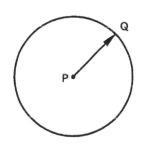

Figure 7.3

codomain $\mathscr{C}$ is the volume occupied by the body in the final state. We thus have a vector field of displacements, which can be projected onto a fixed XYZ reference system (Figure 7.1)

$$\{\eta\} = u(x, y, z)\, \bar{i} + v(x, y, z)\, \bar{j} + w(x, y, z)\, \bar{k} \tag{7.2}$$

Each component u, v, w, is a function of the three cartesian coordinates x, y, z. As will be shown, the function $\{\eta\}$ is the primary unknown in structural problems.

For the moment we formulate a hypothesis of regularity of the function $f: R^3 \rightarrow R^3$, such as to rule out **fracture** and **overlapping** (Figure 7.2). In Chapter 20, we shall then consider the phenomenon of fracture explicitly and thus the discontinuity of the displacement function. The function f and its inverse function f^{-1} must be continuous and bijective (bi-univocal). These requirements can be summarized under a single term: f and f^{-1} must be **homeomorphisms**.

Consider an arbitrary point P, within a deformable body, and a point Q belonging to the infinitesimal volume surrounding it (Figure 7.3), such that

$$PQ = \{dr\} = (x_Q - x_P)\, \bar{i} + (y_Q - y_P)\, \bar{j} + (z_Q - z_P)\, \bar{k} \tag{7.3}$$

If the function f is sufficiently regular, i.e. continuous together with its first partial derivatives, it is possible to expand this function applying Taylor series up to terms of the first order:

$$u_Q = u_P + \left(\frac{\partial u}{\partial x}\right)_P dx + \left(\frac{\partial u}{\partial y}\right)_P dy + \left(\frac{\partial u}{\partial z}\right)_P dz \tag{7.4a}$$

$$v_Q = v_P + \left(\frac{\partial v}{\partial x}\right)_P dx + \left(\frac{\partial v}{\partial y}\right)_P dy + \left(\frac{\partial v}{\partial z}\right)_P dz \tag{7.4b}$$

$$w_Q = w_P + \left(\frac{\partial w}{\partial x}\right)_P dx + \left(\frac{\partial w}{\partial y}\right)_P dy + \left(\frac{\partial w}{\partial z}\right)_P dz \tag{7.4c}$$

The scalar expressions (7.4) may be summarized in a single matrix expression:

$$\{\eta_Q\} = \{\eta_P\} + [J_P]\{dr\} \tag{7.5}$$

where $[J_P]$ is the Jacobian matrix of the dependent variables u, v, w with respect to the independent variables x, y, z.

On the other hand, if the displacements were caused exclusively by a rigid motion, equation (7.5) would be reduced to equation (3.6), which, in compact form, can be represented thus:

$$\{\eta_Q\} = \{\eta_P\} + [\varphi_P]\{dr\} \tag{7.6}$$

In this particular case, the Jacobian matrix is represented by the rotation matrix, already defined in Chapter 3:

$$[\varphi_P] = \begin{bmatrix} 0 & -\varphi_z & \varphi_y \\ \varphi_z & 0 & -\varphi_x \\ -\varphi_y & \varphi_{x'} & 0 \end{bmatrix} \tag{7.7}$$

Since the Jacobian matrix must be the sum of a contribution of rigid motion and a contribution of deformation, where the former consists of the skew-

175

symmetric matrix (7.7), it may be inferred that the purely deforming contribution is equal to the difference between the Jacobian matrix and the rotation matrix. On the other hand, each square matrix consists of the sum of a symmetric matrix and a skew-symmetric matrix. The following identity in fact holds good:

$$[J_P] = \frac{1}{2}\left([J_P] - [J_P]^T\right) + \frac{1}{2}\left([J_P] + [J_P]^T\right) \tag{7.8}$$

Of course the first term represents the rotational contribution (skew-symmetric matrix $[\varphi_P]$), whereas the latter represents the contribution of deformation (symmetric matrix $[\varepsilon_P]$).

Finally, equation (7.5) expands as follows:

$$\{\eta_Q\} = \{\eta_P\} + [\varphi_P]\{dr\} + [\varepsilon_P]\{dr\} \tag{7.9}$$

where the first term is the translational contribution, the second one is the rotational contribution and the third is the contribution of deformation. The matrices of rotation and strain, on the basis of the identity (7.8), can be rendered explicit as functions of the first partial derivatives of the components of the displacement vector, u, v, w:

$$[\varphi_P] = \begin{bmatrix} 0 & \frac{1}{2}\left(\frac{\partial u}{\partial y} - \frac{\partial v}{\partial x}\right) & \frac{1}{2}\left(\frac{\partial u}{\partial z} - \frac{\partial w}{\partial x}\right) \\ \frac{1}{2}\left(\frac{\partial v}{\partial x} - \frac{\partial u}{\partial y}\right) & 0 & \frac{1}{2}\left(\frac{\partial v}{\partial z} - \frac{\partial w}{\partial y}\right) \\ \frac{1}{2}\left(\frac{\partial w}{\partial x} - \frac{\partial u}{\partial z}\right) & \frac{1}{2}\left(\frac{\partial w}{\partial y} - \frac{\partial v}{\partial z}\right) & 0 \end{bmatrix}_P \tag{7.10a}$$

$$[\varepsilon_P] = \begin{bmatrix} \frac{\partial u}{\partial x} & \frac{1}{2}\left(\frac{\partial u}{\partial y} + \frac{\partial v}{\partial x}\right) & \frac{1}{2}\left(\frac{\partial u}{\partial z} + \frac{\partial w}{\partial x}\right) \\ \frac{1}{2}\left(\frac{\partial v}{\partial x} + \frac{\partial u}{\partial y}\right) & \frac{\partial v}{\partial y} & \frac{1}{2}\left(\frac{\partial v}{\partial z} + \frac{\partial w}{\partial y}\right) \\ \frac{1}{2}\left(\frac{\partial w}{\partial x} + \frac{\partial u}{\partial z}\right) & \frac{1}{2}\left(\frac{\partial w}{\partial y} + \frac{\partial v}{\partial z}\right) & \frac{\partial w}{\partial z} \end{bmatrix}_P \tag{7.10b}$$

The strain matrix will henceforth be represented thus:

$$[\varepsilon] = \begin{bmatrix} \varepsilon_x & \frac{1}{2}\gamma_{yx} & \frac{1}{2}\gamma_{zx} \\ \frac{1}{2}\gamma_{xy} & \varepsilon_y & \frac{1}{2}\gamma_{zy} \\ \frac{1}{2}\gamma_{xz} & \frac{1}{2}\gamma_{yz} & \varepsilon_z \end{bmatrix} \tag{7.11}$$

where the ε terms represent the partial derivatives of the components of the displacement vector in the corresponding directions, while the γ terms represent the sums of the cross partial derivatives. The elements of the strain matrix are pure numbers, to which we shall give a precise physical interpretation in the next section. In the framework of the hypotheses of small displacements and of regularity of the function f, the parameters ε and γ are small compared with unity.

7.3 Dilations and shearing strains

Consider two orthogonal segments PQ and PR of infinitesimal length within the body in the initial position and in the undeformed configuration. Choose the reference system XYZ so that the X and Y axes are parallel to the segments PQ and PR respectively. When the body is in the final position and in the deformed configuration, the two transformed segments $P'Q'$ and $P'R'$, in addition to their having undergone a rototranslation, appear to be of different length from the initial one and no longer form a right angle (Figure 7.4). Applying equations (7.4) to the particular case described above, for point Q we have

$$u_Q = u_P + \left(\frac{\partial u}{\partial x}\right)_P dx \qquad (7.12a)$$

$$v_Q = v_P + \left(\frac{\partial v}{\partial x}\right)_P dx \qquad (7.12b)$$

$$w_Q = w_P + \left(\frac{\partial w}{\partial x}\right)_P dx \qquad (7.12c)$$

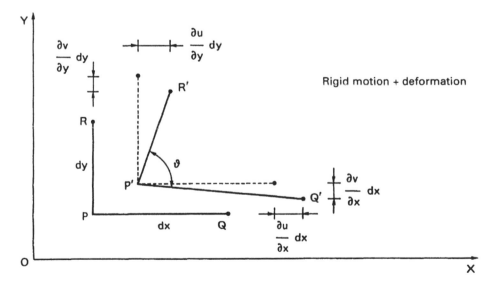

Figure 7.4

ANALYSIS OF STRAIN AND STRESS

and for point R

$$u_R = u_P + \left(\frac{\partial u}{\partial y}\right)_P dy \qquad (7.13a)$$

$$v_R = v_P + \left(\frac{\partial v}{\partial y}\right)_P dy \qquad (7.13b)$$

$$w_R = w_P + \left(\frac{\partial w}{\partial y}\right)_P dy \qquad (7.13c)$$

The projection on the XY plane of the two rototranslated and distorted segments is represented in Figure 7.4, on the basis of the relations (7.12 a, b) and (7.13 a, b).

Neglecting infinitesimals of a higher order, the **specific dilation** in the direction of the X axis equals

$$\frac{\left(dx + \frac{\partial u}{\partial x}dx\right) - dx}{dx} = \frac{\partial u}{\partial x} = \varepsilon_x \qquad (7.14)$$

and thus coincides with the first diagonal term of the strain matrix. Likewise, the specific dilations in the Y and Z directions will be represented by the remaining diagonal elements ε_y and ε_z.

Next, as regards the **shearing strains** $\gamma_{xy} = \gamma_{yx}$, $\gamma_{xz} = \gamma_{zx}$, $\gamma_{yz} = \gamma_{zy}$, these represent the decreases (or negative variations) that the right angles, formed by the initial directions, undergo as a result of the deformation. For the X and Y directions, and neglecting infinitesimals of a higher order, we have (Figure 7.4)

$$\frac{\pi}{2} - \vartheta = \frac{\frac{\partial u}{\partial y}dy}{dy} + \frac{\frac{\partial v}{\partial x}dx}{dx} = \gamma_{xy} \qquad (7.15)$$

where ϑ indicates the new angle formed by the above-mentioned axes. Note that, in the diagram of Figure 7.4, the term $\partial u/\partial y$ is positive, while $\partial v/\partial x$ is negative.

We shall now check the physical meaning of the elements of the strain matrix, considering directly only the contributions of deformation (Figure 7.5). The initial segments PQ and PR in this case only undergo a variation in length and a distortion, while the contributions of rotation and translation are obliterated. On the basis of equation (7.9), and, precisely, the third term of its second member, we can write

$$u_Q = \left(\frac{\partial u}{\partial x}\right)_P dx \qquad (7.16a)$$

$$v_Q = \frac{1}{2}\left(\frac{\partial v}{\partial x} + \frac{\partial u}{\partial y}\right)_P dx \qquad (7.16b)$$

$$w_Q = \frac{1}{2}\left(\frac{\partial w}{\partial x} + \frac{\partial u}{\partial z}\right)_P dx \qquad (7.16c)$$

178

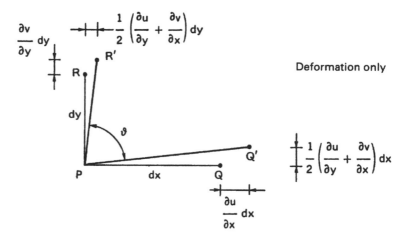

Figure 7.5

and likewise

$$u_R = \frac{1}{2}\left(\frac{\partial u}{\partial y} + \frac{\partial v}{\partial x}\right)_P dy \qquad (7.17a)$$

$$v_R = \left(\frac{\partial v}{\partial y}\right)_P dy \qquad (7.17b)$$

$$w_R = \frac{1}{2}\left(\frac{\partial w}{\partial y} + \frac{\partial v}{\partial z}\right)_P dy \qquad (7.17c)$$

The projection on the XY plane of the two purely distorted (and not rototranslated) segments is represented in Figure 7.5, on the basis of the relations (7.16a, b) and (7.17a, b). Also in this case, and neglecting infinitesimals of a higher order, the specific dilations appear to be equal to the diagonal terms of the strain matrix, just as the shearing strains coincide with the decrease in the angles formed by the straight lines passing through point P and parallel to the coordinate axes. For the X and Y directions, we have in fact (Figure 7.5)

$$\frac{\pi}{2} - \vartheta = 2 \times \frac{1}{2}\left(\frac{\partial u}{\partial y} + \frac{\partial v}{\partial x}\right) = \gamma_{xy} \qquad (7.18)$$

7.4 Law of transformation of the strain tensor for rotations of the reference system

Consider an infinitesimal sphere of unit radius with its centre in point P (Figure 7.6). The unit vector $\{n\}$ identifies point Q on this sphere. The

179

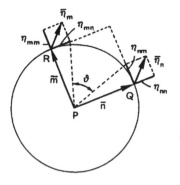

Figure 7.6

displacement vector of point Q, once the components of rototranslation have been obliterated, is given by the third term on the right-hand side of equation (7.9):

$$\{\eta_n\} = [\varepsilon]\{n\} \tag{7.19}$$

The projection of the displacement $\{\eta_n\}$ on the same direction n therefore equals

$$\eta_{nn} = \{n\}^T[\varepsilon]\{n\} \tag{7.20}$$

whilst the projection on another generic direction m, identified by a point R of the sphere, equals (Figure 7.6)

$$\eta_{nm} = \{m\}^T[\varepsilon]\{n\} \tag{7.21}$$

On the other hand, the displacement of the point R is

$$\{\eta_m\} = [\varepsilon]\{m\} \tag{7.22}$$

and thus its projection on the direction n is

$$\eta_{mn} = \{n\}^T[\varepsilon]\{m\} \tag{7.23}$$

The expressions (7.21) and (7.23) can be shown to coincide. Rendering the matrix products explicit, we obtain the following **bilinear form**:

$$\eta_{nm} = \eta_{mn} = \varepsilon_x n_x m_x + \varepsilon_y n_y m_y + \varepsilon_z n_z m_z + \tag{7.24}$$
$$\frac{1}{2}\gamma_{xy}(n_x m_y + n_y m_x) +$$
$$\frac{1}{2}\gamma_{xz}(n_x m_z + n_z m_x) +$$
$$\frac{1}{2}\gamma_{yz}(n_y m_z + n_z m_y)$$

where n_x, n_y, n_z are the direction cosines of the n direction, and m_x, m_y, m_z are the direction cosines of the m direction. The equality (7.24) expresses the **law of reciprocity** for the projections of the displacement vector.

Having made the radius of the sphere of Figure 7.6 equal to unity, we can note how the projection η_{nn} also represents the specific dilation in the direction n. Hence, by equation (7.20), we obtain

$$\varepsilon_n = \{n\}^T[\varepsilon]\{n\} \tag{7.25}$$

If then the directions n and m are assumed to be orthogonal, we obtain the corresponding shearing strain

$$\gamma_{nm} = \gamma_{mn} = \eta_{nm} + \eta_{mn} \tag{7.26}$$

The law of reciprocity gives us

$$\gamma_{nm} = \gamma_{mn} = 2\eta_{nm} = 2\eta_{mn} \tag{7.27}$$

180

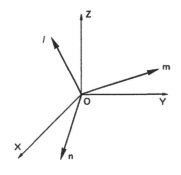

Figure 7.7

and hence, from equations (7.21) and (7.23), we obtain finally

$$\frac{1}{2}\gamma_{nm} = \frac{1}{2}\gamma_{mn} = \{m\}^T[\varepsilon]\{n\} \tag{7.28}$$
$$= \{n\}^T[\varepsilon]\{m\}$$

Considering three mutually orthogonal directions, n, m, l, rotated with respect to the initial reference system, X, Y, Z, it is now possible to express the law of transformation of the strain matrix for rotations of the reference system (Figure 7.7). This matrix in the rotated reference system nml is indicated by $[\varepsilon^*]$, where the asterisk implies the operation of rotation:

$$[\varepsilon^*] = \begin{bmatrix} \varepsilon_n & \frac{1}{2}\gamma_{mn} & \frac{1}{2}\gamma_{ln} \\ \frac{1}{2}\gamma_{nm} & \varepsilon_m & \frac{1}{2}\gamma_{lm} \\ \frac{1}{2}\gamma_{nl} & \frac{1}{2}\gamma_{ml} & \varepsilon_l \end{bmatrix} \tag{7.29}$$

From relations (7.25) and (7.28) we obtain

$$[\varepsilon^*] = \begin{bmatrix} \{n\}^T[\varepsilon]\{n\} & \{n\}^T[\varepsilon]\{m\} & \{n\}^T[\varepsilon]\{l\} \\ \{m\}^T[\varepsilon]\{n\} & \{m\}^T[\varepsilon]\{m\} & \{m\}^T[\varepsilon]\{l\} \\ \{l\}^T[\varepsilon]\{n\} & \{l\}^T[\varepsilon]\{m\} & \{l\}^T[\varepsilon]\{l\} \end{bmatrix} \tag{7.30}$$
$$= \begin{bmatrix} \{n\}^T \\ \{m\}^T \\ \{l\}^T \end{bmatrix} [\varepsilon] \left[\{n\}\{m\}\{l\}\right]$$

This last matrix product only apparently and formally relates a column matrix, a scalar quantity and a row matrix. Actually all three matrices are square (3×3), and the law (7.30) can be put in an even more compact form *

$$[\varepsilon^*] = [N][\varepsilon][N]^T \tag{7.31}$$

*The law of transformation (7.31) may be readily obtained by considering the relation (7.19) in the rotated reference system

$$[N]\{\eta_n\} = [\varepsilon^*][N]\{n\}$$

Premultiplying both sides by $[N]^T$, we obtain

$$\{\eta_n\} = [N]^T[\varepsilon^*][N]\{n\}$$

This equation coincides with equation (7.19) if

$$[\varepsilon] = [N]^T[\varepsilon^*][N]$$

and thus in the case where equation (7.31) holds.

where the matrix $[N]$ is orthogonal and represents the rotation which changes the reference system from XYZ to nml (Figure 7.7)

$$[N] = \begin{bmatrix} n_x & n_y & n_z \\ m_x & m_y & m_z \\ l_x & l_y & l_z \end{bmatrix} \tag{7.32}$$

A law similar to that of equation (7.31) has already been met with in Chapter 2, in the case of the moment of inertia tensor. It is then possible to conclude that the strain matrix in actual fact is a **tensor**, by virtue of the form that its law of transformation for rotations assumes. Also strain, like inertia, is a physical quantity that may be described solely in tensor terms.

7.5 Principal directions of strain

There is now posed the problem of determining, if they exist, the points of the infinitesimal sphere of centre P that undergo only radial displacements (Figure 7.6). The task will be to define the directions along which only dilations, and not shearing strains, occur. This means that the unit vector $\{n\}$ of such a direction must be parallel to the corresponding displacement vector $\{\eta_n\}$

$$\{\eta_n\} = \varepsilon_n \{n\} \tag{7.33}$$

In general, on the other hand, the relation (7.19) holds good, so that, by virtue of the transitive law, we obtain the characteristic equation (**eigenvalue equation**) which governs the problem

$$\big([\varepsilon] - [1]\varepsilon_n\big)\{n\} = \{0\} \tag{7.34}$$

where [1] indicates the (3×3) identity matrix. In explicit terms, equation (7.34) is presented thus:

$$\begin{bmatrix} (\varepsilon_x - \varepsilon_n) & \frac{1}{2}\gamma_{yx} & \frac{1}{2}\gamma_{zx} \\ \frac{1}{2}\gamma_{xy} & (\varepsilon_y - \varepsilon_n) & \frac{1}{2}\gamma_{zy} \\ \frac{1}{2}\gamma_{xz} & \frac{1}{2}\gamma_{yz} & (\varepsilon_z - \varepsilon_n) \end{bmatrix} \begin{bmatrix} n_x \\ n_y \\ n_z \end{bmatrix} = \begin{bmatrix} 0 \\ 0 \\ 0 \end{bmatrix} \tag{7.35}$$

The trivial solution of the system of linear algebraic equations (7.35) is without physical meaning, as the direction cosines must obey the relation of normality

$$n_x^2 + n_y^2 + n_z^2 = 1 \tag{7.36}$$

The solution is different from the trivial one and represents a **principal direction**, if and only if the determinant of the matrix of coefficients vanishes. This last condition gives a third order algebraic equation in the unknown ε_n

$$\varepsilon_n^3 - J_I \varepsilon_n^2 - J_{II} \varepsilon_n - J_{III} = 0 \tag{7.37}$$

182

where the coefficients are the so-called **scalar invariants of strain**, since they remain constant as the reference system varies:

$$J_I = \varepsilon_x + \varepsilon_y + \varepsilon_z \tag{7.38a}$$

$$J_{II} = -\begin{vmatrix} \varepsilon_x & \frac{1}{2}\gamma_{yx} \\ \frac{1}{2}\gamma_{xy} & \varepsilon_y \end{vmatrix} - \begin{vmatrix} \varepsilon_x & \frac{1}{2}\gamma_{zx} \\ \frac{1}{2}\gamma_{xz} & \varepsilon_z \end{vmatrix} - \begin{vmatrix} \varepsilon_y & \frac{1}{2}\gamma_{zy} \\ \frac{1}{2}\gamma_{yz} & \varepsilon_z \end{vmatrix} \tag{7.38b}$$

$$J_{III} = \det[\varepsilon] \tag{7.38c}$$

The first invariant is referred to as the **trace** of the tensor and is equal to the sum of the diagonal elements. The second invariant is equal to the sum of the opposites of the determinants of the principal minors. The third invariant is equal to the determinant of the strain tensor. If these coefficients varied as the reference system varied, the solution of the physical problem would also vary as the reference system varied, which would be absurd.

Equation (7.37) possesses three roots ε_1, ε_2, ε_3, referred to as the **eigenvalues** of the problem, so that the system (7.35) possesses three different solutions, $\{n_1\}$, $\{n_2\}$, $\{n_3\}$, called **eigenvectors** of the problem. The eigenvalues are real, since the tensor $[\varepsilon]$ is symmetric, and represent the three **principal dilations**, while the eigenvectors, if $\varepsilon_1 \neq \varepsilon_2 \neq \varepsilon_3$, are mutually orthogonal and represent the three **principal directions**. If we consider, in fact, the two principal directions i and j, the law of reciprocity gives the equality

$$\varepsilon_i \cos \vartheta_{ij} = \varepsilon_j \cos \vartheta_{ij} \tag{7.39}$$

where the angle ϑ_{ij} is that contained between the given directions. If $\varepsilon_i \neq \varepsilon_j$, we must have $\vartheta_{ij} = \pi/2$, whereas when $\varepsilon_i = \varepsilon_j$, ϑ_{ij} can assume any value. Three cases may thus present themselves:

1. $\varepsilon_1 \neq \varepsilon_2 \neq \varepsilon_3$: the three principal directions are mutually orthogonal;
2. $\varepsilon_1 = \varepsilon_2 \neq \varepsilon_3$: the direction $\{n_3\}$ is principal together with the ∞^1 directions orthogonal to it (principal plane 12);
3. $\varepsilon_1 = \varepsilon_2 = \varepsilon_3$: the ∞^2 directions are all principal.

Of course, the strain tensor in the principal reference system 123 is diagonal (Figure 7.8):

$$\begin{bmatrix} \varepsilon_1 & 0 & 0 \\ 0 & \varepsilon_2 & 0 \\ 0 & 0 & \varepsilon_3 \end{bmatrix} \tag{7.40}$$

as the shearing strains are zero. The invariants of the strain may therefore be expressed as functions of the principal dilations

$$J_I = \varepsilon_1 + \varepsilon_2 + \varepsilon_3 \tag{7.41a}$$

$$J_{II} = -(\varepsilon_1\varepsilon_2 + \varepsilon_1\varepsilon_3 + \varepsilon_2\varepsilon_3) \tag{7.41b}$$

$$J_{III} = \varepsilon_1\varepsilon_2\varepsilon_3 \tag{7.41c}$$

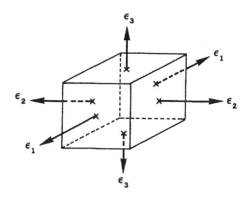

Figure 7.8

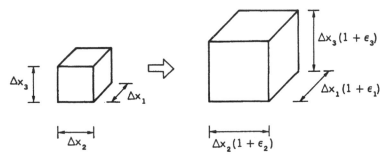

Figure 7.9

The first invariant assumes the physical meaning of **cubic dilation** (or **volumetric dilation**). An elemental parallelepiped of sides Δx_1, Δx_2, Δx_3, oriented according to the principal directions, is in fact transformed, once deformation has come about, into the parallelepiped of sides $\Delta x_1 (1 + \varepsilon_1)$, $\Delta x_2 (1 + \varepsilon_2)$, $\Delta x_3 (1 + \varepsilon_3)$ (Figure 7.9). The volume of the dilated element is thus

$$V' = V(1+\varepsilon_1)(1+\varepsilon_2)(1+\varepsilon_3) \tag{7.42}$$

V being the volume in the undeformed configuration. Neglecting the infinitesimals of a higher order, we have

$$V' = V(1+\varepsilon_1+\varepsilon_2+\varepsilon_3) \tag{7.43}$$

from which we obtain the volumetric dilation

$$\frac{\Delta V}{V} = \frac{V'-V}{V} = \varepsilon_1 + \varepsilon_2 + \varepsilon_3 = J_I \tag{7.44}$$

7.6 Equations of compatibility

As we have seen, the displacement vector is a function $f: R^3 \rightarrow R^3$, which associates with each position vector the ordered triad u, v, w of the components of the corresponding displacement. On the other hand, the strain tensor, defined in the foregoing sections, can be considered as a function $g: R^3 \rightarrow R^6$, which associates with each position vector the ordered set ε_x, ε_y, ε_z, γ_{xy}, γ_{xz}, γ_{yz} of the dilations and shearing strains. The function g may be derived from the function f, on the basis of the relations (7.10b). It thus follows that not all the continuous and derivable tensor fields produce, on integration, displacement fields that are bicontinuous functions. The six components of strain must hence be connected by three differential relations, which limit their mutual independence.

These relations, known as **equations of compatibility**, are obtained by deriving the shearing strains with respect to both the corresponding variables and noting that the third order partial derivatives that are obtained in the displacements correspond to those of the second order in the dilations

$$\frac{\partial^2 \gamma_{xy}}{\partial x \partial y} = \frac{\partial^3 u}{\partial x \partial y^2} + \frac{\partial^3 v}{\partial x^2 \partial y} = \frac{\partial^2 \varepsilon_x}{\partial y^2} + \frac{\partial^2 \varepsilon_y}{\partial x^2} \tag{7.45a}$$

$$\frac{\partial^2 \gamma_{xz}}{\partial x \partial z} = \frac{\partial^3 u}{\partial x \partial z^2} + \frac{\partial^3 w}{\partial x^2 \partial z} = \frac{\partial^2 \varepsilon_x}{\partial z^2} + \frac{\partial^2 \varepsilon_z}{\partial x^2} \tag{7.45b}$$

$$\frac{\partial^2 \gamma_{yz}}{\partial y \partial z} = \frac{\partial^3 v}{\partial y \partial z^2} + \frac{\partial^3 w}{\partial y^2 \partial z} = \frac{\partial^2 \varepsilon_y}{\partial z^2} + \frac{\partial^2 \varepsilon_z}{\partial y^2} \tag{7.45c}$$

7.7 Stress tensor

Let us consider a body in equilibrium under the action of forces distributed over the unit external surface, $\{p\}$, and in the unit volume, $\{\mathcal{F}\}$ (Figure 7.10). The cardinal equations of statics impose

$$\int_S \{p\}\, dS + \int_V \{\mathcal{F}\}\, dV = \{0\} \tag{7.46a}$$

$$\int_S \{r\} \wedge \{p\}\, dS + \int_V \{r\} \wedge \{\mathcal{F}\}\, dV = \{0\} \tag{7.46b}$$

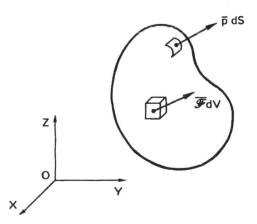

Figure 7.10

185

where S represents the boundary of the body, V the volume occupied by it and $\{r\}$ the position vector of its points. In the case where the body is deformable, let us assume that $\{r\}$ can be confused with the final position vector $\{r\} + \{\eta\}$.

Now imagine sectioning the body with a plane A (Figure 7.11(a)), passing through one of its generic points P. Each of the two portions into which the solid has been ideally subdivided will be in equilibrium under the action of the surface forces (as well as the body forces): both those corresponding to the external partial surface S_A and those corresponding to the surface of the section Ω_A and transmitted by the complementary portion of the body. Expressed in symbols

$$\int_{S_A} \{p\}\, \mathrm{d}S + \int_{\Omega_A} \{t\}\, \mathrm{d}\Omega + \int_{V_A} \{\mathscr{F}\}\, \mathrm{d}V = \{0\} \qquad (7.47a)$$

$$\int_{S_A} \{r\} \wedge \{p\}\, \mathrm{d}S + \int_{\Omega_A} \{r\} \wedge \{t\}\, \mathrm{d}\Omega + \int_{V_A} \{r\} \wedge \{\mathscr{F}\}\, \mathrm{d}V = \{0\} \qquad (7.47b)$$

where $\{t\}$ is the **tension vector**, i.e. the force transmitted to the elementary area $\mathrm{d}\Omega_A$, which constitutes the area surrounding point P on the plane A. This vector is not in general orthogonal to the plane A as occurs for fluids under pressure, and it is a function both of point P and of the secant plane A.

In fact, if we consider a different section of the body, obtained by a plane B passing again through point P, we find a different tension vector acting on the elementary area $\mathrm{d}\Omega_B$, the surrounding area of point P on the plane B (Figure 7.11(b)). The cardinal equations of statics for this new portion of the solid will prove to be similar to equations (7.47), if we replace the subscript A with B.

Once the position vector $\{r\}$ and the unit vector $\{n\}$ normal to the elementary area $\mathrm{d}\Omega$ are known, we are able to define the tension vector

$$\{t\} = \{t(\{r\},\{n\})\} = \{t_n\} \qquad (7.48)$$

Of this vector, it is possible to consider the components with respect to the external reference system (Figure 7.10),

$$\{t_n\} = t_{nx}\bar{i} + t_{ny}\bar{j} + t_{nz}\bar{k} \qquad (7.49)$$

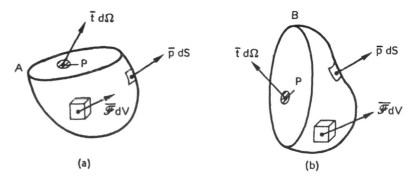

(a)   (b)

Figure 7.11

186

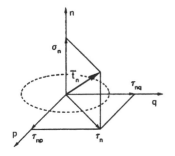

Figure 7.12

or the components with respect to a local system, with one of the coordinate axes coinciding with the normal to the elementary area $d\Omega$ and the other two lying on the section plane (Figure 7.12)

$$\{t_n\} = \sigma_n \bar{n} + \tau_{np} \bar{p} + \tau_{nq} \bar{q} \tag{7.50}$$

where $\bar{n}, \bar{p}, \bar{q}$ are the unit vectors of the above axes, σ_n is the **normal component**, while τ_{np} and τ_{nq} are the shearing stress components on the axes p and q. The resultant of τ_{np} and τ_{nq} on the section plane is referred to as the **total shearing stress component**

$$\tau_n = \left(\tau_{np}^2 + \tau_{nq}^2 \right)^{\frac{1}{2}} \tag{7.51}$$

Having established the point P inside the body, we now propose to determine the law of variation of the tension vector, as the plane of the elementary area $d\Omega$ varies. For this purpose consider a volume surrounding point P having the form of a tetrahedron with three sides parallel to the coordinate planes and the oblique face with normal unit vector $\{n\}$ (Figure 7.13(a)). Let this infinitesimal tetrahedron be subjected to the action of the tension vectors – $\{t_x\}$, $-\{t_y\}$, $-\{t_z\}$, $\{t_n\}$, and at the same time let the body force be negligible. By virtue of equilibrium with regard to translation, we have

$$\{t_n\}\, d\Omega_n - \{t_x\}\, d\Omega_x - \{t_y\}\, d\Omega_y - \{t_z\}\, d\Omega_z = \{0\} \tag{7.52}$$

where the areas of the projections of the triangular surface $d\Omega_n$ on the coordinate planes are equal to

$$d\Omega_x = n_x\, d\Omega_n \tag{7.53a}$$

$$d\Omega_y = n_y\, d\Omega_n \tag{7.53b}$$

$$d\Omega_z = n_z\, d\Omega_n \tag{7.53c}$$

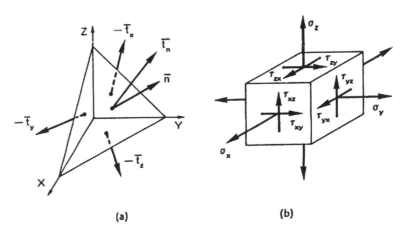

(a)   (b)

Figure 7.13

187

ANALYSIS OF STRAIN AND STRESS

Dividing the expression (7.52) by $d\Omega_n$ we then obtain

$$\{t_n\} = \{t_x\}n_x + \{t_y\}n_y + \{t_z\}n_z \tag{7.54}$$

The foregoing expression, due to Cauchy, can be put in the matrix form considering the so-called **special stress components**, i.e. the components of the vectors $\{t_x\}$, $\{t_y\}$, $\{t_z\}$ on the X, Y, Z axes

$$\begin{bmatrix} t_{nx} \\ t_{ny} \\ t_{nz} \end{bmatrix} = \begin{bmatrix} t_{xx} & t_{yx} & t_{zx} \\ t_{xy} & t_{yy} & t_{zy} \\ t_{xz} & t_{yz} & t_{zz} \end{bmatrix} \begin{bmatrix} n_x \\ n_y \\ n_z \end{bmatrix} \tag{7.55}$$

or using the traditional notation

$$\begin{bmatrix} t_{nx} \\ t_{ny} \\ t_{nz} \end{bmatrix} = \begin{bmatrix} \sigma_x & \tau_{yx} & \tau_{zx} \\ \tau_{xy} & \sigma_y & \tau_{zy} \\ \tau_{xz} & \tau_{yz} & \sigma_z \end{bmatrix} \begin{bmatrix} n_x \\ n_y \\ n_z \end{bmatrix} \tag{7.56}$$

where the σ terms are normal stress components and the τ terms are shearing stress components (Figure 7.13(b)).

The matrix relation (7.56) may be represented in compact form as follows:

$$\{t_n\} = [\sigma]\{n\} \tag{7.57}$$

which interprets the **stress matrix** $[\sigma]$ as a matrix of transformation of the normal unit vector $\{n\}$ into the corresponding tension vector $\{t_n\}$. The analogy with relation (7.19) is evident. In the latter, the strain matrix $[\varepsilon]$ may be interpreted as the matrix of transformation of the normal unit vector $\{n\}$ into the corresponding displacement vector $\{\eta_n\}$.

Considering then the equilibrium with regard to rotation of the tetrahedron of Figure 7.14, the matrix $[\sigma]$ is shown to be symmetric. The centroid G of the triangular elementary area $d\Omega_n$ has as projections on the coordinate planes the centroids of the elementary areas $d\Omega_x$, $d\Omega_y$, $d\Omega_z$. Let the tension vectors be applied to these centroids G_x, G_y, G_z, and let the conditions of equilibrium with regard to rotation of the tetrahedron with respect to the axes GG_x, GG_y, GG_z be expressed. In the case, for instance, of the axis GG_x, the five special components σ_x, τ_{xy}, τ_{xz}, σ_y, σ_z, present a zero arm, while τ_{yx} and τ_{zx} are parallel to the axis. The only two components that contribute to the moment with respect to the axis GG_x are τ_{zy} and τ_{yz},

$$\tau_{zy}\ d\Omega_z\ \frac{dz}{3} - \tau_{yz}\ d\Omega_y\ \frac{dy}{3} = 0 \tag{7.58}$$

where the product of a stress component by the corresponding elementary area represents the infinitesimal resultant force, whilst dz/3 and dy/3 represent the arms of these two forces. Noting that

$$\frac{1}{3}d\Omega_y\ dy = \frac{1}{3}d\Omega_z\ dz = dV \tag{7.59}$$

188

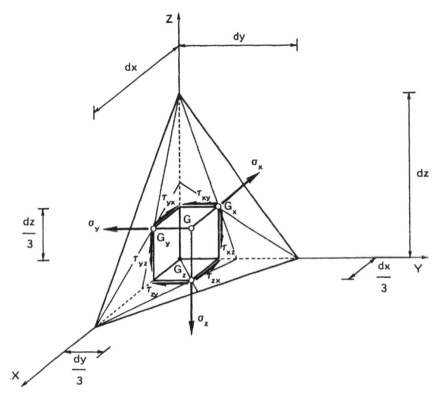

Figure 7.14

where dV is the volume of the elementary tetrahedron, we obtain that $\tau_{yz} = \tau_{zy}$, just as $\tau_{xz} = \tau_{zx}$ and $\tau_{xy} = \tau_{yx}$. The stress matrix $[\sigma]$ is thus symmetric, with only six significant components.

7.8 Law of transformation of the stress tensor for rotations of the reference system

As we have already done in the case of the displacement vector in Section 7.4, we shall now express the projection of the tension vector on a generic axis. Recalling the law of transformation expressed by equation (7.57), the component normal to the elementary area of normal unit vector $\{n\}$ equals

$$\sigma_n = t_{nn} = \{n\}^{\mathrm{T}}\{t_n\} = \{n\}^{\mathrm{T}}[\sigma]\{n\} \tag{7.60}$$

More generally, the projection of the tension vector $\{t_n\}$ on a generic direction of unit vector $\{m\}$ equals

$$t_{nm} = \{m\}^{\mathrm{T}}[\sigma]\{n\} \tag{7.61}$$

189

On the other hand, the projection of the tension vector $\{t_m\}$ on the direction of unit vector $\{n\}$ equals

$$t_{mn} = \{n\}^{\mathrm{T}}[\sigma]\{m\} \qquad (7.62)$$

Expanding the expressions (7.61) and (7.62), we obtain the **law of reciprocity** for the projections of the tension vector

$$t_{nm} = t_{mn} = \sigma_x n_x m_x + \sigma_y n_y m_y + \sigma_z n_z m_z + \qquad (7.63)$$

$$\tau_{xy}(n_x m_y + n_y m_x) +$$

$$\tau_{xz}(n_x m_z + n_z m_x) +$$

$$\tau_{yz}(n_y m_z + n_z m_y)$$

where n_x, n_y, n_z and m_x, m_y, m_z are the components of the unit vectors $\{n\}$ and $\{m\}$, respectively.

If then the unit vectors $\{n\}$ and $\{m\}$ are assumed to be orthogonal, the projections t_{nm} and t_{mn} become special shearing stress components:

$$\tau_{nm} = \tau_{mn} = \{m\}^{\mathrm{T}}[\sigma]\{n\} \qquad (7.64)$$

$$= \{n\}^{\mathrm{T}}[\sigma]\{m\}$$

The foregoing equality expresses the **law of reciprocity of shearing stresses**, of which the symmetry of the matrix $[\sigma]$ is a clear example (Figure 7.13(b)).

If n, m, l are three mutually orthogonal directions, rotated with respect to the initial reference directions X, Y, Z, on the basis of the foregoing laws of projection, it is possible to express the law of transformation of the stress matrix for rotations of the reference system (Figure 7.7). The transformed matrix is marked with an asterisk:

$$[\sigma^*] = \begin{bmatrix} \sigma_n & \tau_{mn} & \tau_{ln} \\ \tau_{nm} & \sigma_m & \tau_{lm} \\ \tau_{nl} & \tau_{ml} & \sigma_l \end{bmatrix} \qquad (7.65)$$

From relations (7.60) and (7.64) we obtain

$$[\sigma^*] = \begin{bmatrix} \{n\}^{\mathrm{T}}[\sigma]\{n\} & \{n\}^{\mathrm{T}}[\sigma]\{m\} & \{n\}^{\mathrm{T}}[\sigma]\{l\} \\ \{m\}^{\mathrm{T}}[\sigma]\{n\} & \{m\}^{\mathrm{T}}[\sigma]\{m\} & \{m\}^{\mathrm{T}}[\sigma]\{l\} \\ \{l\}^{\mathrm{T}}[\sigma]\{n\} & \{l\}^{\mathrm{T}}[\sigma]\{m\} & \{l\}^{\mathrm{T}}[\sigma]\{l\} \end{bmatrix} \qquad (7.66)$$

$$= \begin{bmatrix} \{n\}^{\mathrm{T}} \\ \{m\}^{\mathrm{T}} \\ \{l\}^{\mathrm{T}} \end{bmatrix} [\sigma] \big[\{n\} \, \{m\} \, \{l\}\big]$$

As has already been noted in the analysis of strain, the three matrices highlighted in the foregoing product are square (3×3), and the law of transformation sought can be put in the form

190

$$[\sigma^*] = [N][\sigma][N]^T \tag{7.67}$$

where $[N]$ is the orthogonal matrix (7.32). The form of the law (7.67) makes it possible to recognize a tensor entity in the matrix $[\sigma]$, referred to as **stress tensor**. Thus, in addition to inertia (Chapter 2) and strain (Section 7.4), stress also proves to be a physical quantity of a tensor nature.

7.9 Principal directions of stress

The problem now is to determine, if they exist, planes with respect to which only normal stresses are present (Figure 7.15). This means that the unit vector $\{n\}$, normal to such a plane, must be parallel to the corresponding tension vector $\{t_n\}$:

$$\{t_n\} = \sigma_n \{n\} \tag{7.68}$$

In general the relation (7.57) holds, from which, by the transitive law, we obtain the **characteristic equation** which governs the problem

$$([\sigma] - [1]\sigma_n)\{n\} = \{0\} \tag{7.69}$$

where $[1]$ indicates the (3×3) identity matrix. Note the perfect formal identity of equations (7.34) and (7.69). In explicit terms, equation (7.69) can be presented as follows:

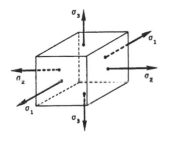

Figure 7.15

$$\begin{bmatrix} (\sigma_x - \sigma_n) & \tau_{yx} & \tau_{zx} \\ \tau_{xy} & (\sigma_y - \sigma_n) & \tau_{zy} \\ \tau_{xz} & \tau_{yz} & (\sigma_z - \sigma_n) \end{bmatrix} \begin{bmatrix} n_x \\ n_y \\ n_z \end{bmatrix} = \begin{bmatrix} 0 \\ 0 \\ 0 \end{bmatrix} \tag{7.70}$$

The trivial solution of the system (7.70) is without physical meaning, as the direction cosines must obey the relation of normality (7.36). The solution is different from the trivial one and represents a **principal direction**, if and only if the determinant of the matrix of coefficients vanishes. This condition gives a third order algebraic equation in the unknown σ_n, formally identical to equation (7.37),

$$\sigma_n^3 - J_I\sigma_n^2 - J_{II}\sigma_n - J_{III} = 0 \tag{7.71}$$

where the coefficients are the **scalar invariants of stress**

$$J_I = \sigma_x + \sigma_y + \sigma_z \tag{7.72a}$$

$$J_{II} = -\begin{vmatrix} \sigma_x & \tau_{yx} \\ \tau_{xy} & \sigma_y \end{vmatrix} - \begin{vmatrix} \sigma_x & \tau_{zx} \\ \tau_{xz} & \sigma_z \end{vmatrix} - \begin{vmatrix} \sigma_y & \tau_{zy} \\ \tau_{yz} & \sigma_z \end{vmatrix} \tag{7.72b}$$

$$J_{III} = \det[\sigma] \tag{7.72c}$$

The first invariant, or trace of the tensor, is equal to the sum of the diagonal elements. The second invariant is equal to the sum of the opposites of the determinants of the principal minors. The third invariant is equal to the determinant of the stress tensor. If these coefficients varied as the reference system

191

varied, the solution of the physical problem would also vary, and this would be absurd.

Equation (7.71) possesses three roots σ_1, σ_2, σ_3, referred to as **eigenvalues** of the problem. Consequently, the system (7.70) admits of three different solutions, $\{n_1\}$, $\{n_2\}$, $\{n_3\}$, called **eigenvectors** of the problem. The eigenvalues are real, since the tensor $[\sigma]$ is symmetric, and represent the three **principal stresses**, while the eigenvectors are mutually orthogonal and represent the three **principal directions**, if the eigenvalues are all distinct (Figure 7.15). This follows from the law of reciprocity, as has already been illustrated in the case of strain. Also in the case of stress, if two eigenvalues coincide, then there exists one principal direction and one principal plane which are mutually orthogonal. Thus when all three eigenvalues are equal, the directions are all principal.

The stress tensor in the principal reference system 123 is of course diagonal (Figure 7.15),

$$\begin{bmatrix} \sigma_1 & 0 & 0 \\ 0 & \sigma_2 & 0 \\ 0 & 0 & \sigma_3 \end{bmatrix} \tag{7.73}$$

since the shearing stress components are all zero by definition. The invariants of stress may be expressed therefore as functions of the principal stresses:

$$J_I = \sigma_1 + \sigma_2 + \sigma_3 \tag{7.74a}$$

$$J_{II} = -(\sigma_1\sigma_2 + \sigma_1\sigma_3 + \sigma_2\sigma_3) \tag{7.74b}$$

$$J_{III} = \sigma_1\sigma_2\sigma_3 \tag{7.74c}$$

The first invariant assumes the physical meaning of **mean normal stress**, but for the factor 3:

$$J_I = 3\frac{\sigma_1 + \sigma_2 + \sigma_3}{3} \tag{7.75}$$

$$= 3\frac{\sigma_x + \sigma_y + \sigma_z}{3} = 3\overline{\sigma}$$

Each stress tensor, corresponding to a generic reference system, may thus be represented as the sum of two components,

$$[\sigma] = [\sigma^i] + [\sigma^d] \tag{7.76}$$

where the first is referred to as the **hydrostatic tensor**,

$$[\sigma^i] = \begin{bmatrix} \overline{\sigma} & 0 & 0 \\ 0 & \overline{\sigma} & 0 \\ 0 & 0 & \overline{\sigma} \end{bmatrix} \tag{7.77a}$$

and the second is called the **deviatoric tensor,**

$$[\sigma^d] = \begin{bmatrix} \sigma_x - \bar{\sigma} & \tau_{yx} & \tau_{zx} \\ \tau_{xy} & \sigma_y - \bar{\sigma} & \tau_{zy} \\ \tau_{xz} & \tau_{yz} & \sigma_z - \bar{\sigma} \end{bmatrix} \qquad (7.77b)$$

Whereas the hydrostatic tensor does not depend on the reference system, since it is a function only of the trace, the deviatoric tensor varies as the orientation of the reference system varies.

The component (7.77a) is given the name hydrostatic, because liquids under pressure exchange internal stresses of this sort. The principal stresses are all three equal and thus all the directions are principal (see gyroscopic areas, section 2.5). In fluids under pressure the stress vector is always normal to any elementary area $d\Omega$; it is compressive and its magnitude is equal to the pressure of the fluid. Perfect fluids, in fact, do not transmit shearing stresses internally, just as ropes do not transmit shearing force (and bending moment), but only axial force.

The problem of principal stresses will now be given a graphical interpretation based on the method of **Mohr's circles**. This method has already been introduced in Chapter 2, in the framework of the geometry of areas, in order to seek the principal directions of inertia.

In the principal reference system 123, the equations (7.57) and (7.60) are particularized in the following way:

$$\{t_n\} = \begin{bmatrix} \sigma_1 & 0 & 0 \\ 0 & \sigma_2 & 0 \\ 0 & 0 & \sigma_3 \end{bmatrix} \begin{bmatrix} n_1 \\ n_2 \\ n_3 \end{bmatrix} \qquad (7.78)$$

$$\sigma_n = [\, n_1 \; n_2 \; n_3 \,] \begin{bmatrix} \sigma_1 & 0 & 0 \\ 0 & \sigma_2 & 0 \\ 0 & 0 & \sigma_3 \end{bmatrix} \begin{bmatrix} n_1 \\ n_2 \\ n_3 \end{bmatrix} \qquad (7.79)$$

where n_1, n_2, n_3 are the direction cosines of the generic direction n in the principal system.

Equation (7.79) may be developed by working out the matrix products

$$\sigma_n = \sigma_1 n_1^2 + \sigma_2 n_2^2 + \sigma_3 n_3^2 \qquad (7.80a)$$

On the other hand, the magnitude squared of the tension vector (7.78) is

$$\sigma_n^2 + \tau_n^2 = \sigma_1^2 n_1^2 + \sigma_2^2 n_2^2 + \sigma_3^2 n_3^2 \qquad (7.80b)$$

while for the direction cosines we have the condition of normality

$$n_1^2 + n_2^2 + n_3^2 = 1 \qquad (7.80c)$$

Equations (7.80) constitute a system of three linear algebraic equations in the three unknowns n_1^2, n_2^2, n_3^2. The solution of the system is the following:

$$n_1^2 = \frac{\tau_n^2 + (\sigma_n - \sigma_2)(\sigma_n - \sigma_3)}{(\sigma_1 - \sigma_2)(\sigma_1 - \sigma_3)} \tag{7.81a}$$

$$n_2^2 = \frac{\tau_n^2 + (\sigma_n - \sigma_1)(\sigma_n - \sigma_3)}{(\sigma_2 - \sigma_1)(\sigma_2 - \sigma_3)} \tag{7.81b}$$

$$n_3^2 = \frac{\tau_n^2 + (\sigma_n - \sigma_1)(\sigma_n - \sigma_2)}{(\sigma_3 - \sigma_1)(\sigma_3 - \sigma_2)} \tag{7.81c}$$

Let us assume that between the principal stresses there exists the order relation: $\sigma_1 \geqslant \sigma_2 \geqslant \sigma_3$. As the expressions on the right-hand sides of relations (7.81) should be positive, the following inequalities are obtained:

$$\tau_n^2 + (\sigma_n - \sigma_2)(\sigma_n - \sigma_3) \geqslant 0 \tag{7.82a}$$

$$\tau_n^2 + (\sigma_n - \sigma_1)(\sigma_n - \sigma_3) \leqslant 0 \tag{7.82b}$$

$$\tau_n^2 + (\sigma_n - \sigma_1)(\sigma_n - \sigma_2) \geqslant 0 \tag{7.82c}$$

It is easy to verify that these inequalities are equivalent to the ones given below:

$$\tau_n^2 + \left(\sigma_n - \frac{\sigma_2 + \sigma_3}{2}\right)^2 \geqslant \left(\frac{\sigma_2 - \sigma_3}{2}\right)^2 \tag{7.83a}$$

$$\tau_n^2 + \left(\sigma_n - \frac{\sigma_1 + \sigma_3}{2}\right)^2 \leqslant \left(\frac{\sigma_1 - \sigma_3}{2}\right)^2 \tag{7.83b}$$

$$\tau_n^2 + \left(\sigma_n - \frac{\sigma_1 + \sigma_2}{2}\right)^2 \geqslant \left(\frac{\sigma_1 - \sigma_2}{2}\right)^2 \tag{7.83c}$$

On **Mohr's plane** (Figure 7.16) all the pairs of components, consisting of the normal stress σ_n and the shearing stress τ_n, which are obtained as the unit vector $\{n\}$ varies, are represented by the intersection of the three domains (7.83). The first domain is the one external to the circumference that has its centre on the axis σ_n in the point C_1 $[\frac{1}{2}(\sigma_2 + \sigma_3),0]$ and radius $R_1 = \frac{1}{2}(\sigma_2 - \sigma_3)$. The second domain is the one internal to the circumference of centre C_2 $[\frac{1}{2}(\sigma_1 + \sigma_3),0]$ and radius $R_2 = \frac{1}{2}(\sigma_1 - \sigma_3)$. Finally, the third domain is the one external to the circumference of centre C_3 $[\frac{1}{2}(\sigma_1 + \sigma_2),0]$ and radius $R_3 = \frac{1}{2}(\sigma_1 - \sigma_2)$. Note that the possible pairs (σ_n, τ_n) are ∞^2, just as the directions n issuing from a point are ∞^2. There exists a bi-univocal relation which links each unit vector $\{n\}$ to each point of the hatched area of Figure 7.16. For reasons of brevity, we shall not enter into further details of this relation.

In the case where one, or two, of the principal stresses are zero, the graphical construction described previously will present one, or two, of the intersections of the circumferences with the axis σ_n, coincident with the origin. The five possible cases are represented in Figure 7.17.

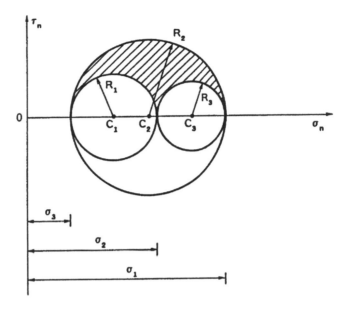

Figure 7.16

7.10 Plane stress condition

The stress condition in a point is said to be **plane** if the tension vector belongs in every case to one and the same plane, independently of the section chosen. A necessary and sufficient condition for the stress state to be plane is that one of the three principal stresses should vanish. If, for example, $\sigma_1 \neq 0$, $\sigma_2 \neq 0$, $\sigma_3 = 0$ (Figure 7.17(c)), it can easily be shown that the tension vector $\{t_n\}$ always belongs to the **plane of stresses** σ_1–σ_2, whatever the orientation of the elementary area $d\Omega_n$ (Figure 7.18). Equation (7.78) becomes

$$\begin{bmatrix} t_{n1} \\ t_{n2} \\ t_{n3} \end{bmatrix} = \begin{bmatrix} \sigma_1 & 0 & 0 \\ 0 & \sigma_2 & 0 \\ 0 & 0 & 0 \end{bmatrix} \begin{bmatrix} n_1 \\ n_2 \\ n_3 \end{bmatrix} \tag{7.84}$$

from which we obtain

$$t_{n3} = 0 \tag{7.85}$$

Since the tension vector has always zero components in the direction 3, the stress tensor with respect to a generic system of axes $XY3$, will present the third row, and thus by symmetry the third column, identically equal to zero

$$\begin{bmatrix} \sigma_x & \tau_{yx} & 0 \\ \tau_{xy} & \sigma_y & 0 \\ 0 & 0 & 0 \end{bmatrix} \tag{7.86}$$

195

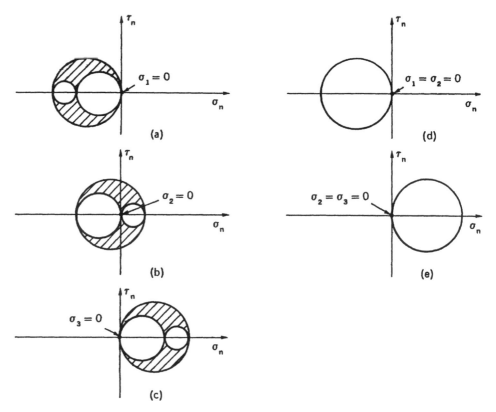

Figure 7.17

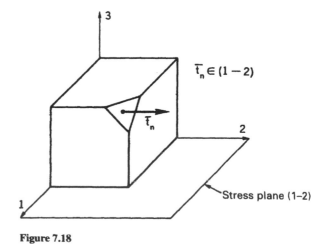

Figure 7.18

The eigenvalue problem is resolved then by equating to zero the determinant of the following matrix:

$$\det \begin{bmatrix} \sigma_x - \sigma_n & \tau_{yx} & 0 \\ \tau_{xy} & \sigma_y - \sigma_n & 0 \\ 0 & 0 & -\sigma_n \end{bmatrix} = 0 \qquad (7.87)$$

The three roots are obtained from the two conditions

$$\sigma_n = 0 \qquad (7.88a)$$

$$\sigma_n^2 - (\sigma_x + \sigma_y)\sigma_n + (\sigma_x \sigma_y - \tau_{xy}^2) = 0 \qquad (7.88b)$$

Whereas the first equation gives a result already known, because by hypothesis $\sigma_3 = 0$, the second gives the two principal stresses different from zero, σ_1 and σ_2. Note that the first coefficient $(\sigma_x + \sigma_y)$ is the trace of the significant principal minor of the stress tensor (7.86), whilst the second coefficient $(\sigma_x \sigma_y - \tau_{xy}^2)$ is the determinant of this minor. Resolving equation (7.88b), we obtain the pair of roots

$$\sigma_1 = \frac{\sigma_x + \sigma_y}{2} + \frac{1}{2}\left[(\sigma_x - \sigma_y)^2 + 4\tau_{xy}^2\right]^{\frac{1}{2}} \qquad (7.89a)$$

$$\sigma_2 = \frac{\sigma_x + \sigma_y}{2} - \frac{1}{2}\left[(\sigma_x - \sigma_y)^2 + 4\tau_{xy}^2\right]^{\frac{1}{2}} \qquad (7.89b)$$

It is possible to obtain the same result by imposing that the significant principal minor of the tensor (7.86) should be diagonal (this approach has already been adopted in Chapter 2 for seeking the principal directions of inertia)

$$[\sigma^*] = \begin{bmatrix} \cos\vartheta & \sin\vartheta \\ -\sin\vartheta & \cos\vartheta \end{bmatrix} \begin{bmatrix} \sigma_x & \tau_{yx} \\ \tau_{xy} & \sigma_y \end{bmatrix} \begin{bmatrix} \cos\vartheta & -\sin\vartheta \\ \sin\vartheta & \cos\vartheta \end{bmatrix} \qquad (7.90)$$

Equating the off-diagonal term to zero, we have

$$\tau_{xy}^* = \tau_{xy}\cos 2\vartheta - \frac{1}{2}(\sigma_x - \sigma_y)\sin 2\vartheta = 0 \qquad (7.91)$$

and thus the angle by which the XY system must turn to reach the principal system is

$$\vartheta_0 = \frac{1}{2}\arctan\left(\frac{2\tau_{xy}}{\sigma_x - \sigma_y}\right) \qquad (7.92)$$

with $-\pi/4 < \vartheta_0 < \pi/4$.

The graphical construction of Mohr's circle, representing all the pairs of normal stresses σ_n and shearing stresses τ_n, which we have as the orientation of the elementary area $d\Omega_n$ varies in such a way that the unit vector $\{n\}$ should belong to the plane of stresses, is made in exactly the same way as that already shown in Chapter 2 for the geometry of areas. Let us imagine that the stresses σ_x, σ_y, τ_{xy}, acting on an elementary parallelepiped having sides parallel to the axes $XY3$, are known (Figure 7.19(a)). Let the shearing stress τ_{xy} be considered positive if it tends to rotate the element in a clockwise direction (and *vice versa* for τ_{yx}). Let there be fixed on Mohr's plane (Figure 7.19(b)) the two notable points: P (σ_x, τ_{xy}), P' $(\sigma_y, -\tau_{xy})$. The intersection of the segment PP' with the axis σ_n gives the centre C of Mohr's circle, while the segments CP and CP' represent two opposite radii of this circle. The line parallel to the axis σ_n is then drawn through P and the line parallel to the axis τ_n is drawn through P'. These two lines meet at the pole P^*. The straight lines joining P^* with the points M and N of the axis σ_n, which are the intersections of the circumference with the axis, give the directions of the two principal axes. Of course the points M and N each have for their abscissae the value of a principal stress. In particular, in Figure 7.19(b), the abscissa of M is σ_2 and the abscissa of N is σ_1, as we have assumed $\sigma_x > \sigma_y$ and the order relations are maintained:

$$\sigma_x > \sigma_y \Rightarrow \sigma_1 > \sigma_2 \tag{7.93a}$$

$$\sigma_x < \sigma_y \Rightarrow \sigma_1 < \sigma_2 \tag{7.93b}$$

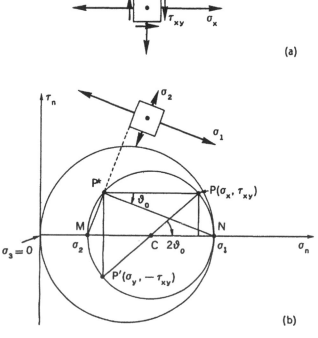

(a)

(b)

Figure 7.19

198

When $\sigma_x = \sigma_y$ and $\tau_{xy} \neq 0$, it makes no difference whether the XY reference system is rotated $\pi/4$ clockwise or counterclockwise to obtain the principal directions. When then $\sigma_x = \sigma_y$ and $\tau_{xy} = 0$, Mohr's circle degenerates into a point. Note, on the other hand, that assuming $\sigma_3 = 0$, as the unit vector $\{n\}$ also varies outside the plane of stresses, the set of pairs (σ_n, τ_n) is represented by the circumference with centre at the point C $(\sigma_x/2, 0)$ and radius $R = \sigma_x/2$. The problem, in other words, always remains three-dimensional.

It is easy to verify that the graphical construction outlined above reflects both the analytical solutions (7.89) and (7.92). To justify the former, consider in fact the abscissa of the centre C and the right triangle PP^*P', which has as its hypotenuse the diameter of Mohr's circle (Figure 7.19(b)). To justify the latter, note that PP^*N is a circumferential angle corresponding to the central angle PCN, and that the latter has an amplitude equal to arctan $2\tau_{xy}/(\sigma_x-\sigma_y)$.

8 Theory of elasticity

8.1 Introduction

In this chapter the general problem of the non-homogeneous and anisotropic three-dimensional linear elastic body is formulated. The properties of linearity, homogeneity and isotropy concern exclusively the constitutive relations which link stresses and strains; they do not in any way affect the kinematic relations, which define dilations and shearing strains, nor the static relations provided by the indefinite equations of equilibrium. An intimate correlation exists between the static and kinematic relations, in that the two corresponding matrix operators are each the transpose of the other. The same correlation is present, at a finite level, in the case of rigid systems, as has been seen in Chapter 3.

Static–kinematic duality leads to an extremely direct demonstration of the Principle of Virtual Work for deformable bodies, just as it enables a representation of the elastic problem in a symmetrical manner by combining the three above-mentioned relations in a single operator equation which has as its unknown the displacement vector.

Having demonstrated the classical theorems of Clapeyron and Betti, which hold good also in the case of anisotropic material, we then proceed to the analysis of isotropic material, which is of particular importance for its practical engineering applications, and to the definition of the corresponding Young's modulus and Poisson's ratio. The chapter closes with the strength criteria for biaxial and triaxial stress conditions, where the modes of rupture of ductile materials are distinguished from those of brittle materials.

8.2 Indefinite equations of equilibrium

In the last chapter the stress tensor was defined as the matrix of transformation of the unit vector into the corresponding tension vector. It was seen at the same time how the elements of the tensor represent the special components of stress on the coordinate planes. We then studied the law of variation of the stress tensor with the variation in orientation of the reference system and identified the principal reference system, with respect to which the tensor becomes diagonal and the shearing stresses thus vanish. We shall now determine the system of differential equations that govern the variations of the stress tensor as the point under consideration varies. Having so far limited our investigation to examining what happens at point P of the body, in the present section we shall define the differential relation that links the stresses that develop in points of the body that are very close to one another.

To this end let us consider an elementary parallelepiped with the sides parallel to the coordinate planes, of length dx, dy, dz, respectively (Figure 8.1). On the opposite faces of the parallelepiped there act components of stress

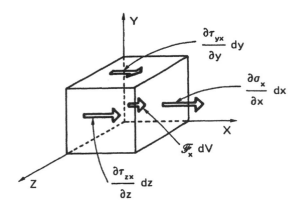

Figure 8.1

which, but for an infinitesimal increment, are equal to one another. Equilibrium to translation in the X direction, for instance, imposes

$$\frac{\partial \sigma_x}{\partial x} dx(\, dy \, dz) + \frac{\partial \tau_{yx}}{\partial y} \, dy(\, dx \, dz) + \frac{\partial \tau_{zx}}{\partial z} dz(\, dx \, dy) + \mathcal{F}_x(\, dx \, dy \, dz) = 0 \quad (8.1)$$

where only the increments of stress, multiplied by the elementary areas on which they act, and the body force, multiplied by the elementary volume in which it acts, are present. Dividing equation (8.1) by the elementary volume $dV = dx \, dy \, dz$, we obtain the first of the **indefinite equations of equilibrium**:

$$\frac{\partial \sigma_x}{\partial x} + \frac{\partial \tau_{yx}}{\partial y} + \frac{\partial \tau_{zx}}{\partial z} + \mathcal{F}_x = 0 \quad (8.2a)$$

The analogous equations of equilibrium in the Y and Z directions appear as follows:

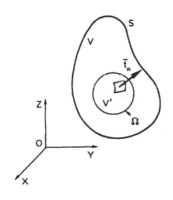

Figure 8.2

$$\frac{\partial \tau_{xy}}{\partial x} + \frac{\partial \sigma_y}{\partial y} + \frac{\partial \tau_{zy}}{\partial z} + \mathcal{F}_y = 0 \quad (8.2b)$$

$$\frac{\partial \tau_{xz}}{\partial x} + \frac{\partial \tau_{yz}}{\partial y} + \frac{\partial \sigma_z}{\partial z} + \mathcal{F}_z = 0 \quad (8.2c)$$

Equations (8.2) may also be obtained by integration. Consider a domain V', contained in the domain V and having for boundary a closed and regular surface Ω (Figure 8.2). The equation of equilibrium to translation in vector form is written

$$\int_\Omega \{t_n\} \, d\Omega + \int_{V'} \{\mathcal{F}\} \, dV = \{0\} \quad (8.3)$$

201

This vector equation is equivalent to three scalar equations, of which the first is

$$\int_{\Omega} t_{nx} \, d\Omega + \int_{V'} \mathscr{F}_x \, dV = 0 \tag{8.4}$$

Applying the first of the Cauchy relations (7.56), the foregoing equation becomes

$$\int_{\Omega} (\sigma_x n_x + \tau_{yx} n_y + \tau_{zx} n_z) \, d\Omega + \int_{V'} \mathscr{F}_x \, dV = 0 \tag{8.5}$$

The application of Green's Theorem to the surface integral transforms it into a volume integral,

$$\int_{V'} \left(\frac{\partial \sigma_x}{\partial x} + \frac{\partial \tau_{yx}}{\partial y} + \frac{\partial \tau_{zx}}{\partial z} + \mathscr{F}_x \right) dV = 0 \tag{8.6}$$

and this equation must hold good for any V' subdomain. The integrand must therefore be identically equal to zero, thus verifying the first of the indefinite equations of equilibrium (8.2a).

Note that equations (8.2) constitute a system of three differential equations with partial derivatives, in the six unknown functions σ_x, σ_y, σ_z, τ_{xy}, τ_{xz}, τ_{yz}. This system is hence three times indeterminate, and, using the terminology introduced for beam systems, it is possible to state that the problem of a three-dimensional solid is three times hyperstatic. We shall presently see how, by adding the kinematic equations and the constitutive equations to the static equations, the problem as a whole becomes determinate.

On the boundary of the domain V (Figure 8.3), the tension vector must on the other hand coincide with the surface force $\{p\}$, externally applied:

$$t_{nx} = \sigma_x n_x + \tau_{yx} n_y + \tau_{zx} n_z = p_x \tag{8.7a}$$

$$t_{ny} = \tau_{xy} n_x + \sigma_y n_y + \tau_{zy} n_z = p_y \tag{8.7b}$$

$$t_{nz} = \tau_{xz} n_x + \tau_{yz} n_y + \sigma_z n_z = p_z \tag{8.7c}$$

The above relations are known as **boundary conditions of equivalence**. They represent one of the two boundary conditions for the general problem of the mechanics of elastic solids, which will be introduced in the ensuing sections.

8.3 Static–kinematic duality

We are now able to express in matrix form the systems of differential equations that govern, on the one hand, **congruence** and, on the other, **equilibrium**. These two systems have intimately connected formal structures, as we shall show in this section.

As regards congruence, we recall the relations (7.10b) which define the elements of the strain tensor. The six independent components can be ordered in

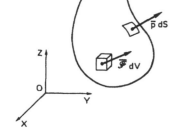

Figure 8.3

a **strain vector**, which can be obtained by premultiplying formally the displacement vector by a (6×3) matrix operator:

$$
\begin{bmatrix} \varepsilon_x \\ \varepsilon_y \\ \varepsilon_z \\ \gamma_{xy} \\ \gamma_{xz} \\ \gamma_{yz} \end{bmatrix}
=
\begin{bmatrix}
\frac{\partial}{\partial x} & 0 & 0 \\
0 & \frac{\partial}{\partial y} & 0 \\
0 & 0 & \frac{\partial}{\partial z} \\
\frac{\partial}{\partial y} & \frac{\partial}{\partial x} & 0 \\
\frac{\partial}{\partial z} & 0 & \frac{\partial}{\partial x} \\
0 & \frac{\partial}{\partial z} & \frac{\partial}{\partial y}
\end{bmatrix}
\begin{bmatrix} u \\ v \\ w \end{bmatrix}
\tag{8.8}
$$

The relations (8.8) can also be written in compact form,

$$
\{\varepsilon\} = [\partial]\{\eta\}
\tag{8.9}
$$

and are called **kinematic equations**.

On the other hand, as regards equilibrium, the indefinite equations of equilibrium may be reproposed, also in matrix form and with the stress components ordered in the **stress vector**:

$$
\begin{bmatrix}
\frac{\partial}{\partial x} & 0 & 0 & \frac{\partial}{\partial y} & \frac{\partial}{\partial z} & 0 \\
0 & \frac{\partial}{\partial y} & 0 & \frac{\partial}{\partial x} & 0 & \frac{\partial}{\partial z} \\
0 & 0 & \frac{\partial}{\partial z} & 0 & \frac{\partial}{\partial x} & \frac{\partial}{\partial y}
\end{bmatrix}
\begin{bmatrix} \sigma_x \\ \sigma_y \\ \sigma_z \\ \tau_{xy} \\ \tau_{xz} \\ \tau_{yz} \end{bmatrix}
+
\begin{bmatrix} \mathscr{F}_x \\ \mathscr{F}_y \\ \mathscr{F}_z \end{bmatrix}
=
\begin{bmatrix} 0 \\ 0 \\ 0 \end{bmatrix}
\tag{8.10}
$$

The **static equations** are written in compact form

$$
[\partial]^{\mathrm{T}}\{\sigma\} + \{\mathscr{F}\} = \{0\}
\tag{8.11}
$$

the static differential operator being the exact transpose of the kinematic one appearing in equation (8.8).

In this way, just as in the mechanics of rigid bodies the static matrix is the transpose of the kinematic one, also in the mechanics of deformable bodies there exists the same profound interconnection between the two matrix operators. In the case of the mechanics of rigid bodies, we saw how this interconnection implies the validity of the Principle of Virtual Work (Section 3.9). The validity of this Principle will, on the other hand, be extended to the case of deformable bodies, precisely on the basis of static–kinematic duality.

To conclude, it is also possible to give an explicit matrix form to the **boundary equations of equivalence** (8.7),

$$
\begin{bmatrix} n_x & 0 & 0 & n_y & n_z & 0 \\ 0 & n_y & 0 & n_x & 0 & n_z \\ 0 & 0 & n_z & 0 & n_x & n_y \end{bmatrix}
\begin{bmatrix} \sigma_x \\ \sigma_y \\ \sigma_z \\ \tau_{xy} \\ \tau_{xz} \\ \tau_{yz} \end{bmatrix}
= \begin{bmatrix} p_x \\ p_y \\ p_z \end{bmatrix}
\tag{8.12}
$$

or in compact form

$$
[\mathcal{N}]^T \{\sigma\} = \{p\}
\tag{8.13}
$$

The reader's attention is drawn to the perfect correspondence that exists between the matrix operator $[\partial]^T$ and the algebraic matrix $[\mathcal{N}]^T$: the partial derivatives of the one are matched by the corresponding direction cosines of the other, in the spirit of Green's Theorem.

8.4 Principle of Virtual Work

The Principle of Virtual Work is the fundamental identity in the ambit of the mechanics of **deformable bodies**. It states the equality between external virtual work (forces multiplied by corresponding displacements) and internal virtual work (stresses multiplied by corresponding strains). More precisely, the Principle of Virtual Work may be said to constitute the very definition of strain energy. From Rational Mechanics, the concept of work as scalar product of the force vector and the displacement vector is well-known. However, it is not obvious that strain energy is expressible as the scalar product of the stress vector and strain vector, for the very reason that the intimate nature of these latter quantities is tensorial. On the other hand, when the body is rigid, the strains are zero and the internal virtual work vanishes, as we have already assumed in Section 3.9.

A system a of external forces (of volume $\{\mathcal{F}_a\}$ and surface $\{p_a\}$) and stresses $\{\sigma_a\}$ is said to be **statically admissible** when these forces satisfy the equations of statics (8.11) and the boundary conditions expressed by equation (8.13)

$$
[\partial]^T \{\sigma_a\} = -\{\mathcal{F}_a\}, \qquad \forall P \in V \tag{8.14a}
$$

$$
[\mathcal{N}]^T \{\sigma_a\} = \{p_a\}, \qquad \forall P \in S \tag{8.14b}
$$

where V is the three-dimensional domain occupied by the body and S is the boundary of that domain, on which the external forces $\{p_a\}$ are applied; the latter may, however, be zero over the entire boundary S or over a subset of this.

On the other hand, a system b of displacements $\{\eta_b\}$ and strains $\{\varepsilon_b\}$ is said to be **kinematically admissible** when the equations of kinematics (8.9) are satisfied:

$$[\partial]\{\eta_b\} = \{\varepsilon_b\}, \qquad \forall P \in V \tag{8.15}$$

At this point a digression is called for to demonstrate the rule of integration by parts on a three-dimensional domain. This rule is nothing other than an extension of Green's Theorem, and will be used for demonstrating the **Principle of Virtual Work for Deformable Bodies.**

Consider two functions of the three cartesian coordinates, $f(x, y, z)$ and $g(x, y, z)$, defined on a three-dimensional domain V. Perform the partial derivation of the product with respect, for instance, to the x coordinate

$$\frac{\partial}{\partial x}(fg) = \frac{\partial f}{\partial x}g + f\frac{\partial g}{\partial x} \tag{8.16}$$

and integrate both members on a generic chord parallel to the X axis and belonging to a generic section A of the domain, $z =$ constant (Figure 8.4)

$$\int_{\alpha(y)}^{\beta(y)} \frac{\partial}{\partial x}(fg)\,dx = \int_{\alpha(y)}^{\beta(y)} \frac{\partial f}{\partial x}g\,dx + \int_{\alpha(y)}^{\beta(y)} f\frac{\partial g}{\partial x}\,dx \tag{8.17}$$

The integral of the derivative of the product is equal to the difference of the values that the product presents at the extremes of the interval of integration

$$[fg]_{\alpha(y)}^{\beta(y)} = \int_{\alpha(y)}^{\beta(y)} \frac{\partial f}{\partial x}g\,dx + \int_{\alpha(y)}^{\beta(y)} f\frac{\partial g}{\partial x}\,dx \tag{8.18}$$

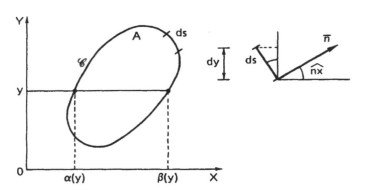

Figure 8.4

205

Integrating both sides of equation (8.18) with respect to the variable y, we obtain

$$\oint_{\mathscr{C}} fg \, dy = \int_A \frac{\partial f}{\partial x} g \, dx \, dy + \int_A f \frac{\partial g}{\partial x} \, dx \, dy \tag{8.19}$$

where on the left-hand side there appears the counterclockwise integration around a closed path of the product performed on the boundary $\mathscr{C}$ of the two-dimensional domain A. If n_x indicates the direction cosine with respect to the X axis of the normal to the curve $\mathscr{C}$, and ds indicates the increment of the curvilinear coordinate, from the foregoing equation we obtain

$$\oint_{\mathscr{C}} fg n_x \, ds = \int_A \frac{\partial f}{\partial x} g \, dA + \int_A f \frac{\partial g}{\partial x} \, dA \tag{8.20}$$

Finally, integrating both members with respect to the third coordinate z, we have

$$\int_S fg n_x \, dS = \int_V \frac{\partial f}{\partial x} g \, dV + \int_V f \frac{\partial g}{\partial x} \, dV \tag{8.21}$$

S being the boundary of the entire three-dimensional domain V. Equation (8.21) represents an extension of Green's Theorem, as is evident if we put $f(x, y, z) = 1$, or $g(x, y, z) = 1$.

On the basis of equation (8.21), and using the compact matrix formulation, it is not difficult to obtain the equation of virtual work. Consider the virtual work performed by the body forces $\{\mathscr{F}_a\}$ times the displacements $\{\eta_b\}$, these two fields belonging to two altogether independent systems, the first being statically admissible and the second kinematically admissible:

$$L_F = \int_V \{\mathscr{F}_a\}^{\mathrm{T}} \{\eta_b\} \, dV \tag{8.22}$$

Applying the equations of statics (8.14a), we have

$$L_F = -\int_V \left([\partial]^{\mathrm{T}} \{\sigma_a\}\right)^{\mathrm{T}} \{\eta_b\} \, dV \tag{8.23}$$

Noting that, under the sign of transposition, there is not an algebraic matrix product, but instead a differential operator which transforms a vector function, and having recourse to equation (8.21), we obtain

$$L_F = \int_V \{\sigma_a\}^{\mathrm{T}} [\partial] \{\eta_b\} \, dV - \int_S \{\sigma_a\}^{\mathrm{T}} [\mathscr{N}] \{\eta_b\} \, dS \tag{8.24}$$

Applying the equations of kinematics (8.15) and the boundary conditions of equivalence (8.14b), we have then

$$L_F = \int_V \{\sigma_a\}^{\mathrm{T}} \{\varepsilon_b\} \, dV - \int_S \{p_a\}^{\mathrm{T}} \{\eta_b\} \, dS \tag{8.25}$$

and thus

$$\int_V \{\sigma_a\}^T \{\varepsilon_b\} \, dV = \int_V \{\mathscr{F}_a\}^T \{\eta_b\} \, dV + \int_S \{p_a\}^T \{\eta_b\} \, dS \qquad (8.26)$$

which constitutes the final form of the **Principle of Virtual Work for Deformable Bodies**. Whilst the right-hand side represents the external virtual work L_{ve}, the left-hand side represents and defines the internal virtual work L_{vi}, as the scalar product of the stress and strain vectors:

$$L_{vi} = L_{ve} \qquad (8.27)$$

Note that up to this point we have not framed any hypothesis on the nature of the material. This means that the Principle of Virtual Work is of general application, whatever the constitutive law of the material.

8.5 Elastic constitutive law

We now introduce the concept of the **elastic body**. As will be expressed more rigorously later, a deformable body is elastic when its strain energy, i.e. the work performed from outside to bring it into a certain **strain condition** $\{\varepsilon\}$, or into a certain **stress condition** $\{\sigma\}$, does not depend on the loading process (i.e. on the previous events), but only on the final condition. It is usually said that the strain energy is in this case a **state function**.

Consider a deformable body, in equilibrium under the action of the body $\{\mathscr{F}\}$ and surface $\{p\}$ forces. There is generated within it a displacement field $\{\eta\}$, different from zero, except on a constrained part of the external surface (Figure 8.5). Now imagine increasing the external forces by elementary quantities. Let the incremental fields be $\{d\mathscr{F}\}$ and $\{dp\}$ and let them generate an incremental displacement field $\{d\eta\}$ from which there follows an incremental strain field $\{d\varepsilon\}$.

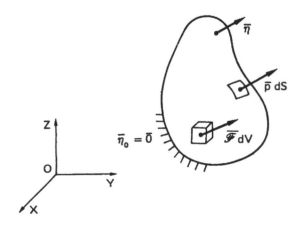

Figure 8.5

Apply the Principle of Virtual Work, considering as a statically admissible system that of the external forces $\{\mathscr{F}\}$ and $\{p\}$, and of the stresses produced by them $\{\sigma\}$, and as a kinematically admissible system that of the incremental displacements $\{\mathrm{d}\eta\}$ and of the incremental strains $\{\mathrm{d}\varepsilon\}$. Note that, in this case, the infinitesimal work is non-virtual, in the sense that the external forces work really by the increments of displacement, while the work of the increments of the forces $\{\mathrm{d}\mathscr{F}\}$ and $\{\mathrm{d}p\}$, by the increments of displacement $\{\mathrm{d}\eta\}$, is an infinitesimal of a higher order and is consequently negligible.

We have therefore

$$\mathrm{d}L_e = \int_S \{p\}^{\mathrm{T}} \{\mathrm{d}\eta\}\, \mathrm{d}S + \int_V \{\mathscr{F}\}^{\mathrm{T}} \{\mathrm{d}\eta\}\, \mathrm{d}V \qquad (8.28\mathrm{a})$$

$$\mathrm{d}L_i = \int_V \{\sigma\}^{\mathrm{T}} \{\mathrm{d}\varepsilon\}\, \mathrm{d}V \qquad (8.28\mathrm{b})$$

and, by the Principle of Virtual Work, the following equality holds:

$$\mathrm{d}L_e = \mathrm{d}L_i \qquad (8.29)$$

On the other hand, a deformable body is defined as **elastic** when the infinitesimal work expressed by equation (8.28b) is an **exact differential**. In particular, for the infinitesimal work $\mathrm{d}L_i$ to be an exact differential, it is necessary for its integrand to be an exact differential:

$$\mathrm{d}\Phi = \{\sigma\}^{\mathrm{T}} \{\mathrm{d}\varepsilon\} \qquad (8.30)$$

The function

$$\Phi = \Phi\,(\varepsilon_x, \varepsilon_y, \varepsilon_z, \gamma_{xy}, \gamma_{xz}, \gamma_{yz}) \qquad (8.31)$$

must, that is, be a state function and it is referred to as **elastic potential**, because it is possible to deduce from it the components of stress by means of partial derivation. The total differential of the function Φ can be expressed as

$$\mathrm{d}\Phi = \frac{\partial \Phi}{\partial \varepsilon_x}\, \mathrm{d}\varepsilon_x + \frac{\partial \Phi}{\partial \varepsilon_y}\, \mathrm{d}\varepsilon_y + \frac{\partial \Phi}{\partial \varepsilon_z}\, \mathrm{d}\varepsilon_z + \qquad (8.32\mathrm{a})$$

$$\frac{\partial \Phi}{\partial \gamma_{xy}}\, \mathrm{d}\gamma_{xy} + \frac{\partial \Phi}{\partial \gamma_{xz}}\, \mathrm{d}\gamma_{xz} + \frac{\partial \Phi}{\partial \gamma_{yz}}\, \mathrm{d}\gamma_{yz}$$

while, rendering equation (8.30) explicit, we have

$$\mathrm{d}\Phi = \sigma_x\, \mathrm{d}\varepsilon_x + \sigma_y\, \mathrm{d}\varepsilon_y + \sigma_z\, \mathrm{d}\varepsilon_z + \tau_{xy}\, \mathrm{d}\gamma_{xy} + \tau_{xz}\, \mathrm{d}\gamma_{xz} + \tau_{yz}\, \mathrm{d}\gamma_{yz} \qquad (8.32\mathrm{b})$$

From equations (8.32a, b) we obtain the components of stress

$$\sigma_x = \frac{\partial \Phi}{\partial \varepsilon_x}, \quad \sigma_y = \frac{\partial \Phi}{\partial \varepsilon_y}, \quad \sigma_z = \frac{\partial \Phi}{\partial \varepsilon_z}, \qquad (8.33)$$

$$\tau_{xy} = \frac{\partial \Phi}{\partial \gamma_{xy}}, \quad \tau_{xz} = \frac{\partial \Phi}{\partial \gamma_{xz}}, \quad \tau_{yz} = \frac{\partial \Phi}{\partial \gamma_{yz}}$$

In the uniaxial case, both Φ and σ_x are functions of the dilation ε_x alone,

$$\Phi = \Phi\left(\varepsilon_x\right), \quad \sigma_x = \sigma_x(\varepsilon_x) \tag{8.34}$$

so that, if we imagine loading and then unloading the one-dimensional body (e.g. a bar in tension), the paths forward and backward in the plane $\varepsilon_x - \sigma_x$ coincide (Figure 8.6(a)). The strain energy, represented by the potential Φ, is equal to the area under the curve $\sigma_x(\varepsilon_x)$. Hence, when the body is unloaded completely, there is no dissipation of energy and the stored elastic energy is fully recovered.

In the case of an **inelastic** one-dimensional body (Figure 8.6(b)), the functions Φ and σ_x no longer present the property of monodromy, i.e. to one value of ε_x there can correspond two or more values of the work and of the force. Unloading the body, we no longer go along the curve $\sigma_x(\varepsilon_x)$ corresponding to the loading, and hence we encounter residual or permanent deformations, with dissipation of energy and only partial recovery of the strain energy. Reversing then the direction of the force and submitting the body to loading cycles, closed or spiral shaped curves will be described in the plane $\varepsilon_x - \sigma_x$. These trajectories will be traversed in a clockwise direction, giving rise to a dissipation of energy by **hysteresis**.

Consider the infinitesimal **virtual** work

$$\mathrm{d}\Psi = \{\mathrm{d}\sigma\}^{\mathrm{T}}\{\varepsilon\} \tag{8.35}$$

Using the definition (8.30), we have

$$\mathrm{d}\Phi + \mathrm{d}\Psi = \mathrm{d}\left(\{\sigma\}^{\mathrm{T}}\{\varepsilon\}\right) \tag{8.36}$$

The elastic potential Φ and the scalar product $\{\sigma\}^{\mathrm{T}}\{\varepsilon\}$ are both state functions, and thus $\mathrm{d}\Phi$ and $\mathrm{d}(\{\sigma\}^{\mathrm{T}}\{\varepsilon\})$ are exact differentials. It follows from equation (8.36) that also $\mathrm{d}\Psi$ is an exact differential and hence that Ψ is a state function. The function

$$\Psi = \Psi\left(\sigma_x, \sigma_y, \sigma_z, \tau_{xy}, \tau_{xz}, \tau_{yz}\right) \tag{8.37}$$

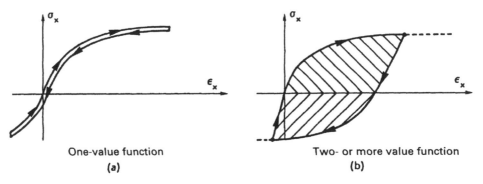

One-value function

(a)

Two- or more value function

(b)

Figure 8.6

is called **complementary elastic potential**, and it is possible to deduce from it the strain components via partial derivation. The total differential of the function Ψ may be expressed as follows:

$$d\Psi = \frac{\partial\Psi}{\partial\sigma_x}d\sigma_x + \frac{\partial\Psi}{\partial\sigma_y}d\sigma_y + \frac{\partial\Psi}{\partial\sigma_z}d\sigma_z + \qquad (8.38\text{a})$$

$$\frac{\partial\Psi}{\partial\tau_{xy}}d\tau_{xy} + \frac{\partial\Psi}{\partial\tau_{xz}}d\tau_{xz} + \frac{\partial\Psi}{\partial\tau_{yz}}d\tau_{yz}$$

whereas rendering equation (8.35) explicit, we have

$$d\Psi = \varepsilon_x d\sigma_x + \varepsilon_y d\sigma_y + \varepsilon_z d\sigma_z + \qquad (8.38\text{b})$$

$$\gamma_{xy}d\tau_{xy} + \gamma_{xz}d\tau_{xz} + \gamma_{yz}d\tau_{yz}$$

From equations (8.38a, b) we obtain the components of strain:

$$\varepsilon_x = \frac{\partial\Psi}{\partial\sigma_x}, \quad \varepsilon_y = \frac{\partial\Psi}{\partial\sigma_y}, \quad \varepsilon_z = \frac{\partial\Psi}{\partial\sigma_z}, \qquad (8.39)$$

$$\gamma_{xy} = \frac{\partial\Psi}{\partial\tau_{xy}}, \quad \gamma_{xz} = \frac{\partial\Psi}{\partial\tau_{xz}}, \quad \gamma_{yz} = \frac{\partial\Psi}{\partial\tau_{yz}}$$

In the uniaxial case it is easy to give a graphical interpretation of Ψ. The elastic complementary energy is the area contained between the curve of loading $\varepsilon_x(\sigma_x)$ and the axis σ_x (Figure 8.7(a)), i.e. it is the area complementary to the one representing the elastic energy Φ, with respect to the rectangle of sides ε_x, σ_x. The latter would represent the work of deformation in the case where, during the entire deformation process, the value of stress was constant and equal to the final value. In the general case, where the relation $\sigma_x = \sigma_x(\varepsilon_x)$ is not linear (Figure 8.7(a)), the deformable body is said to be **nonlinear elastic**. In the particular case, instead, where the relation is linear (Figure 8.7(b)), the body is said to be **linear elastic**. In the linear and one-dimensional case, it is obvious that $\Psi = \Phi$. We shall demon-

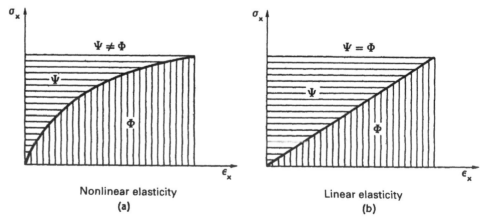

Nonlinear elasticity
(a)

Linear elasticity
(b)

Figure 8.7

strate, however, that equality also exists between elastic potential and complementary elastic potential in the linear and three-dimensional case.

8.6 Linear elasticity

Let us reconsider the elastic potential function (8.31) and expand it using Maclaurin series about the origin, i.e. about the undeformed condition

$$
\begin{aligned}
\Phi\left(\varepsilon_x, \varepsilon_y, \ldots, \gamma_{yz}\right) = \Phi\left(0\right) &+ \left(\frac{\partial \Phi}{\partial \varepsilon_x}\right)_0 \varepsilon_x + \left(\frac{\partial \Phi}{\partial \varepsilon_y}\right)_0 \varepsilon_y + \ldots + \\
\left(\frac{\partial \Phi}{\partial \gamma_{yz}}\right)_0 \gamma_{yz} &+ \frac{1}{2}\left\{\left(\frac{\partial^2 \Phi}{\partial \varepsilon_x^2}\right)_0 \varepsilon_x^2 + \left(\frac{\partial^2 \Phi}{\partial \varepsilon_y^2}\right)_0 \varepsilon_y^2 + \ldots + \right. \\
\left(\frac{\partial^2 \Phi}{\partial \gamma_{yz}^2}\right)_0 \gamma_{yz}^2 &+ 2\left(\frac{\partial^2 \Phi}{\partial \varepsilon_x \partial \varepsilon_y}\right)_0 \varepsilon_x \varepsilon_y + 2\left(\frac{\partial^2 \Phi}{\partial \varepsilon_x \partial \varepsilon_z}\right)_0 \varepsilon_x \varepsilon_z + \ldots + \\
2&\left. \left(\frac{\partial^2 \Phi}{\partial \gamma_{xz} \partial \gamma_{yz}}\right)_0 \gamma_{xz} \gamma_{yz} \right\} + \ldots
\end{aligned}
\tag{8.40}
$$

If the strains are sufficiently small, a good approximation will be achieved by neglecting terms in powers above the second. On the other hand, since the stresses are obtained by derivation of Φ, the value that the function presents at the origin is an arbitrary constant, which we can take to be zero: $\Phi\left(0\right) = 0$. Also the coefficients of the first order terms are zero, since they represent the stresses in the undeformed condition

$$
\left(\frac{\partial \Phi}{\partial \varepsilon_x}\right)_0 = \sigma_x(0) = 0
\tag{8.41a}
$$

$$
\left(\frac{\partial \Phi}{\partial \varepsilon_y}\right)_0 = \sigma_y(0) = 0
\tag{8.41b}
$$

$$
\vdots \qquad\qquad\qquad \vdots
$$

$$
\left(\frac{\partial \Phi}{\partial \gamma_{yz}}\right)_0 = \tau_{yz}(0) = 0
\tag{8.41f}
$$

The analogy with the potential well of the harmonic oscillator is evident (Figure 8.8). Also in that case the derivative of the potential provides the force necessary to remove the material point from the origin (or the opposite of the restoring force)

$$
\Phi = \frac{1}{2} kx^2
\tag{8.42a}
$$

$$
\text{Force} = \frac{d\Phi}{dx} = kx
\tag{8.42b}
$$

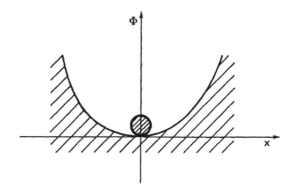

Figure 8.8

where k is the stiffness of the spring. Also in this case, as is well-known, the restoring force is zero in the origin.

The Maclaurin expansion (8.40) then reduces to a **quadratic form**, with 21 coefficients that can be ordered in the (6×6) Hessian matrix

$$[H] = \begin{bmatrix} \left(\dfrac{\partial^2 \Phi}{\partial \varepsilon_x^2}\right)_0 & \left(\dfrac{\partial^2 \Phi}{\partial \varepsilon_x \partial \varepsilon_y}\right)_0 & \cdots & \left(\dfrac{\partial^2 \Phi}{\partial \varepsilon_x \partial \gamma_{yz}}\right)_0 \\[2ex] \left(\dfrac{\partial^2 \Phi}{\partial \varepsilon_y \partial \varepsilon_x}\right)_0 & \left(\dfrac{\partial^2 \Phi}{\partial \varepsilon_y^2}\right)_0 & \cdots & \left(\dfrac{\partial^2 \Phi}{\partial \varepsilon_y \partial \gamma_{yz}}\right)_0 \\[1ex] \vdots & & & \vdots \\[1ex] \left(\dfrac{\partial^2 \Phi}{\partial \gamma_{yz} \partial \varepsilon_x}\right)_0 & \left(\dfrac{\partial^2 \Phi}{\partial \gamma_{yz} \partial \varepsilon_y}\right)_0 & \cdots & \left(\dfrac{\partial^2 \Phi}{\partial \gamma_{yz}^2}\right)_0 \end{bmatrix} \qquad (8.43)$$

Note that the Hessian matrix is the Jacobian of the six first partial derivatives.

It is then possible to write in compact matrix form

$$\Phi = \frac{1}{2} \{\varepsilon\}^{T} [H] \{\varepsilon\} \qquad (8.44a)$$

This relation is analogous to equation (8.42a), once the strain vector $\{\varepsilon\}$ is made to correspond to the elongation x, and the Hessian matrix is made to correspond to the stiffness of the spring. Via partial derivation, it can be readily verified that the stress vector is given by

$$\{\sigma\} = [H] \{\varepsilon\} \qquad (8.44b)$$

This relation is analogous to equation (8.42b), which gives the restoring force of the spring in the case of a one-dimensional problem (harmonic oscillator). From equations (8.44a, b) a new expression of elastic potential is derived:

212

$$\Phi = \frac{1}{2}\{\varepsilon\}^T\{\sigma\} \tag{8.45}$$

Since the undeformed condition must represent an absolute minimum of Φ, and not only a stationary point as equations (8.41) ensure, the Hessian matrix must be **positive definite**, i.e. its determinant and those of its principal minors must be greater than zero. Thus, once more it is possible to discern the analogy with the one-dimensional case, in which the stiffness k of the spring, which is the second derivative of the potential, must be positive.

The positiveness of the Hessian implies the reversibility of the relation (8.44b) and hence of the Hessian matrix itself:

$$\{\varepsilon\} = [H]^{-1}\{\sigma\} \tag{8.46}$$

The relations (8.44b) and (8.46) link together linearly the stress and strain vectors, and constitute the link, hitherto missing, between statics and kinematics.

From the relation (8.36), but for an arbitrary constant, we obtain

$$\Phi + \Psi = \{\sigma\}^T\{\varepsilon\} = \sigma_x\varepsilon_x + \sigma_y\varepsilon_y + \sigma_z\varepsilon_z + \tau_{xy}\gamma_{xy} + \tau_{xz}\gamma_{xz} + \tau_{yz}\gamma_{yz} \tag{8.47}$$

i.e. the sum of the elastic potential and complementary elastic potential is always equal (also in nonlinear cases) to the scalar product of $\{\sigma\}$ and $\{\varepsilon\}$. It has also been demonstrated that, in linear cases, Φ is half of that product (equation (8.45)). Hence, for linear elastic bodies we have

$$\Phi = \Psi = \frac{1}{2}\{\varepsilon\}^T\{\sigma\} = \frac{1}{2}\{\sigma\}^T\{\varepsilon\} \tag{8.48}$$

We can then give the following form to the complementary elastic potential:

$$\Psi = \frac{1}{2}\{\sigma\}^T[H]^{-1}\{\sigma\} \tag{8.49}$$

which corresponds to equation (8.44a) and shows how Ψ is a quadratic form of the components of stress.

8.7 Problem of a linear elastic body

As has previously been noted, the three indefinite equations of equilibrium do not suffice to determine the six components of stress. On the other hand, by adding to them the six elastic constitutive equations (8.44b), we obtain a system of nine differential equations in nine unknowns: σ_x, σ_y, σ_z, τ_{xy}, τ_{xz}, τ_{yz}; u, v, w.

In matrix form it is possible to give a very synthetic and expressive representation of the linear elastic problem, considering as the primary unknown the displacement vector $\{\eta\}$. If in the static equation (8.11) we introduce the constitutive law (8.44b), and then the kinematic equation (8.9), we obtain a matrix equation, called **Lamé's equation in operator form**:

$$\left([\partial]^T[H][\partial]\right)\{\eta\} = -\{\mathcal{F}\} \tag{8.50}$$

213

The matrix and second order differential operator in round brackets is called the **Lamé operator**,

$$[\mathcal{L}] = [\partial]^{\mathrm{T}} \; [H] \; [\partial] \tag{8.51}$$
$$\underset{(3\times3)}{} \quad \underset{(3\times6)}{} \; \underset{(6\times6)}{} \; \underset{(6\times3)}{}$$

It turns out to be a (3×3) matrix and, in non-homogeneous problems, where the matrix $[H]$ is a function of the point, it too is a function of the point.

Recalling the boundary equations of equivalence (8.13) and assuming that they hold good on a portion S_p of the external surface of the body and that, on the complementary portion S_η, there is imposed a congruent field of displacements $\{\eta_0\}$, the three-dimensional elastic problem can be synthesized as follows:

$$[\mathcal{L}]\{\eta\} = -\{\mathcal{F}\}, \quad \forall P \in V \tag{8.52a}$$

$$([\mathcal{V}]^{\mathrm{T}}[H][\partial])\{\eta\} = \{p\}, \quad \forall P \in S_p \tag{8.52b}$$

$$\{\eta\} = \{\eta_0\}, \quad \forall P \in S_\eta \tag{8.52c}$$

For example, in the case of a linear elastic one-dimensional body, restrained at one end, submitted to a tensile force p at the opposite end and to a distribution $\mathcal{F}_x(x)$ of axial forces (Figure 8.9), equations (8.52) take on a notably simplified appearance:

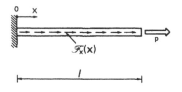

Figure 8.9

$$EA \frac{\mathrm{d}^2 u}{\mathrm{d}x^2} = -\mathcal{F}_x(x), \quad \text{for } 0 \leqslant x \leqslant l \tag{8.53a}$$

$$EA \frac{\mathrm{d}u}{\mathrm{d}x} = p, \quad \text{for } x = l \tag{8.53b}$$

$$u = 0, \quad \text{for } x = 0 \tag{8.53c}$$

where l is the length of the bar and the product EA represents the longitudinal stiffness of the bar.

Once the problem (8.52) is resolved and the displacement field $\{\eta\}$ has been identified, if we reverse the procedure and equations (8.9) and (8.44b) are reapplied, the strain field $\{\varepsilon\}$ and the stress field $\{\sigma\}$ are respectively determined. Since equations (8.52) are linear, the **Principle of Superposition** holds. It means that, if a loading system $\{\mathcal{F}_a\}, \{p_a\}, \{\eta_{0a}\}$ generates a displacement field $\{\eta_a\}$ and thus the strain and stress fields $\{\varepsilon_a\}, \{\sigma_a\}$, and if a different loading system $\{\mathcal{F}_b\}, \{p_b\}, \{\eta_{0b}\}$ generates the fields $\{\eta_b\}$, $\{\varepsilon_b\}, \{\sigma_b\}$, the loading system $\{\mathcal{F}_a\} + \{\mathcal{F}_b\}, \{p_a\} + \{p_b\}, \{\eta_{0a}\} + \{\eta_{0b}\}$, generates displacement, strain and stress fields, which are the sum of the previous ones:

$$\{\eta_a\} + \{\eta_b\}, \{\varepsilon_a\} + \{\varepsilon_b\}, \{\sigma_a\} + \{\sigma_b\}$$

On the basis of the Principle of Superposition, it is possible to demonstrate **Kirchhoff's Theorem** or the **Solution Uniqueness Theorem**: if the solution $\{\eta\}$ exists, it is the only one. The demonstration must be conducted *ab absurdo*. Imagine, that is, that one loading system, $\{\mathcal{F}\}, \{p\}, \{\eta_0\}$, can generate two different responses: $\{\eta_a\}, \{\varepsilon_a\}, \{\sigma_a\}$ or $\{\eta_b\}, \{\varepsilon_b\}, \{\sigma_b\}$. Applying the Principle of Virtual Work to the difference system, we have

$$\int_V \{0\}^T \{\Delta\eta\} \, dV + \int_{S_p} \{0\}^T \{\Delta\eta\} \, dS + \int_{S_\eta} \{\Delta R\}^T \{0\} \, dS \qquad (8.54)$$

$$= \int_V \{\Delta\sigma\}^T \{\Delta\varepsilon\} \, dV$$

$\{\Delta R\}$ being the difference in constraint reactions. The integrand on the right-hand side represents twice the elastic potential, so that we have

$$2 \int_V \Phi \left(\Delta\varepsilon_x, \Delta\varepsilon_y, \ldots, \Delta\gamma_{yz} \right) dV = 0 \qquad (8.55)$$

On the other hand, it is known that Φ is a positive definite quadratic form, so that the integral (8.55) vanishes only when the integrand is zero at each point of the elastic body. This is found to be the case only when, at each point, we have

$$\Delta\varepsilon_x = \Delta\varepsilon_y = \ldots = \Delta\gamma_{yz} = 0 \qquad (8.56)$$

i.e. only when solutions (a) and (b) coincide.

8.8 Clapeyron's Theorem

Consider a linear elastic body subjected to body forces $\{\mathscr{F}\}$ and to surface forces $\{p\}$. Let $\{\eta\}$ be the displacement field that is generated in the body at the end of the loading process that brings the external forces from zero to the aforesaid values. The application of the Principle of Virtual Work gives the following equality:

$$\int_V \{\mathscr{F}\}^T \{\eta\} \, dV + \int_S \{p\}^T \{\eta\} \, dS = \int_V \{\sigma\}^T \{\varepsilon\} \, dV \qquad (8.57)$$

if the corresponding **final** fields are considered as statically and kinematically admissible systems. Multiplying both sides of equation (8.57) by the factor 1/2, we have

$$\frac{1}{2} \int_V \{\mathscr{F}\}^T \{\eta\} \, dV + \frac{1}{2} \int_S \{p\}^T \{\eta\} \, dS = \int_V \Phi \, dV \qquad (8.58)$$

since, for a linear elastic body, the relation (8.45) holds. Equation (8.58) expresses the fact that the work of deformation performed by the external forces to bring the body from the initial undeformed condition to the final deformed condition is equal to half of the work that these forces would perform if they presented their final value during the whole deformation process. The content of **Clapeyron's Theorem** was already implicitly presented, when we considered the one-dimensional case of Figure 8.7(b). If, for example, a linear elastic beam is subjected to the action of a concentrated force in the

215

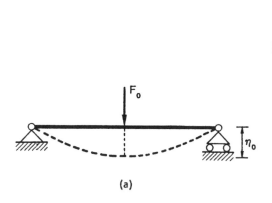
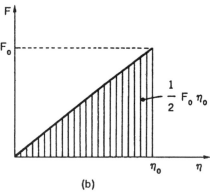

(a)   (b)

Figure 8.10

centre which increases slowly (so as not to induce dynamic phenomena) from zero to the final value F_0, and at the same time the deflection at the centre increases from zero to the final value η_0, then, according to Clapeyron's Theorem, the work of deformation performed on the beam is (Figure 8.10)

$$L_{\text{def.}} = \int_0^{\eta_0} F(\eta)\, d\eta = \frac{1}{2}F_0\eta_0 \qquad (8.59)$$

8.9 Betti's Reciprocal Theorem

Betti's Reciprocal Theorem shows how the Principle of Superposition in linear elasticity holds only for displacements, strain and stress and is not applicable, instead, to the work of deformation.

Consider a linear elastic body and submit it to a quasi-static (i.e. very slow) loading process so that the final forces applied are $\{\mathscr{F}_a\}$, $\{p_a\}$, and the work of deformation performed is L_a. Then proceed loading with the quasi-static application of a second system of forces $\{\mathscr{F}_b\}$, $\{p_b\}$. Let the work performed by these forces be L_b; the work of the forces $\{\mathscr{F}_a\}$, $\{p_a\}$ by the displacements $\{\eta_b\}$ is called **mutual work** L_{ab}. Expressed in formulas,

$$L_{a+b} = L_a + L_b + L_{ab} \qquad (8.60)$$

with

$$L_a = \frac{1}{2}\int_V \{\mathscr{F}_a\}^{\mathrm{T}}\{\eta_a\}\, dV + \frac{1}{2}\int_S \{p_a\}^{\mathrm{T}}\{\eta_a\}\, dS \qquad (8.61\text{a})$$

$$L_b = \frac{1}{2}\int_V \{\mathscr{F}_b\}^{\mathrm{T}}\{\eta_b\}\, dV + \frac{1}{2}\int_S \{p_b\}^{\mathrm{T}}\{\eta_b\}\, dS \qquad (8.61\text{b})$$

$$L_{ab} = \int_V \{\mathscr{F}_a\}^{\mathrm{T}}\{\eta_b\}\, dV + \int_S \{p_a\}^{\mathrm{T}}\{\eta_b\}\, dS \qquad (8.61\text{c})$$

216

Equation (8.60) clearly expresses the non-applicability of the Principle of Superposition to the work of deformation.

Imagine now that the process of loading described above is reversed, i.e. that first the forces $\{\mathscr{F}_b\}$, $\{p_b\}$ are applied, and then the forces $\{\mathscr{F}_a\}$, $\{p_a\}$. The total work of deformation will then be expressible as follows:

$$L_{b+a} = L_b + L_a + L_{ba} \tag{8.62}$$

where the mutual work of the forces $\{\mathscr{F}_b\}$, $\{p_b\}$ acting through the further displacements $\{\eta_a\}$ is

$$L_{ba} = \int_V \{\mathscr{F}_b\}^{\mathrm{T}} \{\eta_a\}\, \mathrm{d}V + \int_S \{p_b\}^{\mathrm{T}} \{\eta_a\}\, \mathrm{d}S \tag{8.63}$$

Comparing expressions (8.60) and (8.62) and noting that the total work of deformation must not depend on the loading path (i.e. on the order in which the external forces are applied), since the body was assumed as being elastic, we obtain the equality of the two mutual work expressions (8.61c) and (8.63):

$$L_{ab} = L_{ba} \tag{8.64}$$

In general, for an elastic body subjected to two systems of surface and body forces, the work done by the first system acting through the displacements resulting from the second and that done by the second system acting through the displacements resulting from the first are equal and different from zero. When in particular these are both zero, the two systems of forces are said to be **energetically orthogonal** and the Principle of Superposition becomes valid also for the work of deformation. Consider, for instance, a supported beam made of linear elastic material, subjected to a concentrated force F in the centre (Figure 8.11(a)) or to a concentrated moment m at one end (Figure 8.11(b)). The two systems, F and m, are not energetically orthogonal, since their mutual work is different from zero,

$$F\eta(m) = m\varphi(F) \neq 0 \tag{8.65}$$

where $\eta(m)$ is the deflection in the centre caused by the moment m, and $\varphi(F)$ is the angle of elastic rotation at the ends caused by the force F. Instead, the

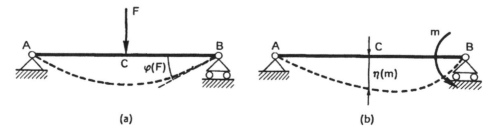

Figure 8.11

217

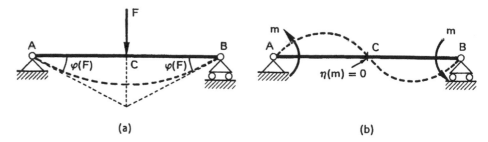

Figure 8.12

two systems of external forces of Figure 8.12, i.e. the concentrated force F in the centre and the two concentrated moments m applied in the same direction at the ends, are energetically orthogonal. In fact, their mutual work is zero,

$$F \times 0 = m\varphi(F) - m\varphi(F) = 0 \tag{8.66}$$

the deflection in the centre being zero in the diagram of Figure 8.12(b), which is skew-symmetrical, and the end elastic rotations being equal and opposite in the diagram of Figure 8.12(a), which is symmetrical.

As will be shown later, the characteristics of the internal reaction of the beams are, except for marginal cases, energetically orthogonal.

8.10 Isotropy

The deformable body is considered in this section also as **isotropic**, as well as linear elastic. This means that the mechanical properties are considered as identical in all directions issuing from the generic point P (the case of anisotropic material is dealt with in Appendix C). As there do not exist preferential directions, the complementary elastic potential Ψ will depend on the values of the three principal stresses, and not on the orientation of the principal reference system

$$\Psi = \Psi(\sigma_1, \sigma_2, \sigma_3) \tag{8.67}$$

As Ψ is a quadratic form of the stress components, it may be cast in the following form:

$$\Psi = \frac{1}{2E}\left\{(\sigma_1^2 + \sigma_2^2 + \sigma_3^2) - 2\nu(\sigma_1\sigma_2 + \sigma_1\sigma_3 + \sigma_2\sigma_3)\right\} \tag{8.68}$$

where $1/2E$ and $-\nu/E$ are two coefficients which multiply the squares and the mutual products, respectively. The coefficients thus reduce from 21 to two, on the hypothesis of isotropy alone. The constants E and ν, as we shall see presently, have a precise physical meaning.

Recalling the expressions (7.74) of the invariants as functions of the principal stresses, equation (8.68) can be written as follows:

$$\Psi = \frac{1}{2E}\left\{J_I^2 + 2J_{II}(1 + \nu)\right\} \tag{8.69}$$

218

Expressing then the invariants as functions of the generic components of the stress vector, as equations (7.72) show, we have

$$\Psi = \frac{1}{2E}(\sigma_x^2 + \sigma_y^2 + \sigma_z^2) - \frac{v}{E}(\sigma_x\sigma_y + \sigma_x\sigma_z + \sigma_y\sigma_z) + \qquad (8.70)$$

$$\frac{1}{2G}(\tau_{xy}^2 + \tau_{xz}^2 + \tau_{yz}^2)$$

where we have set

$$G = \frac{E}{2(1+v)} \qquad (8.71)$$

Equation (8.70) thus represents the complementary elastic potential in the case of linear and isotropic elasticity.

The components of strain are obtained according to equations (8.39), by partial derivation of Ψ:

$$\varepsilon_x = \frac{\partial \Psi}{\partial \sigma_x} = \frac{\sigma_x}{E} - \frac{v}{E}\sigma_y - \frac{v}{E}\sigma_z \qquad (8.72a)$$

$$\varepsilon_y = \frac{\partial \Psi}{\partial \sigma_y} = \frac{\sigma_y}{E} - \frac{v}{E}\sigma_x - \frac{v}{E}\sigma_z \qquad (8.72b)$$

$$\varepsilon_z = \frac{\partial \Psi}{\partial \sigma_z} = \frac{\sigma_z}{E} - \frac{v}{E}\sigma_x - \frac{v}{E}\sigma_y \qquad (8.72c)$$

$$\gamma_{xy} = \frac{\partial \Psi}{\partial \tau_{xy}} = \frac{\tau_{xy}}{G} \qquad (8.72d)$$

$$\gamma_{xz} = \frac{\partial \Psi}{\partial \tau_{xz}} = \frac{\tau_{xz}}{G} \qquad (8.72e)$$

$$\gamma_{yz} = \frac{\partial \Psi}{\partial \tau_{yz}} = \frac{\tau_{yz}}{G} \qquad (8.72f)$$

Note that, whereas the shearing strains are linearly dependent, with a relation of proportionality, only on the respective shearing stresses, the dilations each depend on all three normal stresses. In explicit matrix form we may write

$$\begin{bmatrix} \varepsilon_x \\ \varepsilon_y \\ \varepsilon_z \\ \gamma_{xy} \\ \gamma_{xz} \\ \gamma_{yz} \end{bmatrix} = \begin{bmatrix} \frac{1}{E} & -\frac{v}{E} & -\frac{v}{E} & 0 & 0 & 0 \\ -\frac{v}{E} & \frac{1}{E} & -\frac{v}{E} & 0 & 0 & 0 \\ -\frac{v}{E} & -\frac{v}{E} & \frac{1}{E} & 0 & 0 & 0 \\ 0 & 0 & 0 & \frac{1}{G} & 0 & 0 \\ 0 & 0 & 0 & 0 & \frac{1}{G} & 0 \\ 0 & 0 & 0 & 0 & 0 & \frac{1}{G} \end{bmatrix} \begin{bmatrix} \sigma_x \\ \sigma_y \\ \sigma_z \\ \tau_{xy} \\ \tau_{xz} \\ \tau_{yz} \end{bmatrix} \qquad (8.73)$$

The foregoing matrix relation in compact form is represented by equation (8.46). The inverse of equation (8.73) is as follows:

$$\frac{1}{2G}\begin{bmatrix} \sigma_x \\ \sigma_y \\ \sigma_z \\ \tau_{xy} \\ \tau_{xz} \\ \tau_{yz} \end{bmatrix} = \begin{bmatrix} \frac{1-v}{1-2v} & \frac{v}{1-2v} & \frac{v}{1-2v} & 0 & 0 & 0 \\ \frac{v}{1-2v} & \frac{1-v}{1-2v} & \frac{v}{1-2v} & 0 & 0 & 0 \\ \frac{v}{1-2v} & \frac{v}{1-2v} & \frac{1-v}{1-2v} & 0 & 0 & 0 \\ 0 & 0 & 0 & \frac{1}{2} & 0 & 0 \\ 0 & 0 & 0 & 0 & \frac{1}{2} & 0 \\ 0 & 0 & 0 & 0 & 0 & \frac{1}{2} \end{bmatrix}\begin{bmatrix} \varepsilon_x \\ \varepsilon_y \\ \varepsilon_z \\ \gamma_{xy} \\ \gamma_{xz} \\ \gamma_{yz} \end{bmatrix} \qquad (8.74)$$

Since the Hessian matrix has been assumed as positive definite (the undeformed condition must represent an absolute minimum of the work of deformation), also its inverse is positive definite (i.e. likewise the unstressed condition must represent an absolute minimum of the work of deformation). All the principal minors of the matrix (8.73) must therefore be greater than zero. The following principal minor must therefore be positive definite:

$$\begin{bmatrix} \frac{1}{E} & -\frac{v}{E} & -\frac{v}{E} \\ -\frac{v}{E} & \frac{1}{E} & -\frac{v}{E} \\ -\frac{v}{E} & -\frac{v}{E} & \frac{1}{E} \end{bmatrix} \qquad (8.75a)$$

and the following condition must at the same time hold:

$$\frac{1}{G} > 0 \qquad (8.75b)$$

From equations (8.75) four inequalities are drawn,

$$\frac{1}{E} > 0 \qquad (8.76a)$$

$$\frac{1}{E^2}(1-v^2) > 0 \qquad (8.76b)$$

$$\frac{1}{E^3}(1+v)^2(1-2v) > 0 \qquad (8.76c)$$

$$\frac{E}{2(1+v)} > 0 \qquad (8.76d)$$

which, when resolved, impose

$$E > 0 \qquad (8.77a)$$

$$-1 < v < 1 \qquad (8.77b)$$

$$v < \frac{1}{2} \qquad (8.77c)$$

$$v > -1 \qquad (8.77d)$$

Combining the conditions (8.77b, c) and noting that equation (8.77d) contains equation (8.77b), we obtain finally the following bounds:

$$E > 0 \tag{8.78a}$$

$$-1 < v < \frac{1}{2} \tag{8.78b}$$

On the other hand, physically we note that the coefficient v is never negative. In the sequel we shall discuss the physical meaning of the parameters E, v, G, and their limitations.

Consider an elementary parallelepiped subjected to the normal tensile stress component σ_x only (Figure 8.13(a)). Equations (8.72) in this case take on a particular form

$$\varepsilon_x = \frac{\sigma_x}{E} \tag{8.79a}$$

$$\varepsilon_y = -\frac{v}{E}\sigma_x = -v\varepsilon_x \tag{8.79b}$$

$$\varepsilon_z = -\frac{v}{E}\sigma_x = -v\varepsilon_x \tag{8.79c}$$

$$\gamma_{xy} = \gamma_{xz} = \gamma_{yz} = 0 \tag{8.79d}$$

From equation (8.79a) there derives for E the physical meaning of stiffness of the material. On the plane ε_x–σ_x, E represents in fact the positive slope of the straight line passing through the origin, which describes the process of loading (Figure 8.13(b)). The slope E is known as the **normal elastic modulus** or **Young's modulus**.

The coefficient v, on the other hand, represents the ratio between the dilations induced in the directions perpendicular to that of stress and the dilation in the direction of stress:

$$v = \left| \frac{\varepsilon_y}{\varepsilon_x} \right| = \left| \frac{\varepsilon_z}{\varepsilon_x} \right| \tag{8.80}$$

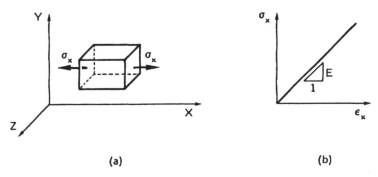

(a)   (b)

Figure 8.13

221

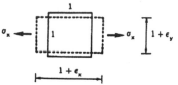

Figure 8.14

Since the former are experimentally always of the opposite algebraic sign with respect to the latter (Figure 8.14), physically v is positive, and thus the limitations (8.78b) corresponding to it are in actual fact more severe:

$$0 < v < \frac{1}{2} \tag{8.81}$$

The constant v is called **ratio of transverse contraction** or **Poisson's ratio**. It is interesting to note that the reversibility of the Hessian matrix implies a volumetric dilation concordant with the normal stress σ_x, and *vice versa*. In fact, from relations (8.79), we have

$$\frac{\Delta V}{V} = \varepsilon_x + \varepsilon_y + \varepsilon_z = \frac{\sigma_x}{E}(1 - 2v) \tag{8.82}$$

Since from equation (8.78a) we have $E > 0$, and from equation (8.78b) $v < 1/2$, equation (8.82) shows how $\Delta V/V$ and σ_x are concordant, i.e. if σ_x is a tensile stress, a positive volumetric dilation is produced, whereas if σ_x is a compressive stress, a negative volumetric dilation (i.e. a volumetric contraction) is produced.

Consider once more the elementary parallelepiped, stressed in this case only by the shearing stress component τ_{xy} (Figure 8.15(a)). Equations (8.72) are particularized as follows:

$$\varepsilon_x = \varepsilon_y = \varepsilon_z = \gamma_{xz} = \gamma_{yz} = 0 \tag{8.83a}$$

$$\gamma_{xy} = \frac{\tau_{xy}}{G} \tag{8.83b}$$

The parameter G thus represents the stiffness that the solid opposes to shearing strain. This is called the **shear elastic modulus**, and graphically is the positive angular coefficient of the line of loading in the plane $\gamma_{xy}-\tau_{xy}$ (Figure 8.15(b)). From the relation (8.83b) and the similar relations which show the proportionality between shearing strains and corresponding shearing stresses we deduce how, in the linear elastic and isotropic body, the principal directions of strain and stress coincide.

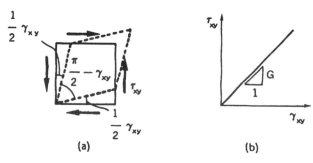

(a)   (b)

Figure 8.15

A linear elastic and isotropic body is characterized by the values that the two parameters E and v assume at each point. If the point functions E and v prove to be constant, then the body is said to be **homogeneous**.

Two typical building materials, steel and concrete, are with good approximation considered linear elastic, isotropic and homogeneous. On the other hand, the hypothesis of linear elasticity is acceptable only in the conditions where the materials are not excessively stressed. Beyond certain threshold values, of which we shall speak presently, the behaviour of the material becomes non-linear and inelastic. When the materials are subjected to increasing stresses, critical conditions are eventually reached, on account of which the body can no longer be considered even a deformable continuum. In other words, fractures form and hence produce discontinuities in the displacement function.

In Table 8.1 indicative values of E, v and σ_{max} for the two materials mentioned above are given, σ_{max} being the value of tensile normal stress which causes yielding or fracturing.

Table 8.1

	E (kg/cm^2)	v	σ_{max}(kg/cm^2)
Steel	2 100 000	0.30	2400
Concrete	250 000	0.15	30

The ratio σ_{max}/E indicates the order of magnitude of the dilation, below which the linear elastic idealization has a physical meaning. It may be noted that this order of magnitude lies between 10^{-3} and 10^{-4}. The strains which are usually found in structural elements are in fact very small.

As regards the **anisotropic elastic constitutive law**, the reader is referred to Appendix C.

8.11 Strength, ductility, fracture energy

Once the elastic stress field of a structural element has been calculated, on the basis of the forces applied $\{\mathscr{F}\}$ and $\{p\}$ and the displacements imposed $\{\eta_0\}$, using the equations of statics and kinematics and the constitutive equations and then resolving Lamé's equation, we are faced with the problem of evaluating whether the theoretically determined stresses exceed, albeit in only one point of the body or in one of its portions, the strength of the material of which the body is made. In fact, as has already been mentioned in the foregoing section, even though the law $\sigma(\varepsilon)$ is linear and elastic in the initial portion of the curve, it loses its linearity in the next portion, giving rise to phenomena of yielding, plastic deformation and ultimately fracture. The stress field, on the other hand, is proportional to the forces applied so far as the structural behaviour is linearly elastic. Consequently, if calculation, in this conventionally elastic case, gives excessively high values of stress, it is possible to consider the external loads reduced by a suitable factor to obtain a really elastic stress field. This factor will therefore be equal to the ratio between the conventionally elastic and excessively high stresses and the really elastic and hence admissible stresses for the material. The latter are chosen by the structural designer on the basis of criteria of resistance and safety which we shall look at later on.

Structural materials are traditionally catalogued, on the basis of the characteristics of the $\sigma(\varepsilon)$ curve, into two distinct categories: **ductile materials** and **brittle materials**. Whereas the former show large portions of the $\sigma(\varepsilon)$ diagram that are not linear, before they reach the fracture point, the latter break suddenly, when the response is still substantially elastic and linear. A second characteristic which distinguishes them clearly is the ratio between tensile strength and compressive strength. Whereas for ductile materials this ratio is close to unity, for brittle materials it is a good deal lower (in some cases, 10^{-1} to 10^{-2}). The differences in behaviour depend to a great extent upon the microscopic mechanisms of damage and fracture, which, in the various structural materials present notable differences. In metal alloys, for instance, sliding takes place between the planes of atoms and crystals which gives rise to a behaviour of a plastic and ductile kind, with considerable permanent deformations. In concrete and rock, on the other hand, the microcracks and debondings between the granular components and the matrix can extend and combine to form a macroscopic crack which splits the structural element suddenly into two parts. This unstable fracturing process causes the material to behave in a brittle manner.

On the other hand, it is not always easy to determine the microscopic magnitude of the mechanisms of damage. It may present very different dimensions according to the nature of the mechanisms and the heterogeneity of the material. In crystals damage occurs at an atomic level, with vacancies and dislocations; in metal alloys cracks spread at an intergranular or transgranular level; and in concrete the cracking occurs at the interface between the aggregates and the cement matrix. It is thus understandable how the scale of damage comes to depend upon the regularity of the solid and hence upon the size of heterogeneities present in it. Alongside the traditional building materials, more recent times have witnessed the advent of a large number of new materials which present highly heterogeneous and anisotropic features on account of their being reinforced with fibres and composed of laminas. These materials, called **composites**, may have a polymer, metal, ceramic or cement matrix. In these there are essentially two mechanisms of damage: fibre pull-out and delamination (i.e. the debonding of the layers).

The distinction between ductile materials and brittle materials is not always so clear in practice, because the ductility of the material also depends upon the ambient temperature and, as we shall see later, upon the size of the structural element. Of the two, the latter is the factor that is harder to grasp, because in this case ductility ceases to be a property of the material and becomes a property of the structure as a whole.

Let us consider a **uniaxial tensile test** carried out on a test specimen of ductile material, for instance, steel (Figure 8.16). Let the test specimen have the usual hourglass shape, to prevent fracture occurring in the vicinity of the ends where the specimen is clamped to the testing equipment. Let A_0 be the area of the initial cross section of the tensile specimen in the middle zone, and l_0 the initial distance between two sensors glued at two distinct points of the middle zone. Let this distance be measured by an electrical device that connects the two points. Let the **nominal stress** σ be defined as the ratio between the force F transmitted by the testing equipment and the initial area A_0:

$$\sigma = F/A_0 \tag{8.84}$$

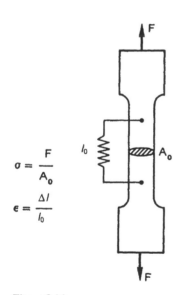

$$\sigma = \frac{F}{A_0}$$

$$\epsilon = \frac{\Delta l}{l_0}$$

Figure 8.16

In this way the elastic and possible plastic transverse contractions are neglected. Then let the **conventional dilation** ε be defined as the ratio between the variation in the distance between the two sensors, Δl, and the initial distance l_0:

$$\varepsilon = \Delta l / l_0 \qquad (8.85)$$

This dilation is the average dilation for the zone being checked. It is very likely that during the test, and especially in the nonlinear regime, dilation is not uniform and consequently at a given point does not coincide with the average.

Now let all the pairs of points recorded during the loading process be plotted on the σ–ε plane (Figure 8.17). Between points O and L the diagram is linear and elastic. From L onwards the response is no longer linear and the material begins to yield. When the specimen is unloaded, there is evidence of permanent deformation ε_{pl}. This means that part of the strain energy has been recovered (triangle ABA'), i.e. that corresponding to the strain ε_{el}, whereas the remainder has dissipated plastically (area $OLAA'$). When the test specimen is again loaded, once more it covers elastically the path $A'A$, which is parallel to the path OL. When it arrives at A, the specimen yields again at a stress $\sigma > \sigma_l$. Virgin material, then, yields at lower levels of stress than does material that has already undergone yielding. This phenomenon is referred to as **hardening**. When the applied force F is further increased, the curve ceases to be linear (portion AU). In this phase the increase in stress per unit increase in dilation (usually called tangential stiffness) continues to diminish, until it vanishes at point U. When the point U is reached, if the loading process is controlled by the external force F, the specimen breaks, because F cannot increase any further.

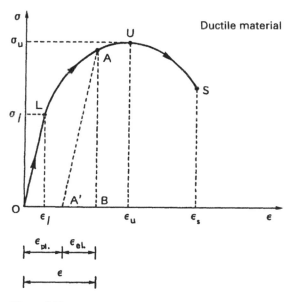

Figure 8.17

225

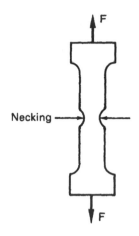

Figure 8.18

On the other hand, if the loading process is controlled by the variation in distance Δl (i.e. if a slope is electronically imposed on that quantity in time), it is possible to investigate the behaviour of the material beyond the point of ultimate strength U. Beyond the point U, in fact, the tangential stiffness becomes negative and, to positive increments of displacement Δl, there correspond negative increments of the force F. This is due to the phenomenon of **plastic transverse contraction** or **necking** (Figure 8.18), whereby the area A of the actual cross section becomes notably less than A_0, in a localized band between the two sensors. Finally, once a terminal point S is reached, the specimen gives suddenly, even though the loading process is deformation-controlled.

In the case of certain metal alloys, such as low carbon steels, a sudden yield follows the proportional limit L, so that dilation increases by a finite quantity under constant loading (Figure 8.19). In these cases it is thus easy to identify the value of **uniaxial yielding stress** σ_P, as this coincides with the proportional limit σ_l. When, instead, the proportional limit is followed by the hardening portion of the curve, it is more difficult to define σ_P. In this case, it is conventional to use the stress value of which the permanent deformation ε_{pl} at unloading is equal to 2‰.

Whereas ductile materials present similar behaviours in tension and compression, brittle materials behave in considerably different ways. Concrete, for instance, is ductile in compression but brittle in tension, and presents an ultimate compressive strength that is about one order of magnitude greater than its ultimate tensile strength. A tensile test on a specimen of concrete, if conducted by applying a load or, as is usually said, under controlled loading conditions, shows an approximately linear elastic response up to a point where the load drops sharply corresponding to the sudden formation of a crack. However, today's electronic techniques allow us to control the strain (*input* = strain ε, *output* = stress σ). By so doing, the post-peak response curve of the cement material is highlighted (Figure 8.20). Only recently has it been realized that there exists an extensive branch of **softening** and that it is possible for concrete to dissipate a considerable amount of energy per unit volume. This energy is represented by the area under the curve $\sigma(\varepsilon)$.

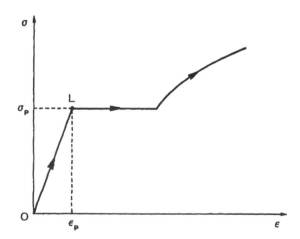

Figure 8.19

226

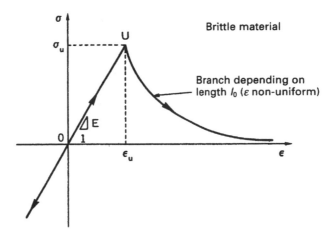

Figure 8.20

Even more recently it has been possible to demonstrate that energy, in actual fact, is not dissipated uniformly in the volume, but rather is dissipated over a localized band, which subsequently becomes a crack (the same phenomenon occurs in ductile materials with necking). In other words, point dilation between the two sensors of Figure 8.16 is not a constant function. On the contrary, it presents a notable peak corresponding to the crack that is forming. Ideally it is possible to imagine that the dilation ε is a Dirac δ function, since the dilation is infinite where a discontinuity occurs in the **axial displacement** function.

As a consequence of the localization of the strain ε, the decreasing branch of the $\sigma(\varepsilon)$ curve comes to depend on the length l_0 of the measurement base. What, instead, emerges as a true characteristic of the material is the $\sigma(w)$ diagram, which represents the stress transmitted through the crack, as a function of the opening (or width) of the crack itself (Figure 8.21). This law of decay indicates, of course, a weakening of the interaction with the increase in the distance w between the faces (or free surfaces) of the crack. When w reaches the limiting value w_c, the interaction ceases totally and the crack becomes a complete disconnection which divides the specimen into two distinct parts. The area under the curve $\sigma(w)$ represents the energy dissipated over the unit

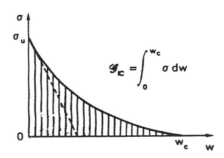

Figure 8.21

227

surface of fracture. Since the **cohesive law** $\sigma(w)$ is a characteristic of the material, which depends on the intimate structure and on the mechanisms of damage of the material, the **fracture energy** $\mathscr{G}_{IC}$ is also an intrinsic property of the material:

$$\mathscr{G}_{IC} = \int_0^{w_c} \sigma(w)\,dw \tag{8.86}$$

The energy dissipated over the surface of the crack is equal to $\mathscr{G}_{IC}A_0$, since $\mathscr{G}_{IC}$ is work per unit surface and thus force per unit length, $[F][L]^{-1}$. Since, however, we have assumed that the dissipation of energy has only occurred on the fracture surface and not in the volume of the undamaged material, the energy dissipated globally in the volume $A_0 l_0$ is still equal to $\mathscr{G}_{IC}A_0$ (this is rigorously valid only in the absence of hardening). If the response curves are then plotted on the plane F–Δl, with the increase in the length l_0 of the specimen, we obtain elastic portions of curve having a decreasing stiffness and softening portions having a growing negative slope and, beyond a certain limit, having a positive slope (Figure 8.22(a)). The area under each curve must in fact be constant and equal to $\mathscr{G}_{IC}A_0$.

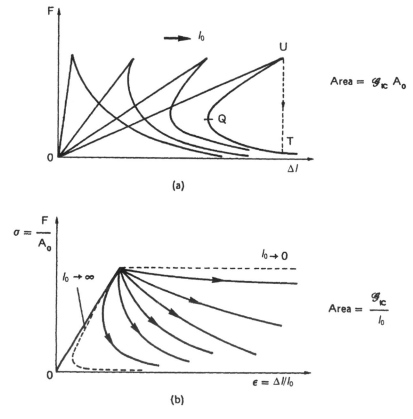

(a)

(b)

Figure 8.22

228

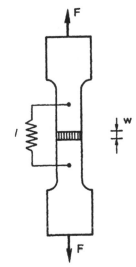

Figure 8.23

On the σ–ε plane (Figure 8.22(b)) the transition just described is represented by a single linear elastic portion of curve and by a fan of softening branches, as the length l_0 varies. The area under the curve, in fact, in this case varies with l_0, as it is equal to $\mathscr{G}_{IC}/l_0$. For $l_0 \to 0$ the softening branch becomes horizontal and represents a **perfectly plastic** structural response. On the other hand, for $l_0 \to \infty$ the area between the $\sigma(\varepsilon)$ curve and the axis ε must tend to zero, and thus the softening branch tends to coincide with the elastic portion (Figure 8.22(b)).

The positive slope of the softening branch may be justified not only, as we have seen, by considering the dissipated energy, but also by analytical derivation of the function $\varepsilon(\sigma)$. In the post-peak regime, we have (Figure 8.23)

$$\varepsilon = \frac{\Delta l}{l_0} = \frac{\varepsilon_{el} l_0 + w}{l_0} \qquad (8.87)$$

where ε_{el} indicates the specific longitudinal dilation of the undamaged zone:

$$\varepsilon_{el} = \sigma/E \qquad (8.88)$$

From equation (8.87) we then draw

$$\varepsilon = \frac{\sigma}{E} + \frac{1}{l_0} w(\sigma) \qquad (8.89)$$

and deriving with respect to σ

$$\frac{d\varepsilon}{d\sigma} = \frac{1}{E} + \frac{1}{l_0} \frac{dw}{d\sigma} \qquad (8.90)$$

This derivative, and consequently also the inverse $d\sigma/d\varepsilon$, is greater than zero for

$$l_0 > E \left| \frac{dw}{d\sigma} \right| \qquad (8.91)$$

It follows that there are portions of softening having a positive slope for

$$l_0 > E \Big/ \left| \frac{d\sigma}{dw} \right|_{max} \qquad (8.92)$$

i.e. when the length of the specimen, or rather the distance l_0, between the points of which the relative displacement is measured, is higher than the ratio between the elastic modulus and the maximum slope of the cohesive law. This is due to the fact that, during the softening phase, the stress σ diminishes and, while the point which represents the fracture zone drops along the curve $\sigma(w)$ (Figure 8.21), the point representing the undamaged zone drops along the straight line $\sigma(\varepsilon)$ (Figure 8.20) and describes an elastic unloading. If the length l_0 is sufficiently great, the elastic contraction prevails over the dilation of the fracture zone, giving rise to the phenomenon described before.

Softening with a positive slope represents a phenomenon that falls within the scope of Catastrophe Theory. If, in fact, the loading process is governed by the conventional dilation ε, or by the elongation Δl, once the point U is reached (Figure 8.22(a)), there is a vertical drop in the load, until the lower softening portion of curve, which has a negative slope, is encountered. The portion UQT is

229

THEORY OF ELASTICITY

thus ignored and becomes virtual. To record this portion experimentally, it is necessary to govern the process of loading via the opening w of the crack, a procedure rendered possible by modern-day electronic techniques. The instability described above is called **snap-back**. All relatively brittle materials (e.g. concrete, cast iron, glass, plexiglass) which possess a low **fracture energy** $\mathcal{G}_{IC}$ with the normal lengths l_0 of the measurement base present a sharp drop in load when the global behaviour of the specimen is still linear elastic.

To conclude this section, it is expedient to note how strength and fracture energy are intrinsic properties of the material, whereas ductility depends on a structural factor, such as the length of the specimen. In Chapter 20 this subject will be taken up again, and we shall see how, among the factors that affect structural ductility, or brittleness, the size scale of the structural element must be included.

8.12 Strength criteria

In the case of bodies subjected to a condition of uniaxial stress, such as ropes or columns, the check on strength is immediate, once the service stress and the yielding or ultimate strength are known. In ductile materials, the material does not reach the critical point, i.e. yielding, in the case where the following relation holds:

$$-\sigma_P < \sigma < \sigma_P \tag{8.93a}$$

as the yielding stress has approximately the same absolute value in tension as it does in compression. In brittle materials, the behaviour in compression is usually different from that in tension and the critical point for the material is avoided if

$$-\sigma_c < \sigma < \sigma_u \tag{8.93b}$$

where σ_c is the ultimate compressive strength.

The relations (8.93) would provide real safety limits if all the quantities involved were known with certainty and without statistical oscillations. However, in the physical world, the quantities will always possess a degree of approximation. For instance, the strength measured in the laboratory, the dimensions of the body, the forces actually applied are not deterministic quantities. **Admissible stress**, to permit conditions of safety, will then be represented by a fraction of the nominal strength, in such a way that the so-called **safety criteria**, for ductile and brittle materials, will be presented as follows:

$$-\frac{\sigma_P}{s} < \sigma < \frac{\sigma_P}{s} \tag{8.94a}$$

$$-\frac{\sigma_c}{s} < \sigma < \frac{\sigma_u}{s} \tag{8.94b}$$

The parameter $s > 1$ is called the **safety factor**. The higher this factor, the less foreseeable is the behaviour of the material and hence the less repeatable are the laboratory results. Whereas for ductile materials it is common practice to take $s = 1.5$, for brittle materials the factor of safety is usually higher (in some cases $s = 6$), because in the case of these materials the mechanisms of damage are more unstable and hence, as has already been shown, there do not exist ductile or plastic reserves beyond the ultimate load.

230

In general, however, the service of structural elements is not restricted to only one axis. Nonetheless, the tests that are usually performed in laboratories are uniaxial, hence the need to correlate biaxial and triaxial stress states to the uniaxial ones. In other words, the procedure must be to define a function of the stress tensor, called **ideal stress** or **equivalent stress**, to compare with the stress of uniaxial yield

$$\sigma_{eq} = \sigma_{eq}(\sigma_x, \sigma_y, \sigma_z, \tau_{xy}, \tau_{xz}, \tau_{yz}) < \sigma_P \qquad (8.95)$$

In the case of isotropy of the material, the ideal stress is a function of the principal components of stress alone:

$$\sigma_{eq} = \sigma_{eq}(\sigma_1, \sigma_2, \sigma_3) < \sigma_P \qquad (8.96)$$

As regards uniaxial tensile and compressive tests on metal specimens, it may be noted that the hydrostatic pressure of the environment in which the tests are conducted does not influence the yield stress value. This important experimental fact has given rise to two strength criteria, which have been widely confirmed by experience: Tresca's criterion and Von Mises' criterion.

Tresca's criterion or the **criterion of maximum shearing stress** considers shearing stress responsible for the yielding of the material when it is subjected to a triaxial stress condition. It is thus implicitly assumed that not only in the uniaxial case but also in the triaxial case the superposition of a hydrostatic condition does not affect the strength of the material. It is very simple and expressive to represent Tresca's criterion on Mohr's plane. The maximum shearing stress τ_{max} is in fact equal to half the difference between the extreme principal stresses (Figure 8.24)

$$\tau_{max} = \frac{1}{2} \max \left\{ |\sigma_1 - \sigma_2| \,, \, |\sigma_1 - \sigma_3| \,, \, |\sigma_2 - \sigma_3| \right\} \qquad (8.97)$$

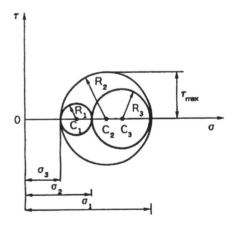

Figure 8.24

and, from the condition of uniaxial yielding, we obtain (Figure 8.25)

$$\tau_P = \frac{1}{2}\sigma_P \qquad (8.98)$$

where τ_P is the yield shearing stress. Tresca's condition $\tau_{max} < \tau_P$, applying equations (8.97) and (8.98), is translated into the following:

$$\max\left\{|\sigma_1 - \sigma_2|, |\sigma_1 - \sigma_3|, |\sigma_2 - \sigma_3|\right\} < \sigma_P \qquad (8.99)$$

where the term on the left-hand side is called **Tresca's equivalent stress**. Equation (8.99) is thus a particular case of equation (8.96). On Mohr's plane, the original condition $\tau_{max} < \tau_P$ is represented by an infinite strip, bounded by the two parallel straight lines $\tau = \pm\tau_P$ (Figure 8.25).

In the case of a **plane stress condition**, one of the three principal stresses vanishes, for example $\sigma_3 = 0$, and the condition (8.99) becomes

$$\max\left\{|\sigma_1 - \sigma_2|, |\sigma_1|, |\sigma_2|\right\} < \sigma_P \qquad (8.100)$$

The inequality (8.100) may be interpreted as the intersection of three different inequalities:

$$|\sigma_1| < \sigma_P \qquad (8.101a)$$

$$|\sigma_2| < \sigma_P \qquad (8.101b)$$

$$|\sigma_1 - \sigma_2| < \sigma_P \qquad (8.101c)$$

The first two, on the plane σ_1–σ_2, represent **Rankine's criterion**, or the **criterion of maximum normal stress**. The strength domain would in this case be a square (Figure 8.26), but, in actual fact, such a simple criterion has been proved inadequate by the experimental evidence. The inequality (8.101c) in fact further sections the square domain and finally furnishes a hexagonal domain, called **Tresca's hexagon** (Figure 8.26). Note how the four points of intersection of Tresca's hexagon with the axes σ_1 and σ_2 represent the uniaxial critical points, of tension and compression, in the two principal directions.

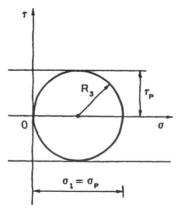

Figure 8.25

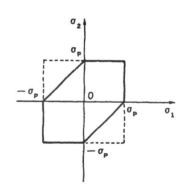

Figure 8.26

232

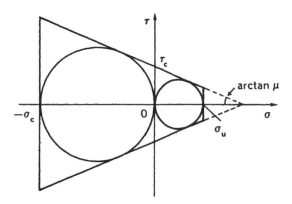

Figure 8.27

It is worthwhile mentioning here the extrapolation of Tresca's criterion for brittle materials, which present different values for tensile and compressive strength. Instead of considering the critical shearing stress as a constant independent of the state of stress, the **Mohr–Coulomb criterion** proposes a limit τ_P that is a function of the corresponding normal stress

$$\tau_{\max} < \tau_c - \mu\sigma \tag{8.102}$$

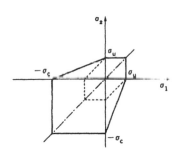

Figure 8.28

where τ_c represents the **cohesiveness** of the material and μ the **coefficient of internal friction**. The strength increases with the increase in normal compression, so that the strength domain on Mohr's plane is represented by a strip that broadens in the direction of the negative stresses σ (Figure 8.27). Cutting this domain with the two vertical straight lines that represent the uniaxial critical states of tension and compression, we obtain a trapezoidal domain which reproduces satisfactorily the critical states of brittle or incoherent materials (e.g. concrete, rock, soil). On the plane of the principal stresses σ_1–σ_2, the Mohr–Coulomb domain is represented by a symmetrical hexagon with respect to the bisector of the first and third quadrants (Figure 8.28), which reduces to Tresca's hexagon for $\mu = 0$ (zero internal friction).

Von Mises' criterion, or the **criterion of maximum energy of distortion**, considers the strain energy corresponding to the **deviatoric tensor** (7.77b) responsible for the critical condition of the material. Since, in general, the strain energy per unit volume (or strain energy density) may be expressed as a function of the first two invariants of stress, as is shown by equation (8.69),

$$\Psi = \Phi = \frac{1}{2E}\left\{ J_I^2 + 2J_{II}(1+v) \right\} \tag{8.103}$$

this expression can be particularized to the case of the deviatoric tensor. The first invariant of the deviatoric tensor is zero by definition, so that the energy Φ_d associated to this tensor is

$$\Phi_d = \frac{1+v}{E} J_{IId} \tag{8.104}$$

233

where J_{IId} indicates the second deviatoric invariant. This equals

$$J_{IId} = -\left[(\sigma_1 - \bar{\sigma})(\sigma_2 - \bar{\sigma}) + (\sigma_1 - \bar{\sigma})(\sigma_3 - \bar{\sigma}) + (\sigma_2 - \bar{\sigma})(\sigma_3 - \bar{\sigma})\right] \quad (8.105)$$

where $\bar{\sigma}$ indicates the mean stress:

$$\bar{\sigma} = \frac{1}{3}(\sigma_1 + \sigma_2 + \sigma_3) \quad (8.106)$$

Substituting equation (8.106) in equation (8.105) we obtain

$$J_{IId} = -\frac{1}{9}\left[(2\sigma_1 - \sigma_2 - \sigma_3)(2\sigma_2 - \sigma_1 - \sigma_3) + \ldots\right] \quad (8.107)$$

and then, multiplying the three pairs of trinomials and ordering the terms that result:

$$\begin{aligned} J_{IId} &= -\frac{1}{9}\left[\sigma_3^2 - 2\sigma_1^2 - 2\sigma_2^2 + 5\sigma_1\sigma_2 - \sigma_1\sigma_3 - \sigma_2\sigma_3 + \right. \\ &\quad \sigma_2^2 - 2\sigma_1^2 - 2\sigma_3^2 + 5\sigma_1\sigma_3 - \sigma_1\sigma_2 - \sigma_3\sigma_2 + \\ &\quad \left. \sigma_1^2 - 2\sigma_2^2 - 2\sigma_3^2 + 5\sigma_2\sigma_3 - \sigma_2\sigma_1 - \sigma_3\sigma_1\right] \\ &= \frac{1}{3}\left[(\sigma_1^2 + \sigma_2^2 + \sigma_3^2) - (\sigma_1\sigma_2 + \sigma_1\sigma_3 + \sigma_2\sigma_3)\right] \end{aligned} \quad (8.108)$$

The distortion energy is then

$$\Phi_d = \frac{1+\nu}{3E}\left[(\sigma_1^2 + \sigma_2^2 + \sigma_3^2) - (\sigma_1\sigma_2 + \sigma_1\sigma_3 + \sigma_2\sigma_3)\right] \quad (8.109)$$

In the uniaxial critical condition, the distortion energy reaches its limit value:

$$\Phi_{dP} = \frac{1+\nu}{3E}\sigma_P^2 \quad (8.110)$$

The original condition of Von Mises,

$$\Phi_d < \Phi_{dP} \quad (8.111)$$

therefore translates into the following:

$$\left[(\sigma_1^2 + \sigma_2^2 + \sigma_3^2) - (\sigma_1\sigma_2 + \sigma_1\sigma_3 + \sigma_2\sigma_3)\right] < \sigma_P^2 \quad (8.112)$$

where the term on the left-hand side represents the square of **Von Mises'
equivalent stress**. Equation (8.112) is thus a particular case of equation (8.96).

We now intend to express the condition of Von Mises as a function of the special components of stress, instead of as a function of the principal stresses. Noting that the inequality (8.112) can be expressed as a function of the invariants

$$(J_I^2 + 3J_{II}) < \sigma_P^2 \quad (8.113)$$

and using expressions (7.72a,b), we obtain

$$\left[(\sigma_x^2 + \sigma_y^2 + \sigma_z^2) - (\sigma_x\sigma_y + \sigma_x\sigma_z + \sigma_y\sigma_z) + 3(\tau_{xy}^2 + \tau_{xz}^2 + \tau_{yz}^2)\right] < \sigma_P^2 \quad (8.114)$$

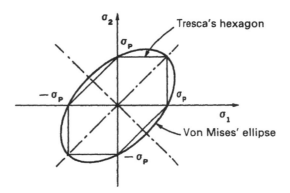

Figure 8.29

As regards **plane stress conditions**, for example $\sigma_3 = 0$, the condition (8.112) becomes

$$(\sigma_1^2 + \sigma_2^2 - \sigma_1\sigma_2) < \sigma_P^2 \qquad (8.115)$$

The inequality (8.115) represents all the points inside an ellipse which has its major axis coincident with the bisector of the first and third quadrants (Figure 8.29). **Von Mises' ellipse** is circumscribed to Tresca's hexagon and intersects the axes σ_1 and σ_2 at the same four notable points. The two boundaries have in common also the two points of intersection with the bisector of the first and third quadrants.

Von Mises' criterion is less conservative than Tresca's criterion, since the elliptical domain is wider than the hexagonal one. On the other hand, Von Mises' criterion is simpler to use than is Tresca's, since it imposes only one inequality (8.115) as against the three of equation (8.101).

Finally, it should be noted how the triaxial strength domains, both according to Tresca and according to Von Mises, are represented in the space σ_1–σ_2–σ_3 by a cylinder with the generators parallel to the trisector of the first octant. The inequality (8.99) represents in fact the cylinder that has as its directrix on the plane σ_1–σ_2 the hexagon of Figure 8.26, while the inequality (8.112) represents the cylinder that has as its directrix on the plane σ_1–σ_2 the ellipse of Figure 8.29. That the trisector of the first octant should be entirely contained in the above-mentioned strength domains is a result consistent with the basic experimental observation, i.e. with the independence of the conditions of yielding from a superimposed hydrostatic state. This means that, according to these criteria, even severe hydrostatic conditions of loading do not cause in any case the failure of the material.

As regards the **strength criteria for anisotropic materials**, the reader is referred to Appendix C.

235

9 The Saint Venant problem

9.1 Introduction

This chapter deals with the particular case of a cylindrical, homogeneous and isotropic, linearly elastic solid loaded exclusively on its end planes. This solid, known as the Saint Venant solid, represents a relatively simple and highly useful model in the case of beams. All the fundamental loadings are studied, which correspond to the internal reaction characteristics already introduced in Chapter 5: axial force, shearing force, twisting moment, and bending moment. For each loading both the stress condition produced and the corresponding deformation characteristic are obtained: axial dilation, mean shearing strain, unit angle of torsion, and curvature. It is shown that, in the case of symmetrical sections, the fundamental reactions are all mutually energetically orthogonal, and that in this case, the strain energy is a diagonal quadratic form of the reactions themselves.

The chapter further deals with combined loadings consisting of eccentric axial force, with the definition of the central core of inertia, and of shear-torsion, with the definition of the corresponding centre of shear or of torsion. As regards the case of torsion of thin-walled sections, the considerable difference existing between closed sections (tubular sections) and open sections is emphasized. The latter are in fact subjected to stresses and torsional rotations that are far higher.

The chapter closes with a number of examples of strength tests, carried out on beams having as sections the areas considered in the closing part of Chapter 2.

9.2 Fundamental hypotheses

The **Saint Venant problem** constitutes a particular elastic problem regarding a cylindrical solid, loaded at its ends.

Consider a generic area with its central (i.e. centroidal and principal) XY reference system. Imagine translating this area perpendicularly to its own plane, so as to cause its centroid G to describe a rectilinear trajectory, normal to the XY coordinate plane and of length l (Figure 9.1). Let the oriented straight line of this trajectory constitute the third reference axis Z. The cylindrical volume thus described constitutes the domain of the Saint Venant solid.

Having thus described the geometry of the Saint Venant solid, it is necessary to specify the material of which it is made and the forces by which it is loaded. As regards the material, it is assumed to be linear elastic, isotropic and homogeneous. The last of these assumptions may be omitted, as is shown in Appendix D. As far as the external forces applied are concerned, on the other hand, these are assumed to be only surface forces, which act exclusively on the end planes (Figure 9.1). Body forces are hence excluded, as are displacements imposed on the boundary. Of course the surface forces acting on both

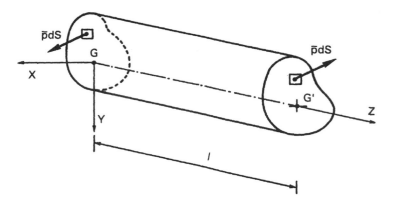

Figure 9.1

ends, A and A', must make up a balanced system and consequently satisfy the cardinal equations of statics:

$$\int_{A \cup A'} \{p\}\, dA = \{0\} \tag{9.1a}$$

$$\int_{A \cup A'} (\{r\} \wedge \{p\})\, dA = \{0\} \tag{9.1b}$$

The equations governing the Saint Venant problem are the same as those governing the elastic problem, the former being merely a particular case in the framework of the latter. These equations are those of statics (8.2), kinematics (8.8) and the constitutive equations (8.73). As shall be noted in the sequel, in the case in point, some of these are identically satisfied, precisely on account of the simplifying hypotheses introduced.

On the other hand, the boundary conditions of equivalence on the lateral surface, where $n_z = 0$, are

$$\sigma_x n_x + \tau_{yx} n_y = 0 \tag{9.2a}$$

$$\tau_{xy} n_x + \sigma_y n_y = 0 \tag{9.2b}$$

$$\tau_{xz} n_x + \tau_{yz} n_y = 0 \tag{9.2c}$$

while those on the end planes, where $n_x = 0$, $n_y = 0$, $n_z = 1$, take the following form:

$$\tau_{zx} = p_x \tag{9.3a}$$

$$\tau_{zy} = p_y \tag{9.3b}$$

$$\sigma_z = p_z \tag{9.3c}$$

We may now state Saint Venant's fundamental hypothesis, one which has been amply borne out both experimentally and theoretically (Figure 9.2):

At a sufficient distance from each end plane, the strain and stress fields depend only upon the resultant $\{R\}$ of the forces acting on the end itself and upon the resultant moment $\{M\}$ of the forces with respect to the centroid of the end considered.

237

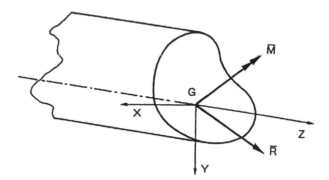

Figure 9.2

In line with this hypothesis, the conditions of equivalence at the ends can be global, and not local conditions as in the case of equations (9.3):

$$\int_A \{t_z\}\, dA = \int_A \{p\}\, dA = \{R\} \tag{9.4a}$$

$$\int_A \left(\{r\} \wedge \{t_z\}\right) dA = \int_A \left(\{r\} \wedge \{p\}\right) dA = \{M\} \tag{9.4b}$$

Consider, for instance, a rectilinear beam having rectangular cross section, of base b and depth h, subjected on each end plane, in one case to a couple made up of two forces F with arm h (Figure 9.3(a)), in the other to a couple consisting of two forces $2F$ with arm $h/2$ (Figure 9.3(b)). The strain and stress fields are approximately the same at distances from the ends greater than the depth h. Reversing one of the two systems of forces and applying the Principle of Superposition, two self-balancing systems acting on the end planes are obtained, which generate approximately zero strain and stress fields, except in the end regions. The damping of the perturbation created by the self-balancing systems of forces occurs at distances from the ends approximately greater than the maximum dimension of the cross section, when the cross section itself is compact and of a regular shape. When, instead, the cross section is thin-walled, the distance of damping may prove far greater.

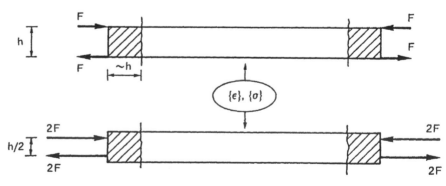

Figure 9.3

238

As we have already said in Chapter 5, the components of the resultant force $\{R\}$ and the resultant moment $\{M\}$ are called **elementary internal reactions**:

$R_x = T_x =$ shearing force along the X axis;
$R_y = T_y =$ shearing force along the Y axis;
$R_z = N =$ axial force;
$M_x =$ bending moment along the X axis;
$M_y =$ bending moment along the Y axis;
$M_z =$ twisting moment.

The cases of **elementary loading** for the Saint Venant solid are thus the following:

1. centred axial force N;
2. flexure M_x ;
3. flexure M_y ;
4. shear T_y (and flexure M_x);
5. shear T_x (and flexure M_y);
6. torsion M_z.

It should be noted that, since the shear is the derivative of the moment function, its presence also presupposes that of the corresponding flexure, whereas the reverse is not true.

In the sequel we shall see how the three elementary loadings, N, M_x, M_y, produce an axial one-dimensional stress field, where only normal stress σ_z is present. The remaining three elementary loadings, T_y, T_x, M_z, produce instead, on each cross section, a field of shearing stress with the presence of the components τ_{zx}, τ_{zy}, and thus of τ_z only. Also for this reason, it is customary to combine the elementary loadings, so as to obtain **complex loadings** which produce only normal stresses, σ_z, or only shearing stresses, τ_z:

1. biaxial flexure: M_x, M_y;
2. eccentric axial force: N, M_x, M_y;
3. shear-torsion: T_x, T_y, M_z.

It should be noted that the eccentric axial force is equivalent to an axial force N, exerted at a point of the XY plane which does not coincide with the centroid. On the other hand, the shear-torsion is equivalent to a force T, with the line of action belonging to the XY plane and not passing necessarily through the centroid. In the last analysis, then, the resultant force $\{R\}$ and the resultant moment $\{M\}$ are a system equivalent to that of the two skew and orthogonal forces N and T (Figure 9.4).

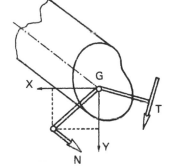

Figure 9.4

9.3 Centred axial force

As regards centred axial force and, as we shall see, also the other elementary loadings that produce normal stress σ_z, we start by assuming a stress solution and then verify that this solution satisfies all the equations of the elastic problem, including the boundary conditions.

Let us assume then

$$\sigma_x = \sigma_y = \tau_{xy} = \tau_{xz} = \tau_{yz} = 0 \qquad (9.5a)$$

$$\sigma_z = c \qquad (9.5b)$$

where c is a constant that can be determined on the basis of the global boundary conditions of equivalence (9.4). These conditions are all identically satisfied, except for the projection of the first on the Z axis

$$\int_A \sigma_z \, dA = N \tag{9.6}$$

whence we obtain

$$cA = N \tag{9.7}$$

and hence the constant stress on the cross section

$$\sigma_z = \frac{N}{A} \tag{9.8}$$

On the other hand, the indefinite equations of equilibrium (8.2), as well as the boundary conditions of equivalence (9.2), are all identically satisfied by the solution given by equations (9.5a) and (9.8).

The elastic constitutive equations (8.73) then give the strain field

$$\varepsilon_x = \varepsilon_y = -v \frac{N}{EA} \tag{9.9a}$$

$$\varepsilon_z = \frac{N}{EA} \tag{9.9b}$$

$$\gamma_{xy} = \gamma_{xz} = \gamma_{yz} = 0 \tag{9.9c}$$

If the axial force N is tensile, there is thus a uniform dilation in the axial direction and contractions that are all equal to one another in the transverse directions, while the shearing strains are zero.

Integrating the strain field, it is possible finally to obtain the displacement field, save for components of rigid rototranslation. We have in fact

$$\frac{\partial u}{\partial x} = -v \frac{N}{EA} \tag{9.10a}$$

$$\frac{\partial v}{\partial y} = -v \frac{N}{EA} \tag{9.10b}$$

$$\frac{\partial w}{\partial z} = \frac{N}{EA} \tag{9.10c}$$

$$\frac{\partial u}{\partial y} + \frac{\partial v}{\partial x} = 0 \tag{9.10d}$$

$$\frac{\partial u}{\partial z} + \frac{\partial w}{\partial x} = 0 \tag{9.10e}$$

$$\frac{\partial v}{\partial z} + \frac{\partial w}{\partial y} = 0 \tag{9.10f}$$

Integrating the first three equations, we obtain

$$u = -v \frac{N}{EA} x + u_0(y, z) \tag{9.11a}$$

240

$$v = -v\frac{N}{EA}y + v_0(x,z) \qquad (9.11b)$$

$$w = \frac{N}{EA}z + w_0(x,y) \qquad (9.11c)$$

Substituting equations (9.11) into equations (9.10 d, e, f), we have

$$\frac{\partial u_0}{\partial y} + \frac{\partial v_0}{\partial x} = 0 \qquad (9.12a)$$

$$\frac{\partial u_0}{\partial z} + \frac{\partial w_0}{\partial x} = 0 \qquad (9.12b)$$

$$\frac{\partial v_0}{\partial z} + \frac{\partial w_0}{\partial y} = 0 \qquad (9.12c)$$

as well as, of course

$$\frac{\partial u_0}{\partial x} = 0 \qquad (9.12d)$$

$$\frac{\partial v_0}{\partial y} = 0 \qquad (9.12e)$$

$$\frac{\partial w_0}{\partial z} = 0 \qquad (9.12f)$$

From equations (9.12) it follows that the field of displacements u_0, v_0, w_0 does not have strain components. It, therefore, represents a generic rigid rototranslation.

The solution obtained consists of stresses (9.5a) and (9.8) and displacements (9.11). It is the only one possible, by virtue of Kirchhoff's Solution Uniqueness Theorem.

The elementary work of deformation, for an infinitesimal segment of length dz of the Saint Venant solid, may be obtained by applying Clapeyron's Theorem (Figure 9.5)

$$dL = \frac{1}{2}N\,dw = \frac{1}{2}N\varepsilon_z\,dz \qquad (9.13)$$

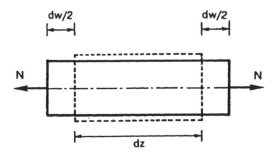

Figure 9.5

241

Substituting the expression (9.9b) in equation (9.13), we have then

$$\frac{dL}{dz} = \frac{1}{2}\frac{N^2}{EA} \qquad (9.14)$$

The factor 1/2 (characteristic of linear elasticity) is therefore multiplied by the square of the static characteristic N, and divided by the product of the elastic characteristic E and the geometric characteristic A. It will be noted in the sequel how the structure of the formula (9.14) is also conserved in the case of the other elementary loadings.

9.4 Flexure

Also in the case of flexure a uniaxial stress field σ_z is assumed, in this case linearly variable on the cross section

$$\sigma_x = \sigma_y = \tau_{xy} = \tau_{xz} = \tau_{yz} = 0 \qquad (9.15a)$$

$$\sigma_z = ax + by + c \qquad (9.15b)$$

The constants a, b, c may be determined on the basis of the boundary conditions expressed by equations (9.4):

$$\int_A \sigma_z \, dA = N = 0 \qquad (9.16a)$$

$$\int_A \tau_{zx} \, dA = T_x = 0 \qquad (9.16b)$$

$$\int_A \tau_{zy} \, dA = T_y = 0 \qquad (9.16c)$$

$$\int_A \sigma_z y \, dA = M_x \neq 0 \qquad (9.16d)$$

$$\int_A \sigma_z x \, dA = -M_y = 0 \qquad (9.16e)$$

$$\int_A (\tau_{zy}x - \tau_{zx}y) \, dA = M_z = 0 \qquad (9.16f)$$

Whilst the conditions (9.16b, c, f) are identically satisfied, the conditions (9.16 a, d, e) give the constants a, b, c. From equations (9.16a) and (9.15b) we have in fact

$$\int_A (ax + by + c) \, dA = aS_y + bS_x + cA = 0 \qquad (9.17)$$

and, since the static moments S_x and S_y are zero as the XY system is a centroidal reference system, we obtain

$$c = 0 \qquad (9.18)$$

From equation (9.16 e) we have then

$$\int_A (ax + by)x \, dA = aI_y + bI_{xy} = 0 \qquad (9.19)$$

and, since the product of inertia I_{xy} is zero as XY is a principal reference system, we obtain

$$a = 0 \qquad (9.20)$$

Finally, from equation (9.16 d) we have

$$\int_A by^2 \, dA = bI_x = M_x \qquad (9.21)$$

from which we obtain

$$b = \frac{M_x}{I_x} \qquad (9.22)$$

and thus

$$\sigma_z = \frac{M_x}{I_x} y \qquad (9.23)$$

If only the elementary load M_x is present, the stress σ_z depends upon the y coordinate alone, and increases in absolute value as it moves away from the X axis (Figure 9.6). It is directly proportional to the applied moment and inversely proportional to the relative moment of inertia. The stress σ_z is zero for $y = 0$, i.e. on the X axis which, for this reason, is called the **neutral axis**. The Y axis is called the **loading axis**, because the couple M_x belongs to the plane YZ, which in turn is called the **loading plane**. In this case the loading axis and the neutral axis are mutually orthogonal. The case of the **heterogeneous beam in flexure** is dealt with in Appendix D.

The stress solution, represented by equations (9.15a) and (9.23), identically satisfies the indefinite equations of equilibrium (8.2) and equivalence (9.2), and, via the elastic constitutive equations (8.73), gives the following strain field:

$$\varepsilon_x = \varepsilon_y = -v\frac{M_x}{EI_x} y \qquad (9.24a)$$

$$\varepsilon_z = \frac{M_x}{EI_x} y \qquad (9.24b)$$

$$\gamma_{xy} = \gamma_{rz} = \gamma_{yz} = 0 \qquad (9.24c)$$

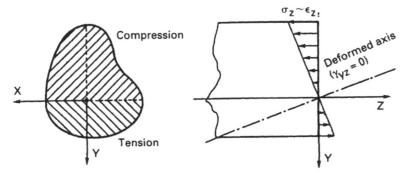

Figure 9.6

243

Integrating the strains of equations (9.24), it is possible to obtain the displacement field, except for the components of rigid rotation. From equations (9.24) we have

$$\frac{\partial u}{\partial x} = -v \frac{M_x}{EI_x} y \qquad (9.25a)$$

$$\frac{\partial v}{\partial y} = -v \frac{M_x}{EI_x} y \qquad (9.25b)$$

$$\frac{\partial w}{\partial z} = \frac{M_x}{EI_x} y \qquad (9.25c)$$

$$\frac{\partial u}{\partial y} + \frac{\partial v}{\partial x} = 0 \qquad (9.25d)$$

$$\frac{\partial u}{\partial z} + \frac{\partial w}{\partial x} = 0 \qquad (9.25e)$$

$$\frac{\partial v}{\partial z} + \frac{\partial w}{\partial y} = 0 \qquad (9.25f)$$

Integrating equations (9.25a, b, c), we obtain

$$u = -v \frac{M_x}{EI_x} xy + u_0(y, z) \qquad (9.26a)$$

$$v = -v \frac{M_x}{EI_x} \frac{y^2}{2} + v_0(x, z) \qquad (9.26b)$$

$$w = \frac{M_x}{EI_x} yz + w_0(x, y) \qquad (9.26c)$$

Substituting equations (9.26) in equations (9.25d, e, f), we have

$$\frac{\partial u_0}{\partial y} + \frac{\partial v_0}{\partial x} = v \frac{M_x}{EI_x} x \qquad (9.27a)$$

$$\frac{\partial u_0}{\partial z} + \frac{\partial w_0}{\partial x} = 0 \qquad (9.27b)$$

$$\frac{\partial v_0}{\partial z} + \frac{\partial w_0}{\partial y} = -\frac{M_x}{EI_x} z \qquad (9.27c)$$

Integrating equation (9.27a) with respect to x, we obtain

$$v_0(x, z) = v \frac{M_x}{EI_x} \frac{x^2}{2} + v_1(x, z) \qquad (9.28)$$

while from equation (9.27c) we draw

$$\frac{\partial v_1}{\partial z} + \frac{\partial w_0}{\partial y} = -\frac{M_x}{EI_x} z \qquad (9.29)$$

244

and hence

$$v_1(x, z) = -\frac{M_x}{EI_x}\frac{z^2}{2} + v_2(x, z) \tag{9.30}$$

In conclusion, we can express the displacement field as follows:

$$u = -v\frac{M_x}{EI_x}xy + u_0(y, z) \tag{9.31a}$$

$$v = -\frac{M_x}{2EI_x}\left[z^2 + v(y^2 - x^2)\right] + v_2(x, z) \tag{9.31b}$$

$$w = \frac{M_x}{EI_x}yz + w_0(x, y) \tag{9.31c}$$

Substituting the foregoing equations (9.31) in equations (9.25d, e, f), we have

$$\frac{\partial u_0}{\partial y} + \frac{\partial v_2}{\partial x} = 0 \tag{9.32a}$$

$$\frac{\partial u_0}{\partial z} + \frac{\partial w_0}{\partial x} = 0 \tag{9.32b}$$

$$\frac{\partial v_2}{\partial z} + \frac{\partial w_0}{\partial y} = 0 \tag{9.32c}$$

as well as, of course

$$\frac{\partial u_0}{\partial x} = 0 \tag{9.32d}$$

$$\frac{\partial v_2}{\partial y} = 0 \tag{9.32e}$$

$$\frac{\partial w_0}{\partial z} = 0 \tag{9.32f}$$

From equations (9.32) it follows that the field of displacements u_0, v_2, w_0 does not have any strain components. It therefore represents a generic rigid rototranslation, which shall henceforth be neglected.

The points of the Z axis, of coordinates $P(0, 0, \bar{z})$, are transformed, once deformation has occurred, into the points $P'(0, v_P, \bar{z})$, since by equation (9.31b) we have

$$v_P = v(0, 0, \bar{z}) = -\frac{M_x}{2EI_x}\bar{z}^2 \tag{9.33}$$

In the foregoing equation the term of rigid rototranslation v_2 has been omitted (Figure 9.7).

245

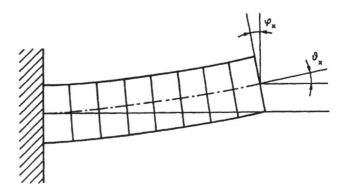

Figure 9.7

On the other hand, the points of a generic cross section, of coordinates $Q(x, y, \bar{z})$, are transformed, as a consequence of the deformation, into the points $Q'(x + u_Q, y + v_Q, \bar{z} + w_Q)$, equations (9.31) yielding

$$u_Q = u(x, y, \bar{z}) = -v \frac{M_x}{EI_x} xy \qquad (9.34a)$$

$$v_Q = v(x, y, \bar{z}) = -\frac{M_x}{2EI_x} \left[\bar{z}^2 + v(y^2 - x^2) \right] \qquad (9.34b)$$

$$w_Q = w(x, y, \bar{z}) = \frac{M_x}{EI_x} y\bar{z} \qquad (9.34c)$$

Hence the transformation $Q \to Q'$ will be, in general, nonlinear:

$$x' = x - v \frac{M_x}{EI_x} xy \qquad (9.35a)$$

$$y' = y - \frac{M_x}{2EI_x} \left[\bar{z}^2 + v(y^2 - x^2) \right] \qquad (9.35b)$$

$$z' = \bar{z} + \frac{M_x}{EI_x} y\bar{z} \qquad (9.35c)$$

If, on the other hand, we consider the Saint Venant solid sufficiently slender to allow us to consider the coordinates x and y as always being infinitesimal compared to the $\bar{z}$ coordinate, equations (9.35) may be rewritten as follows:

$$x' = x \qquad (9.36a)$$

$$y' = y - \frac{M_x}{2EI_x} \bar{z}^2 \qquad (9.36b)$$

$$z' = \bar{z} \left(1 + \frac{M_x}{EI_x} y \right) \qquad (9.36c)$$

In equations (9.36a, b) the second order infinitesimal terms, which represent the phenomenon of transverse contraction, have been omitted.

246

In the framework of the above hypotheses the **principle of conservation of plane sections** holds, according to which each individual cross section rotates rigidly by the angle φ_x about the X axis, and translates by the quantity v_P in the Y direction (Figure 9.7). Furthermore, each individual cross section, after deformation, remains perpendicular to the deformed axis, since $\gamma_{yz} = 0$, whilst the latter undergoes bending but not variations in length, in that $\varepsilon_z (x = 0, y = 0) = 0$.

A further confirmation of what is stated above comes from the relation (9.36c):

$$\varphi_x = \frac{z' - \bar{z}}{y} = \frac{M_x}{EI_x} \bar{z} \tag{9.37}$$

In fact, the angle ϑ_x which the geometrical tangent to the deformed configuration of the axis expressed by equation (9.33) forms with the Z axis, is approximately equal to (Figure 9.7)

$$\vartheta'_x \quad -\frac{\mathrm{d}v_P}{\mathrm{d}\bar{z}} = \frac{M_x}{EI_x} \bar{z} \tag{9.38}$$

Comparing expressions (9.37) and (9.38), which represent the rotation of the section and the rotation of the axis, respectively, we find what has been previously stated, viz. that $\varphi_x = \vartheta_x$.

Referring to equation (9.37), the differential of the angle φ_x equals

$$\mathrm{d}\varphi_x = \frac{M_x}{EI_x}\,\mathrm{d}z \tag{9.39}$$

On the other hand, if we denote by R_x the radius of curvature of the deformed axis and by $\chi_x = 1/R_x$ the curvature due to flexure, we have (Figure 9.8)

$$\mathrm{d}\varphi_x = \frac{\mathrm{d}z}{R_x} = \chi_x\,\mathrm{d}z \tag{9.40}$$

From the comparison of the two equations (9.39) and (9.40), we obtain the **curvature**,

$$\chi_x = \frac{M_x}{EI_x} \tag{9.41}$$

Likewise, the curvature produced by the flexure M_y equals

$$\chi_y = \frac{M_y}{EI_y} \tag{9.42}$$

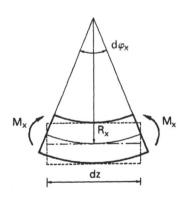

Figure 9.8

The foregoing formula presents the same structure as equation (9.9b), which gives the axial dilation produced by the centred axial force. In both cases we have the static characteristic divided by the product of the modulus of elasticity and the geometrical characteristic.

By then applying Clapeyron's Theorem to the infinitesimal beam segment of Figure 9.8, we obtain the elementary work of deformation,

$$\mathrm{d}L = \frac{1}{2} M_x\,\mathrm{d}\varphi_x = \frac{1}{2} M_x \chi_x\,\mathrm{d}z \tag{9.43}$$

Substituting relation (9.41) into equation (9.43), we have finally

$$\frac{dL}{dz} = \frac{1}{2}\frac{M_x^2}{EI_x} \tag{9.44}$$

9.5 Eccentric axial force and biaxial flexure

Consider the combined loading of **eccentric axial force** consisting of an axial force N, having eccentricity e_y with respect to the X axis and e_x with respect to the Y axis, and thus eccentricity $e = (e_x^2 + e_y^2)^{1/2}$ with respect to the centroidal axis (Figure 9.9). This force is equivalent, on the other hand, to the system of elementary loadings made up of the centred axial force N and the flexures

$$M_x = Ne_y \tag{9.45a}$$

$$M_y = -Ne_x \tag{9.45b}$$

Summing up the uniaxial stress fields that correspond to the aforesaid elementary loadings, we obtain

$$\sigma_z = \frac{N}{A} + \frac{M_x}{I_x}y - \frac{M_y}{I_y}x \tag{9.46}$$

The negative sign of the last term is due to the fact that the moment M_y, if positive, stretches the longitudinal fibres of the half-plane $x < 0$.

Substituting equations (9.45) in equation (9.46), we deduce the following relation:

$$\sigma_z = \frac{N}{A} + \frac{Ne_y}{I_x}y + \frac{Ne_x}{I_y}x \tag{9.47}$$

which, expressing the moments of inertia as functions of the area A and of the respective central radii of gyration, becomes

$$\sigma_z = \frac{N}{A}\left(1 + \frac{e_y}{\rho_x^2}y + \frac{e_x}{\rho_y^2}x\right) \tag{9.48}$$

Defining the **neutral axis**, also in the case of combined loading, as the straight line on which the stress σ_z vanishes, we obtain by simply equating the expression (9.48) to zero,

$$1 + \frac{e_y}{\rho_x^2}y + \frac{e_x}{\rho_y^2}x = 0 \tag{9.49}$$

The neutral axis may cut the cross section or not, according to the eccentricity e and the angle α (Figure 9.9). For small eccentricities the neutral axis does not intersect the cross section, and the stresses σ_z all have the same sign, whereas, for large eccentricities, the neutral axis intersects the cross section, and the stresses σ_z change sign on it. The term **central core of inertia** is given to the area within which the eccentric axial force falls, in such a way that the neutral axis does not intersect the cross section of the beam. This concept is of particular importance in reference to compressed beams made of brittle or

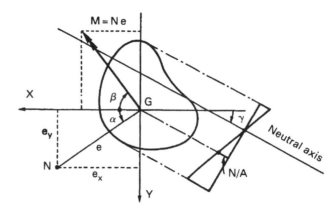

Figure 9.9

non-traction-bearing materials (concrete, masonry, etc.). We shall shortly see two particularly significant examples.

The linear variation of the stresses σ_z on the cross section can be represented by drawing a reference line perpendicular to the neutral axis (Figure 9.9) and by marking in the two values corresponding to the neutral axis, $\sigma_z = 0$, and to the centroid, $\sigma_z = N/A$. The diagram of stresses σ_z will therefore be butterfly-shaped in the case where the neutral axis intersects the cross section (Figure 9.9), or trapezoidal in the case where the neutral axis is external to the cross section (Figure 9.10).

As the eccentricities e_x, e_y, vary, three particular cases may arise.

1. $e_x = e_y = 0$: centred axial force. The neutral axis degenerates into the straight line at infinity.

2. (a) $e_x = 0$: centred axial force plus flexure. The neutral axis is parallel to the X axis:

$$y = -\frac{\rho_x^2}{e_y} \tag{9.50a}$$

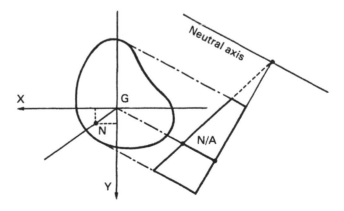

Figure 9.10

249

(b) $e_y = 0$: centred axial force plus flexure. The neutral axis is parallel to the Y axis:

$$x = -\frac{\rho_y^2}{e_x} \qquad (9.50b)$$

Note that the centre of pressure and the centroid always belong to the same one of the two half-planes into which the neutral axis divides the XY plane. In this regard, see also the general equation (9.49).

3. $e_y/e_x = \tan \alpha$; $e_x \to \infty$, $e_y \to \infty$. The latter is the case of **biaxial flexure**. The general equation (9.49) then becomes

$$\frac{y}{\rho_x^2} \tan \alpha + \frac{x}{\rho_y^2} = 0 \qquad (9.51)$$

which represents a centroidal straight line. That is, when only the flexures M_x, M_y are present, the neutral axis is centroidal; however, in general, this is not perpendicular to the loading axis.

The **loading axis** is the straight line NG (Figure 9.9) represented by the equation

$$y = x \tan \alpha \qquad (9.52)$$

On the other hand, the **neutral axis** can be represented by the equation

$$y = -\frac{1}{\tan \alpha} \left(\frac{\rho_x}{\rho_y} \right)^2 x - \frac{\rho_x^2}{e_y} \qquad (9.53)$$

It is readily recognizable that the lines (9.52) and (9.53) are orthogonal, not only for $e_x = 0$ or for $e_y = 0$, but also when the central radii of gyration are equal: $\rho_x = \rho_y$. In this case we have $I_x = I_y$ and thus the section is of a gyroscopic nature (Chapter 2).

We shall now determine the central cores of inertia of two cross sections that are very frequently encountered in building practice: the rectangle and the circle.

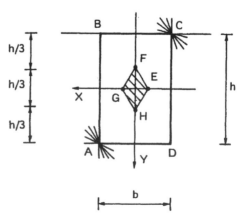

Figure 9.11

In the case of the rectangular cross section of Figure 9.11, the neutral axis can be assumed to coincide with the straight line to which the upper side BC belongs. Using equation (9.50a), we obtain

$$-\frac{h}{2} = \frac{\left(-h^2/12\right)}{e_y} \tag{9.54}$$

from which

$$e_y = \frac{h}{6} \tag{9.55}$$

The eccentric axial force corresponding to the neutral axis BC is applied to the point H, belonging to the Y axis at a distance $h/6$ from the X axis. Similarly, the centres of pressure corresponding to the straight lines CD, DA, AB are G, F, E, respectively (Figure 9.11).

It is easy to show that the central core of inertia is represented by the rhombus $EFGH$. The neutral axis corresponding to each point of the segment GH,

$$\frac{e_x}{(b/6)} + \frac{e_y}{(h/6)} = 1 \tag{9.56}$$

is given by the equation (9.53), once we insert the ratio

$$\tan \alpha = \frac{e_y}{e_x} = \frac{h(b - 6e_x)}{6be_x} \tag{9.57}$$

We thus have

$$y = \frac{6e_x b}{h(6e_x - b)} \left(\frac{h}{b}\right)^2 x - \frac{h^2}{12} \frac{1}{e_y} \tag{9.58}$$

Equation (9.58) is satisfied by the coordinates of the point $C(-b/2, -h/2)$ for any pair of values (e_x, e_y) which satisfies the relation (9.56).

The cross section is traditionally said to be entirely in compression if the centre of pressure is within the **middle third**. In fact, in the case of compressive axial force and flexure, when the force N acts between F and H ($\overline{FH} = h/3$), the neutral axis is external to the cross section, and we have $\sigma_z < 0$ in each point (Figure 9.11).

A similar line of reasoning, though simpler on account of polar symmetry, leads us to conclude that the central core of inertia of a circular cross section of radius R is the concentric circle of radius $R/4$. Equation (9.50a), in the case of the neutral axis being tangent in point A (Figure 9.12), gives in fact

$$-R = -\frac{(R^2/4)}{e_y} \tag{9.59}$$

from which there follows

$$e_y = \frac{R}{4} \tag{9.60}$$

As regards deformation, eccentric axial force produces a rigid rotation of each individual cross section about the neutral axis, as well as an axial

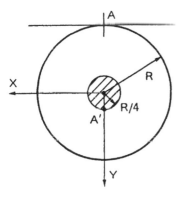

Figure 9.12

translation of the section. Referring to Figure 9.9 and to the relation (9.37), we can see that the moment $M_x = Ne_y$ produces the rotation

$$\varphi_x = \frac{M_x}{EI_x} z = \frac{Nz}{EA} \frac{e_y}{\rho_x^2} \tag{9.61a}$$

just as the moment $M_y = -Ne_x$ produces the rotation

$$\varphi_y = \frac{M_y}{EI_y} z = -\frac{Nz}{EA} \frac{e_x}{\rho_y^2} \tag{9.61b}$$

Since the rotations φ_x and φ_y are infinitesimal, the vector summation of them may be made and will give the total rotation vector

$$\{\varphi\} = \varphi_x \bar{i} + \varphi_y \bar{j} \tag{9.62}$$

where $\bar{i}$ and $\bar{j}$ are the unit vectors of the X axis and the Y axis, respectively. The angular coefficient of the axis of rotation is thus equal to

$$\frac{\varphi_y}{\varphi_x} = -\frac{e_x}{e_y}\left(\frac{\rho_x}{\rho_y}\right)^2 = -\frac{1}{\tan\alpha}\left(\frac{\rho_x}{\rho_y}\right)^2 \tag{9.63}$$

and coincides with that of the neutral axis (9.53).

Finally, it may be shown that the three elementary loadings, N, M_x, M_y, are energetically orthogonal, and the work associated with the eccentric axial force is therefore equal to the sum of the work of each of the individual characteristics. The elementary work, corresponding to an infinitesimal segment of beam, is equal to

$$dL = \int_A \Psi \, dA \, dz \tag{9.64}$$

where Ψ represents the work per unit volume. Recalling expression (8.70), which furnishes the complementary elastic potential as a function of the stress condition, we have

$$dL = \int_A \frac{\sigma_z^2}{2E} \, dA \, dz \tag{9.65}$$

If we use the solution (9.46), we obtain

$$\frac{dL}{dz} = \frac{1}{2E}\int_A \left(\frac{N}{A} + \frac{M_x}{I_x} y - \frac{M_y}{I_y} x\right)^2 dA \tag{9.66}$$

Expanding the square of the trinomial under the integral sign, we split the elementary work into two parts; in the first part there appear the squares of the characteristics, whilst in the second there appear the mutual products

$$\frac{dL}{dz} = \frac{1}{2}\left[\frac{N^2}{EA} + \frac{M_x^2}{EI_x} + \frac{M_y^2}{EI_y}\right] + \tag{9.67}$$

$$\frac{1}{E}\left[\frac{N M_x}{A I_x} S_x - \frac{N M_y}{A I_y} S_y - \frac{M_x M_y}{I_x I_y} I_{xy}\right]$$

Since the X and Y axes are central, the corresponding static moments vanish, as does the product of inertia, and hence the second part of expression (9.67) vanishes

$$\frac{dL}{dz} = \frac{1}{2}\left[\frac{N^2}{EA} + \frac{M_x^2}{EI_x} + \frac{M_y^2}{EI_y}\right] \qquad (9.68)$$

The relation (9.68) proves that the three characteristics so far considered are energetically orthogonal.

9.6 Torsion in beams of circular cross section

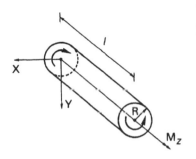

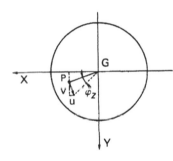

Let us consider a beam of circular cross section, subjected at its ends to the action of two twisting moments M_z, of equal magnitude and opposite direction (Figure 9.13). Let l denote the length of the beam and R the radius of the cross section.

Whereas in the case of axial force and flexure we have assumed the stresses σ_z (static hypothesis), in the case of torsion the hypothesis regards the displacement field (kinematic hypothesis). We assume in fact that each cross section rotates rigidly about the longitudinal axis, remaining at the same time plane. Expressed in formulas (Figure 9.14),

$$u = -\varphi_z y \qquad (9.69a)$$
$$v = \varphi_z x \qquad (9.69b)$$
$$w = 0 \qquad (9.69c)$$

Figure 9.13

Figure 9.14

where φ_z is the angle of infinitesimal rotation. For this purpose we have had recourse to equations (3.8). Since, if we exclude the end regions, the relative rotation per unit length of the beam must be constant and equal to Θ, we shall have for example

$$\varphi_z = \Theta z \qquad (9.70)$$

in the case where the beam is built-in in the cross section $z = 0$. The characteristic of deformation Θ is called the **unit angle of torsion**. Equations (9.69) then become

$$u = -\Theta yz \qquad (9.71a)$$
$$v = \Theta xz \qquad (9.71b)$$
$$w = 0 \qquad (9.71c)$$

From the displacement field expressed by equations (9.71) the strain field may be immediately derived via the **kinematic equations**

$$\varepsilon_x = \varepsilon_y = \varepsilon_z = \gamma_{xy} = 0 \qquad (9.72a)$$
$$\gamma_{zx} = -\Theta y \qquad (9.72b)$$
$$\gamma_{zy} = \Theta x \qquad (9.72c)$$

Applying then the **elastic constitutive equations** (8.74), from the strain field (9.72) it is possible to obtain the corresponding stress field:

$$\sigma_x = \sigma_y = \sigma_z = \tau_{xy} = 0 \qquad (9.73a)$$
$$\tau_{zx} = -G\Theta y \qquad (9.73b)$$
$$\tau_{zy} = G\Theta x \qquad (9.73c)$$

From the shearing stress components (9.73b, c) the magnitude of the shearing stress vector is obtained,

$$\tau_z = G\Theta\, r \qquad (9.74)$$

where r is the radial distance of the generic point from the centre of the circular cross section.

The **indefinite equations of equilibrium** (8.2) are identically satisfied by the stress field (9.73), just as are the **equations of equivalence on the lateral surface** (9.2). Equation (9.2c) in fact becomes

$$G\Theta\,(-yn_x + xn_y) = 0 \qquad (9.75)$$

from which relation we obtain the proportion

$$\frac{x}{y} = \frac{n_x}{n_y} \qquad (9.76)$$

which is identically satisfied on the basis of the similarity of the right-angled triangles ABC and $A'\,B'\,B$ in Figure 9.15. Note that equation (9.2c) is equivalent to equating to zero the scalar product of the stress vector and the unit vector normal to the lateral surface

$$\{\tau_z\}^T\{n\} = 0 \qquad (9.77)$$

whereby the stress vector $\{\tau_z\}$ is always tangential to the lateral surface and thus, in the case of a circular cross section, it is also perpendicular to the radius vector. On the other hand, not only on the lateral surface, but indeed at each point of the circular cross section the vector $\{\tau_z\}$ is perpendicular to the radius vector (Figure 9.15). From equations (9.73b, c), we have in fact

$$\{\tau_z\}^T\{r\} = -G\Theta\,yx + G\Theta\,xy = 0 \qquad (9.78)$$

Since $\sigma_z = 0$, the **conditions of equivalence on the end planes** $N = M_x = M_y = 0$, are identically satisfied. On the other hand, the shear along the X axis vanishes,

$$T_x = \int_A \tau_{zx}\, dA = -G\Theta \int_A y\, dA = -G\Theta S_x = 0 \qquad (9.79)$$

since the static moment of the cross section with respect to the centroidal axis X vanishes. The same applies for shear along the Y axis,

$$T_y = \int_A \tau_{zy}\, dA = G\Theta \int_A x\, dA = G\Theta S_y = 0 \qquad (9.80)$$

The only significant condition of equivalence remains the one corresponding to the twisting moment M_z

$$M_z = \int_A \{r\} \wedge \{\tau_z\}\, dA = \int_A (x\tau_{zy} - y\tau_{zx})\, dA \qquad (9.81)$$

which, on the basis of equations (9.73b, c), gives

$$M_z = G\Theta \int_A r^2\, dA = G\Theta I_p \qquad (9.82)$$

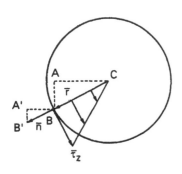

Figure 9.15

254

where I_p is the polar moment of inertia of the circular cross section. From the foregoing relation (9.82), we obtain the unit angle of torsion

$$\Theta = \frac{M_{\tilde{z}}}{GI_p} \qquad (9.83)$$

which is the characteristic of deformation corresponding to the twisting moment. As in the case of longitudinal dilation (9.9b) produced by the centred axial force, and in that of the curvature (9.41) produced by flexure, the angle Θ is directly proportional to the static characteristic $M_{\tilde{z}}$ and inversely proportional to the elastic characteristic G and the geometric characteristic I_p.

Reconsidering equation (9.74), it is now possible to formulate the definitive expression of **global shearing stress**

$$\tau_{\tilde{z}} = \frac{M_{\tilde{z}}}{I_p} r \qquad (9.84)$$

This expression has a structure similar to equation (9.23) for flexure. If the twisting moment is counterclockwise (positive), the vector $\{\tau_{\tilde{z}}\}$ will always give positive moment with respect to the centre and will follow concentric circular lines of flux. It will increase linearly from zero, at the centre, to its maximum value, at the boundary (Figure 9.16(a))

$$\tau_{\max} = \frac{M_{\tilde{z}}}{I_p} R \qquad (9.85)$$

Since the polar moment of inertia is equal to $\pi R^4/2$, the expression (9.85) may be rewritten as a function of the radius R alone, as well as of the static characteristic $M_{\tilde{z}}$

$$\tau_{\max} = \frac{2 M_{\tilde{z}}}{\pi R^3} \qquad (9.86)$$

In the case of a cross section with the form of an annulus (Figure 9.16(b)), it is possible to repeat the entire line of reasoning so far developed for the case

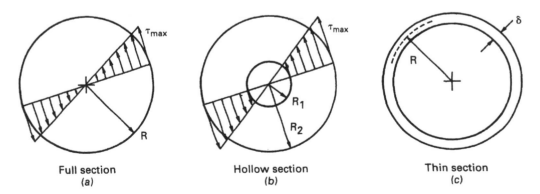

Full section
(a)

Hollow section
(b)

Thin section
(c)

Figure 9.16

of a circular cross section. It will be sufficient to take into account the different polar moment of inertia

$$I_p = \frac{\pi}{2}(R_2^4 - R_1^4) \tag{9.87}$$

in the expression of stress (9.84)

$$\tau_z = \frac{2\,M_z}{\pi(R_2^4 - R_1^4)}\,r \tag{9.88}$$

from which we obtain the maximum shearing stress

$$\tau_{max} = \frac{2\,M_z R_2}{\pi(R_2^4 - R_1^4)} \tag{9.89}$$

Hence, given an equal maximum radius R_2, the hollow cross section is subjected to a higher stress.

Finally, let us consider a thin-walled circular cross section of thickness δ (Figure 9.16(c)), which may be understood as a limit case of the preceding example:

$$R_1 \, ' \; R_2 = R \tag{9.90a}$$

$$R_2 - R_1 = \delta \tag{9.90b}$$

Expanding the difference of the fourth powers, the relation (9.89) becomes

$$\tau_{max} = \frac{2\,M_z R_2}{\pi(R_2^2 + R_1^2)\,(R_2 + R_1)\,(R_2 - R_1)} \tag{9.91}$$

and applying the approximations (9.90)

$$\tau_{max} = \frac{M_z}{2\,\pi R^2 \delta} \tag{9.92}$$

The foregoing formula may also be deduced from equation (9.85), setting $I_p = (2\pi R\delta)R^2$.

On Mohr's plane, the stress condition is represented by a circumference with its centre in the origin. This stress condition is referred to as the state of **pure shear** and implies two principal directions, one of tension and the other of compression, rotated by 45° with respect to the Z axis (Figure 9.17(a)). It is thus understandable why cylindrical elements of concrete subjected to torsion are reinforced with spiral-shaped bars, inclined at an angle of 45° with respect to the axis and disposed in such a way as to stand up to the principal tensile stress (Figure 9.17(b)).

The work of deformation of an element of the beam (Figure 9.13) is obtained once more by applying Clapeyron's Theorem

$$dL = \frac{1}{2}M_z \Theta\, dz \tag{9.93}$$

and using the expression (9.83) of the unit angle of torsion

$$\frac{dL}{dz} = \frac{1}{2}\frac{M_z^2}{GI_p} \tag{9.94}$$

The quadratic expression (9.94) is altogether analogous to equations (9.14) and (9.44), obtained for the centred axial force and for the flexure, respectively.

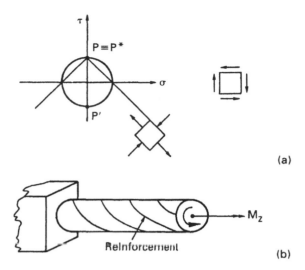

(a)

M_z

Reinforcement

(b)

Figure 9.17

9.7 Torsion in beams of generic cross section

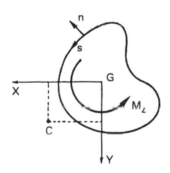

Figure 9.18

In the case of beams of generic cross section, each cross section is assumed to rotate about a longitudinal axis (which may not coincide with the centroidal one), called the **axis of torsion**, at the same time not remaining plane. Expressed in formulas and analogously to equations (9.71), we have (Figure 9.18)

$$u = -\Theta z(y - y_C) \tag{9.95a}$$

$$v = \Theta z(x - x_C) \tag{9.95b}$$

$$w = \Theta \omega(x,y) \tag{9.95c}$$

where x_C, y_C are the coordinates of the **centre of torsion** C, while the function $\omega(x, y)$, called the **warping function**, represents the axial displacements of the points of the cross section.

From the kinematic equations (8.8), we obtain the strain field

$$\varepsilon_x = \varepsilon_y = \varepsilon_z = \gamma_{xy} = 0 \tag{9.96a}$$

$$\gamma_{zx} = -\Theta(y - y_C) + \Theta\frac{\partial\omega}{\partial x} \tag{9.96b}$$

$$\gamma_{zy} = \Theta(x - x_C) + \Theta\frac{\partial\omega}{\partial y} \tag{9.96c}$$

and consequently, applying the elastic constitutive equations (8.74), we obtain the stress field

$$\sigma_x = \sigma_y = \sigma_z = \tau_{xy} = 0 \tag{9.97a}$$

$$\tau_{zx} = G\Theta\left[\frac{\partial\omega}{\partial x} - (y - y_C)\right] \tag{9.97b}$$

$$\tau_{zy} = G\Theta\left[\frac{\partial\omega}{\partial y} + (x - x_C)\right] \tag{9.97c}$$

257

In this case, the third of the indefinite equations of equilibrium (8.2c) is satisfied if and only if

$$\frac{\partial^2 \omega}{\partial x^2} + \frac{\partial^2 \omega}{\partial y^2} = 0 \qquad (9.98)$$

i.e. if and only if ω is a harmonic function. On the other hand, the third of the equations of equivalence on the lateral surface (9.2c) is satisfied if and only if the following boundary condition is satisfied:

$$\left[\frac{\partial \omega}{\partial x} - (y - y_C) \right] n_x + \left[\frac{\partial \omega}{\partial y} + (x - x_C) \right] n_y = 0 \qquad (9.99)$$

Laplace's equation (9.98) associated with the boundary condition (9.99) constitutes a problem for which the solution exists and is the unique one, but for an arbitrary additional constant (Neumann problem). It should be noted, however, that up to this point the coordinates x_C, y_C of the centre of torsion are still unknown.

The conditions of equivalence on the end planes

$$N = M_x = M_y = 0 \qquad (9.100)$$

are identically satisfied, since $\sigma_z = 0$, whilst from the annihilation of the shear we obtain the coordinates of the centre of torsion C. We have in fact

$$T_x = \int_A \tau_{zx} \, dA = G\Theta \int_A \left[\frac{\partial \omega}{\partial x} - (y - y_C) \right] dA = 0 \qquad (9.101a)$$

$$T_y = \int_A \tau_{zy} \, dA = G\Theta \int_A \left[\frac{\partial \omega}{\partial y} + (x - x_C) \right] dA = 0 \qquad (9.101b)$$

whence we obtain

$$\int_A \frac{\partial \omega}{\partial x} \, dA - S_x + y_C A = 0 \qquad (9.102a)$$

$$\int_A \frac{\partial \omega}{\partial y} \, dA + S_y - x_C A = 0 \qquad (9.102b)$$

and thus, since the centroidal static moments are zero

$$x_C = \frac{1}{A} \int_A \frac{\partial \omega}{\partial y} \, dA \qquad (9.103a)$$

$$y_C = -\frac{1}{A} \int_A \frac{\partial \omega}{\partial x} \, dA \qquad (9.103b)$$

Applying Green's Theorem, the coordinates of the centre of torsion finally appear as follows:

$$x_C = \frac{1}{A} \oint_{\mathscr{C}} \omega n_y \, ds \qquad (9.104a)$$

$$y_C = -\frac{1}{A} \oint_{\mathscr{C}} \omega n_x \, ds \qquad (9.104b)$$

where $\mathscr{C}$ indicates the boundary of the cross section; or, in the alternative form

$$x_C = -\frac{1}{A} \oint_{\mathscr{C}} \omega \, \mathrm{d}x \qquad (9.105a)$$

$$y_C = -\frac{1}{A} \oint_{\mathscr{C}} \omega \, \mathrm{d}y \qquad (9.105b)$$

The boundary condition expressed by equation (9.99) for the warping function then transforms into the following integrodifferential equation:

$$\left[\frac{\partial \omega}{\partial x} - y - \frac{1}{A} \oint_{\mathscr{C}} \omega \, \mathrm{d}y \right] n_x + \left[\frac{\partial \omega}{\partial y} + x + \frac{1}{A} \oint_{\mathscr{C}} \omega \, \mathrm{d}x \right] n_y = 0 \qquad (9.106)$$

The equations (9.98) and (9.106) give the warping function, and consequently, from equations (9.105) we obtain the coordinates of the centre of torsion. For a numerical solution to this problem, the reader is referred to Appendix F.

There remains to be imposed the last condition of equivalence on the end planes, the one corresponding to twisting moment

$$M_z = \int_A \{r\} \wedge \{\tau_z\} \, \mathrm{d}A = \int_A (x\tau_{zy} - y\tau_{zx}) \, \mathrm{d}A \qquad (9.107)$$

which, on the basis of equations (9.97b, c), gives

$$M_z = G\Theta \int_A \left(x^2 + y^2 + x\frac{\partial \omega}{\partial y} - y\frac{\partial \omega}{\partial x} \right) \mathrm{d}A \qquad (9.108)$$

The integral represents the so-called **factor of torsional rigidity** I_t, which is a quantity that is always less than or, at the most, equal to the polar moment of inertia I_p. The **unit angle of torsion** can thus be expressed in general as follows:

$$\Theta = \frac{M_z}{GI_t} \qquad (9.109)$$

with

$$I_t = \int_A \left(x^2 + y^2 + x\frac{\partial \omega}{\partial y} - y\frac{\partial \omega}{\partial x} \right) \mathrm{d}A \qquad (9.110)$$

Note how I_t is an exclusively geometrical characteristic of the cross section.

From equations (9.97b, c) and (9.109) we obtain the shearing stress vector

$$\tau_{zx} = \frac{M_z}{I_t}\left[\frac{\partial \omega}{\partial x} - (y - y_C) \right] \qquad (9.111a)$$

$$\tau_{zy} = \frac{M_z}{I_t}\left[\frac{\partial \omega}{\partial y} + (x - x_C) \right] \qquad (9.111b)$$

Finally, Clapeyron's Theorem gives the elementary work of deformation

$$\frac{\mathrm{d}L}{\mathrm{d}z} = \frac{1}{2}\frac{M_z^2}{GI_t} \qquad (9.112)$$

259

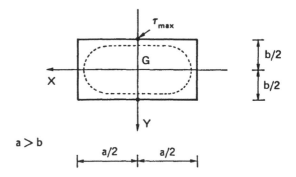

Figure 9.19

9.8 Torsion in open thin-walled sections

The solution previously outlined regarding a cross section of generic shape can be applied to the case of the **rectangular** cross section. It may thus be deduced how the maximum stress is reached in the intermediate point of the larger side (Figure 9.19)

$$\tau_{max} = \alpha \frac{M_z}{ab^2} \tag{9.113}$$

where α is a dimensionless coefficient which depends only on the ratio a/b between the sides of the cross section. At the same time, it is possible to obtain the factor of torsional rigidity to be inserted into equation (9.109)

$$I_t = \beta ab^3 \tag{9.114}$$

where β is another dimensionless coefficient which depends only on the ratio a/b between the sides of the cross section. The two coefficients, α and β, are given in Table 9.1 as functions of a/b.

In the case of a **thin rectangular cross section** ($a/b \rightarrow \infty$), the lines of flux of the vector $\{\tau_z\}$ are closed curves with two portions practically parallel to the larger sides (Figure 9.20). These portions reverse their course only in the end regions, so that the shearing stress vector $\{\tau_z\}$ presents the sole component τ_{zx} for the most part of the cross section. The shearing stress component τ_{zx} is shown to have a linear distribution over the thickness b,

$$\tau_{zx} = -\frac{6 M_z}{ab^3} y \tag{9.115}$$

with a maximum absolute value at the boundary,

$$\tau_{max} = \frac{3 M_z}{ab^2} \tag{9.116}$$

Table 9.1

a/b	1	1.5	2	3	10	∞
α	4.80	4.33	4.06	3.74	3.20	3
β	0.141	0.196	0.229	0.263	0.312	1/3

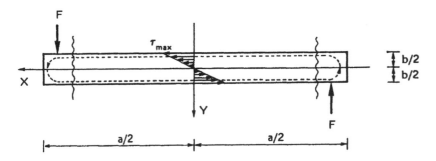

Figure 9.20

Since on the basis of equation (9.114) and of the tabulated value we have

$$I_t = \frac{1}{3} ab^3 \qquad (9.117)$$

we also obtain

$$\tau_{\max} = \frac{M_z}{I_t} b \qquad (9.118)$$

Note that the stresses parallel to the larger sides are not, however, sufficient of themselves to ensure equivalence with the applied twisting moment M_z. In fact, on the basis of equation (9.16f), we obtain

$$M_z(\tau_{zx}) = -\int_A y\tau_{zx} \, dA \qquad (9.119)$$

Substituting the expression (9.115) in equation (9.119), we obtain

$$M_z(\tau_{zx}) = \frac{6 M_z}{ab^3} \int_{-a/2}^{a/2} dx \int_{-b/2}^{b/2} y^2 \, dy \qquad (9.120)$$

from which we have

$$M_z(\tau_{zx}) = \frac{M_z}{2} \qquad (9.121)$$

The other half of the moment M_z is furnished by the components τ_{zy}, which are important only in the end regions of the cross section, where they present an arm with respect to the centroid (or centre of torsion) which is much greater than that of the components τ_{zx} $(a \gg b)$.

Let us consider the case of an **open thin-walled** cross section made up of a number of thin rectangles welded together so as not to create any closed path (Figure 9.21), and let us allow the applied moment M_z to be distributed in such a way that the ith section takes up the amount M_z^i. The maximum stress on the ith section will be given by equation (9.118),

$$\tau_{\max}^i = \frac{M_z^i}{I_t^i} b_i \qquad (9.122)$$

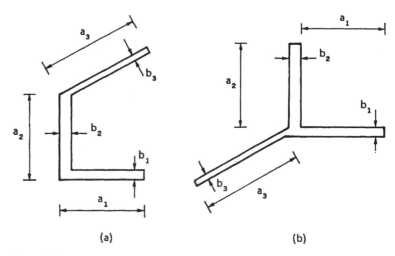

(a)   (b)

Figure 9.21

where b_i is the thickness of the ith section and

$$I_t^i = \frac{1}{3} a_i b_i^3 \qquad (9.123)$$

from equation (9.117). On the other hand, for reasons of congruence, since each section must rotate by the same angle Θ, we have

$$\Theta^i = \frac{M_z^i}{GI_t^i} = \Theta = \frac{M_z}{GI_t} \qquad (9.124)$$

where I_t is the factor of torsional rigidity of the entire section. From equations (9.122) and (9.124), we obtain

$$\tau_{max}^i = \frac{M_z}{I_t} b_i \qquad (9.125)$$

The factor I_t is obtained from considerations of equilibrium, or rather of equivalence. Since the sum of the partial moments M_z^i must equal the applied moment M_z, we have

$$M_z = G\Theta I_t = \sum_i M_z^i = G\Theta \sum_i I_t^i \qquad (9.126)$$

from which we obtain

$$I_t = \sum_i I_t^i \qquad (9.127)$$

The global factor of torsional rigidity is thus equal to the sum of the partial factors, as is the case when the elements are in parallel. From equation (9.124) we then have

262

$$M_{\underline{z}}^i = M_{\underline{z}} \frac{I_t^i}{\sum\limits_i I_t^i} \tag{9.128}$$

where the ratio $I_t^i / \sum_i I_t^i$ is called **coefficient of distribution**. Recalling equation (9.127), the global factor may be expressed as follows:

$$I_t = \frac{1}{3} \sum_i a_i b_i^3 \tag{9.129}$$

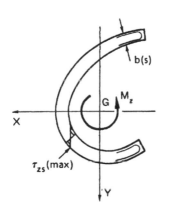

If the open thin-walled section has as its mid-line a regular curve (Figure 9.22), and not a broken line as in the previous case, equations (9.125) and (9.129) will be replaced by the following:

$$\tau_{zs}(\max) = \frac{M_{\underline{z}}}{I_t} b(s) \tag{9.130}$$

$$I_t = \frac{1}{3} \int_{\mathscr{C}} b^3(s)\, ds \tag{9.131}$$

where s is the curvilinear coordinate on the mid-line $\mathscr{C}$.

The distribution of the stress τ_{zs} is linear on the thickness and vanishes on the mid-line. The maximum absolute value (9.130), which is recorded on the boundary, is in turn maximum in the points in which the thickness $b(s)$ is maximum. The flux lines of the vector $\{\tau_z\}$ are thus parallel to the mid-line and reverse their course only in the end regions (Figure 9.22).

Figure 9.22

9.9   Torsion in closed thin-walled sections

Closed thin-walled sections are also called **tubular sections**. These are frequently used in building applications, because, as we shall see shortly, they present notable strength and stiffness. Their disparity of behaviour in regard to open thin-walled sections is due substantially to the flux of the vector $\{\tau_z\}$, which in the case of closed sections manages to develop more conveniently, i.e. with the shearing stresses τ_{zs} that are approximately constant on each chord that is perpendicular to the mid-line $\mathscr{C}$ (Figure 9.23(a)). In addition, even though the thickness $b(s)$ is not constant, the product of shearing stress τ_{zs} and thickness must be constant:

$$\tau_{zs}(s)\, b(s) = \text{constant} \tag{9.132}$$

This follows from the simple application of the law of reciprocity of shearing stresses to an element of the thin wall, obtained by sectioning the beam with two planes perpendicular to the axis of the beam, located at a distance dz apart, and with two planes perpendicular to the mid-line of the cross-section (Figure 9.23(b)). If we denote as b_1, b_2 the chords of intersection of the longitudinal and the transverse sections, and as τ_1, τ_2 their respective shearing stresses, by virtue of the axial equilibrium of the element, we must have

$$\tau_1 b_1\, dz = \tau_2 b_2\, dz \tag{9.133}$$

from which there follows the constancy of the product (9.132), since equation (9.133) holds good for any pair of planes which cut the mid-line at right angles to it.

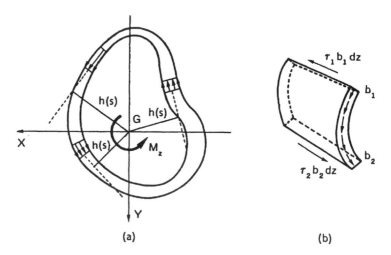

(a) (b)

Figure 9.23

From equivalence we must find

$$M_z = \oint_{\mathscr{C}} \tau_{zs}(s)\, b(s)\, h(s)\, \mathrm{d}s \qquad (9.134)$$

where $h(s)$ represents the arm of the elementary force $(\tau_{zs}\, b\, \mathrm{d}s)$ with respect to the centroid (Figure 9.23(a)).

From equations (9.132) and (9.134) we deduce

$$M_z = \tau_{zs}(s)\, b(s) \oint_{\mathscr{C}} h(s)\, \mathrm{d}s \qquad (9.135)$$

and thus, noting that the integral given above represents twice the area Ω enclosed by the mid-line $\mathscr{C}$, we obtain

$$\tau_{zs} = \frac{M_z}{2\,\Omega b(s)} \qquad (9.136)$$

The expression (9.136), called **Bredt's formula,** underlines how the maximum stress is produced where the thickness $b(s)$ is minimum, in sharp contrast with what occurs in the case of open thin-walled sections, as described by equation (9.130).

An application of Clapeyron's Theorem then furnishes the **factor of torsional rigidity**. The elementary work of deformation is equal to

$$\mathrm{d}L = \frac{1}{2} \frac{M_z^2}{G I_t}\, \mathrm{d}z = \int_V \Psi\, \mathrm{d}V \qquad (9.137)$$

where Ψ is the complementary elastic potential expressed by equation (8.70). Since the stress component τ_{zs} alone is different from zero, we have

$$\mathrm{d}L = \frac{\mathrm{d}z}{2G} \int_A \tau_{zs}^2\, \mathrm{d}A \qquad (9.138)$$

264

and thus, substituting equation (9.136)

$$dL = \frac{dz}{2G} \frac{M_z^2}{4\Omega^2} \oint_{\mathscr{C}} \frac{1}{b^2(s)} \, (b(s) \, ds) \qquad (9.139)$$

From equations (9.137) and (9.139) we obtain the factor of torsional rigidity

$$I_t = 4\Omega^2 \bigg/ \oint_{\mathscr{C}} \frac{ds}{b(s)} \qquad (9.140)$$

This formula simplifies when the thickness of the section is constant,

$$I_t = \frac{4\Omega^2 b}{s} \qquad (9.141)$$

where s is the length of the mid-line. It should be noted that, whereas the thickness b appears raised to the first power in equation (9.141), it appears raised to the third in equation (9.117), thus jeopardizing the stiffness of the open sections.

In the case of a thin circular section of thickness δ we find again equation (9.92), applying Bredt's formula

$$\tau_{zs} = \frac{M_z}{2\pi R^2 \delta} \qquad (9.142)$$

while equation (9.141) gives

$$I_t = \frac{4(\pi R^2)^2 \delta}{2\pi R} \qquad (9.143)$$

which proves to be the polar moment of inertia of the cross section

$$I_t = 2\pi R^3 \delta = I_p \qquad (9.144)$$

In certain cases it may happen that a thin section is made up of tubular parts and open parts (flanges). It may readily be shown that practically the whole of the applied moment M_z is sustained by the tubular part, while the remaining open thin parts are subjected to much lower amounts.

Consider, for instance, the **box section** of Figure 9.24. Cross sections of this sort are frequently used for beams of road bridges. It consists of a thin rectangular cross section (1) and of two flanges (2).

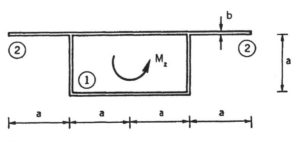

Figure 9.24

265

Angular congruence gives

$$\Theta_1 = \frac{M_1}{GI_1} = \Theta_2 = \frac{M_2}{GI_2} \qquad (9.145)$$

where the symbols have their obvious meanings. On the other hand, from equivalence we have

$$M_1 + M_2 = M_z \qquad (9.146)$$

Equations (9.145) and (9.146) constitute a system of linear equations in the two unknowns M_1, M_2. Once solved, these give the distribution that was sought:

$$M_1 = M_z \frac{I_1}{I_1 + I_2} \qquad (9.147a)$$

$$M_2 = M_z \frac{I_2}{I_1 + I_2} \qquad (9.147b)$$

Using relations (9.141) and (9.129), we then have

$$I_1 = \frac{8}{3} a^3 b \qquad (9.148a)$$

$$I_2 = \frac{2}{3} ab^3 \qquad (9.148b)$$

whence the ratio between rigidity factors becomes

$$\frac{I_1}{I_2} = 4 \left(\frac{a}{b} \right)^2 \qquad (9.149)$$

If, for example, $a/b \simeq 10$, we obtain $I_1/I_2 \simeq 400$, and thus $M_1 \simeq M_z$ and $M_2 \simeq 0$.
Using equations (9.136) and (9.130), we obtain the shearing stresses

$$\tau_1 = M_z \frac{I_1}{I_1 + I_2} \frac{1}{4\, a^2 b} \qquad (9.150a)$$

$$\tau_2 = M_z \frac{I_2}{I_1 + I_2} \frac{b}{\frac{2}{3} ab^3} \qquad (9.150b)$$

whence the ratio between the shearing stresses is

$$\frac{\tau_1}{\tau_2} = 4 \left(\frac{a}{b} \right)^2 \times \frac{1}{6} \left(\frac{b}{a} \right) = \frac{2}{3} \left(\frac{a}{b} \right) \qquad (9.151)$$

Note that, while the ratio (9.149) between the stiffnesses is proportional to the square of the ratio (a/b), the ratio (9.151) between the stresses is proportional to the first power of (a/b). The flanges are in any case under a slight stress.

As regards multiply-connected thin-walled sections, the reader is referred to Appendix G.

9.10 Combined shearing and torsional loading

We intend to show how shear is energetically orthogonal to twisting moment, only if applied to the centre of torsion. To do so, let us consider a beam built-

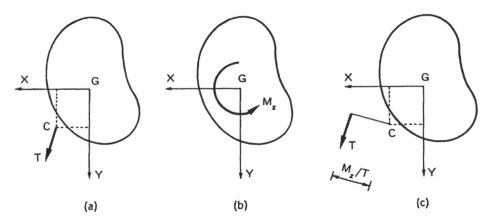

Figure 9.25

in at one of its ends and loaded on the other, in one case by a force T passing through the centre of torsion C (Figure 9.25(a)), in the other by a twisting moment M_z (Figure 9.25(b)). If we denote as $\{\eta_C(M_z)\}$ the displacement in the XY plane of the point C, caused by the moment M_z, and as $\varphi_z(T)$ the rotation of the cross section in the XY plane, caused by the shear T, the application of Betti's Reciprocal Theorem expresses the equality of the work performed by either force system acting through the displacement resulting from the other

$$\{T\}^{\mathrm{T}}\{\eta_C(M_z)\} = M_z\varphi_z(T) \tag{9.152}$$

On the other hand the displacement $\{\eta_C(M_z)\}$ of the centre of torsion is zero by definition, so that the work done by either system is likewise zero and in particular the rotation $\varphi_z(T)$ caused by the shear acting in point C is zero. For this reason, the **centre of torsion** is also called the **centre of shear**.

When the shearing force, then, is applied to the centre of torsion, it causes only translations and not rotations of the cross section in the XY plane. As there is no torsional deformation, i.e. $\Theta = 0$, from equations (9.97) it may be deduced that the torsional stresses must also be zero.

Following the same line of reasoning, it is evident that the combined loading of **shear-torsion** (Figures 9.25(a), (b)) is equivalent to a single force, parallel to the assigned shear, that presents a moment M_z with respect to the centre of torsion (Figure 9.25(c)). In other words, the global twisting moment M_z is evaluated as the moment of the force tangent to the cross section with respect to the centre of torsion, and not with respect to the centroid, as one might have been erroneously led to think previously.

9.11 Shearing force

Consider a Saint Venant solid loaded on the end planes by two equal and opposite shearing forces T_y, whose lines of action pass through the corresponding centres of torsion and are parallel to the central direction of inertia Y, and by a bending moment M_x which counterbalances the couple $T_y l$, where l is the length of the cylindrical solid (Figure 9.26). Whereas the shear is constant, the bending moment varies linearly, from zero to the value $-T_y l$, along the axis of the beam. A typical case is that of a cantilever beam loaded by a force T_y at the end. As has already been mentioned, it is not possible to isolate the

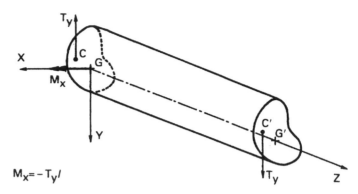

$$M_x = -T_y l$$

Figure 9.26

shearing characteristic, as this is the derivative of the bending moment function.

Let us now set ourselves the problem of determining the mean shearing stress acting orthogonally to a generic chord BB' (Figure 9.27(a)). For this purpose let us consider a portion of the solid, bounded by two cross sections a distance dz apart and by the plane parallel to the axis of the beam, the projection of which on the cross section is represented by the chord BB' (Figure 9.27(b)). If we indicate by s the direction perpendicular to the chord BB' on the XY plane, the distribution of the shearing stresses τ_{sz} will be given by overturning the reciprocal shearing stresses τ_{zs} about the edge BB' (Figure 9.27(b)). From the equilibrium with regard to axial translation of the element of solid considered, we can then put

$$\int_B^{B'} \tau_{zs} \, dz \, d\xi = \int_{A'} \frac{\partial \sigma_z}{\partial z} \, dz \, dA \qquad (9.153)$$

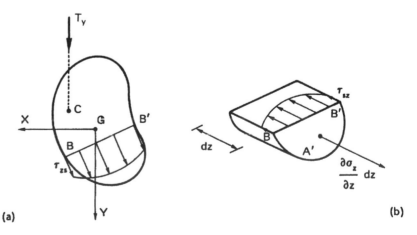

(a)

(b)

Figure 9.27

where A' is the lower portion of the cross section, and where the integrand on the right-hand side of the equation represents the increment of normal stress σ_z produced by the bending moment $M_x(z)$. From equation (9.23) we obtain

$$\frac{\partial \sigma_z}{\partial z} = \frac{\partial M_x}{\partial z}\frac{y}{I_x} \qquad (9.154)$$

where M_x is the only function of z. Recalling then that the derivative of the moment is equal to the corresponding shear, we have

$$\frac{\partial \sigma_z}{\partial z} = T_y \frac{y}{I_x} \qquad (9.155)$$

The relation (9.155), introduced into equation (9.153), gives the mean shearing stresses acting orthogonally to the chord BB',

$$\bar{\tau}_{zs} = \frac{T_y S_x^{A'}}{I_x b} \qquad (9.156)$$

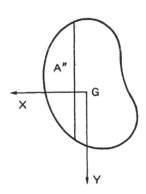

Figure 9.28

where $S_x^{A'}$ is the static moment of the area A' with respect to the X axis, and b is the length of the chord. The relation (9.156) is known as the **Jourawski formula**. The components of $\{\tau_z\}$ parallel to the chord BB' are negligible in the cases in which the chord BB' is parallel to the central axis X. The mean values of these stresses, on a chord parallel to the central axis Y, are in fact proportional to the static moment $S_x^{A''}$, where A'' is one of the two portions of the section separated by the vertical chord (Figure 9.28).

From the applications point of view, the Jourawski formula proves to be particularly useful, because, in addition to the fact that the above-mentioned components are negligible, the local values τ_{zs} are not sensitive to variation from the mean value given by the relation (9.156). A typical example of a cross section for which the Jourawski formula provides a highly reliable estimation of the maximum shearing stress is the rectangular one (Figure 9.29(a)). Consider a generic horizontal chord BB' for which the static moment of the area A' with respect to the X axis is

$$S_x^{A'} = b\left(\frac{h}{2}-y\right)\left[y+\frac{1}{2}\left(\frac{h}{2}-y\right)\right] = \frac{b}{2}\left(\frac{h^2}{4}-y^2\right) \qquad (9.157)$$

Recalling that the moment of inertia of a rectangular area is $I_x = bh^3/12$, and applying equation (9.156), we obtain

$$\bar{\tau}_{zy} = \frac{6\,T_y}{bh^3}\left(\frac{h^2}{4}-y^2\right) \qquad (9.158)$$

The stress τ_{zy}, or rather its mean value on the horizontal chords, varies parabolically, vanishing on the horizontal sides and presenting one maximum on the centroidal chord (Figure 9.29(b)):

$$\tau_{max} = \frac{3}{2}\frac{T_y}{bh} \qquad (9.159)$$

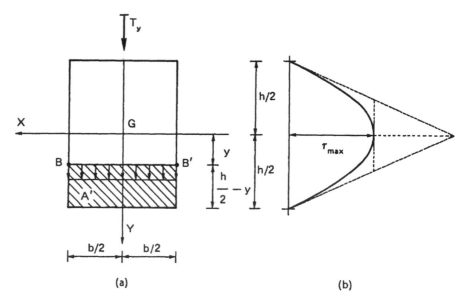

Figure 9.29

More generally, the following relation will hold good:

$$\tau_{max} = s_y \frac{T_y}{A} \tag{9.160}$$

where A stands for the area of the cross section, while s_y is a factor always greater than unity, which depends on the geometrical shape of the cross section. For the circular cross section, for instance, we have $s_y = 4/3$.

In slender beams, the shearing stresses are usually far smaller than the axial stresses due to bending moment. In the case, for instance, of a cantilever beam having rectangular cross section and loaded by a force T_y at the free end, at the built-in constraint we have

$$\sigma_{max} = \frac{T_y l}{bh^3/12} \frac{h}{2} = \frac{6 T_y l}{bh^2} \tag{9.161}$$

so that the ratio

$$\frac{\tau_{max}}{\sigma_{max}} = \frac{1}{4}\left(\frac{h}{l}\right) \tag{9.162}$$

tends to zero for $l \to \infty$.

In the case of stubby beams, the shearing stresses can be considerable, even if compared with axial stresses due to bending moment. It should be noted, however, that below certain ratios of slenderness ($l/h \lesssim 3$) the Saint Venant theory fails to apply, since the end regions of the solid must then be neglected (the length of these regions is approximately equal to the beam depth h). Nor must it be forgotten that by the law of reciprocity of shearing stresses, longitudinal shearing stresses τ_{yz} are also present, as well as the stresses τ_{zy} acting upon the cross section (Figure 9.30(a)). The existence of the former stresses τ_{yz} may be inferred if we consider two beams resting one on top of the other

270

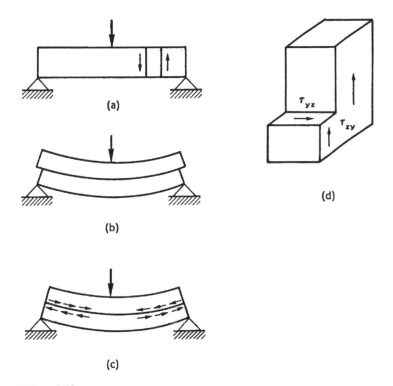

Figure 9.30

without friction (Figure 9.30(b)) If the lower beam is supported at its ends and the upper one is loaded with a force, then each beam will bend, presenting fibres in tension below and fibres in compression above. At the interface between the two beams we shall then have dilations in the upper beam and contractions in the lower one. If we now imagine that there is some form of friction acting between the two free surfaces in contact, the two beams will develop interactions that will tend to contract the upper one and dilate the lower one. These interactions, in the case where relative sliding is prevented by the presence of a bonding agent, are represented by the shearing stresses τ_{yz} (Figures 9.30(c), (d)). In laminate composite materials, the mechanism of **delamination** is caused by the stresses τ_{yz} exceeding the limits of bonding resistance.

As regards deformation, the only strain component substantially different from zero is γ_{zy}. In the case of a rectangular cross section, γ_{zy} will show a parabolic variation equal, but for the factor $1/G$, to that of τ_{zy} (Figure 9.31):

$$\gamma_{zy} = \frac{\tau_{zy}}{G} = \frac{6T_y}{Gbh^3}\left(\frac{h^2}{4} - y^2\right) \qquad (9.163)$$

Shearing strain is thus maximum on the centroidal plane $y = 0$, and zero on the outermost planes $y = \pm h/2$ (Figure 9.31). The result then is an inflection or warping of the cross sections out of their original planes, so as to maintain orthogonality only between the deformed section and the outermost planes.

271

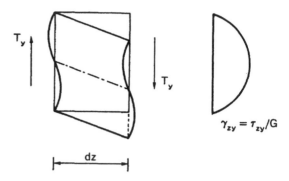

Figure 9.31

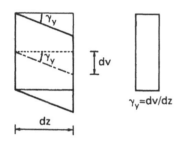

Figure 9.32

The relative sliding between the end sections of the element of length dz (Figure 9.32) may be obtained via the application of Clapeyron's Theorem

$$dL = \frac{1}{2} T_y \, dv = \frac{1}{2} T_y \gamma_y \, dz \qquad (9.164)$$

where γ_y is called **mean shearing strain** and represents the characteristic of deformation corresponding to the shearing force T_y. The elementary work dL is expressible via the complementary elastic potential,

$$dL = \int_V \Psi \, dV = \frac{dz}{2G} \int_A \tau_{zy}^2 \, dA \qquad (9.165)$$

where the component τ_{zx} has been neglected, while the component τ_{zy} has been assumed to be given locally by its mean on the corresponding horizontal chord.

Substituting the expression (9.158) in equation (9.165), we obtain

$$dL = \frac{dz}{2G} \int_{-b/2}^{b/2} dx \int_{-h/2}^{h/2} \frac{36 T_y^2}{b^2 h^6} \left(\frac{h^2}{4} - y^2 \right)^2 dy \qquad (9.166)$$

from which

$$dL = \frac{dz}{2G} \frac{6}{5} \frac{T_y^2}{bh} \qquad (9.167)$$

A comparison between the foregoing expression and equation (9.164) yields

$$\gamma_y = \frac{6}{5} \frac{T_y}{Gbh} \qquad (9.168)$$

More generally the following relation will hold:

$$\gamma_y = t_y \frac{T_y}{GA} \qquad (9.169)$$

where t_y, called the **shear factor**, is always greater than unity and depends on the geometrical shape of the cross section. For the circular cross section, for instance, we have $t_y = 32/27$.

272

From equation (9.164) we thus obtain

$$\frac{dL}{dz} = \frac{1}{2}t_y\frac{T_y^2}{GA} \tag{9.170}$$

which shows how the work per unit length of the beam is a quadratic function also of the shearing force.

9.12 Biaxial shearing force

In Section 9.10 it has been shown how the shear passing through the centre of torsion causes only translation and not rotation of the cross section in its plane. We did not, however, then specify the relation that links this translation to the shearing force components T_x, T_y. On the other hand, in the last section we implicitly assumed that, where the axis Y is central and one of symmetry, the shearing force T_y causes exclusively a shearing strain γ_y and hence a translation of the cross section in its own Y direction. However, it is not possible to rule out *a priori* that, where the Y axis is central but not one of symmetry, the shearing force T_y may cause also a shearing strain γ_x and hence also a translation of the cross section in a direction different from its own.

To analyse the problem in a rigorous way, it is necessary to consider (open or closed) thin-walled cross sections, for which the shearing stress τ_{zs} can be considered constant on each chord orthogonal to the mid-line $\mathscr{C}$, while the shearing stress component orthogonal to τ_{zs} can be neglected (Figure 9.33). In the case of **biaxial shear**, i.e. where there is the simultaneous presence of the two shears T_x, T_y, the stress τ_{zs} is equal to the algebraic sum of the corresponding contributions

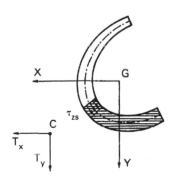

Figure 9.33

$$\tau_{zs}(s) = \frac{T_y}{I_x}\frac{S_x(s)}{b(s)} + \frac{T_x}{I_y}\frac{S_y(s)}{b(s)} \tag{9.171}$$

The elementary work of an element of beam is always equal to the integral of the complementary elastic energy

$$dL = \frac{dz}{2G}\int_{\mathscr{C}} \tau_{zs}^2 b \, ds \tag{9.172}$$

From equation (9.171) we obtain the relation

$$dL = \frac{dz}{2G}\left[\frac{T_y^2}{I_x^2}\int_{\mathscr{C}}\frac{S_x^2}{b}\,ds + \frac{T_x^2}{I_y^2}\int_{\mathscr{C}}\frac{S_y^2}{b}\,ds + 2\frac{T_xT_y}{I_xI_y}\int_{\mathscr{C}}\frac{S_xS_y}{b}\,ds\right] \tag{9.173}$$

which may be cast in the form

$$dL = \frac{dz}{2GA}(t_xT_x^2 + t_yT_y^2 + 2t_{xy}T_xT_y) \tag{9.174}$$

where

$$t_x = \frac{A}{I_y^2}\int_{\mathscr{C}}\frac{S_y^2}{b}\,ds \tag{9.175a}$$

$$t_y = \frac{A}{I_x^2}\int_{\mathscr{C}}\frac{S_x^2}{b}\,ds \tag{9.175b}$$

273

are the shear factors, while

$$t_{xy} = \frac{A}{I_x I_y} \int_{\mathscr{C}} \frac{S_x S_y}{b} \, ds \qquad (9.175c)$$

is the **mutual shear factor**.

When in general $t_{xy} \neq 0$, the **mutual work**

$$dL \,(\text{mutual}) = t_{xy} \frac{T_x T_y}{GA} \, dz \qquad (9.176)$$

is different from zero, and thus the shears T_x and T_y are **not energetically orthogonal**. On the other hand, t_{xy} is null when there is even only one axis of symmetry, as may be deduced from the integral expression (9.175c). Hence, only for cross sections with one or more axes of symmetry are the shears T_x and T_y energetically orthogonal.

Applying Clapeyron's Theorem, we also have

$$dL = \frac{1}{2} T_x \, du + \frac{1}{2} T_y \, dv \qquad (9.177)$$

and therefore, introducing the mean shearing strains

$$dL = \frac{1}{2} (T_x \, \gamma_x + T_y \gamma_y) \, dz \qquad (9.178)$$

On the other hand, from the linearity of the elastic problem, the mean shearing strains are homogeneous linear functions of the shears

$$\gamma_x = a_{xx} T_x + a_{xy} T_y \qquad (9.179a)$$

$$\gamma_y = a_{yx} T_x + a_{yy} T_y \qquad (9.179b)$$

Substituting equations (9.179) into equation (9.178), and taking into account that we must have $a_{xy} = a_{yx}$ according to Betti's Reciprocal Theorem, we have

$$dL = \frac{dz}{2} (a_{xx} T_x^2 + a_{yy} T_y^2 + 2 a_{xy} T_x T_y) \qquad (9.180)$$

A comparison between the preceding expression and equation (9.174) furnishes, via the law of identity of polynomials, the linear relation that links the shearing force vector to the shearing strain (or translation) vector of the cross section:

$$\gamma_x = \frac{1}{GA} (t_x T_x + t_{xy} T_y) \qquad (9.181a)$$

$$\gamma_y = \frac{1}{GA} (t_{xy} T_x + t_y T_y) \qquad (9.181b)$$

The relations (9.181) hold good also for compact sections, in which case, however, the factors t_x, t_y, t_{xy} are difficult to determine.

9.13 Thin-walled cross sections subjected to shear

As has been mentioned in the foregoing section, the Jourawski formula furnishes the exact shearing stresses only in the case of thin-walled sections, with thickness b tending to zero. In this case, in fact, it is legitimate to equate the local stress with the mean stress on the chord. On the other hand, only the

274

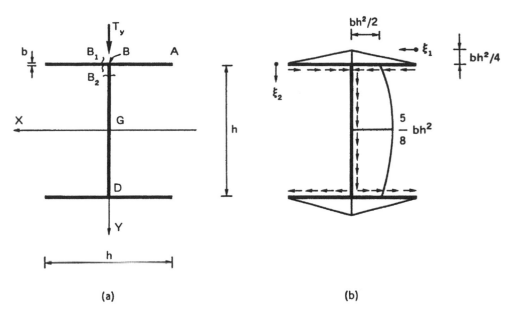

(a)   (b)

Figure 9.34

shearing stress component τ_{zs} (parallel to the mid-line) is present, by virtue of the condition of equivalence on the lateral surface of the beam.

In the sequel we shall examine a number of typical cases, namely the I-section, the box section and the C-section, both from the standpoint of stress and from that of deformation.

Consider the I-section of Figure 9.34(a). This type of cross section is frequently used in metal constructions, because it allows the maximum moment of inertia to be achieved with the minimum use of material. A high moment of inertia implies, in fact, low axial stresses σ_z due to bending.

The static moment $S_x^{A'}$ may be represented graphically on the cross section, if we take as our reference line the mid-line (Figure 9.34(b)). The static moment varies linearly on the flanges and parabolically on the web. In the points where the flanges and the web converge, it undergoes a discontinuity, in such a way that the flux of the vector $\{\tau_{zs}\}$ is preserved. If we denote by B_1 and B_2 the points which immediately precede and follow the node B, in absolute value we have (Figure 9.34(b))

$$S_x^{A'}(B_1) = b\left(\frac{h}{2}\right)\left(\frac{h}{2}\right) = \frac{1}{4}bh^2 \tag{9.182a}$$

$$S_x^{A'}(B_2) = 2\,S_x^{A'}(B_1) = \frac{1}{2}bh^2 \tag{9.182b}$$

The static moment in the centroid is then

$$S_x^{A'}(G) = \frac{1}{2}bh^2 + b\left(\frac{h}{2}\right)\left(\frac{h}{4}\right) = \frac{5}{8}bh^2 \tag{9.182c}$$

The maximum stress is in the centroid. Applying the Jourawski formula and using the moment of inertia

275

$$I_x = \frac{1}{12} bh^3 + 2(bh) \left(\frac{h}{2}\right)^2 = \frac{7}{12} bh^3 \qquad (9.183a)$$

we find that the maximum stress is

$$\tau_{max} = \frac{T_y\left(\frac{5}{8} bh^2\right)}{\left(\frac{7}{12} bh^3\right) b} = \frac{15}{14} \frac{T_y}{bh} \qquad (9.183b)$$

More particularly, considering the two coordinates ξ_1 and ξ_2 along the midline, we have (Figure 9.34(b))

$$S_x^{A'}(\xi_1) = \frac{1}{2} bh\xi_1 \qquad (9.184a)$$

$$S_x^{A'}(\xi_2) = \frac{1}{2} bh^2 + b\xi_2 \left(\frac{h}{2} - \frac{\xi_2}{2}\right) \qquad (9.184b)$$

It is thus possible to verify that it is the web BD alone that withstands the shearing force. We have in fact

$$\int_B^D \tau_{zs} b \, ds = 2 \int_0^{h/2} \frac{T_y S_x^{A'}(\xi_2)}{I_x b} b \, d\xi_2 \qquad (9.185)$$

Substituting expressions (9.183a) and (9.184b) into equation (9.185), we obtain

$$\int_B^D \tau_{zs} b \, ds = \frac{2 T_y}{\frac{7}{12} bh^3} \int_0^{h/2} \left(\frac{1}{2} bh^2 + \frac{1}{2} bh\xi_2 - \frac{1}{2} b\xi_2^2\right) d\xi_2 \qquad (9.186)$$

and thus we verify that

$$\int_B^D \tau_{zs} b \, ds = T_y \qquad (9.187)$$

As regards the shear deformation, the expression (9.175b) in the case of the cross section under examination becomes

$$t_y = \frac{3 h}{\left(\frac{7}{12} bh^3\right)^2} \left\{4 \int_0^{h/2} [S_x^{A'}(\xi_1)]^2 \, d\xi_1 + 2 \int_0^{h/2} [S_x^{A'}(\xi_2)]^2 \, d\xi_2\right\} \qquad (9.188)$$

whence we obtain

$$t_y \simeq 3.38$$

Figure 9.35 gives qualitatively the diagram of shearing stresses for the box section already considered.

Finally let us consider the C-section of Figure 9.36(a). It is symmetrical only with respect to the central axis X, so that the centre of torsion will be found on that axis in a position that is unknown beforehand. In the case where there is a shear T_y passing through the centre of torsion (Figure 9.36(a)), the shearing stresses produced must constitute a system of forces equivalent to the only characteristic present, T_y (in addition to the bending moment M_x). Resolving then a simple problem of static equivalence, it is possible to identify the position of the centre of torsion in the case of thinwalled cross sections.

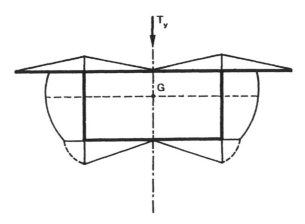

Figure 9.35

With the two reference systems ξ_1 and ξ_2 fixed on the mid-line, we have (Figure 9.36(a))

$$S_x^{A'}(\xi_1) = bh\xi_1 \tag{9.189a}$$

$$S_x^{A'}(\xi_2) = 2\,bh^2 + b\xi_2\left(h - \frac{\xi_2}{2}\right) \tag{9.189b}$$

for which the application of the Jourawski formula gives the following shearing stresses:

$$\tau_{zs}(\xi_1) = \tau_{zx} = \frac{T_y}{I_x b}\,bh\xi_1 \tag{9.190a}$$

$$\tau_{zs}(\xi_2) = \tau_{zy} = \frac{T_y}{I_x b}\left(2\,bh^2 + bh\xi_2 - \frac{1}{2}b\xi_2^{\,2}\right) \tag{9.190b}$$

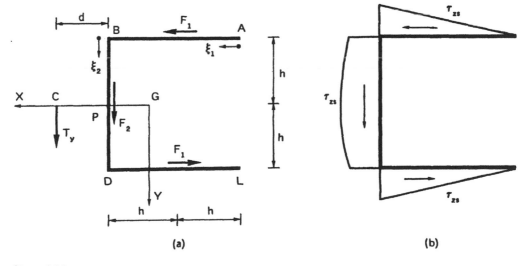

(a)

(b)

Figure 9.36

277

where $I_x = \frac{14}{3} bh^3$. The variation of the stress τ_{zs} is thus linear on the two horizontal plates and parabolic on the vertical plate (Figure 9.36(b)). The stresses τ_{zs} create a flux which runs from the upper end A to the lower one L. The magnitude of the resultant force of each horizontal distribution τ_{zs} is

$$F_1 = \int_0^{2h} \tau_{zs}(\xi_1)b\,d\xi_1 = \frac{3}{14}\frac{T_y}{h^2}\left[\frac{\xi_1^2}{2}\right]_0^{2h} = \frac{3}{7}T_y \qquad (9.191a)$$

whilst the magnitude of the vertical resultant is equal to the shear T_y:

$$F_2 = \int_0^{2h} \tau_{zs}(\xi_2)b\,d\xi_2 = \frac{3}{14}\frac{T_y}{h^3}\left[2h^2\xi_2 + h\frac{\xi_2^2}{2} - \frac{1}{2}\frac{\xi_2^3}{3}\right]_0^{2h} = T_y \qquad (9.191b)$$

The two horizontal resultants therefore create a counterclockwise couple $2hF_1$, and hence the shear T_y is equivalent to the system formed by this couple and by the force F_2, when it acts to the left of the vertical plate, at a distance d, such that

$$2\,h\,F_1 = d\,F_2 \qquad (9.192)$$

from which we obtain the position of the centre of torsion C:

$$d = 2\,h\frac{F_1}{F_2} = \frac{6}{7}h \qquad (9.193)$$

From the foregoing considerations, it may be deduced that, for thin-walled sections formed by thin rectangular elements converging to a single common point (Figure 9.37), the centre of torsion coincides with that point. In fact, the problem to be solved is a static one of equivalence for two or more forces all passing through a single pole. Evidently the resultant will also pass through the pole.

Also in the case of oblique polar symmetry (Figure 9.38), considerations similar to the preceding ones apply, which make it possible to identify the position of the centre of torsion in the centre of symmetry.

However, in the case of non-symmetrical sections (Figure 9.39), the task of locating the centre of torsion is far more complex and calls for the resolution of two static problems of equivalence, with respect to the shears T_x and T_y. The intersection of the lines of action of the two resultants furnishes the centre of torsion.

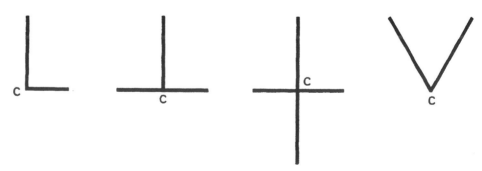

Figure 9.37

278

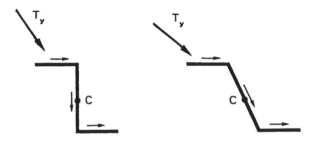

Figure 9.38

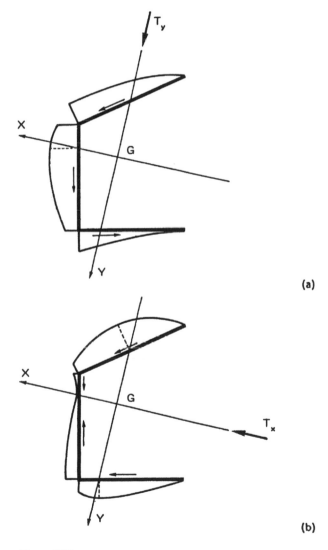

(a)

(b)

Figure 9.39

279

9.14 Beam strength analysis

Combining all the elementary loadings acting on the Saint Venant solid, we obtain two orthogonal forces that are generally skew forces (Figure 9.40(a)): the axial force N, eccentric with respect to the centroid G, and the shearing force T, eccentric with respect to the centre of torsion C. The first combined load is equal to the sum of a centred axial force N and the two bending moments M_x, M_y, and generates only the axial component of stress σ_z. The second combined load is equal to the sum of a twisting moment M_z and the two shears T_x, T_y, and generates only the shearing stress component τ_z. Therefore, in each point of any given cross section, there will in general be present both the components σ_z and τ_z (Figure 9.40(a)), thus in every case giving rise to a plane state of stress (Figure 9.40(b)). It is evident that, in general, the plane of stresses varies from point to point in the cross section.

The stress tensor for the Saint Venant solid thus takes the following form:

$$[\sigma] = \begin{bmatrix} 0 & 0 & \tau_{xz} \\ 0 & 0 & \tau_{yz} \\ \tau_{zx} & \tau_{zy} & \sigma_z \end{bmatrix} \tag{9.194}$$

three of the six significant components always being zero. It should be noted that the case of the tensor (9.194) is complementary to that of the cylindrical solid of thickness l tending to zero, loaded exclusively on the lateral surface by forces contained in the middle plane. This solid, called the **Clebsch solid**, idealizes a plane plate of small thickness, loaded by forces contained in its own middle plane and thus not subject to bending. This problem will be dealt with more extensively in Chapter 19.

The representation of the state of stress in a point is obtained graphically on Mohr's plane (Figure 9.41). It is sufficient to identify the two notable points

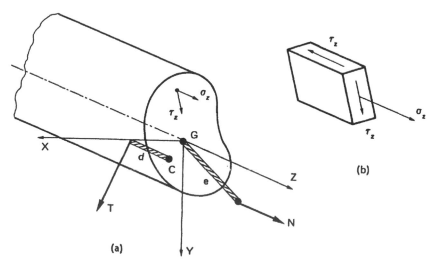

Figure 9.40

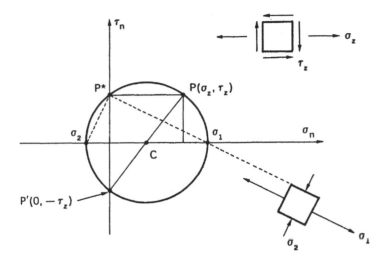

Figure 9.41

$P(\sigma_z, \tau_z)$ and $P'(0, -\tau_z)$, where the stress $\tau_z > 0$ if, when applied to the plane normal to the axis Z, it tends to cause the element to rotate in a clockwise direction. The intersection of the diameter PP' with the axis σ_n gives the centre of Mohr's circle (Figure 9.41). The intersection of the horizontal through P with the vertical through P' defines the pole P^*, while the lines joining the pole P^* with the points of intersection of the circumference with the axis σ_n define the principal directions of stress. From the graphical construction, it may be noted how one of the two principal stresses is always negative and thus compressive, while the other is always positive and thus tensile.

The application of Tresca's criterion (criterion of maximum shearing stress), gives the following condition (Figure 9.41):

$$\tau_{\max} = \overline{PC} = \sqrt{\left(\frac{\sigma_z}{2}\right)^2 + \tau_z^2} < \frac{1}{2}\sigma_P \qquad (9.195)$$

where σ_P is the uniaxial yielding stress. From equation (9.195) we obtain then the following strength condition for the beams:

$$\sqrt{\sigma_z^2 + 4\,\tau_z^2} < \sigma_P \qquad (9.196)$$

On the other hand, Von Mises' criterion (criterion of the maximum energy of distortion), via the general equation (8.114) gives

$$\sqrt{\sigma_z^2 + 3\,\tau_z^2} < \sigma_P \qquad (9.197)$$

From a comparison between the inequalities (9.196) and (9.197), we verify again how Tresca's criterion is more conservative than that of Von Mises, in that it presents the factor 4 as against the factor 3.

There follow a number of strength analyses for cross sections already introduced in Chapter 2.

281

EXAMPLE 1

Let us reconsider the cross section of Figure 2.18, subjected to an eccentric axial force of compression (combined compression and bending). Since the eccentricities in the central reference system are (Figure 9.42)

$$e_x = 43.54 \text{ mm}, \ e_y = 29.30 \text{ mm}$$

and the corresponding central radii of gyration

$$\rho_x^2 = I_x/A = 423.16 \text{ mm}^2$$
$$\rho_y^2 = I_y/A = 96.16 \text{ mm}^2$$

equation (9.49) for the neutral axis is as follows:

$$1 + \frac{29.30}{423.16}y + \frac{43.54}{96.16}x = 0$$

or, in segmentary form

$$\frac{y}{14.44} + \frac{x}{2.21} = -1$$

Thus on the basis of equation (9.48) point A, having coordinates $x_A = 15.84$ mm, $y_A = 41.08$ mm, is subjected to the compressive stress $\sigma_A = -17.88$ kg/mm^2,

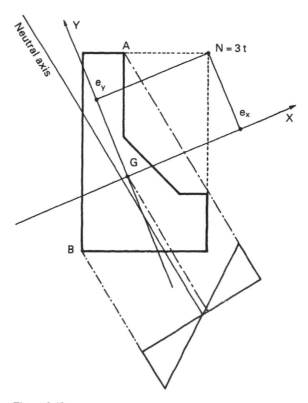

Figure 9.42

282

whilst point B, having coordinates $x_B = -25.00$ mm, $y_B = -17.32$ mm, is subjected to the tensile stress $\sigma_B = 18.67$ kg/mm^2. Both stresses are lower in absolute value than the yield stress of steel, $\sigma_P = 24$ kg/mm^2.

EXAMPLE 2

Let us reconsider the cross section of Figure 2.19, subjected to a combination of compression and bending. Since

$e_x = 0$, $e_y = 35.76$ mm

$\rho_x^2 = I_x/A = 267$ mm^2

equation (9.50a) of the neutral axis is

$y = -7.5$ mm

The maximum stress in absolute value is thus

$$\sigma_{max} = \frac{5000\,kg}{1996\,mm^2} \times \frac{35.76 + 7.5}{7.5} = 14.42\,kg/mm^2$$

and proves to be less than $\sigma_P = 24$ kg/mm^2.

EXAMPLE 3

Let us reconsider the cross section of Figure 2.20, subjected to axial force and twisting moment. The axial force produces a uniform normal stress

$$\sigma_z = \frac{N}{A} = \frac{1500 \times 10^3\,kg}{1260\,cm^2} = 12\,kg/mm^2$$

while the twisting moment, if we neglect the contribution of the flanges, produces a uniform shearing stress

$$\tau_z = \frac{M_t}{2\Omega\delta} = \frac{40 \times 10^3\,kg \times 10^2\,cm}{12\,cm \times 1082\,cm^2} = 3\,kg/mm^2$$

furnished by the Bredt formula (9.136).

Applying Tresca's criterion (9.196), we obtain

$$\sqrt{\sigma_z^2 + 4\tau_z^2} = \sqrt{12^2 + 4 \times 3^2} = 13.4\,kg/mm^2 < 24\,kg/mm^2$$

The yield strength test thus gives a positive response.

EXAMPLE 4

Let us consider again the cross section of Figure 2.21, subjected to a combined loading of shear-torsion. The twisting moment is $M_z = 400\,t \times 0.4\,m = 160$ tm, and produces a counterclockwise uniform shearing stress equal to

$$\tau_z(M_z) = \frac{160 \times 10^3\,kg \times 10^2\,cm}{10\,cm \times (2513 + 1600)\,cm^2} = 3.89\,kg/mm^2$$

On the other hand, for the calculation of stresses due to shear, it is necessary to evaluate the static moment of the part of cross section which remains above the central axis ξ.

283

In Chapter 2 the static moment of the circular segment (4) with respect to the X axis has already been evaluated:

$$S_X^{(4)} = 22\,600 \text{ cm}^3$$

so that

$$S_\xi^{(4)} = S_x^{(4)} - A^{(4)}y_G = 22\,600 - 444 \times 24.61 = 11\,673 \text{ cm}^3$$

The two rectilinear segments above the axis ξ present the following static moment:

$$S_\xi^{(2')} = S_\xi^{(3')} = \frac{1}{2}\,(40 - 24.61)^2 \times 5 = 592 \text{ cm}^3.$$

Altogether we have

$$S_\xi^{\max} = (11\,673 + 2 \times 592) \text{ cm}^3 = 12\,857 \text{ cm}^3.$$

The maximum shearing stress due to shear is therefore

$$\tau_z(T) = \frac{TS_\xi^{\max}}{2I_\xi\delta} = \frac{400 \times 10^3 \text{ kg} \times 12\,857\,\text{cm}^3}{624\,000\,\text{cm}^4 \times 10\,\text{cm}} = 8.24 \text{ kg/mm}^2$$

The maximum shearing stress is found on the left-hand vertical segment, in correspondence with the global centroid

$$\tau_{\max} = \tau_z(M_z) + \tau_z(T) = 12.13 \text{ kg/mm}^2$$

Applying Tresca's criterion (9.196), we find

$$\sqrt{4\tau_{\max}^2} = \sqrt{4 \times 12.13^2} = 24.26 \text{ kg/mm}^2 > \sigma_P$$

whereas Von Mises' criterion (9.197) yields a positive result:

$$\sqrt{3\tau_{\max}^2} = \sqrt{3 \times 12.13^2} = 21 \text{ kg/mm}^2 < \sigma_P$$

EXAMPLE 5

Let us consider again the cross section of Figure 2.22, subjected to biaxial shear. The corresponding components are equal to

$$T_\eta \simeq 193 \text{ t}, \, T_\xi \simeq 52.4 \text{ t}$$

The maximum static moment with respect to the ξ axis is (Figure 9.43(a))

$$S_\xi^{\max} = (160 \times 20 \times 0.96 + 120 \times 13.58) \text{ cm}^3 = 4701 \text{ cm}^3$$

so that the maximum shearing stress due to T_η develops in the centre and is equal to

$$\tau_z(T_\eta) = \frac{T_\eta S_\xi^{\max}}{I_\xi\delta} = \frac{193 \times 10^3 \text{ kg} \times 4701\,\text{cm}^3}{204\,000\,\text{cm}^4 \times 4\,\text{cm}} = 11.12 \text{ kg/mm}^2$$

The shearing stress in the central point due to T_ξ, on the other hand, is in the opposite sense (Figure 9.43(b)), and tends to mitigate the previous contribution. The test is thus positive.

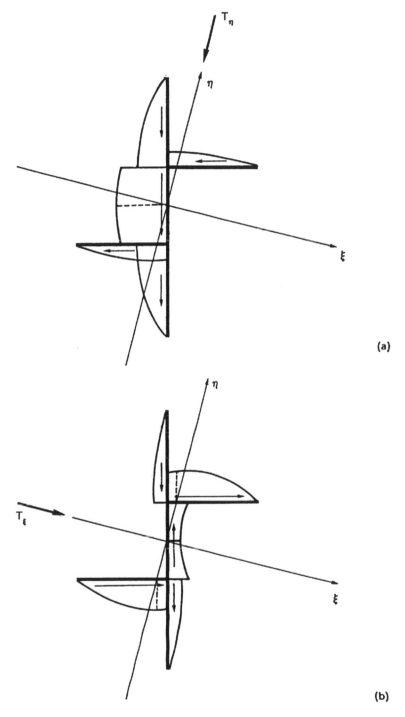

(a)

(b)

Figure 9.43

285

10 Beams and plates in flexure

10.1 Introduction

In this chapter deflected beams and plates are studied. It is shown that, for both elements, it is possible to define the deformation characteristics by derivation of the components of the generalized displacement vector, which, in addition to displacements in the strict sense, also presents rotations. Analogously with the three-dimensional solid encountered in Chapter 8, it may be noted that, also for the one- or two-dimensional solid, the static operator is, but for the algebraic sign, the transpose, or rather the adjoint, of the kinematic operator. This property of duality proves of great utility in the case of discretization into finite elements (Chapter 11).

The cases of curved plane beams and shells having double curvature present notable analogies with the cases of rectilinear plane beams and plates respectively, once the rotation matrix, which converts the external reference system to the local reference system, is included in the analysis. A further noteworthy analogy is found between the differential equation of the elastic line for rectilinear beams and the equation of the elastic plane for plates. Both equations, in fact, neglect the shearing deformability and turn out to be of the fourth order in the single unknown function consisting of deflection or transverse displacement.

Two particular cases of application of the equation of the elastic line are represented by the beam on an elastic foundation and by the beam subjected to steady free oscillations.

10.2 Technical theory of beams

The results obtained for the Saint Venant solid on the basis of restrictive hypotheses, as regards both the geometry of the solid (rectilinear axis, constant cross section) and external loads (lateral surface not loaded, zero body forces), are usually extended in technical applications to cases in which these hypotheses are not satisfied. This means that the stresses and strains in the cross sections of beams are usually calculated using formulas obtained in the foregoing chapter, by introducing, instead of the loads acting on the end planes of the Saint Venant solid, the internal beam reactions acting in the cross sections considered. This extrapolation is also made in the case of

1. beams having non-rectilinear axes;
2. beams of variable cross section;
3. beams loaded on the lateral surface.

It is required, in any case, that the radius of curvature of the geometrical axis of the beam should be much greater than the characteristic dimensions of the cross section, and that the cross section should, at most, be only slightly variable.

10.3 Beams with rectilinear axes

Let us consider an elementary portion of a beam with rectilinear axis and a cross section that is symmetrical with respect to the Y axis. Let this portion be subjected to bending moment M_x and to shear T_y. Deformations due to these two characteristics will produce relative displacements between the centroids of the two extreme cross sections of the beam portion, exclusively in the direction of the Y axis. In the case of the shear T_y, we have (Figure 10.1 (a))

$$dv^T = \gamma_y \, dz \tag{10.1a}$$

where dv^T is the relative displacement in the Y direction due to the shear, γ_y is the shearing strain dual of the shearing force and dz is the length of the infinitesimal element of beam. In the case of bending moment, and considering the rotation φ_x of the element, we have (Figure 10.1 (b))

$$dv^M = -\varphi_x \, dz \tag{10.1b}$$

having neglected the infinitesimals of a higher order due to the curvature, i.e. to the slope variation $d\varphi_x$ (Figure 10.1 (c)),

$$d(dv^M) = -\frac{1}{2} d\varphi_x \, dz \tag{10.1c}$$

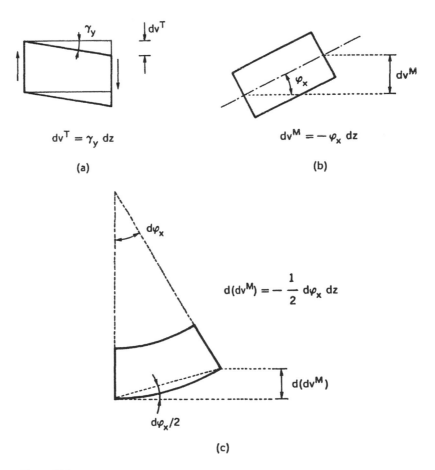

$$dv^T = \gamma_y \, dz$$

(a)

$$dv^M = -\varphi_x \, dz$$

(b)

$$d(dv^M) = -\frac{1}{2} d\varphi_x \, dz$$

(c)

Figure 10.1

287

Summing up the two significant contributions of the shear, equation (10.1a), and of the bending moment, equation (10.1b), we obtain

$$dv = dv^T + dv^M = \gamma_y \, dz - \varphi_x \, dz \tag{10.2}$$

from which

$$\frac{dv}{dz} = \gamma_y - \varphi_x \tag{10.3}$$

At this point we are able to formulate the fundamental equations of the elastic problem for one-dimensional solids with a rectilinear axis. The kinematic equations constitute, as in the case of the three-dimensional solid, the definition of the characteristics of deformation as functions of the generalized displacements

$$
\begin{bmatrix}
\gamma_x \\
\gamma_y \\
\varepsilon_z \\
\chi_x \\
\chi_y \\
\Theta
\end{bmatrix}
=
\begin{bmatrix}
\dfrac{d}{dz} & 0 & 0 & 0 & -1 & 0 \\
0 & \dfrac{d}{dz} & 0 & +1 & 0 & 0 \\
0 & 0 & \dfrac{d}{dz} & 0 & 0 & 0 \\
0 & 0 & 0 & \dfrac{d}{dz} & 0 & 0 \\
0 & 0 & 0 & 0 & \dfrac{d}{dz} & 0 \\
0 & 0 & 0 & 0 & 0 & \dfrac{d}{dz}
\end{bmatrix}
\begin{bmatrix}
u \\
v \\
w \\
\varphi_x \\
\varphi_y \\
\varphi_z
\end{bmatrix}
\tag{10.4}
$$

where among the components of the deformation vector appear the shearing strains, γ_x, γ_y, the axial dilation ε_z, the curvatures, χ_x, χ_y, and the unit angle of torsion, Θ, whilst among the components of the displacement vector appear, in addition to the ordinary components, u, v, w, also the generalized components, φ_x, φ_y, φ_z, i.e. the angles of rotation about the reference axes. The transformation matrix is differential and shows on the diagonal the total derivative d/dz, while the off-diagonal terms are all zero except for two, which have absolute values of unity and derive from relation (10.3) and from its analogue

$$\frac{du}{dz} = \gamma_x + \varphi_y \tag{10.5}$$

Also relation (10.4), like relation (8.8), may be written in compact form,

$$\{q\} = [\partial] \{\eta\} \tag{10.6}$$

where $\{q\}$ indicates the vector of the deformation characteristics, $\{\eta\}$ the vector of generalized displacements, and $[\partial]$ the matrix operator of relation (10.4).

288

On the other hand, also the indefinite equations of equilibrium, if relations (5.12) are taken into account, can be presented in matrix form,

$$
\begin{bmatrix}
\dfrac{d}{dz} & 0 & 0 & 0 & 0 & 0 \\
0 & \dfrac{d}{dz} & 0 & 0 & 0 & 0 \\
0 & 0 & \dfrac{d}{dz} & 0 & 0 & 0 \\
0 & -1 & 0 & \dfrac{d}{dz} & 0 & 0 \\
+1 & 0 & 0 & 0 & \dfrac{d}{dz} & 0 \\
0 & 0 & 0 & 0 & 0 & \dfrac{d}{dz}
\end{bmatrix}
\begin{bmatrix}
T_x \\ T_y \\ N \\ M_x \\ M_y \\ M_z
\end{bmatrix}
+
\begin{bmatrix}
q_x \\ q_y \\ p \\ m_x \\ m_y \\ m_z
\end{bmatrix}
=
\begin{bmatrix}
0 \\ 0 \\ 0 \\ 0 \\ 0 \\ 0
\end{bmatrix}
\tag{10.7}
$$

where among the components of the vector of static characteristics appear the shearing forces, T_x, T_y, the axial force N, the bending moments, M_x, M_y, and the twisting moment, M_z, whilst among the components of the vector of external generalized forces appear, in addition to the transverse distributed loads, q_x, q_y, and the axial distributed load, p, also the bending distributed moments, m_x, m_y and the twisting distributed moment, m_z. The matrix operator presents the total derivative d/dz in all diagonal positions, while the off-diagonal terms are all zero, except for two which are equal to unity and express the equality of the shear with the derivative of the corresponding bending moment (neglecting the distributed moments, m_x, m_y). The matrix operator (10.7) is equal to the transpose of the operator (10.4), but for the finite terms which change algebraic sign. This is said to be the **adjoint operator** of the previous one, and *vice versa*. The reason why the finite terms change sign will emerge clearly in the next chapter. In compact form we can write

$$
[\partial]^*\{Q\} + \{\mathscr{F}\} = \{0\}
\tag{10.8}
$$

where $\{Q\}$ is the vector of static characteristics, and $\{\mathscr{F}\}$ is the vector of external forces.

It should be noted that, unlike in the case of the three-dimensional solid, the matrices $[\partial]$ and $[\partial]^*$ are square (6×6). This means, as we already know, that it is possible to determine the internal reactions of a statically determinate beam by using only the relations of equilibrium. This, unfortunately, does not occur in the case of a three-dimensional body constrained isostatically, the stress field of which may be determined only by applying, in addition to the static equations, also the constitutive and kinematic equations.

On the other hand, also in the case of beams constrained in a redundant manner (i.e. statically indeterminate beams), it is necessary to have recourse to the constitutive equations to define their static characteristics, in addition, of course, to the deformations and the displacements. The relations which link the static characteristics with the dual characteristics of deformation may be presented in matrix form:

$$
\begin{bmatrix} \gamma_x \\ \gamma_y \\ \varepsilon_z \\ \chi_x \\ \chi_y \\ \Theta \end{bmatrix} =
\begin{bmatrix}
\frac{t_x}{GA} & \frac{t_{xy}}{GA} & 0 & 0 & 0 & 0 \\
\frac{t_{xy}}{GA} & \frac{t_y}{GA} & 0 & 0 & 0 & 0 \\
0 & 0 & \frac{1}{EA} & 0 & 0 & 0 \\
0 & 0 & 0 & \frac{1}{EI_x} & 0 & 0 \\
0 & 0 & 0 & 0 & \frac{1}{EI_y} & 0 \\
0 & 0 & 0 & 0 & 0 & \frac{1}{GI_t}
\end{bmatrix}
\begin{bmatrix} T_x \\ T_y \\ N \\ M_x \\ M_y \\ M_z \end{bmatrix}
\tag{10.9}
$$

Expressed in compact form, they are

$$
\{q\} = [H]^{-1}\{Q\} \tag{10.10a}
$$

where $[H]^{-1}$ represents the inverse of the Hessian matrix of the elastic potential of the beam. On the other hand, the inverse relation of equation (10.10a) also holds:

$$
\{Q\} = [H]\{q\} \tag{10.10b}
$$

Applying Clapeyron's Theorem, we obtain the work of deformation per unit length of the beam, i.e. the elastic potential of the beam,

$$
\frac{dL}{dz} = \frac{1}{2}\{Q\}^T\{q\} \tag{10.11}
$$

which, on the basis of equation (10.10a), becomes

$$
\frac{dL}{dz} = \frac{1}{2}\{Q\}^T[H]^{-1}\{Q\} \tag{10.12a}
$$

or, on the basis of equation (10.10b)

$$
\frac{dL}{dz} = \frac{1}{2}\{q\}^T[H]\{q\} \tag{10.12b}
$$

where we have used the relation of symmetry $[H]^T = [H]$. Rendering relation (10.12a) explicit, we obtain

$$
\frac{dL}{dz} = \frac{1}{2}\left(t_x\frac{T_x^2}{GA} + t_y\frac{T_y^2}{GA} + 2t_{xy}\frac{T_xT_y}{GA} + \frac{N^2}{EA} + \frac{M_x^2}{EI_x} + \frac{M_y^2}{EI_y} + \frac{M_z^2}{GI_t} \right) \tag{10.13}
$$

which is a quadratic form of the static characteristics. In the case where the cross section of the beam presents at least one axis of symmetry, the mutual factor of shear, t_{xy}, vanishes and the total work is equal to the sum of the contributions of the single characteristics, the Principle of Superposition being applicable in this case.

Having now at our disposal the equations of kinematics and statics and the constitutive equations for the beam, we can obtain **Lamé's equation in operator form**

$$([\partial]^*[H][\partial])\{\eta\} = -\{\mathcal{F}\} \qquad (10.14)$$

The matrix and differential operator of the second order in round brackets can be called the **Lamé operator**

$$[\mathcal{L}] = [\partial]^* \ [H] \ [\partial] \qquad (10.15)$$
$$\underset{(6 \times 6)}{} \quad \underset{(6 \times 6)(6 \times 6)(6 \times 6)}{}$$

It turns out to be a (6×6) matrix and, in non-homogeneous problems, in which the matrix $[H]$ is a function of the axial coordinate z, is also a function of z.

Finally, the boundary conditions may be conditions of equivalence at the ends,

$$[\mathcal{N}]^{\mathrm{T}}\{Q\} = \{Q_0\} \qquad (10.16)$$

where the matrix $[\mathcal{N}]^{\mathrm{T}}$ coincides with the identity matrix [1], and causes a value of unity to correspond to each differential term of the matrix $[\partial]^*$. The boundary conditions can, on the other hand, also represent displacements imposed at the ends:

$$\{\eta\} = \{\eta_0\} \qquad (10.17)$$

In conclusion, the **elastic problem of the rectilinear beam** can be summarized as follows:

$$[\mathcal{L}]\{\eta\} = -\{\mathcal{F}\}, \qquad \text{for } 0 < z < l \qquad (10.18a)$$

$$([H][\partial])\{\eta\} = \{Q_0\}, \qquad \text{for } z = 0, l \qquad (10.18b)$$

$$\{\eta\} = \{\eta_0\}, \qquad \text{for } z = 0, l \qquad (10.18c)$$

where the static boundary condition (10.18b) holds good in the end point or points that are subjected to loading, as does the kinematic boundary condition (10.18c) in the constrained end point or points. In the case where there are no conditions on the displacements, the loads $\{Q_0\}$ must constitute a self-balancing system.

The previously developed formulation is notably simplified in the case where the beam is **loaded in the plane**. If we assume a cross section symmetrical with respect to the Y axis, the characteristics of deformation reduce to the shearing strain γ_y, the axial dilation ε_z and the curvature χ_x, so that of the relations (10.4), (10.7) and (10.9) only those corresponding to the second, third and fourth rows and columns remain significant. In particular, the **kinematic equations** simplify as follows:

$$\begin{bmatrix} \gamma_y \\ \varepsilon_z \\ \chi_x \end{bmatrix} = \begin{bmatrix} \dfrac{d}{dz} & 0 & +1 \\ 0 & \dfrac{d}{dz} & 0 \\ 0 & 0 & \dfrac{d}{dz} \end{bmatrix} \begin{bmatrix} v \\ w \\ \varphi_x \end{bmatrix} \qquad (10.19)$$

291

The **static equations** likewise simplify thus:

$$\begin{bmatrix} \dfrac{d}{dz} & 0 & 0 \\[2mm] 0 & \dfrac{d}{dz} & 0 \\[2mm] -1 & 0 & \dfrac{d}{dz} \end{bmatrix} \begin{bmatrix} T_y \\ N \\ M_x \end{bmatrix} + \begin{bmatrix} q \\ p \\ m \end{bmatrix} = \begin{bmatrix} 0 \\ 0 \\ 0 \end{bmatrix} \qquad (10.20)$$

The **constitutive equations** reduce then to a diagonal relation

$$\begin{bmatrix} \gamma_y \\ \varepsilon_z \\ \chi_x \end{bmatrix} = \begin{bmatrix} \dfrac{t_y}{GA} & 0 & 0 \\[2mm] 0 & \dfrac{1}{EA} & 0 \\[2mm] 0 & 0 & \dfrac{1}{EI_x} \end{bmatrix} \begin{bmatrix} T_y \\ N \\ M_x \end{bmatrix} \qquad (10.21)$$

10.4 Plane beams with curvilinear axes

Let us consider once more the element of beam with curvilinear axis of Figure 5.3. The curvilinear coordinate s is considered as increasing as we proceed from left to right along the beam, while the angle $d\vartheta$ is considered to be positive if it is counterclockwise. In accordance with the above conventions, also the radius of curvature r acquires an algebraic sign on the basis of the relation

$$ds = r \, d\vartheta \qquad (10.22)$$

As regards the generalized displacements of the generic cross section, the radial displacement v is positive if it is in the positive direction of the Y^* axis (where Y^*Z^* is a system of right-handed axes travelling along the axis of the beam), the axial displacement w is positive if it is in the positive direction of the curvilinear coordinate s, and finally the **variation of the angle** φ is positive if it is counterclockwise (Figure 10.2 (a)).

We shall now show how the kinematic equation (10.19) must be modified to take into account the **intrinsic curvature** of the beam. The axial displacement w produces, in fact, a slope variation $\varphi(w)$, which, in accordance with the

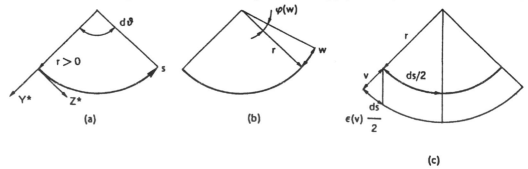

(a)   (b)   (c)

Figure 10.2

scheme of Figure 10.2(b), and neglecting infinitesimals of a higher order, equals

$$\varphi(w) = \frac{w}{r} \tag{10.23}$$

On the other hand, the radial displacement v produces an axial dilation $\varepsilon(v)$ which, in accordance with the scheme of Figure 10.2(c), and neglecting infinitesimals of a higher order, is given by

$$\varepsilon(v) = \frac{v}{r} \tag{10.24}$$

As a consequence of an infinitesimal relative rotation $d\varphi$ of the extreme cross sections of the beam element of Figure 5.3, the angle between the sections can be obtained as the sum $(d\vartheta + d\varphi)$ of the initial and intrinsic relative rotation and the elastic and flexural relative rotation. The new curvature is then

$$\chi_{\text{total}} = \frac{(d\vartheta + d\varphi)}{ds} \tag{10.25}$$

so that the **variation of curvature** is

$$\chi = \chi_{\text{total}} - \frac{1}{r} = \frac{d\varphi}{ds} \tag{10.26}$$

Substituting the relations (10.23), (10.24) and (10.26), which furnish respectively the slope variation to be deducted, the additional axial dilation and the variation of curvature, the **kinematic equations** (10.19) transform as follows:

$$\begin{bmatrix} \gamma \\ \varepsilon \\ \chi \end{bmatrix} = \begin{bmatrix} \dfrac{d}{ds} & -\dfrac{1}{r} & 1 \\ \dfrac{1}{r} & \dfrac{d}{ds} & 0 \\ 0 & 0 & \dfrac{d}{ds} \end{bmatrix} \begin{bmatrix} v \\ w \\ \varphi \end{bmatrix} \tag{10.27}$$

The **indefinite equations of equilibrium**, or **static equations**, have previously been derived in Section 5.2. Equations (5.4) may be reproposed in matrix form:

$$\begin{bmatrix} \dfrac{d}{ds} & -\dfrac{1}{r} & 0 \\ \dfrac{1}{r} & \dfrac{d}{ds} & 0 \\ -1 & 0 & \dfrac{d}{ds} \end{bmatrix} \begin{bmatrix} T \\ N \\ M \end{bmatrix} + \begin{bmatrix} q \\ p \\ m \end{bmatrix} = \begin{bmatrix} 0 \\ 0 \\ 0 \end{bmatrix} \tag{10.28}$$

It should be noted that, but for the algebraic signs of the non-differential terms, the static matrix is the transpose of the kinematic one.

Finally, as regards the **constitutive equations**, if the radius of curvature is much greater than the characteristic dimensions of the cross section, equation (10.21) can be used to very good approximation.

The rotation matrix which transforms the global reference system YZ into the local reference system Y^*Z^* is the following:

$$[N] = \begin{bmatrix} \cos\vartheta & \sin\vartheta & 0 \\ -\sin\vartheta & \cos\vartheta & 0 \\ 0 & 0 & 1 \end{bmatrix} \qquad (10.29)$$

so that the vectors of the external forces and of the generalized displacements in the local reference system may be expressed by premultiplying the respective vectors evaluated in the global reference system by the matrix $[N]$

$$\{\mathscr{F}^*\} = [N]\{\mathscr{F}\} \qquad (10.30a)$$

$$\{\eta^*\} = [N]\{\eta\} \qquad (10.30b)$$

The static and kinematic equations can thus be expressed as follows:

$$[\partial]^*\{Q\} + \{\mathscr{F}^*\} = \{0\} \qquad (10.31a)$$

$$\{q\} = [\partial]\{\eta^*\} \qquad (10.31b)$$

which, on the basis of equations (10.30), become

$$[\partial]^*\{Q\} + [N]\{\mathscr{F}\} = \{0\} \qquad (10.32a)$$

$$\{q\} = [\partial][N]\{\eta\} \qquad (10.32b)$$

Substituting equations (10.10b) and (10.32b) in equation (10.32a), we have

$$[\partial]^*[H][\partial][N]\{\eta\} + [N]\{\mathscr{F}\} = \{0\} \qquad (10.33)$$

Premultiplying both sides of equation (10.33) by $[N]^\mathrm{T}$, we obtain finally

$$\left([N]^\mathrm{T}[\partial]^*[H][\partial][N]\right)\{\eta\} = -\{\mathscr{F}\} \qquad (10.34)$$

which is Lamé's equation for curved beams and arches.

The **elastic problem for curved beams and arches** can then be summarized as follows:

$$[\mathscr{L}]\{\eta\} = -\{\mathscr{F}\}, \qquad \text{for } 0 < s < l \qquad (10.35a)$$

$$\left([N]^\mathrm{T}[H][\partial][N]\right)\{\eta\} = \{Q_0\}, \qquad \text{for } s = 0, l \qquad (10.35b)$$

$$\{\eta\} = \{\eta_0\}, \qquad \text{for } s = 0, l \qquad (10.35c)$$

10.5 Differential equation of the elastic line

As regards beams with rectilinear axes and cross sections that are symmetrical with respect to the Y axis, loaded in the plane of symmetry YZ, we shall arrive at a differential equation in the unknown function $v(z)$, called **deflection** or **transverse displacement**, by neglecting the contributions of deformation due to shear, which, for sufficiently slender beams, are much less than the contributions of deformation due to bending moment.

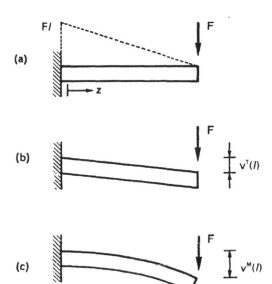

Figure 10.3

Let us consider a cantilever beam of length l, loaded at the free end by a concentrated force F (Figure 10.3 (a)). The increment of vertical displacement due to shear is (Figure 10.3 (b))

$$dv^T = \gamma_y \, dz \tag{10.36}$$

so that the vertical displacement at the free end due to shear is

$$v^T(l) = \gamma_y l \tag{10.37}$$

On the other hand, the increment of vertical displacement due to bending moment is (Figure 10.3 (c))

$$dv^M = -\varphi_x \, dz \tag{10.38}$$

Differentiating both sides of equation (10.38) with respect to z, we obtain

$$\frac{d^2 v^M}{dz^2} = -\frac{d\varphi_x}{dz} \tag{10.39}$$

Recalling the expressions of elastic curvature (9.40) and (9.41), we obtain a second order differential equation

$$\frac{d^2 v^M}{dz^2} = -\frac{M_x}{EI_x} \tag{10.40}$$

The bending moment M_x, as well as the elastic characteristic E and the geometrical characteristic I_x, can be functions of the z coordinate. Usually, however, cases are considered where the material and the cross section do not vary

295

along the axis of the beam. The bending moment, in the case of the cantilever beam of Figure 10.3(a), instead varies linearly,

$$M_x = F(z - l) \tag{10.41}$$

so that the differential equation (10.40) becomes

$$\frac{d^2 v^M}{dz^2} = \frac{F}{EI_x}(l - z) \tag{10.42}$$

and produces the complete integral

$$v^M(z) = -\frac{F}{6\,EI_x}z^3 + \frac{Fl}{2\,EI_x}z^2 + C_1 z + C_2 \tag{10.43}$$

The two constants C_1 and C_2 are obtained by applying the boundary conditions. The built-in constraint does not allow any translations or rotations of the fixed-end section of the beam, so that

$$v^M(0) = C_2 = 0 \tag{10.44a}$$

$$\frac{dv^M}{dz}(0) = C_1 = 0 \tag{10.44b}$$

and the vertical displacement at the free end is therefore (Figure 10.3 (c))

$$v^M(l) = \frac{Fl^3}{3\,EI_x} \tag{10.45}$$

From relations (10.37) and (10.45) it is already evident that the vertical displacement due to shear is a linear function of the length l of the beam, while the vertical displacement due to bending moment is proportional to the third power of l. More particularly, it is possible to sum up the two contributions and write

$$v(l) = \frac{Fl^3}{3\,EI_x}\left[1 + 6(1 + v)t_y \left(\frac{\rho_x}{l}\right)^2\right] \tag{10.46}$$

where ρ_x indicates the radius of gyration of the cross section with respect to the X axis. It is hence possible to understand how, since we normally have $\rho_x \ll l$, the contribution of shear is negligible.

The **differential equation of the elastic line** thus coincides substantially with equation (10.40)

$$\frac{d^2 v}{dz^2} = -\frac{M_x}{EI_x} \tag{10.47}$$

If in a cross section within the beam the bending moment vanishes, also the curvature vanishes and, in accordance with equation (10.47), the second derivative of the vertical displacement v vanishes. In correspondence with that cross section the elastic deformed configuration (or elastic line) of the axis of the beam will thus show a point of inflection.

Differentiating both sides of equation (10.47) with respect to z, we obtain

296

$$\frac{d^3v}{dz^3} = -\frac{T_y}{EI_x} \qquad (10.48)$$

which is a third order differential equation that has as its known term the shear function. Differentiating again we finally obtain

$$\frac{d^4v}{dz^4} = \frac{q}{EI_x} \qquad (10.49)$$

which is the alternative version of the equation of the elastic line, of the fourth order, with the known term proportional to the transverse distributed load $q(z)$. This version requires a more laborious integration, with the identification of four arbitrary constants, on the basis of the static and kinematic boundary conditions. To compensate for this, on the other hand, it is not necessary to determine beforehand the bending moment function $M_x(z)$.

From equation (10.49) it may be deduced how a discontinuity in the transverse distributed load function $q(z)$ causes a discontinuity on the fourth derivative of the transverse displacement $v(z)$, and thus a relatively minor discontinuity. In like manner and considering equation (10.48), we may deduce how a discontinuity in the shear function $T_y(z)$, and thus a concentrated transverse load, causes a discontinuity on the third derivative of the transverse displacement $v(z)$. From equation (10.47) we can deduce how a discontinuity in the bending moment function $M_x(z)$, and thus a concentrated moment, causes a discontinuity on the second derivative of the transverse displacement $v(z)$. To have, instead, the discontinuities more pronounced in the transverse displacement $v(z)$, i.e. the discontinuity on the first derivative, or relative rotation, and the discontinuity on the function $v(z)$ itself, or relative sliding, the appropriate disconnections are necessary, represented in one case by the hinge and in the other by the double rod parallel to the axis of the beam. Figure 10.4 provides a synthesis of the various cases presented previously.

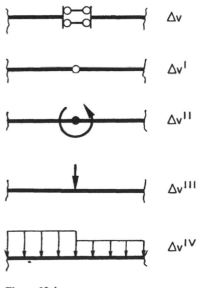

Figure 10.4

297

10.6 Notable displacements and rotations in elementary schemes

Here we shall discuss displacements and rotations of notable cross sections in elementary structural schemes. To this end, we shall proceed to integrate the differential equation of the elastic line introduced previously, considering at the same time the specific boundary conditions of constraint.

As a first case, let us examine the cantilever beam *AB*, built in at *A* and subjected to a concentrated moment at point B (Figure 10.5(a)). Since the bending moment is constant along the axis of the cantilever and also negative, equation (10.47) takes the following form:

$$\frac{d^2v}{dz^2} = \frac{m}{EI} \tag{10.50}$$

the flexural rigidity *EI* also being considered constant. The complete integral of equation (10.50) is thus

$$v(z) = \frac{m}{EI}\frac{z^2}{2} + C_1 z + C_2 \tag{10.51}$$

the two constants C_1 and C_2 being obtainable by means of the two boundary conditions

$$v(0) = v'(0) = 0 \tag{10.52}$$

which express the vanishing of the vertical displacement and of the rotation at the built-in constraint, respectively. We obtain

$$C_1 = C_2 = 0 \tag{10.53}$$

and thus the vertical displacement at the end *B* is

$$v_B = v(l) = \frac{ml^2}{2\,EI} \tag{10.54a}$$

while the rotation at the same end is

$$\varphi_B = -v'(l) = -\frac{ml}{EI} \tag{10.54b}$$

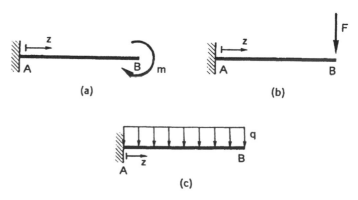

(a)　(b)

(c)

Figure 10.5

A slightly more complex case is that of the same cantilever beam, loaded by a force perpendicular to the axis at the end B (Figure 10.5 (b)). In this case the moment is a function of the axial coordinate z, and therefore the differential equation of the elastic line (10.47) is particularized as follows:

$$\frac{d^2v}{dz^2} = \frac{F(l-z)}{EI} \tag{10.55}$$

A first integration gives

$$v'(z) = -\frac{Fz^2}{2\,EI} + \frac{Fl}{EI}z + C_1 \tag{10.56}$$

and a second one

$$v(z) = -\frac{Fz^3}{6\,EI} + \frac{Fl}{2\,EI}z^2 + C_1 z + C_2 \tag{10.57}$$

As the boundary conditions of equation (10.52) still hold, we obtain finally

$$v_B = \frac{Fl^3}{3\,EI} \tag{10.58a}$$

$$\varphi_B = -\frac{Fl^2}{2\,EI} \tag{10.58b}$$

As our last elementary case for the cantilever beam, let us take a uniform distributed load q (Figure 10.5 (c)). In this case the differential equation is

$$\frac{d^2v}{dz^2} = \frac{q(l-z)^2}{2\,EI} \tag{10.59}$$

which, once integrated, yields

$$v'(z) = \frac{q}{2\,EI}\left(\frac{z^3}{3} - lz^2 + l^2 z + C_1\right) \tag{10.60}$$

$$v(z) = \frac{q}{2\,EI}\left(\frac{z^4}{12} - l\frac{z^3}{3} + l^2\frac{z^2}{2} + C_1 z + C_2\right) \tag{10.61}$$

The application of the conditions of equations (10.52) confirms equations (10.53), and thus

$$v_B = \frac{ql^4}{8\,EI} \tag{10.62a}$$

$$\varphi_B = -\frac{ql^3}{6\,EI} \tag{10.62b}$$

Let us now look at the case of a beam supported at both ends subjected to a concentrated moment at one of the two ends (Figure 10.6 (a)). The equation of the elastic line is

$$\frac{d^2v}{dz^2} = -\frac{mz}{EIl} \tag{10.63}$$

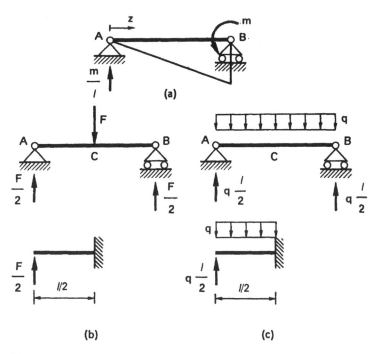

Figure 10.6

with the boundary conditions

$$v(0) = v(l) = 0 \qquad (10.64)$$

Integrating equation (10.63), we obtain

$$v'(z) = \frac{1}{EI}\left(-\frac{m}{2l}z^2 + C_1\right) \qquad (10.65a)$$

$$v(z) = \frac{1}{EI}\left(-\frac{m}{6l}z^3 + C_1 z + C_2\right) \qquad (10.65b)$$

from which, applying equations (10.64), there follows

$$C_1 = \frac{ml}{6} \qquad (10.66a)$$

$$C_2 = 0 \qquad (10.66b)$$

The rotation function (10.65a) thus takes on the following final aspect:

$$v'(z) = \frac{1}{EI}\left(-\frac{m}{2l}z^2 + \frac{ml}{6}\right) \qquad (10.67)$$

whence it is possible to determine the rotations of the end sections:

$$\varphi_A = -v'(0) = -\frac{ml}{6\,EI} \qquad (10.68a)$$

$$\varphi_B = -v'(l) = \frac{ml}{3\,EI} \qquad (10.68b)$$

When the supported beam is loaded symmetrically, it is possible to consider only one half of it, with a built-in constraint which represents kinematically the condition of symmetry and the reactive force in place of the support. This is the case, for instance, of the vertical force in the centre (Figure 10.6 (b)). In this way we find ourselves once again faced with the case of the cantilever beam of Figure 10.5 (b), for which equations (10.58) provide, respectively, the vertical displacement in the centre and the rotations at the ends:

$$v_C = \frac{\dfrac{F}{2}\left(\dfrac{l}{2}\right)^3}{3\,EI} = \frac{Fl^3}{48\,EI} \qquad (10.69a)$$

$$\varphi_B = -\varphi_A = \frac{\dfrac{F}{2}\left(\dfrac{l}{2}\right)^2}{2\,EI} = \frac{Fl^2}{16\,EI} \qquad (10.69b)$$

In like manner, in the case of uniform distributed load (Figure 10.6 (c)), using equations (10.58) and (10.62), we obtain

$$v_C = \frac{q\dfrac{l}{2}\left(\dfrac{l}{2}\right)^3}{3\,EI} - \frac{q\left(\dfrac{l}{2}\right)^4}{8\,EI} = \frac{5}{384}\frac{ql^4}{EI} \qquad (10.70a)$$

$$\varphi_B = -\varphi_A = \frac{q\dfrac{l}{2}\left(\dfrac{l}{2}\right)^2}{2\,EI} - \frac{q\left(\dfrac{l}{2}\right)^3}{6\,EI} = \frac{ql^3}{24\,EI} \qquad (10.70b)$$

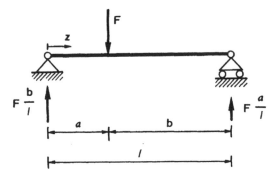

Figure 10.7

301

If the vertical force F is not applied in the centre (Figure 10.7), it is not possible to exploit the properties of symmetry. The bending moment function is piecewise linear:

$$M(z) = \frac{Fb}{l} z, \qquad \text{for } 0 \leqslant z \leqslant a \qquad (10.71a)$$

$$M(z) = \frac{Fb}{l} z - F(z - a), \qquad \text{for } a \leqslant z \leqslant l \qquad (10.71b)$$

Two distinct differential equations must therefore be considered:

$$\frac{d^2 v_1}{dz^2} = -\frac{Fb}{EIl} z, \qquad \text{for } 0 \leqslant z \leqslant a \qquad (10.72a)$$

$$\frac{d^2 v_2}{dz^2} = -\frac{Fb}{EIl} z + \frac{F}{EI}(z - a), \qquad \text{for } a \leqslant z \leqslant l \qquad (10.72b)$$

Integrating equation (10.72a), we obtain

$$\frac{dv_1}{dz} = -\frac{Fb}{2\,EIl} z^2 + C_1 \qquad (10.73a)$$

$$v_1 = -\frac{Fb}{6\,EIl} z^3 + C_1 z + C_2 \qquad (10.73b)$$

whilst, integrating equation (10.72b), we get

$$\frac{dv_2}{dz} = -\frac{Fb}{2\,EIl} z^2 + \frac{F}{2\,EI}(z - a)^2 + C_3 \qquad (10.74a)$$

$$v_2 = -\frac{Fb}{6\,EIl} z^3 + \frac{F}{6\,EI}(z - a)^3 + C_3 z + C_4 \qquad (10.74b)$$

The two conditions of continuity for the rotation and for the displacement in $z = a$,

$$\left(\frac{dv_1}{dz}\right)_{z=a} = \left(\frac{dv_2}{dz}\right)_{z=a} \qquad (10.75a)$$

$$v_1(z = a) = v_2(z = a) \qquad (10.75b)$$

give

$$C_1 = C_3 \qquad (10.76a)$$

$$C_2 = C_4 \qquad (10.76b)$$

while the conditions at the ends,

$$v_1(0) = 0 \qquad (10.77a)$$

$$v_2(l) = 0 \qquad (10.77b)$$

determine the values

$$C_1 = \frac{Fb}{6\,EIl}(l^2 - b^2) \qquad (10.78a)$$

$$C_2 = 0 \qquad (10.78b)$$

302

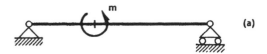

(a)

(b)

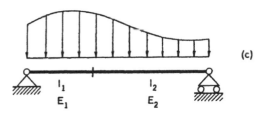

(c)

Figure 10.8

In general it is necessary to resort to the conditions of continuity expressed by equations (10.75) whenever there is a discontinuity or non-regularity in the moment function (Figures 10.8 (a), (b)), or a discontinuity in the flexural rigidity EI of the beam (Figure 10.8 (c)). The latter eventually occurs in the case of an abrupt variation of the cross section, or in the case of a sharp variation in the material of which the beam is made. When these variations occur with continuity, it is instead necessary to integrate the differential equation of the elastic line, considering the continuous functions $E(z)$, $I(z)$.

Let us consider finally the beam with overhanging end of Figure 10.9 (a). Also in this case the bending moment is piecewise linear, and thus it is expedient to consider two unknown functions v_1 (portion AB) and v_2 (portion BC). The differential equations of the elastic line, corresponding to the two portions, are (Figure 10.9 (b))

$$\frac{d^2 v_1}{dz^2} = \frac{Fz}{EI}, \qquad\qquad \text{for } 0 \leqslant z \leqslant l \qquad (10.79\text{a})$$

$$\frac{d^2 v_2}{dz^2} = \frac{Fz}{EI} - \frac{2\,F(z-l)}{EI}, \qquad \text{for } l \leqslant z \leqslant 2l \qquad (10.79\text{b})$$

with the boundary conditions

$$v_1(0) \; = 0 \qquad\qquad\qquad (10.80\text{a})$$
$$v_1(l) \; = 0 \qquad\qquad\qquad (10.80\text{b})$$
$$v_2(l) \; = 0 \qquad\qquad\qquad (10.80\text{c})$$
$$v_1'(l) \; = v_2'(l) \qquad\qquad (10.80\text{d})$$

303

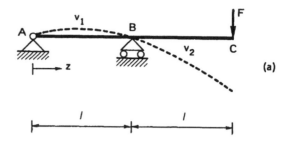

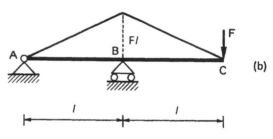

Figure 10.9

Integrating equation (10.79a), we obtain

$$\frac{dv_1}{dz} = \frac{F}{2\,EI}z^2 + C_1 \tag{10.81a}$$

$$v_1 = \frac{F}{6\,EI}z^3 + C_1 z + C_2 \tag{10.81b}$$

while integrating equation (10.79b), we find

$$\frac{dv_2}{dz} = \frac{F}{2\,EI}z^2 - \frac{F}{EI}(z-l)^2 + C_3 \tag{10.82a}$$

$$v_2 = \frac{F}{6\,EI}z^3 - \frac{F}{3\,EI}(z-l)^3 + C_3 z + C_4 \tag{10.82b}$$

By applying equations (10.80), we obtain four equations in the four arbitrary constants C_i, $i = 1, 2, 3, 4$:

$$C_2 = 0 \tag{10.83a}$$

$$\frac{F}{6\,EI}l^3 + C_1 l + C_2 = 0 \tag{10.83b}$$

$$\frac{F}{6\,EI}l^3 + C_3 l + C_4 = 0 \tag{10.83c}$$

304

$$\frac{F}{2\,EI}\,l^2 + C_1 = \frac{F}{2\,EI}\,l^2 + C_3 \qquad\qquad (10.83\text{d})$$

The solution is readily obtained:

$$C_1 = C_3 = -\frac{Fl^2}{6\,EI} \qquad\qquad (10.84\text{a})$$

$$C_2 = C_4 = 0 \qquad\qquad (10.84\text{b})$$

The rotation and the vertical displacement at the end of the overhang can be found on the basis of equations (10.82) and (10.84):

$$\varphi_C = -v_2'(2\,l) = -\frac{5}{6}\frac{Fl^2}{EI} \qquad\qquad (10.85\text{a})$$

$$v_C = v_2(2\,l) = \frac{2}{3}\frac{Fl^3}{EI} \qquad\qquad (10.85\text{b})$$

10.7 Composition of rotations and displacements

As can be seen in the last section, the resolution of the equation of the elastic line is not always immediate, and indeed often involves very laborious calculations. In some cases, as in that of Figure 10.9, it is certainly more convenient to apply the Principle of Superposition, considering separately the effects of the force on the cantilever beam *BC* (Figure 10.10 (a)) and the effects of the

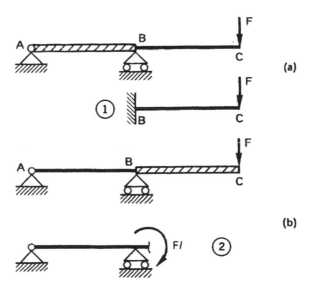

Figure 10.10

internal reactions on the supported beam AB (Figure 10.10 (b)). As regards the cantilever beam BC, we have

$$\varphi_A^{(1)} = 0 \tag{10.86a}$$

$$\varphi_B^{(1)} = 0 \tag{10.86b}$$

$$\varphi_C^{(1)} = -\frac{Fl^2}{2\,EI} \tag{10.86c}$$

$$v_C^{(1)} = \frac{Fl^3}{3\,EI} \tag{10.86d}$$

while for the supported beam, subjected to the moment Fl at the end B, we have

$$\varphi_A^{(2)} = \frac{(Fl)l}{6\,EI} \tag{10.87a}$$

$$\varphi_B^{(2)} = -\frac{(Fl)l}{3\,EI} \tag{10.87b}$$

$$\varphi_C^{(2)} = \varphi_B^{(2)} \tag{10.87c}$$

$$v_C^{(2)} = \left| \varphi_B^{(2)} \right| l \tag{10.87d}$$

whereby finally, summing up the contributions of equations (10.86) and (10.87), we obtain

$$\varphi_A = \frac{Fl^2}{6\,EI} \tag{10.88a}$$

$$\varphi_B = -\frac{Fl^2}{3\,EI} \tag{10.88b}$$

$$\varphi_C = -\frac{5}{6}\frac{Fl^2}{EI} \tag{10.88c}$$

$$v_C = \frac{2}{3}\frac{Fl^3}{EI} \tag{10.88d}$$

Note how equations (10.88c, d) coincide with equations (10.85a, b).

If we intend to determine the rotation and displacement of the midpoint of a cantilever beam of length $2l$, loaded by a force at the end (Figure 10.11 (a)), it

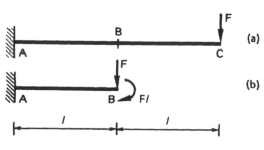

Figure 10.11

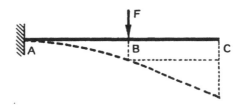

Figure 10.12

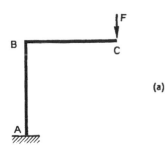

(a)

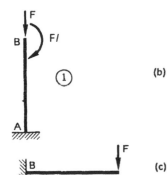

(b)

(c)

Figure 10.13

is sufficient to consider a cantilever beam halved and loaded by the internal characteristics of shear and bending moment (Figure 10.11 (b)). Summation of the contributions yields

$$\varphi_B = -\frac{Fl^2}{2\,EI} - \frac{(Fl)l}{EI} = -\frac{3}{2}\frac{Fl^2}{EI} \tag{10.89a}$$

$$v_B = \frac{Fl^3}{3\,EI} + \frac{(Fl)l^2}{2\,EI} = \frac{5}{6}\frac{Fl^3}{EI} \tag{10.89b}$$

If, instead, the same cantilever beam is loaded in its midpoint (Figure 10.12), and we wish to determine the rotation and displacement of the end cross section, taking into account that the portion BC is not subjected to bending but only to rigid rotation, we have

$$\varphi_C = \varphi_B = -\frac{Fl^2}{2\,EI} \tag{10.90a}$$

$$v_C = v_B + |\varphi_B|l = \frac{Fl^3}{3\,EI} + \frac{Fl^2}{2\,EI}l = \frac{5}{6}\frac{Fl^3}{EI} \tag{10.90b}$$

The displacements (10.89b) and (10.90b) are identical on the basis of Betti's Reciprocal Theorem.

Let us consider the L-shaped cantilever beam of Figure 10.13 (a), subjected to a force acting at the end C, and let us seek to determine the translation and rotation of this cross section. We shall therefore make the summation of the contributions that result from the scheme of Figure 10.13(b), in which the characteristics of axial force and bending moment act on the vertical cantilever AB, with the contributions that emerge from the scheme of Figure 10.13(c), consisting of the overhang BC, built-in at B and loaded with the external force F at C

$$u_B^{(1)} = \frac{(Fl)l^2}{2\,EI} \tag{10.91a}$$

$$\varphi_B^{(1)} = -\frac{(Fl)l}{EI} \tag{10.91b}$$

$$\varphi_C^{(1)} = \varphi_B^{(1)} \tag{10.91c}$$

$$v_C^{(1)} = \left| \varphi_B^{(1)} \right| l \tag{10.91d}$$

$$u_C^{(1)} = u_B^{(1)} \tag{10.91e}$$

$$v_C^{(2)} = \frac{Fl^3}{3\,EI} \tag{10.91f}$$

307

$$\varphi_C^{(2)} = -\frac{Fl^2}{2\,EI} \tag{10.91g}$$

Summing up the corresponding terms, we obtain

$$\varphi_C = \varphi_C^{(1)} + \varphi_C^{(2)} = -\frac{3}{2}\frac{Fl^2}{EI} \tag{10.92a}$$

$$v_C = v_C^{(1)} + v_C^{(2)} = \frac{4}{3}\frac{Fl^3}{EI} \tag{10.92b}$$

$$u_C = u_C^{(1)} = \frac{Fl^3}{2\,EI} \tag{10.92c}$$

As our final example, let us now examine the beam of Figure 10.14 subjected to a concentrated moment at A. The overhang is devoid of loads and hence is not deflected, but only rotates rigidly, under the action of the remainder of the structure. On the hypothesis of small displacements, and applying the rules of kinematics of rigid bodies, we have

$$\varphi_D = \varphi_B = -\frac{ml}{6\,EI} \tag{10.93a}$$

$$v_D = v_C = |\varphi_B|\frac{l}{2} = \frac{ml^2}{12\,EI} \tag{10.93b}$$

$$u_D = |\varphi_B|h = \frac{mlh}{6\,EI} \tag{10.93c}$$

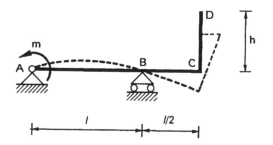

Figure 10.14

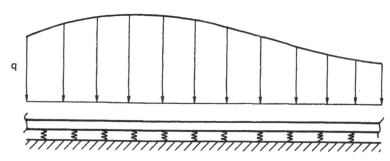

Figure 10.15

10.8 Beam on elastic foundation

We shall now examine the case of a beam supported on a foundation which reacts elastically and bilaterally. Referring to the fact that the reaction of the foundation is assumed to be proportional to the vertical displacement v, we represent this foundation usually as a bed of springs (Figure 10.15). This model provides a significant representation of the case of a beam set in the ground, or of a rail fastened to sleepers. We shall see later in Chapter 12 how this can be of aid in studying cylindrical shells.

If we denote by K the elastic rigidity of the foundation, the differential equation (10.49), which holds in the absence of the foundation, is modified as follows:

$$\frac{d^4v}{dz^4} = \frac{q}{EI} - \frac{K}{EI}\,v \qquad (10.94)$$

where the index x of the moment of inertia has been omitted. In the case where $q = 0$, we have

$$\frac{d^4v}{dz^4} + 4\,\beta^4 v = 0 \qquad (10.95)$$

with

$$\beta = \sqrt[4]{\frac{K}{4\,EI}} \qquad (10.96)$$

The complete integral of equation (10.95) is

$$v(z) = e^{\beta z}(C_1 \cos \beta z + C_2 \sin \beta z) + e^{-\beta z}(C_3 \cos \beta z + C_4 \sin \beta z) \qquad (10.97)$$

where the constants C_i, with $i = 1, 2, 3, 4$, are to be identified via the boundary conditions.

When instead, q is constant and different from zero, the complete integral (10.97) must be supplemented by the particular solution $v = q/K$.

Once the analytical expression of the displacement v is known, it is possible to obtain by derivation the rotation φ, the bending moment M, and the shear T.

In the case of an infinitely long beam resting on an elastic foundation, loaded by a concentrated force F (Figure 10.16), the terms of equation (10.97) that contain the factor $e^{\beta z}$ must vanish because at infinity the displacement vanishes. We shall therefore have $C_1 = C_2 = 0$, so that

$$v(z) = e^{-\beta z}(C_3 \cos \beta z + C_4 \sin \beta z) \qquad (10.98)$$

At the point of application of the force, from symmetry we have then

$$\left(\frac{dv}{dz}\right)_{z=0} = -\beta(C_3 - C_4) = 0 \qquad (10.99)$$

and hence the displacement will be determined but for a single factor

$$v(z) = Ce^{-\beta z}(\cos \beta z + \sin \beta z) \qquad (10.100)$$

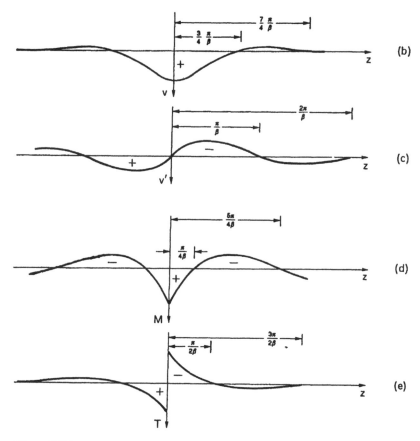

Figure 10.16

This factor C can finally be determined on the basis of the condition of equilibrium to the vertical translation of the entire beam

$$F = 2\int_0^\infty K\upsilon \; dz \tag{10.101}$$

Substituting equation (10.100) in equation (10.101), we obtain

$$F = \frac{2\,KC}{\beta} \tag{10.102}$$

and hence the elastic line is represented by the function

$$\upsilon(z) = \frac{F\beta}{2\,K}\,e^{-\beta z}(\cos \beta z + \sin \beta z) = \frac{F\beta}{2\,K}\,A_{\beta z} \tag{10.103a}$$

By derivation we obtain

$$\varphi = -\frac{dv}{dz} = \frac{F\beta^2}{K}e^{-\beta z}\sin\beta z = \frac{F\beta^2}{K}B_{\beta z} \qquad (10.103b)$$

$$M = -EI\frac{d^2v}{dz^2} = \frac{F}{4\beta}e^{-\beta z}(\cos\beta z - \sin\beta z) = \frac{F}{4\beta}C_{\beta z} \qquad (10.103c)$$

$$T = -EI\frac{d^3v}{dz^3} = -\frac{F}{2}e^{-\beta z}\cos\beta z = -\frac{F}{2}D_{\beta z} \qquad (10.103d)$$

Table 10.1 gives the values of the functions $A_{\beta z}, B_{\beta z}, C_{\beta z}, D_{\beta z}$, as the argument βz varies. The maximum values attained by the transverse displacement, the bending moment and the shear, are found for $z = 0$ (Figure 10.16). The functions (10.103) represent all exponentially smoothing sinusoids. The wavelength λ is defined by the relation

$$\beta\lambda = 2\pi \qquad (10.104)$$

from which we deduce

$$\lambda = \frac{2\pi}{\beta} = 2\pi\sqrt[4]{\frac{4EI}{K}} \qquad (10.105)$$

In the case of multiple concentrated loads, the Principle of Superposition can furnish the displacement and the internal characteristics at each point. For example, when two equal forces are applied, at a distance apart of 1500 mm, on a beam with $\beta = 10^{-3}\text{mm}^{-1}$, we have (Table 10.1)

$$A_{\beta z}(1.5) \simeq 0.23$$
$$C_{\beta z}(1.5) \simeq -0.20$$

whereby the total displacement under each force is increased by approximately 23%,

$$v = \frac{F\beta}{2K}(1 + 0.23)$$

while the total bending moment under each force is reduced by approximately 20%,

$$M = \frac{F}{4\beta}(1 - 0.20)$$

Let us consider finally a semi-infinite beam, loaded at its end by a force F and a moment m (Figure 10.17). Since also in this case the displacement at infinity must be zero, the relation (10.98) will hold with the boundary conditions

$$\left(\frac{d^2v}{dz^2}\right)_{z=0} = -\frac{m}{EI} \qquad (10.106a)$$

$$\left(\frac{d^3v}{dz^3}\right)_{z=0} = \frac{F}{EI} \qquad (10.106b)$$

311

Table 10.1

βz	$A_{\beta z}$	$B_{\beta z}$	$C_{\beta z}$	$D_{\beta z}$	βz	$A_{\beta z}$	$B_{\beta z}$	$C_{\beta z}$	$D_{\beta z}$
0	1.0000	0	1.0000	1.0000	3.6	−0.0366	−0.0121	−0.0124	−0.0245
0.1	0.9907	0.0903	0.8100	0.9003	3.7	−0.0341	−0.0131	−0.0079	−0.0210
0.2	0.9651	0.1627	0.6398	0.8024	3.8	−0.0314	−0.0137	−0.0040	−0.0177
0.3	0.9267	0.2189	0.4888	0.7077	3.9	−0.0286	−0.0140	−0.0008	−0.0147
0.4	0.8784	0.2610	0.3564	0.6174	$5\pi/4$	−0.0278	−0.0140	0	−0.0139
0.5	0.8231	0.2908	0.2415	0.5323	4.0	−0.0258	−0.0139	0.0019	−0.0120
0.6	0.7628	0.3099	0.1431	0.4530	4.1	−0.0231	−0.0136	0.0040	−0.0095
0.7	0.6997	0.3199	0.0599	0.3798	4.2	−0.0204	−0.0131	0.0057	−0.0074
$\pi/4$	0.6448	0.3224	0	0.3224	4.3	−0.0179	−0.0125	0.0070	−0.0054
0.8	0.6354	0.3223	−0.0093	0.3131	4.4	−0.0155	−0.0117	0.0079	−0.0038
0.9	0.5712	0.3185	−0.0657	0.2527	4.5	−0.0132	−0.0108	0.0085	−0.0023
1.0	0.5083	0.3096	−0.1108	0.1988	4.6	−0.0111	−0.0100	0.0089	−0.0011
1.1	0.4476	0.2967	−0.1457	0.1510	4.7	−0.0092	−0.0091	0.0090	−0.0001
1.2	0.3899	0.2807	−0.1716	0.1091	$3\pi/2$	−0.0090	−0.0090	0.0090	0
1.3	0.3355	0.2626	−0.1897	0.0729	4.8	−0.0075	−0.0082	0.0089	0.0007
1.4	0.2849	0.2430	−0.2011	0.0419	4.9	−0.0059	−0.0073	0.0087	0.0014
1.5	0.2384	0.2226	−0.2068	0.0158	5.0	−0.0046	−0.0065	0.0084	0.0019
$\pi/2$	0.2079	0.2079	−0.2079	0	5.1	−0.0033	−0.0057	0.0080	0.0023
1.6	0.1959	0.2018	−0.2077	−0.0059	5.2	−0.0023	−0.0049	0.0075	0.0026
1.7	0.1576	0.1812	−0.2047	−0.0235	5.3	−0.0014	−0.0042	0.0069	0.0028
1.8	0.1234	0.1610	−0.1985	−0.0376	5.4	−0.0006	−0.0035	0.0064	0.0029
1.9	0.0932	0.1415	−0.1899	−0.0484	$7\pi/4$	0	−0.0029	0.0058	0.0029
2.0	0.0667	0.1230	−0.1794	−0.0563	5.5	0.0000	−0.0029	0.0058	0.0029
2.1	0.0439	0.1057	−0.1675	−0.0618	5.6	0.0005	−0.0023	0.0052	0.0029
2.2	0.0244	0.0895	−0.1548	−0.0652	5.7	0.0010	−0.0018	0.0046	0.0028
2.3	0.0080	0.0748	−0.1416	−0.0668	5.8	0.0013	−0.0014	0.0041	0.0027
$3\pi/4$	0	0.0671	−0.1342	−0.0671	5.9	0.0015	−0.0010	0.0036	0.0026
2.4	−0.0056	0.0613	−0.1282	−0.0669	6.0	0.0017	−0.0007	0.0031	0.0024
2.5	−0.0166	0.0492	−0.1149	−0.0658	6.1	0.0018	−0.0004	0.0026	0.0022
2.6	−0.0254	0.0383	−0.1019	−0.0636	6.2	0.0019	−0.0002	0.0022	0.0020
2.7	−0.0320	0.0287	−0.0895	−0.0608	2π	0.0019	0	0.0019	0.0019
2.8	−0.0369	0.0204	−0.0777	−0.0573	6.3	0.0019	+0.0001	0.0018	0.0018
2.9	−0.0403	0.0132	−0.0666	−0.0534	6.4	0.0018	0.0003	0.0015	0.0017
3.0	−0.0423	0.0070	−0.0563	−0.0493	6.5	0.0018	0.0004	0.0012	0.0015
3.1	−0.0431	0.0019	−0.0469	−0.0450	6.6	0.0017	0.0005	0.0009	0.0013
π	−0.0432	0	−0.0432	−0.0432	6.7	0.0016	0.0006	0.0006	0.0011
3.2	−0.0431	−0.0024	−0.0383	−0.0407	6.8	0.0015	0.0006	0.0004	0.0010
3.3	−0.0422	−0.0058	−0.0306	−0.0364	6.9	0.0014	0.0006	0.0002	0.0008
3.4	−0.0408	−0.0085	−0.0237	−0.0323	7.0	0.0013	0.0006	0.0001	0.0007
3.5	−0.0389	−0.0106	−0.0177	−0.0283	$9\pi/4$	0.0012	0.0006	0	0.0006

Figure 10.17

312

From equations (10.98) and (10.106) we obtain

$$C_3 = \frac{F - \beta m}{2\,\beta^3 EI} \qquad\qquad (10.107a)$$

$$C_4 = \frac{m}{2\,\beta^2\, EI} \qquad\qquad (10.107b)$$

Equation (10.98) therefore transforms as follows:

$$v(z) = \frac{e^{-\beta z}}{2\,\beta^3 EI}\big[F\cos\beta z - \beta m(\cos\beta z - \sin\beta z)\big] \qquad (10.108a)$$

$$= \frac{2\,F\beta}{K}\,D_{\beta z} - \frac{2\,m\beta^2}{K}\,C_{\beta z}$$

By derivation we then obtain

$$\varphi(z) = \frac{2\,F\beta^2}{K}\,A_{\beta z} - \frac{4\,m\beta^3}{K}\,D_{\beta z} \qquad\qquad (10.108b)$$

The vertical displacement and the rotation at the end of the beam are therefore

$$v(0) = \frac{2\,\beta}{K}(F - m\beta) \qquad\qquad (10.109a)$$

$$\varphi(0) = \frac{2\,\beta^2}{K}(F - 2\,m\beta) \qquad\qquad (10.109b)$$

The latter result will be used in the study of cylindrical shells which have their ends fastened to plane or hemispherical bases (Section 12.11).

10.9 Dynamics of deflected beams

With the purpose of analysing the free flexural oscillations of beams, let us consider the differential equation of the elastic line (10.49), replacing the distributed load $q(z)$ with the force of inertia,

$$\frac{\partial^4 v}{\partial z^4} = -\frac{\mu}{EI}\frac{\partial^2 v}{\partial t^2} \qquad\qquad (10.110)$$

where μ denotes the linear density of the beam (mass per unit length).

Equation (10.110) is an equation with separable variables, the solution being representable as the product of two different functions, each one having a single variable:

$$v(z,t) = \eta(z)f(t) \qquad\qquad (10.111)$$

Substituting equation (10.111) in equation (10.110), we obtain

$$\frac{d^4\eta}{dz^4}f + \frac{\mu}{EI}\eta\frac{d^2 f}{dt^2} = 0 \qquad\qquad (10.112)$$

313

Dividing equation (10.112) by the product ηf, we have

$$-\frac{\left(\dfrac{d^2 f}{dt^2}\right)}{f} = \frac{EI}{\mu}\frac{\left(\dfrac{d^4 \eta}{dz^4}\right)}{\eta} = \omega^2 \tag{10.113}$$

where ω^2 represents a positive constant, the two first terms of equation (10.113) being at the most functions of the time t and the coordinate z, respectively.

From equation (10.113) there follow two ordinary differential equations

$$\frac{d^2 f}{dt^2} + \omega^2 f = 0 \tag{10.114a}$$

$$\frac{d^4 \eta}{dz^4} - \alpha^4 \eta = 0 \tag{10.114b}$$

with

$$\alpha = \sqrt[4]{\frac{\mu\omega^2}{EI}} \tag{10.115}$$

Whereas equation (10.114a) is the equation of the harmonic oscillator, with the well-known complete integral

$$f(t) = A\cos\omega t + B\sin\omega t \tag{10.116a}$$

equation (10.114b) has the complete integral

$$\eta(z) = C\cos\alpha z + D\sin\alpha z + E\cosh\alpha z + F\sinh\alpha z \tag{10.116b}$$

As we shall see later on, the constants A, B may be determined on the basis of the initial conditions, while the constants C, D, E, F may be determined on the basis of the boundary conditions. However, the parameter ω remains for the moment undetermined, and so also the parameter α according to equation (10.115). This represents the eigenvalue of the problem, from the mathematical standpoint, or the angular frequency of the system, from the mechanical point of view. In the sequel we shall see how the angular frequency ω may also be obtained on the basis of the boundary conditions. We shall obtain in fact an infinite number of eigenvalues ω_i, and thus α_i, just as also an infinite number of eigenfunctions f_i, and thus η_i. The complete integral of the differential equation (10.110) may therefore be given the following form, on the basis of the Principle of Superposition:

$$v(z,t) = \sum_{i=1}^{\infty} \eta_i(z) f_i(t) \tag{10.117}$$

with

$$f_i(t) = A_i\cos\omega_i t + B_i\sin\omega_i t \tag{10.118a}$$
$$\eta_i(z) = C_i\cos\alpha_i z + D_i\sin\alpha_i z + E_i\cosh\alpha_i z + F_i\sinh\alpha_i z \tag{10.118b}$$

The eigenfunctions η_i are **orthonormal functions**. We may in fact write equation (10.114b) for two different eigensolutions

$$\eta_j^{IV} = \alpha_j^4 \eta_j \tag{10.119a}$$

314

$$\eta_k^{IV} = \alpha_k^4 \eta_k \qquad (10.119b)$$

Multiplying the first of equations (10.119) by η_k, the second by η_j, and integrating over the length of the beam, we obtain

$$\int_0^l \eta_k \eta_j^{IV} \, dz = \alpha_j^4 \int_0^l \eta_k \eta_j \, dz \qquad (10.120a)$$

$$\int_0^l \eta_j \eta_k^{IV} \, dz = \alpha_k^4 \int_0^l \eta_j \eta_k \, dz \qquad (10.120b)$$

Integrating by parts the left-hand sides, the foregoing equations transform as follows:

$$[\eta_k \eta_j''']_0^l - [\eta_k' \eta_j'']_0^l + \int_0^l \eta_k'' \eta_j'' \, dz = \alpha_j^4 \int_0^l \eta_k \eta_j \, dz \qquad (10.121a)$$

$$[\eta_j \eta_k''']_0^l - [\eta_j' \eta_k'']_0^l + \int_0^l \eta_j'' \eta_k'' \, dz = \alpha_k^4 \int_0^l \eta_j \eta_k \, dz \qquad (10.121b)$$

When each of the two ends of the beam is constrained by a built-in support ($\eta = \eta' = 0$), or by a hinge ($\eta = \eta'' = 0$), or by a double rod ($\eta''' = \eta' = 0$), or yet again is unconstrained ($\eta''' = \eta'' = 0$), in this last case the remaining end of the beam being built-in, the quantities in square brackets vanish. Subtracting member by member we thus have

$$(\alpha_j^4 - \alpha_k^4) \int_0^l \eta_j \eta_k \, dz = 0 \qquad (10.122)$$

from which there follows the condition of orthonormality,

$$\int_0^l \eta_j \eta_k \, dz = \delta_{jk} \qquad (10.123)$$

where δ_{jk} is the **Kronecker delta**. Thus when the eigenvalues are distinct, the integral of the product of the corresponding eigenfunctions vanishes. When, instead, the indices j and k coincide, the condition of normality reminds us that the eigenfunctions are defined but for a factor of proportionality, as follows from the homogeneity of equation (10.114b).

As we have already had occasion to mention, the constants A_i, B_i of equation (10.116a) are determined via the initial conditions

$$v(z, 0) = v_0(z) \qquad (10.124a)$$

$$\frac{\partial v}{\partial t}(z, 0) = \dot{v}_0(z) \qquad (10.124b)$$

which, on the basis of equations (10.117) and (10.118a), become

$$\sum_{i=1}^\infty A_i \eta_i(z) = v_0(z) \qquad (10.125a)$$

$$\sum_{i=1}^\infty \omega_i B_i \eta_i(z) = \dot{v}_0(z) \qquad (10.125b)$$

315

Multiplying by any desired eigenfunction η_j, and integrating over the length of the beam, we obtain

$$\sum_{i=1}^{\infty} A_i \int_0^l \eta_i \eta_j \, dz = \int_0^l \eta_j v_0 \, dz \qquad (10.126a)$$

$$\sum_{i=1}^{\infty} \omega_i B_i \int_0^l \eta_i \eta_j \, dz = \int_0^l \eta_j \dot{v}_0 \, dz \qquad (10.126b)$$

Taking into account the condition of orthonormality, equation (10.123), finally we have

$$A_j = \int_0^l \eta_j v_0 \, dz \qquad (10.127a)$$

$$B_j = \frac{1}{\omega_j} \int_0^l \eta_j \dot{v}_0 \, dz \qquad (10.127b)$$

When the system is initially perturbed, by assigning to the beam a deformed configuration proportional to one of the eigenfunctions, with an initial zero velocity, the beam, once left free to oscillate, continues to do so in proportion to the initial deformed configuration. In this case we have

$$v_0(z) = a\eta_i(z) \qquad (10.128a)$$
$$\dot{v}_0(z) = 0 \qquad (10.128b)$$

where a is an arbitrary constant of proportionality. Equations (10.127) then furnish

$$A_j = a\delta_{ij} \qquad (10.129a)$$
$$B_j = 0 \qquad (10.129b)$$

and hence the complete integral (10.117) takes the following form:

$$v(z,t) = a\eta_i(z) \cos \omega_i t = v_0(z) \cos \omega_i t \qquad (10.130)$$

The beam therefore oscillates in proportion to the initial deformation and with an angular frequency that corresponds to the same eigenfunction. These oscillations are called the **natural modes of vibration of the system.**

More particularly, as regards a beam supported at both ends, of length l (Figure 10.18(a)), the boundary conditions imposed on the expression (10.118b) in correspondence to the end A yield

$$\eta(0) = C + E = 0 \qquad (10.131a)$$
$$\eta''(0) = -\alpha^2(C - E) = 0 \qquad (10.131b)$$

whence we obtain

$$C = E = 0 \qquad (10.132)$$

316

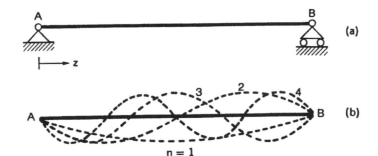

Figure 10.18

The boundary conditions corresponding to the end B yield, on the other hand

$$\eta(l) = D \sin \alpha l + F \sinh \alpha l = 0 \tag{10.133a}$$

$$\eta''(l) = -\alpha^2 (D \sin \alpha l - F \sinh \alpha l) = 0 \tag{10.133b}$$

from which it follows that

$$D \sin \alpha l = 0 \tag{10.134a}$$

$$F \sinh \alpha l = 0 \tag{10.134b}$$

From equation (10.134b) we obtain

$$F = 0 \tag{10.135}$$

whilst from equation (10.134a), once the trivial solution $D = 0$ has been ruled out, it follows that

$$\alpha = n \frac{\pi}{l} \tag{10.136}$$

where n is a natural number.

From relation (10.115) we deduce

$$\alpha_n^4 = n^4 \frac{\pi^4}{l^4} = \frac{\mu \omega_n^2}{EI} \tag{10.137}$$

whereby we obtain the **natural angular frequencies** of the system

$$\omega_n = n^2 \frac{\pi^2}{l^2} \sqrt{\frac{EI}{\mu}} \tag{10.138}$$

and hence the proper periods thereof:

$$T_n = \frac{2\pi}{\omega_n} = \frac{2 l^2}{\pi n^2} \sqrt{\frac{\mu}{EI}} \tag{10.139}$$

317

The normalized eigenfunctions are thus represented by the following succession (Figure 10.18(b)):

$$\eta_n(z) = \sqrt{\frac{2}{l}} \sin n\pi \frac{z}{l} \tag{10.140}$$

In the case of the cantilever beam of Figure 10.19, the boundary conditions imposed on the expression (10.118b) are

$$\eta(0) = C + E = 0 \tag{10.141a}$$

$$\eta'(0) = \alpha(D + F) = 0 \tag{10.141b}$$

$$\eta''(l) = -\alpha^2 (C \cos \alpha l + D \sin \alpha l - E \cosh \alpha l - F \sinh \alpha l) = 0 \tag{10.141c}$$

$$\eta'''(l) = \alpha^3 (C \sin \alpha l - D \cos \alpha l + E \sinh \alpha l + F \cosh \alpha l) = 0 \tag{10.141d}$$

Whilst from the first two equations we obtain

$$E = -C, \qquad F = -D \tag{10.142}$$

from the last two there follows

$$C(\cos \alpha l + \cosh \alpha l) + D(\sin \alpha l + \sinh \alpha l) = 0 \tag{10.143a}$$

$$C(\sin \alpha l - \sinh \alpha l) - D(\cos \alpha l + \cosh \alpha l) = 0 \tag{10.143b}$$

The system of algebraic equations (10.143) gives, on the other hand, a solution different from the trivial one, if and only if the determinant of the coefficients is zero:

$$(\cos \alpha l + \cosh \alpha l)^2 + (\sin^2 \alpha l - \sinh^2 \alpha l) = 0 \tag{10.144}$$

Computing expression (10.144), we obtain the trigonometric equation which provides the set or **spectrum** of eigenvalues

$$\cos \alpha_n l \cosh \alpha_n l = -1 \tag{10.145}$$

The first three roots of equation (10.145) are

$$\alpha_1 l = 1.875, \ \alpha_2 l = 4.694, \ \alpha_3 l = 7.885$$

The angular frequencies and the proper periods of the cantilever beam are given by

$$\omega_n = \alpha_n^2 \sqrt{\frac{EI}{\mu}} \tag{10.146a}$$

$$T_n = \frac{2\pi}{\alpha_n^2} \sqrt{\frac{\mu}{EI}} \tag{10.146b}$$

The fundamental period is therefore

$$T_1 = 1.79 \, l^2 \sqrt{\frac{\mu}{EI}} \tag{10.147}$$

and is approximately three times as long as that of the supported beam given by equation (10.139):

$$T_1 = 0.64 \, l^2 \sqrt{\frac{\mu}{EI}} \qquad (10.148)$$

The first three eigenfunctions have the aspect shown in Figure 10.19.

In the case of a rope in tension (Figure 10.18), the flexural rigidity EI is vanishingly small, so that the bending moment, in the case of large displacements, is given by the product of the axial force and the transverse displacement

$$M = -Nv \qquad (10.149)$$

Applying the relation (4.25), it is possible to obtain the equivalent transverse load, so that the equation of equilibrium to vertical translation is the following:

$$N\frac{\partial^2 v}{\partial z^2} = \mu \frac{\partial^2 v}{\partial t^2} \qquad (10.150)$$

Equation (10.150) transforms into the wave equation

$$\frac{\partial^2 v}{\partial t^2} = c^2 \frac{\partial^2 v}{\partial z^2} \qquad (10.151)$$

where

$$c^2 = \frac{N}{\mu} \qquad (10.152)$$

is the square of the velocity of the transverse wave in the rope in tension.

Equation (10.151) is formally identical to the equation of longitudinal waves in elastic bars. If in fact we replace the distributed longitudinal force $F_x(x)$ in the static equation (8.53a) with the force of inertia $-\mu(\partial^2 u/\partial t^2)$, we obtain

$$EA\frac{\partial^2 u}{\partial x^2} = \mu \frac{\partial^2 u}{\partial t^2} \qquad (10.153)$$

and thus

$$\frac{\partial^2 u}{\partial t^2} = c^2 \frac{\partial^2 u}{\partial x^2} \qquad (10.154)$$

where in this case

$$c^2 = \frac{EA}{\mu} \qquad (10.155)$$

is the square of the velocity of the longitudinal wave in the elastic bar.

10.10 Plates in flexure

Plates are structural elements where one dimension is negligible in comparison with the other two. This dimension is termed **thickness**. **Plane plates**, in

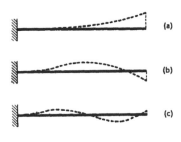

Figure 10.19

(a)

(b)

(c)

319

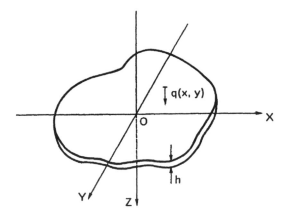

Figure 10.20

particular, are cylindrical solids whose generators are at least one order of magnitude smaller than the dimensions of the faces (the reverse of the situation we have in the case of the Saint Venant solid).

Let us consider a plate of thickness h, loaded by distributed forces orthogonal to the faces and constrained at the edge (Figure 10.20). Let XY be the middle plane of the plate and Z the orthogonal axis. The so-called **Kirchhoff kinematic hypothesis** assumes that the segments orthogonal to the middle plane, after deformation has occurred, remain orthogonal to the deformed middle plane (Figure 10.21). Denoting then as φ_x the angle of rotation about the Y axis and as φ_y the angle of rotation about the X axis, the displacement of a generic point P of coordinates x, y, z will present the following three components:

$$u = \varphi_x z = -\frac{\partial w}{\partial x} z \qquad (10.156a)$$

$$v = \varphi_y z = -\frac{\partial w}{\partial y} z \qquad (10.156b)$$

$$w = w(x, y) \qquad (10.156c)$$

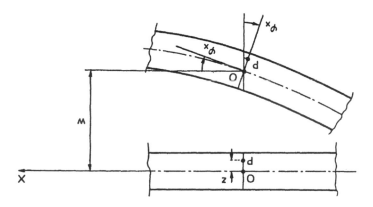

Figure 10.21

320

The relation (10.156c) indicates that all the points belonging to one and the same segment orthogonal to the middle plane are displaced in that direction by the same quantity.

From the kinematic hypothesis (10.156) it follows, by simple derivation, that the strain field is

$$\varepsilon_x = \frac{\partial u}{\partial x} = \frac{\partial \varphi_x}{\partial x} z = -\frac{\partial^2 w}{\partial x^2} z \qquad (10.157a)$$

$$\varepsilon_y = \frac{\partial v}{\partial y} = \frac{\partial \varphi_y}{\partial y} z = -\frac{\partial^2 w}{\partial y^2} z \qquad (10.157b)$$

$$\varepsilon_z = \frac{\partial w}{\partial z} = 0 \qquad (10.157c)$$

$$\gamma_{xy} = \frac{\partial u}{\partial y} + \frac{\partial v}{\partial x} = \left(\frac{\partial \varphi_x}{\partial y} + \frac{\partial \varphi_y}{\partial x}\right) z = -2\frac{\partial^2 w}{\partial x \partial y} z \qquad (10.157d)$$

$$\gamma_{xz} = \frac{\partial u}{\partial z} + \frac{\partial w}{\partial x} = 0 \qquad (10.157e)$$

$$\gamma_{yz} = \frac{\partial v}{\partial z} + \frac{\partial w}{\partial y} = 0 \qquad (10.157f)$$

Kirchhoff's kinematic hypothesis generates therefore a condition of plane strain. The three significant components of strain may be expressed as follows:

$$\varepsilon_x = \chi_x z \qquad (10.158a)$$

$$\varepsilon_y = \chi_y z \qquad (10.158b)$$

$$\gamma_{xy} = \chi_{xy} z \qquad (10.158c)$$

where χ_x and χ_y are the flexural curvatures of the middle plane in the respective directions, and χ_{xy} is twice the unit angle of torsion of the middle plane in the X and Y directions.

For the condition of plane stress, the constitutive relations (8.73) become

$$\varepsilon_x = \frac{1}{E}(\sigma_x - v\sigma_y) \qquad (10.159a)$$

$$\varepsilon_y = \frac{1}{E}(\sigma_y - v\sigma_x) \qquad (10.159b)$$

$$\gamma_{xy} = \frac{1}{G}\tau_{xy} \qquad (10.159c)$$

It is important, however, to note that a condition cannot, at the same time, be both one of plane strain and one of plane stress. The thickness h is, on the other hand, assumed to be so small as to enable very low, and consequently negligible, stresses σ_z to develop. This is an assumption that we shall take up and discuss in greater depth in Chapter 19. From equations (10.159a, b) we find

$$\sigma_x - v\sigma_y = E\varepsilon_x \qquad (10.160a)$$

$$v\sigma_y - v^2\sigma_x = Ev\varepsilon_y \qquad (10.160b)$$

321

whence, by simple addition, we obtain the expressions

$$\sigma_x = \frac{E}{1-v^2}(\varepsilon_x + v\varepsilon_y) \tag{10.161a}$$

$$\sigma_y = \frac{E}{1-v^2}(\varepsilon_y + v\varepsilon_x) \tag{10.161b}$$

$$\tau_{xy} = \frac{E}{2(1+v)}\gamma_{xy} \tag{10.161c}$$

From equations (10.158) there thus follows the stress field of the plate

$$\sigma_x = \frac{E}{1-v^2}(\chi_x + v\chi_y)z \tag{10.162a}$$

$$\sigma_y = \frac{E}{1-v^2}(\chi_y + v\chi_x)z \tag{10.162b}$$

$$\tau_{xy} = \frac{E}{2(1+v)}\chi_{xy}z \tag{10.162c}$$

Integrating, over the thickness, the stresses expressed by equations (10.162), we obtain the **characteristics of the internal reaction**, which are bending and twisting moments per unit length (Figure 10.22)

$$M_x = \int_{-h/2}^{h/2} \sigma_x z \, dz \tag{10.163a}$$

$$M_y = \int_{-h/2}^{h/2} \sigma_y z \, dz \tag{10.163b}$$

$$M_{xy} = M_{yx} = \int_{-h/2}^{h/2} \tau_{xy} z \, dz \tag{10.163c}$$

Substituting equations (10.162) in equations (10.163), we obtain finally the **constitutive equations** of the plane plate,

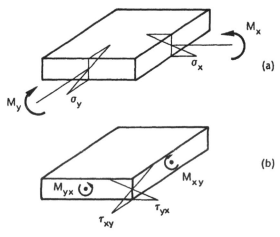

Figure 10.22

$$M_x = D(\chi_x + v\chi_y) \qquad (10.164\text{a})$$

$$M_y = D(\chi_y + v\chi_x) \qquad (10.164\text{b})$$

$$M_{xy} = M_{yx} = \frac{1-v}{2} D\chi_{xy} \qquad (10.164\text{c})$$

where

$$D = \frac{Eh^3}{12(1-v^2)} \qquad (10.165)$$

is the flexural rigidity of the plate.

Let us then determine the **indefinite equations of equilibrium**, considering an infinitesimal element of the plate submitted to the external load and to the static characteristics. The condition of equilibrium with regard to rotation about the Y axis (Figure 10.23(a)) yields

$$\left(\frac{\partial M_x}{\partial x} dx\right) dy + \left(\frac{\partial M_{xy}}{\partial y} dy\right) dx - (T_x\, dy)\, dx = 0 \qquad (10.166)$$

from which we deduce

$$\frac{\partial M_x}{\partial x} + \frac{\partial M_{xy}}{\partial y} - T_x = 0 \qquad (10.167\text{a})$$

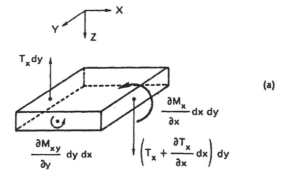

(a)

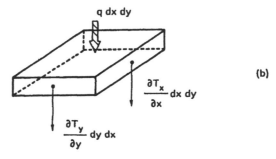

(b)

Figure 10.23

Likewise, we have

$$\frac{\partial M_{xy}}{\partial x} + \frac{\partial M_y}{\partial y} - T_y = 0 \qquad (10.167b)$$

The condition of equilibrium with regard to translation in the direction of the Z axis (Figure 10.23(b)) gives

$$\left(\frac{\partial T_x}{\partial x} dx\right) dy + \left(\frac{\partial T_y}{\partial y} dy\right) dx + q\, dx\, dy = 0 \qquad (10.168)$$

from which we deduce

$$\frac{\partial T_x}{\partial x} + \frac{\partial T_y}{\partial y} + q = 0 \qquad (10.169)$$

The remaining three conditions of equilibrium, that with regard to rotation about the Z axis and those with regard to translation in the X and Y directions, are identically satisfied, in that the plate has been assumed to be loaded by forces not exerted on the middle plane.

If γ_x and γ_y denote the shearing strains due to the shearing forces, T_x and T_y, respectively, the kinematic equations define the characteristics of deformation as functions of the generalized displacements, in the following way:

$$\begin{bmatrix} \gamma_x \\ \gamma_y \\ \chi_x \\ \chi_y \\ \chi_{xy} \end{bmatrix} = \begin{bmatrix} \dfrac{\partial}{\partial x} & +1 & 0 \\ \dfrac{\partial}{\partial y} & 0 & +1 \\ 0 & \dfrac{\partial}{\partial x} & 0 \\ 0 & 0 & \dfrac{\partial}{\partial y} \\ 0 & \dfrac{\partial}{\partial y} & \dfrac{\partial}{\partial x} \end{bmatrix} \begin{bmatrix} w \\ \varphi_x \\ \varphi_y \end{bmatrix} \qquad (10.170)$$

It is to be noted that the shearing strains γ_x and γ_y have so far been neglected, starting from equations (10.156a, b), as also in equations (10.157e, f).

The **static equations** (10.167) and (10.169), on the other hand, in matrix form are presented as follows:

$$\begin{bmatrix} \dfrac{\partial}{\partial x} & \dfrac{\partial}{\partial y} & 0 & 0 & 0 \\ -1 & 0 & \dfrac{\partial}{\partial x} & 0 & \dfrac{\partial}{\partial y} \\ 0 & -1 & 0 & \dfrac{\partial}{\partial y} & \dfrac{\partial}{\partial x} \end{bmatrix} \begin{bmatrix} T_x \\ T_y \\ M_x \\ M_y \\ M_{xy} \end{bmatrix} + \begin{bmatrix} q \\ 0 \\ 0 \end{bmatrix} = \begin{bmatrix} 0 \\ 0 \\ 0 \end{bmatrix} \qquad (10.171)$$

Also in the case of plates, **static–kinematic duality** is expressed by the fact that the static matrix, neglecting the algebraic sign of the unity terms, is the transpose of the kinematic matrix, and *vice versa*.

324

The **constitutive equations** (10.164), finally, can also be cast in matrix form,

$$
\begin{bmatrix} T_x \\ T_y \\ M_x \\ M_y \\ M_{xy} \end{bmatrix} = \begin{bmatrix} \frac{5}{6}Gh & 0 & 0 & 0 & 0 \\ 0 & \frac{5}{6}Gh & 0 & 0 & 0 \\ 0 & 0 & D & \nu D & 0 \\ 0 & 0 & \nu D & D & 0 \\ 0 & 0 & 0 & 0 & \frac{1-\nu}{2}D \end{bmatrix} \begin{bmatrix} \gamma_x \\ \gamma_y \\ \chi_x \\ \chi_y \\ \chi_{xy} \end{bmatrix}
\qquad (10.172)
$$

where the factor 5/6 is the inverse of the shear factor corresponding to a rectangular cross section of unit base and thickness h. Note that, while the thickness h appears in the first two rows raised to the first power, in the remaining rows it appears raised to the third power, in agreement with equation (10.165). The shearing stiffness appears therefore more important than the flexural stiffness, by as much as two orders of magnitude, and this explains why the shearing strains are often neglected.

The equations of kinematics (10.170) and statics (10.171) and the constitutive equations (10.172) may be cast in compact form,

$$\{q\} = [\partial]\{\eta\} \qquad (10.173\text{a})$$

$$[\partial]^*\{Q\} + \{\mathscr{F}\} = \{0\} \qquad (10.173\text{b})$$

$$\{Q\} = [H]\{q\} \qquad (10.173\text{c})$$

so that if, as we have already done in the case of the three-dimensional solid and the beam, we denote by

$$\underset{(3\times3)}{[\mathscr{L}]} = [\partial]^* \underset{(3\times5)(5\times5)(5\times3)}{[H]} [\partial] \qquad (10.174)$$

the Lamé's matrix operator, the elastic problem of the deflected plane plate is represented by the following operator equation furnished with the corresponding boundary conditions:

$$[\mathscr{L}]\{\eta\} = -\{\mathscr{F}\}, \qquad \forall P \in S \qquad (10.175\text{a})$$

$$[\mathscr{N}]^T\{Q\} = \{p\}, \qquad \forall P \in \mathscr{C}_p \qquad (10.175\text{b})$$

$$\{\eta\} = \{\eta_0\}, \qquad \forall P \in \mathscr{C}_\eta \qquad (10.175\text{c})$$

In the above, C_p denotes the portion of the edge C on which the static conditions are assigned, while C_η denotes the complementary portion on which the kinematic (or constraint) conditions are assigned.

Designating as $\{n\}$ the unit vector normal to the boundary portion $\mathscr{C}_p$ (Figure 10.24), it is simple to render explicit the boundary condition of equivalence (10.175b).

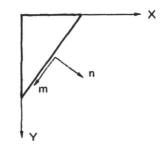

Figure 10.24

$$
\begin{bmatrix} n_x & n_y & 0 & 0 & 0 \\ 0 & 0 & n_x & 0 & n_y \\ 0 & 0 & 0 & n_y & n_x \end{bmatrix} \begin{bmatrix} T_x \\ T_y \\ M_x \\ M_y \\ M_{xy} \end{bmatrix} = \begin{bmatrix} T_n \\ M_{nx} \\ M_{ny} \end{bmatrix}
\qquad (10.176)
$$

325

M_{nx} and M_{ny} being the components of the moment vector acting on the section of normal n. As in the case of the three-dimensional solid, also in the case of the plate, the matrix $[\mathcal{N}]^T$ is linked to the matrix operator $[\partial]^*$ of the static equation (10.171), matching each partial derivative with the corresponding direction cosine of the normal to the boundary.

Finally, the **elastic problem of the deflected plane plate** can be summarized as follows:

$$[\mathscr{L}]\{\eta\} = -\{\mathscr{F}\}, \qquad \forall P \in S \qquad (10.177a)$$

$$([\mathcal{N}]^T[H][\partial])\{\eta\} = \{p\}, \qquad \forall P \in \mathscr{C}_p \qquad (10.177b)$$

$$\{\eta\} = \{\eta_0\}, \qquad \forall P \in \mathscr{C}_\eta \qquad (10.177c)$$

Equations (10.177) are formally identical to equations (8.52), once the three-dimensional domain V has been replaced by the two-dimensional one S of the plate, and the external surface S by the boundary $\mathscr{C}$ of the plate.

10.11 Sophie Germain equation

Neglecting the shearing deformability of the plate, it is possible to arrive at a differential equation in the single kinematic unknown w. Deriving equations (10.167), we find

$$\frac{\partial T_x}{\partial x} = \frac{\partial^2 M_x}{\partial x^2} + \frac{\partial^2 M_{xy}}{\partial x \partial y} \qquad (10.178a)$$

$$\frac{\partial T_y}{\partial y} = \frac{\partial^2 M_{xy}}{\partial x \partial y} + \frac{\partial^2 M_y}{\partial y^2} \qquad (10.178b)$$

and substituting the foregoing equations (10.178) in equation (10.169), we obtain

$$\frac{\partial^2 M_x}{\partial x^2} + 2\frac{\partial^2 M_{xy}}{\partial x \partial y} + \frac{\partial^2 M_y}{\partial y^2} + q = 0 \qquad (10.179)$$

The constitutive equations (10.164), if we neglect the shearing strains, become

$$M_x = -D\left(\frac{\partial^2 w}{\partial x^2} + v\frac{\partial^2 w}{\partial y^2}\right) \qquad (10.180a)$$

$$M_y = -D\left(\frac{\partial^2 w}{\partial y^2} + v\frac{\partial^2 w}{\partial x^2}\right) \qquad (10.180b)$$

$$M_{xy} = -D(1-v)\frac{\partial^2 w}{\partial x \partial y} \qquad (10.180c)$$

Substituting equations (10.180) in equation (10.179), we deduce finally the **Sophie Germain equation**,

$$\frac{\partial^4 w}{\partial x^4} + 2\frac{\partial^4 w}{\partial x^2 \partial y^2} + \frac{\partial^4 w}{\partial y^4} = \frac{q}{D} \qquad (10.181)$$

which is the fourth-order differential equation corresponding to the **elastic plane**. If we indicate by

$$\nabla^2 = \frac{\partial^2}{\partial x^2} + \frac{\partial^2}{\partial y^2} \qquad (10.182)$$

the Laplacian, equation (10.181) can also be written as

$$\nabla^2(\nabla^2 w) = \frac{q}{D} \qquad (10.183)$$

or, even more synthetically

$$\nabla^4 w = \frac{q}{D} \qquad (10.184)$$

Note the formal analogy between equation (10.184) and the equation of the elastic line (10.49).

The term **principal directions of moment** relative to a point of the deflected plate is given to the two orthogonal directions along which the twisting moment M_{xy} vanishes, and consequently the shearing stresses τ_{xy} likewise vanish. These directions thus coincide with the principal ones of stress. The term **principal directions of curvature** is applied to the two orthogonal directions along which the unit angle of torsion $\chi_{xy}/2$ vanishes. In the case where the material is assumed to be isotropic, the constitutive equation (10.164c) shows how the principal directions of moment and the principal directions of curvature must coincide.

The **Finite Difference Method** for the approximate numerical solution of the Sophie Germain equation is proposed in Appendix F. Deflected plates having polar symmetry are dealt with in Chapter 12, while the multilayer plates of **composite materials** are discussed in Appendix E.

10.12 Shells with double curvature

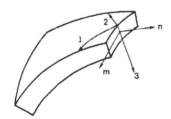

Figure 10.25

Let us consider a shell of thickness h, the middle surface of which has a double curvature. On this surface there exists a system of principal curvilinear coordinates $s_1 s_2$ (Figure 10.25), in correspondence with which the middle surface presents the minimum and maximum curvature. The **membrane regime** consists of the normal forces N_1, N_2, and the shearing force N_{12} contained in the plane tangential to the middle surface (Figure 10.26(a)), as well as of the dilations ε_1, ε_2 and the shearing strain ε_{12} between the principal directions of curvature. The **flexural regime** consists of the shearing forces T_1, T_2, perpendicular to the tangent plane, of the bending moments M_1, M_2, and of the twisting moment M_{12} (Figure 10.26(b)), as well as of the shearing strains γ_1, γ_2 between each principal direction of curvature and the direction normal to the tangent plane, of the flexural curvatures χ_1, χ_2, and of twice the unit angle of torsion χ_{12}.

The **kinematic equations**, which define the characteristics of deformation as functions of the generalized displacements, may be put in matrix form,

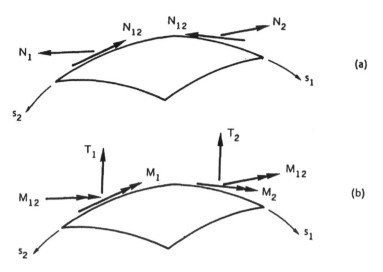

Figure 10.26

$$
\begin{bmatrix} \varepsilon_1 \\ \varepsilon_2 \\ \varepsilon_{12} \\ \gamma_1 \\ \gamma_2 \\ \chi_1 \\ \chi_2 \\ \chi_{12} \end{bmatrix}
=
\begin{bmatrix}
\frac{\partial}{\partial s_1} & +\frac{R_2}{R_1(R_2-R_1)}\frac{\partial R_1}{\partial s_2} & +\frac{1}{R_1} & 0 & 0 \\[2mm]
+\frac{R_1}{R_2(R_1-R_2)}\frac{\partial R_2}{\partial s_1} & \frac{\partial}{\partial s_2} & +\frac{1}{R_2} & 0 & 0 \\[2mm]
\frac{\partial}{\partial s_2}-\frac{R_2}{R_1(R_2-R_1)}\frac{\partial R_1}{\partial s_2} & \frac{\partial}{\partial s_1}-\frac{R_1}{R_2(R_1-R_2)}\frac{\partial R_2}{\partial s_1} & 0 & 0 & 0 \\[2mm]
-\frac{1}{R_1} & 0 & \frac{\partial}{\partial s_1} & +1 & 0 \\[2mm]
0 & -\frac{1}{R_2} & \frac{\partial}{\partial s_2} & 0 & +1 \\[2mm]
0 & 0 & 0 & \frac{\partial}{\partial s_1} & +\frac{R_2}{R_1(R_2-R_1)}\frac{\partial R_1}{\partial s_2} \\[2mm]
0 & 0 & 0 & +\frac{R_1}{R_2(R_1-R_2)}\frac{\partial R_2}{\partial s_1} & \frac{\partial}{\partial s_2} \\[2mm]
0 & 0 & 0 & \frac{\partial}{\partial s_2}-\frac{R_2}{R_1(R_2-R_1)}\frac{\partial R_1}{\partial s_2} & \frac{\partial}{\partial s_1}-\frac{R_1}{R_2(R_1-R_2)}\frac{\partial R_2}{\partial s_1}
\end{bmatrix}
\begin{bmatrix} u_1 \\ u_2 \\ u_3 \\ \varphi_1 \\ \varphi_2 \end{bmatrix}
$$

$$(10.185)$$

where u_1, u_2, u_3 are the components of the displacement on the two principal axes of curvature and on the normal to the tangent plane, φ_1, φ_2 are the rotations about the principal directions of curvature 2 and 1, respectively, and R_1, R_2 are the two principal radii of curvature. Equation (10.185) may be rewritten in compact form,

$$\{q\} = [\partial]\{\eta^*\} \tag{10.186}$$

where $\{\eta^*\}$ is the displacement vector in the rotated reference system 123.

The **indefinite equations of equilibrium**, on the other hand, are five and express the equilibrium with regard to translation in the three directions 1,2, 3, and the equilibrium with regard to rotation about the tangential axes 1 and 2

$$
\begin{bmatrix}
\dfrac{\partial}{\partial c_1}+\dfrac{R_1}{R_2(R_1-R_2)}\dfrac{\partial R_2}{\partial c_2} & -\dfrac{R_1}{R_2(R_1-R_2)}\dfrac{\partial R_2}{\partial c_1} & \dfrac{\partial}{\partial c_2}+\dfrac{2R_2}{R_1(R_2-R_1)}\dfrac{\partial R_1}{\partial c_2} & \dfrac{\partial}{\partial c_1}+\dfrac{R_1}{R_2(R_1-R_2)}\dfrac{\partial R_2}{\partial c_2} & 0 & \dfrac{\partial}{\partial c_1}+\dfrac{R_1}{R_2(R_1-R_2)}\dfrac{\partial R_2}{\partial c_2} & 0 & 0 \\[2em]
-\dfrac{R_2}{R_1(R_2-R_1)}\dfrac{\partial R_1}{\partial c_2} & \dfrac{\partial}{\partial c_2}+\dfrac{R_2}{R_1(R_2-R_1)}\dfrac{\partial R_1}{\partial c_1} & \dfrac{\partial}{\partial c_1}+\dfrac{2R_1}{R_2(R_1-R_2)}\dfrac{\partial R_2}{\partial c_1} & 0 & \dfrac{R_2}{R_1(R_2-R_1)}\dfrac{\partial R_1}{\partial c_2} & 0 & -\dfrac{R_1}{R_2(R_1-R_2)}\dfrac{\partial R_2}{\partial c_2} & \dfrac{\partial}{\partial c_2}+\dfrac{2R_2}{R_1(R_2-R_1)}\dfrac{\partial R_1}{\partial c_2} \\[2em]
-\dfrac{1}{R_1} & -\dfrac{1}{R_2} & 0 & \dfrac{\partial}{\partial c_1}+\dfrac{R_1}{R_2(R_1-R_2)}\dfrac{\partial R_2}{\partial c_2} & \dfrac{\partial}{\partial c_2}+\dfrac{R_2}{R_1(R_2-R_1)}\dfrac{\partial R_1}{\partial c_1} & -\dfrac{R_1}{R_2(R_1-R_2)}\dfrac{\partial R_2}{\partial c_1} & \dfrac{R_2}{R_1(R_2-R_1)}\dfrac{\partial R_1}{\partial c_2} & \dfrac{\partial}{\partial c_1}+\dfrac{2R_1}{R_2(R_1-R_2)}\dfrac{\partial R_2}{\partial c_1} \\[2em]
0 & 0 & 0 & -1 & 0 & -1 & \dfrac{\partial}{\partial c_2}+\dfrac{R_2}{R_1(R_2-R_1)}\dfrac{\partial R_1}{\partial c_1} & \dfrac{\partial}{\partial c_1}+\dfrac{R_1}{R_2(R_1-R_2)}\dfrac{\partial R_2}{\partial c_2} \\[2em]
0 & 0 & 0 & 0 & +\dfrac{1}{R_2} & \dfrac{R_2}{R_1(R_2-R_1)}\dfrac{\partial R_1}{\partial c_2} & +\dfrac{R_1}{R_2(R_1-R_2)}\dfrac{\partial R_2}{\partial c_1} & \dfrac{R_2}{R_1(R_2-R_1)}\dfrac{\partial R_1}{\partial c_2} \\
\end{bmatrix}
\begin{Bmatrix} N_1 \\ N_2 \\ N_{12} \\ T_1 \\ T_2 \\ M_1 \\ M_2 \\ M_{12} \end{Bmatrix}
+
\begin{Bmatrix} 0 \\ 0 \\ 0 \\ p_1 \\ p_2 \\ q \\ 0 \\ 0 \end{Bmatrix}
=
\begin{Bmatrix} 0 \\ 0 \\ 0 \\ 0 \\ 0 \\ 0 \\ 0 \\ 0 \end{Bmatrix}
$$

(10.187)

Equation (10.187) may be written synthetically

$$[\partial]^*\{Q\} + \{\mathscr{F}^*\} = \{0\} \tag{10.188}$$

where $\{F^*\}$ is the vector of the external forces in the rotated reference system 123.

It should be noted that the kinematic equations represent a more complete version than those proposed by Novozhilov. On the other hand, they have been obtained heuristically, i.e. considering the adjoint of the static operator.

The constitutive equations are the following:

$$
\begin{bmatrix} N_1 \\ N_2 \\ N_{12} \\ T_1 \\ T_2 \\ M_1 \\ M_2 \\ M_{12} \end{bmatrix}
=
\begin{bmatrix}
\frac{12D}{h^2} & v\frac{12D}{h^2} & 0 & 0 & 0 & 0 & 0 & 0 \\
v\frac{12D}{h^2} & \frac{12D}{h^2} & 0 & 0 & 0 & 0 & 0 & 0 \\
0 & 0 & \frac{1-v}{2}\frac{12D}{h^2} & 0 & 0 & 0 & 0 & 0 \\
0 & 0 & 0 & (1-v)\frac{5D}{h^2} & 0 & 0 & 0 & 0 \\
0 & 0 & 0 & \, & (1-v)\frac{5D}{h^2} & 0 & 0 & 0 \\
0 & 0 & 0 & 0 & 0 & D & vD & 0 \\
0 & 0 & 0 & 0 & 0 & vD & D & 0 \\
0 & 0 & 0 & 0 & 0 & 0 & 0 & \frac{1-v}{2}D
\end{bmatrix}
\begin{bmatrix} \varepsilon_1 \\ \varepsilon_2 \\ \varepsilon_{12} \\ \gamma_1 \\ \gamma_2 \\ \chi_1 \\ \chi_2 \\ \chi_{12} \end{bmatrix}
\tag{10.189}
$$

or, in compact form

$$\{Q\} = [H]\{q\} \tag{10.190}$$

The vectors of the external forces and of the displacements in the local system 123 may be obtained by premultiplying the corresponding vectors in the global reference system XYZ by the orthogonal matrix of rotation $[N]$

$$\{\mathscr{F}^*\} = [N]\{\mathscr{F}\} \tag{10.191a}$$

$$\{\eta^*\} = [N]\{\eta\} \tag{10.191b}$$

where $[N]$ is the following 5×6 matrix:

$$
[N] =
\begin{bmatrix}
\cos \hat{1x} & \cos \hat{1y} & \cos \hat{1z} & 0 & 0 & 0 \\
\cos \hat{2x} & \cos \hat{2y} & \cos \hat{2z} & 0 & 0 & 0 \\
\cos \hat{3x} & \cos \hat{3y} & \cos \hat{3z} & 0 & 0 & 0 \\
0 & 0 & 0 & \cos \hat{1x} & \cos \hat{1y} & \cos \hat{1z} \\
0 & 0 & 0 & -\cos \hat{2x} & -\cos \hat{2y} & -\cos \hat{2z}
\end{bmatrix}
\tag{10.19}
$$

and $\{F\}$, $\{\eta\}$ are the following six-component vectors:

$$\{\mathscr{F}\}^{\mathrm{T}} = [q_x, \quad q_y, \quad q_z, \quad 0, \quad 0, \quad 0] \tag{10.193a}$$

$$\{\eta\}^{\mathrm{T}} = [u, \quad v, \quad w, \quad \varphi_x, \quad \varphi_y, \quad \varphi_z] \tag{10.193b}$$

The matrix $[N]$ is not square, since the moment and rotation vectors, which are always contained in the tangent plane, in the global XYZ system generally possess three components.

Substituting equations (10.191) in equations (10.186) and (10.188), we obtain

$$[\mathscr{L}]\{\eta\} = -\{\mathscr{F}\} \qquad (10.194)$$

the **Lamé operator** being given by the following matrix product:

$$[\mathscr{L}] = \underset{(6 \times 6)}{[N]^{\mathrm{T}}} \underset{(6 \times 5)}{[\partial]^{*}} \underset{(5 \times 8)}{[H]} \underset{(8 \times 8)}{[\partial]} \underset{(8 \times 5)}{[N]} \qquad (10.195)$$

On the other hand, the boundary condition of equivalence takes the form:

$$\underset{(5 \times 8)}{[\mathscr{N}]^{\mathrm{T}}} \underset{(8 \times 1)}{\{Q\}} = \underset{(5 \times 6)}{[N]} \underset{(6 \times 1)}{\{p\}} \qquad (10.196)$$

where $\{p\}$ is the vector of the forces and of the moments applied to the unit length of the boundary and referred to the global system XYZ, whilst $[\mathscr{N}]^{\mathrm{T}}$ is the matrix which transforms the static characteristics into the aforesaid loadings, referred to the local system 123. Designating as n the axis belonging to the tangent plane and orthogonal to the boundary (Figure 10.25), and as m the axis tangential to the boundary, in such a way that the reference system $nm3$ is right-handed, we have

$$[\mathscr{N}]^{\mathrm{T}} = \begin{bmatrix} n_1 & 0 & n_2 & 0 & 0 & 0 & 0 & 0 \\ 0 & n_2 & n_1 & 0 & 0 & 0 & 0 & 0 \\ 0 & 0 & 0 & n_1 & n_2 & 0 & 0 & 0 \\ 0 & 0 & 0 & 0 & 0 & n_1 & 0 & n_2 \\ 0 & 0 & 0 & 0 & 0 & 0 & n_2 & n_1 \end{bmatrix} \qquad (10.197)$$

where the submatrix formed by the last three rows and the last five columns replicates equation (10.176), obtained for the plate. The submatrix obtained from the first two rows and the first three columns reproduces equation (8.12) in the case of plane stress condition. It is therefore apparent how, also in the framework of boundary condition, the membrane and flexural regimes remain separate and are not interacting. As in all other cases so far considered, the matrix $[\mathscr{N}]^{\mathrm{T}}$ is directly correlated to the operator $[\partial]^{*}$.

The **elastic problem of the shell with double curvature** is thus summarized in the following equations:

$$\begin{array}{lll} [\mathscr{L}]\{\eta\} = -\{\mathscr{F}\}, & \forall P \in S & (10.198a) \\[4pt] \left([N]^{\mathrm{T}}[\mathscr{N}]^{\mathrm{T}}[H][\partial][N]\right)\{\eta\} = \{p\}, & \forall P \in \mathscr{C}_p & (10.198b) \\[4pt] \{\eta\} = \{\eta_0\}, & \forall P \in \mathscr{C}_{\eta} & (10.198c) \end{array}$$

11 Finite element method

11.1 Introduction

The Finite Element Method is illustrated here as a method of discretization and interpolation for the approximate solution of elastic problems. This method is introduced in an altogether general manner, without specifying the structural element to which it is applied, whether it is of one, two, or three dimensions, and in the first two cases, whether it does or does not have an intrinsic curvature. On the other hand, the two dimensions that characterize the element are brought into the forefront: that of the generalized displacement vector and that common to the two vectors of static and deformation characteristics.

Applying the Principle of Minimum Total Potential Energy and the Ritz–Galerkin numerical approximation, we arrive at the analytical and variational definition of the Finite Element Method. Furthermore, applying the Principle of Virtual Work, also the alternative definition of the method is given, the one more widely known in the engineering field, viz. the mechanical and matrix one. Via the definition of shape functions, we arrive at the notion of local stiffness matrix of the individual element. This matrix is thus expanded and assembled, i.e. added to all the other similar matrices, to provide, finally, the global stiffness matrix. In this context an explanation is given of the change of algebraic sign shown by the unity terms of the static and kinematic matrix operators, which, in the case of beams and plates, are each the adjoint of the other.

The chapter closes with a reference to the dynamics of elastic solids. On the basis of results already reached in the case of the dynamics of deflected beams (Chapter 10), the so-called **modal analysis** is developed, a topic which will be taken up again in Chapter 14, where we shall study the dynamics of beam systems and, more particularly, of orthogonal multistorey frames.

11.2 Single-degree-of-freedom system

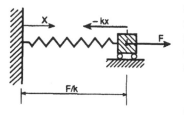

Figure 11.1

Consider a material point subjected to an external force F and to the elastic restoring force of a linear spring having stiffness k (Figure 11.1). Since the restoring force is proportional to the elongation x of the spring and is acting in the opposite direction to that of the external force, the condition of static equilibrium will be expressed by the following equation:

$$F - kx = 0 \tag{11.1}$$

from which the abscissa of the **position of equilibrium** is deduced:

$$x = F/k \tag{11.2}$$

This simple result may be obtained in principle by also considering the **total potential energy** of the system, which is equal to the sum of the potential energy of the spring and the potential energy of the non-positional force F:

$$W(x) = \frac{1}{2}kx^2 - Fx \tag{11.3}$$

332

As is known from Rational Mechanics, the derivative of the total potential energy, with change of sign, yields the total force acting on the system,

$$\text{Total force} = -\frac{dW}{dx} = -kx + F \tag{11.4}$$

from which we deduce that the system is in equilibrium when equation (11.2) is satisfied. The condition

$$-\frac{dW}{dx} = 0 \tag{11.5}$$

also defines a point of stationarity, viz. the **minimum total potential energy**, which can be represented by a parabolic curve as a function of the elongation x (Figure 11.2). This parabola is also known as the **potential well**.

The position of equilibrium (11.2) may also be obtained by applying the Principle of Virtual Work: if we impose a virtual displacement Δx on the system in a condition of equilibrium, the work thus produced must be zero,

$$F\Delta x - kx\,\Delta x = 0 \tag{11.6}$$

from which, by cancelling Δx, the already known result of relation (11.2) follows.

The Finite Element Method, even though it concerns multiple-degree-of-freedom systems, can be introduced by following the two paths indicated for the single-degree-of-freedom system: (1) **Principle of Minimum Total Potential Energy**; (2) **Principle of Virtual Work**. The Finite Element Method is basically a **discretization method**, in the sense that, instead of the continuous function of the displacements $\{\eta\}$, it considers as unknowns only the displacements $\{\delta\}$ of a discrete number n of points called **nodes**. It is at the same time an **interpolation method**, in the sense that, once the displacements $\{\delta\}$ are determined, it connects them with sufficiently regular functions. The problem is thus reduced to the determination of the **equilibrium configuration** $\{\delta\}$, from which we then obtain, by interpolation, the displacement field $\{\eta\}$, by derivation, the strain field $\{\varepsilon\}$ or the deformation characteristics field $\{q\}$, and, via the constitutive equations, the stress field $\{\sigma\}$ or the static characteristics field $\{Q\}$.

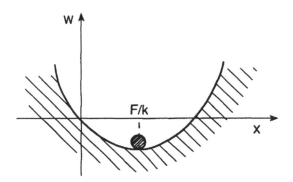

Figure 11.2

333

For each of the structural geometries considered hitherto, there are two numbers that characterize the system: (1) the **degrees of freedom** g, which are represented by the dimension of the displacement vector $\{\eta\}$; (2) the dimension d of the strain characteristics vector, $\{\varepsilon\}$ or $\{q\}$, and the static characteristics vector, $\{\sigma\}$ or $\{Q\}$. Table 11.1 gives the characteristic numbers g and d for all the one-, two- and three-dimensional elastic solids so far analysed.

Table 11.1

Structural element	g	d
Beam in the plane	3	3
Beam in space	6	6
Deflected plate	3	5
Shell with double curvature	5	8
Two-dimensional solid	2	3
Three-dimensional solid	3	6

Note that only in the case of the beam do the degrees of freedom g coincide with the dimension d of the characteristics vectors. In all other cases we have $g < d$.

11.3 Principle of minimum total potential energy

In the foregoing chapters we have seen how every elastic problem, in one, two or three dimensions, can be referred to **Lamé's equation** of the sort

$$\underset{(g\times g)(g\times 1)}{[\mathscr{L}]\ \{\eta\}} = -\underset{(g\times 1)}{\{\mathscr{F}\}} \tag{11.7}$$

where $[\mathscr{L}]$ is the differential and matrix operator corresponding to the geometry in question, $\{\eta\}$ is the vector of the generalized displacements, $\{\mathscr{F}\}$ is the vector of the external forces acting in the elastic domain.

In the case where the structural element does not present an intrinsic curvature (arches and shells), the **boundary conditions of equivalence** assume the following form:

$$\underset{(g\times d)\ (d\times d)(d\times g)\ (g\times 1)}{([\mathscr{N}]^{\mathrm{T}}\ [H]\ [\partial])\ \{\eta\}} = \underset{(g\times 1)}{\{p\}} \tag{11.8}$$

where $[\mathscr{N}]^{\mathrm{T}}$ is the matrix which transforms the static characteristics vector into the vector of the external forces acting on the boundary, $[H]$ is the Hessian matrix of the elastic potential Φ, $[\partial]$ is the kinematic operator, and $\{p\}$ is the vector of the external forces acting on the boundary S of the elastic domain.

In the case where the external forces $\{\mathscr{F}\}$ and $\{p\}$ are not self-balanced, constraint conditions of a kinematic type are necessary,

$$\{\eta\} = \{\eta_0\} \qquad (11.9)$$
$$\scriptstyle (g\times 1) \quad (g\times 1)$$

valid over a part of the boundary, or else over the entire boundary.

In the sequel it will be shown how the **operator formulation** of the elastic problem, i.e. equations (11.7) and (11.8), imply the Principle of Minimum Total Potential Energy, and *vice versa*.

The total potential energy, in the case of an elastic structural element, is defined as follows:

$$W(\eta) = \int_V \Phi(q)dV - \int_V \{\eta\}^T \{\mathscr{F}\}dV - \int_S \{\eta\}^T \{p\}dS \qquad (11.10)$$

Applying Clapeyron's Theorem, we have

$$W(\eta) = \frac{1}{2}\left(\int_V \{\eta\}^T \{\mathscr{F}\}dV + \int_S \{\eta\}^T \{p\}dS \right) - \qquad (11.11)$$
$$\left(\int_V \{\eta\}^T \{\mathscr{F}\}dV + \int_S \{\eta\}^T \{p\}dS \right)$$

where the terms in brackets are identical and both represent twice the strain energy. Denoting by $[\mathscr{L}_0]$ the operator for the boundary conditions of equivalence (11.8)

$$[\mathscr{L}_0] \{\eta\} = \{p\} \qquad (11.12)$$
$$\scriptstyle (g\times g)(g\times 1) \quad (g\times 1)$$

and substituting equation (11.7) and the preceding equation (11.12), only in the first term of equation (11.11), we have

$$W(\eta) = \frac{1}{2}\left(-\int_V \{\eta\}^T [\mathscr{L}]\{\eta\}dV + \int_S \{\eta\}^T [\mathscr{L}_0]\{\eta\}dS \right) - \quad (11.13)$$
$$\left(\int_V \{\eta\}^T \{\mathscr{F}\}dV + \int_S \{\eta\}^T \{p\}dS \right)$$

The total potential energy for the external forces $\{\mathscr{F}\}$, $\{p\}$, and for an incremented displacement vector $\{\eta + \Delta\eta\}$, will be written as follows:

$$W(\eta + \Delta\eta) = \frac{1}{2}\left(-\int_V \{\eta + \Delta\eta\}^T [\mathscr{L}]\{\eta + \Delta\eta\}dV + \qquad (11.14)\right.$$
$$\left.\int_S \{\eta + \Delta\eta\}^T [\mathscr{L}_0]\{\eta + \Delta\eta\}dS \right) -$$
$$\left(\int_V \{\eta + \Delta\eta\}^T \{\mathscr{F}\}dV + \int_S \{\eta + \Delta\eta\}^T \{p\}dS \right)$$

335

The associative property allows us to dismember each integral into a sum of several integrals:

$$W(\eta + \Delta\eta) = \frac{1}{2}\left(-\int_V \{\eta\}^T[\mathscr{L}]\{\eta\}dV - \int_V \{\Delta\eta\}^T[\mathscr{L}]\{\eta\}dV - \right. \qquad (11.15)$$

$$\int_V \{\eta\}^T[\mathscr{L}]\{\Delta\eta\}dV - \int_V \{\Delta\eta\}^T[\mathscr{L}]\{\Delta\eta\}dV +$$

$$\int_S \{\eta\}^T[\mathscr{L}_0]\{\eta\}dS + \int_S \{\Delta\eta\}^T[\mathscr{L}_0]\{\eta\}dS +$$

$$\int_S \{\eta\}^T[\mathscr{L}_0]\{\Delta\eta\}dS + \int_S \{\Delta\eta\}^T[\mathscr{L}_0]\{\Delta\eta\}dS \Bigg) -$$

$$\left(\int_V \{\eta\}^T\{\mathscr{F}\}dV + \int_V \{\Delta\eta\}^T\{\mathscr{F}\}dV + \right.$$

$$\int_S \{\eta\}^T\{p\}dS + \int_S \{\Delta\eta\}^T\{p\}dS \Bigg)$$

Application of Betti's Reciprocal Theorem yields

$$W(\eta + \Delta\eta) = W(\eta) - \qquad (11.16)$$

$$\int_V \{\Delta\eta\}^T[\mathscr{L}]\{\eta\}dV + \int_S \{\Delta\eta\}^T[\mathscr{L}_0]\{\eta\}dS +$$

$$\frac{1}{2}\left(-\int_V \{\Delta\eta\}^T[\mathscr{L}]\{\Delta\eta\}dV + \int_S \{\Delta\eta\}^T[\mathscr{L}_0]\{\Delta\eta\}dS \right) -$$

$$\int_V \{\Delta\eta\}^T\{\mathscr{F}\}dV - \int_S \{\Delta\eta\}^T\{p\}dS$$

On the basis of the field equations (11.7) and the boundary equations (11.12), four of the six integrals of equation (11.16) cancel each other out, thus yielding

$$W(\eta + \Delta\eta) = W(\eta) + \frac{1}{2}\left(-\int_V \{\Delta\eta\}^T[\mathscr{L}]\{\Delta\eta\}dV + \right. \qquad (11.17)$$

$$\int_S \{\Delta\eta\}^T[\mathscr{L}_0]\{\Delta\eta\}dS \Bigg)$$

The integrals in parentheses represent the work that the body forces $\{\Delta\mathscr{F}\}$ and surface forces $\{\Delta p\}$ perform by the displacements caused by them, where

$$[\mathscr{L}]\{\Delta\eta\} = -\{\Delta\mathscr{F}\} \qquad (11.18a)$$

$$[\mathscr{L}_0]\{\Delta\eta\} = \{\Delta p\} \qquad (11.18b)$$

Hence, by virtue of Clapeyron's Theorem, we obtain

$$W(\eta + \Delta\eta) = W(\eta) + \int_V \Phi(\Delta q)dV \qquad (11.19)$$

336

The total potential energy, for the displacement field $\{\eta\}$, which resolves the elastic problem, is thus the minimum, with respect to any other arbitrarily chosen field $\{\eta + \Delta\eta\}$. The elastic potential Φ is in fact a positive definite quadratic form of the deformation characteristics $\{\Delta q\}$. We have thus shown how the operator formulation expressed by equations (11.7) and (11.8) implies the so-called **variational formulation**

$$W(\eta) = \text{minimum} \tag{11.20}$$

On the other hand, by virtue of the arbitrariness of the incremental vector $\{\Delta\eta\}$, also the reverse holds. In fact, taking the foregoing formulation in the inverse direction, we arrive at the implications of orthogonality (cf. equation (11.16)),

$$([\mathscr{L}]\{\eta\} + \{\mathscr{F}\}) \perp \{\Delta\eta\} \tag{11.21a}$$

$$([\mathscr{L}_0]\{\eta\} - \{p\}) \perp \{\Delta\eta\} \tag{11.21b}$$

which hold for any incremental vector $\{\Delta\eta\}$. From this it follows that the vectors on the left-hand sides of equations (11.21) vanish and hence the operator formulation holds good.

11.4   Ritz–Galerkin method

When the Ritz–Galerkin numeric approximation method is used, the **functional** $W(\eta)$ is assumed to be stationary, expressing the unknown function $\{\eta\}$ as the sum of known and linearly independent functions $\{\eta_i\}$, with $i = 1,2, ...,$ $(g \times n)$:

$$\{\eta\} = \sum_{i=1}^{g \times n} \alpha_i \{\eta_i\} \tag{11.22}$$

To express it using the customary language of Functional Analysis, the functional $W(\eta)$ is rendered stationary on a subspace of finite dimension, subtended by a set of known linearly independent functions. The problem thus emerges as discretized, since, instead of the vector function $\{\eta\}$, the new unknowns are now the $(g \times n)$ coefficients α_i, where n is the number of the nodes and g the degrees of freedom of each node.

Inserting the linear combination (11.12) in the expression of total potential energy (11.13), and applying the associative property, we obtain

$$W(\eta) = \frac{1}{2}\left(-\sum_{i=1}^{g \times n}\sum_{j=1}^{g \times n} \alpha_i\alpha_j \int_V \{\eta_i\}^{\mathrm{T}}[\mathscr{L}]\{\eta_j\}\mathrm{d}V + \right. \tag{11.23}$$

$$\left. \sum_{i=1}^{g \times n}\sum_{j=1}^{g \times n} \alpha_i\alpha_j \int_S \{\eta_i\}^{\mathrm{T}}[\mathscr{L}_0]\{\eta_j\}\mathrm{d}S \right) -$$

$$\left(\sum_{i=1}^{g \times n} \alpha_i \int_V \{\eta_i\}^{\mathrm{T}}\{\mathscr{F}\}\mathrm{d}V + \sum_{i=1}^{g \times n} \alpha_i \int_S \{\eta_i\}^{\mathrm{T}}\{p\}\mathrm{d}S \right)$$

In a more synthetic form, we have

$$W(\alpha) = \frac{1}{2}\{\alpha\}^{\mathrm{T}}[L]\{\alpha\} - \{\alpha\}^{\mathrm{T}}\{F\} \tag{11.24}$$

where $\{\alpha\}$ is the vector of the unknown coefficients of the linear combination. The square matrix $[L]$ has the dimension $(g \times n)$ and as elements the following integrals:

$$L_{ij} = -\int_V \{\eta_i\}^{\mathrm{T}}[\mathscr{L}]\{\eta_j\}\mathrm{d}V + \int_S \{\eta_i\}^{\mathrm{T}}[\mathscr{L}_0]\{\eta_j\}\mathrm{d}S \tag{11.25}$$

while the vector $\{F\}$ has the dimension $(g \times n)$ and as elements

$$F_{i=} \int_V \{\eta_i\}^{\mathrm{T}}\{\mathscr{F}\}\mathrm{d}V + \int_S \{\eta_i\}^{\mathrm{T}}\{p\}\mathrm{d}S \tag{11.26}$$

The matrix $[L]$ is symmetrical by virtue of Betti's Reciprocal Theorem, and is called the **Ritz–Galerkin matrix**.

The minimum of the total potential energy is obtained by deriving the expression (11.23) with respect to each coefficient α_i and equating the result to zero:

$$\sum_{j=1}^{g \times n} L_{ij}\alpha_j - F_i = 0, \;\; \text{for } i = 1, 2, ..., (g \times n) \tag{11.27}$$

We have therefore arrived at a system of $(g \times n)$ linear algebraic equations in the $(g \times n)$ unknowns α_j, which in synthetic form may be written thus:

$$[L]\{\alpha\} = \{F\} \tag{11.28}$$

That the condition of stationarity (11.28) is also a condition of minimum is guaranteed by the fact that the quadratic form present in equation (11.23) is positive definite, representing as it does the strain energy of the solid in a discretized form.

In the case where the functions $\{\eta_i\}$ are defined over the entire domain V, the matrix $[L]$ is ill-conditioned, and thus the resolving numerical algorithm presents problems of instability. With the **Isoparametric Finite Element Method**, the so-called **splines** are used as $\{\eta_i\}$ functions. These are functions defined only on subsets of the domain V (whence the term 'finite elements'), which present a value of unity in one node and zero values in all the other nodes which belong to their own domain of definition. The splines can be linear or of a higher order. A number of examples are shown in Figure 11.3. The simplest are of course the linear splines. To each node k there corresponds a spline η_k and hence, if the degrees of freedom are g, then there correspond g vectors of dimension g:

$$
\begin{array}{c}
1- \\ 2- \\ \vdots \\ \vdots \\ g-
\end{array}
\begin{bmatrix} \eta_k \\ 0 \\ 0 \\ \vdots \\ 0 \end{bmatrix}
\begin{bmatrix} 0 \\ \eta_k \\ 0 \\ \vdots \\ 0 \end{bmatrix}, ...,
\begin{bmatrix} 0 \\ 0 \\ 0 \\ \vdots \\ \eta_k \end{bmatrix}, \;\; k = 1, 2, ..., n \tag{11.29}
$$
$$
\begin{array}{ccc} 1- & 2- & \quad g- \end{array}
$$

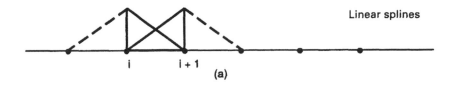

Linear splines

(a)

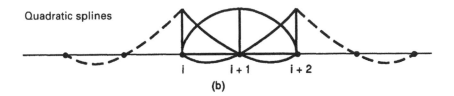

Quadratic splines

(b)

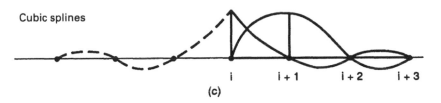

Cubic splines

(c)

Figure 11.3

It is therefore evident that, with splines, the coefficients α_i of the linear combination (11.22) coincide with the nodal values of the generalized displacements. Ordering these values in the vector $\{\delta\}$, of dimension $(g \times n)$, the resolving equation (11.28) becomes more expressive, no longer presenting as unknowns simple coefficients, but rather the nodal displacements themselves:

$$[L]\{\delta\} = \{F\} \tag{11.30}$$

11.5 Principle of Virtual Work

In this section we shall define again the Finite Element Method on the basis of the Principle of Virtual Work, and we shall show how this is equivalent to the definition proposed in the previous section and based on the Principle of Minimum Total Potential Energy.

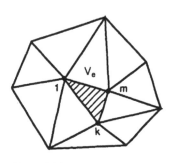

Figure 11.4

Let the elastic domain V be divided into subdomains V_e, called **finite elements** of the domain V, and let each element contain m nodal points (Figure 11.4). Usually in the two-dimensional cases (plane stress or strain conditions, plates or shells, axisymmetrical solids, etc.), the elements are triangular or quadrangular, with the nodes at the vertices, on the sides and, in some cases, inside. In three-dimensional cases, the elements are usually tetrahedrons or prisms with quadrangular sides. A number of specific examples are presented in Appendix H.

To each of the nodal points of the element V_e let there correspond a spline, defined on the sole element V_e if the node is internal, also on the adjacent element if the node is on one side, and also on all the other elements to which the

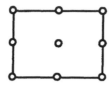
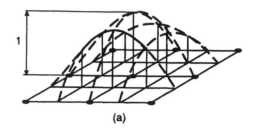

(a)

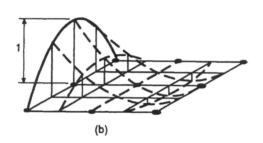
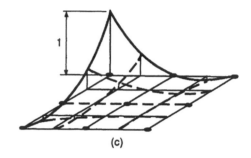

(b)

(c)

Figure 11.5

node belongs if this coincides with a vertex (Figure 11.5). To each node k of the element V_e let there then correspond a diagonal matrix made up of the g vectors (11.29)

$$\underset{(g \times g)}{[\eta_k]} = \begin{bmatrix} \eta_k & & & \\ & \eta_k & & \\ & & \ddots & \\ & & & \eta_k \end{bmatrix}, \quad k = 1, 2, \ldots, m \tag{11.31}$$

These matrices are referred to as **shape functions**, and have the following properties:

$$[\eta_k]_k = [1] \tag{11.32a}$$

$$[\eta_k]_j = [0], \quad k \neq j \tag{11.32b}$$

Using the Kronecker symbol, we can write more synthetically

$$[\eta_k]_j = [\delta_{kj}] \tag{11.33}$$

The displacement vector can be expressed by interpolation, via the shape functions and on the basis of the nodal displacements:

$$\underset{(g \times 1)}{\{\eta\}} = \overset{(g \times g)}{[\eta_1]} \ldots \overset{(g \times g)}{[\eta_k]} \ldots \overset{(g \times g)}{[\eta_m]} \begin{bmatrix} \delta_1 \\ \vdots \\ \delta_k \\ \vdots \\ \delta_m \end{bmatrix} \begin{matrix} (g \times 1) \\ \\ (g \times 1) \\ \\ (g \times 1) \end{matrix} \tag{11.34}$$

340

In compact form, the **displacement vector field** defined on the element V_e may be represented as

$$\underset{(g\times 1)}{\{\eta_e\}} = \underset{g\times(g\times m)}{[\eta_e]} \ \underset{(g\times m)\times 1}{\{\delta_e\}} \tag{11.35}$$

The **deformation characteristics vector** is obtained by derivation

$$\underset{(d\times 1)}{\{q_e\}} = \underset{(d\times g)(g\times 1)}{[\partial]\ \{\eta_e\}} \tag{11.36}$$

whence, applying relation (11.35), we obtain

$$\underset{(d\times 1)}{\{q_e\}} = \underset{(d\times g)}{[\partial]} \ \underset{g\times(g\times m)}{[\eta_e]} \ \underset{(g\times m)\times 1}{\{\delta_e\}} \tag{11.37}$$

or, in synthetic form

$$\underset{(d\times 1)}{\{q_e\}} = \underset{d\times(g\times m)(g\times m)\times 1}{[B_e] \ \{\delta_e\}} \tag{11.38}$$

where the matrix

$$\underset{d\times(g\times m)}{[B_e]} = \underset{(d\times g)}{[\partial]} \ \underset{g\times(g\times m)}{[\eta_e]} \tag{11.39}$$

is calculated by derivation of the splines. If the splines are linear, $[B_e]$ is a constant matrix on the element V_e.

The **static characteristics vector** is obtained by premultiplying the deformation characteristics vector by the Hessian matrix of strain energy

$$\underset{(d\times 1)}{\{Q_e\}} = \underset{(d\times d)}{[H]} \ \underset{d\times(g\times m)}{[B_e]} \ \underset{(g\times m)\times 1}{\{\delta_e\}} \tag{11.40}$$

Let the Principle of Virtual Work be now applied to the element V_e. This fundamental principle has been demonstrated, for the three-dimensional solid, in Section 8.4. The same formal demonstration also holds in the cases considered in Table 11.1: rectilinear or curved beams, plates or shells, plane stress or plane strain conditions, etc. In the cases where there is the presence of an intrinsic curvature, it is sufficient to substitute the operators $[\partial\ ]$, $[\partial\ ]^T$, $[\mathscr{N}\ ]$ and $[\mathscr{N}\ ]^T$, respectively, with

$$[\partial][N],\ [N]^T[\partial]^*,\ [\mathscr{N}][N],\ [N]^T[\mathscr{N}]^T \tag{11.41}$$

Let a field of virtual displacements $\{\Delta\eta\}$ be imposed on the element V_e. The Principle of Virtual Work implies the following equality:

$$\int_{V_e} \{\Delta q\}^T \{Q_e\} \mathrm{d}V = \int_{V_e} \{\Delta\eta\}^T \{\mathscr{F}\}\, \mathrm{d}V + \int_{S_e} \{\Delta\eta\}^T \{p\}\mathrm{d}S \tag{11.42}$$

On the basis of equations (11.35), (11.38) and (11.40), we deduce

$$\int_{V_e} \{\Delta\delta\}^T [B_e]^T [H][B_e]\{\delta_e\}\mathrm{d}V \tag{11.43}$$

$$= \int_{V_e} \{\Delta\delta\}^T [\eta_e]^T \{\mathscr{F}\}\mathrm{d}V + \int_{S_e} \{\Delta\delta\}^T [\eta_e]^T \{p\}\mathrm{d}S$$

341

Cancelling on both sides the virtual nodal displacement $\{\Delta\delta\}^T$, we obtain

$$\int_{V_e \; (g\times m)\times d \; (d\times d) \; d\times(g\times m)} [B_e]^T \; [H] \; [B_e] \; \mathrm{d}V \cdot \{\delta_e\}_{(g\times m)\times 1} \qquad (11.44)$$

$$= \int_{V_e \; (g\times m)\times g \, (g\times 1)} [\eta_e]^T \; \{\mathcal{F}\}\mathrm{d}V + \int_{S_e \; (g\times m)\times g \, (g\times 1)} [\eta_e]^T \; \{p\}\mathrm{d}S$$

The **vector of the nodal displacements of the element** V_e, $\{\delta_e\}$, has been carried out from under the integral sign since it is constant. The integral on the left-hand side is called the **local stiffness matrix**:

$$\underset{(g\times m)(g\times m)}{[K_e]} \; = \int_{V_e \; (g\times m)\times d \, (d\times d) \, d\times(g\times m)} [B_e]^T \; [H] \; [B_e] \; \mathrm{d}V \qquad (11.45)$$

Equation (11.44) therefore takes on the following form:

$$\underset{(g\times m)(g\times m)}{[K_e]} \quad \underset{(g\times m)\times 1}{\{\delta_e\}} \; = \; \underset{(g\times m)\times 1}{\{F_e\}} \; + \; \underset{(g\times m)\times 1}{\{p_e\}} \qquad (11.46)$$

The two vectors on the right-hand side are the **vectors of the equivalent nodal forces**, and represent the integrated effect of the forces distributed in the domain and on the boundary of the element V_e. Once the local stiffness matrix is calculated, it would be possible to determine the vector of the nodal displacements $\{\delta_e\}$ on the basis of the local relation (11.46), only if the forces $\{p\}$ acting on the boundary of the element were known beforehand and hence the vector of the equivalent forces $\{p_e\}$ were obtained by integration. Whereas, that is, the body forces $\{\mathcal{F}\}$ are a datum of the problem, the forces $\{p\}$, which exchange between them the elements at the reciprocal boundaries, are *a priori* unknown.

To get round this obstacle and, at the same time, to resolve the general problem of the determination of the vector of all the nodal displacements $\{\delta\}$ of the solid, one must add the relation (11.46), valid for the element V_e, to all the similar relations valid for the other elements of the mesh. In this way, the surface contributions $\{p_e\}$ all cancel out, except for those that do not belong to interfaces between elements, but which belong to the outer boundary. This operation is called **assemblage**, and involves a prior **expansion** of the vectors $\{\delta_e\}$, $\{F_e\}$, $\{p_e\}$, from the local dimension $(g \times m)$ to the global dimension $(g \times n)$, where n is the global number of nodal points of the mesh. The procedure will therefore be to order all the nodes of the mesh of finite elements, so as to be able to insert the nodes of the generic element V_e in the positions that they should have. This may be achieved by premultiplying the vector of the local nodal displacements $\{\delta_e\}$ by a suitable assemblage matrix $[A_e]^T$, of dimension $(g \times n) \times (g \times m)$, where all the elements are zero, except for $(g \times m)$ elements having the value of unity set in the $(g \times m)$ different rows to be filled, and corresponding to the $(g \times m)$ columns

$$\{\delta^e\} \; = \; [A_e]^T \quad \{\delta_e\} \qquad (11.47\text{a})$$

$$\{F^e\} \; = \; [A_e]^T \quad \{F_e\} \qquad (11.47\text{b})$$

$$\underset{(g\times n)\times 1}{\{p^e\}} \; = \; \underset{(g\times n)(g\times m)}{[A_e]^T} \quad \underset{(g\times m)\times 1}{\{p_e\}} \qquad (11.47\text{c})$$

342

Substituting the inverse relations in equation (11.46), we obtain

$$[K_e][A_e]\{\delta^e\} = [A_e]\{F^e\} + [A_e]\{p^e\} \tag{11.48}$$

which, premultiplied by $[A_e]^T$, yields

$$\underset{(g\times n)(g\times n)}{([A_e]^T[K_e][A_e])} \underset{(g\times n)\times 1}{\{\delta^e\}} = \underset{(g\times n)\times 1}{\{F^e\}} + \underset{(g\times n)\times 1}{\{p^e\}} \tag{11.49}$$

The relation (11.49) remains valid even if the expanded vector of local displacements $\{\delta^e\}$ is substituted with the **global vector of nodal displacements** $\{\delta\}$

$$[K^e]\{\delta\} = \{F^e\} + \{p^e\} \tag{11.50}$$

where $[K^e]$ is the local stiffness matrix in expanded form:

$$\underset{(g\times n)(g\times n)}{[K^e]} = \underset{(g\times n)(g\times m)}{[A_e]^T} \underset{(g\times m)(g\times m)}{[K_e]} \underset{(g\times m)(g\times n)}{[A_e]} \tag{11.51}$$

The local relation, but in expanded form (11.50), may be added to the similar relations for the other finite elements:

$$\underset{(g\times n)(g\times n)}{\left(\sum_e [K^e]\right)} \underset{(g\times n)\times 1}{\{\delta\}} = \underset{(g\times n)\times 1}{\{F\}} \tag{11.52}$$

having gathered to a common factor the vector of the nodal displacements $\{\delta\}$, and where

$$\{F\} = \sum_e (\{F^e\} + \{p^e\}) \tag{11.53}$$

From equations (11.44) and (11.47) we deduce

$$\{F\} = \int_V [A_e]^T [\eta_e]^T \{\mathcal{F}\} dV + \int_S [A_e]^T [\eta_e]^T \{p\} dS \tag{11.54}$$

where the integrals extended to the boundaries of the elements cancel out two by two, since the forces that the interfaces of the elements exchange are equal and opposite. It is easy to verify that the vector (11.54) has equations (11.26) as its components.

Finally, we thus derive the equation

$$[K]\{\delta\} = \{F\} \tag{11.55}$$

which coincides with equation (11.30), once the equality of the **global stiffness matrix** $[K]$ with the **Ritz–Galerkin matrix** $[L]$ has been demonstrated. However, this is possible on the basis of relations (11.45) and (11.39):

$$[K_e] = \int_{V_e} [B_e]^T [H][B_e] dV = \int_{V_e} ([\partial][\eta_e])^T [H][\partial][\eta_e] dV \tag{11.56}$$

Applying the rule of integration by parts on a three-dimensional domain (already used in the demonstration of the Principle of Virtual Work in Section 8.4), we have

$$[K_e] = -\int_{V_e} [\eta_e]^{\mathrm{T}} [\partial]^*[H][\partial][\eta_e]\mathrm{d}V + \qquad (11.57)$$

$$\int_{S_e} [\eta_e]^{\mathrm{T}} [\mathcal{N}]^{\mathrm{T}} [H][\partial][\eta_e]\mathrm{d}S$$

The minus sign in front of the first integral, which derives from the rule of integration by parts and is necessary for the terms of the matrix $[\partial]$ which are differential operators, constitutes the reason why the non-differential terms of the matrix $[\partial]$ change their algebraic sign in the adjoint matrix $[\partial]^*$.

In equation (11.57) the presence of the Lamé operators $[\mathcal{L}]$ and $[\mathcal{L}_0]$ can be recognized:

$$[K_e] = -\int_{V_e} [\eta_e]^{\mathrm{T}} [\mathcal{L}][\eta_e]\mathrm{d}V + \qquad (11.58)$$

$$\int_{S_e} [\eta_e]^{\mathrm{T}} [\mathcal{L}_0][\eta_e]\mathrm{d}S$$

The global stiffness matrix is therefore obtained by summing up all the contributions (11.58), after premultiplying them by the matrices $[A_e]^{\mathrm{T}}$ and postmultiplying them by the matrices $[A_e]$:

$$[K] = \sum_e [K^e] = -\int_V [A_e]^{\mathrm{T}} [\eta_e]^{\mathrm{T}} [\mathcal{L}][\eta_e][A_e]\mathrm{d}V + \qquad (11.59)$$

$$\int_S [A_e]^{\mathrm{T}} [\eta_e]^{\mathrm{T}} [\mathcal{L}_0][\eta_e][A_e]\mathrm{d}S$$

The contributions corresponding to the interface between elements cancel each other out. It is easy to verify that the matrix (11.59) has equations (11.25) as its elements, and hence the identity between the global stiffness matrix and the Ritz–Galerkin matrix holds:

$$[K] = [L] \qquad (11.60)$$

On the basis of equations (11.24) and (11.60), the total potential energy can thus be expressed as follows:

$$W(\delta) = \frac{1}{2}\{\delta\}^{\mathrm{T}}[K]\{\delta\} - \{\delta\}^{\mathrm{T}}\{F\} \qquad (11.61)$$

where the first term represents the strain energy of the discretized elastic solid, and the second term represents the potential energy of the external (body and surface) forces.

11.6 Kinematic boundary conditions

So far we have not considered the boundary conditions of a kinematic type, as given by equation (11.9). However, the Principle of Minimum Total Potential Energy can be reposed in the case where the external forces do not constitute a self-balanced system, so that we arrive at the same resolving equation (11.55). Some of the elements of the vector $\{\delta\}$ are in this case known, rather than unknown, terms, just as the constraint reactions now play a role of unknowns, and no longer of known terms, as hitherto assumed.

Partitioning the vectors and the stiffness matrix in such a way as to separate the free displacements from the constrained ones, we obtain

$$\begin{bmatrix} K_{LL} & K_{LV} \\ K_{VL} & K_{VV} \end{bmatrix} \begin{bmatrix} \delta_L \\ \delta_V \end{bmatrix} = \begin{bmatrix} F_L \\ F_V \end{bmatrix} \qquad (11.62)$$

While the constrained displacements $\{\delta_V\}$ are zero, or anyway predetermined in the case of imposed displacements, the free displacements $\{\delta_L\}$ represent the unknowns of the problem

$$[K_{LL}]\{\delta_L\} = \{F_L\} - [K_{LV}]\{\delta_V\} \qquad (11.63)$$

from which we obtain

$$\{\delta_L\} = [K_{LL}]^{-1}\left(\{F_L\} - [K_{LV}]\{\delta_V\}\right) \qquad (11.64)$$

The external constraint reactions are thus expressible as follows:

$$\{Q_V\} = \{F_V\} - \{F_V^0\} \qquad (11.65)$$

where

$$\{F_V\} = [K_{VL}]\{\delta_L\} + [K_{VV}]\{\delta_V\} \qquad (11.66)$$

represents the vector of the equivalent nodal forces, acting in the constrained nodes, while

$$\{F_V^0\} = \sum_e \{F_V^e\} \qquad (11.67)$$

represents the vector of the nodal forces, equivalent to the body forces.

11.7 Dynamics of elastic solids

If the body force $\{F\}$ in the operator equation (11.7) is substituted by the force of inertia,

$$-[\rho]\frac{\partial^2}{\partial t^2}\{\eta\} \qquad (11.68)$$

we obtain the equation of free oscillations for the elastic solid under examination

$$\left([\mathscr{L}] - [\rho]\frac{\partial^2}{\partial t^2}\right)\{\eta\} = \{0\} \qquad (11.69)$$

Here $[\rho]$ denotes the **density matrix**, which is a diagonal matrix of dimension $(g \times g)$, where the density ρ of the material corresponds to the translations, and the moment of inertia $(1/12)\,\rho h^2$ corresponds to the rotations, h being the thickness of the beam or plate.

In the absence of body and surface forces of a static type and in the presence of inertial forces, relation (11.58) becomes

$$[K_{ed}] = -\int_{V_e} [\eta_e]^T[\mathscr{L}][\eta_e]\mathrm{d}V + \int_{V_e} [\eta_e]^T[\rho][\eta_e]\mathrm{d}V \cdot \frac{\partial^2}{\partial t^2} + \qquad (11.70)$$

$$\int_{S_e} [\eta_e]^T[\mathscr{L}_0][\eta_e]\mathrm{d}S$$

FINITE ELEMENT METHOD

from which we obtain the **dynamic stiffness matrix**

$$[K_{ed}] = [K_e] + [M_e]\frac{\partial^2}{\partial t^2} \tag{11.71}$$

with

$$[M_e] = \int_{V_e} [\eta_e]^T [\rho][\eta_e] dV \tag{11.72}$$

which represents the **local matrix of masses**.

In equivalent manner, replacing in equation (11.44) the body forces $\{\mathcal{F}\}$ with the inertial forces (11.68), we obtain

$$[K_e]\{\delta_e\} = -\int_{V_e} [\eta_e]^T [\rho][\eta_e] dV \cdot \frac{\partial^2}{\partial t^2}\{\delta_e\} + \int_{S_e} [\eta_e]^T \{p\} dS \tag{11.73}$$

At this point, expanding and assembling the local dynamic stiffness matrices, we obtain

$$\sum_e [K^{ed}] = \sum_e [A_e]^T [K_{ed}][A_e] \tag{11.74}$$

a relation which is analogous to equations (11.51) and (11.52), and furnishes the **global matrix of masses**

$$[M] = \sum_e [A_e]^T [M_e][A_e] \tag{11.75}$$

Finally we obtain then the equation

$$[K]\{\delta\} + [M]\{\ddot{\delta}\} = \{0\} \tag{11.76}$$

which is formally analogous to the equation of a harmonic oscillator with one degree of freedom, devoid of viscous forces and forcing loads. It should be noted that $[M]$ is not in general a diagonal matrix.

It would have been possible to arrive at the same equation by considering the inertial forces as static body forces, and by applying sequentially equations (11.53), (11.54), (11.68), (11.35) and (11.47a).

In relation to equation (11.76), let a solution be chosen of the form

$$\{\delta(t)\} = \{\delta\} f(t) \tag{11.77}$$

separating the temporal variable t and considering the same oscillatory law for all the generalized coordinates of the system.

Substituting equation (11.77) into equation (11.76), we obtain

$$-\frac{\ddot{f}}{f} = \frac{\{\delta\}^T [K]\{\delta\}}{\{\delta\}^T [M]\{\delta\}} = \lambda \tag{11.78}$$

and hence the separation of the temporal problem from the spatial one:

$$\ddot{f} + \lambda f = 0 \tag{11.79a}$$

$$\{\delta\}^T ([K] - \lambda[M])\{\delta\} = 0 \tag{11.79b}$$

346

DYNAMICS OF ELASTIC SOLIDS

From equation (11.79b) we obtain

$$\det\ ([K] - \lambda[M]) = 0 \qquad (11.80)$$

which is an algebraic equation in the unknown λ, of a degree equal to the number of degrees of freedom of the system. This equation is called the **characteristic equation**, and its solutions are the **eigenvalues** of the problem. The **eigenvectors** are obtained, but for a factor of proportionality, from the equation

$$([K] - \lambda[M])\{\delta\} = \{0\} \qquad (11.81)$$

The eigenvectors have the property of orthonormality. Let equation (11.81) be written for two different eigenvectors

$$[K]\{\delta_j\} = \lambda_j[M]\{\delta_j\} \qquad (11.82a)$$

$$[K]\{\delta_k\} = \lambda_k[M]\{\delta_k\} \qquad (11.82b)$$

Premultiplying the former by $\{\delta_k\}^T$ and the latter by $\{\delta_j\}^T$, taking into account the symmetry of $[K]$ and $[M]$ and subtracting member from member, we obtain

$$0 = (\lambda_j - \lambda_k)\{\delta_k\}^T[M]\{\delta_j\} \qquad (11.83)$$

and hence, in normal form

$$\{\delta_k\}^T[M]\{\delta_j\} = \delta_{jk} \qquad (11.84a)$$

where the term on the right-hand side of the equation is the Kronecker symbol. Equations (11.82a) and (11.84a) also imply

$$\{\delta_k\}^T[K]\{\delta_j\} = \lambda_j\delta_{jk} \qquad (11.84b)$$

Since the matrices $[M]$ and $[K]$ are symmetrical and positive definite, it is possible to demonstrate how the eigenvalues λ_j, $j = 1, 2, ..., (g \times n)$ are all real and positive. Equation (11.79a) is thus written

$$\ddot{f}_i + \omega_i^2 f_i = 0, \ \ i = 1, 2, ..., (g \times n) \qquad (11.85)$$

and has the following integral:

$$f_i(t) = A_i \cos \omega_i t + B_i \sin \omega_i t \qquad (11.86)$$

The complete integral (11.77) can therefore be put in the following form:

$$\{\delta(t)\} = \sum_{i=1}^{g \times n} \{\delta_i\}(A_i \cos \omega_i t + B_i \sin \omega_i t) \qquad (11.87)$$

The $2 \times (g \times n)$ constants A_i, B_i are determined by imposing the initial conditions, in a manner similar to that adopted in Chapter 10 in the case of deflected beams:

$$\{\delta(0)\} = \{\delta_0\} \qquad (11.88a)$$

$$\{\dot{\delta}(0)\} = \{\dot{\delta}_0\} \qquad (11.88b)$$

From equations (11.87) and (11.88) we deduce in fact that

347

$$\sum_{i=1}^{g \times n} A_i \{\delta_i\} = \{\delta_0\} \qquad (11.89a)$$

$$\sum_{i=1}^{g \times n} B_i \omega_i \{\delta_i\} = \{\dot{\delta}_0\} \qquad (11.89b)$$

We thus obtain two distinct systems of equations in the unknowns A_i and B_i, respectively, which can be resolved by transposition of the individual members

$$\sum_{i=1}^{g \times n} A_i \{\delta_i\}^T = \{\delta_0\}^T \qquad (11.90a)$$

$$\sum_{i=1}^{g \times n} B_i \omega_i \{\delta_i\}^T = \{\dot{\delta}_0\}^T \qquad (11.90b)$$

Postmultiplying by $[M]\{\delta_j\}$ and exploiting the property of orthonormality, we obtain

$$A_j = \{\delta_0\}^T [M]\{\delta_j\} \qquad (11.91a)$$

$$B_j = \frac{1}{\omega_j} \{\dot{\delta}_0\}^T [M]\{\delta_j\} \qquad (11.91b)$$

for $j = 1, 2, ..., (g \times n)$.

As in the case of deflected beams, also in the more general framework of the Finite Element Method, a system perturbed initially according to an eigenvector, with zero initial velocity, then continues to oscillate indefinitely in proportion to that deformed configuration. Assume that

$$\{\delta_0\} = a\{\delta_i\} \qquad (11.92a)$$

$$\{\dot{\delta}_0\} = \{0\} \qquad (11.92b)$$

From equations (11.91) we derive

$$A_j = a\delta_{ij} \qquad (11.93a)$$

$$B_j = 0 \qquad (11.93b)$$

and hence the complete integral is

$$\{\delta(t)\} = a\{\delta_i\} \cos \omega_i t = \{\delta_0\} \cos \omega_i t \qquad (11.94)$$

The eigenvectors being known, it is possible to consider as generalized coordinates of the system the temporal functions f_i. Then ordering these functions, called **normal coordinates**, in the vector $\{f\}$, we perform the following coordinate transformation:

$$\{\delta(t)\} = [\Delta]\{f\} \qquad (11.95)$$

which is an alternative way of writing equation (11.87), $[\Delta]$ being the **modal matrix**, which has as its columns the eigenvectors

348

$$[\Delta] = [\delta_1 \mid \delta_2 \mid \ldots \mid \delta_{g \times n}] \qquad (11.96)$$

Substituting equation (11.95) in equation (11.76), and premultiplying by $[\Delta]^T$, we obtain

$$([\Delta]^T [M][\Delta])\{\ddot{f}\} + ([\Delta]^T [K][\Delta])\{f\} = \{0\} \qquad (11.97)$$

Taking into account equations (11.84a,b), we have on the other hand

$$[\Delta]^T [M][\Delta] = [1] \qquad (11.98a)$$

$$[\Delta]^T [K][\Delta] = [\Lambda] \qquad (11.98b)$$

where $[1]$ is the unit matrix, and $[\Lambda]$ is the diagonal matrix of the eigenvalues. Equation (11.97) leads therefore to the vector form

$$\{\ddot{f}\} + [\Lambda]\{f\} = \{0\} \qquad (11.99)$$

and thus to the scalar form (11.85). The equations of motion are thus decoupled, each containing a single unknown, made up of one of the normal coordinates. The transformation of coordinates (11.95) allows the expressions of the elastic potential and kinetic energy to be reduced to the so-called **canonical form**:

$$W = \frac{1}{2}\{\delta\}^T [K]\{\delta\} = \frac{1}{2}\{f\}^T [\Lambda]\{f\} \qquad (11.100a)$$

$$\mathscr{T} = \frac{1}{2}\{\dot{\delta}\}^T [M]\{\dot{\delta}\} = \frac{1}{2}\{\dot{f}\}^T [1]\{\dot{f}\} \qquad (11.100b)$$

The application of the Finite Element Method to physical problems different from those examined in this chapter is illustrated in Appendix I. A number of complementary topics are also dealt with in the appendices; these include the problem of initial strains and residual stresses (Appendix J), the dynamic behaviour of elastic solids with linear damping (Appendix K) and plane elasticity with couple stresses (Appendix L).

12 Structural symmetry

12.1 Introduction

In this chapter we shall consider one-, two- and three-dimensional solids, with properties of geometrical, constraint and static symmetry. These are structures which present axes or centres of symmetry about which the distribution of matter, the constraints and the loads applied are symmetrical quantities. As we shall be able to investigate more fully in the sequel, the most notable consequence deriving from the properties of symmetry of a statically indeterminate structure is the reduction of its effective degree of indeterminacy. The increase in regularity with respect to structures devoid of symmetry is in general the cause of a decrease in static indeterminacy.

As regards beam systems, both axial symmetry and polar symmetry will be considered, as well as the corresponding skew symmetries. The study of shells having double curvature will be restricted to the case of shells of revolution, loaded both symmetrically and otherwise with respect to the axis of symmetry. This study will then be particularized to the specific, but, from the technical standpoint, highly significant, cases of membranes of revolution, thin shells of revolution, circular plates and cylindrical shells. In this context, reference will also be made to the problem of pressurized vessels having a cylindrical shape and flat or spherical bases. Finally, the problem of axi-symmetrical three-dimensional solids, loaded either symmetrically or otherwise with respect to the axis of symmetry, will be dealt with.

12.2 Beam systems with axial symmetry

A beam system is said to be symmetrical with respect to an axis when one of the two halves into which the structure is subdivided by the axis comes to superpose itself on the other, if it is made to rotate by 180° about the axis itself (Figure 12.1(a)). A beam system with axial symmetry is said to be symmetrically loaded if, in the above-mentioned rotation, the loads that act on one half also come to superpose themselves on those acting on the other half. In addition to the beams and the loads, the constraints, both external and internal, must of course also respect the condition of symmetry so that the structural behaviour should be specularly symmetrical.

In a beam system with axial symmetry, the structural response, whether static or kinematic, must logically prove symmetrical. This means that the characteristics, both static and deformation, must be specular. Whereas then the axial force and the bending moment are equal and have the same sign in the pairs of symmetrical points, shearing force is equal but has an opposite sign, with a skew-symmetrical diagram. Likewise, the rotations and elastic displacements perpendicular to the axis of symmetry are equal and opposite, while the elastic displacements in the direction of the axis are equal and of the same sign.

If the conditions described above are to be maintained on the axis of symmetry, it will be necessary for the shearing force, as well as the rotation and the component of the displacement orthogonal to the axis of symmetry, to

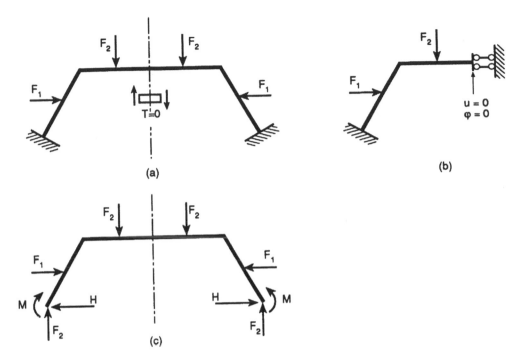

Figure 12.1

vanish. Thus, whereas the shearing force T vanishes in order to satisfy symmetry and equilibrium simultaneously, the displacement u and rotation φ vanish to satisfy symmetry and congruence simultaneously (Figure 12.1(a)). These conditions, both static and kinematic, at the points where the axis of symmetry encounters the structure, are realized by a double rod perpendicular to the axis itself (Figure 12.1(b)). It is therefore possible to reduce the study of the entire structure to that of one half, constrained at a point corresponding to its axis of symmetry by a double rod. Consider then that the axial force and bending moment diagrams are symmetrical, whereas the shearing force diagram is skew-symmetrical.

The reduced structure of Figure 12.1(b) has two degrees of indeterminacy, while the original structure apparently has three (Figure 12.1(a)). This means that the structure of Figure 12.1(a) actually has two degrees of indeterminacy for reasons of symmetry. Whereas in fact the vertical reactions are each equal to one half of the vertical load, the fixed-end moments and the horizontal reactions are represented by equal and symmetrical loadings, which remain, however, statically indeterminate (Figure 12.1(c)).

In the case where, instead of the internal fixed joint, there is a weaker constraint at the axis of symmetry, it is possible to apply once again what has already been said, but excluding *a priori* from the conditions of symmetry the characteristics not transmitted by the constraint itself, and at the same time including the relative displacements permitted by it. If, for example, the two symmetrical parts of a structure are connected by a hinge (Figure 12.2(a)), the moment in the centre will vanish by definition of the hinge constraint, while the shear will vanish by virtue of symmetry. Hence the only remaining static characteristic transmitted by the hinge will be the axial force. The existence of

351

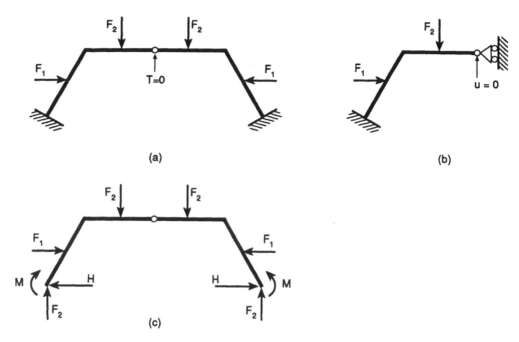

Figure 12.2

the hinge allows, on the other hand, relative rotations between the two parts, just as symmetry allows displacements of the centre in the direction of the axis. However, displacements of the centre are not possible in either direction perpendicularly to the axis, for reasons of symmetry, nor are detachment and overlapping possible for reasons of congruence.

These conditions, both static and kinematic, are realized by a vertically moving rolling support or by a horizontal connecting rod. The equivalent structure of Figure 12.2(b) has one degree of indeterminacy, whilst the original structure of Figure 12.2(a) apparently has two. As before, the vertical reactions are statically determinate and each equal to one half of the vertical load, while the horizontal reactions and the fixed-end moments are linked together by the equation of equilibrium to rotation of each part about the hinge (Figure 12.2(c)).

In the case where there are columns or uprights on the axis of symmetry (Figure 12.3(a)), it is necessary to consider, in addition to the conditions of symmetry, the conditions of equilibrium of the central fixing-node (Figure 12.3(b)). It is simple to conclude that the upright is loaded by an axial force which, in absolute value, is twice the shear transmitted by each of the two horizontal beams, while the characteristics of shear and bending moment are zero on the upright for reasons of symmetry. If the upright is considered as axially undeformable, the equivalent structure is reduced to that of Figure 12.3(c), where the centre is constrained with a perfect fixed joint. This structure is thus indeterminate to the second degree, whereas the original structure is apparently indeterminate to the third degree (Figure 12.3(a)). If, instead, we wish to take into account the axial compliance of the central upright, it is necessary to consider a fixed joint elastically compliant to vertical translation, having a stiffness of $EA/2h$, where h denotes the height of the upright.

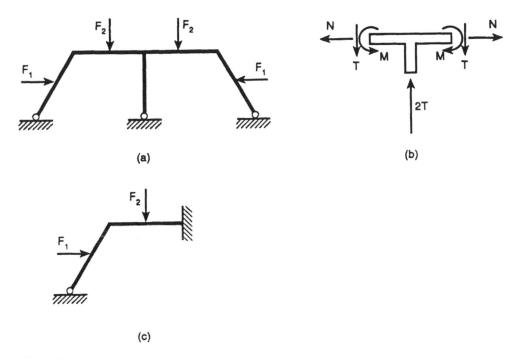

(a)

(b)

(c)

Figure 12.3

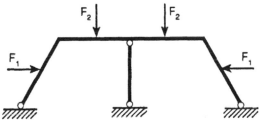

Figure 12.4

Nothing changes with respect to the previous case, if the central upright, instead of being fixed, is only hinged to the horizontal beam, and thus consists of a simple vertical connecting rod (Figure 12.4). As before, it will transmit only a vertical force to the overlying beam. The equivalent scheme is thus yet again that of Figure 12.3(c).

Finally, let us take the case where a concentrated force is applied in the centre (Figure 12.5(a)). By reason of the equilibrium of the central beam element and from symmetry, it is possible to refer to the equivalent scheme of Figure 12.5(b), where the end constrained by the double rod is also loaded by a force equal to one half of the total.

Applying in inverse manner the considerations so far made, it is possible to calculate the elastic rotation at the hinged end of a beam, constrained at the opposite end by a double rod (Figure 12.6(a)). From symmetry, this scheme is

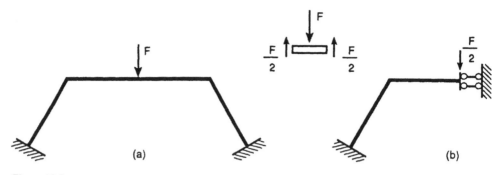

(a)

(b)

Figure 12.5

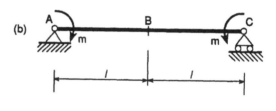

(a)

(b)

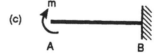

(c)

Figure 12.6

in fact equivalent to a beam of twice the length, hinged at the ends and loaded by two specular loads (Figure 12.6(b)). In the specific case of the moment applied to the hinged end (Figure 12.6(a)), the scheme of Figure 12.6(b) yields the rotation

$$\varphi_A = -\varphi_C = -\frac{m(2l)}{3EI} - \frac{m(2l)}{6EI} = -\frac{ml}{EI} \tag{12.1}$$

The elastic line of the beam of Figure 12.6(a) is, on the other hand, equal, but for an additional constant, to that of the cantilever of Figure 12.6(c), where the rotation of the free end is given by relation (10.54b).

12.3 Beam systems with axial skew-symmetry

A symmetrical beam system is said to be loaded in a skew-symmetrical way when the loads acting on one of the halves are the opposite of, and symmetrical to, the loads acting on the remaining half (Figure 12.7(a)).

354

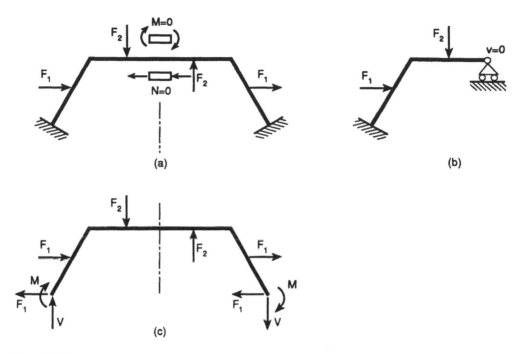

Figure 12.7

In a beam system with axial skew-symmetry, the structural response, both static and kinematic, must logically be skew-symmetrical. This means that the characteristics, both static and deformation, must be opposite to those produced at the symmetrical points.

Whereas, then, the so-called symmetrical characteristics – axial force and bending moment – will present a skew-symmetrical diagram, the skew-symmetrical characteristic – the shearing force – will present a symmetrical diagram. Likewise, the rotations and the elastic displacements orthogonal to the axis of symmetry are equal and have the same sign, whilst the elastic displacements in the direction of the axis are equal and opposite.

If the above conditions are also to be respected on the axis of symmetry, both the axial force and the bending moment must vanish, as must the component of the displacement in the direction of the axis of symmetry. Thus, whereas the axial force N and the bending moment M vanish in order to satisfy skew-symmetry and equilibrium simultaneously, the displacement v vanishes to satisfy skew-symmetry and congruence simultaneously (Figure 12.7(a)). These conditions, which are static and kinematic, at the points where the axis of symmetry encounters the structure, are realized by a roller support moving orthogonally to the axis itself (Figure 12.7(b)). The study of the original structure is thus reduced to that of one of its halves, constrained in the centre with a roller support. The reduced structure of Figure 12.7(b) thus has one degree of redundancy, whereas the original structure apparently has three (Figure 12.7(a)). While in fact the horizontal reactions are each equal to one half of the horizontal load, the fixed-end moments and the vertical reactions are represented by equal and skew-symmetrical loads, which, respecting the condition of equilibrium to rotation, remain only once indeterminate (Figure 12.7(c)).

355

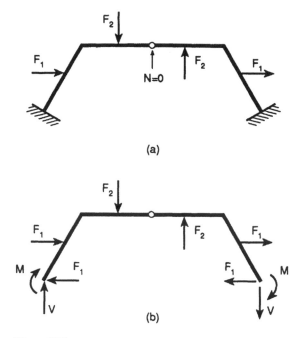

Figure 12.8

In the case where, instead of the internal fixed joint, there is a weaker constraint at the axis of symmetry, the considerations made above may be repeated, taking into account, however, only the reactions transmitted by the constraint and adding the relative displacements allowed by this. If, for instance, the two symmetrical parts of a structure loaded skew-symmetrically are connected by a hinge (Figure 12.8(a)), the shear, as before, is the only characteristic transmitted, and hence the reduced scheme is again that of Figure 12.7(b). This scheme is indeterminate to the first degree, whilst the original structure of Figure 12.8(a) is apparently indeterminate to the second degree. The vertical reactions V and the fixed-end moments M (Figure 12.8(b)) are in fact linked by a condition of equilibrium to rotation, and are thus once indeterminate.

If, instead, the two symmetrical parts are connected by a double rod (Figure 12.9(a)), the constraint cannot transmit either of the two symmetrical characteristics and the equivalent scheme will be represented by the cantilever beam of Figure 12.9(b). The original structure of Figure 12.9(a) is therefore substantially statically determinate, since the horizontal reactions are each equal to one half of the horizontal load, just as the vertical reactions are each equal to the vertical load acting on the corresponding part (Figure 12.9(c)). The fixed-end moments may, on the other hand, be determined via an equation of global equilibrium with regard to rotation.

In the case where there are columns or uprights on the axis of symmetry (Figure 12.10(a)), it is necessary to consider, in addition to the conditions of skew-symmetry, the conditions of equilibrium of the central fixing-node (Figure 12.10(b)). Unlike in the case of symmetry, the upright is subjected to

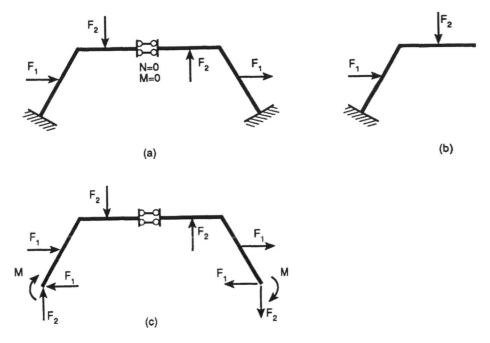

Figure 12.9

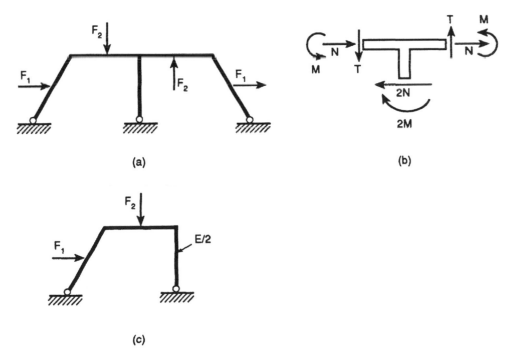

Figure 12.10

357

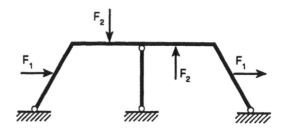

Figure 12.11

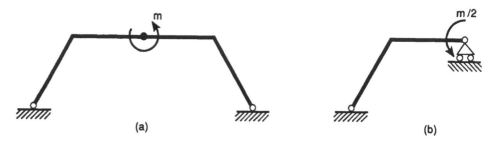

(a)   (b)

Figure 12.12

moment and shear, while the axial force exerted on it is zero. The bending moment and shearing force, in the immediate vicinity of the node, are equal, respectively, to twice the moment and twice the axial force acting on the horizontal beams. It is thus possible to consider the reduced scheme of Figure 12.10(c), where the material of the upright is considered with its elastic modulus halved.

In the case where, in place of the upright, there is a simple connecting rod (Figure 12.11), it is obvious that the reduced scheme which must be referred to remains that of Figure 12.7(b), with a hinge instead of the built-in support.

When a concentrated moment m is applied in the centre (Figure 12.12(a)), it is possible to consider this as a skew-symmetrical load consisting of two moments equal to $m/2$ and having the same sign. The equivalent scheme is that of Figure 12.12(b). In the case, therefore, of a simply supported beam (Figure 12.13(a)), we revert to the supported beam of halved length (Figure 12.13(b)). The elastic rotation of the ends thus equals

$$\varphi_A = \varphi_B = -\frac{\left(\dfrac{m}{2}\right)\left(\dfrac{l}{2}\right)}{6EI} = -\frac{ml}{24EI} \tag{12.2}$$

while the elastic rotation of the centre section is

$$\varphi_C = \frac{\left(\dfrac{m}{2}\right)\left(\dfrac{l}{2}\right)}{3EI} = \frac{ml}{12EI} \tag{12.3}$$

358

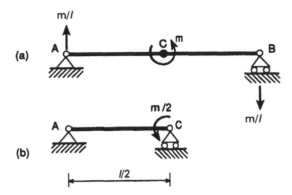

Figure 12.13

12.4 Beam systems with polar symmetry

A beam system is said to be symmetrical with respect to a pole when one of the two halves into which the structure is subdivided by the pole may be superposed on the other half, if made to rotate by 180° about the pole itself (Figure 12.14(a)). A system of beams with polar symmetry is said to be

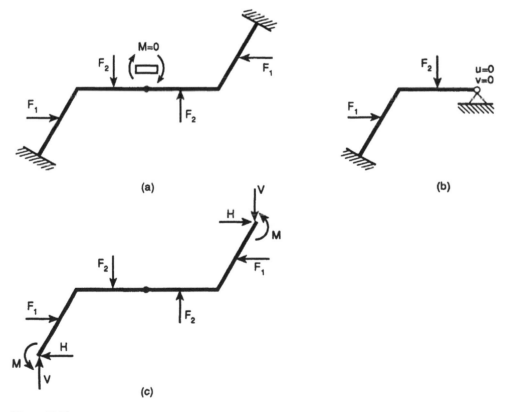

Figure 12.14

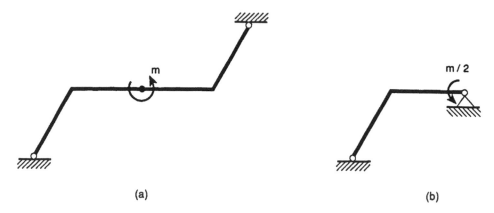

(a)   (b)

Figure 12.15

symmetrically loaded if, in the above rotation, the loads acting on one half are also superposed on those acting on the remaining half.

If the structure presents polar symmetry, it is logical that from polar-symmetrical causes there follow polar-symmetrical effects, which are both static and kinematic. In particular, at the pole the moment vanishes in order to satisfy polar symmetry and equilibrium simultaneously, just as the displacement vanishes to satisfy polar symmetry and congruence simultaneously. These conditions are realized by a hinge, so that the equivalent reduced structure appears as in Figure 12.14(b). This scheme thus proves to have two degrees of indeterminacy. The original structure also has two degrees of indeterminacy (Figure 12.14(c)). The reactions H, V, M are in fact linked only by the global equation of equilibrium with regard to rotation, the equations of equilibrium with regard to translation already being identically satisfied.

Of course, in the case where there is originally a hinge at the pole, the considerations outlined above again all apply.

Finally, in the case where a concentrated moment m is applied at the pole (Figure 12.15(a)), this load can be considered as polar-symmetrical and as consisting of two moments equal to $m/2$ and having the same sign. The reduced scheme is thus that of Figure 12.15(b).

12.5 Beam systems with polar skew-symmetry

A polar-symmetrical beam system is said to be loaded skew-symmetrically when the loads that act on one of the halves are the opposite of, and symmetrical to those acting on the remaining half (Figure 12.16(a)).

At the pole, the axial force and the shearing force vanish in order to satisfy polar skew-symmetry and equilibrium simultaneously, just as the elastic rotation vanishes to satisfy polar skew-symmetry and congruence simultaneously. These conditions are realized by a double articulated parallelogram (Figure 12.16(b)). The equivalent scheme appears to be indeterminate to the first degree, whereas the original structure is apparently indeterminate to the third degree. The degree of residual redundancy is due to the indeterminacy of the fixed-end moment M (Figure 12.16(c)).

In the case where there is a hinge at the pole (Figure 12.17(a)), the equivalent scheme reduces to the cantilever beam of Figure 12.17(b). The structure is therefore substantially statically determinate. The fixed-end moment M in this

360

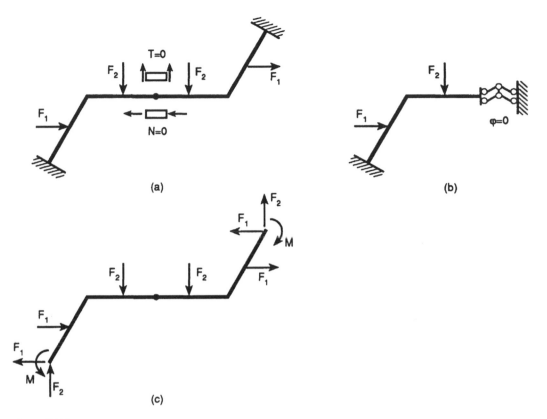

Figure 12.16

case is in fact determined via a condition of partial equilibrium to rotation about the hinge (Figure 12.17(c)).

Finally, in the case where a concentrated force F is applied at the pole (Figure 12.18(a)), this load can be considered polar skew-symmetrical and consisting of two forces equal to $F/2$, having the same sign. The reduced scheme is thus that shown in Figure 12.18(b).

12.6 Non-symmetrically loaded shells of revolution

The term **shell of revolution** refers to a shell, generally of double curvature, generated by the complete rotation of a plane curve $r(z)$ about the axis of symmetry Z (Figure 12.19). The set of the infinite configurations which the generating curve $r(z)$ assumes in its rotation are called **meridians**. The set of the infinite circular trajectories described by the individual points of the curve are called **parallels**. The meridians and the parallels represent the so-called lines of curvature, on which a system of principal curvilinear coordinates s_1 s_2 can be defined.

Denoting by s_1 the curvilinear coordinate along the meridians, and by s_2 the curvilinear coordinate along the parallels, we have (Figure 12.19)

$$ds_1 = ds = \frac{dz}{\cos \alpha} \tag{12.4a}$$

$$ds_2 = r\, d\vartheta \tag{12.4b}$$

361

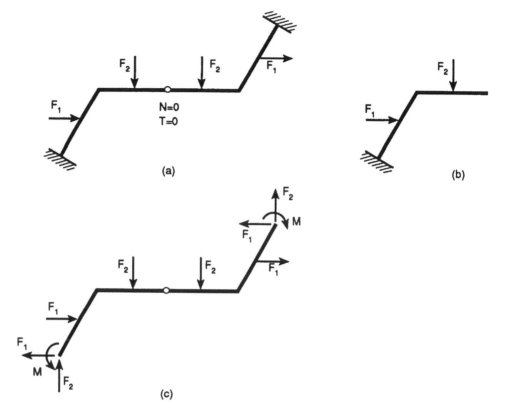

Figure 12.17

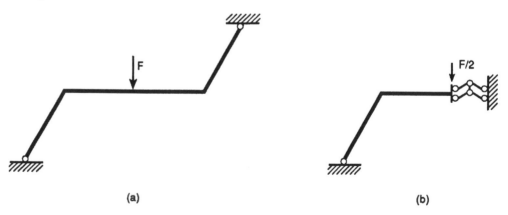

Figure 12.18

where α is the angle which the tangent to the meridian forms with the axis of symmetry, which is equal also to the angle that the normal to the surface forms with the radius r (Figure 12.19), and where ϑ represents the longitude. On the basis of **Meusnier's Theorem**, the radius r and the principal radius of curvature R_2 are linked by the following relation:

$$r = R_2 \cos \alpha \qquad (12.5)$$

362

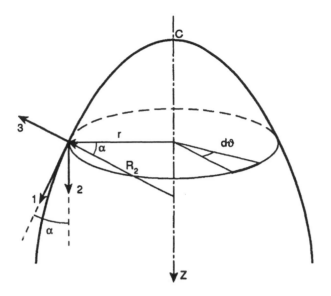

Figure 12.19

Whereas the variations of the radius of curvature R_1 are equal to zero with respect to the coordinate s_2 (i.e. along the parallels), the variations of the radius of curvature R_2 with respect to the coordinate s_1 (i.e. along the meridians) are:

$$\frac{\partial R_2}{\partial s_1} = \frac{\partial}{\partial s}\left(\frac{r}{\cos \alpha}\right)$$
$$= \frac{\partial r}{\partial s}\left(\frac{1}{\cos \alpha}\right) + r\frac{\sin \alpha}{\cos^2 \alpha}\frac{\partial \alpha}{\partial s}. \tag{12.6}$$

By the definition of curvature $\partial \alpha / \partial s = -1/R_1$, and thus from equation (12.6) we have:

$$\frac{\partial R_2}{\partial s_1} = \frac{\sin \alpha}{\cos \alpha} - \frac{R_2}{R_1}\frac{\sin \alpha}{\cos \alpha}$$
$$= \tan \alpha \left(1 - \frac{R_2}{R_1}\right). \tag{12.7}$$

The term $1/\rho_2$, recurring in the kinematic (10.185) and static (10.187) matrices, can be expressed as follows:

$$\frac{1}{\rho_2} = \frac{R_1}{R_2(R_1 - R_2)}\frac{\partial R_2}{\partial s_1} = \frac{\tan \alpha}{R_2} = \frac{\sin \alpha}{r}. \tag{12.8}$$

363

The kinematic equations for the shells of revolution non-symmetrically loaded are thus:

$$
\begin{bmatrix} \varepsilon_s \\ \varepsilon_\vartheta \\ \gamma_{s\vartheta} \\ \gamma_s \\ \gamma_\vartheta \\ \chi_s \\ \chi_\vartheta \\ \chi_{s\vartheta} \end{bmatrix} =
\begin{bmatrix}
\dfrac{\partial}{\partial s} & 0 & +\dfrac{1}{R_1} & 0 & 0 \\[2mm]
+\dfrac{\sin\alpha}{r} & \dfrac{1}{r}\dfrac{\partial}{\partial\vartheta} & +\dfrac{1}{R_2} & 0 & 0 \\[2mm]
\dfrac{1}{r}\dfrac{\partial}{\partial\vartheta} & \left(\dfrac{\partial}{\partial s}-\dfrac{\sin\alpha}{r}\right) & 0 & 0 & 0 \\[2mm]
-\dfrac{1}{R_1} & 0 & \dfrac{\partial}{\partial s} & +1 & 0 \\[2mm]
0 & -\dfrac{1}{R_2} & \dfrac{1}{r}\dfrac{\partial}{\partial\vartheta} & 0 & +1 \\[2mm]
0 & 0 & 0 & \dfrac{\partial}{\partial s} & 0 \\[2mm]
0 & 0 & 0 & +\dfrac{\sin\alpha}{r} & \dfrac{1}{r}\dfrac{\partial}{\partial\vartheta} \\[2mm]
0 & 0 & 0 & \dfrac{1}{r}\dfrac{\partial}{\partial\vartheta} & \left(\dfrac{\partial}{\partial s}-\dfrac{\sin\alpha}{r}\right)
\end{bmatrix}
\begin{bmatrix} u \\ v \\ w \\ \varphi_s \\ \varphi_\vartheta \end{bmatrix}.
\tag{12.9}
$$

The static equations are dual with respect to equations (12.9):

$$
\begin{bmatrix}
\left(\dfrac{\partial}{\partial s}+\dfrac{\sin\alpha}{r}\right) & -\dfrac{\sin\alpha}{r} & \dfrac{1}{r}\dfrac{\partial}{\partial\vartheta} & +\dfrac{1}{R_1} & 0 & 0 & 0 & 0 \\[2mm]
0 & \dfrac{1}{r}\dfrac{\partial}{\partial\vartheta} & \left(\dfrac{\partial}{\partial s}+\dfrac{2\sin\alpha}{r}\right) & 0 & +\dfrac{1}{R_2} & 0 & 0 & 0 \\[2mm]
-\dfrac{1}{R_1} & -\dfrac{1}{R_2} & 0 & \left(\dfrac{\partial}{\partial s}+\dfrac{\sin\alpha}{r}\right) & \dfrac{1}{r}\dfrac{\partial}{\partial\vartheta} & 0 & 0 & 0 \\[2mm]
0 & 0 & 0 & -1 & 0 & \left(\dfrac{\partial}{\partial s}+\dfrac{\sin\alpha}{r}\right) & -\dfrac{\sin\alpha}{r} & \dfrac{1}{r}\dfrac{\partial}{\partial\vartheta} \\[2mm]
0 & 0 & 0 & 0 & -1 & 0 & \dfrac{1}{r}\dfrac{\partial}{\partial\vartheta} & \left(\dfrac{\partial}{\partial s}+\dfrac{2\sin\alpha}{r}\right)
\end{bmatrix}
\begin{bmatrix} N_s \\ N_\vartheta \\ N_{s\vartheta} \\ T_s \\ T_\vartheta \\ M_s \\ M_\vartheta \\ M_{s\vartheta} \end{bmatrix}
+
\begin{bmatrix} p_s \\ p_\vartheta \\ q \\ 0 \\ 0 \end{bmatrix}
=
\begin{bmatrix} 0 \\ 0 \\ 0 \\ 0 \\ 0 \end{bmatrix}
\tag{12.10}
$$

12.7 Symmetrically loaded shells of revolution

When a shell of revolution is loaded symmetrically with respect to axis Z, relations (12.9) and (12.10) simplify, since only the curvilinear coordinate s is present as an independent variable, while the displacement v along the parallels vanishes, as well as the deformations $\gamma_{s\vartheta}$, γ_ϑ, $\chi_{s\vartheta}$, and the corresponding internal reactions $N_{s\vartheta}$, T_ϑ, $M_{s\vartheta}$:

364

$$
\begin{bmatrix} \varepsilon_s \\ \varepsilon_\vartheta \\ \gamma_s \\ \chi_s \\ \chi_\vartheta \end{bmatrix} = \begin{bmatrix} \dfrac{d}{ds} & \dfrac{1}{R_1} & 0 \\[2mm] +\dfrac{\sin\alpha}{r} & \dfrac{1}{R_2} & 0 \\[2mm] -\dfrac{1}{R_1} & \dfrac{d}{ds} & +1 \\[2mm] 0 & 0 & \dfrac{d}{ds} \\[2mm] 0 & 0 & +\dfrac{\sin\alpha}{r} \end{bmatrix} \begin{bmatrix} u \\ w \\ \varphi_s \end{bmatrix} \qquad (12.11)
$$

$$
\begin{bmatrix} \left(\dfrac{d}{ds}+\dfrac{\sin\alpha}{r}\right) & -\dfrac{\sin\alpha}{r} & \dfrac{1}{R_1} & 0 & 0 \\[2mm] -\dfrac{1}{R_1} & -\dfrac{1}{R_2} & \left(\dfrac{d}{ds}+\dfrac{\sin\alpha}{r}\right) & 0 & 0 \\[2mm] 0 & 0 & -1 & \left(\dfrac{d}{ds}+\dfrac{\sin\alpha}{r}\right) & -\dfrac{\sin\alpha}{r} \end{bmatrix} \begin{bmatrix} N_s \\ N_\vartheta \\ T_s \\ M_s \\ M_\vartheta \end{bmatrix} + \begin{bmatrix} \mathcal{F}_s \\ \mathcal{F}_n \\ 0 \end{bmatrix} = \begin{bmatrix} 0 \\ 0 \\ 0 \end{bmatrix}. \quad (12.12)
$$

Observe that, again for reasons of symmetry, the conditions of equilibrium to translation along the parallels and to rotation around the meridians are identically satisfied and thus do not appear in equation (12.12). Finally, we have three equations of equilibrium (respectively, with regard to translation along the meridians, to translation along the normal n, and to rotation about the parallels) in the five static unknowns N_s, N_ϑ, T_s, M_s, M_ϑ (Figure 12.20). The elastic problem for shells of revolution thus has two degrees of internal redundancy, while the more general problem of shells with double curvature appears to have three degrees of redundancy. Just as for beam systems then, symmetry reduces the degree of statical indeterminacy of the elastic problem also for shells.

Equations (12.12) are verified by imposing the above three conditions of equilibrium on an infinitesimal shell element, bounded by two meridians located at an infinitesimal distance $ds_2 = r\,d\vartheta$ and by two parallels located at an infinitesimal distance $ds_1 = ds$ (Figure 12.20).

The condition of equilibrium with regard to translation along the meridians yields the equation (Figures 12.20(a), (c))

$$
dN_s r\,d\vartheta + N_s dr\,d\vartheta - N_\vartheta \sin\alpha\,ds\,d\vartheta + T_s \frac{ds}{R_1} r\,d\vartheta + \mathcal{F}_s r\,ds\,d\vartheta = 0 \qquad (12.13a)
$$

which, divided by $rdsd\vartheta$, coincides with the first of equations (12.12).

The condition of equilibrium with regard to translation along the normal n furnishes the equation (Figures 12.20(c), (d), (e))

365

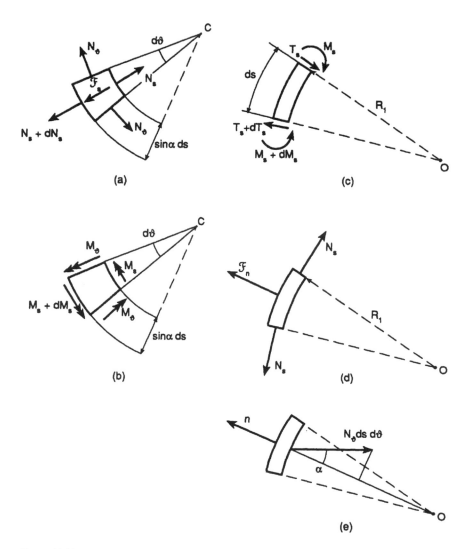

Figure 12.20

$$-N_s \frac{\mathrm{d}s}{R_1} r \, \mathrm{d}\vartheta - N_\vartheta \, \mathrm{d}s \, \mathrm{d}\vartheta \cos\alpha + \mathrm{d}T_s r \, \mathrm{d}\vartheta + T_s \, \mathrm{d}r \, \mathrm{d}\vartheta + \mathscr{F}_n r \, \mathrm{d}s \, \mathrm{d}\vartheta = 0 \quad (12.13\mathrm{b})$$

which, divided by $r \, \mathrm{d}s \, \mathrm{d}\vartheta$, coincides with the second of equations (12.12).

Finally, the condition of equilibrium with regard to rotation about the parallels furnishes the equation (Figures 12.20(b), (c))

$$-T_s \, r \, \mathrm{d}\vartheta \, \mathrm{d}s + \mathrm{d}M_s \, r \, \mathrm{d}\vartheta + M_s \mathrm{d}r \, \mathrm{d}\vartheta - M_\vartheta \sin\alpha \, \mathrm{d}s \, \mathrm{d}\vartheta = 0 \qquad (12.13\mathrm{c})$$

366

which, divided by $r\,\mathrm{d}s\,\mathrm{d}\vartheta$, coincides with the third of equations (12.12).

Notice that, in the indefinite equations of equilibrium (12.12), some contributions have been enclosed which, for example in the particular case of a circular plane plate, will not be negligible. These contributions are due to the fact that the parallel curvilinear sides of the shell element of Figures 12.20(a), (b) differ by the amount $\mathrm{d}r\,\mathrm{d}\vartheta$. In the static matrix (10.187) they appear in the first element of the first row, in the fourth element of the third row and in the sixth element of the fourth row.

12.8 Membranes and thin shells

Membranes are two-dimensional structural elements without flexural rigidity. These elements can sustain only tensile forces contained in the tangent plane. A similar but opposite case is provided by **thin shells**, which are shells of such small thickness that they present an altogether negligible flexural rigidity. These elements can sustain only compressive forces contained in the tangent plane. In the case of membranes, therefore, a zero compressive stiffness is assumed whereas in the case of thin shells a zero tensile stiffness is assumed. Both hypotheses imply a zero flexural rigidity.

As regards membranes and thin shells of revolution, the kinematic and static equations simplify notably compared with equations (12.11) and (12.12), since only forces along the meridians and the parallels, N_s and N_ϑ, are present, as well as the displacements along the meridians and those perpendicular to the middle surface, u and w, respectively

$$\begin{bmatrix} \varepsilon_s \\ \\ \varepsilon_\vartheta \end{bmatrix} = \begin{bmatrix} \dfrac{\mathrm{d}}{\mathrm{d}s} & \dfrac{1}{R_1} \\ \\ \dfrac{\sin\alpha}{r} & \dfrac{1}{R_2} \end{bmatrix} \begin{bmatrix} u \\ \\ w \end{bmatrix} \tag{12.14a}$$

$$\begin{bmatrix} \dfrac{\mathrm{d}}{\mathrm{d}s} & -\dfrac{\sin\alpha}{r} \\ \\ -\dfrac{1}{R_1} & -\dfrac{1}{R_2} \end{bmatrix} \begin{bmatrix} N_s \\ \\ N_\vartheta \end{bmatrix} + \begin{bmatrix} 0 \\ \\ \mathscr{F}_n \end{bmatrix} = \begin{bmatrix} 0 \\ \\ 0 \end{bmatrix} \tag{12.14b}$$

From the second of equations (12.14b) we obtain the fundamental algebraic relation which links the forces N_s and N_ϑ

$$\frac{N_s}{R_1} + \frac{N_\vartheta}{R_2} = \mathscr{F}_n \tag{12.15a}$$

while from the first we obtain the following differential equation:

$$\frac{\mathrm{d}N_s}{\mathrm{d}s} - \frac{\sin\alpha}{r} N_\vartheta = 0 \tag{12.15b}$$

On the other hand, by means of equation (12.15a) we can express N_ϑ as a function of N_s,

$$N_\vartheta = R_2\left(\mathscr{F}_n - \frac{N_s}{R_1}\right) \tag{12.16}$$

and this expression, inserted in equation (12.15b), gives

$$\frac{dN_s}{ds} + \frac{\tan\alpha}{R_1}N_s = \mathscr{F}_n \tan\alpha \tag{12.17}$$

which is a differential equation with ordinary derivatives in the unknown function $N_s(s)$.

   Instead of resolving the foregoing differential equation, alternatively we can consider equilibrium to translation in the Z direction of the portion of a thin shell (or membrane) which remains above a generic parallel (Figure 12.21)

$$Q = N_s \cos\alpha(2\pi r) \tag{12.18}$$

where Q is the integral of the vertical loads acting on that portion. From equation (12.18) we obtain immediately

$$N_s = \frac{Q}{2\pi r \cos\alpha} \tag{12.19}$$

Via equation (12.16) we then obtain the corresponding force along the parallel.

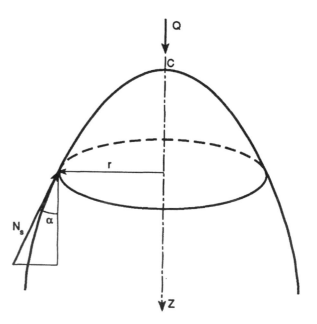

Figure 12.21

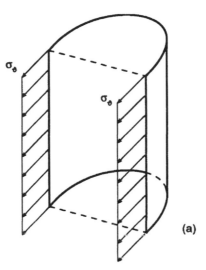

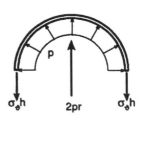

(b)

(a)

Figure 12.22

If σ_s and σ_ϑ denote the internal forces transmitted per unit area of the cross section (N_s and N_ϑ are forces per unit length), equation (12.15a) is transformed as follows:

$$\frac{\sigma_s}{R_1} + \frac{\sigma_\vartheta}{R_2} = \frac{p}{h} \tag{12.20}$$

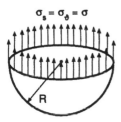

$\sigma_s = \sigma_\vartheta = \sigma$

$\sigma\,(2\pi\,Rh) = p\,(\pi\,R^2)$

Figure 12.23

where p denotes the pressure acting normally to the middle surface and h denotes the thickness of the thin shell (or membrane).

In the case of an indefinitely long **cylindrical membrane** subjected to the internal pressure p (Figure 12.22), we have $R_1 \to \infty$, $R_2 = r$, and thus the circumferential stress is

$$\sigma_\vartheta = \frac{pr}{h} \tag{12.21}$$

This internal reaction increases naturally with the increase in the pressure p and the radius r, and with the decrease in the thickness h.

In the case of a **spherical membrane** subjected to the internal pressure p (Figure 12.23), we have $R_1 = R_2 = R$, and thus the state of stress is isotropic:

$$\sigma_s = \sigma_\vartheta = \frac{pR}{2h} \tag{12.22}$$

Also in this case the internal reaction increases with pressure and radius, and decreases with the increase in thickness.

In the case of an indefinite **conical membrane** subjected to the internal pressure p (Figure 12.24), we have $R_1 \to \infty$, $R_2 = r/\cos\alpha$, and hence the circumferential stress is

369

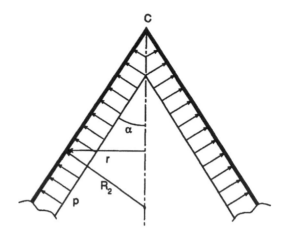

Figure 12.24

$$\sigma_\vartheta = \frac{pr}{h \cos \alpha} \tag{12.23}$$

The conditions of internal loading in the case of finite cylinders and cones are equal to the above ones obtained only at sufficiently large distances from the externally constrained zones. This amount is, however, to be evaluated in relation to the thickness h of the shell.

In the case of a **toroidal membrane** under pressure (Figure 12.25), equations (12.20) and (12.19) become

$$\frac{\sigma_s}{R_1} + \frac{\sigma_\vartheta}{r} \cos \alpha = \frac{p}{h} \tag{12.24a}$$

$$\sigma_s h = \frac{\pi\left(r^2 - r_0^2\right)p}{2\pi r \cos \alpha} \tag{12.24b}$$

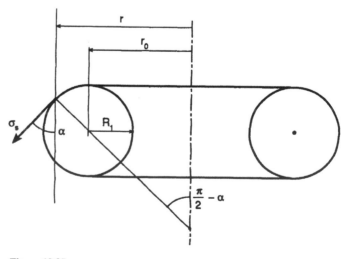

Figure 12.25

Since $r = r_0 + R_1 \cos\alpha$ (Figure 12.25), equation (12.24b) offers

$$\sigma_s = \frac{pR_1}{2h} \frac{2r_0 + R_1 \cos\alpha}{r_0 + R_1 \cos\alpha} \qquad (12.25)$$

On the crown of the toroidal surface we have $\alpha = \pi/2$ and hence

$$\sigma_s = \frac{pR_1}{h} \qquad (12.26)$$

This stress corresponds to the circumferential stress of a pressurized cylinder of radius R_1. The minimum stress σ_s occurs on the maximum parallel of the torus for $\alpha = 0$,

$$\sigma_s = \frac{pR_1}{2h} \frac{2r_0 + R_1}{r_0 + R_1} \qquad (12.27)$$

while the maximum occurs on the minimum parallel of the torus for $\alpha = \pi$,

$$\sigma_s = \frac{pR_1}{2h} \frac{2r_0 - R_1}{r_0 - R_1} \qquad (12.28)$$

The stress along the parallels is obtained from relation (12.24a) and in each point of the torus amounts to

$$\sigma_\vartheta = \frac{pR_1}{2h} \qquad (12.29)$$

It is equal to the longitudinal stress of a pressurized cylinder of radius R_1, having, for example, hemispherical ends (Figure 12.23).

12.9 Circular plates

The case of shells of revolution loaded symmetrically reduces to the particular case of **circular plates**, for $R_1 \rightarrow \infty$, $\alpha = \pi/2$. The curvilinear coordinate along the meridian, s, coincides with the radial coordinate r, so that the kinematic equation (12.11) transforms as follows:

$$
\begin{bmatrix} \varepsilon_r \\ \varepsilon_\vartheta \\ \gamma_r \\ \chi_r \\ \chi_\vartheta \end{bmatrix} = \begin{bmatrix} \dfrac{d}{dr} & 0 & 0 \\ \dfrac{1}{r} & 0 & 0 \\ 0 & \dfrac{d}{dr} & +1 \\ 0 & 0 & \dfrac{d}{dr} \\ 0 & 0 & \dfrac{1}{r} \end{bmatrix} \begin{bmatrix} u \\ w \\ \varphi_r \end{bmatrix} \qquad (12.30)
$$

The static equation (12.12), on the other hand, becomes:

$$
\begin{bmatrix}
\left(\dfrac{d}{dr}+\dfrac{1}{r}\right) & -\dfrac{1}{r} & 0 & 0 & 0 \\[2ex]
0 & 0 & \left(\dfrac{d}{dr}+\dfrac{1}{r}\right) & 0 & 0 \\[2ex]
0 & 0 & -1 & \left(\dfrac{d}{dr}+\dfrac{1}{r}\right) & -\dfrac{1}{r}
\end{bmatrix}
\begin{bmatrix} N_r \\ N_\vartheta \\ T_r \\ M_r \\ M_\vartheta \end{bmatrix}
+
\begin{bmatrix} \mathscr{F}_r \\ \mathscr{F}_n \\ 0 \end{bmatrix}
=
\begin{bmatrix} 0 \\ 0 \\ 0 \end{bmatrix}
\quad (12.31)
$$

Restricting the analysis to the flexural regime only, we obtain the following equations, which are kinematic and static, respectively:

$$
\begin{bmatrix} \gamma_r \\[1.5ex] \chi_r \\[1.5ex] \chi_\vartheta \end{bmatrix}
=
\begin{bmatrix}
\dfrac{d}{dr} & +1 \\[2ex]
0 & \dfrac{d}{dr} \\[2ex]
0 & \dfrac{1}{r}
\end{bmatrix}
\begin{bmatrix} w \\[2ex] \varphi_r \end{bmatrix}
\quad (12.32)
$$

$$
\begin{bmatrix}
\left(\dfrac{d}{dr}+\dfrac{1}{r}\right) & 0 & 0 \\[2ex]
-1 & \left(\dfrac{d}{dr}+\dfrac{1}{r}\right) & -\dfrac{1}{r}
\end{bmatrix}
\begin{bmatrix} T_r \\ M_r \\ M_\vartheta \end{bmatrix}
+
\begin{bmatrix} -q \\ 0 \end{bmatrix}
=
\begin{bmatrix} 0 \\ 0 \end{bmatrix}
\quad (12.33)
$$

The indefinite equation of equilibrium (12.33) represents a system of two differential equations in the three unknowns T_r, M_r, M_ϑ. The polar symmetry thus reduces the degree of static indeterminacy of the deflected plane plates from two to one.

The first of equations (12.33) represents the condition of equilibrium with regard to the vertical translation of a plate element identified by two radii forming the angle $d\vartheta$, and by two circumferences of radius r and $r + dr$ (Figure 12.26(a))

$$
dT_r r\, d\vartheta + T_r\, dr\, d\vartheta - qr\, dr\, d\vartheta = 0 \quad (12.34a)
$$

The second term of the foregoing equation is due to the greater length presented by the outermost arc of circumference. Dividing by the elementary area $rdrd\vartheta$, we once more obtain equation (12.33).

The second of equations (12.33) represents the condition of equilibrium with regard to rotation of the same plate element about the circumference of radius r (Figure 12.26(b)):

$$
-T_r\, r\, d\vartheta\, dr + dM_r\, r\, d\vartheta + M_r\, dr\, d\vartheta - M_\vartheta\, dr\, d\vartheta = 0 \quad (12.34b)
$$

372

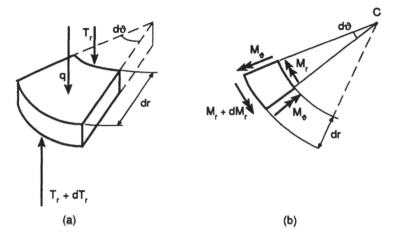

Figure 12.26

Also in this case the contribution of the third term is due to the greater length of the outermost arc of circumference.

Using the elastic constitutive equations which link bending moments and curvatures:

$$M_r = D(\chi_r + v\chi_\vartheta) \tag{12.35a}$$

$$M_\vartheta = D(\chi_\vartheta + v\chi_r) \tag{12.35b}$$

and assuming zero shearing strain γ_r

$$\gamma_r = \frac{dw}{dr} + \varphi_r = 0 \tag{12.36a}$$

whereby

$$\chi_r = \frac{d\varphi_r}{dr} = -\frac{d^2w}{dr^2} \tag{12.36b}$$

$$\chi_\vartheta = \frac{\varphi_r}{r} = -\frac{1}{r}\frac{dw}{dr} \tag{12.36c}$$

the second of the indefinite equations of equilibrium (12.33) transforms into a third-order differential equation in the unknown function w:

$$-T_r - D\frac{d}{dr}\left(\frac{d^2w}{dr^2} + \frac{v}{r}\frac{dw}{dr}\right) - \frac{D}{r}\left(\frac{d^2w}{dr^2} + \frac{v}{r}\frac{dw}{dr}\right) + \frac{D}{r}\left(\frac{1}{r}\frac{dw}{dr} + v\frac{d^2w}{dr^2}\right) = 0 \tag{12.37}$$

Reordering the terms, we obtain

$$\frac{d^3w}{dr^3} + \frac{1}{r}\frac{d^2w}{dr^2} - \frac{1}{r^2}\frac{dw}{dr} = -\frac{T_r}{D} \tag{12.38}$$

373

The foregoing equation is equivalent to the following one, in which the unknown function is the radial rotation φ_r

$$\frac{d^2\varphi_r}{dr^2} + \frac{1}{r}\frac{d\varphi_r}{dr} - \frac{1}{r^2}\varphi_r = \frac{T_r}{D} \tag{12.39}$$

If $Q(r)$ denotes the integral of the vertical loads acting on the plate within the circumference of radius r, from equilibrium we have

$$2\pi r T_r = Q(r) \tag{12.40}$$

whereby equation (12.39) can be cast in the following form:

$$\frac{d}{dr}\left[\frac{1}{r}\frac{d}{dr}(r\varphi_r)\right] = \frac{Q(r)}{2\pi Dr} \tag{12.41}$$

A first integration yields

$$\frac{1}{r}\frac{d}{dr}(r\varphi_r) = \int_0^r \frac{Q(r)}{2\pi Dr}dr + C_1 \tag{12.42}$$

Multiplying by r and integrating again, we obtain

$$r\varphi_r = \int_0^r\left[r\int_0^r \frac{Q(r)}{2\pi Dr}dr\right]dr + C_1\frac{r^2}{2} + C_2 \tag{12.43}$$

from which, on further integration, we find the equation of the elastic deformed configuration $w(r)$.

Consider, for instance, a circular plate of radius R clamped at the boundary and uniformly loaded with a pressure p. In this case we have $Q(r) = p\pi r^2$, so that equation (12.43) becomes

$$r\varphi_r = \int_0^r \frac{pr^3}{4D}dr + C_1\frac{r^2}{2} + C_2 \tag{12.44}$$

and hence

$$\varphi_r = \frac{pr^3}{16D} + C_1\frac{r}{2} + \frac{C_2}{r} \tag{12.45}$$

For reasons of symmetry we must have $\varphi_r(0) = 0$, and hence the constant C_2 is zero. The condition of a built-in constraint at the edge, on the other hand, furnishes the relation

$$\varphi_r(R) = \frac{pR^3}{16D} + C_1\frac{R}{2} = 0 \tag{12.46}$$

from which we obtain the constant C_1:

$$C_1 = -\frac{pR^2}{8D} \tag{12.47}$$

The equation of the elastic deformed configuration is thus drawn from the integration of the following equation:

$$-\frac{dw}{dr} = \frac{pr^3}{16D} - \frac{pR^2}{16D}r \tag{12.48}$$

374

The displacement w is thus defined but for a constant C_3,

$$w = -\frac{pr^4}{64D} + \frac{pR^2}{32D}r^2 + C_3 \qquad (12.49)$$

which may be determined by imposing the annihilation of the displacement at the built-in constraint

$$C_3 = -\frac{pR^4}{64D} \qquad (12.50)$$

Finally, therefore, the displacement orthogonal to the middle plane and the radial rotation are expressible as follows:

$$w = -\frac{p}{64D}\left(R^2 - r^2\right)^2 \qquad (12.51a)$$

$$\varphi_r = -\frac{pr}{16D}\left(R^2 - r^2\right) \qquad (12.51b)$$

The vertical displacement at the centre of the plate is therefore equal to

$$f = |w(0)| = \frac{pR^4}{64D} \qquad (12.52)$$

The bending moments (12.35) are obtained taking into account equations (12.36b, c) and (12.51)

$$M_r = -\frac{p}{16}\left[(1+v)R^2 - (3+v)r^2\right] \qquad (12.53a)$$

$$M_\vartheta = -\frac{p}{16}\left[(1+v)R^2 - (1+3v)r^2\right] \qquad (12.53b)$$

In the centre the two moments, the radial one and the circumferential one, are equal to one another

$$M_r(0) = M_\vartheta(0) = -(1+v)\frac{pR^2}{16} \qquad (12.54)$$

On the clamped edge we have

$$M_r(R) = \frac{pR^2}{8}, \quad M_\vartheta(R) = v\frac{pR^2}{8} \qquad (12.55)$$

The maximum moment is the radial one at the built-in constraint. Figure 12.27 shows the elastic deformed configuration and the internal reactions M_r and M_ϑ.

If the above plate is loaded by the concentrated force Q, equation (12.43) becomes

$$r\varphi_r = \frac{Q}{2\pi D}\int_0^r r\log r \ dr + C_1\frac{r^2}{2} + C_2 \qquad (12.56)$$

and hence

$$\varphi_r = \frac{Qr}{8\pi D}(2\log r - 1) + C_1\frac{r}{2} + \frac{C_2}{r} \qquad (12.57)$$

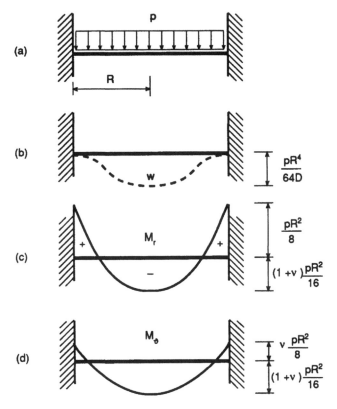

Figure 12.27

The symmetry condition and the boundary condition, respectively

$$\varphi_r(0) = 0 \qquad (12.58a)$$

$$\varphi_r(R) = 0 \qquad (12.58b)$$

furnish the corresponding values of the two constants

$$C_2 = 0 \qquad (12.59a)$$

$$C_1 = -\frac{Q}{4\pi D}(2\log R - 1) \qquad (12.59b)$$

We thus find

$$\varphi_r = -\frac{Q}{4\pi D} r \, \log \frac{R}{r} \qquad (12.60a)$$

$$w = -\frac{Q}{16\pi D}\left(R^2 - r^2 - 2r^2 \log \frac{R}{r}\right) \qquad (12.60b)$$

The vertical displacement at the centre of the plate is therefore

$$f = |w(0)| = \frac{QR^2}{16\pi D} \qquad (12.61)$$

376

In the case where $Q = p\pi R^2$, this latter vertical displacement is four times that of equation (12.52). The bending moments are given by

$$M_r = -\frac{Q}{4\pi}\left[(1+v)\log\frac{R}{r}-1\right] \qquad (12.62a)$$

$$M_\vartheta = -\frac{Q}{4\pi}\left[(1+v)\log\frac{R}{r}-v\right] \qquad (12.62b)$$

In the centre they are theoretically infinite, whilst at the clamped edge they are equal to

$$M_r(R) = \frac{Q}{4\pi}, \quad M_\vartheta(R) = v\frac{Q}{4\pi} \qquad (12.63)$$

Figure 12.28 depicts the elastic deformed configuration and presents the M_r and M_ϑ bending moment diagrams.

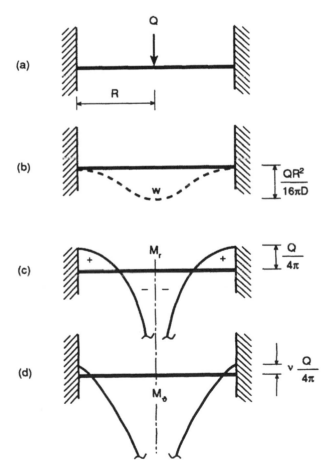

Figure 12.28

377

12.10   Cylindrical shells

In **cylindrical shells** the principal radius of curvature $R_1 \to \infty$, while the angle α between the generatrix and the axis of symmetry vanishes. Furthermore, the curvilinear coordinate along the meridian, s, coincides with the longitudinal coordinate x, and the second principal radius of curvature $R_2 = r$ coincides with the radius R of the circular directrix.

The kinematic equation (12.11) is thus transformed as follows (Figure 12.29):

$$
\begin{bmatrix} \varepsilon_x \\ \varepsilon_\vartheta \\ \gamma_x \\ \chi_x \\ \chi_\vartheta \end{bmatrix} = \begin{bmatrix} \dfrac{\mathrm{d}}{\mathrm{d}x} & 0 & 0 \\ 0 & \dfrac{1}{R} & 0 \\ 0 & \dfrac{\mathrm{d}}{\mathrm{d}x} & +1 \\ 0 & 0 & \dfrac{\mathrm{d}}{\mathrm{d}x} \\ 0 & 0 & 0 \end{bmatrix} \begin{bmatrix} u \\ w \\ \varphi_x \end{bmatrix}
\tag{12.64}
$$

whilst the static equation (12.12) becomes

$$
\begin{bmatrix} \dfrac{\mathrm{d}}{\mathrm{d}x} & 0 & 0 & 0 & 0 \\ 0 & -\dfrac{1}{R} & \dfrac{\mathrm{d}}{\mathrm{d}x} & 0 & 0 \\ 0 & 0 & -1 & \dfrac{\mathrm{d}}{\mathrm{d}x} & 0 \end{bmatrix} \begin{bmatrix} N_x \\ N_\vartheta \\ T_x \\ M_x \\ M_\vartheta \end{bmatrix} + \begin{bmatrix} 0 \\ q \\ 0 \end{bmatrix} = \begin{bmatrix} 0 \\ 0 \\ 0 \end{bmatrix}
\tag{12.65}
$$

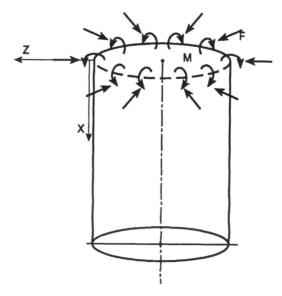

Figure 12.29

378

in the case where the only external force acting on the shell is a load $q(x)$, normal to the middle surface. Notice that the variation in curvature χ_ϑ vanishes, just as the moment M_ϑ is not involved in any of the three equations of equilibrium.

The first of equations (12.65) is the equation of equilibrium with regard to longitudinal translation,

$$\frac{dN_x}{dx} = 0 \tag{12.66a}$$

which gives N_x = constant. The second of equations (12.65) is the equation of equilibrium with regard to normal translation,

$$-\frac{N_\vartheta}{R} + \frac{dT_x}{dx} = -q \tag{12.66b}$$

while the third is the equation of equilibrium with regard to rotation about the parallel,

$$T_x = \frac{dM_x}{dx} \tag{12.66c}$$

Substituting equation (12.66c) into equation (12.66b), we obtain

$$-\frac{N_\vartheta}{R} + \frac{d^2 M_x}{dx^2} = -q \tag{12.67}$$

In the case where the longitudinal dilation ε_x is zero, we have

$$N_\vartheta = Eh\varepsilon_\vartheta = \frac{Eh}{R} w \tag{12.68}$$

The moment M_x is, on the other hand, proportional to the variation in curvature χ_x, as $\chi_\vartheta = 0$:

$$M_x = D\chi_x = D\frac{d\varphi_x}{dx} \tag{12.69}$$

If we disregard the shearing strain

$$\gamma_x = \frac{dw}{dx} + \varphi_x = 0 \tag{12.70}$$

we get

$$M_x = -D\frac{d^2 w}{dx^2} \tag{12.71}$$

Substituting relations (12.68) and (12.71) into equation (12.67), we obtain the following differential equation in the unknown function w:

$$D\frac{d^4 w}{dx^4} + \frac{Eh}{R^2} w = q \tag{12.72}$$

Equation (12.72) is formally identical to the differential equation of the beam on an elastic foundation (10.94). In the case where $q = 0$, equation (12.72) can be cast in the form

$$\frac{d^4 w}{dx^4} + 4\beta^4 w = 0 \tag{12.73}$$

379

where β denotes the parameter

$$\beta = \sqrt[4]{\frac{Eh}{4DR^2}} \qquad (12.74)$$

From equation (10.165) we obtain

$$\beta = \sqrt[4]{\frac{3(1-v^2)}{h^2R^2}} \qquad (12.75)$$

For the solution of equation (12.73), the reader is referred to Section 10.8.

In the case where the force N_x is zero, note that equation (12.73) is still valid. From equations (10.161a, b) we have

$$N_x = \frac{Eh}{1-v^2}\left(\varepsilon_x + v\varepsilon_\vartheta\right) = 0 \qquad (12.76a)$$

$$N_\vartheta = \frac{Eh}{1-v^2}\left(\varepsilon_\vartheta + v\varepsilon_x\right) \qquad (12.76b)$$

Equation (12.76a) gives $\varepsilon_x = -v\varepsilon_\vartheta$, which, substituted into equation (12.76b), yields again equation (12.68).

12.11 Cylindrical vessels with faces subjected to internal pressure

The displacement and rotation at the edge of a cylindrical shell, produced by forces and moments distributed along the edge itself (Figure 12.29) can be obtained on the basis of the analogy between cylindrical shells and beams on an elastic foundation (Figure 10.17). Equations (10.109) can be expressed in the following form:

$$w(0) = -\lambda_{FF}F + \lambda_{FM}M \qquad (12.77a)$$

$$\varphi_x(0) = -\lambda_{MF}F + \lambda_{MM}M \qquad (12.77b)$$

where the elastic coefficients λ_{ij}, which for the beam on an elastic foundation were found to be equal to

$$\lambda_{FF} = \frac{2\beta}{K} = \frac{1}{2\beta^3 EI} \qquad (12.78a)$$

$$\lambda_{FM} = \lambda_{MF} = \frac{2\beta^2}{K} = \frac{1}{2\beta^2 EI} \qquad (12.78b)$$

$$\lambda_{MM} = \frac{4\beta^3}{K} = \frac{1}{\beta EI} \qquad (12.78c)$$

for the semi-infinite cylinder are likewise equal to

$$\lambda_{FF} = \frac{1}{2\beta^3 D} \qquad (12.79a)$$

$$\lambda_{FM} = \lambda_{MF} = \frac{1}{2\beta^2 D} \qquad (12.79b)$$

$$\lambda_{MM} = \frac{1}{\beta D} \qquad (12.79c)$$

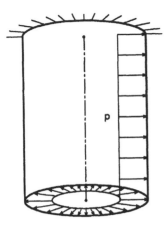

Figure 12.30

If the edges of a pressurized cylinder were free, we should have exclusively the circumferential stress given by equation (12.21) and the radial displacement

$$w = \frac{pR^2}{Eh} \tag{12.80}$$

If, instead, the edges of the cylinder are clamped (Figure 12.30), and the cylinder is assumed as being sufficiently long, a localized flexural regime will be produced, which will be superposed in the surroundings of the edge on the aforementioned membrane regime. Since the constraint prevents both the radial displacement and the rotation of the edge, we have the following two equations of congruence (Figures 12.29 and 12.30):

$$w(0) = -\lambda_{FF}F + \lambda_{FM}M + \frac{pR^2}{Eh} = 0 \tag{12.81a}$$

$$\varphi_x(0) = -\lambda_{MF}F + \lambda_{MM}M = 0 \tag{12.81b}$$

where F and M are the statically indeterminate reactions which act along the edge. Using equations (12.79) and (12.75), we obtain

$$M = \frac{p}{2\beta^2} \tag{12.82a}$$

$$F = \frac{p}{\beta} \tag{12.82b}$$

For a strength test, it will be possible to consider, at the edge, the circumferential stress

$$\sigma_\vartheta = \frac{pR}{h} \pm v \frac{6}{h^2} M \tag{12.83a}$$

and the longitudinal stress

$$\sigma_x = \pm \frac{6}{h^2} M \tag{12.83b}$$

as well as the shearing stress

$$\tau = \frac{F}{h} \tag{12.83c}$$

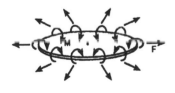

Figure 12.31

In the case of a pressurized flat-faced cylinder, the edges of the cylinder and of the circular plate exchange a distributed force and a distributed moment (Figures 12.29, 12.31), the plate system having two degrees of static indeterminacy, as in the previous case.

The equation of angular congruence takes the form

$$-\lambda_{MF}F + \lambda_{MM}M = -\frac{MR}{(1+v)D} + \frac{pR^3}{8(1+v)D} \tag{12.84}$$

where the left-hand side of the equation represents the rotation of the edge of the cylinder, whilst the two terms on the right represent the rotations of the edge of the cylinder face, due, respectively, to the redundant moment and to the internal pressure. The first rotation is deduced in the case of uniform

381

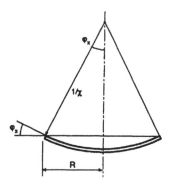

Figure 12.32

bending. If $M_x = M_y = M$, also the curvatures equal one another, $\chi_x = \chi_y = \chi$, so that the relation (10.164a) gives

$$\chi = \frac{M}{(1+v)D} \qquad (12.85)$$

and thus the angle of rotation at the edge (Figure 12.32)

$$\varphi_x = R\chi \qquad (12.86)$$

which appears in equation (12.84). The second rotation can then be obtained from the foregoing one by substituting, in place of M, the radial moment at the built-in constraint (12.55).

The second equation of congruence is the one corresponding to the radial displacement

$$-\lambda_{FF}F + \lambda_{FM}M + \frac{pR^2}{2Eh}(2-v) = \frac{FR}{Eh}(1-v) \qquad (12.87)$$

The third term on the left-hand side represents the radial displacement of the cylinder, which, when pressurized, is subject to a biaxial stress condition

$$\sigma_\vartheta = \frac{pR}{h}, \quad \sigma_x = \frac{pR}{2h} \qquad (12.88)$$

whereby the circumferential dilation

$$\varepsilon_\vartheta = \frac{1}{E}(\sigma_\vartheta - v\sigma_x) \qquad (12.89)$$

produces the radial displacement

$$w = \varepsilon_\vartheta R = \frac{pR^2}{2Eh}(2-v) \qquad (12.90)$$

The right-hand side of equation (12.87) represents, on the other hand, the radial displacement of the edge of the circular plate, since this is in a condition of uniform stress $\sigma = F/h$.

In the case of a pressurized cylinder with hemispherical faces (Figures 12.29, 12.33), the equation of angular congruence takes the following form:

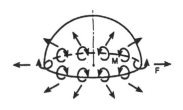

Figure 12.33

$$-\lambda_{MF}F + \lambda_{MM}M = -\lambda_{MF}F - \lambda_{MM}M \qquad (12.91)$$

if we assume, as is approximately the case, that the elastic coefficients of the cylinder and the hemisphere are equal. From equation (12.91) it follows that the redundant moment M vanishes.

On the other hand, the equation of congruence for the radial displacement is

$$-\lambda_{FF}F + \frac{pR^2}{2Eh}(2-v) = \lambda_{FF}F + \frac{pR^2}{2Eh}(1-v) \qquad (12.92)$$

where the second terms on both sides of the equation take ino account the biaxial stress condition of the shells, which are cylindrical and spherical, respectively. From equation (12.92) we obtain

$$2\lambda_{FF}F = \frac{pR^2}{2Eh} \qquad (12.93)$$

382

and thus, applying equations (12.79a) and (12.74), we have

$$F = \frac{pR^2\beta^3 D}{2Eh} = \frac{p}{8\beta} \tag{12.94}$$

12.12 Three-dimensional solids of revolution

In the case of a three-dimensional solid of revolution, not loaded symmetrically, the kinematic and static equations appear as follows (Figure 12.34):

$$
\begin{bmatrix} \varepsilon_r \\ \varepsilon_\vartheta \\ \varepsilon_z \\ \gamma_{r\vartheta} \\ \gamma_{rz} \\ \gamma_{\vartheta z} \end{bmatrix} =
\begin{bmatrix}
\dfrac{\partial}{\partial r} & 0 & 0 \\[2mm]
\dfrac{1}{r} & \dfrac{1}{r}\dfrac{\partial}{\partial\vartheta} & 0 \\[2mm]
0 & 0 & \dfrac{\partial}{\partial z} \\[2mm]
\dfrac{1}{r}\dfrac{\partial}{\partial\vartheta} & \left(\dfrac{\partial}{\partial r}-\dfrac{1}{r}\right) & 0 \\[2mm]
\dfrac{\partial}{\partial z} & 0 & \dfrac{\partial}{\partial r} \\[2mm]
0 & \dfrac{\partial}{\partial z} & \dfrac{1}{r}\dfrac{\partial}{\partial\vartheta}
\end{bmatrix}
\begin{bmatrix} u \\ v \\ w \end{bmatrix} \tag{12.95a}
$$

$$
\begin{bmatrix}
\left(\dfrac{\partial}{\partial r}+\dfrac{1}{r}\right) & -\dfrac{1}{r} & 0 & \dfrac{1}{r}\dfrac{\partial}{\partial\vartheta} & \dfrac{\partial}{\partial z} & 0 \\[2mm]
0 & \dfrac{1}{r}\dfrac{\partial}{\partial\vartheta} & 0 & \left(\dfrac{\partial}{\partial r}+\dfrac{2}{r}\right) & 0 & \dfrac{\partial}{\partial z} \\[2mm]
0 & 0 & \dfrac{\partial}{\partial z} & 0 & \left(\dfrac{\partial}{\partial r}+\dfrac{1}{r}\right) & \dfrac{1}{r}\dfrac{\partial}{\partial\vartheta}
\end{bmatrix}
\begin{bmatrix} \sigma_r \\ \sigma_\vartheta \\ \sigma_z \\ \tau_{r\vartheta} \\ \tau_{rz} \\ \tau_{\vartheta z} \end{bmatrix} +
$$

$$
\begin{bmatrix} \mathscr{F}_r \\ \mathscr{F}_\vartheta \\ \mathscr{F}_z \end{bmatrix} =
\begin{bmatrix} 0 \\ 0 \\ 0 \end{bmatrix} \tag{12.95b}
$$

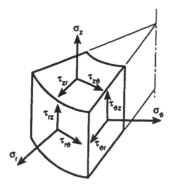

Figure 12.34

The static equations (12.95b) represent the three indefinite equations of equilibrium with regard to translation, in the radial, circumferential and axial directions, respectively (Figure 12.34). Notice that, once again, the terms $1/r$ in the static matrix are due to the difference between the areas of the two parallel curved faces of the element of Figure 12.34, as well as the different action lines of the stresses σ_ϑ and $\tau_{\vartheta r}$ acting on the two opposite faces.

In the case where the solid of revolution is also loaded symmetrically with respect to its axis, the degrees of internal redundancy become two instead of three:

$$
\begin{bmatrix} \varepsilon_r \\ \varepsilon_\vartheta \\ \varepsilon_z \\ \gamma_{rz} \end{bmatrix} = \begin{bmatrix} \dfrac{\partial}{\partial r} & 0 \\ \dfrac{1}{r} & 0 \\ 0 & \dfrac{\partial}{\partial z} \\ \dfrac{\partial}{\partial z} & \dfrac{\partial}{\partial r} \end{bmatrix} \begin{bmatrix} u \\ w \end{bmatrix}
\tag{12.96a}
$$

$$
\begin{bmatrix} \left(\dfrac{\partial}{\partial r}+\dfrac{1}{r}\right) & -\dfrac{1}{r} & 0 & \dfrac{\partial}{\partial z} \\ 0 & 0 & \dfrac{\partial}{\partial z} & \left(\dfrac{\partial}{\partial r}+\dfrac{1}{r}\right) \end{bmatrix} \begin{bmatrix} \sigma_r \\ \sigma_\vartheta \\ \sigma_z \\ \tau_{rz} \end{bmatrix} + \begin{bmatrix} \mathscr{F}_r \\ \mathscr{F}_z \end{bmatrix} = \begin{bmatrix} 0 \\ 0 \end{bmatrix}
\tag{12.96b}
$$

When the problem presents a plane stress condition, we have $\sigma_z = \tau_{rz} = 0$, and hence only the first of equations (12.96b) remains significant:

$$
\frac{d\sigma_r}{dr} + \frac{\sigma_r - \sigma_\vartheta}{r} + \mathscr{F}_r = 0
\tag{12.97}
$$

13 Statically indeterminate structures: method of forces

13.1 Introduction

Redundant beam systems, i.e. ones that contain a surplus number of constraints, are statically indeterminate. As we have seen in Chapter 3, this means that they can be balanced by ∞^{v-g} different sets of reactive forces, $v-g$ being the degree of redundancy. We therefore need to identify the particular single set of reactive forces which, in addition to equilibrium, also implies **congruence**, or rather the respect of the internal and external constraints, notwithstanding the deformations induced in the structural elements.

From the operative viewpoint, the **method of forces** consists of eliminating $v-g$ degrees of constraint, so as to reduce the given structure to a statically determinate beam system, and applying to this system, in addition to the external forces, the unknown constraint reactions exerted by the constraints that have been removed. The $v-g$ equations of congruence will then impose abeyance of the kinematic conditions corresponding to the suppressed constraints, and, once resolved, will yield the $v-g$ elementary reactions exerted by these constraints, which are called **hyperstatic unknowns**.

When resolving a statically indeterminate structure, one is therefore confronted with the problem of finding a suitable way of disconnecting it so as to obtain the statically determinate scheme on which to impose the conditions of congruence. In principle, the disconnection can be performed in an infinite number of different ways, as it is possible to reduce the degrees of constraint both externally and internally, and, among the internal constraints, it is possible to reduce the infinite internal fixed-joint constraints which guarantee the continuity of the beam. Normally, however, it is convenient to reduce or suppress the external constraints, or to interrupt the continuity of the structure, by inserting hinges in points of concurrence of two or more beams (fixed-joint nodes). In the first case the equations of congruence will impose the annihilation of the displacements of the points that are sites of statically indeterminate reactions, whereas in the second case the so-called **angular congruence** will be imposed, i.e. an equal elastic rotation at all the beam ends which converge at the same fixed-joint node.

13.2 Axial indeterminacy

Consider a rectilinear beam of length l, hinged at the ends A and B and subjected to an axial force F, acting at a distance a from the end A and b from the end B (Figure 13.1(a)). Thus loaded, the beam is statically indeterminate, since the pairs of reactions H_A and H_B, which together with the force F make up a balanced system, are infinite. Replacing the hinge B with a roller support having a horizontal plane of movement (Figure 13.1(b)) and applying the hyperstatic unknown X at the same end B, we obtain the equivalent statically determinate scheme. The equation of congruence must express the existence

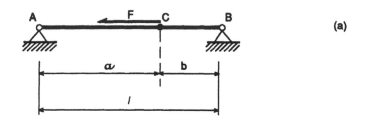
(a)

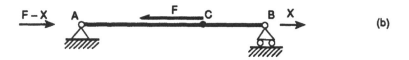
(b)

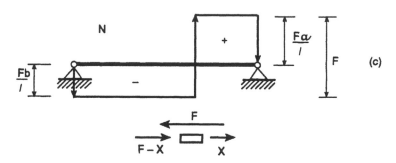
(c)

Figure 13.1

of the suppressed constraint, i.e. that the displacement of the roller support is zero

$$w_B = \frac{X}{EA}l - \frac{F}{EA}a = 0 \qquad (13.1)$$

where the first term represents the contribution of the reaction X, while the second represents the contribution of the external force. Notice that we have implicitly made use of the Principle of Superposition. The force X, in fact, generates a characteristic of tension on the entire beam, while the force F generates a characteristic of compression only on the portion AC. The force F thus contracts the portion AC, while the portion CB is drawn along by a rigid translation.

From equation (13.1) we derive

$$X = F\frac{a}{l} \qquad (13.2a)$$

$$F - X = F\frac{b}{l} \qquad (13.2b)$$

so that the force F is supported by the two end constraints in direct proportion to the reciprocal distances from the point of application. The portion CB is thus subjected to tension while the portion AC is subjected to compression.

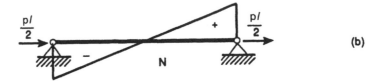

Figure 13.2

The axial force diagram (Figure 13.1(c)) thus shows a discontinuity at the point of application of the force. The element of beam straddling this point is in equilibrium under the action of the external force and of two internal reactions having the same sense.

If the beam considered previously were submitted to a uniform distribution of axial forces p (Figure 13.2(a)), the skew-symmetry of the structural scheme would make it possible to recognize two equal constraint reactions having the same sense, so that, on the basis of equation (5.12a), the axial force diagram would be linear and skew-symmetrical, with a zero in the centre and the extreme values equal to $-\frac{1}{2}pl$ in A and $+\frac{1}{2}pl$ in B (Figure 13.2(b)).

Finally, let us consider a case of double axial redundancy: a beam of length $2l$ hinged at the ends and in the centre, loaded by a concentrated axial force acting in the centre of the left-hand span (Figure 13.3(a)). The equivalent statically determinate scheme may be obtained by transforming two of the three hinges into as many horizontally moving roller supports. Figure 13.3(b) depicts the scheme with the roller supports in B and C and the respective redundant reactions X_1 and X_2. The two equations of congruence express the immovability of the points B and C,

$$w_B = \frac{(X_1 + X_2)}{EA}l - \frac{F}{EA}\frac{l}{2} = 0 \qquad (13.3a)$$

$$w_C = \frac{X_2}{EA}2l + \frac{X_1}{EA}l - \frac{F}{EA}\frac{l}{2} = 0 \qquad (13.3b)$$

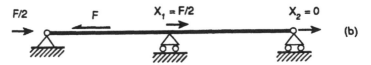

Figure 13.3

from which we obtain

$$X_1 = \frac{F}{2}, \quad X_2 = 0 \tag{13.4}$$

The solution found expresses the fact that only the two hinges between which the external force F is applied react.

13.3 Elementary statically indeterminate schemes

Consider the rectilinear beam of length l, built-in at the end A and supported at the end B (Figure 13.4(a)), subjected to the distributed load q. This structure has one degree of redundancy. The equivalent statically determinate scheme is obtained by eliminating one of the three external constraints (excluding the axial one) and imposing congruence, i.e. abeyance of the constraint that has been suppressed. Alternatively, though only in principle, it would also be possible to disconnect the beam internally, but this would not prove to be a convenient approach in actual operative terms.

A first equivalent statically determinate scheme is obtained by eliminating the roller support in B and subjecting the cantilever beam AB, not only to the distributed load q, but also to the redundant reaction X, which is an unknown vertical force acting at the end B (Figure 13.4(b)). Superposing the effects, the condition of congruence becomes

$$v_B = \frac{ql^4}{8EI} - \frac{Xl^3}{3EI} = 0 \tag{13.5}$$

This equation contains the single unknown X. From equation (13.5) we obtain

$$X = \frac{3}{8}ql \tag{13.6}$$

The reactions at the built-in end are then equal to (Figure 13.4(b))

$$V_A = ql - \frac{3}{8}ql = \frac{5}{8}ql \tag{13.7a}$$

$$M_A = \frac{1}{2}ql^2 - \frac{3}{8}ql^2 = \frac{1}{8}ql^2 \tag{13.7b}$$

The shear diagram is thus linear with extreme values equal to $\frac{5}{8}ql$ in A and $-\frac{3}{8}ql$ in B (Figure 13.4(c)). The bending moment diagram may be plotted by points. The moment at the built-in constraint is equal in fact to $\frac{1}{8}ql^2$, while the moment is zero at the hinge B (Figure 13.4(d)) and at the point where the function

$$M(z) = \frac{3}{8}qlz - \frac{1}{2}qz^2 = -\frac{1}{2}qz\left(z - \frac{3}{4}l\right) \tag{13.8}$$

vanishes, i.e. at a distance $z = \frac{3}{4}l$ from the end B. Another notable value of the moment is the maximum one at the point of zero shear

$$M_{max} = M\left(\frac{3}{8}l\right) = \frac{9}{128}ql^2 \tag{13.9}$$

388

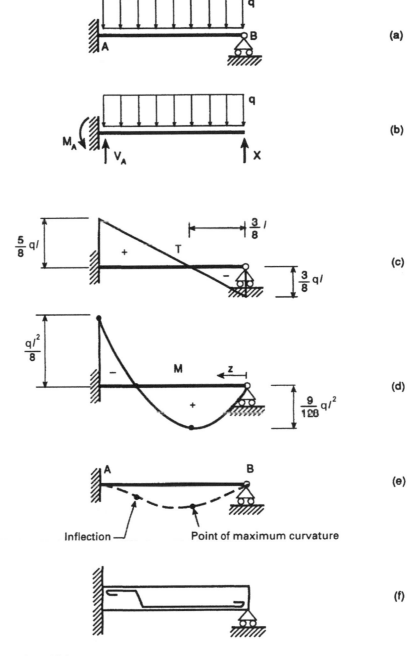

Figure 13.4

On the basis of the four notable points discussed above, the diagram can be immediately plotted (Figure 13.4(d)). The elastic deformed configuration of the beam can be plotted qualitatively, with respect to the constraints and the moment diagram (Figure 13.4(e)). The built-in constraint A imposes a deformed configuration with zero vertical displacement and zero rotation at A, just as the roller support B imposes a deformed configuration with zero vertical displacement at B, while it allows rotation of the end section B. The deformed configuration of course presents a point of inflection where the moment becomes zero and then undergoes a change in sign. There will be extended fibres in the upper portion between the end A and the inflection, and, *vice versa*, in the lower portion between the inflection and the end B. Summarizing, we can say that the deformed configuration will present the point of inflection at a distance $\frac{3}{4} l$ from the end B, and the point of maximum curvature at the distance $\frac{3}{8} l$ once again from point B (Figure 13.4(e)). In the case of the beam made of reinforced concrete, the reinforcement must follow the path of the stretched fibres (Figure 13.4(f)).

A second equivalent statically determinate scheme is obtained by eliminating the degree of constraint with regard to rotation of the built-in end A; i.e. by replacing the built-in constraint with a hinge, and by applying an unknown redundant moment X at the end A itself (Figure 13.5(a)). Summing up the elastic rotations of the end section A, due both to the external load and to the redundant reaction X, we obtain the equation of congruence

$$\varphi_A = -\frac{ql^3}{24EI} + \frac{Xl}{3EI} = 0 \qquad (13.10)$$

from which there follows

$$X = \frac{1}{8} ql^2 \qquad (13.11)$$

This result coincides with the moment at the built-in constraint, deduced in the previous solution.

The moment diagram, in the framework of the present scheme of resolution, may be obtained graphically (Figure 13.5(b)). The partial diagram due to the redundant reaction is linear with extreme values equal to $-\frac{1}{8} ql^2$ in A and zero in B. On the other hand, the partial diagram due to the distributed load is parabolic with the maximum which again equals $\frac{1}{8} ql^2$. The graphical sum of the two partial diagrams may be obtained by following the usual procedure, outlined in Chapters 4, 5 and 6, based on the properties of the arcs of parabola. From the mid-point of the triangular diagram two consecutive vertical segments are drawn, each having a length of $\frac{1}{8} ql^2$, so as to obtain the third point of the parabola and the point of intersection of the end tangents, respectively. The third tangent is parallel to the line joining the end points. The total diagram thus obtained coincides with that of Figure 13.4(d).

As regards the shear diagram, the equivalent statically determinate scheme must be balanced by two vertical forces at either end, directed upwards and equal to $\frac{1}{2} ql$, and by a clockwise couple of vertical forces $X/l = \frac{1}{8} ql$ (Figure 13.5(c)). Making the vector summation of the partial reactions, we obtain once more the extreme values of shearing force shown in Figure 13.4(c).

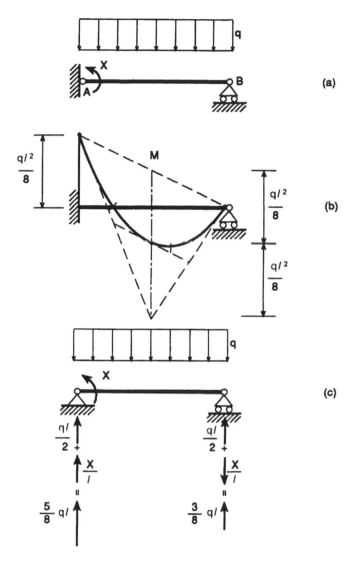

Figure 13.5

Let us now consider the same beam, built-in at one end and supported at the other, loaded in this case by a couple m at the end B (Figure 13.6(a)). Even though it is possible to accommodate the case within the equivalent statically determinate scheme consisting of the cantilever beam and obtained by eliminating the roller support, in the ensuing treatment we shall consider the second scheme used previously: that of the beam supported at either end (Figure 13.6(b)). The condition of congruence is

$$\varphi_A = \frac{Xl}{3EI} - \frac{ml}{6EI} = 0 \qquad (13.12)$$

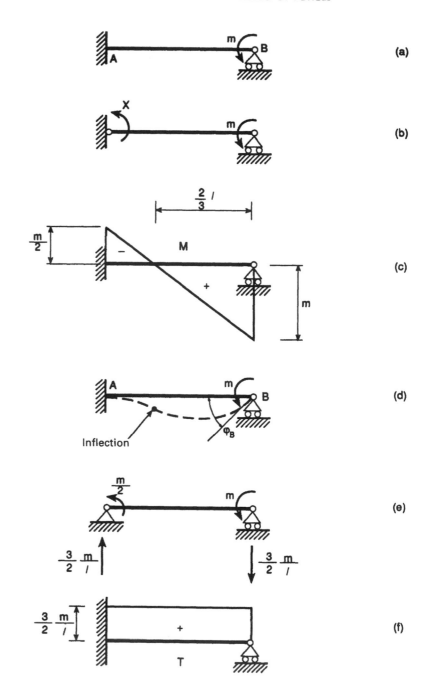

Figure 13.6

from which we obtain

$$X = \frac{m}{2} \tag{13.13}$$

The moment at the built-in constraint is then half that applied. The moment diagram is linear and hence vanishes at the distance $\frac{2}{3}l$ from the support (Figure 13.6(c)). At that point the elastic deformed configuration of the beam undergoes an inflection (Figure 13.6(d)). It is, moreover, possible to calculate the elastic rotation of the end section B:

$$\varphi_B = \frac{ml}{3EI} - \frac{(m/2)l}{6EI} = \frac{ml}{4EI} \tag{13.14}$$

As will emerge more clearly in the sequel, it is important to introduce the concept of rotational stiffness of the beam built-in at one end and supported at the other, for moments applied at the supported end:

$$k = \frac{m}{\varphi_B} = \frac{4EI}{l} \tag{13.15}$$

This stiffness is directly proportional to the elastic modulus of the material and to the moment of inertia of the cross section, and inversely proportional to the length of the beam.

The vertical reactions must balance the two moments having the same sense, m and $X = m/2$. They are thus equal and opposite forces of magnitude $(3/2)m/l$ (Figure 13.6(e)). The shear diagram is thus constant and positive (Figure 13.6(f)).

Note that, if in the cases hitherto examined we were to replace the roller support with a hinge, the solutions would not vary at all, on account of the absence of the axial force. Even if a redundant reaction were supposed, this would be zero, thus yielding the only contribution to the equation of axial congruence $w_B = 0$ (Figure 13.1(b)).

In the case of a rectilinear beam built in at both ends, loaded in any manner whatsoever (Figure 13.7(a)), the degree of static indeterminacy is three and can be eliminated by removing one of the two built-in constraints (Figure 13.7(b)). The three redundant unknowns, consisting of the elementary built-in constraint reactions, may be determined using the three equations of congruence with regard to horizontal translation, to vertical translation and to rotation, respectively:

$$w_B = 0, \quad v_B = 0, \quad \varphi_B = 0 \tag{13.16}$$

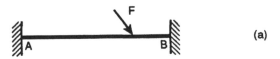

(a)

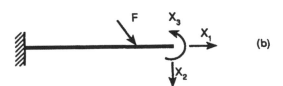

(b)

Figure 13.7

393

In the particular case of a vertical load concentrated in the centre (Figure 13.8(a)), the scheme has only one degree of redundancy. The other two degrees of redundancy, which are potentially present, do not appear owing to symmetry and the lack of horizontal components in the external loading. The equivalent statically determinate scheme is then found by inserting two hinges at the ends and applying two equal and opposite redundant moments X (Figure 13.8(b)). Just as there is only one degree of redundancy, there is also a single equation of congruence with regard to rotation

$$\varphi_A = -\varphi_B = \frac{Xl}{3EI} + \frac{Xl}{6EI} - \frac{Fl^2}{16EI} = 0 \tag{13.17}$$

whence we obtain

$$X = \frac{1}{8} Fl \tag{13.18}$$

The bending moment diagram can be constructed graphically by superposition (Figure 13.8(c)). The two redundant moments furnish a constant partial

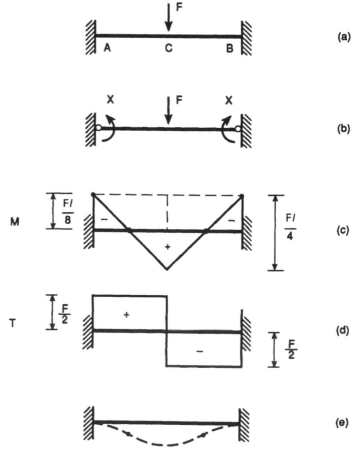

Figure 13.8

394

diagram which is equal in absolute value to $\frac{1}{8}Fl$, whilst the central force F produces a bilinear partial diagram with the maximum equal to $\frac{1}{4}Fl$. The total diagram thus intersects the axis of the beam at two symmetrical points at a distance $l/4$ from the built-in constraints.

The shear diagram is that produced exclusively by the force F, the two redundant moments constituting a self-balanced system (Figure 13.8(d)).

Since the elastic deformed configuration of the beam has to satisfy the constraint conditions as well as the deflections suggested by the moment diagram, it will appear as in Figure 13.8(e), with two inflections corresponding to the points where the bending moment becomes zero.

It is interesting to see how, by using the properties of symmetry of the structure, it is possible to reduce it even to a statically determinate scheme, and so resolve it with the use of equilibrium equations alone. One may consider just the half-beam on the left, once the centre C is constrained with a double connecting rod (Figure 13.9(a)). The vertical force $F/2$ is countered by the built-in constraint, so that it is possible to transform the scheme of Figure 13.9(a) into that of Figure 13.9(b), where the built-in constraint has been replaced by a second double rod loaded by the vertical reaction $F/2$. This latter scheme is skew-symmetrical and can be reduced to that of Figure 13.9(c), which presents a roller support in the new centre and thus emerges as statically determinate. It is clear that the reactive moment exerted by the double rod A, in the scheme of Figure 13.9(c), is

$$M_A = \frac{F}{2} \times \frac{l}{4} = \frac{1}{8}Fl$$

and represents the fixed-end moment already defined following another procedure.

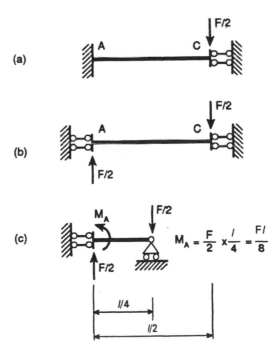

Figure 13.9

395

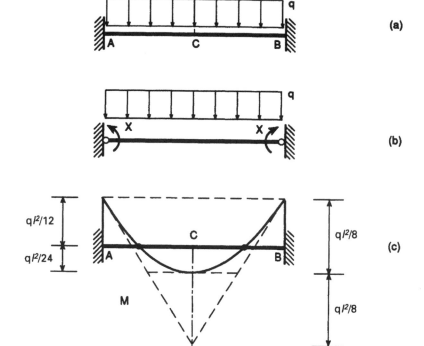

Figure 13.10

Adopting the same synthetic approach seen hitherto, it is not possible, on the other hand, to resolve the case of a beam built in at both ends and subjected to a distributed load q (Figure 13.10(a)). The equivalent statically determinate scheme is that obtained, as we have already seen, by replacing the built-in constraints with hinges and by applying two equal and opposite redundant moments (Figure 13.10(b)). The equation of congruence is

$$\varphi_A = -\varphi_B = \frac{Xl}{3EI} + \frac{Xl}{6EI} - \frac{ql^3}{24EI} = 0 \qquad (13.19)$$

from which we obtain

$$X = \frac{ql^2}{12} \qquad (13.20)$$

Whereas then the shear diagram is equal to the one of the scheme of a beam supported at both ends and subjected to a distributed load q (Figure 5.20(c)), the moment diagram is obtained from the graphical addition of a constant diagram with value equal to $-\frac{1}{12}ql^2$, and a parabolic diagram with a maximum of $\frac{1}{8}ql^2$ (Figure 13.10(c)). The moment in the centre is thus $M_C = \frac{1}{24}ql^2$.

The scheme of a beam constrained by a built-in support and a double rod, loaded by a vertical force F applied to the double rod (Figure 13.11(a)), from symmetry is equivalent to the beam of twice the length, i.e. $2l$, built in at both ends and subjected to twice the force, $2F$, in the centre (Figure 13.11(b)), or to

396

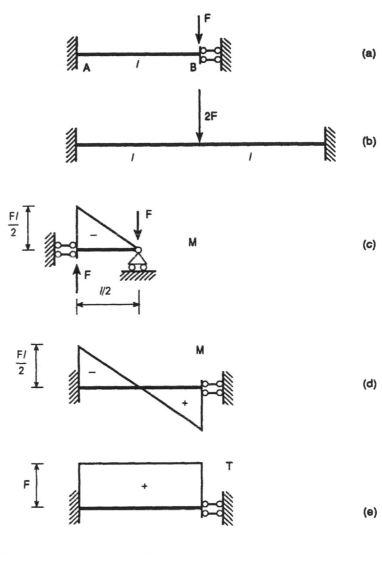

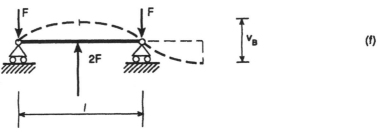

Figure 13.11

the beam of length $l/2$ constrained by a double rod and a roller support (Figure 13.11(c)). In either case the bending moment and shearing force diagrams are those represented in Figure 13.11(d) and Figure 13.11(e), respectively. The elastic displacement of point B can be determined in various ways. The simplest is, however, the one based on the scheme of Figure 13.11(c), or rather, by symmetry, on the scheme of Figure 13.11(f)

$$v_B = 2\left(\frac{2Fl^3}{48EI}\right) = \frac{Fl^3}{12EI} \tag{13.21}$$

Likewise, the scheme of a beam constrained by a hinge and a double rod, acted upon by a vertical force F applied at the double rod (Figure 13.12(a)), from symmetry is equivalent to the beam of twice the length, $2l$, supported at both ends and subjected in the centre to twice the force, $2F$ (Figure 13.12(b)). The bending moment and shear diagrams are those represented in Figures 13.12(c), (d). The vertical displacement of the point B constrained by the double rod is

$$v_B = \frac{2F(2l)^3}{48EI} = \frac{Fl^3}{3EI} \tag{13.22}$$

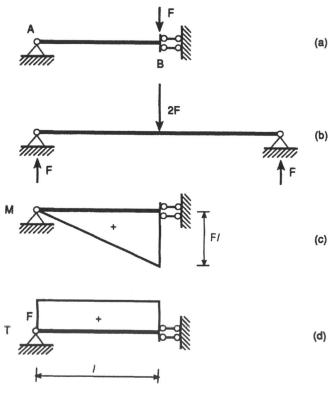

Figure 13.12

398

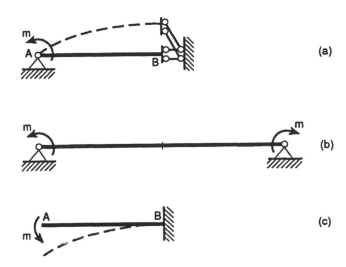

Figure 13.13

The above structure, loaded by a moment at the hinged end (Figure 13.13(a)) is equivalent to the beam of twice the length, $2l$, supported at both ends and subjected to two opposite moments at the ends (Figure 13.13(b)). The moment is thus constant and the shear absent. The rotation at the extreme section A is

$$\varphi_A = \frac{m(2l)}{3EI} + \frac{m(2l)}{6EI} = \frac{ml}{EI} \qquad (13.23)$$

Notice that the rotation φ_A is the same as that undergone by the end section of a cantilever beam, built in at B and loaded by the moment m at A (Figure 13.13(c)). In fact the moment diagrams of the schemes of Figures 13.13(a), (c), coincide, just as the same boundary condition, $v'_B = 0$, applies in both cases. The deformed configuration, on the other hand, remains the same but for one additional constant.

13.4 Elastic constraints

Up to now the constraints, whether internal or external, have been considered as rigid, i.e. as conditions of congruence, where the displacements or rotations vanish. In practice, however, the constraints cannot always be treated simply as rigid. They are said to settle elastically when the reaction of the constraint is proportional to the displacement undergone by the constraint itself. In what follows we shall compare the results for rigidly constrained redundant structures with those for the same structures constrained elastically. In statically determinate structures, on the other hand, the constraint reactions and the diagrams of characteristics do not depend on the stiffness of the constraints, since in any case the equilibrium equations are the same.

Let us consider the continuous beam on three supports shown in Figure 13.14(a), subjected to the moment m acting at the end C. There exists an infinite number of triads of constraint reactions V_A, V_B, V_C, equilibrants of the

399

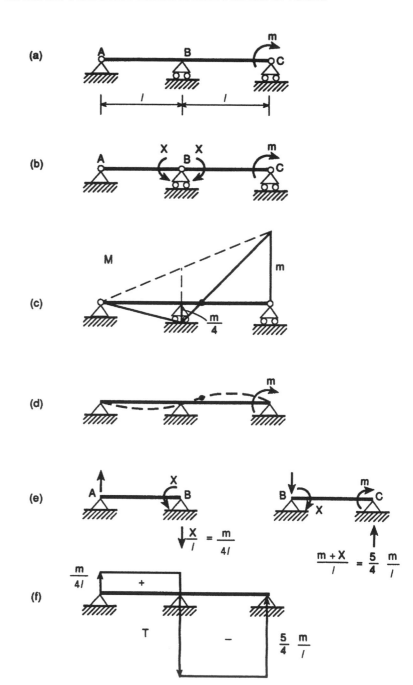

Figure 13.14

moment m. The one that also ensures congruence could be obtained by eliminating any one of the three supports, by applying the corresponding unknown redundant reaction and by imposing the condition that the corresponding vertical displacement should become zero. An even faster approach is that of interrupting the continuity of the beam by inserting a hinge on the support B (Figure 13.14(b)) and applying the redundant reactions transmitted by the removed constraint, viz. two equal and opposite moments. The condition of congruence at this point will concern the continuity of the elastic line, which will not be able to present cusps in B. That is, the section B thought of as belonging to the beam AB must rotate by the same amount as that by which the same cross section, thought of as belonging to the beam BC, rotates:

$$\varphi_{BA} = \varphi_{BC} \tag{13.24}$$

Rendering both sides explicit, we have

$$\frac{Xl}{3EI} = -\frac{Xl}{3EI} + \frac{ml}{6EI} \tag{13.25}$$

from which we obtain

$$X = \frac{m}{4} \tag{13.26}$$

Knowing now the moments at the three supports, the moment diagram can at once be drawn (Figure 13.14(c)). It will suffice to lay out to scale a segment of length m above the support C, a segment of length $m/4$ beneath the support B, and then join with straight line segments the notable points of the diagram thus defined. A point of annihilation of the moment in the right-hand span is then identified at the distance $l/5$ from the central support. At this point the elastic deformed configuration possesses an inflection (Figure 13.14(d)).

The constraint reactions, and hence the shear diagram, may be determined by isolating the supported beams AB and BC, acted upon by the external and the redundant loads (Figure 13.14(e)). The two end reactions are directed upwards and are $V_A = m/4l$, $V_C = 5m/4l$, while the reaction of the intermediate support is the sum of the reactions that apply to the two schemes of Figure 13.14(e), $V_B = 3m/2l$. It is a force directed downwards, which produces a discontinuity of the first kind in the shear diagram (Figure 13.14(f)).

Consider again the foregoing scheme, assuming, however, an intermediate elastically compliant support (Figure 13.15(a)). This compliance can, for instance, represent the axial compliance of a connecting rod. In this case the stiffness of the equivalent spring is $k = EA/l$ (Figure 13.15(b)). When the external constraints are elastically compliant, it is necessary to consider, not only, as in the usual case, the action of the constraint on the structure, but also the action of the structure on the compliant constraint. The scheme of Figure 13.15(c) shows how the spring is loaded by a force X/l transmitted by the left-hand beam and by a force $(m+X)/l$ transmitted by the right-hand beam, both directed upwards in accordance with the conventions assumed. The support will then rise by the quantity

$$\delta = \frac{m + 2X}{kl} \tag{13.27}$$

401

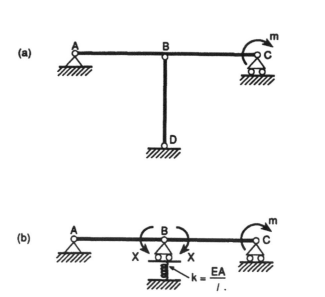

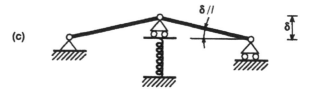

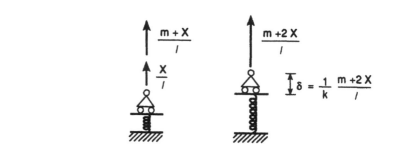

Figure 13.15

402

which is a function of the redundant unknown X, and will induce a rigid rotation in both beams, with an absolute value equal to δ/l (Figure 13.15(c)).

The equation of congruence, in implicit form, is once more equation (13.24), where, on the other hand, also the contributions of rigid rotation must appear, as well as those of elastic rotation previously considered. We shall therefore have

$$\frac{Xl}{3EI} + \frac{m+2X}{kl^2} = -\frac{Xl}{3EI} + \frac{ml}{6EI} - \frac{m+2X}{kl^2} \tag{13.28}$$

whence we obtain

$$X = m \frac{\left(\dfrac{l}{6EI} - \dfrac{2}{kl^2}\right)}{\left(\dfrac{2l}{3EI} + \dfrac{4}{kl^2}\right)} \tag{13.29}$$

When the stiffness of the spring tends to infinity, we find again the previous result (Figure 13.14(c))

$$\lim_{k \to \infty} X = \frac{m}{4} \tag{13.30}$$

On the other hand, when the stiffness of the spring tends to zero, we find again the value of moment in the centre for a beam supported at both ends, having a length of $2l$, loaded by a moment at the end

$$\lim_{k \to 0} X = -\frac{m}{2} \tag{13.31}$$

Note that the moment X vanishes when

$$k = \frac{12EI}{l^3} \tag{13.32}$$

Hence, for smaller stiffness the inflection disappears in the elastic line, and only the upper fibres are stretched. For $k = 12EI/l^3$, the moment is zero in the left-hand span, since only the supports B and C react, and the deformed configuration of the beam is rigid between the supports A and B (Figure 13.15(d)), i.e. the left-hand span rotates rigidly counterclockwise, drawn along by the deflection of the right-hand span.

As a second example of elastic constraint, consider again the beam built in at one end and supported at the other, acted upon by a distributed load q (Figure 13.16(a)). In this case, assume that the built-in constraint A is angularly compliant. The equivalent statically determinate scheme is the same as that of Figure 13.5(a), but, in the equation of congruence, the action of the beam on the built-in constraint must be taken into account (Figure 13.16(b)). Whereas, that is, the built-in constraint acts on the beam with a counterclockwise moment X, the beam will act on the built-in constraint with a clockwise moment X, and will cause it to rotate by an angle $-X/k$. The equation of congruence is thus modified as follows:

$$\varphi_A = \frac{Xl}{3EI} - \frac{ql^3}{24EI} = -\frac{X}{k} \tag{13.33}$$

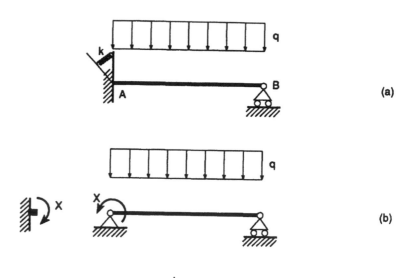

(a)

(b)

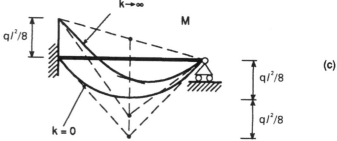

(c)

Figure 13.16

from which we obtain the redundant unknown

$$X = \frac{\left(\dfrac{ql^3}{24EI}\right)}{\left(\dfrac{l}{3EI} + \dfrac{1}{k}\right)}$$

(13.34)

When the angular stiffness of the built-in constraint tends to infinity, we find again the **rigid joint moment**

$$\lim_{k \to \infty} X = \frac{1}{8}ql^2$$

(13.35a)

On the other hand, when the angular stiffness of the built-in constraint tends to zero, the moment vanishes because the constraint A turns into a hinge:

$$\lim_{k \to 0} X = 0$$

(13.35b)

The moment diagram will thus in general be contained between the two limit diagrams of Figure 13.16(c).

13.5 Inelastic constraints (imposed displacements)

The cases of inelastic constraint settlements, which will be dealt with in the ensuing discussion, can be more appropriately termed **imposed displacements**. In fact, it is not a question of modifying in some way the reactive properties of the constraint, but rather of imposing a predetermined displacement on the constraint itself. Such a displacement will therefore not be a function of the loads and the redundant unknowns, but will itself perform the function of an external load, it being a datum of the problem.

In the case where infinitesimal displacements are imposed on the constraints of a statically determinate beam system, the system adapts by undergoing only rigid rototranslations. Consequently no external or internal reactions develop. On the other hand, with the exception of particular cases, it may be stated that the displacements imposed on a statically indeterminate beam system generate reactions and deflections in the beams. Evidently this is due to the redundant degree of constraint, which may be said to oppose the deformation of the system. A rational explanation of this may be found in the matrix treatment of the kinematics of rigid systems, presented in Chapter 3.

The kinematic matrix of a statically determinate beam system is square, and hence the augmented matrix, containing the imposed displacements, will have the same rank, whatever the displacements imposed on the constraints. By virtue of the Rouché-Capelli Theorem, it is thus possible to state that there exists a single kinematic solution, represented by the rigid deformed configuration of the system.

The kinematic matrix of a statically indeterminate beam system is rectangular and is augmented by the imposed displacements on its longer side. This allows the augmented matrix to have a rank greater than that of the kinematic matrix, unless the imposed displacements are so particular as to produce a column of known terms linearly dependent on the columns of the kinematic matrix. If we exclude the latter case, the Rouché-Capelli Theorem makes it possible to assert that there does not exist a solution in the framework of the kinematics of the rigid body. The beam system will thus have to adapt by undergoing deformations (dilations and deflections). The solution is determinable only in the context of the kinematics of the deformable body.

Consider a beam built in at one end and supported at the other and let the vertical displacement η_0 of the built-in constraint be imposed (Figure 13.17(a)). Let the equivalent statically determinate scheme be that of a beam supported at both ends (Figure 13.17(b)). This scheme rotates only rigidly, after the imposition of the displacement η_0 (Figure 13.17(c)). The equation of congruence will consider both the elastic rotation induced by the redundant moment X, and the rigid rotation η_0/l,

$$\varphi_A = -\frac{Xl}{3EI} + \frac{\eta_0}{l} = 0 \qquad (13.36)$$

from which we obtain

$$X = \frac{3EI}{l^2}\eta_0 \qquad (13.37)$$

The reactive moment of the built-in constraint is thus proportional to the magnitude η_0 of the inelastic settlement. The bending moment diagram is thus

405

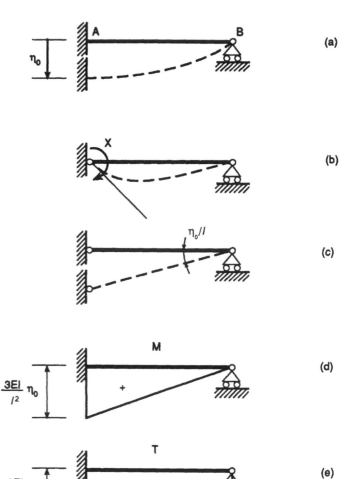

Figure 13.17

linear with a maximum at the built-in constraint (Figure 13.17(d)), whilst the shear is constant and equal to $-X/l$ (Figure 13.17(e)).

A different equivalent statically determinate scheme could be that of a cantilever beam, with the reaction V_B at the free end determinable using the equation of congruence

$$v_B = \eta_0 - \frac{V_B l^3}{3EI} = 0 \tag{13.38}$$

whence

$$V_B = \frac{3EI}{l^3}\eta_0 \tag{13.39}$$

406

On the other hand, the reaction V_B may be obtained also by inverting the formula (13.22) corresponding to the scheme of a beam with a hinge and a double rod (Figure 13.12(a)).

Let a rotation φ_0 of the built-in constraint be imposed on the scheme of the beam built in at one end and supported at the other (Figure 13.18(a)). Let the equivalent statically determinate scheme consist of a beam supported at either end, loaded by the unknown moment X (Figure 13.18(b)). The condition of angular congruence thus reads

$$\varphi_A = \frac{Xl}{3EI} = \varphi_0 \tag{13.40}$$

from which we deduce

$$X = \frac{3EI}{l}\varphi_0 \tag{13.41}$$

Also in this case the moment diagram is linear (Figure 13.18(c)) with a maximum at the built-in constraint, while the shear is constant and equal to X/l (Figure 13.18(d)).

Let us now pass on to the beam built in at both ends, and let a vertical displacement and a rotation be imposed separately to one of the two built-in constraints. In the case of the imposed displacement η_0 (Figure 13.19(a)), the

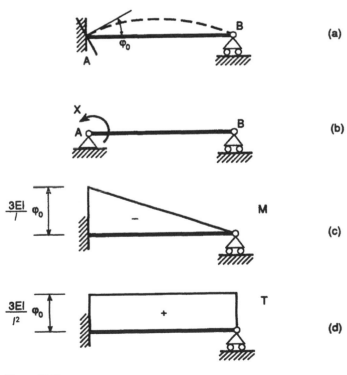

Figure 13.18

407

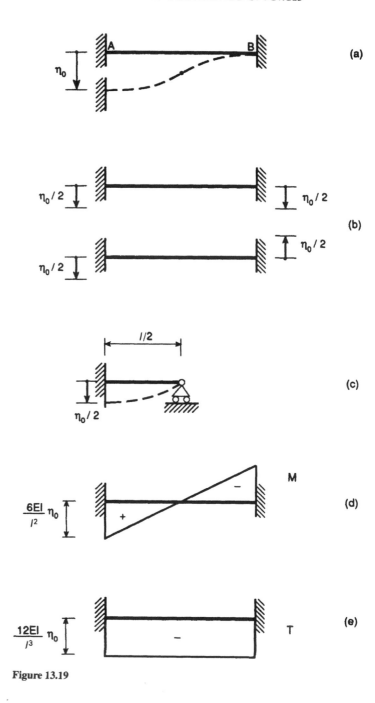

Figure 13.19

scheme can be split up into one symmetrical scheme plus one skew-symmetrical scheme (Figure 13.19(b)). Whereas the symmetrical scheme represents a simple rigid translation $\eta_0/2$ downwards of the entire beam, the skew-symmetrical scheme can be brought back to the one previously studied of a beam built in at

408

one end and supported at the other (Figure 13.19(c)), for which the moment at the built-in constraint is

$$X = \frac{3EI}{(l/2)^2}\frac{\eta_0}{2} = \frac{6EI}{l^2}\eta_0 \qquad (13.42)$$

The moment diagram is thus linear and skew-symmetrical (Figure 13.19(d)), while the shear diagram is constant and equal to (Figure 13.19(e))

$$T = -\frac{2X}{l} = -\frac{12EI}{l^3}\eta_0 \qquad (13.43)$$

In the case of imposed rotation φ_0 (Figure 13.20(a)), the equivalent statically determinate scheme is that of a beam supported at either end with unknown redundant moments at the ends (Figure 13.20(b)). The equations of congruence will impose the rotations φ_0 in A and zero in B

$$\varphi_A = \frac{X_1 l}{3EI} - \frac{X_2 l}{6EI} = \varphi_0 \qquad (13.44a)$$

$$\varphi_B = \frac{X_2 l}{3EI} - \frac{X_1 l}{6EI} = 0 \qquad (13.44b)$$

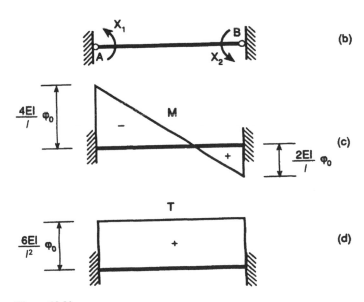

Figure 13.20

409

From the foregoing equations we can deduce the two redundant unknowns

$$X_1 = \frac{4EI}{l}\varphi_0 \qquad (13.45a)$$

$$X_2 = \frac{2EI}{l}\varphi_0 \qquad (13.45b)$$

The moment diagram is thus linear and asymmetrical, with a value at the end at which the built-in constraint rotates which is twice that at the other end (Figure 13.20(c)). The shear is constant (Figure 13.20(d)), having a value of

$$T = \frac{X_1 + X_2}{l} = \frac{6EI}{l^2}\varphi_0 \qquad (13.46)$$

Note that it would have been possible to disconnect just the built-in constraint A, in which case the equation of congruence would have been expressible on the basis of equation (13.14):

$$\varphi_A = \frac{X_1 l}{4EI} = \varphi_0 \qquad (13.47)$$

In this way the solution outlined previously is once more obtained.

Finally, let us consider the scheme of the beam built in at one end and constrained with a double rod at the other. Whereas the vertical displacement of the built-in constraint does not generate reactions and deflections (it is one of those cases in which the augmented matrix has the same rank as the kinematic matrix), an imposed rotation φ_0 generates the elastic deformation of Figure 13.21(a). The equivalent statically determinate scheme consists of the hinged beam constrained by a double rod, loaded by the redundant moment X (Figure

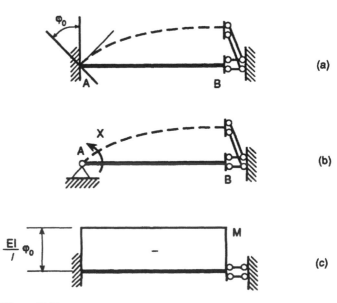

Figure 13.21

13.21(b)). This scheme has already been studied in Section 13.3 and, on the basis of equation (13.23), yields the following equation of congruence:

$$\varphi_A = \frac{Xl}{EI} = \varphi_0 \qquad (13.48)$$

from which we deduce

$$X = \frac{EI}{l}\varphi_0 \qquad (13.49)$$

The moment diagram is constant (Figure 13.21(c)), whereas the shear is zero.

13.6 Thermal distortions

Thermal distortions are the deformations induced in beams by variations in temperature within the depth h. In the present section only linear variations of temperature within the depth h will be considered. Each **linear variation of temperature** can then be divided into a **uniform thermal variation** and a **butterfly-shaped thermal variation** (Figure 13.22). In the context of small displacements, it will be possible to apply the Principle of Superposition.

A uniform thermal variation ΔT, acting on an infinitesimal beam element of length dz (Figure 13.23), induces an elongation of the element given by

$$dw = \alpha\,\Delta T\,dz \qquad (13.50)$$

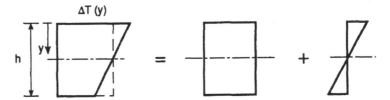

Figure 13.22

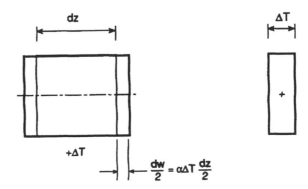

Figure 13.23

where α is the **coefficient of thermal expansion** of the material. Thus a **thermal dilation** will be produced:

$$\varepsilon_T = \frac{\mathrm{d}w}{\mathrm{d}z} = \alpha\,\Delta T \qquad (13.51)$$

The butterfly-shaped thermal variation of Figure 13.24 brings about lengthening of the lower fibres and shortening of the upper ones, so that, if the cross sections are assumed to be orthogonal to the axis of the beam even after thermal deformation has occurred, a relative rotation is produced between the end sections of the element,

$$\mathrm{d}\varphi = 2\,\frac{\alpha\,\Delta T\,\dfrac{\mathrm{d}z}{2}}{\dfrac{h}{2}} = 2\alpha\,\Delta T\,\frac{\mathrm{d}z}{h} \qquad (13.52)$$

and hence a **thermal curvature**

$$\chi_T = \frac{\mathrm{d}\varphi}{\mathrm{d}z} = 2\alpha\,\frac{\Delta T}{h} \qquad (13.53)$$

Since, on the other hand, the rotation is equal, but for the algebraic sign, to the derivative of the vertical displacement v,

$$\varphi = -\frac{\mathrm{d}v}{\mathrm{d}z} \qquad (13.54)$$

the equation (13.53) becomes

$$\frac{\mathrm{d}^2 v}{\mathrm{d}z^2} = -2\alpha\,\frac{\Delta T}{h} \qquad (13.55)$$

The differential equation (13.55), which governs the thermal deflection of the beam, is formally identical to equation (10.47), which governs the elastic deflection and, in place of the temperature variation ΔT, presents the bending

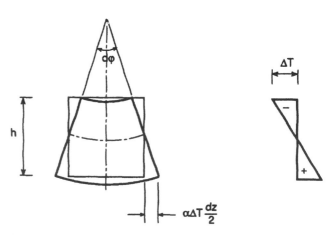

Figure 13.24

412

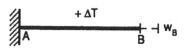

Figure 13.25

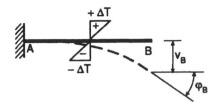

Figure 13.26

moment M. Putting together the two contributions, it is possible to write the **equation of the thermoelastic line**:

$$\frac{\mathrm{d}^2 v}{\mathrm{d}z^2} = -\left(\frac{M}{EI} + 2\alpha \frac{\Delta T}{h}\right) \tag{13.56}$$

Let us now consider some elementary statically determinate schemes subject to thermal distortions. For the same reasons pointed out in the case of imposed displacements, the thermal distortions do not generate additional reactions and elastic deformations in statically determinate structures. The latter are thus freely deformed without forcing of any sort. The cantilever beam of Figure 13.25, subjected to a uniform thermal variation, is lengthened, for example, by the amount

$$w_B = \int_A^B \varepsilon_T \, \mathrm{d}z = \alpha \Delta T \, l \tag{13.57}$$

The same cantilever beam of Figure 13.25, subjected to a butterfly-shaped thermal variation (Figure 13.26), will undergo a rotation of thermal origin at the end section B,

$$\varphi_B = \int_A^B \chi_T \, \mathrm{d}z = -2\alpha \frac{\Delta T}{h} l \tag{13.58a}$$

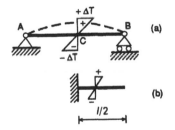

(a)

(b)

Figure 13.27

and a deflection,

$$v_B = \int_A^B \chi_T z \, \mathrm{d}z = \alpha \frac{\Delta T}{h} l^2 \tag{13.58b}$$

which can be determined also by resolving the differential equation (13.55), with the boundary conditions: $v(A) = v'(A) = 0$.

The supported beam subjected to a butterfly-shaped thermal variation (Figure 13.27(a)) constitutes a symmetrical structural scheme, whereby, using the previous results for the cantilever beam, we have (Figure 13.27(b))

$$\varphi_A = -\varphi_B = \alpha \frac{\Delta T}{h} l \tag{13.59a}$$

$$v_C = -\alpha \frac{\Delta T}{h}\left(\frac{l}{2}\right)^2 = -\alpha \frac{\Delta T}{h} \frac{l^2}{4} \tag{13.59b}$$

In the case of the beam built in at both ends, subjected to a uniform thermal variation (Figure 13.28(a)), the equivalent statically determinate scheme is

413

Figure 13.28

that of a cantilever beam with an unknown axial reaction X (Figure 13.28(b)). The equation of congruence

$$w_B = \alpha\,\Delta T\,l - \frac{X}{EA}l = 0 \qquad (13.60)$$

yields the axial force

$$X = EA\alpha\Delta T \qquad (13.61)$$

In the case of a beam built in at both ends, subjected to a butterfly-shaped thermal variation (Figure 13.29(a)), the equivalent statically determinate scheme is that of a beam supported at either end with redundant reactions X at the ends (Figure 13.29(b)). The equation of congruence

$$\varphi_A = -\varphi_B = \alpha\frac{\Delta T}{h}l - \frac{Xl}{2EI} = 0 \qquad (13.62)$$

yields the built-in constraint moment, which is equal to the bending moment acting on all the sections of the beam

$$X = 2\frac{\alpha\,\Delta TEI}{h} \qquad (13.63)$$

Note that since the elastic curvature is equal and opposite to the thermal curvature,

$$\chi_e = \frac{X}{EI} = 2\alpha\frac{\Delta T}{h} \qquad (13.64a)$$

$$\chi_T = -2\alpha\frac{\Delta T}{h} \qquad (13.64b)$$

the global thermoelastic deformed configuration is zero, and the beam remains rectilinear.

In conclusion, let us take a beam built in at one end and constrained by a vertical connecting rod at the other (Figure 13.30(a)). Let the beam be subjected to a butterfly-shaped thermal variation and the connecting rod to a

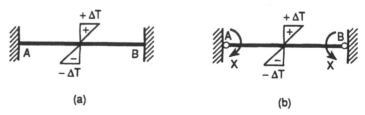

(a)   (b)

Figure 13.29

414

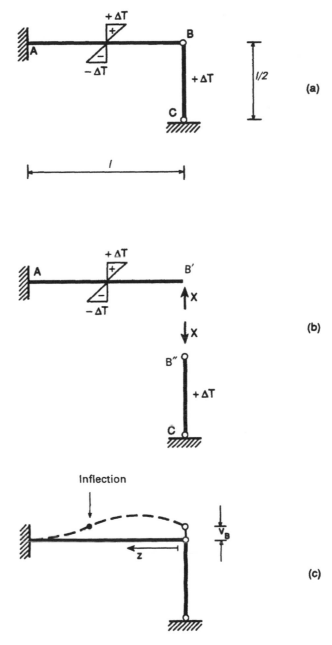

Figure 13.30

uniform thermal variation. If we isolate the cantilever beam and the connecting rod, which exchange the redundant reaction X (Figure 13.30(b)), it is possible to formulate the equation of congruence, which must express the equality of the deflection of point B', belonging to the cantilever beam, and point B'', belonging to the connecting rod,

415

$$v_B = v_{B''} \qquad (13.65)$$

with

$$v_B = \alpha \frac{\Delta T}{h} l^2 - \frac{X l^3}{3EI} \qquad (13.66a)$$

$$v_{B''} = -\alpha \Delta T \frac{l}{2} + \frac{X\left(\dfrac{l}{2}\right)}{EA} \qquad (13.66b)$$

With expressions (13.66) substituted in equation (13.65), we obtain

$$X = \frac{\alpha \Delta T E \left(\dfrac{l}{h} + \dfrac{1}{2}\right)}{\left(\dfrac{1}{2A} + \dfrac{l^2}{3I}\right)} \qquad (13.67)$$

The connecting rod therefore functions as a strut, as was assumed conventionally.

The beam presents a thermoelastic inflection where the elastic curvature equals, in absolute value, the thermal one,

$$\frac{Xz}{EI} = 2\alpha \frac{\Delta T}{h} \qquad (13.68)$$

from which we obtain the coordinate of the inflection (Figure 13.30(c)),

$$z = \frac{2\alpha \Delta T E I}{Xh} \qquad (13.69)$$

where X is given by equation (13.67). Finally

$$z = \frac{\left(\dfrac{I}{A} + \dfrac{2}{3}l^2\right)}{\left(\dfrac{h}{2} + l\right)} \qquad (13.70)$$

and, in the case of a rectangular section and a slender beam ($l/h \to \infty$),

$$z \simeq \frac{2}{3}l \qquad (13.71)$$

Remaining within the restrictive hypotheses made above, the vertical displacement of the end B is

$$v_B = -\frac{1}{2}\alpha \Delta T l \qquad (13.72)$$

and thus, since it is negative, the point B will rise. The thermoelastic deformed configuration of the beam is depicted in Figure 13.30 (c).

13.7 Continuous beams

The term **continuous beams** refers to rectilinear beams devoid of internal disconnections, constrained to the foundation by a series of supports and, in some cases, by built-in supports at the ends. The equivalent statically determinate structure is obtained by inserting a hinge in each support or in each external built-in support. There will thus be as many redundant unknowns and as many equations of congruence as there are disconnected nodes. The equation of angular congruence corresponding to each node will contain the redundant moment acting on the same node, as well as the redundant moments acting on the two adjacent nodes; it is for this reason that it is called **equation of three moments**. In the case of nodes adjacent to the ends or end nodes, the number of unknowns present in the corresponding equation may prove to be less than three.

Consider the three-span continuous beam of Figure 13.31(a). It is constrained by four supports (a hinge and three roller supports) and loaded by a moment m at the right-hand end D. If the two intermediate supports are disconnected with two hinges and the redundant moments X_1 and X_2 are applied (Figure 13.31(b)), the equations of congruence take on the following form:

$$\varphi_{BA} = \varphi_{BC} \tag{13.73a}$$

$$\varphi_{CB} = \varphi_{CD} \tag{13.73b}$$

and thus

$$-\frac{X_1 l}{3EI} = \frac{X_1 l}{3EI} + \frac{X_2 l}{6EI} \tag{13.74a}$$

$$-\frac{X_2 l}{3EI} - \frac{X_1 l}{6EI} = \frac{X_2 l}{3EI} + \frac{ml}{6EI} \tag{13.74b}$$

Once resolved, equations (13.74) yield the solution $X_1 = \frac{1}{15}m$, $X_2 = -\frac{4}{15}m$, and therefore the moment diagram of Figure 13.31 (c). This diagram consists of a broken line which intersects the axis of the beam in two points. The points at which the bending moment becomes zero correspond to inflection points in the elastic deformed configuration (Figure 13.31(d)). Finally, the shear diagram is represented in Figure 13.31(e). The discontinuities or jumps which the function T undergoes, proceeding from left to right, represent the vertical reactions of the supports:

$$V_A = -\frac{1}{15}\frac{m}{l}, \quad V_B = \frac{2}{5}\frac{m}{l}, \quad V_C = -\frac{8}{5}\frac{m}{l}, \quad V_D = \frac{19}{15}\frac{m}{l}$$

Let us now assume that the continuous beam considered previously undergoes an inelastic settlement of the support B. After disconnecting the beam as shown in Figure 13.31(b), it will be necessary to consider the contributions to the rotations of the sections B and C, due both to the elastic deformability of the beams and to the rigid mechanism caused by the deflection η_0

$$-\frac{X_1 l}{3EI} - \frac{\eta_0}{l} = \frac{X_1 l}{3EI} + \frac{X_2 l}{6EI} + \frac{\eta_0}{l} \tag{13.75a}$$

$$-\frac{X_2 l}{3EI} - \frac{X_1 l}{6EI} + \frac{\eta_0}{l} = \frac{X_2 l}{3EI} \tag{13.75b}$$

417

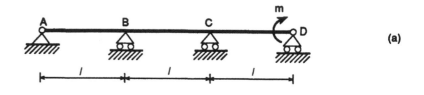

(a)

(b)

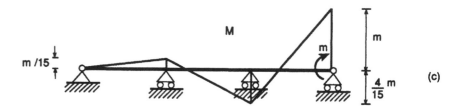

(c)

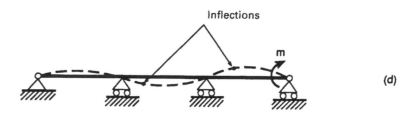

(d)

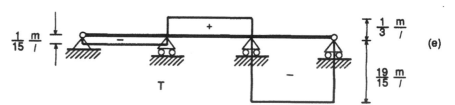

(e)

Figure 13.31

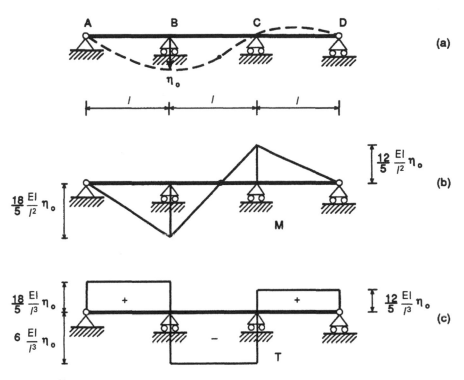

Figure 13.32

The solution to the system (13.75) is as follows:

$$X_1 = -\frac{18}{5}\frac{EI}{l^2}\eta_0 \tag{13.76a}$$

$$X_2 = \frac{12}{5}\frac{EI}{l^2}\eta_0 \tag{13.76b}$$

The moment diagram (Figure 13.32(b)) intersects the axis of the beam at a single point, so that the deformed configuration presents the fibres stretched at the lower edge in the left-hand part and at the upper edge in the right-hand part (Figure 13.32(a)). The shear diagram is depicted in Figure 13.32(c), and also in this case the discontinuities measure the values of the constraint reactions.

As our last example of a continuous beam, we shall examine the one depicted in Figure 13.33(a), which consists of two spans with a built-in constraint in A and supports in B and C. Since the distributed load q acts only on the span BC, it is possible to reduce the scheme to that of Figure 13.16(a), with the elastic compliance of the beam AB represented by the rotational spring. The elementary schemes for resolving the problem are given in Figure 13.33(b), where the stiffness k of the spring is equal to $4EI/l$. From equation (13.34) we obtain $X = ql^2/14$, and hence the moment diagram of Figure 13.33(c). The shear emerges as constant in the beam AB and linear in the beam BC (Figure 13.33(d)). The point of zero shear corresponds to the maximum bending moment (Figure 13.33(c)), just as the points where the moment

419

becomes zero correspond to the inflections of the elastic line (Figure 13.33(e)).

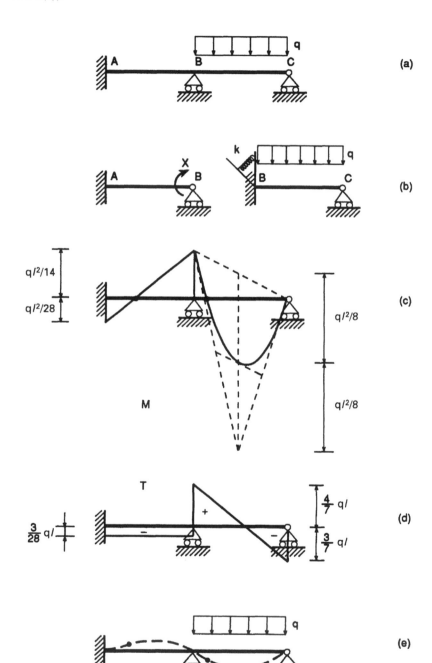

Figure 13.33

420

14 Statically indeterminate structures: method of displacements

14.1 Introduction

The **method of displacements** is dual of the method of forces, introduced in the previous chapter. Thus it is a process of identifying a single set of kinematic parameters which, in addition to congruence, implies also equilibrium.

From the operative standpoint, the **method of displacements** consists of imposing certain displacements or rotations, characteristic of the system, in such a way that the $v - g$ redundant reactions satisfy $v - g$ relations of equilibrium. In the sequel this method will be applied to beam systems of various types. In illustrating the method, we shall start from simple cases of parallel-arranged elements and close the chapter by presenting fundamental concepts on which the **automatic computation** of beam systems with multiple degrees of indeterminacy (trusses, plane frames, plane grids, space frames) is based, in the static as well as the dynamic regime. Some elements of the seismic analysis of multi-storey buildings are also given.

14.2 Parallel-arranged bar systems

Consider a rigid cross member of symmetrical shape constrained with a symmetrical system of parallel connecting rods, which may present different lengths and different cross-sectional areas, and may consist of materials with different elastic moduli (Figure 14.1(a)). If this symmetrical system is loaded symmetrically, for instance by a vertical force F acting in the centre of the cross member, the resulting deformation will be symmetrical and can be globally described by a single datum: the vertical translation δ of the cross member. Each connecting rod will in fact undergo the same elongation δ, since it is hinged to the cross member. For the congruence of the system it is therefore possible to set

$$\delta_i = \frac{X_i l_i}{E_i A_i} = \delta \qquad (14.1)$$

where δ_i is the elongation of the ith connecting rod, X_i is the corresponding axial tensile force, l_i, E_i, A_i are, respectively, the length, the elastic modulus and the cross-sectional area of the connecting rod. Congruence is thus implicitly assumed, while the kinematic unknown δ is determined by considering the equilibrium of the cross member with regard to vertical translation (Figure 14.1(b))

$$F = \sum_i X_i \qquad (14.2)$$

421

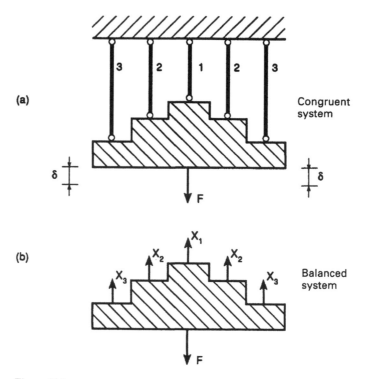

Figure 14.1

From equation (14.1) we obtain in fact the value of each redundant unknown X_i as a function of the primary unknown δ,

$$X_i = \frac{E_i A_i}{l_i} \delta \qquad (14.3)$$

and thus the equation (14.2) yields the displacement that is sought:

$$\delta = \frac{F}{\sum\limits_i \frac{E_i A_i}{l_i}} \qquad (14.4)$$

The foregoing expression may also be cast in the form

$$\delta = \frac{F}{K} \qquad (14.5)$$

where

$$K = \sum\limits_i K_i = \sum\limits_i \frac{E_i A_i}{l_i} \qquad (14.6)$$

denotes the total stiffness of the system. This turns out to be equal to the summation of the partial stiffnesses K_i of the parallel elements.

422

Finally, the expression (14.3) furnishes the reactions of the individual connecting rods

$$X_i = F\left(\frac{E_i A_i}{l_i}\right)\bigg/\left(\sum_i \frac{E_i A_i}{l_i}\right) = F\frac{K_i}{K} \tag{14.7}$$

where the ratio K_i/K, between partial stiffness and total stiffness, is called **coefficient of distribution** and indicates the fraction of the total load supported by the ith element. This fraction is proportional to the partial stiffness of the element itself.

Hence, in this particular case of parallel elements, the convenience of the method of displacements, compared with the method of forces, emerges clearly. In fact, if we were to apply the latter method, we should have to resolve $(n-1)$ equations of congruence, n being the number of unknown redundant reactions (in the example of Figure 14.1 we have $n = 3$).

As a second example of parallel-arranged elements, consider a rigid cross member of asymmetrical shape, constrained by a system of parallel connecting rods of varying lengths and cross sections and of different constitutive materials (Figure 14.2(a)). Let this system then be loaded in an altogether generic manner, on condition that no components of horizontal force are present, with respect to which the system is free. Consider, for instance, a generic vertical force applied to the cross member at a distance d from the centre (Figure 14.2(a)). With respect to this force the system presents $(n-2)$ degrees of indeterminacy, n being the total number of connecting rods.

Once deformation has taken place, the cross member will be rotated by the angle φ with respect to the undeformed configuration, so that there will be two kinematic unknowns: for example, the vertical translation δ of the mid-point of the cross member together with the angle of rotation φ. The congruence of the system thus imposes

$$\delta_i = \frac{X_i l_i}{E_i A_i} = \delta + \varphi x_i \tag{14.8}$$

where x_i denotes the abscissa of the ith connecting rod.

Just as there are two primary unknowns, there are also two equations for resolving them which are represented by the equations of equilibrium of the cross member with regard to vertical translation and to rotation about the centre:

$$F = \sum_i X_i \tag{14.9a}$$

$$Fd = \sum_i X_i x_i \tag{14.9b}$$

Since each individual redundant reaction may be expressed as a function of the two kinematic unknowns, via the relation (14.8),

$$X_i = \frac{E_i A_i}{l_i}\left(\delta + \varphi x_i\right) \tag{14.10}$$

423

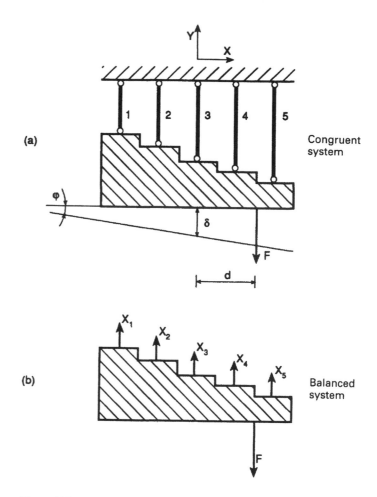

Figure 14.2

equations (14.9) are rendered explicit as follows:

$$\delta\left(\sum_i \frac{E_i A_i}{l_i}\right) + \varphi\left(\sum_i \frac{E_i A_i x_i}{l_i}\right) = F \qquad (14.11a)$$

$$\delta\left(\sum_i \frac{E_i A_i x_i}{l_i}\right) + \varphi\left(\sum_i \frac{E_i A_i x_i^2}{l_i}\right) = Fd \qquad (14.11b)$$

from which it emerges clearly how these represent a linear algebraic system of two equations in the two kinematic unknowns δ, φ. The ordered set of four coefficients represents the stiffness matrix of the system, which is symmetrical by virtue of Betti's Reciprocal Theorem.

Finally consider a set of connecting rods concurrent at a point belonging to the plane (instead of at infinity as in the cases of Figures 14.1 and 14.2). Let the point of concurrence be represented by a hinge-node (Figure 14.3(a)), and let that node be loaded by a generic force F, inclined at the angle β with respect to the horizontal. Once deformation has occurred, the hinge-node will be displaced with respect to its original position, by an amount u in the horizontal direction and by an amount v in the vertical direction (Figure 14.3(a)). Since all the connecting rods are hinged at the end, all sharing the same node, and since the Principle of Superposition holds, the elongation of each individual connecting rod will be given by the two contributions, one corresponding to the displacement u and the other to the displacement v

$$\Delta l_i = \frac{X_i l_i}{E_i A_i} = u \cos \alpha_i + v \sin \alpha_i \tag{14.12}$$

The equations of equilibrium with regard to horizontal and vertical translation of the hinge-node constitute the equations which resolve the problem (Figure 14.3(b))

$$F \cos\beta = \sum_i X_i \cos\alpha_i \tag{14.13a}$$

$$F \sin\beta = \sum_i X_i \sin\alpha_i \tag{14.13b}$$

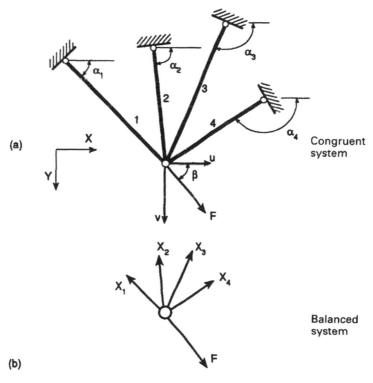

(a)

Congruent system

(b)

Balanced system

Figure 14.3

425

Expressing, according to equation (14.12), each redundant reaction as a function of the two kinematic unknowns u and v

$$X_i = \frac{E_i A_i}{l_i} \left(u \cos \alpha_i + v \sin \alpha_i \right) \qquad (14.14)$$

equations (14.13) transform as follows:

$$u \left(\sum_i \frac{E_i A_i}{l_i} \cos^2 \alpha_i \right) + v \left(\sum_i \frac{E_i A_i}{l_i} \sin \alpha_i \cos \alpha_i \right) = F \cos \beta \quad (14.15a)$$

$$u \left(\sum_i \frac{E_i A_i}{l_i} \cos \alpha_i \sin \alpha_i \right) + v \left(\sum_i \frac{E_i A_i}{l_i} \sin^2 \alpha_i \right) = F \sin \beta \quad (14.15b)$$

Also in this case the stiffness matrix of the system is symmetric, in accordance with Betti's Reciprocal Theorem.

14.3 Parallel-arranged beam systems

Consider n beams connected together by a single fixed joint-node and each constrained at its other end in any manner whatsoever to the foundation (Figure 14.4(a)). Let these beams present different lengths, cross sections and elastic moduli, and let the fixed joint-node be loaded by a counterclockwise moment m. When deformation has taken place, the fixed joint-node will be rotated by an angle φ, as will the end sections of the individual beams which converge at the fixed joint-node itself (Figure 14.4(a))

$$\varphi_i = \frac{X_i l_i}{c_i E_i I_i} = \varphi \qquad (14.16)$$

where φ_i denotes the angle of rotation of the nodal section of the ith beam, X_i is the reactive bending moment due to the above-mentioned distortion, l_i, I_i, E_i are respectively the length, the moment of inertia and the elastic modulus of the beam, and finally c_i is a numerical coefficient dependent on the remaining constraint. From Section 13.5, regarding inelastic constraint settlements and imposed displacements, we can deduce, for instance, the coefficients c_i corresponding to the beams of the scheme of Figure 14.4(a). For beam 1 we have $c_1 = 4$ (Figure 13.20), for beams 2 and 4 we have $c_2 = c_4 = 3$ (Figure 13.18), while for beam 3, $c_3 = 1$ (Figure 13.21).

An alternative and equivalent way of expressing the proportionality relation between the kinematic unknown φ_i and the redundant moment X_i is that of transforming the fixed joint-node into a hinge-node (Figure 14.4(b)) and to apply all the redundant unknowns X_i. It will thus be a question of determining the elastic rotations φ_i on all the various statically indeterminate or statically determinate schemes thus obtained. This time the coefficients c_i are thus derived by assigning to the moments X_i the role of cause, and to the rotations φ_i that of effect, as opposed to the previously adopted procedure. The schemes of Figures 13.6, 10.6(a) and 13.13 confirm, in each case, the corresponding coefficient c_i already deduced following another path.

426

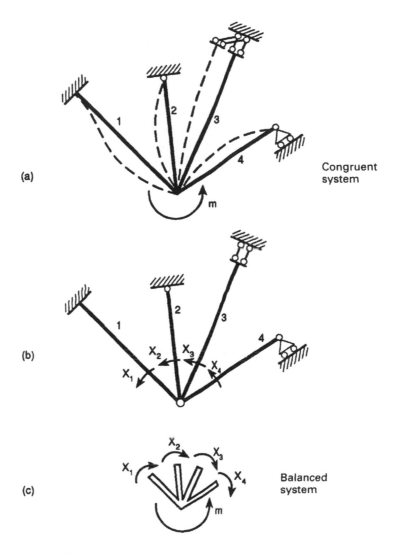

Figure 14.4

The equilibrium of the fixed joint-node, on the other hand, imposes (Figure 14.4(c))

$$m = \sum_i X_i \qquad (14.17)$$

whereby, using equation (14.16), we obtain

$$m = \varphi \sum_i \frac{c_i E_i I_i}{l_i} \qquad (14.18)$$

427

and hence the primary unknown of the problem

$$\varphi = \frac{m}{\displaystyle\sum_i \frac{c_i E_i I_i}{l_i}} = \frac{m}{K} \tag{14.19}$$

where K designates the total stiffness of the beam system, which yet again is equal to the summation of the partial stiffnesses of the individual elements.

From equations (14.16) and (14.19) it is thus possible to derive the individual redundant unknowns,

$$X_i = m\left(\frac{c_i E_i I_i}{l_i}\right) \bigg/ \left(\sum_i \frac{c_i E_i I_i}{l_i}\right) = m\frac{K_i}{K} \tag{14.20}$$

where the ratio K_i/K is the **flexural coefficient of distribution** and represents the fraction of external loading supported by the ith beam.

Consider, as a second fundamental case, a horizontal rigid cross member, constrained to the foundation by a series of uprights of various lengths, having different moments of inertia and consisting of different materials (Figure 14.5(a)). Let the horizontal cross member be loaded by a horizontal force F. When deformation has occurred, the cross member will be translated horizontally by the amount δ, as will the ends of the uprights. From congruence, we therefore have

$$\delta_i = \frac{T_i l_i^3}{c_i E_i I_i} = \delta \tag{14.21}$$

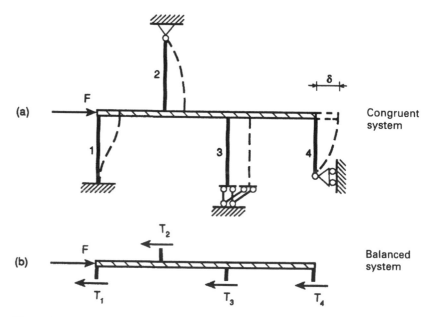

Figure 14.5

where T_i is the shear transmitted to the cross member by the ith upright, l_i, I_i and E_i are as before the characteristics of the upright, while c_i is a numerical coefficient depending on the constraint holding the upright to the foundation. From Section 13.5 we can deduce the coefficients c_i corresponding to the exemplary scheme of Figure 14.5(a). For upright 1 we have $c_1 = 12$ (Figure 13.19), for uprights 2 and 4 we have $c_2 = c_4 = 3$ (Figure 13.17), while for upright 3, $c_3 = 0$, since the horizontal translations of the upright are permitted by the double rod.

The displacement δ_i having been imposed, the redundant shear reaction T_i is thus obtained. Inverting the roles, as has already been seen in the previous fundamental scheme, it is possible to apply the force T_i and find the elastic displacement δ_i. It is sufficient to consider the schemes of Figures 13.11 and 13.12 for the calculation of the coefficients c_1 and $c_2 = c_4$, respectively.

The equation that resolves the problem is thus of equilibrium, with regard to horizontal translation of the cross member (Figure 14.5(b))

$$F = \sum_i T_i \tag{14.22}$$

from which, taking into account equation (14.21), we obtain

$$\delta = \frac{F}{\displaystyle\sum_i \frac{c_i E_i I_i}{l_i^3}} = \frac{F}{K} \tag{14.23a}$$

the total stiffness K being provided by the summation of the partial stiffnesses.

The redundant reactions T_i are instead expressible as follows:

$$T_i = F\left(\frac{c_i E_i I_i}{l_i^3}\right) \bigg/ \left(\sum_i \frac{c_i E_i I_i}{l_i^3}\right) = F\frac{K_i}{K} \tag{14.23b}$$

where the ratio K_i / K is the **shear coefficient of distribution**.

Notice how the uprights, in addition to the shear T_i, generally also transmit to the cross member a moment M_i and an axial force N_i. Equilibrium with regard to vertical translation and to rotation of the cross member will thus be ensured by the combination of these loadings. Whereas, however, the moments M_i are by now known and deducible from schemes already referred to, the axial forces N_i constitute a set of redundant unknowns which may be calculated on the basis of a scheme of parallel connecting rods similar to that of Figure 14.2(a), in which the external load consists of the moments M_i.

In the same way, the shearing forces and axial forces transmitted to the fixed joint-node of Figure 14.4(c) must allow equilibrium with regard to translation of the latter. Of course, in this case the shearing forces are known and their summation constitutes the total force acting on the node, while the axial forces are again unknown but deducible from a scheme of converging connecting rods like that of Figure 14.3(a).

14.4 *Automatic computation of beam systems having multiple degrees of indeterminacy*

Consider a system of rectilinear beams lying in the plane, constrained at the ends, both mutually and externally, by fixed joint-nodes or hinge-nodes

429

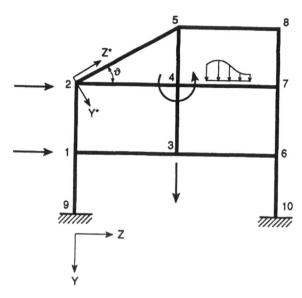

Figure 14.6

(Figure 14.6). Let the fixed joint-nodes and the hinge-nodes be numbered, starting from the internal nodes and ending with the external ones. In the case of fixed joint-nodes, there are three kinematic parameters which characterize their elastic configuration: two orthogonal translations and the rotation (Figure 14.7(a)). In the case of hinge-nodes, the kinematic parameters which characterize their elastic configuration are equal to the number of the beams which converge in the node, augmented by two: it will in fact be necessary to take into account an independent rotation for each end section, and the two orthogonal translations (Figure 14.7(b)). Finally, in the case of connections of a mixed type, fixed joint and hinge (Figure 14.7(c)), there will be three kinematic parameters, plus the number of beams which converge at the hinge.

Further, let the beams be numbered and let each be disposed within a local reference system Y^*Z^*, which will be in general rototranslated with respect to the global reference system YZ (Figure 14.6). Let us then consider a generic loading of the system, consisting of concentrated and distributed loads on the individual beams and of concentrated loads (forces and couples) acting on the internal nodes.

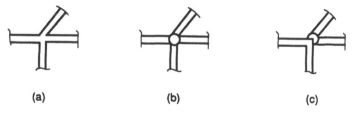

(a)   (b)   (c)

Figure 14.7

430

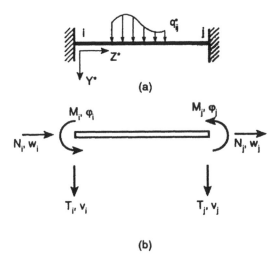

Figure 14.8

At this point let each beam with ends ij be isolated, and let it be considered as constrained by built-in supports in the end sections (Figure 14.8(a)). The procedure will thus be to impose, at each end, the three generalized displacements and find the redundant reactions at the built-in ends. If we designate the reactions at the built-in constraints as M, T and N, and the rotation and the two imposed displacements as φ, v and w, respectively, and if we assume the conventional positive senses indicated in Figure 14.8(b), which are the same at either end, we have the following matrix relation:

$$
\begin{bmatrix} M_i \\ T_i \\ N_i \\ M_j \\ T_j \\ N_j \end{bmatrix} = EI
\begin{bmatrix}
\frac{4}{l} & -\frac{6}{l^2} & 0 & \frac{2}{l} & \frac{6}{l^2} & 0 \\
-\frac{6}{l^2} & \frac{12}{l^3} & 0 & -\frac{6}{l^2} & -\frac{12}{l^3} & 0 \\
0 & 0 & \frac{4}{ll} & 0 & 0 & -\frac{4}{ll} \\
\frac{2}{l} & -\frac{6}{l^2} & 0 & \frac{4}{l} & \frac{6}{l^2} & 0 \\
\frac{6}{l^2} & -\frac{12}{l^3} & 0 & \frac{6}{l^2} & \frac{12}{l^3} & 0 \\
0 & 0 & -\frac{4}{ll} & 0 & 0 & \frac{4}{ll}
\end{bmatrix}
\begin{bmatrix} \varphi_i \\ v_i \\ w_i \\ \varphi_j \\ v_j \\ w_j \end{bmatrix} -
\begin{bmatrix} M_i^0 \\ T_i^0 \\ N_i^0 \\ M_j^0 \\ T_j^0 \\ N_j^0 \end{bmatrix}
\qquad (14.24)
$$

This relation expresses the **constraint reaction vector** as the sum of two contributions: the first deriving from the imposed displacements and the second balancing the external loads acting on the beam. The symmetric (6 × 6) matrix which multiplies the **displacement vector** is referred to as the **stiffness matrix of the element**. Each column of the matrix is obtained by imposing the relative displacement or rotation and by calculating, as was done in Section 13.5, the redundant reactions at the ends. In place of the vector of the

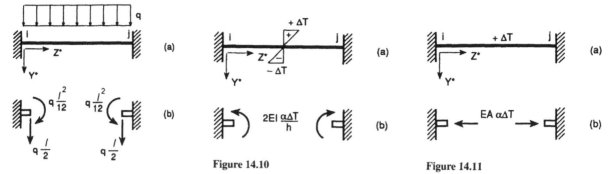

Figure 14.10

Figure 14.11

Figure 14.9

forces balancing the external load acting on the beam, in relation (14.24) appears the **vector of the forces equivalent to the external load**, with a negative algebraic sign. In the case of uniform distributed load (Figure 14.9) we have the vector

$$\left[-\frac{ql^2}{12}, \ \frac{ql}{2}, \ 0, \ \frac{ql^2}{12}, \ \frac{ql}{2}, \ 0 \right]^{\mathrm{T}} \tag{14.25}$$

while, with the butterfly-shaped thermal distortion (Figure 14.10) and the uniform thermal distortion (Figure 14.11), it is necessary to consider the following forces transmitted by the beam to the built-in constraint nodes:

$$\left[2EI\alpha\frac{\Delta T}{h}, 0, 0, -2EI\alpha\frac{\Delta T}{h}, 0, 0 \right]^{\mathrm{T}} \tag{14.26}$$

$$\left[0, 0, -EA\alpha\Delta T, 0, 0, EA\alpha\Delta T \right]^{\mathrm{T}} \tag{14.27}$$

In relation (14.24) it is possible to group together the terms corresponding to the two ends i and j

$$\begin{bmatrix} Q_i^* \\ Q_j^* \end{bmatrix} = \begin{bmatrix} K_{ii} & K_{ij} \\ K_{ji} & K_{jj} \end{bmatrix} \begin{bmatrix} \delta_i^* \\ \delta_j^* \end{bmatrix} - \begin{bmatrix} F_i^* \\ F_j^* \end{bmatrix} \tag{14.28}$$

where the quantities marked by an asterisk are to be understood as referring to the local coordinate axes Y^*Z^*, corresponding to the beam ij (Figure 14.6). In compact form

$$\underset{(6\times6)\ (6\times1)}{[K_e]\{\delta_e^*\}} = \underset{(6\times1)}{\{Q_e^*\}} + \underset{(6\times1)}{\{F_e^*\}} \tag{14.29}$$

The quantities referred to the local system are expressible as functions of the same quantities referred to the global system,

$$\{\delta_e^*\} = [N]\{\delta_e\} \tag{14.30a}$$

432

$$\{Q_e^*\} = [N]\{Q_e\} \tag{14.30b}$$

$$\{F_e^*\} = [N]\{F_e\} \tag{14.30c}$$

[N] being the matrix of rotation which transforms the global reference system YZ into the local reference system Y^*Z^*

$$[N] = \begin{bmatrix} 1 & 0 & 0 & 0 & 0 & 0 \\ 0 & \cos\vartheta & \sin\vartheta & 0 & 0 & 0 \\ 0 & -\sin\vartheta & \cos\vartheta & 0 & 0 & 0 \\ 0 & 0 & 0 & 1 & 0 & 0 \\ 0 & 0 & 0 & 0 & \cos\vartheta & \sin\vartheta \\ 0 & 0 & 0 & 0 & -\sin\vartheta & \cos\vartheta \end{bmatrix} \tag{14.31}$$

Substituting equations (14.30) into relation (14.29), we obtain

$$[K_e][N]\{\delta_e\} = [N](\{Q_e\} + \{F_e\}) \tag{14.32}$$

Finally, premultiplying both sides by $[N]^T$, we obtain

$$([N]^T[K_e][N])\{\delta_e\} = \{Q_e\} + \{F_e\} \tag{14.33}$$

The two vectors on the right-hand side are, respectively, the vector of the constraint reactions and the vector of the equivalent nodal forces, while the matrix in round brackets is the stiffness matrix of the element, reduced to the global reference system.

The **assembly** operation, as in the case of the Finite Element Method, consists of an **expansion** of the vectors $\{\delta_e\}$, $\{Q_e\}$, $\{F_e\}$ from the local dimension (6) to the global dimension n, where n is the total number of the kinematic parameters identifying the deformed configuration of the beam system. Thus the procedure will be to order all the kinematic parameters of the system in one vector, in such a way as to be able to insert the end displacements of the generic element e in the positions that they should occupy. This can be achieved by premultiplying the vector of the local displacements $\{\delta_e\}$ by a suitable assemblage matrix $[A_e]^T$, of dimensions $(n \times 6)$, where all the elements are zero, with the exception of six elements having a value of unity arranged in the six different rows to be filled and corresponding to the six associated columns:

$$\{\delta^e\} = [A_e]^T\{\delta_e\} \tag{14.34a}$$

$$\{Q^e\} = [A_e]^T\{Q_e\} \tag{14.34b}$$

$$\{F^e\} = [A_e]^T\{F_e\} \tag{14.34c}$$

In the case of the beam 2–5 of the frame of Figure 14.6, for example, the matrix $[A_e]^T$ is of dimensions (30×6), where the first element of the fourth row, the second element of the fifth row, the third element of the sixth row, the fourth element of the thirteenth row, the fifth element of the fourteenth row and the sixth element of the fifteenth row are different from zero and have a value of unity:

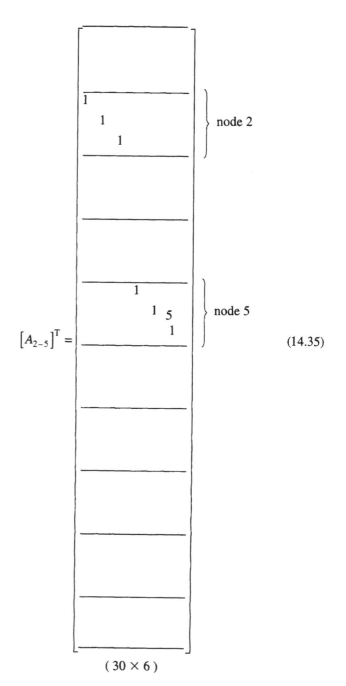

$$[A_{2-5}]^T = \qquad (14.35)$$

Substituting the inverse relations of equations (14.34) into equation (14.33), we obtain

$$\left([N]^T[K_e][N]\right)[A_e]\{\delta^e\} = [A_e]\left(\{Q^e\}+\{F^e\}\right) \qquad (14.36)$$

Premultiplying by $[A_e]^T$, we have

$$\left([A_e]^T[N]^T[K_e][N][A_e]\right)\{\delta^e\} = \{Q^e\}+\{F^e\} \qquad (14.37)$$

The relation (14.37) remains valid even if the expanded vector of local displacements $\{\delta_e\}$ is replaced by the **global vector of nodal displacements** $\{\delta\}$,

$$\underset{(n\times n)\,(n\times1)}{[K^e]}\{\delta\} = \underset{(n\times1)}{\{Q^e\}} + \underset{(n\times1)}{\{F^e\}} \qquad (14.38)$$

where $[K^e]$ denotes the matrix of local stiffness, reduced to the global reference system and expanded to the dimension n.

The local relation in expanded form (14.38) may be summed together with the similar relations corresponding to the other beams:

$$\left(\sum_e [K^e]\right)\{\delta\} = \sum_e \left(\{Q^e\}+\{F^e\}\right) \qquad (14.39)$$

Noting that, for the equilibrium of nodes, we must have (Figure 14.12)

$$-\sum_e \{Q^e\} + \{F\} = \{0\} \qquad (14.40)$$

the equation (14.39) which resolves the problem may be cast in the form

$$[K]\{\delta\} = \{F\}+\{F_{eq}\} \qquad (14.41)$$

where $[K]$ is the **matrix of global stiffness**, $\{F\}$ is the **vector of effective nodal forces** and $\{F_{eq}\}$ is the **vector of equivalent nodal forces**.

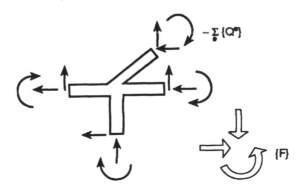

Figure 14.12

435

The vectors of the effective and equivalent nodal forces summarize the external loads acting upon the beam system and are to be considered as known terms or data of the problem. The stiffness matrix of the system is also known, once the geometry and the elastic properties of the beams are known. The unknown of the problem is represented by the vector of nodal displacements $\{\delta\}$. It is evident then how the calculation of a beam system with a multiple degree of statical indeterminacy can be accommodated within the method of displacements and can be performed automatically via computer by means of a pre-ordered succession of elementary operations.

So far we have not taken into account the external constraint conditions. To do this, let us partition the vectors and the matrix which appear in relation (14.41), so as to isolate the free displacements from the constrained displacements:

$$\begin{bmatrix} K_{LL} & K_{LV} \\ K_{VL} & K_{VV} \end{bmatrix} \begin{bmatrix} \delta_L \\ \delta_V \end{bmatrix} = \begin{bmatrix} F_L \\ F_V \end{bmatrix} \tag{14.42}$$

Whereas the constrained displacements $\{\delta_V\}$ are zero (or, at the most, predetermined in the case of inelastic settlement), the free displacements $\{\delta_L\}$ represent the true unknowns of the problem,

$$[K_{LL}]\{\delta_L\} = \{F_L\} - [K_{LV}]\{\delta_V\} \tag{14.43}$$

from which we obtain

$$\{\delta_L\} = [K_{LL}]^{-1} (\{F_L\} - [K_{LV}]\{\delta_V\}) \tag{14.44}$$

On the other hand, from relation (14.42) we have

$$\{F_V\} = [K_{VL}]\{\delta_L\} + [K_{VV}]\{\delta_V\} \tag{14.45}$$

and hence, by virtue of equations (14.44) and (14.45), we obtain the external constraint reactions

$$\underset{(v\times 1)}{\{Q_V\}} = \underset{(v\times l)}{[K_{VL}]} \underset{(l\times l)}{[K_{LL}]^{-1}} \left(\underset{(l\times 1)}{\{F_L\}} - \underset{(l\times v)}{[K_{LV}]} \underset{(v\times 1)}{\{\delta_V\}} \right) + \underset{(v\times v)}{[K_{VV}]} \underset{(v\times 1)}{\{\delta_V\}} - \underset{(v\times 1)}{\{F_V^0\}} \tag{14.46}$$

where $\{F_V^0\}$ is the vector of the equivalent nodal forces corresponding to the constrained nodes. Notice that, beneath the formula (14.46), are given the dimensions of the vectors and matrices that are involved.

Once all the kinematic parameters $\{\delta_L\}$ are known, it is then simple, by applying for each beam the initial relation (14.24), to determine the internal reactions and hence the diagrams of the characteristics.

14.5 Plane trusses

In the case of plane trusses (Figure 14.13), the kinematic parameters of the system reduce to the hinge-node displacements. The local stiffness matrix in this case reduces to a (2×2) matrix,

$$\begin{bmatrix} N_i \\ N_j \end{bmatrix} = EA \begin{bmatrix} \dfrac{1}{l} & -\dfrac{1}{l} \\ -\dfrac{1}{l} & \dfrac{1}{l} \end{bmatrix} \begin{bmatrix} w_i \\ w_j \end{bmatrix} \tag{14.47}$$

436

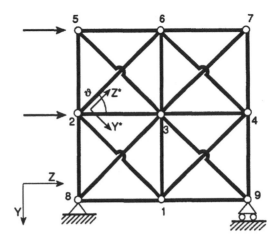

Figure 14.13

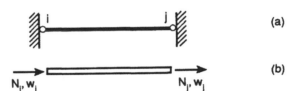

N_i, w_i N_j, w_j

Figure 14.14

where only the axial force and the axial displacement are present (Figure 14.14). The vector of the forces balancing the external load acting on the individual bar is zero, since in general trusses are assumed as being loaded only in the hinge-nodes.

Relation (14.47) can be cast in the compact form

$$\underset{(2\times 2)}{[K_e]}\underset{(2\times 1)}{\{\delta_e^*\}} = \underset{(2\times 1)}{\{Q_e^*\}} \tag{14.48}$$

where the quantities referred to the local system have been marked with an asterisk. On the other hand, both the axial displacement and the axial force can be projected on the axes of the global reference system YZ, giving rise to the four-component vectors $\{\delta_e\}$ and $\{Q_e\}$,

$$\{\delta_e^*\} = [N]\{\delta_e\} \tag{14.49a}$$

$$\{Q_e^*\} = [N]\{Q_e\} \tag{14.49b}$$

where $[N]$ denotes the orthogonal (2×4) matrix:

$$[N] = \begin{bmatrix} -\sin\vartheta & \cos\vartheta & 0 & 0 \\ 0 & 0 & -\sin\vartheta & \cos\vartheta \end{bmatrix} \tag{14.50}$$

437

The assembly operation is based on the relation

$$[K^e]\{\delta\} = \{Q^e\} \qquad (14.51)$$
$$\underset{(n\times n)\,(n\times 1)}{} \qquad \underset{(n\times 1)}{}$$

where

$$[K^e] = [A_e]^T\,[N]^T\,[K_e]\,[N]\,[A_e] \qquad (14.52)$$
$$\underset{(n\times n)}{}\quad\underset{(n\times 4)}{}\quad\underset{(4\times 2)}{}\quad\underset{(2\times 2)(2\times 4)(4\times n)}{}$$

is the local stiffness matrix, reduced to the global reference system and expanded to the dimension n, $\{\delta\}$ is the vector of nodal displacements, which presents two components for each node with a total of n components, and $\{Q^e\}$ is the vector of nodal reactions, this also reduced to the global reference system and expanded to the dimension n. Summing up all the relations (14.51) as the index e characterizing each individual bar varies, we obtain

$$[K]\{\delta\} = \{F\} \qquad (14.53)$$

which is the equation that resolves the problem.

Once all the kinematic parameters $\{\delta\}$ are known, and hence the displacements in the individual local reference systems

$$\{\delta_e^*\} = [N]\,[A_e]\,\{\delta\} \qquad (14.54)$$
$$\underset{(2\times 1)}{}\quad\underset{(2\times 4)(4\times n)(n\times 1)}{}$$

it is then possible to determine the axial force acting in each single bar, by applying the initial relation (14.47).

14.6 Plane frames

Plane frames have already been dealt with extensively in Section 14.4. Here we shall consider two important particular cases:

1. rotating-node frames;
2. translating-node frames with rigid horizontal cross members (also called **shear-type frames**).

In **rotating-node frames** the nodes are not displaced but simply rotate, except for the contributions due to axial deformability of the beams. Figure 14.15 provides an example of a rotating-node frame with four nodes effectively free to rotate (two of these are internal and two external). In these cases, if we neglect the axial deformability of the beams, the local stiffness matrix is, as in the case of trusses, a (2×2) matrix:

$$\begin{bmatrix} M_i \\[1.5em] M_j \end{bmatrix} = EI \begin{bmatrix} \dfrac{4}{l} & \dfrac{2}{l} \\[1em] \dfrac{2}{l} & \dfrac{4}{l} \end{bmatrix} \begin{bmatrix} \varphi_i \\[1.5em] \varphi_j \end{bmatrix} - \begin{bmatrix} M_i^0 \\[1.5em] M_j^0 \end{bmatrix} \qquad (14.55)$$

In this case the **vector of rigid joint moments** $[M_i^0, M_j^0]^T$ is also present.

Relation (14.55) can be put in the synthetic form (14.29). This is one of those cases in which the local and global kinematic parameters coincide, so that relations (14.30) remain valid, even though the orthogonal matrix $[N]$ is the identity matrix of dimensions (2×2). The assemblage matrix $[A_e]$ will

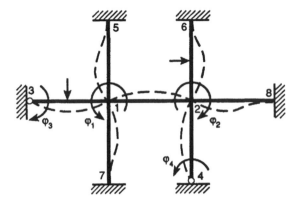

Figure 14.15

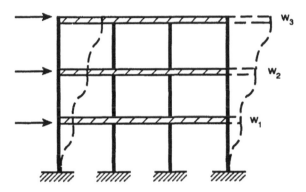

Figure 14.16

therefore have the dimensions $(2 \times n)$, where n is the number of the nodes of the frame.

In **translating-node frames with rigid horizontal cross members**, we have the situation complementary to the preceding one: the nodes do not rotate but are displaced only horizontally, except for the contribution due to the axial deformability of the columns. Figure 14.16 gives an example with three cross members effectively free to translate. Only the vertical beams are thus considered, hence the local stiffness matrix has, as in the foregoing cases (trusses and rotating-node frames), dimension (2×2):

$$\begin{bmatrix} T_i \\ T_j \end{bmatrix} = EI \begin{bmatrix} \frac{12}{l^3} & -\frac{12}{l^3} \\ -\frac{12}{l^3} & \frac{12}{l^3} \end{bmatrix} \begin{bmatrix} v_i \\ v_j \end{bmatrix} - \begin{bmatrix} T_i^0 \\ T_j^0 \end{bmatrix} \qquad (14.56)$$

Also in this case the local kinematic parameters coincide with the global ones, which are precisely the horizontal translations of the cross members. The assemblage matrix $[A_e]$ will therefore have the dimensions $(2 \times n)$, where n is the number of horizontal cross members plus one.

439

14.7 Plane grids

Consider a system of rectilinear beams lying in a plane, mutually constrained via fixed-joint nodes and externally via built-in ends or supports orthogonal to the foundation (Figure 14.17). The kinematic parameters characterizing the deformed configuration of the nodes are three: two rotations with orthogonal axes lying in the plane, and the displacement orthogonal to the plane.

Let each beam be disposed in a local reference system X^*Z^*, which will in general be rototranslated with respect to the global reference system XZ (Figure 14.17) and let a generic loading of the system be considered, consisting of distributed and concentrated loads perpendicular to the plane XZ, acting on the individual beams or on the individual nodes.

Let each beam be isolated, it being considered as built in at the end sections i and j (Figure 14.18(a)). The procedure will be to impose on each end the three generalized displacements and to find the redundant reactions at the built-in constraints. Designating as M, T, M_t the reactions at the built-in constraints, and φ, v, ϑ, respectively, the rotation about the axis X^*, the vertical

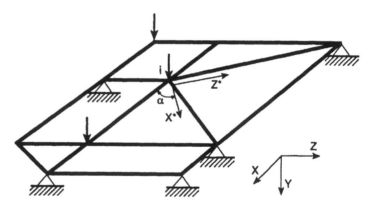

Figure 14.17

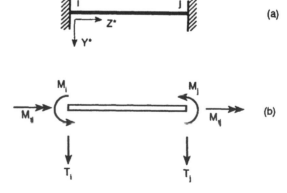

Figure 14.18

440

displacement and the rotation about the axis Z^* (Figure 14.18(b)), we have the following matrix relation:

$$
\begin{bmatrix} M_i \\ T_i \\ M_{ti} \\ M_j \\ T_j \\ M_{tj} \end{bmatrix} = EI
\begin{bmatrix}
\dfrac{4}{l} & -\dfrac{6}{l^2} & 0 & \dfrac{2}{l} & \dfrac{6}{l^2} & 0 \\[2mm]
-\dfrac{6}{l^2} & \dfrac{12}{l^3} & 0 & -\dfrac{6}{l^2} & -\dfrac{12}{l^3} & 0 \\[2mm]
0 & 0 & \dfrac{GI_t}{EIl} & 0 & 0 & -\dfrac{GI_t}{EIl} \\[2mm]
\dfrac{2}{l} & -\dfrac{6}{l^2} & 0 & \dfrac{4}{l} & \dfrac{6}{l^2} & 0 \\[2mm]
\dfrac{6}{l^2} & -\dfrac{12}{l^3} & 0 & \dfrac{6}{l^2} & \dfrac{12}{l^3} & 0 \\[2mm]
0 & 0 & -\dfrac{GI_t}{EIl} & 0 & 0 & \dfrac{GI_t}{EIl}
\end{bmatrix}
\begin{bmatrix} \varphi_i \\ v_i \\ \vartheta_i \\ \varphi_j \\ v_j \\ \vartheta_j \end{bmatrix}
-
\begin{bmatrix} M_i^0 \\ T_i^0 \\ M_{ti}^0 \\ M_j^0 \\ T_j^0 \\ M_{tj}^0 \end{bmatrix}
\qquad (14.57)
$$

where the local stiffness matrix is identical to the one present in relation (14.24) except for the four terms corresponding to torsion, which replace those corresponding to axial force.

The matrix relation (14.57) can be cast in the compact form (14.29). The quantities referred to the local reference system X^*Z^* are expressible as functions of the same quantities referred to the global system XZ. Relations (14.30) therefore continue to hold, $[N]$ being the following rotation matrix:

$$
[N] =
\begin{bmatrix}
\cos\alpha & 0 & \sin\alpha & 0 & 0 & 0 \\
0 & 1 & 0 & 0 & 0 & 0 \\
-\sin\alpha & 0 & \cos\alpha & 0 & 0 & 0 \\
0 & 0 & 0 & \cos\alpha & 0 & \sin\alpha \\
0 & 0 & 0 & 0 & 1 & 0 \\
0 & 0 & 0 & -\sin\alpha & 0 & \cos\alpha
\end{bmatrix}
\qquad (14.58)
$$

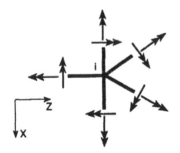

Figure 14.19

Just as the matrix (14.31) reduces shearing force and axial force to the global reference system, so the matrix (14.58) likewise reduces bending moment and twisting moment (Figure 14.19).

As regards calculation of the external constraint reactions, the procedure to follow is yet again that formally outlined in section 14.4.

14.8 Space frames

To conclude, let us consider the most general case, which comprises, as particular cases, both plane frames and plane grids. This is the case of **space frames**, which are three-dimensional systems of rectilinear beams mutually constrained and externally constrained by fixed joints, cylindrical hinges and spherical hinges (Figure 14.20). In the case, for instance, of fixed-joint constraint nodes, there are six kinematic parameters characterizing the deformed configuration: three mutually orthogonal translations and three rotations about three mutually orthogonal axes. In the case of spherical hinges (e.g.

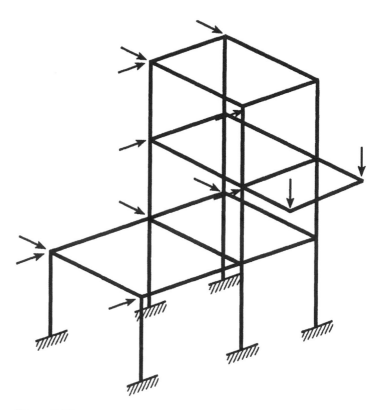

Figure 14.20

three-dimensional trusses), the kinematic parameters are equal to three times the number of beams that converge at the node, augmented by three. In the case of mixed-type connections, the kinematic parameters will be of a lower number. Let each beam be disposed within a local reference system $X^*Y^*Z^*$, which will in general be rototranslated with respect to the global reference system XYZ, and let each be isolated by being considered as built-in at the end sections ij. Following the same procedure as in the previously considered cases, let six generalized displacements be imposed at either end, so as to find the 12 redundant reactions at the built-in supports. Designating as T_x, T_y, N, M_x, M_y and M_z the reactions at the built-in supports, and as u, v, w, φ_x, φ_y and φ_z the imposed displacements, we have the following matrix relation:

$$
\begin{Bmatrix}
T_{xi}^{0}\\[2pt]
T_{yi}^{0}\\[2pt]
N_{i}^{0}\\[2pt]
M_{xi}^{0}\\[2pt]
M_{yi}^{0}\\[2pt]
M_{ji}^{0}\\[2pt]
T_{xj}^{0}\\[2pt]
T_{yj}^{0}\\[2pt]
N_{j}^{0}\\[2pt]
M_{xj}^{0}\\[2pt]
M_{yj}^{0}\\[2pt]
M_{j}^{0}
\end{Bmatrix}
\;-\;
[K]\,
\begin{Bmatrix}
u_{i}\\[2pt]
v_{i}\\[2pt]
w_{i}\\[2pt]
\varphi_{xi}\\[2pt]
\varphi_{yi}\\[2pt]
\varphi_{ji}\\[2pt]
u_{j}\\[2pt]
v_{j}\\[2pt]
w_{j}\\[2pt]
\varphi_{xj}\\[2pt]
\varphi_{yj}\\[2pt]
\varphi_{j}
\end{Bmatrix}
\;=\;
\begin{Bmatrix}
T_{xi}\\[2pt]
T_{yi}\\[2pt]
N_{i}\\[2pt]
M_{xi}\\[2pt]
M_{yi}\\[2pt]
M_{ji}\\[2pt]
T_{xj}\\[2pt]
T_{yj}\\[2pt]
N_{j}\\[2pt]
M_{xj}\\[2pt]
M_{yj}\\[2pt]
M_{j}
\end{Bmatrix}
$$

$$
[K]=
\begin{bmatrix}
\dfrac{12EI_y}{l^3} & 0 & 0 & 0 & -\dfrac{6EI_y}{l^2} & 0 & -\dfrac{12EI_y}{l^3} & 0 & 0 & 0 & -\dfrac{6EI_y}{l^2} & 0\\[6pt]
0 & \dfrac{12EI_x}{l^3} & 0 & \dfrac{6EI_x}{l^2} & 0 & 0 & 0 & -\dfrac{12EI_x}{l^3} & 0 & \dfrac{6EI_x}{l^2} & 0 & 0\\[6pt]
0 & 0 & \dfrac{EA}{l} & 0 & 0 & 0 & 0 & 0 & -\dfrac{EA}{l} & 0 & 0 & 0\\[6pt]
0 & \dfrac{6EI_x}{l^2} & 0 & \dfrac{4EI_x}{l} & 0 & 0 & 0 & -\dfrac{6EI_x}{l^2} & 0 & \dfrac{2EI_x}{l} & 0 & 0\\[6pt]
-\dfrac{6EI_y}{l^2} & 0 & 0 & 0 & \dfrac{4EI_y}{l} & 0 & \dfrac{6EI_y}{l^2} & 0 & 0 & 0 & \dfrac{2EI_y}{l} & 0\\[6pt]
0 & 0 & 0 & 0 & 0 & \dfrac{GI_t}{l} & 0 & 0 & 0 & 0 & 0 & -\dfrac{GI_t}{l}\\[6pt]
-\dfrac{12EI_y}{l^3} & 0 & 0 & 0 & \dfrac{6EI_y}{l^2} & 0 & \dfrac{12EI_y}{l^3} & 0 & 0 & 0 & \dfrac{6EI_y}{l^2} & 0\\[6pt]
0 & -\dfrac{12EI_x}{l^3} & 0 & -\dfrac{6EI_x}{l^2} & 0 & 0 & 0 & \dfrac{12EI_x}{l^3} & 0 & -\dfrac{6EI_x}{l^2} & 0 & 0\\[6pt]
0 & 0 & -\dfrac{EA}{l} & 0 & 0 & 0 & 0 & 0 & \dfrac{EA}{l} & 0 & 0 & 0\\[6pt]
0 & \dfrac{6EI_x}{l^2} & 0 & \dfrac{2EI_x}{l} & 0 & 0 & 0 & -\dfrac{6EI_x}{l^2} & 0 & \dfrac{4EI_x}{l} & 0 & 0\\[6pt]
-\dfrac{6EI_y}{l^2} & 0 & 0 & 0 & \dfrac{2EI_y}{l} & 0 & \dfrac{6EI_y}{l^2} & 0 & 0 & 0 & \dfrac{4EI_y}{l} & 0\\[6pt]
0 & 0 & 0 & 0 & 0 & -\dfrac{GI_t}{l} & 0 & 0 & 0 & 0 & 0 & \dfrac{GI_t}{l}
\end{bmatrix}
$$

The local stiffness matrix in this case has the dimension (12×12).

The rotation matrix $[N]$ also has the dimensions (12×12) and can be partitioned into 16 submatrices of dimensions (3×3), of which 12 are zero and the four diagonal ones are mutually identical,

$$[N] = \begin{bmatrix} N_0 & 0 & 0 & 0 \\ 0 & N_0 & 0 & 0 \\ 0 & 0 & N_0 & 0 \\ 0 & 0 & 0 & N_0 \end{bmatrix} \tag{14.60}$$

where

$$[N_0] = \begin{bmatrix} \cos X^*X & \cos X^*Y & \cos X^*Z \\ \cos Y^*X & \cos Y^*Y & \cos Y^*Z \\ \cos Z^*X & \cos Z^*Y & \cos Z^*Z \end{bmatrix} \tag{14.61}$$

The assembly operation and the determination of the external constraint reactions are formally identical to those corresponding to plane systems, described in Section 14.4.

14.9 Dynamics of beam systems

If the distributed masses of a plane frame are assumed as being concentrated in the nodes, the equilibrium equation (14.41) can be transformed into the **equation of free oscillations**

$$[K]\{\delta\} + [M]\{\ddot{\delta}\} = \{0\} \tag{14.62}$$

where the **mass matrix** $[M]$ is a diagonal matrix which presents an equivalent mass corresponding to each nodal translation and a zero mass corresponding to each nodal rotation. The equivalent mass can, for instance, be calculated by adding the weights, divided by two, of the beams that converge in the node. The formal identity of equation (14.62) and equation (11.76) allows the concepts and formulas developed in Section 11.7 for elastic solids and finite elements to be extrapolated to beam systems.

However, it should be noted that, if one wishes to take into account the real distribution of the masses along beams and columns, it is always possible to apply the Finite Element Method already discussed in Chapter 11, subdividing each beam or column into one or more finite elements. In this way, of course, the mass matrix would not be diagonal.

In practice, the procedure often adopted is to carry out a further simplification and approximation with respect to the two methods just outlined, i.e. the **Finite Element Method** and the **method of the masses concentrated in the nodes**. Since in fact the moment of inertia of horizontal beams is usually much greater than that of columns, the horizontal cross members are considered as rigid and the masses as concentrated in the cross members alone. This scheme, already introduced previously (Figure 14.16) is referred to as the shear-type frame, and the procedure which we are about to describe is known as the **method of rigid cross members**.

444

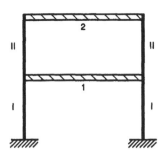

Figure 14.21

Let us take, as an example, the two-storey one-bay frame of Figure 14.21. Denoting the masses of the two cross members by m_1, m_2, and designating the shear rigidities of the pairs of uprights as k_1, k_2,

$$k_i = \frac{24EI_i}{h_i^3} \tag{14.63}$$

the two equations of the motion of the cross members are

$$m_1 \ddot{\delta}_1 = -k_1 \delta_1 + k_2 (\delta_2 - \delta_1) \tag{14.64a}$$

$$m_2 \ddot{\delta}_2 = -k_2 (\delta_2 - \delta_1) \tag{14.64b}$$

where δ_1, δ_2 denote the horizontal translations of the cross members. Equations (14.64) may be written in matrix form

$$\begin{bmatrix} (k_1 + k_2) & -k_2 \\ -k_2 & k_2 \end{bmatrix} \begin{bmatrix} \delta_1 \\ \delta_2 \end{bmatrix} + \begin{bmatrix} m_1 & 0 \\ 0 & m_2 \end{bmatrix} \begin{bmatrix} \ddot{\delta}_1 \\ \ddot{\delta}_2 \end{bmatrix} = \begin{bmatrix} 0 \\ 0 \end{bmatrix} \tag{14.65}$$

To study the **free oscillations** of the system with two degrees of freedom, let us suppose that the coordinates δ_1, δ_2 of the system vary harmonically in time with equal angular frequency and without phase shift,

$$\delta_1(t) = \delta_1 \sin \omega t \tag{14.66a}$$

$$\delta_2(t) = \delta_2 \sin \omega t \tag{14.66b}$$

where the angular frequency ω and the maximum amplitudes δ_1 and δ_2 are to be determined via an eigenvalue problem. Substituting equations (14.66) into equation (14.65), we obtain the following homogeneous algebraic equation:

$$\begin{bmatrix} (k_1 + k_2 - m_1 \omega^2) & -k_2 \\ -k_2 & (k_2 - m_2 \omega^2) \end{bmatrix} \begin{bmatrix} \delta_1 \\ \delta_2 \end{bmatrix} = \begin{bmatrix} 0 \\ 0 \end{bmatrix} \tag{14.67}$$

This equation possesses a solution different from the trivial one if and only if the determinant of the coefficient matrix is zero:

$$\omega^4 - \left(\frac{k_1 + k_2}{m_1} + \frac{k_2}{m_2} \right) \omega^2 + \frac{k_1 k_2}{m_1 m_2} = 0 \tag{14.68}$$

In the case where the columns have the same moment of inertia and the same height, we have $k_1 = k_2 = k$. If we further assume that also the masses of the cross members are equal, $m_1 = m_2 = m$, the characteristic equation (14.68) simplifies as follows:

$$\omega^4 - \frac{3k}{m} \omega^2 + \frac{k^2}{m^2} = 0 \tag{14.69}$$

and yields the following two eigenvalues:

$$\omega^2 = \frac{3 \pm \sqrt{5}}{2} \frac{k}{m} \tag{14.70}$$

445

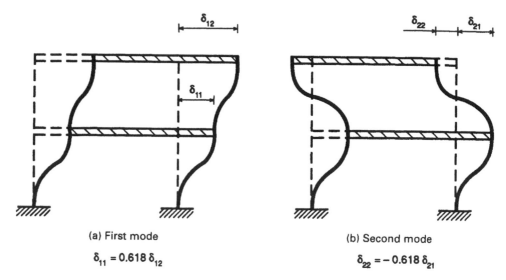

(a) First mode

$\delta_{11} = 0.618\,\delta_{12}$

(b) Second mode

$\delta_{22} = -0.618\,\delta_{21}$

Figure 14.22

whence it follows that

$$\omega_1 = 0.618\sqrt{\frac{k}{m}} \qquad (14.71\text{a})$$

$$\omega_2 = 1.618\sqrt{\frac{k}{m}} \qquad (14.71\text{b})$$

The eigenvectors are obtained by resolving the homogeneous system (14.67), after substituting the corresponding eigenvalues (14.71):

$$\delta_{11} = 0.618\delta_{12} \qquad (14.72\text{a})$$

$$\delta_{22} = -0.618\delta_{21} \qquad (14.72\text{b})$$

The eigenvectors (14.72) are determined but for a factor of proportionality, and are represented in Figure 14.22, normalizing the maximum absolute value coordinate. The second modal deformed configuration presents translations of opposite sign (Figure 14.22(b)).

In general in multi-storey frames, the oscillation modes higher than the first (known as the **fundamental mode**) present reversals of sign in the translations of the cross members. Also for these systems, the equation of motion reduces to the form of equation (14.62), where [*M*] is the diagonal mass matrix of the cross members and [*K*] is the global stiffness matrix, already defined in Section 14.6.

In the sequel we shall examine the problem of **forced oscillations** due to movements of external constraints. Suppose that the base of the columns of a multi-storey frame with rigid cross members is subjected to a seismic movement with displacement $x(t)$, velocity $\dot{x}(t)$ and acceleration $\ddot{x}(t)$. Equation (14.62) will then assume the following form:

$$[K]\{\delta\} + [M]\{\ddot{\delta} + \ddot{x}\} = \{0\} \qquad (14.73)$$

since the elastic restoring forces remain dependent on the translations evaluated in the reference system of the frame, while the forces of inertia are proportional to the accelerations in the absolute system. From equation (14.73) we deduce

$$[K]\{\delta\} + [M]\{\ddot{\delta}\} = -[M]\{\ddot{x}\} \qquad (14.74)$$

This is a non-homogeneous dynamic equation, which presents a forcing load on the right-hand side, which derives from the acceleration $\ddot{x}$ of the foundation.

Introducing the normal coordinates of the system f_i and the eigenvectors $\{\delta_i\}$, and premultiplying by $\{\delta_i\}^T$, we obtain

$$f_i\{\delta_i\}^T[K]\{\delta_i\} + \ddot{f}_i\{\delta_i\}^T[M]\{\delta_i\} = -\{\delta_i\}^T[M]\{\ddot{x}\} \qquad (14.75)$$

Dividing both sides of the equation by the coefficient of $\ddot{f}_i$, we find

$$\ddot{f}_i + f_i \frac{\{\delta_i\}^T[K]\{\delta_i\}}{\{\delta_i\}^T[M]\{\delta_i\}} = -\frac{\{\delta_i\}^T[M]\{\ddot{x}\}}{\{\delta_i\}^T[M]\{\delta_i\}} \qquad (14.76)$$

Finally, recalling equation (11.78) and terming the coefficient of f_i the **Rayleigh ratio**, we obtain the following decoupled equation for each oscillating mode and hence for each normal coordinate f_i :

$$\ddot{f}_i + \omega_i^2 f_i = -g_i \ddot{x}, \qquad \text{for } i = 1, 2, ..., n \qquad (14.77)$$

Since for frames with rigid cross members the mass matrix is diagonal, the factor g_i takes the following form:

$$g_i = \frac{\sum_{j=1}^{n} m_j \delta_{ij}}{\sum_{j=1}^{n} m_j \delta_{ij}^2} \qquad (14.78)$$

and is usually referred to as the **coefficient of participation** of the system to the ith mode of oscillation.

The analysis of the forced oscillations of the system thus comes back to the analysis of n elementary oscillators, each subject to the fraction g_i of excitation at the base. Then once the differential equation (14.77) is resolved, it is possible to retrace the displacements of the cross members in terms of time,

$$\{\delta(t)\} = \sum_{i=1}^{n} f_i\{\delta_i\} \qquad (14.79)$$

via the eigenvectors $\{\delta_i\}$, and hence again via linear transformations, it is possible to deduce the internal characteristics acting on the columns of the frame.

The maximum acceleration that a system with one degree of freedom undergoes as a result of a seismic event is set by current standards according to the formula

$$a = CR(\omega)g \tag{14.80}$$

where C is the **coefficient of seismic intensity**, which depends on the degree of seismicity of the area where the structure is located, $R(\omega)$ is the **coefficient of response** which depends on the fundamental frequency of the system, and g is the acceleration due to gravity.

If the effects of the earthquake on the ith oscillation mode of a frame having multiple degrees of freedom are considered, the maximum acceleration that can be attributed to that elementary oscillator is then

$$\left|\ddot{f}_i\right| = CR(\omega_i)gg_i \tag{14.81}$$

From equation (14.81) we thus obtain the maximum acceleration undergone by the jth storey as a result of the ith mode of oscillation,

$$a_{ij} = CR(\omega_i)gg_i\delta_{ij} \tag{14.82}$$

and, consequently, the maximum force

$$F_{ij} = a_{ij}m_j \tag{14.83}$$

Equation (14.83) can be expressed in the form

$$F_{ij} = CR(\omega_i)W_j\gamma_{ij} \tag{14.84}$$

where W_j is the weight of the jth storey and

$$\gamma_{ij} = g_i\delta_{ij} \tag{14.85}$$

the so-called **coefficient of distribution** of the ith mode on the jth storey.

The maximum storey forces expressed by equation (14.84) act at the same time in the context of the same ith mode, but are out of phase when different modes of oscillation are considered. Then, once the maximum value of a characteristic has been calculated, for instance the bending moment M_i, as the ith mode varies, there is the problem of combining these values. We recall the following empirical formulas for obtaining an equivalent maximum moment:

$$M_{\max} = \sum_{i=1}^{n}|M_i| \tag{14.86a}$$

$$M_{\max} = |M_1| \tag{14.86b}$$

$$M_{\max} = \sqrt{\sum_{i=1}^{n}M_i^2} \tag{14.86c}$$

Equation (14.86a), which corresponds to the hypothesis that the maxima occur simultaneously, proves to be excessively conservative. Equation (14.86b), which considers only the fundamental mode, proves not to be very reliable for slender buildings, where the modes above the first are involved to a far from

448

insignificant extent. Equation (14.86c), on the other hand, meets with wider agreement and is called the formula of **composition in quadrature**.

Finally, it should be noted that the coefficient of participation g_i, given by equation (14.78), and hence the coefficients of distribution γ_{ij}, given by equation (14.85), clearly tend to diminish with the increase in the index i of the oscillation mode, because of the change of sign in the translations δ_{ij}. It is therefore quite easy to understand the reason for the usual limitation of seismic analysis to modes with lower indices only, thus making it possible to avoid the burden of a complete modal analysis.

449

15 Plane frames

15.1 Introduction

As we have already seen in the previous chapter, a frame is a system of beams having many degrees of indeterminacy. In this chapter we shall refer expressly to plane frames loaded in their own plane, with the exception of a brief look at the case of a portal frame loaded by forces perpendicular to its own plane.

In reference to the deformed configuration of their own fixed joint-nodes, plane frames can be subdivided into:

1. **rotating-node frames** (Figure 15.1);
2. **translating-node frames** (Figure 15.2).

In the former, the fixed joint-nodes rotate elastically but do not undergo translation, provided that it is possible to neglect the axial deformability of the individual beams. In the latter, the fixed joint-nodes not only rotate but also undergo considerable translation. In some cases, as in that of the portal frame of Figure 15.1(a), the symmetry of the load condition implies the annihilation of the translations of the fixed joint-nodes. The same portal frame loaded in a non-symmetrical fashion (Figure 15.2(a)) proves to be a translating-node frame. More precisely, the two upper nodes undergo a translation in the hori-

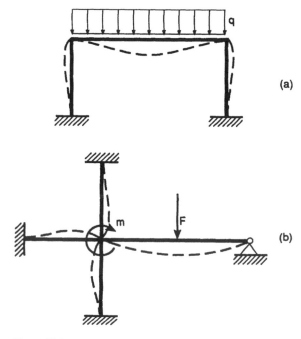

(a)

(b)

Figure 15.1

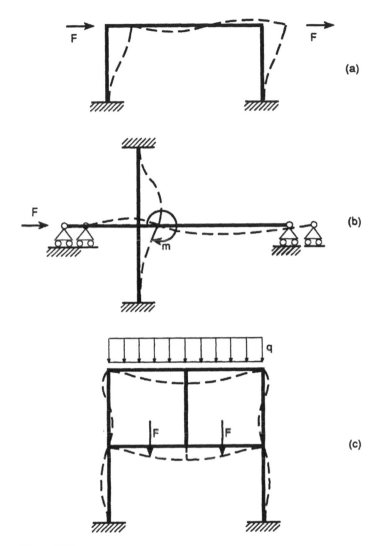

Figure 15.2

zontal direction. In general, it is possible to state that the beam systems sufficiently constrained externally prove to be rotating-node frames (Figure 15.1(b)). By suppressing some of the external constraints, the same frames can transform into translating-node systems (Figure 15.2(b)). Notice, however, how symmetry does not always imply non-translation of nodes. For example, the lack of a central column in the frame of Figure 15.2(c) renders the two central nodes vertically translating.

When, in the last chapter, we dealt with the automatic computation of plane frames, the above-mentioned distinction between rotating-node frames and translating-node frames was not made. In that case, the method of resolution was not susceptible to such a distinction. In the present chapter, we shall propose a method of solution in which the translations of the fixed joint-nodes are among the unknowns of the problem, together with the redundant moments

451

which develop at the fixed joint-nodes themselves. It is, in other words, a hybrid method, half-way between the method of forces and the method of displacements, in which the equations that resolve the problem consist partly of relations of congruence and partly of relations of equilibrium.

The method outlined above consists of disconnecting, with respect to rotation, all the external built-in supports and the internal fixed joint-nodes, putting hinges in them and applying the corresponding redundant moments (Figures 15.3(a),(b)). On account of the equilibrium with regard to rotation of the fixed joint-node, the redundant moments are linked together by the following relation (Figures 15.3(c),(d)):

$$\sum_i X_i = 0$$

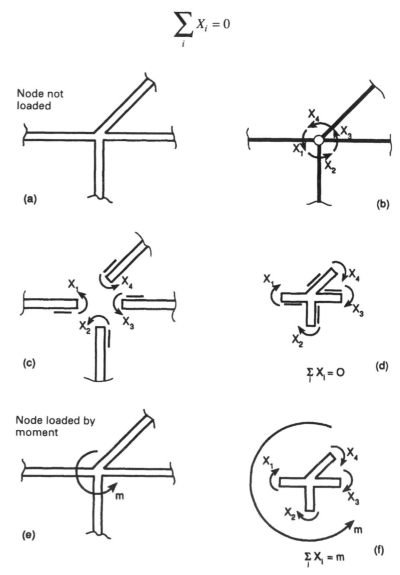

Figure 15.3

452

It is important to distinguish between the moments X_i exerted by the fixed joint-node on the convergent beams and assumed as being counterclockwise in the schemes of Figures 15.3(b),(c), and the moments X_i exerted by the beams on the fixed joint-node, which are opposite to the previous ones (Figure 15.3(d)). In the case where the fixed joint-node is loaded by a concentrated moment m (Figure 15.3(e)), the relation that expresses the equilibrium of the node becomes (Figure 15.3(f))

$$\sum_i X_i = m$$

Once hinges have been inserted in all built-in supports and fixed joint-nodes, we obtain a beam system, called **associated truss structure**, which may be either redundant, isostatic or hypostatic. In the former two cases, the original frame is a rotating-node frame (Figure 15.4), with the unknowns consisting of the redundant moments alone, and the equations for resolving the problem consisting of the relations of angular congruence alone. In the latter case, the original frame is a translating-node frame (Figure 15.5), with the supplementary unknowns consisting of the displacements of the nodes and the supplementary equations consisting of as many relations of equilibrium, in general expressible via the application of the Principle of Virtual Work. In the case, for instance, of the structure of Figure 15.2(a), the associated truss

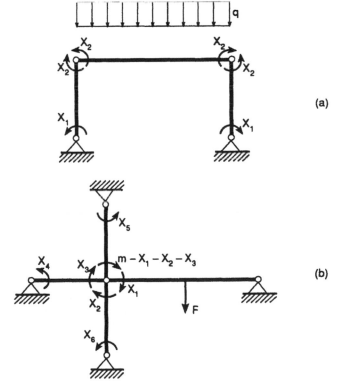

(a)

(b)

Figure 15.4

453

(Figure 15.5(a)) consists of an articulated parallelogram whose only degree of freedom can be described by the horizontal translation of the upper nodes. The same thing can be said for the structure of Figure 15.2(b), whose associated truss (Figure 15.5(b)) presents the possibility of translations of the horizontal cross member. However, the truss associated with the frame of Figure 15.2(c) presents three degrees of freedom, corresponding, respectively, to the horizontal translations of the two cross members and to the vertical translation of the central column (Figures 15.5(c)–(e)).

15.2 Rotating-node frames

The solution of **rotating-node frames** is altogether analogous to the solution of continuous beams, dealt with in Section 13.7, and consists of writing down a number of equations of angular congruence equal to the number of unknown redundant moments. Once these moments are known (they turn out to be end moments for the individual beams), it is easy then to draw the moment diagrams, superposing the diagrams corresponding to external loads on the linear functions that correspond to the redundant moments.

Consider the frame of Figure 15.6(a), consisting of a horizontal continuous beam and a vertical upright built-in at the top in the centre of the beam and, at the bottom, to the foundation. Let two hinges be inserted, respectively, in the fixed joint-node B and in the external built-in support D, applying at the same time the corresponding redundant moments (Figure 15.6(b)). Note that, owing

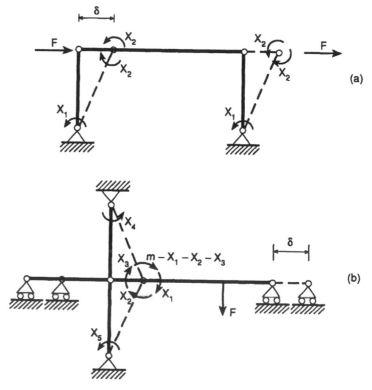

Figure 15.5a, b

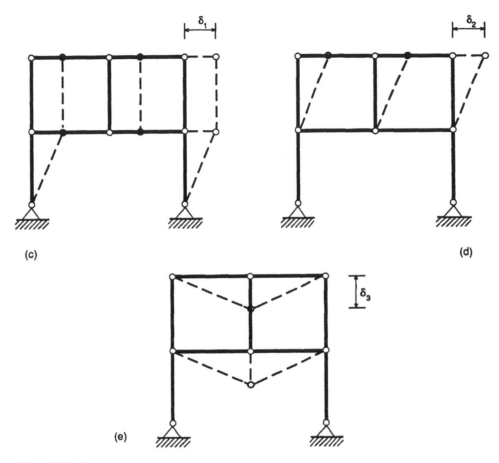

(c)

(d)

(e)

Figure 15.5c, d, e

to the equilibrium of the node B, the redundant moment acting on the end B of the upright DB is a function of the remaining two redundant moments acting on the same node. More precisely, if we designate by X_1 the moment transmitted between the fixed joint-node and the horizontal beam AB, and by X_2 the moment transmitted between the fixed joint-node and the horizontal beam CB, the moment exchanged between the fixed joint-node and the upright is equal to (X_1+X_2) and has a sense opposite to the previous moments. Clearly the senses assumed are altogether conventional and may not be confirmed by calculation, should negative redundant moments be obtained.

There are thus basically three redundant unknowns, X_1, X_2, X_3 (Figure 15.6(b)), just as there are three equations of angular congruence

$$\varphi_{BA} = \varphi_{BC} \tag{15.1a}$$

$$\varphi_{BA} = \varphi_{BD} \tag{15.1b}$$

$$\varphi_{DB} = 0 \tag{15.1c}$$

The first two of equations (15.1) express the fact that all three branches of the fixed joint-node of Figure 15.6(b) rotate by the same amount, and that hence

455

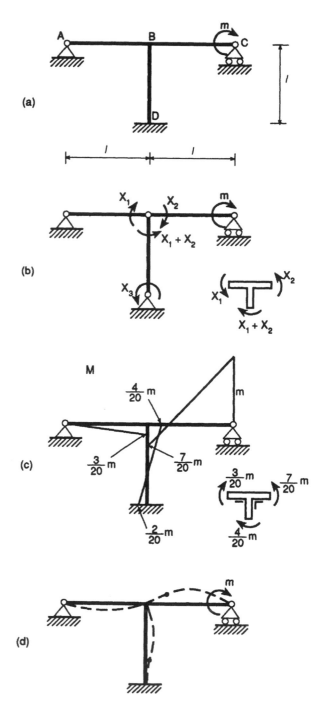

Figure 15.6

the fixed joint-node as a whole rotates rigidly without relative rotations between the various branches. The third equation expresses the existence in D of a built-in support, which prevents any displacement and hence also the absolute rotation of the base section D of the upright. It is then possible to show how the associated truss structure of Figure 15.6(b) constitutes a statically determinate system of beams, and hence the frame under examination may be considered a rotating-node type.

The terms of equations (15.1) can be rendered explicit by taking into account exclusively elastic rotations, which have already been calculated for the elementary schemes of a supported beam (Section 10.6):

$$-\frac{X_1 l}{3EI} = -\frac{X_2 l}{3EI} + \frac{ml}{6EI} \tag{15.2a}$$

$$-\frac{X_1 l}{3EI} = \frac{(X_1 + X_2)l}{3EI} - \frac{X_3 l}{6EI} \tag{15.2b}$$

$$\frac{X_3 l}{3EI} - \frac{(X_1 + X_2)l}{6EI} = 0 \tag{15.2c}$$

Multiplying all three equations by $6EI/l$, we obtain

$$-2X_1 = -2X_2 + m \tag{15.3a}$$

$$-2X_1 = 2X_1 + 2X_2 - X_3 \tag{15.3b}$$

$$2X_3 - X_1 - X_2 = 0 \tag{15.3c}$$

The first expresses X_2 as a function of X_1,

$$X_2 = X_1 + \frac{m}{2} \tag{15.4}$$

and, substituting this expression into equations (15.3b,c), we obtain a system of two linear algebraic equations in the unknowns X_1 and X_3:

$$6X_1 + m = X_3 \tag{15.5a}$$

$$2X_3 - 2X_1 - \frac{m}{2} = 0 \tag{15.5b}$$

Substituting equation (15.5a) into equation (15.5b), we have

$$X_1 = -\frac{3}{20} m \tag{15.6a}$$

so that equation (15.4) and equation (15.5a) yield respectively

$$X_2 = \frac{7}{20} m \tag{15.6b}$$

$$X_3 = \frac{m}{10} \tag{15.6c}$$

The fixed joint-node B is thus in equilibrium under the action of the moments of Figure 15.6(c). The moment diagram is obtained by laying the

457

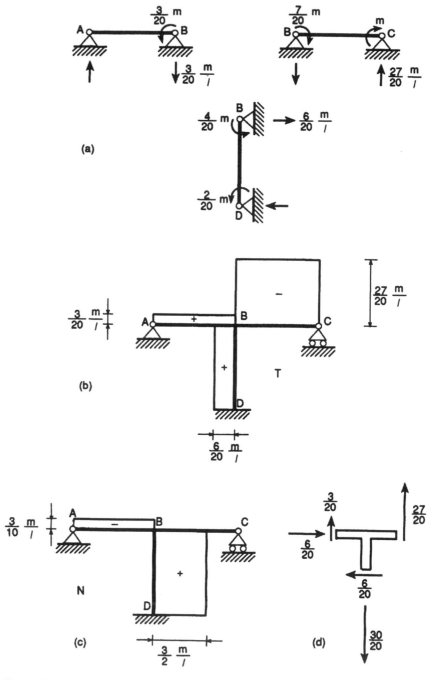

(a)

(b)

(c)

(d)

Figure 15.7a, b, c, d

458

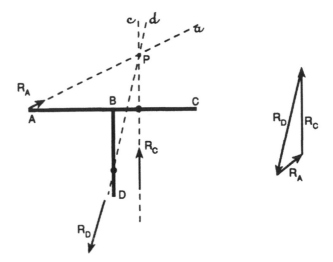

Figure 15.7e

segments out to scale on the side of the stretched fibres, and joining the ends of these with rectilinear segments. The moment diagram is in fact linear, as no transverse distributed loads are present. The deformed configuration of the frame must be consistent with the external constraints and with the moment diagram (Figure 15.6(d)). Notice the two points of inflection corresponding to the sections in which the bending moment vanishes. The fixed joint-node rotates in a counterclockwise direction by the amount $ml/20EI$.

As regards the shear diagram, this is obtainable from the schemes of equilibrium of the individual supported beams to which we are brought back (Figure 15.7(a)), loaded by the external loads and by the redundant moments. This is obviously constant on each individual beam and considerably higher on the beam loaded externally (Figure 15.7(b)). The sign of the shear results from the usual convention. It is possible, however, to deduce the shear as derivative of the moment function, referring the latter to a right-handed system YZ, with origin at one of the two ends of the individual beams.

Finally, the values of the axial force are also constant in each individual beam. The beam BC is not subjected to axial force, as it is constrained by a horizontally moving roller support in C, while the beam AB absorbs an axial force of compression equal in absolute value to the shear of the upright DB (Figure 15.7(c)). In turn, the upright DB absorbs axially the shears of the two horizontal beams AB and CB, being loaded in tension by a total force equal to $\frac{3}{2}\,m/l$. An effective check at this point consists of considering also the equilibrium with regard to translation of the fixed-joint node B (Figure 15.7(d)).

Many of the results just obtained can find a further confirmation from the pressure line. The structure, considered as being completely disconnected externally, is subjected basically to three forces (Figure 15.7(e)). The first is the reaction R_A of the hinge A, which has as its components the axial force $\frac{6}{20}\,m/l$ and the shearing force $\frac{3}{20}\,m/l$. The second is the resultant R_C of the applied moment m and the reaction V_C of the roller support. The third force is the reaction R_D of the built-in support D, which has as its components the axial force

459

$\frac{3}{2}$ *ml* and the shearing force $\frac{3}{10}$ *ml* and passes at a distance X_3/R_D from point *D*. It is possible to verify how these three forces pass through the same pole *P* and constitute a system equivalent to zero.

The pressure line may be defined portion by portion in the following manner:

portion *AB*: straight line *a*;
portion *CB*: straight line *c*;
portion *DB*: straight line *d*.

It may thus be noted that the pressure line passes through those sections in which the bending moment vanishes (Figure 15.6(c)) and in which the elastic deformed configuration presents points of inflection (Figure 15.6(d)).

As a second example, let us consider the square closed configuration of beams shown in Figure 15.8(a), subjected to two equal and opposite forces *F*. The double axial symmetry allows the structure to be considered as a rotating-node frame, and hence there is a single unknown redundant moment *X* (Figure 15.8(b)). The equation that resolves the problem will then be furnished by the condition of angular congruence,

$$\varphi_{AB} = \varphi_{AC} \tag{15.7}$$

which is identical to the other three conditions for the vertices *B,C,D*.

Rendering the condition (15.7) explicit, we have

$$\frac{Xl}{3EI} + \frac{Xl}{6EI} - \frac{Fl^2}{16EI} = -\frac{Xl}{3EI} - \frac{Xl}{6EI} \tag{15.8}$$

whence there results

$$X = \frac{Fl}{16} \tag{15.9}$$

The moment diagram is thus given in Figure 15.8(c). It shows four points of annihilation, symmetrical with respect to both axes of symmetry. The elastic deformed configuration of Figure 15.8(d) presents, of course, points of inflection corresponding to the above-mentioned sections. Whilst the vertical beams are entirely convex outwards, the horizontal beams are convex outwards only in the end regions. The four fixed joint-nodes are not displaced but all rotate by the same amount $Fl^2/32EI$; *A* and *D* rotate clockwise, *B* and *C* counterclockwise.

The shear diagram for the vertical beams is zero, and for the horizontal beams, reproduces that of the scheme of the supported beam (Figure 15.8(e)). On the other hand, the axial force diagram for the horizontal beams is zero, and for the vertical beams is compressive and equal to *F*/2 (Figure 15.8(f)).

At this point it is simple to verify that the pressure line is formed by the two vertical straight lines passing through the points of inflection (Figure 15.8(d)).

Let the symmetrical portal frame of Figure 15.9(a) be loaded by two symmetrical distributions of horizontal forces acting on the upper half of the two uprights. This frame is a rotating-node type by symmetry, and hence an equivalent statically determinate structure can be obtained by disconnecting with regard to rotation the fixed joint-nodes *B* and *C* and eliminating the connecting rod *FG* (Figure 15.9(b)).

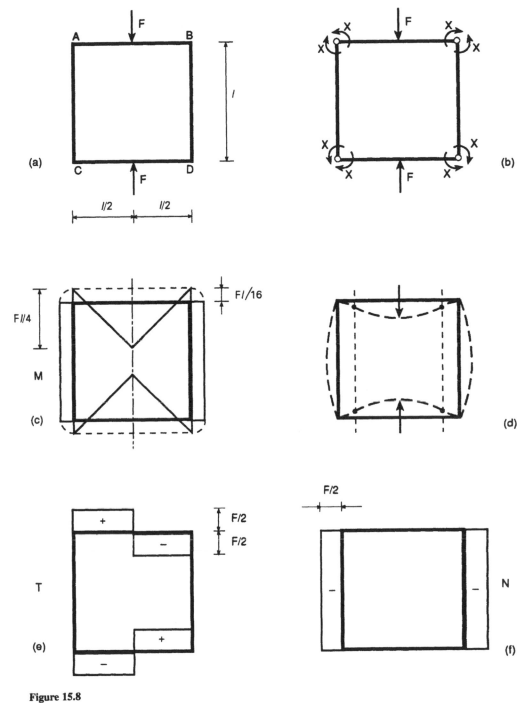

Figure 15.8

In this case the two equations of congruence will have to express the angular congruence of the node B (or C) as well as the annihilation of the horizontal displacement of the point F (or G):

$$\varphi_{BA} = \varphi_{BC} \tag{15.10a}$$

$$u_F = 0 \tag{15.10b}$$

Rendering the foregoing relations explicit as functions of the redundant unknowns X_1 and X_2 (Figure 15.9(b)), we obtain the following two equations that resolve the problem:

$$-\frac{X_1 l}{3EI} - \frac{X_2 l^2}{16EI} + \frac{9}{384}\frac{q l^3}{EI} = \frac{X_1 l}{3EI} + \frac{X_1 l}{6EI} \tag{15.11a}$$

$$\frac{X_1 l^2}{16EI} + \frac{X_2 l^3}{48EI} - \frac{5}{384}\frac{q l^4}{2EI} = 0 \tag{15.11b}$$

The contributions corresponding to the load q acting on the half-beam can be obtained by considering the Principle of Superposition and splitting this load into two components, one symmetrical and the other skew-symmetrical (Figure 15.9(c)). The rotation of the end is

$$\varphi = \frac{\dfrac{q}{2} l^3}{24EI} + \frac{\dfrac{q}{2}\left(\dfrac{l}{2}\right)^3}{24EI} = \frac{9}{384}\frac{q l^3}{EI} \tag{15.12}$$

while the displacement in the centre,

$$v = \frac{5}{384}\frac{\dfrac{q}{2} l^4}{EI} + 0 = \frac{5}{384}\frac{q l^4}{2EI} \tag{15.13}$$

is given by the symmetrical contribution alone.

Reordering equations (15.11) we obtain

$$320X_1 + 24X_2 l = 9q l^2 \tag{15.14a}$$

$$48X_1 + 16X_2 l = 5q l^2 \tag{15.14b}$$

Multiplying the former by two and the latter by −3, and then adding them together, the unknown X_2 is eliminated,

$$X_1 = \frac{3}{496} q l^2 \tag{15.15a}$$

whence we obtain

$$X_2 = \frac{73}{248} q l \tag{15.15b}$$

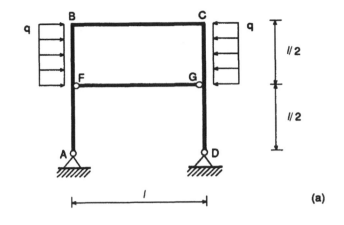

(a)

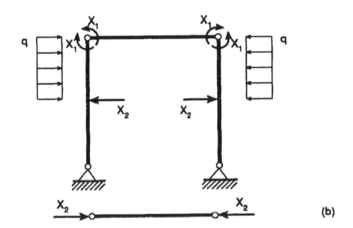

(b)

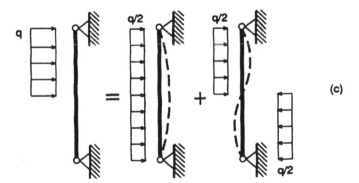

(c)

Figure 15.9

463

Considering the scheme of Figure 15.10(a), rotated by 90° with respect to its actual orientation, and writing the relations of equilibrium with regard to translation and to rotation with respect to the centre,

$$V_A + V_B + X_2 = q\frac{l}{2} \tag{15.16a}$$

$$-V_A\frac{l}{2} + V_B\frac{l}{2} - X_1 - q\frac{l}{2}\frac{l}{4} = 0 \tag{15.16b}$$

we at once obtain the transverse reactions V_A and V_B. Substituting the solutions (15.15) into equations (15.16), we have

$$V_A + V_B = \frac{51}{248}ql \tag{15.17a}$$

$$V_A - V_B = -\frac{65}{248}ql \tag{15.17b}$$

from which there follows

$$V_A = -\frac{7}{248}ql \tag{15.18a}$$

$$V_B = \frac{58}{248}ql \tag{15.18b}$$

The shear diagram on the upright AB is depicted in Figure 15.10(b). The extreme values are equal in absolute value to the reactions V_A and V_B, whilst in the centre there is a positive jump in the function equal to the reaction X_2 of the connecting rod.

As regards bending moment, this presents the following three notable values (Figure 15.10(c)):

$$M(A) = 0$$

$$M(F) = \frac{7}{248}ql \times \frac{l}{2} = \frac{7}{496}ql^2$$

$$M(B) = X_1 = \frac{3}{496}ql^2$$

Whereas in the portion AF the diagram is simply linear (Figure 15.10(c)), in the portion FB to the linear diagram there should be added graphically the parabolic diagram corresponding to the distributed load,

$$M(H) = \frac{M(F) + M(B)}{2} - \frac{1}{8}q\left(\frac{l}{2}\right)^2 = -\frac{21}{992}ql^2$$

A faster way of resolving the twice statically indeterminate structure of Figure 15.9(a) is that of interrupting the continuity of the uprights and then inserting two hinges, one at F (where beams FA, FB and FG converge), and the other at G (where beams GD, GC and GF converge) (Figure 15.11). In this way we have to deal purely with notable schemes of supported beams, with

464

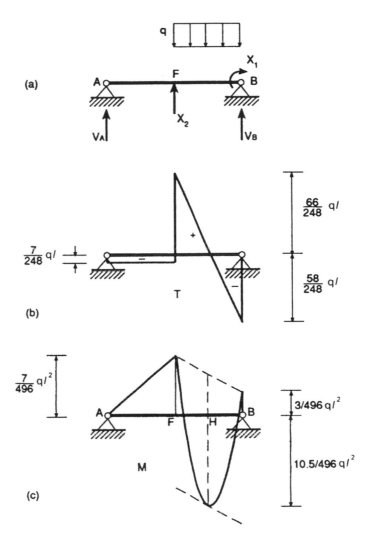

Figure 15.10

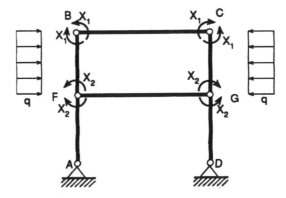

Figure 15.11

465

moments applied at the ends and loads uniformly distributed over the entire span. The two conditions of congruence are in this case both rotational:

$$\varphi_{BC} = \varphi_{BF} \tag{15.19a}$$

$$\varphi_{FB} = \varphi_{FA} \tag{15.19b}$$

Equations (15.19) can be expressed as functions of the two unknown redundant moments X_1, X_2:

$$\frac{X_1 l}{3EI} + \frac{X_1 l}{6EI} = -\frac{X_1\left(\frac{l}{2}\right)}{3EI} - \frac{X_2\left(\frac{l}{2}\right)}{6EI} + \frac{q\left(\frac{l}{2}\right)^3}{24EI} \tag{15.20a}$$

$$\frac{X_2\left(\frac{l}{2}\right)}{3EI} + \frac{X_1\left(\frac{l}{2}\right)}{6EI} - \frac{q\left(\frac{l}{2}\right)^3}{24EI} = -\frac{X_2\left(\frac{l}{2}\right)}{3EI} \tag{15.20b}$$

Multiplying by $6EI/l$, we obtain

$$2X_1 + X_1 = -X_1 - \frac{1}{2}X_2 + \frac{1}{32}ql^2 \tag{15.21a}$$

$$X_2 + \frac{1}{2}X_1 - \frac{1}{32}ql^2 = -X_2 \tag{15.21b}$$

Cross addition of the two sides of equations (15.21) eliminates the term q

$$X_2 = \frac{7}{3}X_1 \tag{15.22}$$

from which we obtain as a verification of the results previously reached otherwise

$$X_1 = \frac{3}{496}ql^2 \tag{15.23a}$$

$$X_2 = \frac{7}{496}ql^2 \tag{15.23b}$$

The moment diagram will thus be constant over the cross member BC and on the two uprights will present the same pattern already defined in Figure 15.10(c).

15.3 Translating-node frames

The solution for **translating-node frames** differs substantially from that for rotating-node frames, starting from the very way in which it is set out. The truss structure associated with the frame is in fact hypostatic and this means that the insertion of the hinges in all the fixed joint-nodes is an excessive operation with respect to the degree of redundancy of the structure. In other words, the associated truss, subjected to the action of external loads and redundant moments, must be in equilibrium for the particular loading condition. In addition to the redundant moments, also the generalized coordinates which define the rigid deformed configuration of the associated truss will be unknown. On the other hand, together with the relations of angular congruence, also a number of equilibrium equations equal to the degrees of freedom of the associated truss will combine to make up the system of equations that provide the solution.

Consider for instance the asymmetrical portal frame of Figure 15.12(a), loaded by a horizontal force F. The associated truss is made up of the articulated parallelogram of Figure 15.12(b), from which the components of rigid rotation of the two uprights are immediately drawn, whilst the cross member undergoes a horizontal translation. From the scheme of Figure 15.12(c), obtained by restraining the displacements of the associated truss, are derived instead the components of elastic rotation at the ends of the individual beams.

The unknowns of the problem are thus the two redundant moments X_1, X_2 and the rigid rotation φ of the left-hand upright (Figure 15.12(b)). There must therefore be three equations that resolve the problem, made up of two equations of angular congruence (taking into account also the rigid rotations) and of an application of the Principle of Virtual Work

$$\varphi_{BA} = \varphi_{BC} \tag{15.24a}$$

$$\varphi_{CB} = \varphi_{CD} \tag{15.24b}$$

$$\text{Principle of Virtual Work} \tag{15.24c}$$

Rendering equations (15.24) explicit, we obtain

$$-\frac{X_1(2l)}{3EI} - \varphi = \frac{X_1(2l)}{3EI} + \frac{X_2(2l)}{6EI} \tag{15.25a}$$

$$-\frac{X_2(2l)}{3EI} - \frac{X_1(2l)}{6EI} = \frac{X_2 l}{3EI} - 2\varphi \tag{15.25b}$$

$$F(2l\varphi) + X_1\varphi - X_2(2\varphi) = 0 \tag{15.25c}$$

Multiplying the first two equation by $3EI/l$ and dividing the third by φ, the system of equation (15.25) transforms as follows:

$$-2X_1 - \frac{3EI}{l}\varphi = 2X_1 + X_2 \tag{15.26a}$$

$$-2X_2 - X_1 = X_2 - \frac{3EI}{l}(2\varphi) \tag{15.26b}$$

$$X_1 = 2X_2 - 2Fl \tag{15.26c}$$

Substituting the expression (15.26c) into the first two, we obtain a system in X_2 and φ:

$$X_2 = \frac{8}{9}Fl - \frac{EI}{3l}\varphi \tag{15.27a}$$

$$X_2 = \frac{2}{5}Fl + \frac{6EI}{5l}\varphi \tag{15.27b}$$

Equating the right-hand sides of equations (15.27), we obtain

$$\varphi = \frac{22}{69}\frac{Fl^2}{EI} \tag{15.28}$$

467

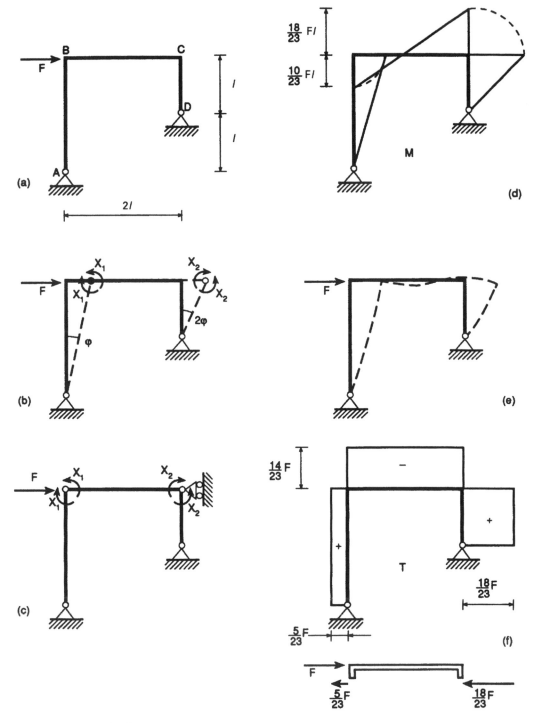

Figure 15.12a, b, c, d, e, f

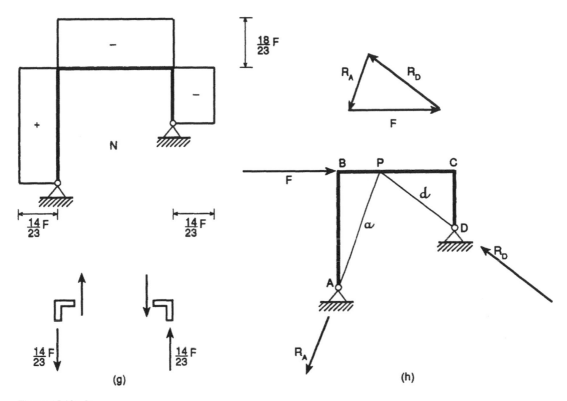

(g)   (h)

Figure 15.12g, h

from which then

$$X_2 = \frac{18}{23} Fl \qquad (15.29a)$$

$$X_1 = -\frac{10}{23} Fl \qquad (15.29b)$$

The redundant moment X_1 turns out to be negative and this means that, contrary to our initial assumption, at the fixed joint-node B the internal fibres are in fact stretched (Figure 15.12(b)). The bending moments at the two fixed joint-nodes thus being known, it is simple to draw the corresponding diagram, taking into account that the moment vanishes at the hinges which are at the feet of the two uprights (Figure 15.12(d)).

The elastic deformed configuration of the translating-node frames must be drawn coherently with the external and internal constraints (angular congruence), with the displacements undergone by the fixed joint-nodes and with the bending moment diagram. The deformed configuration of the asymmetrical portal frame is represented in Figure 15.12(e). Note that the two fixed joint-nodes of the horizontal beam have been translated horizontally by the same amount and rotated clockwise, while the hinges at the foundation allow rotations but not translations of the base sections of the uprights. A point of inflection appears at the point where the moment vanishes on the cross member.

469

The fibres of the beams are thus stretched internally in the left-hand part, and externally in the right-hand part of the portal frame.

The shear is constant on each member and equal, in absolute value, to the slope of the moment diagram (Figure 15.12(f)). The axial force also is constant on each member, being compressive on the horizontal beam and on the right-hand upright, and tensile on the left-hand upright (Figure 15.12(g)). Whereas on the uprights the axial force is equal, in absolute value, to the shear which acts on the horizontal beam, the axial force in the horizontal beam is equal, in absolute value, to the shear acting on the right-hand upright. The schemes of equilibrium to the horizontal translation of the beam and the vertical translation of the nodes B and C are represented in Figure 15.12(f) and Figure 15.12(g), respectively.

The pressure line is made up of two straight lines (Figure 15.12(h)):

portion AB: straight line a;
portion BD: straight line d.

These two straight lines meet at the pole P, which is coincident both with the point of annihilation of the bending moment (Figure 15.12(d)) and with the point of inflection of the elastic deformed configuration (Figure 15.12(e)).

Also the frame of Figure 15.13(a) is a translating-node frame. This consists of a horizontal beam loaded on the overhang, and of two externally hinged uprights. The above-mentioned frame is equivalent to the scheme of Figure 15.13(b), in which the overhang has been eliminated and replaced by the reactions transmitted to the rest of the structure: the vertical force $F = ql$, and the moment $m = ql^2/2$.

The truss structure associated with the statically indeterminate scheme of Figure 15.13(b) has one degree of freedom, and, as in the previous case, the scheme of Figure 15.14(a) provides the components of rigid rotation of the two uprights, with the cross member which undergoes a horizontal translation. The restrained scheme of Figure 15.14(b) in turn provides the components of elastic rotation at the ends of the beams. Note how the moment m, acting on

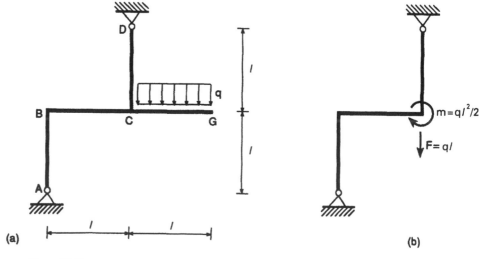

(a)   (b)

Figure 15.13

470

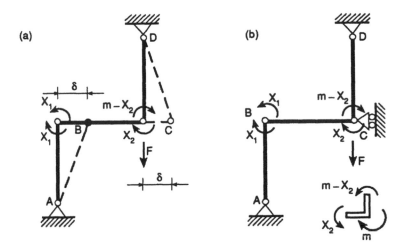

Figure 15.14

the fixed joint-node C (Figure 15.13(b)) has been split into its two components $(m-X_2)$ and X_2, the former acting at the end of the beam CD, and the latter at the end of the beam CB. In this way the equilibrium with regard to rotation of the node C is automatically satisfied (Figure 15.14(b)). The force F, unlike the moment m, does not generate bending moments in the scheme of Figure 15.13(b), but only tensile axial force on the upright CD.

The equations of congruence and the equilibrium relation, which in implicit form appear as follows:

$$\varphi_{BA} = \varphi_{BC} \tag{15.30a}$$

$$\varphi_{CB} = \varphi_{CD} \tag{15.30b}$$

$$\text{Principle of Virtual Work} \tag{15.30c}$$

may be expressed in explicit form, taking into account the rigid translation of the cross member, δ,

$$-\frac{X_1 l}{3EI} - \frac{\delta}{l} = \frac{X_1 l}{3EI} + \frac{X_2 l}{6EI} \tag{15.31a}$$

$$-\frac{X_2 l}{3EI} - \frac{X_1 l}{6EI} = -\frac{(m - X_2)l}{3EI} + \frac{\delta}{l} \tag{15.31b}$$

$$X_1 \frac{\delta}{l} - (m - X_2)\frac{\delta}{l} = 0 \tag{15.31c}$$

Performing the calculations, we obtain

$$X_1 = \frac{m}{6} = \frac{1}{12}ql^2 \tag{15.32a}$$

$$X_2 = \frac{5}{6}m = \frac{5}{12}ql^2 \tag{15.32b}$$

471

$$\delta = -\frac{ml^2}{4EI} = -\frac{ql^4}{8EI} \qquad (15.32c)$$

The positive signs of the redundant moments indicate that the real senses are those assumed, whilst the negative sign of the displacement δ points to a leftward translation of the cross member. The moment diagram is given in Figure 15.15(a), complete with the part that regards the overhang CG. The equilibrium of the node C is guaranteed by the moments that have been determined (Figure 15.15(b)), whilst the lack of points of annihilation in the moment function (except for the external hinges A and D and the end G of the overhang) implies an elastic deformed configuration without inflections (Figure 15.15(c)).

The shear diagram may be obtained very simply from the schemes of equilibrium of the individual supported beams into which the frame has been subdivided (Figure 15.16(a)). The shear is constant on the two uprights and the cross member BC, whereas it is obviously linear on the cantilever CG (Figure 15.16(b)).

The axial force also is constant on all the beams of the frame (Figure 15.16(c)). Whereas it is zero on the overhang, on the uprights it is equal, in absolute value, to the shear of the cross member and *vice versa*. Finally, we

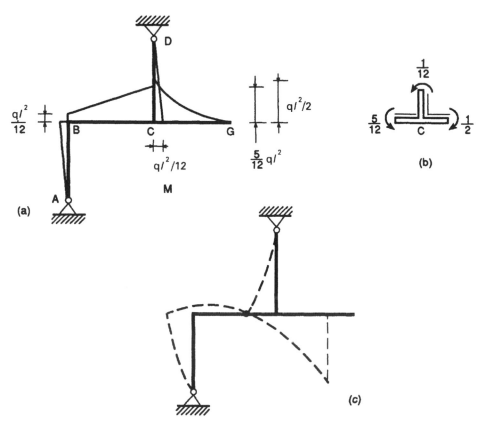

Figure 15.15

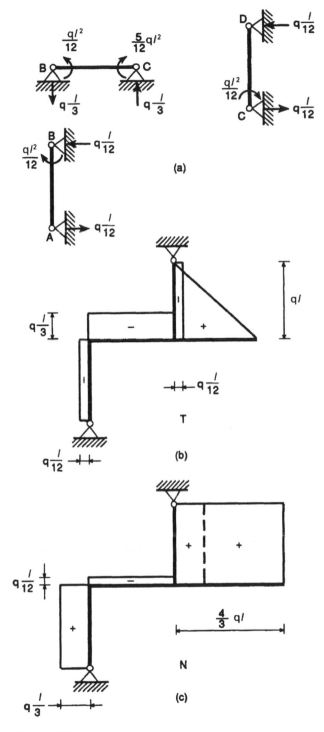

Figure 15.16

473

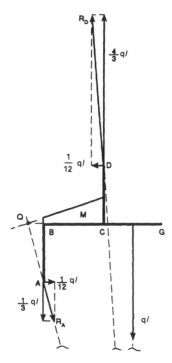

Figure 15.17

have to remember the additional contribution to the axial force on the upright CD made by the force $F = ql$ (Figure 15.13(b)).

The pressure line is represented in Figure 15.17 and basically consists of three straight lines which meet at a common point P. In fact the structure, considered as a single body and completely disconnected externally, is in equilibrium under the action of the external reactions R_A and R_D, as well as the resultant F of the active forces. As may be seen, none of the three straight lines encounters the axis of the beam of which it represents the state of loading. This is consistent with the absence of points of annihilation in the moment diagram (Figure 15.15(a)), and of points of inflection in the elastic deformed configuration (Figure 15.15(c)). The only virtual point at which the moment vanishes (apart from at the two hinges) is the point Q of Figure 15.17, which represents the intersection of the axis of the cross member with the line of action of the reaction R_A.

15.4 Thermal loads and imposed displacements

In the case where a translating-node frame is subjected to thermal dilations or imposed displacements, the components of rigid rotation deriving directly from these anomalous loads must be found from the restrained truss scheme, which, being statically determinate, does not oppose any resistance to such movements.

As an example let us take again the asymmetrical portal frame of Figure 15.18(a), loaded by a uniform thermal variation on the cross member. The associated truss has one degree of freedom, and the generalized coordinate

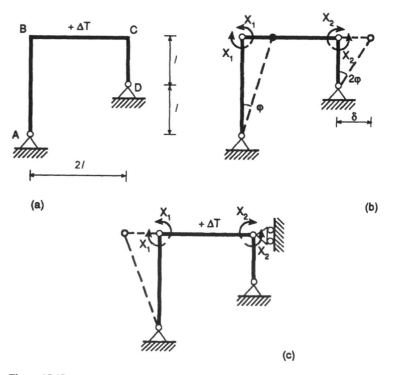

Figure 15.18

474

characterising its rigid deformed configuration is the angle of rotation φ of the upright AB, or the translation δ of the cross member (Figure 15.18(b)). From the restrained truss (Figure 15.18(c)) it is possible to find, as well as the elastic rotations, the rigid rotation of the upright AB, produced directly by the thermal dilation of the cross member. In this case the three equations (15.24) are expressed as follows:

$$-\frac{X_1(2l)}{3EI} - \varphi + \frac{\alpha\Delta T(2l)}{2l} = \frac{X_1(2l)}{3EI} + \frac{X_2(2l)}{6EI} \tag{15.33a}$$

$$-\frac{X_1(2l)}{6EI} - \frac{X_2(2l)}{3EI} = \frac{X_2 l}{3EI} - 2\varphi \tag{15.33b}$$

$$X_1\varphi - X_2(2\varphi) = 0 \tag{15.33c}$$

Multiplying the first two equations by $3EI/l$, and dividing the third by φ, we obtain

$$4X_1 + X_2 = \frac{3\alpha\Delta T EI}{l} - \varphi\frac{3EI}{l} \tag{15.34a}$$

$$X_1 + 3X_2 = 2\varphi\frac{3EI}{l} \tag{15.34b}$$

$$X_1 = 2X_2 \tag{15.34c}$$

Substituting relation (15.34c) into the two previous ones, we have

$$X_2 = \frac{\alpha\Delta T EI}{3l} - \varphi\frac{EI}{3l} \tag{15.35a}$$

$$X_2 = \frac{6}{5}\varphi\frac{EI}{l} \tag{15.35b}$$

from which, by the transitive law, we find

$$\varphi = \frac{5}{23}\alpha\Delta T \tag{15.36a}$$

$$X_1 = \frac{12}{23}\alpha\Delta T\frac{EI}{l} \tag{15.36b}$$

$$X_2 = \frac{6}{23}\alpha\Delta T\frac{EI}{l} \tag{15.36c}$$

The moment diagram (Figure 15.19(a)) envisages the stretched fibres of the beams always outwards, with the absence of points of inflection in the elastic deformed configuration (Figure 15.19(b)). Since the translation of the cross member towards the right is

$$\delta = 2\varphi l = \frac{10}{23}\alpha\Delta T l \tag{15.37}$$

while the thermal elongation of the cross member, neglecting its axial elastic deformability, is

$$\Delta(2l) = 2\alpha\Delta T l \tag{15.38}$$

475

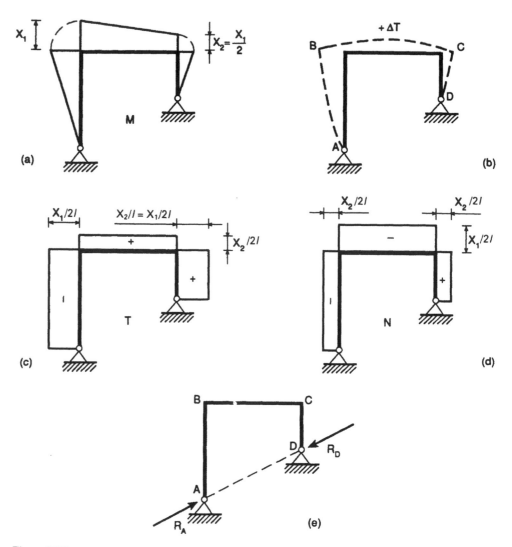

Figure 15.19

in actual fact the node C shifts rightwards by δ, while the node B is displaced leftwards by $[\Delta(2l) - \delta]$.

The shear diagram is constant on each beam (Figure 15.19(c)), as also is the axial force diagram (Figure 15.19(d)), which is compressive on the cross member and on the left-hand upright and tensile on the right-hand upright.

Finally, the pressure line is represented by the straight line AD, joining the two external hinges (Figure 15.19(e)). The frame is in fact in equilibrium under the action of two equal and opposite forces, the components of which are given by the shearing force diagram and the axial force diagram. The pressure line does not intersect the axis of the frame at any point, and this is consistent with the absence of points of moment annihilation (Figure 15.19(a)) and with the absence of inflection points in the elastic deformed configuration (Figure 15.19(b)).

476

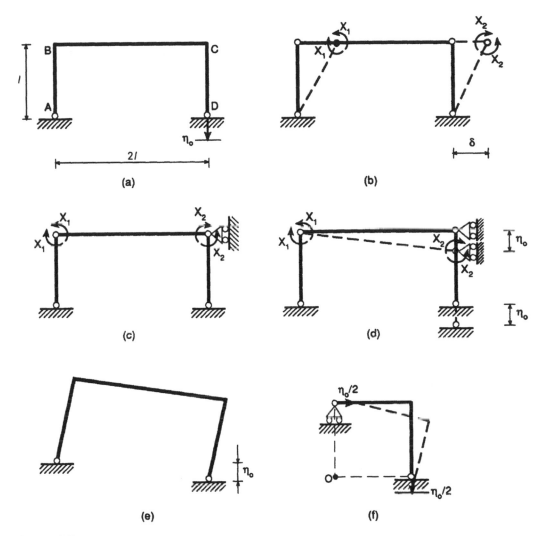

Figure 15.20

As a second example of translating-node frames subjected to distortions, let us look at the symmetrical portal frame of Figure 15.20(a), where a vertical displacement is imposed on the hinge D. The associated truss structure (Figure 15.20(b)) has one degree of freedom and furnishes the rigid rotations, while the restrained truss scheme (Figure 15.20(c)) provides the elastic rotations as well as the rigid rotations deriving directly from the settlement (Figure 15.20(d)). The two equations of angular congruence and the Principle of Virtual Work provide the three equations for resolving the problem:

$$-\frac{X_1 l}{3EI} - \frac{\delta}{l} = \frac{X_1(2l)}{3EI} + \frac{X_2(2l)}{6EI} - \frac{\eta_0}{2l} \qquad (15.39a)$$

$$-\frac{X_2(2l)}{3EI} - \frac{X_1(2l)}{6EI} - \frac{\eta_0}{2l} = \frac{X_2 l}{3EI} - \frac{\delta}{l} \qquad (15.39b)$$

477

$$X_1 \frac{\delta}{l} - X_2 \frac{\delta}{l} = 0 \qquad (15.39c)$$

Note how the Principle of Virtual Work is always to be applied to the associated truss system (Figure 15.20(b)), neglecting the imposed displacements.

From equation (15.39c) we have

$$X_1 = X_2 \qquad (15.40)$$

and multiplying equations (15.39a, b) by $6EI$

$$8X_1 l = -\frac{6EI}{l}\left(\delta - \frac{\eta_0}{2}\right) \qquad (15.41a)$$

$$8X_1 l = -\frac{6EI}{l}\left(\frac{\eta_0}{2} - \delta\right) \qquad (15.41b)$$

Finally we obtain

$$\delta = \eta_0 / 2 \qquad (15.42a)$$

$$X_1 = X_2 = 0 \qquad (15.42b)$$

The resolution of the frame leads us to note that the static characteristics as well as the external reactions are zero. This is due simply to the fact that the imposed displacement η_0 is actually compatible with the constraints of the frame. The hinge A can in fact function as the centre of rotation and allow an infinitesimal rigid rotation of the whole structure in a clockwise direction (Figure 15.20(e)). The trajectory of the point D, as it must be orthogonal to the radius vector AD, is vertical. If we consider the skew-symmetrical component alone of the loading (the symmetrical one producing a trivial vertical translation downwards equal to $\eta_0/2$), the kinematic scheme of Figure 15.20(f) provides an immediate justification for equation (15.42a).

15.5 Frames with non-orthogonal beams

So far we have considered only frames where the individual beams are mutually orthogonal. This is the case which usually concerns the frameworks of buildings, where the columns are vertical and the floors, with their joists, are horizontal. It is not, however, out of place also to consider cases of frames where the individual beams are not mutually orthogonal.

The resolution of frames made up of non-orthogonal beams is accomplished in the same way as that already seen in the foregoing sections. The only differences are represented by an associated truss system which presents more complicated kinematics, since in this case there is no simple translation of the cross members, and by shearing force diagrams and axial force diagrams that are no longer directly derivable from one another by exchanging their components.

Consider, for instance, the portal frame of Figure 15.21(a), which presents the right-hand stanchion inclined at an angle of 45° to the horizontal. The associated truss has one degree of freedom, it being a mechanism whose diagrams of horizontal and vertical displacements are given in Figure 15.21(b). The three unknowns of the problem are the redundant moments X_1, X_2 and the rigid rotation φ of the upright stanchion AB, while the three equations that

478

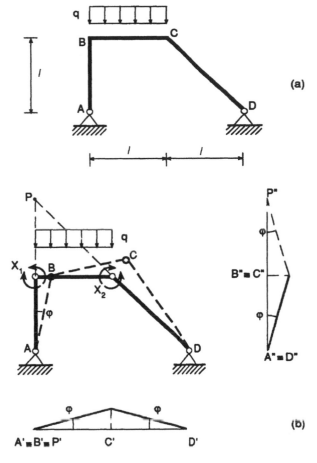

Figure 15.21

resolve the problem are the two equations of angular congruence plus the application of the Principle of Virtual Work:

$$-\frac{X_1 l}{3EI} - \varphi = \frac{X_1 l}{3EI} + \frac{X_2 l}{6EI} + \varphi - \frac{q l^3}{24EI} \qquad (15.43a)$$

$$-\frac{X_2 l}{3EI} - \frac{X_1 l}{6EI} + \varphi + \frac{q l^3}{24EI} = \frac{X_2 l\sqrt{2}}{3EI} - \varphi \qquad (15.43b)$$

$$X_1\varphi + X_1\varphi - X_2\varphi - X_2\varphi - q l\left(\frac{l}{2}\varphi\right) = 0 \qquad (15.43c)$$

Multiplying the first two equations by $6EI/l$ and dividing the third by φ, we have

$$4X_1 + X_2 = -12\varphi\frac{EI}{l} + \frac{q l^2}{4} \qquad (15.44a)$$

$$X_1 + 2X_2(1+\sqrt{2}) = 12\varphi\frac{EI}{l} + \frac{q l^2}{4} \qquad (15.44b)$$

479

$$2(X_1 - X_2) = \frac{1}{2}ql^2 \qquad (15.44c)$$

Resolving, we obtain

$$X_1 = \frac{16 + 3\sqrt{2}}{112}ql^2 \qquad (15.45a)$$

$$X_2 = -\frac{3(4 - \sqrt{2})}{112}ql^2 \qquad (15.45b)$$

$$\varphi = -\frac{8 + 5\sqrt{2}}{448}\frac{ql^3}{EI} \qquad (15.45c)$$

Passing from the irrational expressions above to decimal expressions, we have

$$X_1 \simeq 0.18ql^2 \qquad (15.46a)$$

$$X_2 \simeq -0.07ql^2 \qquad (15.46b)$$

$$\varphi \simeq -0.03\frac{ql^3}{EI} \qquad (15.46c)$$

The unknowns X_2 and φ turn out to be negative, and hence it follows that at the node C, the internal fibres are stretched, and that the rigid rotation of the upright stanchion AB is counterclockwise.

The bending moment diagram (Figure 15.22(a)) is obtained by the graphical summation of the parabolic diagram for the distributed load q and the linear diagram for the redundant moments.

The deformed configuration of the frame (Figure 15.22(b)) is constructed by displacing the two fixed joint-nodes B and C, on the basis of the mechanism of Figure 15.21(b). Both the nodes translate leftwards by the amount φl, while the node C alone translates downwards by the same quantity. The deformed configuration moreover respects the external constraint conditions, the angular congruence at the nodes B and C, and the moment diagram (Figure 15.22(a)). At the point where the moment vanishes there corresponds the inflection of the deformed configuration, which shows the fibres stretched externally in the left-hand part and internally in the right-hand part.

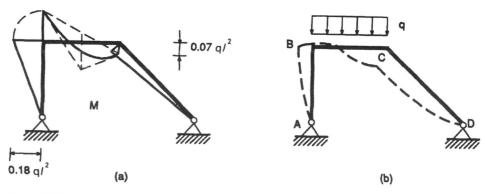

Figure 15.22

480

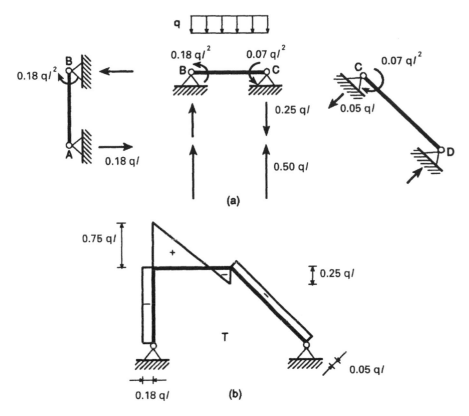

Figure 15.23

The schemes of equilibrium of the individual beams, loaded by the external load q and by the redundant moments (Figure 15.23(a)), immediately furnish the shear diagram (Figure 15.23(b)). The shear is constant on the two stanchions, where the moment is linear, and linear on the cross member, where the moment is parabolic. The point of zero shear corresponds to the section subjected to the maximum moment.

Only on the upright stanchion AB and on the cross member BC do the shearing force and axial force exchange roles. They are both subject to a compression, equal to $0.75ql$ and $0.18ql$, respectively. As regards the stanchion CD, since it is inclined at an angle of $135°$ with respect to the cross member, its axial force can be determined as the sum of the axial components of the horizontal force ($0.18ql$) and the vertical force ($0.25ql$), which the cross member transmits to it (Figure 15.24(a)). The same result may be arrived at by considering the equilibrium with regard to translation of the fixed joint-node C (Figure 15.24(a))

$$N\frac{\sqrt{2}}{2} - T\frac{\sqrt{2}}{2} + 0.18ql = 0 \qquad (15.47a)$$

$$N\frac{\sqrt{2}}{2} + T\frac{\sqrt{2}}{2} + 0.25ql = 0 \qquad (15.47b)$$

481

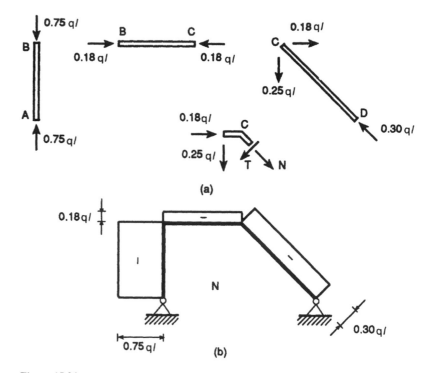

(a)

(b)

Figure 15.24

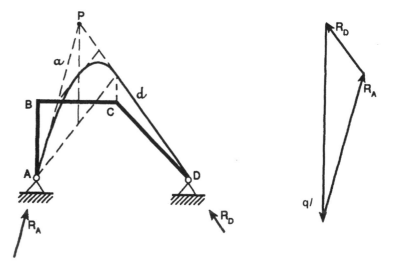

Figure 15.25

Resolving equations (15.47), we obtain

$$N \simeq -0.30ql \qquad (15.48a)$$
$$T \simeq -0.05ql \qquad (15.48b)$$

Equation (15.48b) reconfirms the result already obtained following another approach.

The structure, totally disconnected from the foundation and considered as a single body, is in equilibrium under the action of three forces (Figure 15.25). The reactions of the two hinges and the resultant of the distributed load pass through the pole P. The pressure line consists of the line of action of the external reaction, for each stanchion, and of the arc of parabola which has as its extreme tangents the above-mentioned lines of action, for the cross member. Notice how the arc of parabola corresponds, but for a negative scale factor, to the moment diagram of Figure 15.22(a). Both the curves pass through the section of the cross member which is the site of the deformative inflection point (Figure 15.22(b)).

15.6 Frames loaded out of their own plane

We have already dealt with frames loaded out of their own plane in Sections 14.7 and 14.8, when we considered the automatic computation of **plane grids** and **space frames**. As we have already been able to witness in those cases, the situation is notably complicated with respect to the case of the plane frame, because potentially all six static or kinematic characteristics are involved.

If we wish to proceed along the same lines as those followed in this chapter, we must disconnect the fixed joint-nodes with **spherical hinges**, instead of with the normal **cylindrical hinges**, and apply both bending and twisting redundant moments at the ends of the beams. If the associated truss has n degrees of freedom, it will be necessary to consider n kinematic parameters among the unknowns of the problems, as well as n applications of the Principle of Virtual Work among the equations for resolving the problem. For each spherical node we have, on the other hand, three unknown redundant moments and three equations of angular congruence.

It may prove convenient to apply this method only in the simplest cases, as for example in those of portal frames or balconies. For the solution of the structure of Figure 15.26(a), loaded by a force F on one of the two internal nodes, already as many as ten equations in ten unknowns are required. Two unknowns are represented by the deflections δ_1 and δ_2 of the two internal nodes (Figure 15.26(b)), while the other eight unknowns are represented by the bending and twisting moments which the beams exchange with one another, or which the beams exchange with the built-in constraints in the external wall (Figure 15.26(c)). As regards the equations, there will be two applications of the Principle of Virtual Work and eight conditions of angular congruence

$$\varphi_A = 0 \tag{15.49a}$$

$$\vartheta_A = 0 \tag{15.49b}$$

$$\varphi_{B'} = \vartheta_{B''} \tag{15.49c}$$

$$\vartheta_{B'} = \varphi_{B''} \tag{15.49d}$$

$$\varphi_{C'} = \vartheta_{C''} \tag{15.49e}$$

$$\vartheta_{C'} = \varphi_{C''} \tag{15.49f}$$

$$\varphi_D = 0 \tag{15.49g}$$

$$\vartheta_D = 0 \tag{15.49h}$$

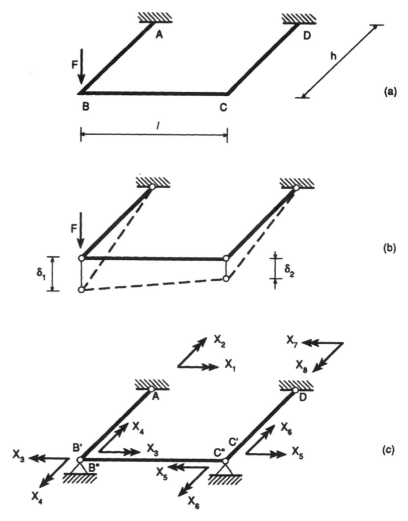

Figure 15.26

where φ denotes the angles of deflection and ϑ those of torsion.

In the case where the previous structure is loaded by a uniform distributed load q on the cross member (Figure 15.27(a)), there is, by symmetry, a single redundant unknown represented by the moment X, which is a bending moment in the case of the cross member and a twisting moment in the case of the two cantilevers (Figure 15.27(b)). The condition of congruence imposes the equality of the angle of deflection of the cross member with that of torsion of the cantilevers,

$$\varphi_{B''} = \vartheta_{B'} \tag{15.49d}$$

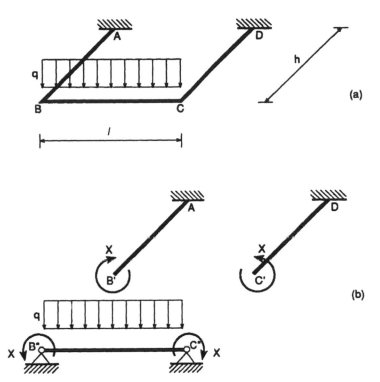

Figure 15.27

and in explicit form

$$\frac{Xl}{3EI} + \frac{Xl}{6EI} - \frac{ql^3}{24EI} = -\frac{Xh}{GI_p} \tag{15.50}$$

where $I_p = 2I$, if the cross section of the beams is assumed to be circular. From equation (15.50) we obtain

$$X = \frac{ql^3}{12[l + 2h(1 + \nu)]} \tag{15.51}$$

Whereas then the cross member is subject to bending moment and shear, the cantilever beams are subject to the constant shear $ql/2$, to the linear bending moment which ranges from zero to a maximum of $qlh/2$ at the built-in support, as well as to the constant twisting moment X.

485

16 Energy methods for the solution of beam systems

16.1 Introduction

The Principle of Virtual Work, proposed in Section 8.4 in the context of three-dimensional solids, may be extrapolated, following a similar line of demonstration, to one- and two-dimensional solids. As regards rectilinear beams, it is sufficient to substitute equation (8.23) with

$$L_F = -\int_0^l ([\partial]^*\{Q_a\})^T \{\eta_b\} dz \tag{16.1}$$

whereby, instead of equation (8.24), we obtain

$$L_F = \int_0^l \{Q_a\}^T [\partial] \{\eta_b\} dz - [\{Q_a\}^T \{\eta_b\}]_0^l \tag{16.2}$$

0 and l being the coordinates of the beam ends.

We then obtain the equation of the Principle of Virtual Work for a rectilinear beam, subjected to loads distributed over the span and to loads concentrated at the ends,

$$\int_0^l \{Q_a\}^T \{q_b\} dz = \int_0^l \{\mathscr{F}_a\}^T \{\eta_b\} dz + [\{Q_a\}^T \{\eta_b\}]_0^l \tag{16.3}$$

where, adopting the same nomenclature used in Section 10.3, $\{Q_a\}$ is the vector of the static characteristics, $\{q_b\}$ is the vector of deformation characteristics, $\{\mathscr{F}_a\}$ is the vector of the distributed forces and $\{\eta_b\}$ is the displacement vector.

The fundamental equation (8.26) can then be shown to be valid also in the case of curvilinear beams, by using the mathematical formalism of Section 10.4 and replacing the operator $[\partial]$ with $[\partial][N]$, where $[N]$ is the rotation matrix, and the operator $[\partial]^*$ with $[N]^T[\partial]^*$.

Finally, equation (8.26) can also be further applied to the case of beam systems, by summing up the contributions of the individual beams. Whereas the integrals of equation (16.3), already extended to the individual beam of length l, must, in this case, be extended to the entire structure S, the second term of the right-hand side of the equation cancels out in all the internal nodes, for obvious reasons of equilibrium. On the other hand, all the contributions corresponding to the ends that are externally constrained or that are subjected to concentrated loads still remain to be accounted for:

$$\int_S \{Q_a\}^T \{q_b\} ds = \int_S \{\mathscr{F}_a\}^T \{\eta_b\} ds + \sum_i \{Q_{ai}\}^T \{\eta_b\} \tag{16.4}$$

Since the aim of the application of Principle of Virtual Work to statically determinate beam systems is to determine the elastic generalized displacements, it is expedient to consider, for each individual beam or structure, two distinct systems:

1. the **real system** or **system of displacements**;
2. the **fictitious system** or **system of forces**.

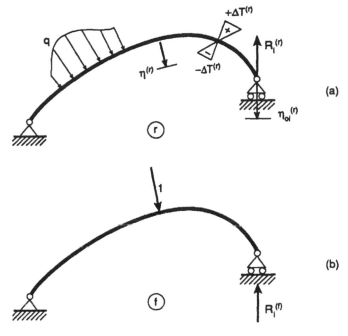

Figure 16.1

The real system consists simply of the structure under examination, subjected to all the external loads, both mechanical and thermal, including the inelastic settlements, and taking into account the elastic settlements (Figure 16.1 (a)). The fictitious system in turn consists of the same structure, loaded in this case by the single unit force, dual of the elastic displacement sought $\eta^{(r)}$ (Figure 16.1 (b)).

An application of equation (16.4) yields

$$\int_S N^{(f)}\left(\frac{N^{(r)}}{EA} + \alpha T_0^{(r)}\right) ds + \int_S T^{(f)}\left(\frac{tT^{(r)}}{GA}\right) ds + \qquad (16.5)$$

$$\int_S M^{(f)}\left(\frac{M^{(r)}}{EI} + 2\frac{\alpha\Delta T^{(r)}}{h}\right) ds$$

$$= 1 \times \eta^{(r)} + \sum_i R_i^{(f)}\left(\eta_{0i}^{(r)} - \frac{R_i^{(r)}}{k_i}\right)$$

where the superscript f denotes the fictitious system or system of forces, the superscript r denotes the real system or system of displacements, $\eta^{(r)}$ designates the real displacement to be determined, R_i indicates the ith constraint reaction, k_i indicates the stiffness and $\eta_{0i}^{(r)}$ denotes the displacement imposed on the corresponding constraint.

With the exclusion of the truss structures, it may be stated that the contribution of the third integral on the left-hand side of equation (16.5), the one corresponding to the bending moment, is usually far greater than that of the two integrals preceding it, which correspond to axial force and shearing

487

force, respectively. Of course, the contributions corresponding to the thermal dilatations and curvatures are absent in the case where there are no thermal loadings ($T_0^{(r)} = \Delta T^{(r)} = 0$). In the same way the summation on the right-hand side of the equation is zero, in the case where there are no elastic or inelastic settlements.

16.2 Determination of elastic displacements in statically determinate structures

We shall now show how it is possible to determine the elastic displacements and rotations in the cross sections belonging to statically determinate beam systems by applying equation (16.5).

As a first elementary example, let us consider the simply supported beam of Figure 16.2 (a), loaded by a concentrated moment at the right-hand end. We intend to calculate the elastic rotation of this end. To do this, let the fictitious system consist of the same beam loaded at the same end by a unit moment, acting in the same direction as the actual moment (Figure 16.2 (b)). We have, therefore

$$M^{(r)}(z) = \frac{m}{l} z \tag{16.6a}$$

$$M^{(f)}(z) = \frac{z}{l} \tag{16.6b}$$

whereby, if we apply equation (16.5), taking into account the absence of axial force, the negligibility of shearing strain and the absence of distortions and settlements, we obtain

$$1 \times \varphi_B = \int_0^l \frac{M^{(f)}M^{(r)}}{EI} \, dz \tag{16.7}$$

Substituting expressions (16.6) into the integral (16.7), we obtain

$$\varphi_B = \frac{ml}{3EI} \tag{16.8}$$

a result already known to us from the treatment of the elastic line (Chapter 10).

We then intend to determine the elastic rotation at the end opposite to the loaded one. In this case we shall have to consider a different fictitious system, loaded by a unit moment at the end A (Figure 16.2 (c)). Here we have

$$M^{(f)}(z) = \frac{l-z}{l} \tag{16.9}$$

whereby application of the Principle of Virtual Work yields

$$1 \times \varphi_A = \int_0^l \frac{M^{(f)}M^{(r)}}{EI} \, dz \tag{16.10}$$

By substituting the functions (16.9) and (16.6a) into equation (16.10), we find that

$$\varphi_A = \frac{ml}{6EI} \tag{16.11}$$

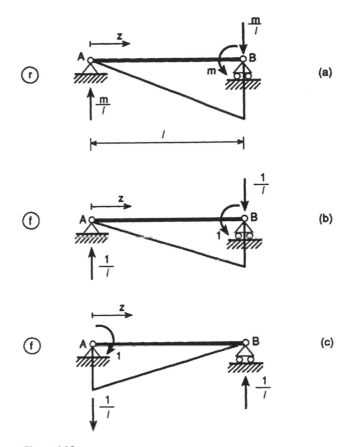

Figure 16.2

In the case of a supported beam subjected to uniform load (Figure 16.3 (a)), the vertical displacement or deflection in the centre can be obtained by means of the fictitious system of Figure 16.3 (b), consisting of the same beam loaded by a vertical unit force acting in the centre. Given the symmetry of the beam under examination and of the displacement sought, i.e. given the symmetry of the two systems, the real one and the fictitious one, it is possible to evaluate the integrals on half the beam and multiply them by two. The two moments, the real one and the fictitious one, can hence be expressed analytically even just on the left-hand span

$$M^{(r)}(z) = \frac{1}{2}qlz - \frac{1}{2}qz^2 \qquad (16.12a)$$

$$M^{(f)}(z) = \frac{z}{2} \qquad (16.12b)$$

for $0 \leqslant z \leqslant l/2$. Applying equation (16.5), we obtain

$$1 \times v_C = \frac{2}{EI}\int_0^{l/2} M^{(f)}M^{(r)}dz \qquad (16.13)$$

489

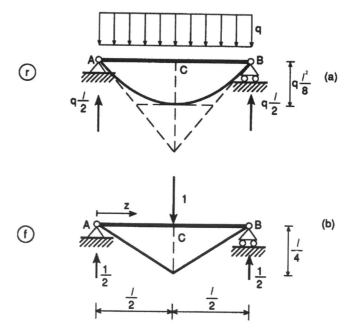

Figure 16.3

and, if we substitute equations (16.12) into equation (16.13), we can easily obtain

$$v_C = \frac{5}{384} \frac{ql^4}{EI} \tag{16.14}$$

Now consider the L-shaped beam of Figure 16.4 (a), uniformly loaded on the cross member. The determination of the elastic displacement of the roller support on which the upright rests may be obtained by means of the fictitious structure of Figure 16.4 (b), on which a unit horizontal force is applied at the end A. Whereas in the fictitious system the upright is subjected to bending moment, this does not occur in the real system, the upright of which is subjected to axial force alone. Hence, taking into account only the contribution of the cross member, we have

$$M^{(r)}(z) = \frac{1}{2}qlz - \frac{1}{2}qz^2 \tag{16.15a}$$

$$M^{(f)}(z) = \frac{h}{l}z \tag{16.15b}$$

whereby we obtain

$$1 \times \delta_A = \frac{1}{EI}\int_0^l \frac{h}{l}z\left(\frac{1}{2}qlz - \frac{1}{2}qz^2\right)dz \tag{16.16}$$

and thus

$$\delta_A = \frac{ql^3}{24EI}h \tag{16.17}$$

490

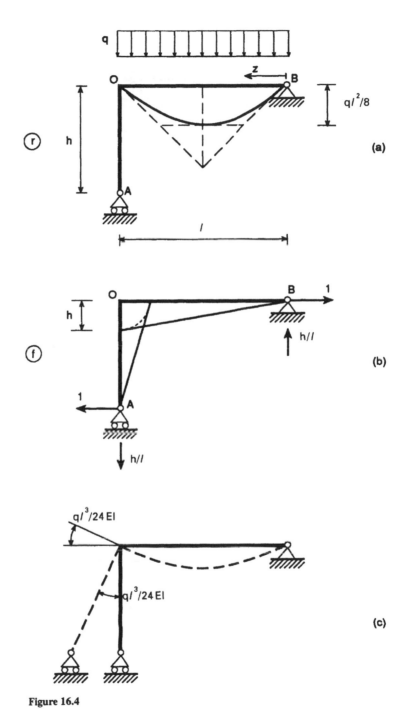

Figure 16.4

Note that this displacement is equal to the product of the angle of elastic rotation of the end of a supported beam by the rigid arm of length h provided by the upright (Figure 16.4 (c)).

491

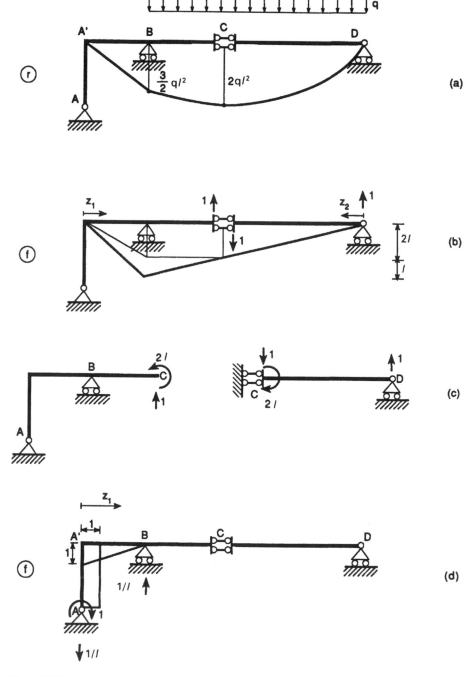

Figure 16.5

Let us reconsider the gerber beam of Figure 16.5 (a), of which we here intend to evaluate the relative vertical displacement at the double rod (Figures 6.3, 6.4 (b)). The fictitious structure consists of the same beam, loaded by two unit vertical, equal and opposite, forces, acting at the ends of the beams connected together by the double rod (Figure 16.5 (b)). Imposing equilibrium first on the portion CD and then on the portion CA (Figure 16.5 (c)), we obtain the fictitious moment functions

$$M^{(f)}(z_1) = 3z_1, \quad 0 \leqslant z_1 \leqslant l \qquad (16.18a)$$

$$M^{(f)}(z_2) = z_2, \quad 0 \leqslant z_2 \leqslant 3l \qquad (16.18b)$$

On the other hand, the real moment functions are equal to

$$M^{(r)}(z_1) = \frac{3}{2}qlz_1, \quad 0 \leqslant z_1 \leqslant l \qquad (16.19a)$$

$$M^{(r)}(z_2) = 2qlz_2 - \frac{1}{2}qz_2^2, \quad 0 \leqslant z_2 \leqslant 3l \qquad (16.19b)$$

Applying the Principle of Virtual Work yields the following equation:

$$1 \times \Delta v_C = \frac{1}{EI} \int_0^l \frac{9}{2}qlz_1^2 dz_1 + \frac{1}{EI} \int_0^{3l} z_2 \left(2qlz_2 - \frac{1}{2}qz_2^2 \right) dz_2 \qquad (16.20)$$

Carrying out the calculations, we obtain

$$\Delta v_C = \frac{75}{8}\frac{ql^4}{EI} \qquad (16.21)$$

If we wish to know by how much one end is raised and by how much the other end is lowered, it is possible to consider the two unit forces of the scheme of Figure 16.5 (b) separately.

In order to define the rigid rotation of the upright, it is necessary to consider the fictitious system of Figure 16.5 (d), in which a unit moment is applied to the end A. In this case the only contribution to the calculation comes from the portion $A'B$, since only here are both the real moment and the fictitious moment different from zero:

$$1 \times \varphi_A = \frac{1}{EI} \int_0^l \left(\frac{3}{2}qlz_1 \right)\left(1 - \frac{z_1}{l} \right) dz_1 \qquad (16.22)$$

Computing, we find the rotation to be positive and hence clockwise as supposed (Figure 6.4 (b)):

$$\varphi_A = \frac{1}{4}\frac{ql^3}{EI} \qquad (16.23)$$

Finally, let us consider the truss of Figure 16.6, and let us determine the elastic displacement of the roller support A. We shall therefore have to reconsider the same truss, loaded by a unit force similar to the actual one. In this case, since the axial force is the only static characteristic present, equation (16.5) reduces as follows:

$$1 \times u_A = \sum_i N_i^{(f)} \frac{N_i^{(r)}}{EA} l_i \qquad (16.24)$$

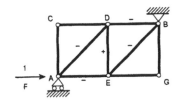

Figure 16.6

493

On the basis of the axial forces obtained and listed in Section 6.3 (Figure 6.9), we have

$$u_A EA = 2\left[\left(-\frac{F}{2}\right)\left(-\frac{1}{2}\right)l\right] +$$ (16.25)

$$2\left[\left(-\frac{F}{2}\sqrt{2}\right)\left(-\frac{1}{2}\sqrt{2}\right)l\sqrt{2}\right] +$$

$$\left[\left(\frac{F}{2}\right)\left(\frac{1}{2}\right)l\right]$$

and thus

$$u_A = \frac{Fl}{EA}\left(\frac{3}{4} + \sqrt{2}\right)$$ (16.26)

16.3   Resolution of structures having one degree of static indeterminacy

In the case where the structure being examined has one degree of static indeterminacy (Figure 16.7), it is possible to write the equation of congruence using the same calculating procedure adopted in the previous section (Figure 16.1). The procedure will be to equate the elastic displacement produced by the external loads and by the redundant reaction to zero or to a quantity different from zero (function of the redundant reaction for elastically compliant constraints), according to whether the suppressed constraint is rigid or not.

From the operative point of view, once the equivalent statically determinate structure has been identified, two schemes are resolved:

Scheme 0, consisting of the equivalent statically determinate structure, subjected to external loads;
Scheme 1, consisting of the equivalent statically determinate structure, subjected to the unit redundant reaction.

At this point, the system of forces consists of Scheme 1, whilst the system of displacements consists of the superposition of Scheme 0 and Scheme 1, which

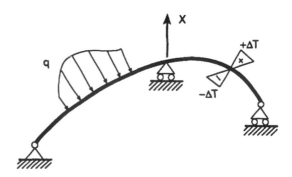

Figure 16.7

is multiplied by the redundant unknown X. Equation (16.5) thus yields the displacement η of the constrained point

$$1 \times \eta = \frac{1}{EI} \int_S M^{(1)} \left(M^{(0)} + X M^{(1)} \right) ds \qquad (16.27)$$

In the case of a rigid or inelastically compliant constraint, the equation of congruence is obtained by equating the right-hand side of equation (16.27) to zero or to η_0, where η_0 is the known entity of settlement. In the case of an elastically compliant constraint, the equation of congruence is obtained by equating the right-hand side of equation (16.27) to $-X/k$, where k is the stiffness of the constraint. In all cases a linear algebraic equation in the single unknown X is obtained.

In the case of an elastically compliant constraint, we have for instance

$$\int_S M^{(0)} M^{(1)} ds + X \int_S (M^{(1)})^2 ds = -X \frac{EI}{k} \qquad (16.28)$$

whence we obtain

$$X = -\frac{\int_S M^{(0)} M^{(1)} ds}{\int_S (M^{(1)})^2 ds + \dfrac{EI}{k}} \qquad (16.29)$$

In the same way, in the case of an inelastically compliant constraint, we obtain

$$X = -\frac{\int_S M^{(0)} M^{(1)} ds - \eta_0 \, EI}{\int_S (M^{(1)})^2 ds} \qquad (16.30)$$

When instead the constraint is rigid ($k \rightarrow \infty$, or $\eta_0 \rightarrow 0$), both expression (16.29) and expression (16.30) reduce to the following:

$$X = -\frac{\int_S M^{(0)} M^{(1)} ds}{\int_S (M^{(1)})^2 ds} \qquad (16.31)$$

We shall now reconsider some of the structures having one degree of static indeterminacy, already studied in Chapter 15 with the method of plane frames. For the very reason that they were not sufficiently constrained, these structures proved to be frames with translating nodes. As will be seen, in these cases the application of the Principle of Virtual Work constitutes a valid alternative to the methods already introduced.

In relation to the frame of Figure 15.13 (a), there are two schemes to be considered to obtain a resolution of the problem using the Principle of Virtual Work. The equivalent statically determinate structure may be obtained, for example, by eliminating the degree of constraint to the horizontal translation of the hinge A, i.e. transforming the hinge into a roller support (Figure 16.8).

Scheme 0 (Figure 16.8 (a)) thus consists of the equivalent statically determinate structure subjected to the distributed load acting on the overhang CG, while Scheme 1 (Figure 16.8 (b)) consists of the same statically determinate structure, loaded in this case by a unit horizontal force applied at point A. The determination of the constraint reactions of the two schemes is immediate, as is the drawing of the respective moment diagrams (Figure 16.8).

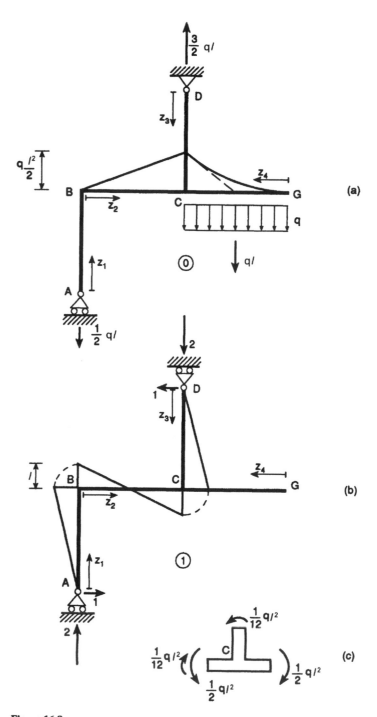

Figure 16.8

496

Table 16.1

Beam	$M^{(0)}$	$M^{(1)}$
AB	0	z_1
BC	$\frac{1}{2}qlz_2$	$l-2z_2$
CD	0	z_3
CG	$\frac{1}{2}qz_4^2$	0

Once a reference system for each beam of the frame has been chosen, it is possible to draw up Table 16.1.

We thus obtain

$$\int_S M^{(0)}M^{(1)}\mathrm{d}s = \int_0^l \frac{1}{2}qlz_2(l-2z_2)\mathrm{d}z_2 = -\frac{1}{12}ql^4 \qquad (16.32a)$$

$$\int_S (M^{(1)})^2\,\mathrm{d}s = \int_0^l z_1^2\mathrm{d}z_1 + \int_0^l (l-2z_2)^2\,\mathrm{d}z_2 + \int_0^l z_3^2\mathrm{d}z_3 = l^3 \qquad (16.32b)$$

From equation (16.31) we obtain the redundant unknown

$$X = \frac{1}{12}ql \qquad (16.33)$$

The equilibrium scheme of the node C, obtained by superposition of Schemes 0 and $1 \times X$, and shown in Figure 16.8 (c), is equivalent to the one obtained with the method of plane frames and shown in Figure 15.15 (b). Applying once more the Principle of Superposition, it is possible to verify the bending moment, shearing force and axial force diagrams, already represented in Figures 15.15 and 15.16.

As a second example, consider again the asymmetrical portal frame of Figure 15.12. As an equivalent statically determinate structure, let us choose, from the infinite range of possibilities, the three-hinged arch ACD (Figure 16.9). In this case, therefore, the structure has been internally disconnected, even though usually, from the point of view of simplicity of calculation, external disconnections are more convenient. Having determined the moment diagrams on Schemes 0 and 1, and having fixed a reference system on each beam, we draw up Table 16.2.

Notice that in the schemes of Figure 16.9, the moment diagram is shown from the side of the stretched fibres, and that where $M^{(0)}$ and $M^{(1)}$ extend opposite fibres, in the table the respective functions must give values of opposite sign. The following two integrals are therefore obtained:

Table 16.2

Beam	$M^{(0)}$	$M^{(1)}$
AB	Fz	$-\dfrac{z}{l}$
BC	Fz	$-\left(1+\dfrac{z}{2l}\right)$
CD	0	$-\dfrac{z}{l}$

$$\int_S M^{(0)}M^{(1)}\mathrm{d}s = \int_0^{2l}\left(-\frac{F}{l}z^2\right)\mathrm{d}z + \int_0^{2l}(-Fz)\left(1+\frac{z}{2l}\right)\mathrm{d}z = -6Fl^2 \quad (16.34a)$$

$$\int_S (M^{(1)})^2\,\mathrm{d}s = \int_0^{2l}\frac{z^2}{l^2}\mathrm{d}z + \int_0^{2l}\left(1+\frac{z^2}{4l^2}+\frac{z}{l}\right)\mathrm{d}z + \int_0^l\frac{z^2}{l^2}\mathrm{d}z = \frac{23}{3}l \quad (16.34b)$$

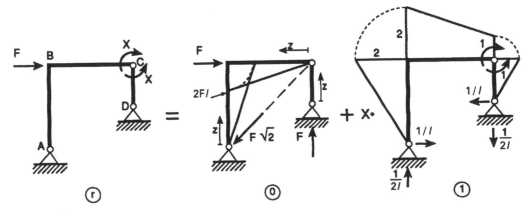

Figure 16.9

497

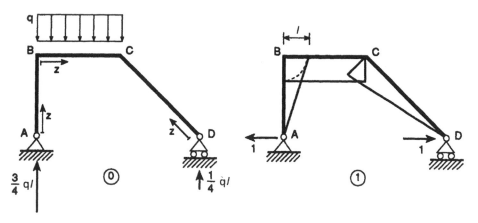

Figure 16.10

From equation (16.31) we obtain the redundant moment

$$X = \frac{18}{23}Fl \qquad (16.35)$$

This calculation is implicitly equivalent to the imposition of angular congruence in the fixed joint-node C:

$$\Delta\varphi_C = 0 \quad \text{or} \qquad (16.36a)$$

$$\varphi_{CD} = \varphi_{CA} \qquad (16.36b)$$

Applying the Principle of Superposition, it is simple to find once more the diagrams of the static characteristics defined in the previous chapter (Figure 15.12).

As a final example of a mechanically loaded structure having one degree of redundancy, let us refer back to the portal frame with oblique stanchion of Figure 15.21 (a). Let the structure be disconnected externally, so as to transform the hinge D into a horizontally moving roller support (Figure 16.10). Once the external reactions have been identified on Schemes 0 and 1 and the reference systems have been defined on the individual beams, the determination of the analytical functions $M^{(0)}$ and $M^{(1)}$ is immediate. It is not necessary at this stage to draw the diagram of these functions, which are given in Table 16.3.

Calculation of the integrals

Table 16.3

Beam	$M^{(0)}$	$M^{(1)}$
AB	0	z
BC	$\frac{3}{4}qlz - \frac{1}{2}qz^2$	l
CD	$\frac{1}{8}qlz\sqrt{2}$	$\frac{\sqrt{2}}{2}z$

$$\int_S M^{(0)}M^{(1)}ds = \int_0^l l\left(\frac{3}{4}qlz - \frac{1}{2}qz^2\right)dz + \qquad (16.37a)$$

$$\int_0^{l\sqrt{2}} \left(\frac{1}{8}qlz\sqrt{2}\right)\left(z\frac{\sqrt{2}}{2}\right)dz = \frac{5+2\sqrt{2}}{24}ql^4$$

$$\int_S (M^{(1)})^2 ds = \int_0^l z^2 dz + \int_0^l l^2 dz + \int_0^{l\sqrt{2}} \frac{1}{2}z^2 dz = \frac{4+\sqrt{2}}{3}l^3 \qquad (16.37b)$$

498

makes it possible, via equation (16.31), to obtain the redundant reaction

$$X = -\frac{16 + 3\sqrt{2}}{112} ql \qquad (16.38)$$

We are thus able to verify that the bending moment in the fixed joint-node C is equal to

$$M_C = Xl + \frac{1}{4}ql^2 = \frac{3(4 - \sqrt{2})}{112} ql^2 \qquad (16.39)$$

which, apart from the sign, coincides with the value given in equation (15.45b) for the solution obtained using the plane frame method.

Before closing this section, we draw attention to the fact that, in general, it is more difficult to construct the moment diagram using the Principle of Virtual Work than it is using the plane frame method, illustrated in the previous chapter. In that case, the nodal values of the bending moment are obtained directly, so that it is simpler to add the partial diagrams for the external loads to the linear diagrams for these values.

16.4 Resolution of structures having two or more degrees of static indeterminacy

In the case of structures having two or more degrees of static indeterminacy, the procedure outlined in the foregoing section can be extended to a pair of fictitious schemes, consisting of the equivalent statically determinate structure, loaded by one redundant unknown at a time. The displacements of the two points in which the disconnection is made can be obtained by applying the Principle of Virtual Work to each fictitious structure.

More precisely, considering as system of displacements the real one (Scheme $0 + X_1 \times$ Scheme $1 + X_2 \times$ Scheme 2) and as system of forces each of the two fictitious systems, we have

$$1 \times \eta_1 = \frac{1}{EI} \int_S M^{(1)}\big(M^{(0)} + X_1 M^{(1)} + X_2 M^{(2)}\big) \ ds \qquad (16.40a)$$

$$1 \times \eta_2 = \frac{1}{EI} \int_S M^{(2)}\big(M^{(0)} + X_1 M^{(1)} + X_2 M^{(2)}\big) \ ds \qquad (16.40b)$$

In the case where all the constraints of the structure are rigid, the two relations of congruence, $\eta_1 = \eta_2 = 0$, yield the following two linear algebraic equations:

$$X_1 \int_S (M^{(1)})^2 \, ds + X_2 \int_S M^{(1)} M^{(2)} ds = -\int_S M^{(1)} M^{(0)} ds \qquad (16.41a)$$

$$X_1 \int_S M^{(2)} M^{(1)} ds + X_2 \int_S (M^{(2)})^2 ds = -\int_S M^{(2)} M^{(0)} ds \qquad (16.41b)$$

If we designate as **coefficient of influence** η_{12} the displacement generated by the redundant reaction $X_2 = 1$ at the point of application and in the direction of the other redundant reaction X_1, and as η_{21} the displacement generated by

499

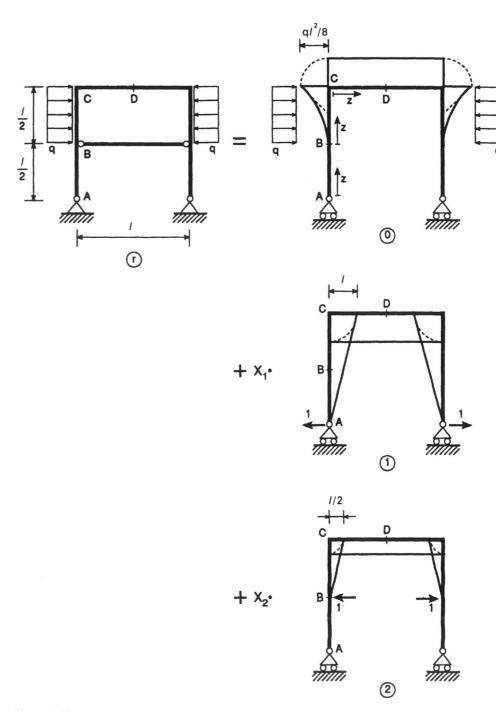

Figure 16.11

$X_1 = 1$ at the point and in the direction of X_2, from Betti's Reciprocal Theorem and the Principle of Virtual Work we can deduce

$$\eta_{12} = \eta_{21} = \frac{1}{EI}\int_S M^{(1)}M^{(2)}ds \qquad (16.42a)$$

while the self-influence coefficients may be expressed as

$$\eta_{11} = \frac{1}{EI}\int_S (M^{(1)})^2\,ds \qquad (16.42b)$$

$$\eta_{22} = \frac{1}{EI}\int_S (M^{(2)})^2\,ds \qquad (16.42c)$$

Equations (16.41) may thus be cast in the form

$$\eta_{11}X_1 + \eta_{12}X_2 = -\eta_{10} \qquad (16.43a)$$
$$\eta_{21}X_1 + \eta_{22}X_2 = -\eta_{20} \qquad (16.43b)$$

η_{10} and η_{20} being the displacements due to the external load. Resolving the system using Cramer's rule, we obtain

$$X_1 = -\frac{\begin{vmatrix}\eta_{10} & \eta_{12}\\ \eta_{20} & \eta_{22}\end{vmatrix}}{\begin{vmatrix}\eta_{11} & \eta_{12}\\ \eta_{21} & \eta_{22}\end{vmatrix}} = -\frac{\eta_{10}\,\eta_{22} - \eta_{12}\,\eta_{20}}{\eta_{11}\,\eta_{22} - \eta_{12}^2} \qquad (16.44a)$$

$$X_2 = -\frac{\begin{vmatrix}\eta_{11} & \eta_{10}\\ \eta_{21} & \eta_{20}\end{vmatrix}}{\begin{vmatrix}\eta_{11} & \eta_{12}\\ \eta_{21} & \eta_{22}\end{vmatrix}} = -\frac{\eta_{11}\,\eta_{20} - \eta_{10}\,\eta_{21}}{\eta_{11}\,\eta_{22} - \eta_{12}^2} \qquad (16.44b)$$

As an example of the application of the above procedure, let us reconsider the frame with two degrees of redundancy of Figure 15.9 (a). As an equivalent statically determinate structure, let us consider the same portal frame, with the connecting rod and the constraint to horizontal translation at the feet of the uprights removed (Figure 16.11). Three schemes are hence to be considered:

Scheme 0, with the external load only (Figure 16.11 (a));
Scheme 1, with two symmetrical and horizontal unit forces acting at the foot of the uprights (Figure 16.11 (b));
Scheme 2, with two symmetrical and horizontal unit forces acting half-way up the uprights (Figure 16.11 (c)).

Figure 16.11 also shows the corresponding moment diagram for each of the three schemes. Taking into account symmetry, there are three portions of the structure on which the integrals of equations (16.42) are to be evaluated. Using a suitable reference system for each portion, we obtain Table 16.4.

Table 16.4

Beam	$M^{(0)}$	$M^{(1)}$	$M^{(2)}$
AB	0	z	0
BC	$-\frac{1}{2}qz^2$	$\frac{l}{2}+z$	z
CD	$-\frac{1}{8}ql^2$	l	$\frac{l}{2}$

There then follows the computation of the coefficients of influence:

$$\int_S (M^{(1)})^2 ds = \int_0^{l/2} z^2 dz + \int_0^{l/2} \left(\frac{l}{2}+z\right)^2 dz + \int_0^{l/2} l^2 dz = \frac{5}{6}l^3 \quad (16.45a)$$

$$\int_S M^{(1)} M^{(2)} ds = \int_0^{l/2} z\left(\frac{l}{2}+z\right) dz + \int_0^{l/2} l\left(\frac{l}{2}\right) dz = \frac{17}{48}l^3 \quad (16.45b)$$

$$\int_S M^{(1)} M^{(0)} ds = \int_0^{l/2}\left(-\frac{1}{2}qz^2\right)\left(\frac{l}{2}+z\right) dz + \int_0^{l/2}\left(-\frac{1}{8}ql^2\right) l dz \quad (16.45c)$$

$$= -\frac{31}{384}ql^4$$

$$\int_S (M^{(2)})^2 ds = \int_0^{l/2} z^2 dz + \int_0^{l/2} \frac{l^2}{4} dz = \frac{l^3}{6} \quad (16.45d)$$

$$\int_S M^{(2)} M^{(0)} ds = \int_0^{l/2}\left(-\frac{1}{2}qz^2\right) z dz + \int_0^{l/2}\left(-\frac{1}{8}ql^2\right)\frac{l}{2} dz \quad (16.45e)$$

$$= -\frac{5}{128}ql^4$$

Considering the equations (16.42) and (16.43), we finally obtain:

$$X_1 = -\frac{7}{248}ql \quad (16.46a)$$

$$X_2 = \frac{73}{248}ql \quad (16.46b)$$

verifying what was already found following another procedure in the previous chapter.

In the case of a structure having three or more degrees of redundancy, the procedure does not substantially change. Once the equivalent statically determinate structure has been identified, $(n+1)$ elementary schemes are considered, where n is the degree of redundancy. Applying the Principle of Virtual Work to each scheme, we arrive at a linear algebraic system of n equations in the n unknowns $X_1, X_2, ..., X_n$

$$\sum_{j=1}^n \eta_{ij} X_j = -\eta_{i0}, \quad \text{for } i = 1, 2, ..., n \quad (16.47)$$

where η_{ij} are the elements of the **influence matrix**

$$\eta_{ij} = \frac{1}{EI}\int_S M^{(i)} M^{(j)} ds, \quad \text{for } i, j = 1, 2, ..., n \quad (16.48)$$

and η_{i0} are the displacements due to the external load

$$\eta_{i0} = \frac{1}{EI}\int_S M^{(i)} M^{(0)} ds, \quad \text{for } i = 1, 2, ..., n \quad (16.49)$$

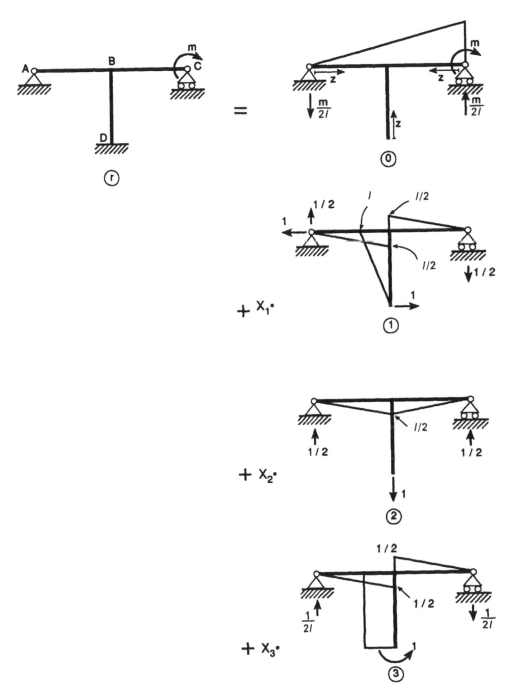

Figure 16.12

503

Table 16.5

Beam	$M^{(0)}$	$M^{(1)}$	$M^{(2)}$	$M^{(3)}$
AB	$-\dfrac{m}{2l}z$	$\dfrac{z}{2}$	$\dfrac{z}{2}$	$\dfrac{z}{2l}$
BC	$-m+\dfrac{m}{2l}z$	$-\dfrac{z}{2}$	$\dfrac{z}{2}$	$-\dfrac{z}{2l}$
BD	0	z	0	1

In the case, for instance, of the frame of Figure 15.6, it is possible to reduce it to an equivalent statically determinate structure by eliminating the built-in support D at the foot of the upright and by applying the three elementary reactions. There will thus be four schemes to be considered (Figures 16.12 (a)–(d)) and Table 16.5, showing the bending moment functions.

The real bending moment may thus be expressed by means of the Principle of Superposition:

$$M_{AB}^{(r)} = M_{AB}^{(0)} + X_1 M_{AB}^{(1)} + X_2 M_{AB}^{(2)} + X_3 M_{AB}^{(3)} \qquad (16.50a)$$

$$= -\frac{m}{2l}z + \frac{1}{2}X_1 z + \frac{1}{2}X_2 z + \frac{1}{2l}X_3 z$$

$$M_{BC}^{(r)} = M_{BC}^{(0)} + X_1 M_{BC}^{(1)} + X_2 M_{BC}^{(2)} + X_3 M_{BC}^{(3)} \qquad (16.50b)$$

$$= -m + \frac{m}{2l}z - \frac{1}{2}X_1 z + \frac{1}{2}X_2 z - \frac{1}{2l}X_3 z$$

$$M_{BD}^{(r)} = M_{BD}^{(0)} + X_1 M_{BD}^{(1)} + X_2 M_{BD}^{(2)} + X_3 M_{BD}^{(3)} \qquad (16.50c)$$

$$= X_1 z + X_3$$

The first equation of congruence will therefore be

$$\int_S M^{(1)} M^{(r)} ds = \qquad (16.51)$$

$$= \int_0^l \left(\frac{z}{2}\right)\left(-\frac{m}{2l}z + \frac{1}{2}X_1 z + \frac{1}{2}X_2 z + \frac{1}{2l}X_3 z\right) dz +$$

$$\int_0^l \left(-\frac{z}{2}\right)\left(-m + \frac{m}{2l}z - \frac{1}{2}X_1 z + \frac{1}{2}X_2 z - \frac{1}{2l}X_3 z\right) dz +$$

$$\int_0^l (z)(X_1 z + X_3) dz = 0$$

which, when the calculations are performed, becomes

$$\frac{1}{4}\left(-\frac{m}{l} + X_1 + X_2 + \frac{X_3}{l}\right)\frac{l^3}{3} - \qquad (16.52)$$

$$\frac{1}{4}\left(\frac{m}{l} - X_1 + X_2 - \frac{X_3}{l}\right)\frac{l^3}{3} + \frac{m}{2}\frac{l^2}{2} +$$

$$X_1 \frac{l^3}{3} + X_3 \frac{l^2}{2} = 0$$

and hence

$$6X_1 l + 8X_3 + m = 0 \qquad (16.53)$$

Equation (16.53) is satisfied by the values

$$X_1 = -\frac{3}{10}\frac{m}{l} \tag{16.54a}$$

$$X_3 = \frac{m}{10} \tag{16.54b}$$

which we obtained following another path in the previous chapter.

In like manner, it is, on the other hand, possible to obtain the remaining two equations of congruence and to verify the results previously obtained. At the same time it may be appreciated how the Principle of Virtual Work is extremely laborious, compared with the method of plane frames, when the degree of redundancy of the structure is equal to or greater than three.

16.5 Thermal distortions and constraint settlements

In the case where the redundant structure undergoes thermal distortions, whether spread uniformly over the entire thickness or butterfly-shaped, equation (16.5) is to be applied, considering the real system as the system of displacements and a statically determinate fictitious system as the system of forces.

Let us take, for instance, the case of the asymmetrical portal frame of Figure 15.18 (a), subjected to a uniform increase in temperature over the cross member. We shall choose as equivalent statically determinate structure the one of Figure 16.13, obtained by replacing the hinge D with a roller support. Table 16.6 gives the bending moment and axial force functions, $M^{(1)}$ and $N^{(1)}$, as $M^{(0)} = N^{(0)} = 0$, on account of the absence of external loads of a mechanical nature.

Equation (16.5) yields

Table 16.6

Beam	$M^{(1)}$	$N^{(1)}$
AB	z	$\dfrac{1}{2}$
BC	$l+\dfrac{z}{2}$	1
CD	z	$-\dfrac{1}{2}$

$$\int_{BC} N^{(1)}\alpha\Delta T\,\mathrm{d}s + \int_{S} M^{(1)}\frac{XM^{(1)}}{EI}\,\mathrm{d}s = 1\times 0 \tag{16.55}$$

and on the basis of Table 16.6

$$2\alpha\Delta Tl + \frac{X}{EI}\left[\int_0^{2l} z^2\mathrm{d}z + \int_0^{2l}\left(l+\frac{z}{2}\right)^2\mathrm{d}z + \int_0^{l} z^2\mathrm{d}z\right] = 0 \tag{16.56}$$

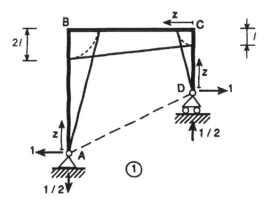

Figure 16.13

505

Evaluating the integrals, we obtain

$$X = -\frac{6}{23}\alpha\Delta T\frac{EI}{l^2}$$ (16.57)

which corresponds exactly to the shear value on the upright CD, obtained in the previous chapter.

As regards statically indeterminate structures having an inelastic constraint settlement, it is necessary to take into account the work performed by the fictitious constraint reaction corresponding to the settlement itself. For example, the portal frame of Figure 15.20 (a) can be rendered statically determinate by replacing the hinge D with a horizontally moving roller support (Figure 16.14). In this case equation (16.5) yields

$$1\times 0 + 0\times\eta_0 = \frac{X}{EI}\int_S (M^{(1)})^2\,ds$$ (16.58)

whence it emerges that $X = 0$; i.e. the structure is not subject to internal reactions, since the displacement η_0 of the point D can be produced by a simple rigid rotation of the portal frame about the hinge A (Figure 15.20 (e)).

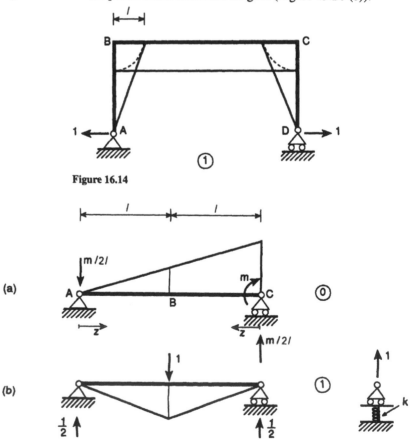

Figure 16.14

(a)

(b)

Figure 16.15

As regards the statically indeterminate structures with an elastic constraint settlement, it was shown in the introduction to this chapter that it is necessary to take into account the work done by the fictitious constraint reaction acting through the settlement caused by the real constraint reaction.

Consider again the continuous beam on an elastic support of Figure 13.15 (a). Scheme 0 consists of the beam AC, loaded by the moment m at the end C (Figure 16.15 (a)), whilst Scheme 1 consists of the same beam with a unit load acting in the centre (Figure 16.15 (b)). The moment functions are given in Table 16.7.

Table 16.7

Beam	$M^{(0)}$	$M^{(1)}$
AB	$-\dfrac{m}{2l}z$	$\dfrac{z}{2}$
BC	$-m+\dfrac{m}{2l}z$	$\dfrac{z}{2}$

Using formula (16.29), we have

$$X = -\frac{\int_S M^{(0)}M^{(1)}\mathrm{d}s}{\int_S (M^{(1)})^2\,\mathrm{d}s + \dfrac{EI}{k}} \qquad (16.59)$$

where $k = EA/h$, and

$$\int_S M^{(0)}M^{(1)}\mathrm{d}s = \int_0^l \left(-\frac{m}{2l}z\right)\left(\frac{z}{2}\right)\mathrm{d}z + \qquad (16.60\text{a})$$

$$\int_0^l \left(-m+\frac{m}{2l}z\right)\frac{z}{2}\,\mathrm{d}z = -\frac{1}{4}ml^2$$

$$\int_S (M^{(1)})^2\,\mathrm{d}s = \int_0^l \frac{z^2}{4}\,\mathrm{d}z + \int_0^l \frac{z^2}{4}\,\mathrm{d}z = \frac{l^3}{6} \qquad (16.60\text{b})$$

Performing the calculation, we obtain

$$X = \frac{\dfrac{1}{4}ml^2}{\dfrac{l^3}{6}+\dfrac{EI}{k}} \qquad (16.61)$$

The two limit cases of an infinitely compliant support and a perfectly rigid support, present, respectively, the following vertical reactions V_B:

$$\lim_{k\to 0} X = 0 \qquad (16.62\text{a})$$

$$\lim_{k\to\infty} X = \frac{3}{2}\frac{m}{l} \qquad (16.62\text{b})$$

Once again we find the reaction of the central support, already determined following another procedure (Figure 13.14).

16.6 Statically indeterminate truss structures

In the case of statically indeterminate truss structures, the application of the Principle of Virtual Work constitutes a highly valid and often rapid method of resolution. The equivalent statically determinate structure is obtained by subtracting a number of bars equal to the degree of redundancy of the structure. These bars, once isolated, must be considered axially compliant under the action of the corresponding redundant reaction.

Let us take as an example the truss structure of Figure 16.16 (a), subjected to a temperature increase ΔT on the bar CE. When the bar has been isolated

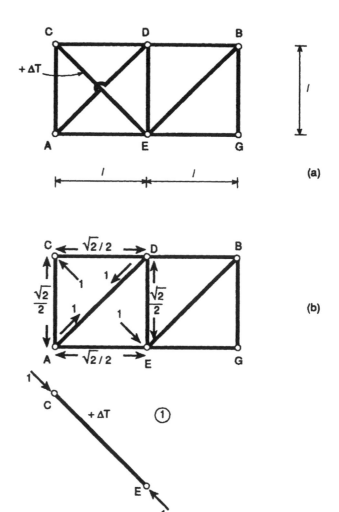

Figure 16.16

and both the bar and the equivalent statically determinate structure have been subjected to the unit fictitious reaction (Figure 16.16 (b)), equation (16.5) becomes

$$1 \times \Delta\eta^{(r)} = \sum_i \frac{N_i^{(f)} N_i^{(r)}}{EA} l_i \qquad (16.63)$$

From the scheme of the bar CE we have, on the other hand,

$$\Delta\eta^{(r)} = \left(\alpha\Delta T - \frac{X}{EA} \right) l\sqrt{2} \qquad (16.64)$$

508

whereby the equation that provides the solution becomes

$$X\left(\sum_i \frac{(N_i^{(1)})^2}{EA} l_i\right) = \left(\alpha\Delta T - \frac{X}{EA}\right) l\sqrt{2} \qquad (16.65)$$

Since

$$N_{AC}^{(1)} = N_{CD}^{(1)} = N_{DE}^{(1)} = N_{AE}^{(1)} = \frac{\sqrt{2}}{2} \qquad (16.66a)$$

$$N_{AD}^{(1)} = -1 \qquad (16.66b)$$

we obtain finally

$$\frac{X}{EA}\left[4\times\left(\frac{\sqrt{2}}{2}\right)^2 \times l + 1 \times l\sqrt{2}\right] = \left(\alpha\Delta T - \frac{X}{EA}\right) l\sqrt{2} \qquad (16.67)$$

and hence

$$X = \frac{2 - \sqrt{2}}{2} \alpha\Delta TEA \qquad (16.68)$$

As our second example, let us examine the truss structure of Figure 16.17(a), where the upper chord is subjected to a uniform temperature increase ΔT. The bars ED and DC are unloaded by virtue of the equilibrium of the node D, so that the scheme providing the solution reduces to that of Figure 16.17(b), where the bar AC has been isolated with respect to the rest of the structure. From the equilibrium of the nodes C and B (Figure 16.17(c)), the following condition results:

$$N_{AB}^{(1)} = N_{BC}^{(1)} = N_{CE}^{(1)} = N_{EA}^{(1)} = -\frac{\sqrt{3}}{3} \qquad (16.69a)$$

$$N_{BE}^{(1)} = \frac{\sqrt{3}}{3} \qquad (16.69b)$$

Since, by virtue of the fact that the load is of thermal origin, we have $N^{(0)} = 0$, applying the Principle of Virtual Work gives

$$-1 \times \frac{Xl\sqrt{3}}{EA} = \sum_i N_i^{(1)}\left(\frac{XN_i^{(1)}l}{EA} + \alpha\Delta T_i l\right) \qquad (16.70)$$

and hence, rendering the terms of the summation explicit

$$-\frac{Xl\sqrt{3}}{EA} = \frac{5}{3}\frac{Xl}{EA} - \frac{\sqrt{3}}{3} \alpha\Delta Tl \qquad (16.71)$$

whence we obtain the unknown reaction N_{AC}

$$X = \frac{9 - 5\sqrt{3}}{2} \alpha\Delta TEA \qquad (16.72)$$

509

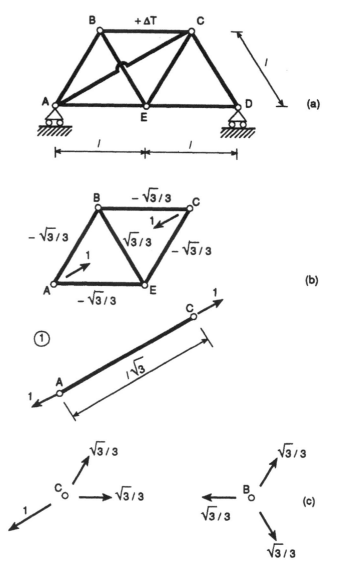

Figure 16.17

The static indeterminacy of the truss structure of Figure 16.18(a) is due to the continuity of the lower chord. The equivalent statically determinate structure is obtained by inserting a complete hinge at D (Figure 16.18(b)) and applying the redundant moment X. The fictitious structure of Scheme 1 proves to be loaded in a symmetrical manner by two unit moments (Figure 16.18(c)). By virtue of the equilibrium of the beams DC and DE, as well as of the nodes A and D (Figure 16.18 (d)), we have

$$N_{AC}^{(1)} = N_{AB}^{(1)} = \frac{2}{l\sqrt{3}} \qquad (16.73\text{a})$$

510

$$N_{AD}^{(1)} = -\frac{2}{l\sqrt{3}} \qquad (16.73b)$$

$$N_{CD}^{(1)} = -\frac{1}{l\sqrt{3}} \qquad (16.73c)$$

Structure of Scheme 0 (Figure 16.18(e)) proves, on the other hand, to be loaded only on the external bars:

$$N_{AC}^{(0)} = -\frac{2F}{\sqrt{3}} \qquad (16.74a)$$

$$N_{CD}^{(0)} = \frac{F}{\sqrt{3}} \qquad (16.74b)$$

$$N_{AB}^{(0)} = -\frac{F}{\sqrt{3}} \qquad (16.74c)$$

$$N_{AD}^{(0)} = 0 \qquad (16.74d)$$

The schemes of equilibrium of the nodes C and A are represented in Figure 16.18(f).

Application of the Principle of Virtual Work leads to the following equation:

$$1 \times 0 = \frac{l}{EA}\left[2N_{AC}^{(0)}N_{AC}^{(1)} + 2N_{CD}^{(0)}N_{CD}^{(1)} + N_{AB}^{(0)}N_{AB}^{(1)}\right] + \qquad (16.75)$$

$$\frac{Xl}{EA}\left[2\left(N_{AC}^{(1)}\right)^2 + 2\left(N_{CD}^{(1)}\right)^2 + \left(N_{AB}^{(1)}\right)^2 + 2\left(N_{AD}^{(1)}\right)^2\right] +$$

$$\frac{X}{EI}\left[2\int_0^l \left(M_{CD}^{(1)}\right)^2 dz\right]$$

which allows the determination of the redundant unknown X. We thus have

$$M_{CD}^{(1)} = \frac{z}{l} \qquad (16.76)$$

whence

$$\int_0^l \left(M_{CD}^{(1)}\right)^2 dz = \frac{l}{3} \qquad (16.77)$$

and hence, substituting equations (16.73), (16.74) and (16.77) into equation (16.75)

$$\frac{l}{EA}\left[-\frac{8F}{3l} - \frac{2F}{3l} - \frac{2F}{3l}\right] + \qquad (16.78)$$

$$\frac{Xl}{EA}\left[\frac{8}{3l^2} + \frac{2}{3l^2} + \frac{4}{3l^2} + \frac{8}{3l^2}\right] +$$

$$\frac{X}{EI}\left[\frac{2}{3}l\right] = 0$$

511

Finally we obtain

$$X = \frac{6Fl}{11 + \left(\dfrac{l}{\rho}\right)^2} \tag{16.79}$$

where ρ denotes the radius of gyration of the cross section of the bars.

Finally, consider the closed structure of Figure 16.19(a), stiffened by a diagonal cross. In order to obtain the equivalent statically determinate structure, let

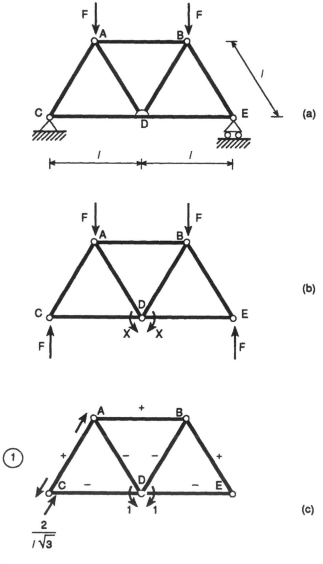

(a)

(b)

(c)

Figure 16.18a, b, c

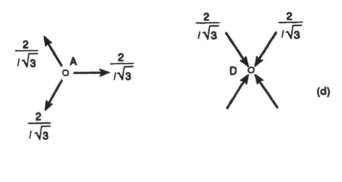

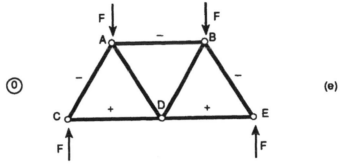

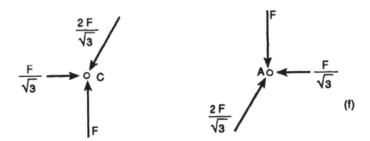

Figure 16.18d, e, f

the cross be isolated and subjected to the action of the unknown axial force X, just as the square framework is subjected to equal and opposite loads (Figure 16.19(b)). The system of Scheme 0 is subjected to a compressive axial force on the beams AB and CD, and to a bending moment on the beams AD and BC (Figure 16.19(c)). The system of Scheme 1 is subjected to a tensile axial force on all the beams (Figure 16.19(d)).

The Principle of Virtual Work yields the following condition:

$$-1 \times \frac{Xl\sqrt{2}/2}{EA} = \int_{S/4} \frac{M^{(f)}M^{(r)}}{EI}\,ds + \int_{S/4} \frac{N^{(f)}N^{(r)}}{EA}\,ds \qquad (16.80)$$

513

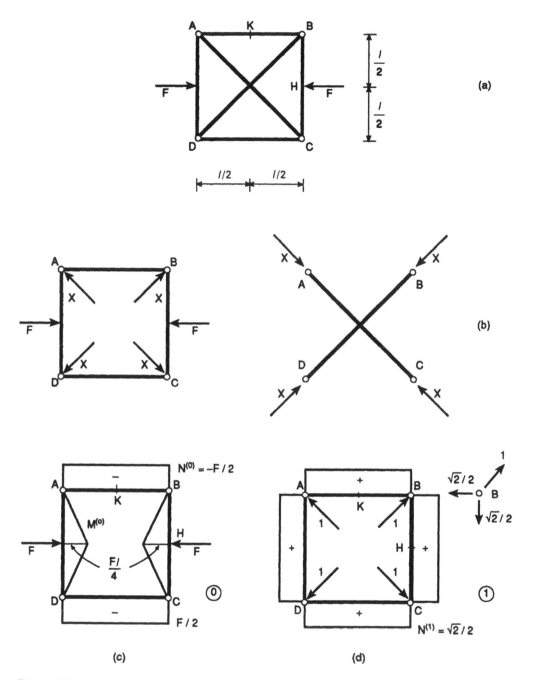

Figure 16.19

514

with

$$M^{(f)} = 0 \tag{16.81a}$$

$$N^{(f)} = N^{(1)} = \frac{\sqrt{2}}{2} \tag{16.81b}$$

$$M_{BH}^{(r)} = \frac{F}{2}z \tag{16.81c}$$

$$M_{BK}^{(r)} = 0 \tag{16.81d}$$

$$N_{BH}^{(r)} = X\frac{\sqrt{2}}{2} \tag{16.81e}$$

$$N_{BK}^{(r)} = -\frac{F}{2} + X\frac{\sqrt{2}}{2} \tag{16.81f}$$

Equation (16.80) thus takes the form

$$-\frac{Xl\sqrt{2}}{2EA} = \frac{l}{2EA}\left(\frac{X}{2} + \frac{X}{2} - \frac{F\sqrt{2}}{4}\right) \tag{16.82}$$

whence it follows that

$$X = \frac{F(2-\sqrt{2})}{4} \tag{16.83}$$

The axial forces in the individual bars are therefore

$$N_{AB} = N_{CD} = -\frac{F}{2} + \frac{F(2-\sqrt{2})}{4}\frac{\sqrt{2}}{2} = F\frac{\sqrt{2}-3}{4} \tag{16.84a}$$

$$N_{AD} = N_{BC} = \frac{F}{4}(\sqrt{2}-1) \tag{16.84b}$$

$$N_{BD} = N_{AC} = -\frac{F}{4}(2-\sqrt{2}) \tag{16.84c}$$

16.7 Arches and rings

In the cases of **arches** and **rings**, and in general of **curvilinear beams**, just as in the previously considered case of truss structures, application of the Principle of Virtual Work proves to be a highly convenient method of solution.

Consider, for instance, the circular cantilever of Figure 16.20(a), subjected to a uniform temperature rise ΔT. Introduce three fictitious schemes, where the cantilever is loaded at the unconstrained end by

Scheme 1: a unit horizontal force (Figure 16.20 (b));
Scheme 2: a unit vertical force (Figure 16.20 (c));
Scheme 3: a unit couple (Figure 16.20 (d)).

515

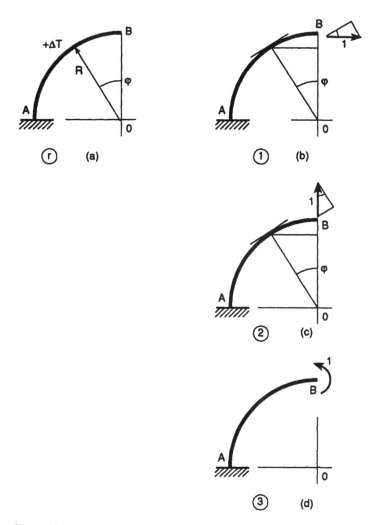

Figure 16.20

It is then possible to calculate the generalized displacements of that end. We have in fact

$$N^{(1)} = \cos\varphi \qquad (16.85a)$$

$$N^{(2)} = \sin\varphi \qquad (16.85b)$$

$$N^{(3)} = 0 \qquad (16.85c)$$

and hence

$$1 \times u_B = \int_S N^{(1)}\varepsilon_T \, ds \qquad (16.86a)$$

$$1 \times v_B = \int_S N^{(2)}\varepsilon_T \, ds \qquad (16.86b)$$

$$1 \times \varphi_B = \int_S N^{(3)}\varepsilon_T \, ds \qquad (16.86c)$$

516

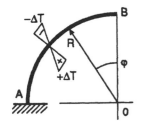

Figure 16.21

Substituting equations (16.85) into equations (16.86), we obtain

$$u_B = v_B = R\alpha\Delta T, \quad \varphi_B = 0 \tag{16.87}$$

If the same cantilever beam is subjected to a butterfly-shaped thermal varia-tion (Figure 16.21), the previously considered fictitious schemes yield the fic-titious moments

$$M^{(1)} = R(\cos\varphi - 1) \tag{16.88a}$$

$$M^{(2)} = R\sin\varphi \tag{16.88b}$$

$$M^{(3)} = 1 \tag{16.88c}$$

and hence the application of the Principle of Virtual Work to each of the three schemes leads to the determination of the displacements of the end B:

$$1 \times u_B = \int_S M^{(1)}\chi_T \, ds \tag{16.89a}$$

$$1 \times v_B = \int_S M^{(2)}\chi_T \, ds \tag{16.89b}$$

$$1 \times \varphi_B = \int_S M^{(3)}\chi_T \, ds \tag{16.89c}$$

From equations (16.88) and (16.89) we obtain

$$u_B = 2\alpha\frac{\Delta T}{h}\int_0^{\pi/2} R^2(\cos\varphi - 1)\,d\varphi \tag{16.90a}$$

$$v_B = 2\alpha\frac{\Delta T}{h}\int_0^{\pi/2} R^2\sin\varphi\,d\varphi \tag{16.90b}$$

$$\varphi_B = 2\alpha\frac{\Delta T}{h}\int_0^{\pi/2} R\,d\varphi \tag{16.90c}$$

and hence

$$u_B = \alpha\Delta T\frac{R^2}{h}(2 - \pi) \tag{16.91a}$$

$$v_B = 2\alpha\Delta T\frac{R^2}{h} \tag{16.91b}$$

$$\varphi_B = \pi\alpha\Delta T\frac{R}{h} \tag{16.91c}$$

To determine the relative horizontal displacement of the ends of the discon-nected ring of Figure 16.22(a), it is sufficient to consider the fictitious scheme of Figure 16.22(b), so that

$$M^{(f)} = M^{(1)} = R(1 - \cos\varphi) \tag{16.92a}$$

$$M^{(r)} = FM^{(1)} = FR(1 - \cos\varphi) \tag{16.92b}$$

Application of the Principle of Virtual Work,

$$1 \times \frac{\Delta u}{2} = \int_{S/2} \frac{M^{(f)}M^{(r)}}{EI}\,ds \tag{16.93}$$

517

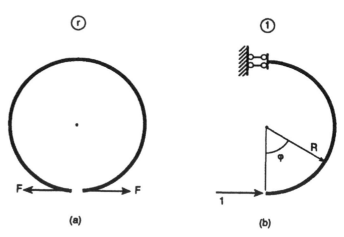

Figure 16.22

on the basis of equations (16.92), yields the following condition:

$$\frac{\Delta u}{2} = \frac{FR^3}{EI} \int_0^\pi (1 - \cos\varphi)^2 \, d\varphi \tag{16.94}$$

from which we find

$$\Delta u = 3\pi \frac{FR^3}{EI} \tag{16.95}$$

To determine the relative displacement of the ends of the disconnected ring of Figure 16.23(a), two fictitious schemes must be used, one with the unit force horizontal (Figure 16.22(b)), and the other with the unit force vertical (Figure 16.23(b)). For the latter scheme we have

$$M^{(f)} = M^{(2)} = R \sin\varphi \tag{16.96a}$$

$$M^{(r)} = FM^{(2)} = FR \sin\varphi \tag{16.96b}$$

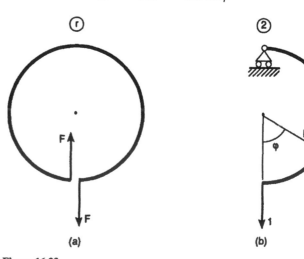

Figure 16.23

and hence the discontinuity of vertical displacement may be deduced from the following equation:

$$1 \times \frac{\Delta v}{2} = \int_{S/2} \frac{M^{(f)}M^{(r)}}{EI} \, ds \qquad (16.97)$$

which, on the basis of equations (16.96), gives

$$\frac{\Delta v}{2} = \frac{FR^3}{EI} \int_0^\pi \sin^2 \varphi \, d\varphi \qquad (16.98)$$

and hence

$$\Delta v = \pi \frac{FR^3}{EI} \qquad (16.99)$$

The discontinuity of horizontal displacement is zero owing to skew symmetry. The absolute displacement may be deduced from the application of the Principle of Virtual Work, considering expression (16.96b) as the real moment and expression (16.92a) as the fictitious moment,

$$1 \times u = \int_{S/2} \frac{FM^{(1)}M^{(2)}}{EI} \, ds \qquad (16.100)$$

from which we obtain

$$u = \frac{FR^3}{EI} \int_0^\pi \sin\varphi(1 - \cos\varphi)d\varphi \qquad (16.101)$$

and hence

$$u = 2\frac{FR^3}{EI} \qquad (16.102)$$

Now consider the statically indeterminate ring of Figure 16.24(a), in which the internal connecting rod undergoes a temperature rise ΔT. By virtue of double symmetry, the ring reduces to the quarter of circumference of Figure 16.24(b), in which the rod has been isolated and subjected to the redundant reaction $2X_1$. The equivalent statically determinate structure can hence appear as in Figure 16.24(c), where the quarter of circumference is subjected also to the second redundant unknown, the moment X_2. The two fictitious structures are represented in Figures 16.24(d), (e). The real moment is equal to

$$M^{(r)} = X_1 M^{(1)} + X_2 M^{(2)} \qquad (16.103a)$$

where

$$M^{(1)} = R \sin \varphi \qquad (16.103b)$$

$$M^{(2)} = 1 \qquad (16.103c)$$

The first equation of congruence for the connecting rod is written

$$1 \times \left(-\frac{2X_1 R}{EA} + \alpha \Delta TR \right) = \int_0^{\pi/2} M^{(1)} \frac{M^{(r)}}{EI} R \, d\varphi \qquad (16.104)$$

519

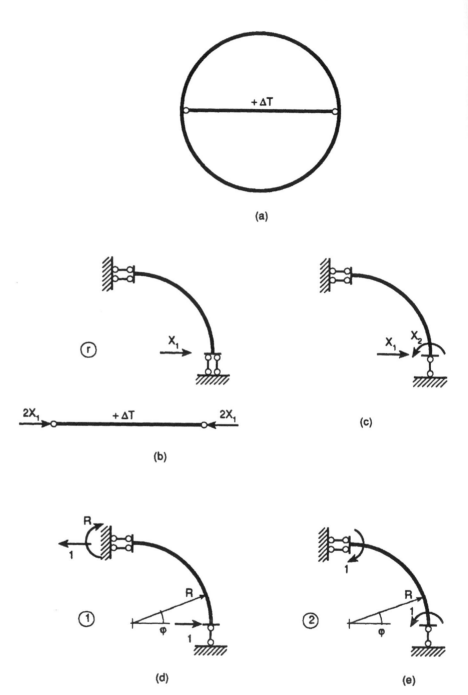

(a)

(b)

(c)

(d)

(e)

Figure 16.24

Performing the calculations, we obtain

$$X_1\left(2\rho^2 + \frac{\pi R^2}{4}\right) + X_2 R = EI\alpha\Delta T \qquad (16.105)$$

Since the radius of gyration of the cross section is much smaller than the dimension R, equation (16.105) simplifies as follows:

$$\pi R^2 X_1 + 4RX_2 = 4EI\alpha\Delta T \qquad (16.106)$$

The second equation of congruence, corresponding to the rotation, is written

$$1 \times 0 = \int_0^{\pi/2} M^{(2)} \frac{M^{(r)}}{EI} R\,\mathrm{d}\varphi \qquad (16.107)$$

From equations (16.103) we have

$$\int_0^{\pi/2} (X_1 R \sin\varphi + X_2)\,\mathrm{d}\varphi = 0 \qquad (16.108)$$

and hence

$$2X_1 R + \pi X_2 = 0 \qquad (16.109)$$

From equations (16.106) and (16.109) we obtain the axial force of the rod

$$X_1 = \frac{4\pi EI\alpha\Delta T}{R^2(\pi^2 - 8)} \qquad (16.110a)$$

as well as the redundant moment

$$X_2 = -\frac{2R}{\pi} X_1 \qquad (16.110b)$$

The statically indeterminate ring of Figure 16.25(a) is studied in the same way as for the preceding one. If we imagine turning the scheme by 90°, it is possible to use the fictitious systems of Schemes 1 and 2 of the previous example (Figures 16.24(d), (e)), whilst Scheme 0 is represented in Figure 16.25(b),

$$M^{(0)} = \frac{FR}{2}(1 - \cos\varphi) \qquad (16.111)$$

521

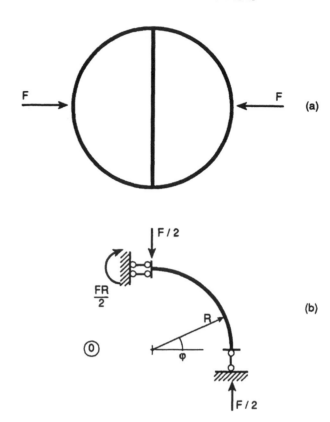

Figure 16.25

The equations of congruence can be written like equations (16.41), with

$$\int_S (M^{(1)})^2 \, ds = R^3 \int_0^{\pi/2} \sin^2 \varphi \, d\varphi = \frac{\pi}{4} R^3 \tag{16.112a}$$

$$\int_S M^{(1)} M^{(2)} ds = R^2 \int_0^{\pi/2} \sin\varphi \, d\varphi = R^2 \tag{16.112b}$$

$$\int_S (M^{(2)})^2 \, ds = R \int_0^{\pi/2} d\varphi = \frac{\pi}{2} R \tag{16.112c}$$

$$\int_S M^{(1)} M^{(0)} ds = \frac{FR^3}{2} \int_0^{\pi/2} \sin \varphi (1 - \cos\varphi) \, d\varphi = \frac{FR^3}{4} \tag{16.112d}$$

$$\int_S M^{(2)} M^{(0)} ds = \frac{FR^2}{2} \int_0^{\pi/2} (1 - \cos\varphi) \, d\varphi = \frac{FR^2}{4}(\pi - 2) \tag{16.112e}$$

We have therefore

$$\frac{\pi}{4} R^3 X_1 + R^2 X_2 = -\frac{FR^3}{4} \tag{16.113a}$$

$$R^2 X_1 + \frac{\pi}{2} RX_2 = -\frac{FR^2}{4}(\pi - 2) \tag{16.113b}$$

522

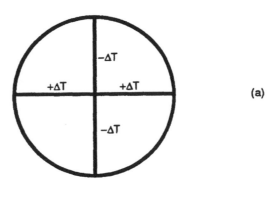

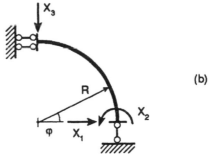

Figure 16.26

From equations (16.44) we have finally

$$X_1 = F\frac{4 - \pi}{8 - \pi^2} \tag{16.114a}$$

$$X_2 = FR\frac{\pi^2 - 2\pi - 4}{2(8 - \pi^2)} \tag{16.114b}$$

Also the statically indeterminate ring of Figure 16.26 (a), the orthogonal diaphragms of which are respectively heated and cooled, can be analysed using the method illustrated previously. In this case (Figure 16.26 (b))

$$M^{(0)} = 0 \tag{16.115a}$$

$$M^{(1)} = R \sin \varphi \tag{16.115b}$$

$$M^{(2)} = 1 \tag{16.115c}$$

$$M^{(3)} = R(1 - \cos\varphi) \tag{16.115d}$$

The ring of Figure 16.27 (a) is subjected to a butterfly-shaped thermal variation. The equivalent statically determinate structure is represented in Figure 16.27 (b), with

$$M^{(0)} = 0, \quad M^{(1)} = 1 \tag{16.116}$$

The real moment is therefore

$$M^{(r)} = M^{(0)} + XM^{(1)} = X \tag{16.117a}$$

523

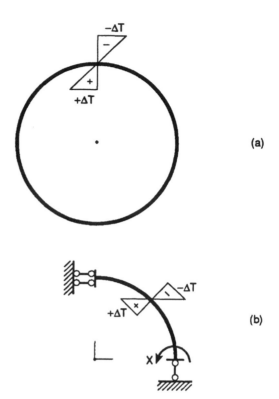

Figure 16.27

so that the real thermoelastic curvature is

$$\chi^{(r)} = \frac{X}{EI} + 2\alpha \frac{\Delta T}{h} \qquad (16.117b)$$

The Principle of Virtual Work yields

$$1 \times 0 = \int_{S/4} M^{(1)} \chi^{(r)} ds \qquad (16.118)$$

from which we obtain

$$\left(\frac{X}{EI} + 2\alpha \frac{\Delta T}{h} \right) \frac{\pi R}{2} = 0 \qquad (16.119)$$

and hence

$$X = -2 \ \alpha \Delta T \frac{EI}{h} \qquad (16.120)$$

Consider again the ring of radius R, loaded by three angularly equidistant radial forces (Figure 16.28 (a)). The equivalent statically determinate structure

524

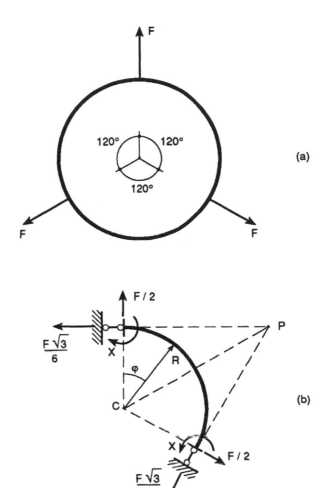

Figure 16.28

is represented in Figure 16.28 (b), so that the moments corresponding to Schemes 0 and 1 are the following:

$$M^{(0)} = \frac{FR}{2} \sin \varphi - \frac{F\sqrt{3}}{6} R(1 - \cos \varphi) \qquad (16.121a)$$

$$M^{(1)} = 1 \qquad (16.121b)$$

Angular congruence imposes

$$1 \times 0 = \int_{S/3} M^{(1)} \frac{M^{(0)} + XM^{(1)}}{EI} \, ds \qquad (16.122)$$

525

and hence

$$\int_0^{2\pi/3} \left[\frac{FR}{2} \sin \varphi - \frac{F\sqrt{3}}{6} R(1 - \cos \varphi) + X \right] R d\varphi = 0 \qquad (16.123)$$

whence we obtain

$$X = -FR \frac{9 - \pi\sqrt{3}}{6\pi} \qquad (16.124)$$

Let the ring of Figure 16.29(a) be subjected to the stresses produced by the thermal dilatation of the two orthogonal diaphragms. The equivalent statically determinate structure is represented in Figure 16.29(b). From symmetry it can be reduced to the fictitious system of Scheme 1 of Figure 16.29(c). We have therefore

$$M^{(0)} = 0 \qquad (16.125a)$$

$$M^{(1)} = \frac{R}{2}(\sin\varphi + \cos\varphi - 1) \qquad (16.125b)$$

Application of the Principle of Virtual Work yields

$$2 \times \frac{1}{2} \times \left(\alpha \Delta T R - \frac{XR}{EA} \right) = \int_{S/4} X \frac{(M^{(1)})^2}{EI} ds \qquad (16.126)$$

from which we obtain

$$\alpha \Delta T R - \frac{XR}{EA} = \frac{XR^3}{4EI} \int_0^{\pi/2} (\sin\varphi + \cos\varphi - 1)^2 d\varphi \qquad (16.127)$$

and hence

$$X = \frac{4 \, \alpha \Delta T E I}{R^2(\pi + 1) + 4\rho^2} \qquad (16.128)$$

Finally let us consider the arch of Figure 16.30(a). The equivalent statically determinate structure of Figure 16.30(b) presents at its lower end an elastically compliant roller support which simulates the lateral cantilever on which the arch rests. Schemes 0 and 1 give the moments (Figure 16.30(c))

$$M^{(0)} = \frac{FR}{2}(1 - \cos\varphi) \qquad (16.129a)$$

$$M^{(1)} = -R \sin\varphi \qquad (16.129b)$$

and thus we have

$$1 \times 0 = \int_{S/2} M^{(1)} \frac{M^{(0)} + XM^{(1)}}{EI} ds \qquad (16.130)$$

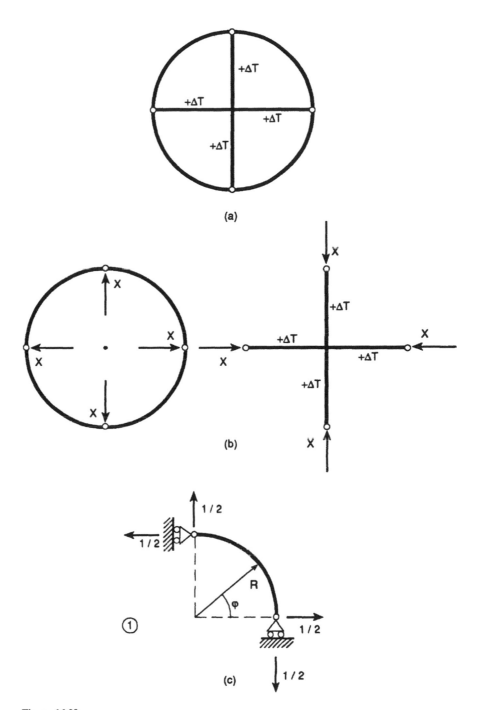

Figure 16.29

527

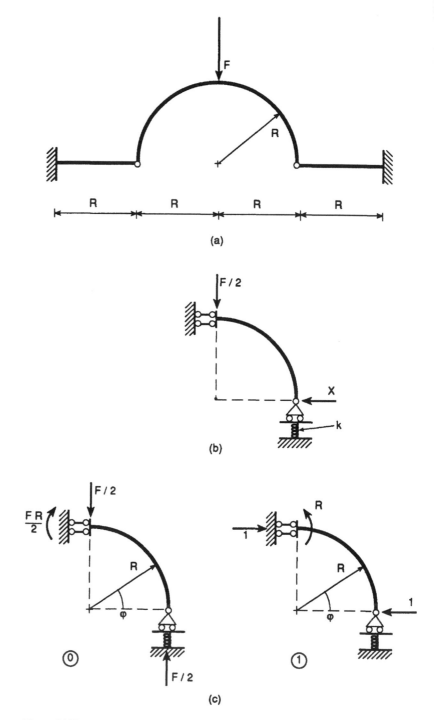

Figure 16.30

Since

$$\int_{S/2} M^{(0)} M^{(1)} \mathrm{d}s = -\frac{FR^3}{4} \qquad (16.131\text{a})$$

$$\int_{S/2} (M^{(1)})^2 \, \mathrm{d}s = R^3 \frac{\pi}{4} \qquad (16.131\text{b})$$

we obtain finally $X = F/\pi$. Note how the stiffness k of the spring does not appear in the solution.

16.8 Castigliano's Theorem

In the case where thermal distortions and constraint settlements are absent, the determination of elastic displacements in statically determinate structures can be made using **Castigliano's Theorem**, as an alternative to the application of the Principle of Virtual Work proposed in Section 16.2.

Consider a statically determinate structure subjected to n different loads F_1, F_2, ..., F_n (Figure 16.31). The Principle of Superposition makes it possible to express the n generalized displacements dual of the forces, η_1, η_2, ..., η_n, via the coefficients of influence η_{ij}:

$$\eta_i = \sum_{j=1}^{n} \eta_{ij} F_j, \qquad \text{for } i = 1, 2, ..., n \qquad (16.132)$$

Clapeyron's Theorem then gives the strain energy of the structure in the form

$$L_{\text{def}} = \frac{1}{2} \sum_{i=1}^{n} F_i \eta_i \qquad (16.133\text{a})$$

which, taking into account equations (16.132), becomes

$$L_{\text{def}} = \frac{1}{2} \sum_{i=1}^{n} \sum_{j=1}^{n} F_i F_j \eta_{ij} \qquad (16.133\text{b})$$

Finally, deriving the strain energy with respect to each force F_i, $i = 1, 2, ..., n$, we have

$$\frac{\partial L_{\text{def}}}{\partial F_i} = \sum_{j=1}^{n} \eta_{ij} F_j, \qquad \text{for } i = 1, 2, ..., n \qquad (16.134)$$

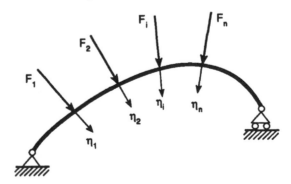

Figure 16.31

529

and hence, on the basis of equation (16.132), we find

$$\frac{\partial L_{\text{def}}}{\partial F_i} = \eta_i, \text{ for } i = 1, 2, \ldots, n \tag{16.135}$$

Relation (16.135) represents the statement of Castigliano's Theorem: **the derivative of the strain energy with respect to the magnitude of an applied force is equal to the global elastic displacement, dual with respect to the same force.**

In the case where one wishes to calculate a generic displacement which does not correspond to an applied force, it is possible to apply a fictitious force, corresponding to the displacement sought, and, once the partial derivative of the work has been obtained, to make the magnitude of the above force tend to zero.

It is possible to demonstrate the perfect equivalence of Castigliano's Theorem with the Principle of Virtual Work. In fact, in the case where axial force and shearing force give a negligible contribution to the strain energy, we have

$$L_{\text{def}} = \frac{1}{2} \int_S \frac{M^2}{EI} \, ds \tag{16.136}$$

and hence, on the basis of equation (16.135)

$$\eta_i = \frac{1}{2EI} \frac{\partial}{\partial F_i} \int_S M^2 ds \tag{16.137}$$

Carrying the differential operator under the integral sign, we obtain

$$\eta_i = \frac{1}{EI} \int_S M \frac{\partial M}{\partial F_i} \, ds \tag{16.138}$$

Since then the real moment M may be interpreted as the sum of n partial moments, each generated by the generic force F_i

$$M = M^{(r)} = \sum_{i=1}^{n} F_i M^{(i)} \tag{16.139}$$

equation (16.138) becomes

$$\eta_i = \frac{1}{EI} \int_S M^{(r)} M^{(i)} ds \tag{16.140}$$

in which relation we can recognize the equation of the Principle of Virtual Work, amply discussed and applied hitherto.

16.9 Menabrea's Theorem

Menabrea's Theorem may be derived directly from Castigliano's Theorem, even though originally the two theorems were demonstrated independently.

Menabrea's Theorem is also called the **Theorem of Minimum Strain Energy** and refers to redundant structures (Figure 16.32). It states in fact that, **given a structure with n degrees of redundancy, the n values of the redundant unknowns make the strain energy of the structure a minimum.**

530

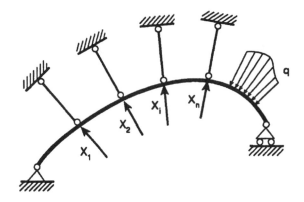

Figure 16.32

Considering the equivalent statically determinate structure and the n redundant unknowns $X_1, X_2, ..., X_n$, Castigliano's Theorem gives

$$\eta_i = \frac{\partial L_{def}}{\partial X_i}, \quad \text{for } i = 1, 2, ..., n \qquad (16.141a)$$

while the conditions of congruence, in the case where all the constraints are rigid, are

$$\eta_i = 0, \quad \text{for } i = 1, 2, ..., n \qquad (16.141b)$$

By the transitive law we obtain therefore

$$\frac{\partial}{\partial X_i} L_{def}(X_1, X_2, ..., X_n) = 0, \quad \text{for } i = 1, 2, ..., n \qquad (16.142)$$

The foregoing equation confirms the statement of Menabrea's Theorem, L_{def} being a positive definite quadratic function in the variables X_i.

531

17 Instability of elastic equilibrium

17.1 Introduction

The hypothesis of small displacements so far advanced considers the cardinal equations of statics (7.46) and (7.47) in relation to the undeformed structural configuration. In other words, the elastic displacements have been hypothesized as being so small as to make it possible for the deformed configuration to be confused with the undeformed one when the static characteristics are to be evaluated. In this chapter this hypothesis will be removed, and it will be shown how the solution of an elastic problem can represent in actual fact a condition of stable, neutral or unstable equilibrium, according to the magnitude of the load applied. Moreover, there exist, around the condition of neutral equilibrium, an infinite number of other similar conditions, characterized by different static parameters (applied loads) and kinematic parameters (configuration of the system).

The instability of elastic equilibrium occurs in general for slender structural elements subjected to compressive loads, such as columns of buildings, machine shafts, struts of trusses, thin arches and shells, and cylindrical and spherical shells subjected to external pressure. But also other cases, which are more complex, as regards both their geometry and the loading conditions, can equally be considered. It will suffice to think of the lateral torsional buckling of beams of thin rectangular cross section, where the disparity between the order of magnitude of the two central moments of inertia can cause, in a deflected beam, a sudden torsional deformation. The instability of elastic equilibrium is, moreover, a critical phenomenon that may affect an entire beam system, before it involves a particular element of the system. This occurs in the case of metal trussed and framed structures, which are frequently made up of extremely slender rods and beams.

The loss of stability of elastic equilibrium is commonly referred to as **buckling**. This is one of the three fundamental phenomena of structural collapse, the other two being **yielding** and **brittle fracturing**, which we shall discuss in the ensuing chapters. These phenomena do not in general occur separately, but interact during the phases of collapse. In this chapter we shall see how yielding can interact with buckling in the context of a transition from one to the other as the slenderness of the structure increases. In Chapter 20 we shall consider, instead, the interaction between yielding and brittle fracturing and the ductile–brittle transition as the size scale increases, the geometrical shape of the structure remaining the same.

17.2 Discrete mechanical systems with one degree of freedom

Let us consider the mechanical system of Figure 17.1(a) consisting of two rigid rods connected by an elastic hinge of rotational rigidity k, and constrained at one end by a hinge and at the other by a roller support. When the system is loaded with a horizontal force N and the absolute rotation φ of the

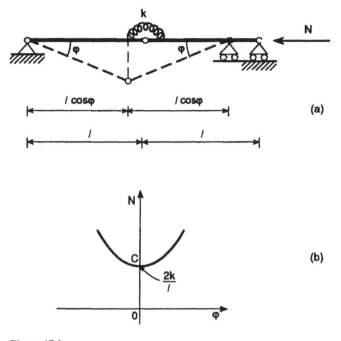

Figure 17.1

two arms is assumed as the generalized coordinate, the total potential energy of the whole system is

$$W(\varphi) = \frac{1}{2}k(2\varphi)^2 - 2Nl(1-\cos\varphi) \qquad (17.1)$$

The conditions of equilibrium are identified by imposing the stationarity of the function (17.1).

$$W'(\varphi) = 4k\varphi - 2Nl\,\sin\varphi = 0 \qquad (17.2)$$

from which we obtain the relation

$$N = \frac{2k\varphi}{l\sin\varphi} \qquad (17.3)$$

which links loading condition and deformed configuration along the branch of equilibrium presented in Figure 17.1(b). The plane N–φ is thus divided into two sectors by the curve of equation (17.3): the points of the upper sector represent conditions of instability, whilst those of the lower sector represent conditions of stability. Starting from the initial condition $\varphi = N = 0$, it will thus be possible to traverse in a stable manner the vertical segment of the axis N up to the point C ($\varphi = 0$, $N = N_c = 2k/l$), then to deviate onto one of the two branches of equilibrium of Figure 17.1(b). Alternatively, it would be possible to proceed along the vertical axis beyond point C of branching, although in this case the equilibrium is of an unstable type.

Note how the global behaviour of the system is then of a hardening type, the increase in deformation requiring a further increase in the external load. A

533

nonlinear and post-critical behaviour of this sort is said to be stable. To verify the stability of the post-critical branch, it is possible to consider the concavity of the potential as expressed by equation (17.1) and hence its derivatives of a higher order, calculated for $\varphi = 0$

$$W'(0) = W''(0) = W'''(0) = 0 \qquad (17.4a)$$

$$W^{IV}(0) = 4k > 0 \qquad (17.4b)$$

The determination of the critical load N_c can be made also via the simple **method of direct equilibrium**, i.e. by equating the destabilizing moment

$$M_i = Nl\sin\,\varphi \simeq Nl\varphi \qquad (17.5a)$$

and the stabilizing moment

$$M_s = 2k\varphi \qquad (17.5b)$$

Note that in relation (17.5a) recourse has been made to the hypothesis of linearized kinematics. This hypothesis simplifies the calculations, even though it prevents the definition of the post-critical behaviour.

Let us now consider the mechanical system of Figure 17.2(a), consisting of two rigid rods connected by a hinge, and constrained externally by a hinge and two roller supports, the intermediate one resting on an elastic foundation of rigidity k. The total potential energy is

$$W(\varphi) = \frac{1}{2}k(l\,\sin\varphi)^2 - 2Nl(1 - \cos\varphi) \qquad (17.6)$$

where the first term represents the potential energy of the spring, while the second represents the potential energy of the horizontal force N. The conditions of equilibrium are obtained by imposing the stationarity of the function (17.6):

$$W'(\varphi) = l\,\sin\varphi\,(kl\,\cos\varphi - 2N) = 0 \qquad (17.7)$$

from which we obtain (Figure 17.2(b))

$$N = \frac{kl}{2}\cos\varphi \qquad (17.8)$$

In this second example, the global behaviour of the system is of a softening type, a decrease in the external load corresponding to an increase in deformation. A post-critical behaviour of this sort is said to be unstable. To verify the instability of the post-critical branch it is possible to consider the convexity of the potential expressed by equation (17.6) and hence its derivatives of a higher order, calculated for $\varphi = 0$:

$$W'(0) = W''(0) = W'''(0) = 0 \qquad (17.9a)$$

$$W^{IV}(0) = -3kl^2 < 0 \qquad (17.9b)$$

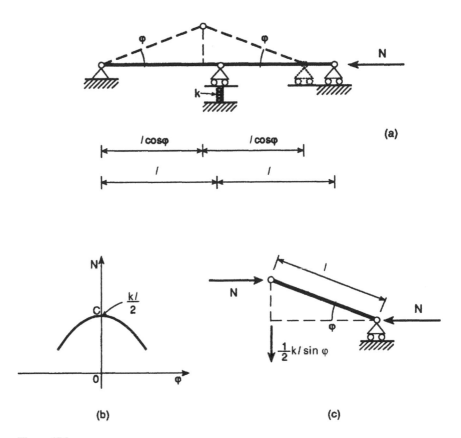

Figure 17.2

Also in this case, the determination of the critical load N_c may be made by equating the destabilizing moment acting on each rod and evaluated around the end supports (Figure 17.2(c))

$$M_i = Nl\sin\,\varphi \simeq Nl\varphi \qquad (17.10a)$$

and the stabilizing moment due to the reaction of the central support (Figure 17.2(c))

$$M_s = \frac{1}{2}kl^2\sin\varphi\,\cos\varphi \simeq \frac{1}{2}kl^2\,\varphi \qquad (17.10b)$$

17.3 Discrete mechanical systems with n degrees of freedom

Let us consider the mechanical system with two degrees of freedom of Figure 17.3(a), consisting of three rigid rods connected by two elastic hinges of rotational rigidity k, and constrained at one end by a hinge and at the other by a roller support. When the system is loaded with a horizontal force N and the vertical

535

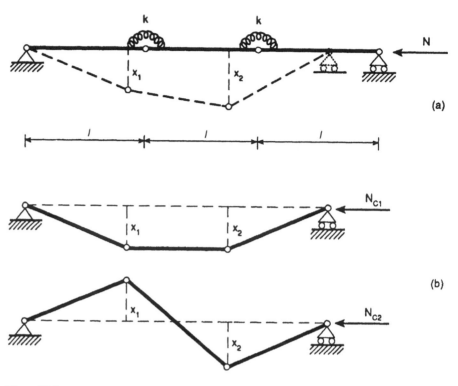

Figure 17.3

displacements x_1 and x_2 of the elastic hinges are assumed to be generalized coordinates, the total potential energy of the entire system is expressible as follows:

$$W(x_1, x_2) = \frac{1}{2}k\left[\left(\arcsin\frac{x_1}{l} - \arcsin\frac{x_2 - x_1}{l}\right)^2 + \right. \qquad (17.11)$$

$$\left(\arcsin\frac{x_2}{l} + \arcsin\frac{x_2 - x_1}{l}\right)^2\right] -$$

$$Nl\left[3 - \cos\left(\arcsin\frac{x_1}{l}\right) - \cos\left(\arcsin\frac{x_2}{l}\right) - \right.$$

$$\left.\cos\left(\arcsin\frac{x_2 - x_1}{l}\right)\right]$$

Performing a Taylor series expansion of equation (17.11) about the origin, we obtain

$$W(x_1, x_2) \simeq \frac{k}{2l^2}(5x_1^2 + 5x_2^2 - 8x_1x_2) - \qquad (17.12)$$

$$\frac{N}{l}(x_1^2 + x_2^2 - x_1x_2)$$

The conditions of equilibrium are identified by imposing the stationarity of the function (17.12):

$$\frac{\partial W}{\partial x_1} = x_1\left(\frac{5k}{l^2} - \frac{2N}{l}\right) - x_2\left(\frac{4k}{l^2} - \frac{N}{l}\right) = 0 \qquad (17.13a)$$

$$\frac{\partial W}{\partial x_2} = -x_1\left(\frac{4k}{l^2} - \frac{N}{l}\right) + x_2\left(\frac{5k}{l^2} - \frac{2N}{l}\right) = 0 \qquad (17.13b)$$

Equations (17.13) constitute a homogeneous system of linear algebraic equations and possess a solution different from the trivial one when the determinant of the coefficients is equal to zero:

$$\begin{vmatrix} \left(\frac{5k}{l^2} - \frac{2N}{l}\right) & -\left(\frac{4k}{l^2} - \frac{N}{l}\right) \\ -\left(\frac{4k}{l^2} - \frac{N}{l}\right) & \left(\frac{5k}{l^2} - \frac{2N}{l}\right) \end{vmatrix} = 0 \qquad (17.14)$$

Evaluating this determinant, we obtain a second-degree algebraic equation in N

$$\left(\frac{5k}{l^2} - \frac{2N}{l}\right)^2 - \left(\frac{4k}{l^2} - \frac{N}{l}\right)^2 = 0 \qquad (17.15)$$

and hence

$$N^2 - \frac{4k}{l}N + 3\frac{k^2}{l^2} = 0 \qquad (17.16)$$

which yields the two eigenvalues

$$N_{c1} = \frac{k}{l} \qquad (17.17a)$$

$$N_{c2} = 3\frac{k}{l} \qquad (17.17b)$$

From the system (17.13) we then obtain the corresponding eigenvectors

$$x_1 = x_2 \qquad (17.18a)$$
$$x_1 = -x_2 \qquad (17.18b)$$

The eigenvectors (17.18) represent the two modes of deformation corresponding to the two critical conditions and are shown, but for a factor of proportionality, in Figure 17.3(b).

To analyse the post-critical branch corresponding to the first eigenvector, the following changes of variable are useful:

$$x_1 = \varepsilon + y_1$$
$$x_2 = \varepsilon + y_2$$

537

where y_1 and y_2 are infinitesimals of a higher order with respect to the displacement ε, which may even be finite. The condition of equilibrium is identified by imposing stationarity of the function $W(\varepsilon+y_1, \varepsilon+y_2)$ and reconsidering the original expression (17.11):

$$N = \frac{k}{\varepsilon}\arcsin\frac{\varepsilon}{l} \qquad (17.19)$$

The expression (17.19), for $\varepsilon \to 0$, tends to the first eigenvalue k/l in accordance with equation (17.17a). Substituting equation (17.19) into equation (17.11) and computing the elements of the Hessian

$$[H] = \begin{bmatrix} \dfrac{\partial^2 W}{\partial y_1^2} & \dfrac{\partial^2 W}{\partial y_1 \partial y_2} \\[2ex] \dfrac{\partial^2 W}{\partial y_2 \partial y_1} & \dfrac{\partial^2 W}{\partial y_2^2} \end{bmatrix} \qquad (17.20)$$

we obtain

$$\frac{\partial^2 W}{\partial y_1^2} = \frac{k}{l^2}\left[3 + \frac{7}{6}\left(\frac{\varepsilon}{l}\right)^2 + \frac{137}{120}\left(\frac{\varepsilon}{l}\right)^4 + \frac{629}{560}\left(\frac{\varepsilon}{l}\right)^6 + \ldots\right] \qquad (17.21a)$$

$$\det[H] = \frac{k^2}{l^4}\left[2\left(\frac{\varepsilon}{l}\right)^2 + \frac{52}{15}\left(\frac{\varepsilon}{l}\right)^4 + \frac{6043}{1260}\left(\frac{\varepsilon}{l}\right)^6 + \ldots\right] \qquad (17.21b)$$

These two series expansions continue with the even powers only of ε/l and with all the coefficients positive. It may thus be concluded that, since the Hessian is positive definite, the post-critical branch is stable analogously to what has already been seen in the example of Figure 17.1.

The eigenvalue problem just illustrated can be solved rapidly, considering directly the equations of equilibrium with regard to rotation about the elastic hinges, in the framework of linearized kinematics (Figure 17.3(a)):

$$Nx_1 = k\left(\frac{x_1}{l} - \frac{x_2 - x_1}{l}\right) \qquad (17.22a)$$

$$Nx_2 = k\left(\frac{x_2}{l} + \frac{x_2 - x_1}{l}\right) \qquad (17.22b)$$

From equations (17.22) we find

$$x_1\left(N - \frac{2k}{l}\right) + \frac{k}{l}x_2 = 0 \qquad (17.23a)$$

$$\frac{k}{l}x_1 + x_2\left(N - \frac{2k}{l}\right) = 0 \qquad (17.23b)$$

538

Also in this case the homogeneous system (17.23) possesses other solutions besides the trivial one, when the determinant of the coefficients becomes zero:

$$\begin{vmatrix} \left(N - \dfrac{2k}{l}\right) & \dfrac{k}{l} \\[2ex] \dfrac{k}{l} & \left(N - \dfrac{2k}{l}\right) \end{vmatrix} = 0 \qquad (17.24)$$

Consequently we again obtain the characteristic equation (17.16).

Note how, for $N = 0$, the matrix (17.14) coincides with the stiffness matrix of the discrete system being considered (Figure 17.3(a)). The columns of the stiffness matrix are obtained in fact by setting one of the two generalized coordinates equal to unity and equating the other to zero (Figure 17.4). The elements of each individual column are furnished by the vertical forces which produce this situation. The vertical forces are, on the other hand, the loads dual with respect to the generalized coordinates chosen.

The physical meaning of the matrix (17.24) is not instead equally evident. For $N = 0$, it in fact represents the stiffness matrix of the system, when the reactive

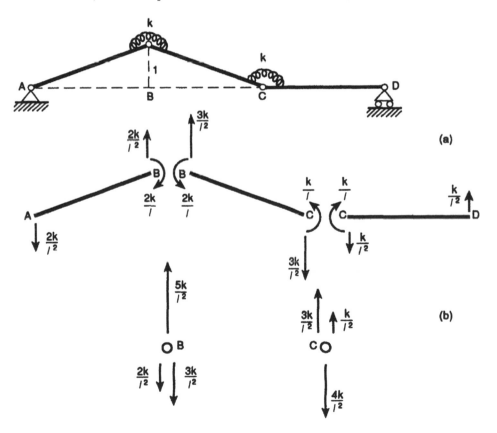

Figure 17.4

539

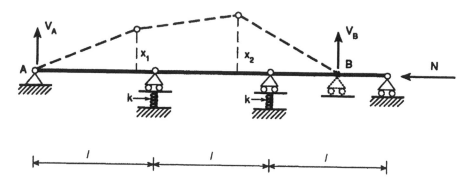

Figure 17.5

moments of the elastic hinges are assumed to be static characteristics. These moments are not, however, the loads dual to the coordinates x_1 and x_2.

As a second example of a system with two degrees of freedom, let us examine that of Figure 17.5, which consists of three rigid rods on four supports, of which the central ones are assumed to be elastically compliant with rigidity k. The total potential energy may be expressed as follows:

$$W(x_1, x_2) = \frac{1}{2}k(x_1^2 + x_2^2) - Nl\left[3 - \cos\left(\arcsin\frac{x_1}{l}\right) - \right. \tag{17.25}$$

$$\left. \cos\left(\arcsin\frac{x_2}{l}\right) - \cos\left(\arcsin\frac{x_2 - x_1}{l}\right)\right]$$

Expanding equation (17.25) into a Taylor series about the origin, we obtain

$$W(x_1, x_2) \simeq \frac{1}{2}k(x_1^2 + x_2^2) - \tag{17.26}$$

$$\frac{N}{l}(x_1^2 + x_2^2 - x_1 x_2)$$

The stationarity of the potential W requires that the two first partial derivatives be equal to zero:

$$\frac{\partial W}{\partial x_1} = x_1\left(k - \frac{2N}{l}\right) + \frac{N}{l}x_2 = 0 \tag{17.27a}$$

$$\frac{\partial W}{\partial x_2} = \frac{N}{l}x_1 + x_2\left(k - \frac{2N}{l}\right) = 0 \tag{17.27b}$$

Making the determinant of the coefficient matrix zero,

$$\begin{vmatrix} \left(k - \frac{2N}{l}\right) & \frac{N}{l} \\ \frac{N}{l} & \left(k - \frac{2N}{l}\right) \end{vmatrix} = 0 \tag{17.28}$$

yields the characteristic equation

$$\frac{3}{l^2} N^2 - \frac{4k}{l} N + k^2 = 0 \tag{17.29}$$

and hence the eigenvalues

$$N_{c1} = \frac{1}{3} kl \tag{17.30a}$$

$$N_{c2} = kl \tag{17.30b}$$

From the system of equations (17.27) the two corresponding eigenvectors are found

$$x_1 = -x_2 \tag{17.31a}$$

$$x_1 = x_2 \tag{17.31b}$$

which are the same, in reverse order, as in the case previously considered (Figure 17.3(b)).

To analyse the post-critical branch corresponding to the first eigenvector, the following changes of variable are useful:

$$x_1 = \varepsilon + y_1$$

$$x_2 = -\varepsilon + y_2$$

where y_1 and y_2 are infinitesimals of a higher order with respect to the displacement ε, which may even be finite. The condition of equilibrium is identified by imposing stationarity of the function $W(\varepsilon+y_1, -\varepsilon+y_2)$ and reconsidering the original expression (17.27):

$$N = kl \frac{\left[1-\left(\frac{\varepsilon}{l}\right)^2\right]^{\frac{1}{2}} \left[1-4\left(\frac{\varepsilon}{l}\right)^2\right]^{\frac{1}{2}}}{2\left[1-\left(\frac{\varepsilon}{l}\right)^2\right]^{\frac{1}{2}} + \left[1-4\left(\frac{\varepsilon}{l}\right)^2\right]^{\frac{1}{2}}} \tag{17.32}$$

The expression (17.32), for $\varepsilon \rightarrow 0$, tends to the first eigenvalue $kl/3$ in accordance with equation (17.30a). Substituting equation (17.32) into equation (17.27) and computing the elements of the Hessian (17.20), we obtain

$$\frac{\partial^2 W}{\partial y_1^2} = k\left[\frac{1}{3} - \frac{3}{2}\left(\frac{\varepsilon}{l}\right)^2 - \frac{45}{8}\left(\frac{\varepsilon}{l}\right)^4 - \frac{381}{16}\left(\frac{\varepsilon}{l}\right)^6 - \ldots\right] \tag{17.33a}$$

$$\det[H] = k^2\left[-2\left(\frac{\varepsilon}{l}\right)^2 - 8\left(\frac{\varepsilon}{l}\right)^4 - \frac{143}{4}\left(\frac{\varepsilon}{l}\right)^6 - \ldots\right] \tag{17.33b}$$

These two series expansions continue with the even powers only of ε/l and with all the coefficients negative. It may thus be concluded that, since the Hessian is not positive definite, the post-critical branch is unstable analogously to what has already been seen in the example of Figure 17.2.

A more rapid solution to the problem may be obtained by imposing the equilibrium with regard to rotation of the two end rods with respect to the intermediate hinges:

$$Nx_1 = V_A l \tag{17.34a}$$

$$Nx_2 = V_B l \tag{17.34b}$$

where V_A and V_B denote the vertical reactions of the end supports (Figure 17.5). The equations of equilibrium with regard to vertical translation and to rotation about the point A of the entire structure,

$$V_A + V_B = k(x_1 + x_2) \tag{17.35a}$$

$$kx_1 l + 2kx_2 l = 3V_B l \tag{17.35b}$$

make it possible to obtain the above reactions as functions of the displacements

$$V_B = \frac{k}{3}(x_1 + 2x_2) \tag{17.36a}$$

$$V_A = \frac{k}{3}(2x_1 + x_2) \tag{17.36b}$$

When equations (17.36) are substituted into equations (17.34), we find

$$\begin{bmatrix} \left(\frac{2}{3}kl - N\right) & \frac{kl}{3} \\ \frac{kl}{3} & \left(\frac{2}{3}kl - N\right) \end{bmatrix} \begin{bmatrix} x_1 \\ x_2 \end{bmatrix} = \begin{bmatrix} 0 \\ 0 \end{bmatrix} \tag{17.37}$$

Making the determinant of the coefficient matrix zero, we obtain once more the characteristic equation (17.29).

Also in this case, for $N = 0$, the matrix (17.28) coincides with the stiffness matrix of the system. In general, when a discrete system with n degrees of freedom is considered, as occurs in the case of the Finite Element Method, the problem of the stability of elastic equilibrium can always be cast in the form

$$\left([K] - \lambda[K_g]\right)\{\delta\} = \{0\} \tag{17.38}$$

where $[K]$ designates the **elastic stiffness matrix**, already defined in Chapter 11, $[K_g]$ designates the **geometric stiffness matrix**, $\{\delta\}$ denotes the nodal displacement vector, and λ indicates a multiplier of the loads, which are assumed to increase proportionally. The eigenvalues of the problem are obtained via the condition

$$\det\left([K] - \lambda[K_g]\right) = 0 \tag{17.39}$$

The minimum eigenvalue λ is said to be the **critical multiplier of the loads**, and represents the load of incipient collapse.

As an example, it is underlined that, in the case of both the systems just considered, the geometric stiffness matrix is the same and takes the following form:

$$[K_g] = \begin{bmatrix} \dfrac{2}{l} & -\dfrac{1}{l} \\[2mm] -\dfrac{1}{l} & \dfrac{2}{l} \end{bmatrix} \tag{17.40}$$

17.4 Rectilinear beams with distributed elasticity

Let us consider a slender beam of constant cross section, inextensible and not deformable in shear, though deformable in bending, constrained at one end by a hinge and at the other by a roller support, loaded by an axial force N and by an orthogonal distributed load $q(z)$ (Figure 17.6(a)).

The total potential energy in a deformed configuration $v(z)$ is

$$W = \frac{1}{2} \int_0^l \frac{M^2}{EI}\, dz - Nw - \int_0^l q(z)\,v(z)dz \tag{17.41}$$

Using the differential equation of the elastic line in the form given by equation (10.47) and noting that the displacement w of the point of application of the force N is (Figure 17.6(b))

$$w = \int_0^l (dl - dz) = \int_0^l (1 - \cos\varphi)\, dl \tag{17.42}$$

and hence, with the expansion of the cosine into a Taylor series

$$w \simeq \frac{1}{2} \int_0^l \varphi^2 dz \simeq \frac{1}{2} \int_0^l v'^2\, dz \tag{17.43}$$

the total potential energy can be expressed as follows:

$$W = \int_0^l \left[\frac{1}{2}(EIv''^2 - Nv'^2) - qv \right] dz \tag{17.44}$$

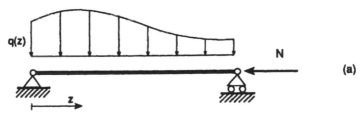

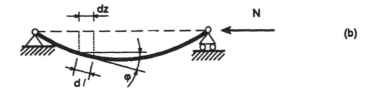

Figure 17.6

Enforcing stationarity of the functional $W(v)$, we obtain

$$\delta W = \int_0^l (EIv''\,\delta v'' - Nv'\,\delta v' - q\delta v)\mathrm{d}z = 0 \qquad (17.45)$$

where δv is referred to as a **perturbation** and indicates a function of infinitesimal values contained in the class of the solutions v. Integration by parts gives

$$-\big[(EIv''' + Nv')\delta v\big]_0^l + \big[EIv''\,\delta v'\big]_0^l + \qquad (17.46)$$

$$\int_0^l (EIv^{IV} + Nv'' - q)\,\delta v\ \mathrm{d}z = 0$$

Since equation (17.46) must hold for any δv, the following equations are identically satisfied:

$$EIv^{IV} + Nv'' - q = 0, \qquad\qquad\qquad (17.47a)$$

$$(EIv''' + Nv')\delta v = 0, \qquad \text{for } z = 0, l \qquad (17.47b)$$

$$(EIv'')\delta v' = 0, \qquad\qquad \text{for } z = 0, l \qquad (17.47c)$$

Equation (17.47a) is called the **equation of the elastic line with second-order effects** and, if we neglect the term Nv'', coincides with equation (10.49). For the simply supported beam (Figure 17.6) we have the boundary conditions $v(0) = v(l) = 0$, which imply $\delta v = 0$ at the ends, and hence that equation (17.47b) is satisfied. On the other hand, v'' is zero at the ends, because the bending moment is zero in the hinges, and hence also equation (17.47c) is satisfied in the specific case considered.

An alternative, and more immediate, mode of obtaining equation (17.47a) is that of considering the equilibrium of a deformed beam element (Figure 17.7). Equilibrium with regard to vertical translation furnishes

$$\frac{\mathrm{d}V}{\mathrm{d}z} = -q \qquad\qquad\qquad (17.48)$$

where V represents the vertical component of the internal reaction, which is not to be confused in this case with the transverse or shearing component.

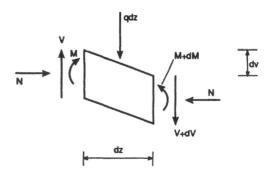

Figure 17.7

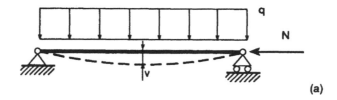

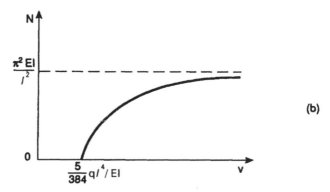

Figure 17.8

On the other hand, equilibrium with regard to rotation furnishes

$$V = \frac{\mathrm{d}M}{\mathrm{d}z} - N\frac{\mathrm{d}v}{\mathrm{d}z} \qquad (17.49)$$

from which, via equation (17.48), we find

$$\frac{\mathrm{d}^2 M}{\mathrm{d}z^2} - N\frac{\mathrm{d}^2 v}{\mathrm{d}z^2} + q = 0 \qquad (17.50)$$

Finally, using equation (10.47), we arrive back at equation (17.47a).

Consider the case of a uniformly distributed load $q(z) = q$ (Figure 17.8(a)). The integral of the equation (17.47a) assumes the form

$$v(z) = A\cos\alpha z + B\sin\alpha z + Cz + D + \frac{qz^2}{2N} \qquad (17.51)$$

where we have set

$$\alpha^2 = \frac{N}{EI} \qquad (17.52)$$

The four constants, A, B, C, D are determined by the boundary conditions

$$v(0) = v(l) = EIv''(0) = EIv''(l) = 0 \qquad (17.53)$$

which yield

$$A = -D = \frac{q}{\alpha^2 N} \qquad (17.54a)$$

545

$$B = \frac{q}{\alpha^2 N} \frac{1 - \cos \alpha l}{\sin \alpha l} \tag{17.54b}$$

$$C = -\frac{ql}{2N} \tag{17.54c}$$

Equation (17.51) therefore becomes

$$v(z) = \frac{q}{N} \left\{ \frac{1}{\alpha^2} \left[(1 - \cos \alpha l) \frac{\sin \alpha z}{\sin \alpha l} - (1 - \cos \alpha z) \right] - \frac{z(l-z)}{2} \right\} \tag{17.55}$$

It is important to note that, for $\alpha l \to \pi$, i.e. for

$$N \to N_c = \pi^2 \frac{EI}{l^2} \tag{17.56}$$

we have $\sin \alpha l \to 0$ and hence a deformed configuration which tends to infinity (Figure 17.8(b)). This means that the flexural stiffness of a compressed beam is less than that of the same beam not loaded in compression, if we take into account the geometrical nonlinearities. This stiffness even becomes zero when the compressive force equals its critical value N_c. The same occurs, for instance, in the case where the supported beam is loaded, not only by the compressive force, N, but also by an end moment m (Figure 17.9).

In the case where the distributed load is absent, i.e. $q = 0$, the equation of the elastic line with geometrical nonlinearities (17.47a) simplifies as follows:

$$EIv^{IV} + Nv'' = 0 \tag{17.57}$$

The integral of equation (17.57) is

$$v(z) = A \cos \alpha z + B \sin \alpha z + Cz + D \tag{17.58}$$

(a)

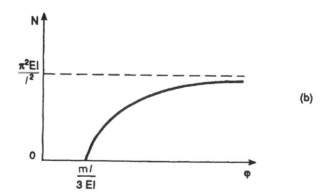

(b)

Figure 17.9

Enforcing the boundary conditions (17.53), we have

$$
\begin{bmatrix}
1 & 0 & 0 & 1 \\
\cos \alpha l & \sin \alpha l & l & 1 \\
-\alpha^2 & 0 & 0 & 0 \\
-\alpha^2\cos \alpha l & -\alpha^2\sin \alpha l & 0 & 0
\end{bmatrix}
\begin{bmatrix}
A \\ B \\ C \\ D
\end{bmatrix}
=
\begin{bmatrix}
0 \\ 0 \\ 0 \\ 0
\end{bmatrix}
\tag{17.59}
$$

The system possesses a solution different from the trivial one if and only if the determinant of the coefficient matrix is zero, and hence when $\sin \alpha l = 0$. This condition coincides with the one that makes the flexural stiffness of the beam zero (equation (17.55)).

It is possible to arrive at the same solution by imposing that, in each section of the beam, the destabilizing moment

$$
M_i = Nv \tag{17.60a}
$$

should be equal to the stabilizing moment

$$
M_s = -EI\frac{d^2v}{dz^2} \tag{17.60b}
$$

Letting therefore $M_i = M_s$, we obtain the differential equation

$$
v'' + \alpha^2 v = 0 \tag{17.61}
$$

the second derivative of which coincides with equation (17.57). The complete integral of equation (17.61) is

$$
v(z) = A\cos \alpha z + B\sin \alpha z \tag{17.62}
$$

and, since we must have $v(0) = v(l) = 0$, it follows that

$$
A = 0, \quad \sin \alpha l = 0 \tag{17.63}
$$

and the coefficient B can assume any value.

From the second of equations (17.63), we obtain the succession of the eigenvalues of the problem

$$
\alpha_n = \frac{n\pi}{l}, \quad n = \text{natural number} \tag{17.64}
$$

and hence from equation (17.52)

$$
N_{cn} = n^2\pi^2\frac{EI}{l^2} \tag{17.65}
$$

To each eigenvalue N_{cn} there corresponds an eigenfunction

$$
v_n(z) = B\sin \alpha_n z \tag{17.66}
$$

which represents the critical mode of deformation for that force. This deformed configuration consists of a number n of sinusoidal half-waves (Figure 17.10). Of course, if there are no further constraints on the beam apart from the two end supports, the critical load is that corresponding to $n = 1$:

$$
N_{c1} = \pi^2\frac{EI}{l^2} \tag{17.67}
$$

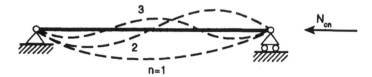

Figure 17.10

This force, called **Euler's critical load**, is the force which determines the buckling of the beam. For $N < N_{c1}$ the equilibrium is stable, for $N = N_{c1}$, the equilibrium is neutral, while for $N > N_{c1}$ the equilibrium is unstable. It should be noted that Euler's critical load increases in proportion to the rigidity EI of the beam, and decreases in inverse proportion to the square of the length of the beam.

Euler's formula shows, on the other hand, limits of validity in the case of insufficiently slender beams, for which the inelastic behaviour of the material can come to interact with the mechanism of buckling.

Let us indicate by

$$\sigma_c = \frac{N_{c1}}{A} \tag{17.68}$$

Euler's critical pressure, which on the basis of equation (17.67) can be cast in the form

$$\sigma_c = \pi^2 \frac{EI}{l^2 A} = \pi^2 E \frac{\rho^2}{l^2} \tag{17.69}$$

where ρ denotes the radius of gyration of the cross section in the direction of the bending axis. If λ designates the slenderness l/ρ, it is possible to express equation (17.69) in the following form:

$$\sigma_c = \frac{\pi^2 E}{\lambda^2} \tag{17.70}$$

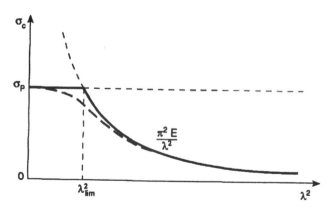

Figure 17.11

Drawing the diagram of equation (17.70) on the plane σ_c–λ^2, we obtain the so-called **Euler's hyperbola** (Figure 17.11). This hyperbola envisages critical loads tending to zero as the slenderness tends to infinity and, conversely, critical loads tending to infinity as the slenderness tends to zero. The latter tendency is unlikely since for stubby beams the failure due to yielding

$$\sigma_c = \sigma_P \tag{17.71}$$

can precede, even markedly, that due to buckling (equation (17.70)). If there were no interaction between the two critical phenomena, there would be a point of discontinuity in the passage from one to the other, corresponding to the **limit slenderness**

$$\lambda_{\lim} = \pi \left(\frac{E}{\sigma_P} \right)^{\frac{1}{2}} \tag{17.72}$$

which proves to be a function of the elastic modulus E and of the yielding stress σ_P of the material. For steel, $E/\sigma_P \sim 10^3$ and hence $\lambda_{\lim} \sim 10^2$.

In actual fact, the two critical phenomena interact and hence there is a gradual transition from one to the other as the slenderness of the beam varies. The critical pressure is thus furnished by the dashed curve of Figure 17.11, which connects the two critical curves corresponding to equations (17.70) and (17.71), rounding off the cusp that these form at their point of intersection. This curve joining the two is normally given in tabulated form, putting

$$\sigma < \sigma_P / \omega \tag{17.73}$$

where ω is a safety factor greater than unity that depends on the material and on the slenderness of the beam.

So far we have examined only the case of a beam that is constrained by a hinge and a roller support. Equation (17.57) represents, on the other hand, the equilibrium equation of a beam, whatever the means of constraint. The boundary conditions instead vary according to the constraints at the ends. Since there are four degrees of freedom – free or restrained – at the two ends (two deflections and two rotations), there are likewise four boundary conditions. These are partly **kinematic (or essential) conditions** and partly **static (or natural) conditions**. Table 17.1 illustrates the different possible cases: a beam supported at either end, a cantilever beam, a beam built-in at one end and supported at the other, a beam with one end built-in and the other constrained with a transverse double rod, a beam with one end built-in and the other constrained by an axial double rod, and a beam supported at one end and constrained at the other by an axial double rod. For each of these cases the kinematic and static boundary conditions are given, with the reminder that the second derivative of the deflection v'' is proportional to the bending moment, while the third derivative v''' is proportional to the shearing force. In the case of the cantilever, the static condition

$$EIv'''(l) + Nv'(l) = 0 \quad \text{or} \tag{17.74}$$

$$T(l) = Nv'(l) \tag{17.75}$$

yields the shear at the end as the transverse component of the horizontal force N (Figure 17.12).

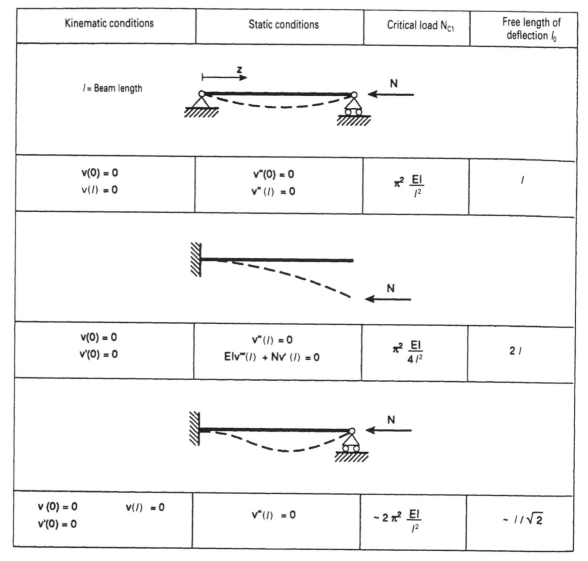

Kinematic conditions	Static conditions	Critical load N_{c1}	Free length of deflection l_0
l = Beam length			
$v(0) = 0$ $v(l) = 0$	$v''(0) = 0$ $v''(l) = 0$	$\pi^2 \dfrac{EI}{l^2}$	l
$v(0) = 0$ $v'(0) = 0$	$v''(l) = 0$ $EIv'''(l) + Nv'(l) = 0$	$\pi^2 \dfrac{EI}{4l^2}$	$2l$
$v(0) = 0 \quad v(l) = 0$ $v'(0) = 0$	$v''(l) = 0$	$\sim 2\pi^2 \dfrac{EI}{l^2}$	$\sim l/\sqrt{2}$

Table 17.1

For each case Table 17.1 then gives the critical load, which is always expressible in the form

$$N_{c1} = \pi^2 \frac{EI}{l_0^2} \tag{17.76}$$

The dimension l_0 is the so-called **free length of deflection**, which represents the distance between two successive points of inflection in the critical deformed configuration.

Finally, notice how the static (or natural) conditions may be deduced also from the boundary conditions (17.47b, c), once the kinematic (or geometrical) conditions are applied to the perturbation δv and to its derivative $\delta v'$.

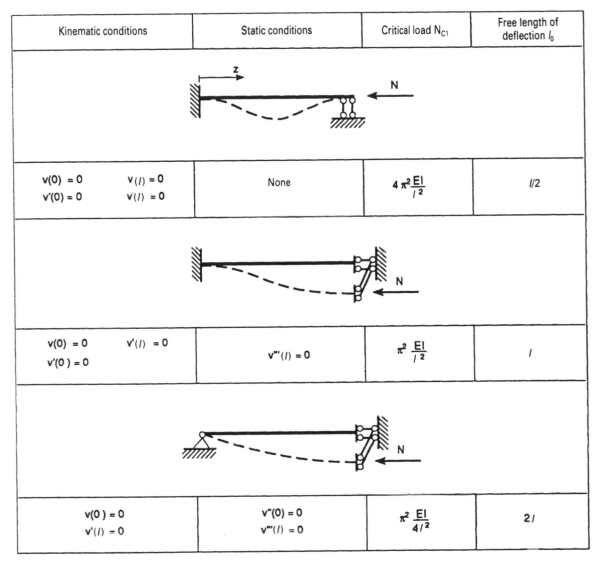

Kinematic conditions	Static conditions	Critical load N_{C1}	Free length of deflection l_0
$v(0) = 0$ $\quad$ $v(l) = 0$ $v'(0) = 0$ $\quad$ $v(l) = 0$	None	$4\pi^2\dfrac{EI}{l^2}$	$l/2$
$v(0) = 0$ $\quad$ $v'(l) = 0$ $v'(0) = 0$	$v'''(l) = 0$	$\pi^2\dfrac{EI}{l^2}$	l
$v(0) = 0$ $v'(l) = 0$	$v''(0) = 0$ $v'''(l) = 0$	$\pi^2\dfrac{EI}{4l^2}$	$2l$

Table 17.1 (continued)

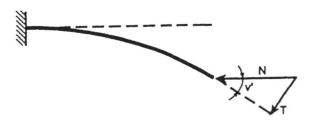

Figure 17.12

551

INSTABILITY OF ELASTIC EQUILIBRIUM

17.5 Beam systems

In some cases beam systems, owing to their simplicity, can be accommodated within the elementary schemes of Table 17.1. In particular, portal frames with rigid cross members can be referred directly to the last four cases, according to whether or not the windbracing is present, and to whether the feet of the columns are hinged or built-in (Figure 17.13).

In other cases, the axial redundant reactions, obtainable with the usual equations of congruence, can cause instability of equilibrium. A classic case is that of bars hinged or built in at the ends (Figure 17.14), subjected to an increase in temperature and hence to a prevented dilation. If the bar is only hinged at the ends, the critical temperature increase is (Figure 17.14(a))

$$\Delta T_c = \pi^2 / \alpha \lambda^2 \tag{17.77}$$

while it is quadrupled if the bar is built-in (Figure 17.14(b)).

When the beam system cannot be reduced to the schemes already seen, it is possible to apply the Finite Element Method, considering the elastic and geometrical stiffness matrices, already introduced in Section 17.3.

For the ith beam we can assume

$$v_i(z) = \{\eta_i\}^T \{\delta_i\} \tag{17.78}$$

$$\underset{(1\times4)\ (4\times1)}{}$$

where v_i represents the transverse displacement, $\{\eta_i\}$ denotes the shape function vector and $\{\delta_i\}$ indicates the nodal displacement vector (two transverse displacements and two rotations).

The shape functions $\{\eta_i\}$ must be chosen in such a way that

$$v_i'(0) = -\delta_{i1},\ v_i(0) = \delta_{i2},\ v_i'(l_i) = -\delta_{i3},\ v_i(l_i) = \delta_{i4}$$

For beams of constant cross section, the shape functions $\{\eta_i\}$ are cubic, and are obtained by imposing, in turn, one of the nodal displacements $\delta_{ij} = 1$, and leaving the others zero (Figure 17.15).

The total potential energy of the ith beam in a generic deformed configuration $v_i(z)$ is equal to

$$W(v_i) = \int_0^{l_i} \left[\frac{1}{2}(EI_i v_i''^2 - N_i v_i'^2) - q_i v_i \right] dz \tag{17.79}$$

From equation (17.78) we have

$$v_i'(z) = \{\eta_i'\}^T \{\delta_i\} \tag{17.80a}$$

$$v_i''(z) = \{\eta_i''\}^T \{\delta_i\} \tag{17.80b}$$

and hence

$$v_i'^2 = \{\delta_i\}^T \{\eta_i'\} \{\eta_i'\}^T \{\delta_i\} \tag{17.81a}$$

$$v_i''^2 = \{\delta_i\}^T \{\eta_i''\} \{\eta_i''\}^T \{\delta_i\} \tag{17.81b}$$

552

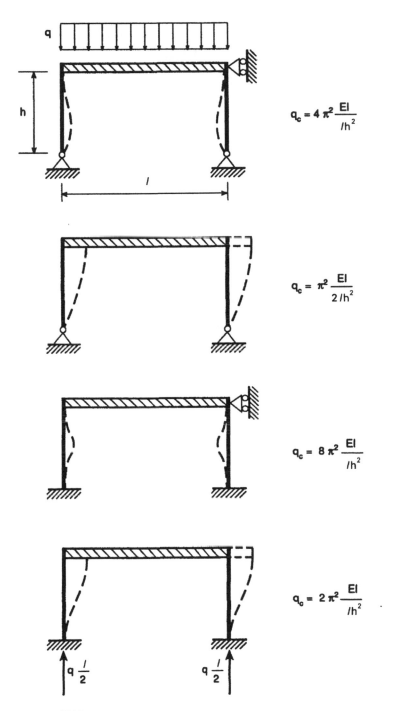

Figure 17.13

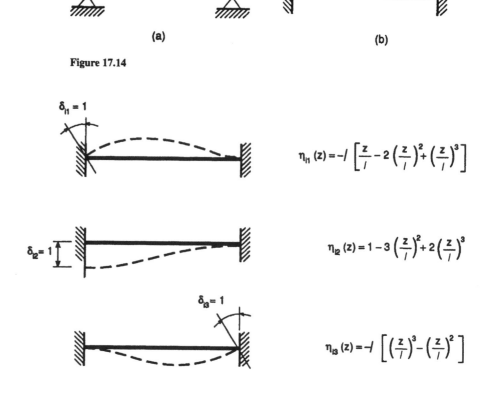

Figure 17.14

$$\eta_{l1}(z) = -l\left[\frac{z}{l} - 2\left(\frac{z}{l}\right)^2 + \left(\frac{z}{l}\right)^3\right]$$

$$\eta_{l2}(z) = 1 - 3\left(\frac{z}{l}\right)^2 + 2\left(\frac{z}{l}\right)^3$$

$$\eta_{l3}(z) = -l\left[\left(\frac{z}{l}\right)^3 - \left(\frac{z}{l}\right)^2\right]$$

$$\eta_{l4}(z) = 3\left(\frac{z}{l}\right)^2 - 2\left(\frac{z}{l}\right)^3$$

Figure 17.15

Substituting equations (17.81) into equation (17.79), we obtain

$$W(v_i) = \frac{1}{2}\{\delta_i\}^{\mathrm{T}}\left[\int_0^{l_i} EI_i\{\eta_i''\}\{\eta_i''\}^{\mathrm{T}}\,\mathrm{d}z\right]\{\delta_i\} - \tag{17.82}$$

$$\frac{1}{2}N_i\{\delta_i\}^{\mathrm{T}}\left[\int_0^{l_i}\{\eta_i'\}\{\eta_i'\}^{\mathrm{T}}\,\mathrm{d}z\right]\{\delta_i\} -$$

$$\left[\int_0^{l_i}q_i\{\eta_i\}^{\mathrm{T}}\,\mathrm{d}z\right]\{\delta_i\}$$

554

BEAM SYSTEMS

Equation (17.82) may be cast in the form

$$W(v_i) = \frac{1}{2}\{\delta_i\}^T\big([K_i] - N_i[K_{gi}]\big)\{\delta_i\} - \{F_i\}^T\{\delta_i\} \qquad (17.83)$$

which, compared against equation (11.61), highlights the **elastic stiffness matrix**

$$[K_i] = \int_0^{l_i} EI_i\{\eta_i''\}\{\eta_i''\}^T\,dz \qquad (17.84a)$$

and the **geometrical stiffness matrix** of the ith beam

$$[K_{gi}] = \int_0^{l_i} \{\eta_i'\}\{\eta_i'\}^T\,dz \qquad (17.84b)$$

as well as the **equivalent nodal force vector**

$$\{F_i\} = \int_0^{l_i} q_i\{\eta_i\}\,dz \qquad (17.84c)$$

Computing with the shape functions given in Figure 17.15, we obtain

$$[K_i] = EI_i
\begin{bmatrix}
\dfrac{4}{l_i} & -\dfrac{6}{l_i^2} & \dfrac{2}{l_i} & \dfrac{6}{l_i^2} \\[2mm]
-\dfrac{6}{l_i^2} & \dfrac{12}{l_i^3} & -\dfrac{6}{l_i^2} & -\dfrac{12}{l_i^3} \\[2mm]
\dfrac{2}{l_i} & -\dfrac{6}{l_i^2} & \dfrac{4}{l_i} & \dfrac{6}{l_i^2} \\[2mm]
\dfrac{6}{l_i^2} & -\dfrac{12}{l_i^3} & \dfrac{6}{l_i^2} & \dfrac{12}{l_i^3}
\end{bmatrix}
\qquad (17.85)$$

which corresponds to equation (14.24), and in addition

$$[K_{gi}] = \frac{1}{l_i}
\begin{bmatrix}
\dfrac{2}{15}l_i^2 & -\dfrac{1}{10}l_i & -\dfrac{1}{30}l_i^2 & \dfrac{1}{10}l_i \\[2mm]
-\dfrac{1}{10}l_i & \dfrac{6}{5} & -\dfrac{1}{10}l_i & -\dfrac{6}{5} \\[2mm]
-\dfrac{1}{30}l_i^2 & -\dfrac{1}{10}l_i & \dfrac{2}{15}l_i^2 & \dfrac{1}{10}l_i \\[2mm]
\dfrac{1}{10}l_i & -\dfrac{6}{5} & \dfrac{1}{10}l_i & \dfrac{6}{5}
\end{bmatrix}
\qquad (17.86)$$

Basically then, the presence of the axial force N_i decreases the stiffness of the ith element.

As regards the subsequent operations of the Finite Element Method, the procedure is exactly as outlined in Chapter 11, with the **rotation** and the **expansion** of the local stiffness matrices. Finally the assembly operation

555

provides the global stiffness matrix, so that the eigenvalue problem for seeking the critical loads is formulated as follows:

$$\det\left([K] - \lambda[K_g]\right) = 0 \tag{17.87}$$

where λ represents the multiplier of the external loads.

In actual fact, the axial forces in the individual beams do not increase in proportion to the loads. It may be assumed, on the other hand, in first approximation, that the axial forces are maintained proportional to the values that are obtained from a geometrically linear analysis

$$N_i(\lambda) = \lambda\, N_i(\lambda = 1) \tag{17.88}$$

17.6 Curvilinear beams: arches and rings

Let us consider a beam with curvilinear axis which is inextensible and not deformable in shear. The kinematic equations (10.27), on the hypothesis that $\gamma = \varepsilon = 0$, yield

$$\varphi = -\frac{dv}{ds} + \frac{w}{r} \tag{17.89a}$$

$$\frac{dw}{ds} = -\frac{v}{r} \tag{17.89b}$$

$$\chi = \frac{d\varphi}{ds} \tag{17.89c}$$

Substituting equation (17.89a) into equation (17.89c), we obtain

$$\chi = -\frac{d^2v}{ds^2} + \frac{d}{ds}\left(\frac{w}{r}\right) \tag{17.90}$$

and hence, neglecting the variation of intrinsic curvature and applying equation (17.89b),

$$\chi = -\frac{d^2v}{ds^2} - \frac{v}{r^2} \tag{17.91}$$

Finally, recalling the relation which links variation of curvature χ and bending moment M, we derive the **equation of the elastic line for curvilinear beams** (Boussinesq gives an analogous treatment)

$$\frac{d^2v}{ds^2} + \frac{v}{r^2} = -\frac{M}{EI} \tag{17.92}$$

Consider a cylindrical shell of radius R, subjected to an external pressure q, in a deformed configuration, symmetrical with respect to two orthogonal diameters (Figure 17.16). If the static characteristics in the point A' are $N_{A'} = -q\overline{A'O}$ and $M_{A'}$, the bending moment in the generic point B' is equal to

$$M_{B'} = M_{A'} + N_{A'} \cdot \overline{A'C} + \frac{1}{2}q\overline{A'B'}^2 \tag{17.93}$$

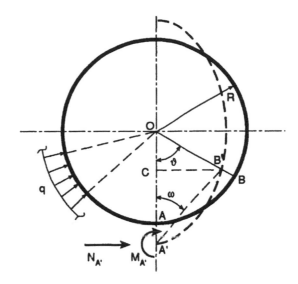

Figure 17.16

and hence

$$M_{B'} = M_{A'} - q \left(\overline{A'O} \cdot \overline{A'C} - \frac{1}{2} \overline{A'B'}^2 \right)$$ (17.94)

Between the sides of the triangle $OA'B'$ there holds the relation

$$\overline{B'O}^2 = \overline{A'O}^2 + \overline{A'B'}^2 - 2 \overline{A'O} \; \overline{A'B'} \cos \omega$$ (17.95)

whence

$$\overline{A'O} \; \overline{A'C} = \frac{1}{2} \left(\overline{A'O}^2 + \overline{A'B'}^2 - \overline{B'O}^2 \right)$$ (17.96)

Having substituted equation (17.96) into equation (17.94), we obtain

$$M_{B'} = M_{A'} - \frac{1}{2} q \left(\overline{A'O}^2 - \overline{B'O}^2 \right)$$ (17.97)

Since we have

$$\overline{A'O} = R + v_0$$ (17.98a)

$$\overline{B'O} = R + v$$ (17.98b)

equation (17.97), once infinitesimals of a higher order have been neglected, becomes

$$M_{B'} = M_{A'} + qR(v - v_0)$$ (17.99)

Taking into account equation (17.99) and that $ds = R \, d\vartheta$, equation (17.92) is transformed as follows:

$$\frac{d^2 v}{d\vartheta^2} + v = -\frac{R^2}{EI} \left[M_{A'} + qR(v - v_0) \right]$$ (17.100)

557

or

$$\frac{d^2v}{d\vartheta^2} + v\left(\frac{qR^3}{EI} + 1\right) = \frac{R^2}{EI}(qRv_0 - M_{A'})$$

(17.101)

Setting

$$\alpha^2 = \frac{qR^3}{EI} + 1$$

(17.102)

the eigenvalue equation becomes

$$\frac{d^2v}{d\vartheta^2} + \alpha^2 v = \frac{R^2}{EI}(qRv_0 - M_{A'})$$

(17.103)

The integral of equation (17.103) is

$$v(\vartheta) = A\sin\alpha\vartheta + B\cos\alpha\vartheta + \frac{qR^3v_0 - M_{A'}R^2}{qR^3 + EI}$$

(17.104)

Enforcing the two conditions of symmetry,

$$\frac{dv}{d\vartheta} = \alpha A\cos\alpha\vartheta - \alpha B\sin\alpha\vartheta = 0, \quad \text{for } \vartheta = 0, \frac{\pi}{2}$$

(17.105)

we obtain the two equations

$$\alpha A = 0$$

(17.106a)

$$\alpha B\sin\frac{\alpha\pi}{2} = 0$$

(17.106b)

which give

$$A = 0$$

(17.107a)

$$\alpha\frac{\pi}{2} = n\pi, \quad n = \text{natural number}$$

(17.107b)

Equation (17.107b) yields the succession of eigenvalues

$$\alpha_n = 2n, \quad n = \text{natural number}$$

(17.108)

For $n = 1$, from equation (17.102) we obtain the first critical load

$$q_c = \frac{3EI}{R^3}$$

(17.109)

and the deformed configuration

$$v(\vartheta) = \frac{M_{A'}R^2 + v_0 EI}{qR^3 + EI}\cos 2\vartheta + \frac{qR^3v_0 - M_{A'}R^2}{qR^3 + EI}$$

(17.110)

On the other hand, from the condition of inextensibility (17.89b) we obtain

$$w(\vartheta) = \frac{M_{A'}R^2 + v_0 EI}{qR^3 + EI}\frac{1}{2}\sin 2\vartheta + \frac{qR^3v_0 - M_{A'}R^2}{qR^3 + EI}\vartheta$$

(17.111)

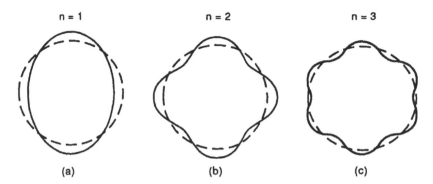

n = 1   n = 2   n = 3

(a)   (b)   (c)

Figure 17.17

For $\vartheta = 0$, $\pi/2$, the axial displacement w must vanish by symmetry, and hence

$$M_{A'} = qRv_0 \qquad (17.112)$$

Substituting equation (17.112) into equation (17.110), we obtain finally the following deformed configuration:

$$v(\vartheta) = v_0 \cos 2\vartheta \qquad (17.113)$$

which represents an **ovalization** of the tube (Figure 17.17(a)). It is possible then to demonstrate that the second configuration of neutral equilibrium consists of the four-lobed buckling of Figure 17.17(b). In general, the nth configuration of neutral equilibrium will present $2n$ lobes.

17.7 Lateral torsional buckling

Consider a beam of thin rectangular cross section, constrained at the ends so that rotation about the longitudinal axis Z is prevented. Let this beam be subjected to uniform bending by means of the application at the ends of two moments m contained in the plane YZ of greater flexural rigidity (Figure 17.18(a)).

Consider a deformed configuration of the beam, with deflection thereof in the XZ plane of smaller flexural rigidity, and simultaneous torsion about the axis Z (Figure 17.18(b)). The deflection $u(z)$ and the torsional rotation $\varphi_z(z)$ generate components of the external moment m in the axial direction Z (Figure 17.18(c)) and in the transverse direction Y (Figure 17.18(d)), respectively:

$$M_{zi} = m\frac{du}{dz} \qquad (17.114a)$$

$$M_{yi} = -m\varphi_z \qquad (17.114b)$$

Both the loads M_{zi} and M_{yi} are destabilizing, because they tend to increase the torsional rotation φ_z and the flexural deflection u, respectively. On the other hand, as for Euler's rod, the corresponding stabilizing loads are present

$$M_{zs} = GI_t\frac{d\varphi_z}{dz} \qquad (17.115a)$$

$$M_{ys} = EI_y\frac{d^2u}{dz^2} \qquad (17.115b)$$

559

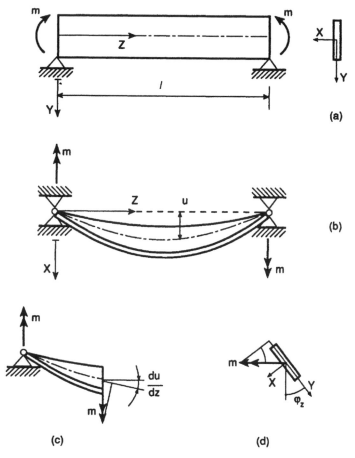

Figure 17.18

The equilibrium is neutral when equations (17.114) are respectively equal to equations (17.115)

$$GI_t \frac{d\varphi_z}{dz} = m \frac{du}{dz} \qquad (17.116a)$$

$$EI_y \frac{d^2u}{dz^2} = -m\varphi_z \qquad (17.116b)$$

Differentiating equation (17.116a) with respect to z and substituting the result into equation (17.116b), we obtain

$$\frac{d^2\varphi_z}{dz^2} + \frac{m^2}{EGI_yI_t}\varphi_z = 0 \qquad (17.117)$$

If we put

$$\alpha^2 = \frac{m^2}{EGI_yI_t} \qquad (17.118)$$

560

the equation

$$\varphi_z'' + \alpha^2 \varphi_z = 0 \qquad (17.119)$$

assumes the same form as equation (17.61), corresponding to the problem of the axially compressed slender rod. Usually, the complete integral of equation (17.119),

$$\varphi_z(z) = A \cos \alpha z + B \sin \alpha z \qquad (17.120)$$

satisfies the boundary conditions

$$\varphi_z(0) = \varphi_z(l) = 0 \qquad (17.121)$$

for $A = 0$ and $\sin \alpha l = 0$. The eigenvalues of the problem are thus

$$\alpha_n = n\frac{\pi}{l}, \quad n = \text{natural number} \qquad (17.122)$$

and the first, $\alpha_1 = \pi/l$, yields the critical load

$$m_c = \frac{\pi}{l}\sqrt{EGI_y I_t} \qquad (17.123)$$

Equation (17.123) is commonly known as **Prandtl's formula**.

The phenomenon of lateral torsional buckling is especially relevant to deep beams, while it is virtually present for beams of compact cross section, for which the critical moment, expressed by equation (17.123), is so high as to exceed the plastic moment of the cross section (see Chapter 18). It is possible to note, on the other hand, how beams of compact cross section can also undergo lateral torsional buckling, in the case where they are particularly slender ($l \to \infty$).

17.8 Plates subjected to compression

On the basis of relations (10.157) and (10.158), the strain energy per unit surface of a deflected plate is

$$\Phi = -\frac{1}{2}\left(M_x \frac{\partial^2 w}{\partial x^2} + M_y \frac{\partial^2 w}{\partial y^2} + 2M_{xy}\frac{\partial^2 w}{\partial x \partial y}\right) \qquad (17.124)$$

Using relations (10.180), we obtain

$$\Phi = \frac{1}{2}D\left[\left(\frac{\partial^2 w}{\partial x^2}\right)^2 + \left(\frac{\partial^2 w}{\partial y^2}\right)^2 + 2v\left(\frac{\partial^2 w}{\partial x^2}\right)\left(\frac{\partial^2 w}{\partial y^2}\right)\right] + \qquad (17.125)$$

$$D(1-v)\left(\frac{\partial^2 w}{\partial x \partial y}\right)^2$$

If, in addition to being considered undeformable in shear, the plate is considered also as being inextensible and subjected to a membrane regime, N_x, N_y, N_{xy}, the potential energy of these loads in a deflected configuration is

$$\Phi_N = \frac{1}{2}\left[N_x\left(\frac{\partial w}{\partial x}\right)^2 + N_y\left(\frac{\partial w}{\partial y}\right)^2 + 2N_{xy}\left(\frac{\partial w}{\partial x}\right)\left(\frac{\partial w}{\partial y}\right)\right] \qquad (17.126)$$

561

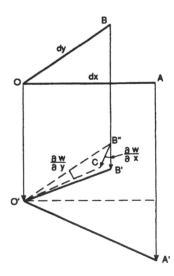

Figure 17.19

As regards the first two terms of equation (17.126), these contributions are analogous to those calculated for the rectilinear beam (see equation (17.44)), while the third term represents the work of the shearing stresses acting through the shearing strains due to the deflection w, and can be justified as follows. Let us consider two infinitesimal segments OA and OB in the directions of the two coordinate axes X and Y (Figure 17.19). Because of the deflection w, these segments are transformed into $O'A'$ and $O'B'$. The difference between the angle $A'O'B'$ and $\pi/2$ represents the shearing strain sought. For the purpose of determining this difference, let us consider the right angle $B''O'A'$. Rotating this angle about the side $O'A'$ by the amount $\partial w/\partial y$, the plane $B''O'A'$ comes to coincide with the plane $B'O'A'$, the point B'' assuming the position C. The displacement $B''C$ is equal to $(\partial w/\partial y)$ dy and is inclined with respect to the vertical $B''B'$ by the angle $\partial w/\partial x$. Consequently the segment CB' is equal to $(\partial w/\partial x)(\partial w/\partial y)dy$, and the angle $CO'B'$, which represents the shearing strain due to the deflection w, is equal to $(\partial w/\partial x)(\partial w/\partial y)$.

The total potential energy of the deflected plate is therefore equal to the sum of the integrals of the strain energy expressed by equation (17.125) and of the potential energy of the membrane forces expressed by equation (17.126)

$$W = \frac{1}{2}D\int_A \left\{ \left(\frac{\partial^2 w}{\partial x^2} + \frac{\partial^2 w}{\partial y^2}\right)^2 - \right. \tag{17.127}$$

$$\left. 2(1-v)\left[\left(\frac{\partial^2 w}{\partial x^2}\right)\left(\frac{\partial^2 w}{\partial y^2}\right) - \left(\frac{\partial^2 w}{\partial x \partial y}\right)^2\right]\right\}dx\,dy\ +$$

$$\frac{1}{2}\int_A \left[N_x\left(\frac{\partial w}{\partial x}\right)^2 + N_y\left(\frac{\partial w}{\partial y}\right)^2 + 2N_{xy}\left(\frac{\partial w}{\partial x}\right)\left(\frac{\partial w}{\partial y}\right)\right]dx\,dy$$

The **variational equation of equilibrium** can be obtained by imposing the stationarity of W, in a way similar to that already seen for the rectilinear beam:

$$D\left(\frac{\partial^4 w}{\partial x^4} + 2\frac{\partial^4 w}{\partial x^2 \partial y^2} + \frac{\partial^4 w}{\partial y^4}\right) \tag{17.128}$$

$$= N_x\frac{\partial^2 w}{\partial x^2} + N_y\frac{\partial^2 w}{\partial y^2} + 2N_{xy}\frac{\partial^2 w}{\partial x \partial y}$$

Note that equation (17.128) is formally analogous to equation (17.57).

In the case of a plane rectangular plate of sides a, b, supported on the four sides and compressed by a force N per unit length of the edge, acting orthogonally to the side b, equation (17.128) assumes the following form:

$$\left(\frac{\partial^4 w}{\partial x^4} + 2\frac{\partial^4 w}{\partial x^2 \partial y^2} + \frac{\partial^4 w}{\partial y^4}\right) = -\frac{N}{D}\left(\frac{\partial^2 w}{\partial x^2}\right) \tag{17.129}$$

The constraints impose $w = 0$ on the four sides, and the annihilation of the bending moment on the edge:

$$w = 0, \quad \left(\frac{\partial^2 w}{\partial y^2}\right) + v\left(\frac{\partial^2 w}{\partial x^2}\right) = 0, \quad \text{for } y = 0, b \qquad (17.130a)$$

$$w = 0, \quad \left(\frac{\partial^2 w}{\partial x^2}\right) + v\left(\frac{\partial^2 w}{\partial y^2}\right) = 0, \quad \text{for } x = 0, a \qquad (17.130b)$$

Each function

$$w(x, y) = A_{nm}\sin n\pi \frac{x}{a}\sin m\pi \frac{y}{b} \qquad (17.131)$$

satisfies the preceding boundary conditions for n, m = natural numbers. Substituting equation (17.131) into equation (17.129) and dividing by the common factor $A_{nm} \sin (n\pi x/a) \sin (m \pi y/b)$, we obtain

$$\left(\frac{n\pi}{a}\right)^4 + 2\left(\frac{n\pi}{a}\right)^2\left(\frac{m\pi}{b}\right)^2 + \left(\frac{m\pi}{b}\right)^4 = \frac{N}{D}\left(\frac{n\pi}{a}\right)^2 \qquad (17.132)$$

and hence

$$N_c^{nm} = \pi^2 D\frac{a^2}{n^2}\left(\frac{n^2}{a^2} + \frac{m^2}{b^2}\right)^2 \qquad (17.133)$$

The smallest value of N_c^{nm} is to be considered the critical load for instability of the elastic equilibrium of the plate. This value is obtained for $m = 1$, since m appears only in the numerator in equation (17.133)

$$N_c^{n1} = \pi^2 \frac{D}{b^2}\left(n\frac{b}{a} + \frac{1}{n}\frac{a}{b}\right)^2 \qquad (17.134)$$

and corresponds to a deformed configuration with only one half-wave along side b and n half-waves along side a.

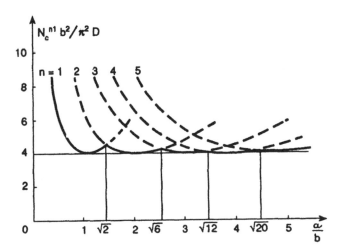

Figure 17.20

563

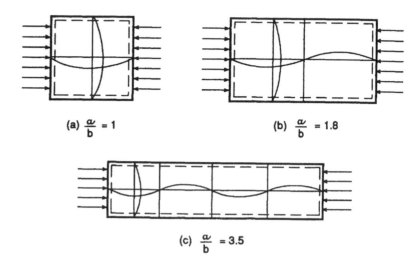

(a) $\dfrac{a}{b} = 1$ (b) $\dfrac{a}{b} = 1.8$

(c) $\dfrac{a}{b} = 3.5$

Figure 17.21

Figure 17.20 presents the diagram of the critical load non-dimensionalized as a function of the ratio a/b between the sides of the rectangle. In actual fact, a succession of curves is obtained as n varies, but, for each value a/b, we have a certain value of n for which N_c^{nl} is a minimum. For $a/b < \sqrt{2}$, the minimum is obtained for $n = 1$. The critical deformed configuration thus presents a half-wave in each direction (Figure 17.21(a)). For $\sqrt{2} < a/b < \sqrt{6}$, we have $n = 2$, and the critical deformed configuration presents two half-waves in the X direction and one half-wave in the Y direction (Figure 17.21(b)). For $a/b > \sqrt{6}$, we have $N_c^{nl} \simeq 4\pi^2 D/b^2$ and n is such as to give rise to half-waves of comparable amplitude along X and along Y (Figure 17.21(c)).

The behaviour of the plate previously analysed is analogous to that of a beam on an elastic foundation. In fact it may be assimilated to that of a system of longitudinal beams constrained to a system of transverse beams. This prevents the value of N_c^{nl} from dropping below the value $4 \pi^2 D/b^2$, whatever the value of a may be.

The total potential energy of a beam on an elastic foundation is (Figure 17.22).

$$W = \frac{1}{2} \int_0^l (EIv''^2 - Nv'^2 + Kv^2)\,\mathrm{d}z \qquad (17.135)$$

where K is the elastic modulus of the foundation.

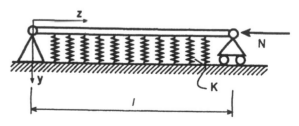

Figure 17.22

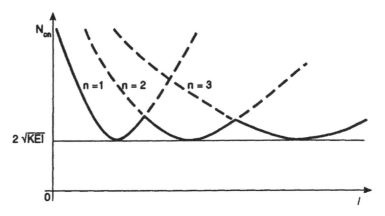

Figure 17.23

In the framework of the Ritz–Galerkin method, let us assume for the deflection v the following series expansion:

$$v(z) = \sum_n A_n \sin n\pi \frac{z}{l} \qquad (17.136)$$

Substituting equation (17.136) into equation (17.135) and recalling the orthonormality of the trigonometric functions

$$\int_0^l \sin n\pi \frac{z}{l} \sin m\pi \frac{z}{l} \, dz \qquad (17.137)$$

$$= \int_0^l \cos n\pi \frac{z}{l} \cos m\pi \frac{z}{l} \, dz = \frac{l}{2} \delta_{nm}$$

where δ_{nm} is the Kronecker symbol, we obtain

$$W = \frac{l}{4} \sum_n A_n^2 \left(EI \frac{n^4 \pi^4}{l^4} - N \frac{n^2 \pi^2}{l^2} + K \right) \qquad (17.138)$$

Equation (17.138) is a diagonal quadratic form in the coefficients A_n, which ceases to be positive definite as soon as N is such as to cause one of the terms in round brackets to vanish:

$$N_{cn} = EI \frac{n^2 \pi^2}{l^2} + K \frac{l^2}{n^2 \pi^2} \qquad (17.139)$$

Figure 17.23 presents the diagram of the critical load as a function of the length l of the beam. As in the case of the plate, we have a succession of curves according to the variation in n, but, for each value of l, we have a certain value of n for which N_{cn} is a minimum. The above curves present local minima for values of l equal to

$$l = n^2 \pi^2 \left(\frac{EI}{K} \right)^{\frac{1}{2}} \qquad (17.140)$$

and these minima are all equal to

565

$$N_{cn} = 2(KEI)^{\frac{1}{2}}$$

(17.141)

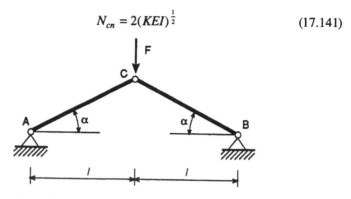

Figure 17.24

17.9 Flat arches

Let us consider the flat arch of Figure 17.24, consisting of two axially deformable rods of stiffness K, hinged together both in the crown, as well as in the foundation. Let the distance between the two springers of the arch be $2l$, and the angle that the two rods AC and BC form initially with the horizontal be α. Under the action of the force F, let this angle diminish by the infinitesimal quantity φ.

If only symmetrical deformations are considered, the system will have only one degree of freedom, and the strain energy of the arch will then be expressible as follows:

$$\Phi(\varphi) = K \left[\frac{l}{\cos \alpha} - \frac{l}{\cos (\alpha - \varphi)} \right]^2$$

(17.142)

In the hypothesis of a flat arch, we can set

$$\cos \alpha \simeq 1 - \frac{\alpha^2}{2}$$

(17.143a)

$$\cos(\alpha - \varphi) \simeq 1 - \frac{1}{2}(\alpha - \varphi)^2$$

(17.143b)

and hence, with a further application of the Taylor series expansion,

$$\frac{1}{\cos \alpha} \simeq 1 + \frac{\alpha^2}{2}$$

(17.144a)

$$\frac{1}{\cos(\alpha - \varphi)} \simeq 1 + \frac{1}{2}(\alpha - \varphi)^2$$

(17.144b)

Substituting equations (17.144) into equation (17.142), we obtain

$$\Phi(\varphi) = \frac{1}{4} K l^2 \varphi^2 (2\alpha - \varphi)^2$$

(17.145)

The deflection caused by the load F is equal, on the other hand, to

$$\eta(\varphi) = l \tan \alpha - l \tan(\alpha - \varphi)$$

(17.146)

whereby, to a first approximation

566

$$\eta(\varphi) \simeq l\varphi \qquad (17.147)$$

Finally, the total potential energy of the system is found from equations (17.145) and (17.147),

$$W(\varphi) = \Phi(\varphi) - F\eta(\varphi) \qquad (17.148)$$

whereby it is equal to

$$W(\varphi) = \frac{1}{4} Kl^2\varphi^2(2\alpha - \varphi)^2 - Fl\varphi \qquad (17.149)$$

The conditions of equilibrium are all those and only those for which the function (17.149) is stationary

$$W'(\varphi) = Kl^2\,\varphi(2\alpha^2 + \varphi^2 - 3\alpha\varphi) - Fl = 0 \qquad (17.150)$$

from which we obtain

$$F = Kl\varphi(\varphi - \alpha)(\varphi - 2\alpha) \qquad (17.151)$$

Relation (17.151) is displayed in Figure 17.25. There thus exist three positions of equilibrium with $F = 0$, when $\varphi = 0$, α, 2α. Whilst the first and the last represent conditions of stable equilibrium with the connecting rods unloaded, the intermediate one is the condition of unstable equilibrium represented by the configuration with the connecting rods aligned and compressed. A rigorous study of the stability is conducted by examining the second derivative of the total potential energy

$$W''(\varphi) = Kl^2(3\varphi^2 - 6\alpha\varphi + 2\alpha^2) \qquad (17.152)$$

which is greater than zero for

$$\varphi < \alpha\left(1 - \frac{\sqrt{3}}{3}\right) \quad \text{or} \quad \varphi > \alpha\left(1 + \frac{\sqrt{3}}{3}\right) \qquad (17.153)$$

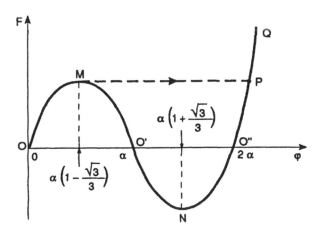

Figure 17.25

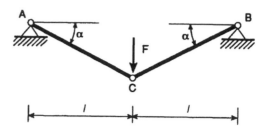

Figure 17.26

The function (17.151) is hence stationary for $\varphi = \alpha(1 - \frac{1}{3}\sqrt{3})$, where it presents a maximum, and for $\varphi = \alpha(1 + \frac{1}{3}\sqrt{3})$, where it presents a minimum (Figure 17.25). Since the third derivative of the total potential energy

$$W'''(\varphi) = 6Kl^2(\varphi - \alpha) \qquad (17.154)$$

is different from zero for $\varphi \neq \alpha$, it may be concluded that both the maximum and the minimum of the curve $F(\varphi)$ represent states of unstable equilibrium.

Therefore, if the flat arch ACB of Figure 17.24 is loaded, the portion OM of the curve $F(\varphi)$ of Figure 17.25 is traversed in a stable manner until the stationary point M is reached. If the load F continues to be increased, there is an abrupt jump on the stable branch PQ, which, with the force F being equal, presents an angle φ, which is much greater, and a configuration of the system which is inverted with respect to the initial one (Figure 17.26).

If, instead, we wish to go along the virtual branch MNP, it is necessary to control the phenomenon by imposing an angle φ, which presents a continuous growth. In this case the force F can be interpreted as a constraint reaction, which between M and N decreases, becoming even negative beyond the point O'. This means that, beyond the aligned connecting rod configuration, a force is necessary in an upward direction so as to proceed along the curve $F(\varphi)$ in a controlled manner.

The energy recovered by the system in the jump MP (Figure 17.25) is equal to the area $MO'NO''P$ multiplied by the length l. This energy will thus be transformed into vibrational kinetic energy of the system about the condition represented by the point P. The instability phenomenon just described, and in particular the jump MP at constant load, is termed **snap-through**, and is thus analogous to the phenomenon of **snap-back** described in Section 8.11.

Also in the more complex cases of flat arches and shells, which are also flexurally compliant, the phenomenon of snap-through can develop, so giving rise to a sudden change of configuration. Figure 17.27 shows the load *vs.* deflection curves corresponding to spherical thin shells built in at the edge, loaded by a uniform pressure q. The dashed curves correspond to a linear analysis, while the continuous line curves correspond to a nonlinear analysis. On the linear response curves the instability loads are also marked, corresponding, respectively, to a complete spherical shell, of equal radius and thickness, and to the same built-in spherical shell. The parameter λ represents the slenderness of the shell

$$\lambda = 2[3(1 - v^2)]^{1/4}(H/h)^{1/2} \qquad (17.155)$$

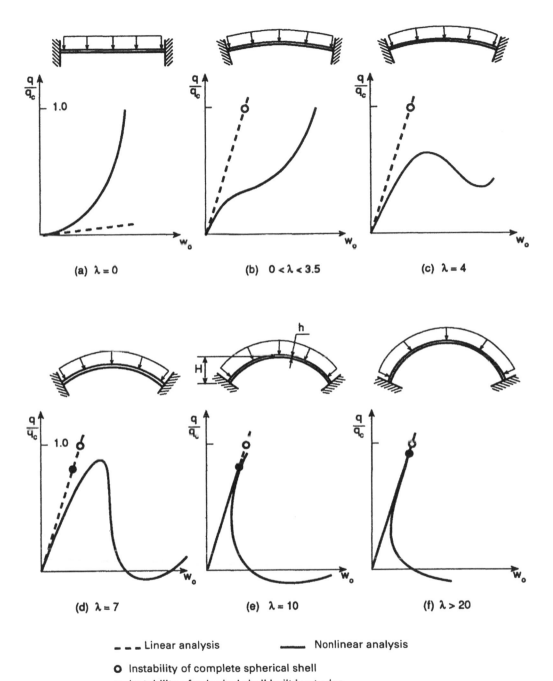

(a) $\lambda = 0$

(b) $0 < \lambda < 3.5$

(c) $\lambda = 4$

(d) $\lambda = 7$

(e) $\lambda = 10$

(f) $\lambda > 20$

- - - Linear analysis ——— Nonlinear analysis

○ Instability of complete spherical shell
● Instability of spherical shell built in at edge

Figure 17.27

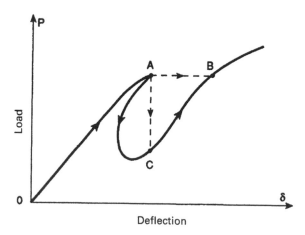

Figure 17.28

where H is the rise of the shell with respect to the edge plane, and h is the thickness.

For $\lambda \leqslant 3.5$, the behaviour of the shell does not present the phenomenon of snap-through. For $\lambda = 0$, the load *vs.* deflection curve presents a stiffening due to the intervention of the tensile force as the plate is deflected. For $3.5 \leqslant \lambda \leqslant 7$, in the curves of Figure 17.27 the phenomenon of snap-through emerges. Finally, for $\lambda \gtrsim 7$, both the phenomenon of snap-back and that of snap-through present themselves. Note that, in this latter interval, the behaviour of the shell prior to instability tends to be increasingly linear as λ increases.

Similar snap-back and snap-through phenomena occur during the phases of cracking of high-strength concrete reinforced beams (Figure 17.28). Whereas snap-back is due basically to brittle fracturing of the concrete, snap-through is due to pulling-out, yielding and hardening of the steel reinforcing bars.

Finally, it is worthwhile recalling how the phenomena of snap-back and snap-through are theoretically predicted both for complete spherical shells subjected to external pressure (Figure 17.29(a)) and for cylindrical shells subjected to axial compression (Figure 17.29(b)). On the other hand, it is difficult to bring out such phenomena experimentally, in view of the considerable sensitivity to initial imperfections displayed by the above-mentioned geometries. With the increase in the initial inherent imperfections, the structural response tends to become less unstable, the phenomenon of snap-back disappearing. Figure 17.30 represents the load *vs.* axial contraction response in the case of an axially loaded cylindrical shell, as the eccentricity of the cross section varies. For particularly high eccentricities, also the phenomenon of snap-through disappears.

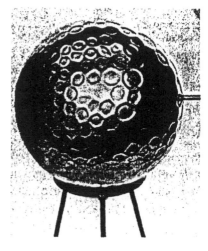

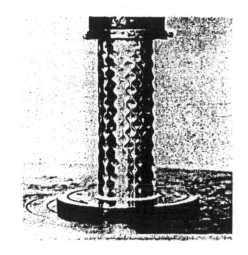

(a)

(b)

Figure 17.29

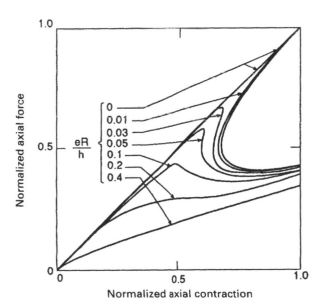

Figure 17.30

571

18 Theory of plasticity

18.1 Introduction

On the basis of the hypotheses of the linear elastic behaviour of the material and of small displacements, the problem of the elastic solid may be resolved, as we have found in Chapters 8, 9 and 10, by means of Lamé's equation, where the operator $[\mathscr{L}]$ is always **linear**. If, that is, $\{\mathscr{F}\}$ is the vector of the external forces and $\{\eta\}$ is the corresponding displacement vector, obtained by resolving equation (8.52(a)), should the loads be multiplied by a constant c, also the displacements, and hence the deformation and the static characteristics will be multiplied by the same constant

$$[\mathscr{L}]\{c\eta\} = -\{c\mathscr{F}\} \tag{18.1}$$

Furthermore, if $\{\mathscr{F}_a\}$, $\{\mathscr{F}_b\}$ are two different vectors of the external forces and $\{\eta_a\}$, $\{\eta_b\}$ the corresponding displacement fields, in the case of superposition of the forces the Principle of Superposition will hold good also for displacements:

$$[\mathscr{L}]\{\eta_a + \eta_b\} = -\{\mathscr{F}_a + \mathscr{F}_b\} \tag{18.2}$$

A first case of nonlinearity was examined in the previous chapter, where it was shown how the external loads do not always increase proportionally to the induced displacements (Figure 17.1(b), 17.2(b), 17.8(b), 17.9(b)) in the case where such displacements cannot be considered small. A second case of nonlinearity will be examined in this chapter, where the ductile behaviour of the material will be considered, as already introduced in Section 8.11. The first is the case of **geometrical nonlinearity**, while the second is the case of **constitutive nonlinearity of the material**.

A first simple example of nonlinear structural behaviour may be offered by the system of parallel connecting rods of Figure 18.1(a), if we assume that they follow a law of elastic–perfectly plastic behaviour (Figure 18.1(b)). This case has already been considered in the elastic regime in Section 14.2, where

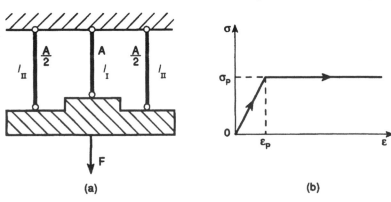

(a)   (b)

Figure 18.1

572

the reactions of the individual connecting rods have been determined, once the behaviour of the cross member is assumed to be rigid. Applying the relations (14.3) and (14.4) and considering each of the two lateral connecting rods as having a cross section equal to half of the cross section of the central connecting rod, the two reactions in the elastic field are obtained

$$X_I = F \frac{\dfrac{A}{l_I}}{\dfrac{A}{l_I} + \dfrac{A}{l_{II}}} = F \frac{l_{II}}{l_I + l_{II}} \qquad (18.3a)$$

$$X_{II} = F \frac{\dfrac{A/2}{l_{II}}}{\dfrac{A}{l_I} + \dfrac{A}{l_{II}}} = \frac{F}{2} \frac{l_I}{l_I + l_{II}} \qquad (18.3b)$$

If $l_I < l_{II}$, the higher tension is developed in the central connecting rod; therefore, if the external force F is increased, this element is the one which first undergoes plastic deformation. Yielding of the central connecting rod occurs for

$$F_1 = \sigma_P A \left(1 + \frac{l_I}{l_{II}}\right) \qquad (18.4a)$$

$$\delta_1 = \frac{\sigma_P l_I}{E} \qquad (18.4b)$$

where the subscript 1 denotes the characteristics of first yielding (force applied to the cross member and vertical displacement thereof).

Yielding of the lateral connecting rods, on the other hand, occurs for

$$F_2 = 2\sigma_P A \qquad (18.5a)$$

$$\delta_2 = \frac{\sigma_P l_{II}}{E} \qquad (18.5b)$$

where the subscript 2 indicates the characteristics of second and ultimate yielding. For $\delta > \delta_2$, in fact, the reaction of the connecting rods cannot increase and remains stationary at the value of ultimate plasticity F_2.

Hence, recapitulating, we obtain a global elastic behaviour for $0 < \delta < \delta_1$

$$F = EA \left(\frac{1}{l_I} + \frac{1}{l_{II}}\right)\delta \qquad (18.6a)$$

a globally strain-hardening behaviour for $\delta_1 < \delta < \delta_2$

$$F = \sigma_P A + \frac{EA}{l_{II}}\delta \qquad (18.6b)$$

and a perfectly plastic behaviour (plastic flow) for $\delta > \delta_2$

$$F = 2\sigma_P A \qquad (18.6c)$$

Figure 18.2 presents in non-dimensional form the force–displacement curves for different values of the ratio l_I/l_{II}. For $l_I \to 0$, the force is sustained

573

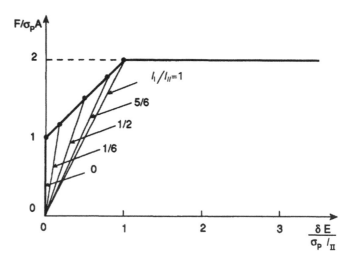

Figure 18.2

in the elastic phase entirely by the central connecting rod, which then yields for $F = \sigma_P A$. When $l_I = l_{II}$, on the other hand, the hardening phase is not present because all three connecting rods yield at the same time. It may be noted that the straight line to which the hardening portion of the diagram of Figure 18.2 belongs does not depend on the ratio l_I/l_{II}. This is due to the fact that, once the central rod has yielded, its length l_I no longer enters into the analysis.

Although the example just considered is particularly simple, since it contains only three elements subjected to axial force, it reflects conceptually the mechanical behaviour of the more complex beam systems, where the prevalent characteristic is bending moment. In such cases the local plastic flow will be represented by a localized rotation and, as the number of such rotations increases, the degree of redundancy of the frame will diminish simultaneously.

In the sequel various examples of **incremental plastic analysis** of beam systems will be shown. In these will be determined, step by step and as the external load increases, the position of the cross sections in which the localized plastic rotation occurs. To this type of incremental analysis of the process of plastic deformation there corresponds the so-called **plastic limit analysis**, which, on the basis of two specific theorems, directly identifies an interval (one that is generally restricted) within which the ultimate load of global plastic flow must necessarily fall. This load is called the **load of plastic collapse**. Once this load is reached, the structure is reduced to a mechanism, i.e. it is hypostatic, even though in equilibrium on account of the particular load condition, and it is not able to sustain further increments of load. On the basis of the aforementioned theorems it is possible to identify also the mechanism of collapse, that is, the positions of the centres of plastic relative rotation.

More particularly, beam systems loaded by concentrated forces will be distinguished from those loaded by distributed forces. In the latter, in fact, identification of the mechanism of collapse is generally more complex. The closing sections of this chapter will deal briefly with problems of nonproportional loading and problems of repeated loading (**shake-down**), as well as with the problem of plastic collapse of deflected plates.

574

18.2 Elastic–plastic flexure

Let us consider the rectangular cross section, of base b and depth h (Figure 18.3), of a beam made of elastic–perfectly plastic material, with equal elastic modulus E and yield stress σ_P both in tension and in compression (Figure 18.1(b)).

Let us assume that as the applied bending moment increases, the cross section of the beam remains plane, even though part of the beam undergoes plastic deformation. As has already been noted in Section 9.4, this is equivalent to considering linear variations of the axial dilation ε_z through the depth of the beam (Figure 18.3(a)). The axial stress σ_z, on the other hand, will not be able to exceed its limit value σ_P, and, once the moment of first plastic deformation M_e has been overstepped, will hence present a linear variation in the central part of the cross section and two plateaus in the outermost parts (Figures 18.3(b), (c)). The diagrams of Figure 18.3 present the succession of the patterns that both ε_z and σ_z follow through the depth, rendering the scales uniform with the values at yielding ε_P and σ_P, respectively. Therefore it clearly emerges from these diagrams that the maximum dilation ε_{max}, which is reached at the outermost edges of the beam, exceeds the dilation ε_P, which is the one that corresponds to yielding. When $\varepsilon_{max} \to \infty$, and hence when plastic flow has occurred, the variation of the stress is bi-rectangular, while the extension $2d$ of the elastic core of the beam vanishes (Figure 18.3(d)).

The **moment of first plastic deformation** (which is the maximum moment in the elastic regime) is readily obtainable via relation (9.23),

$$M_e = \sigma_P \frac{I_x}{h/2} = \sigma_P b \frac{h^2}{6} \qquad (18.7a)$$

while the **moment of ultimate plastic deformation** or **plastic moment** may be evaluated using the diagram of Figure 18.3 (d),

$$M_P = \sigma_P \left(b \frac{h}{2} \right) \frac{h}{2} = \sigma_P b \frac{h^2}{4} \qquad (18.7b)$$

this being equal to that of a couple $\sigma_P(bh/2)$ with moment arm $h/2$. The plastic moment M_P is therefore equal to $\frac{3}{2} M_e$, a fact that enables further exploitation

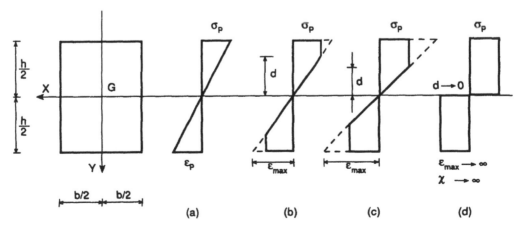

Figure 18.3

of the load-bearing capacity of metal materials, with loads substantially higher than those that meet the **criterion of admissible stresses**.

On the basis of equations (9.23) and (9.41), it is possible to put the axial dilations ε_z in relation with the curvature χ_x of the beam element straddling the section under consideration,

$$\varepsilon_z = \frac{M_x}{EI_x} y = \chi_x y \qquad (18.8)$$

from which we obtain (Figure 18.3)

$$\chi_x = \frac{\varepsilon_{max}}{h/2} = \frac{\varepsilon_P}{d} \qquad (18.9)$$

Equation (18.9) warns us that, for $\varepsilon_{max} \to \infty$, or for $d \to 0$, the beam curvature tends to infinity, giving rise to a localized rotation in the cross section under examination.

We now intend to determine the moment–curvature law, M_x–χ_x, corresponding to the plastic evolution of the cross section (Figure 18.3). At each step of this evolution the applied moment may be evaluated on the basis of the known distribution of the forces:

$$M_x = 2\int_0^d \left(\sigma_P \frac{y}{d}\right) y b \, dy + 2\int_d^{h/2} \sigma_P \, y b \, dy \qquad (18.10)$$

Substituting the half-depth d of the elastic zone with the expression deriving from equation (18.9), we have

$$M_x = 2\sigma_P b \left[\frac{\chi_x}{\varepsilon_P} \int_0^{\varepsilon_P/\chi_x} y^2 dy + \int_{\varepsilon_P/\chi_x}^{h/2} y \, dy \right] \qquad (18.11)$$

from which, evaluating the integrals, we obtain

$$M_x = \sigma_P b \frac{h^2}{4} \left[1 - \frac{1}{3} \frac{\left(\varepsilon_P / \frac{h}{2}\right)^2}{\chi_x^2} \right] \qquad (18.12)$$

Via equation (18.7a) the foregoing function can be cast in a particularly expressive form,

$$\frac{M_x}{M_e} = \frac{3}{2} - \frac{1}{2} \left(\frac{\chi_e}{\chi_x}\right)^2 \qquad (18.13)$$

where χ_e denotes the beam curvature when first plastic deformation occurs. The diagram of Figure 18.4 thus represents a linear law for $\chi_x < \chi_e$, or $M_x < M_e$, and the hyperbolic law of equation (18.13) for $\chi_x > \chi_e$, or $M_x > M_e$. This strain-hardening law is replaced in practice by the elastic–perfectly plastic law represented by the dashed line in the same figure.

When the cross section of the beam presents double symmetry (Figure 18.5), while at the same time not being rectangular, the foregoing reasoning to

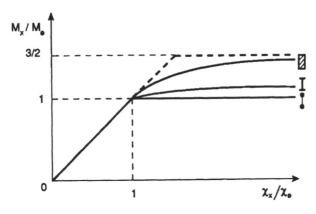

Figure 18.4

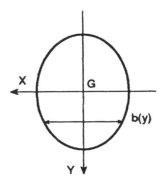

Figure 18.5

a large extent still applies. In particular, relation (18.11) must present the width $b(y)$, which in general is a function of y, under the integral sign

$$M_x = 2\sigma_P\left[\frac{\chi_x}{\varepsilon_P}\int_0^{\varepsilon_P/\chi_x} y^2 b(y)dy + \int_{\varepsilon_P/\chi_x}^{h/2} yb(y)dy\right] \qquad (18.14)$$

The plastic moment is therefore

$$M_P = \lim_{\chi_x \to \infty} M_x = 2\sigma_P\int_0^{h/2} yb(y)dy \qquad (18.15)$$

where the integral represents the half-section static moment $S_x^{A/2}$ with respect to the X axis. The ratio

$$\frac{M_P}{M_e} = \frac{2S_x^{A/2}}{\left(I_x\bigg/\dfrac{h}{2}\right)} \qquad (18.16)$$

is, as has been stated, equal to 1.5 in the case of a rectangular section, while, in the limit case of a section consisting of two concentrated areas set at a distance h apart, it is equal to unity. This means that in such a case the moments of first and ultimate plastic deformation coincide. In the technically highly recurrent case of an I–section, there is no great departure from the limit case just considered, and the ratio given by equation (18.16) is approximately equal to 1.15 (Figure 18.4). The I–sections are thus the most convenient in the elastic regime, while, in the plastic regime, they reveal poor reserves of flexural bearing capacity.

Consider a cross section with a single axis of symmetry, which coincides with the axis of flexure (Figure 18.6). The neutral axis remains orthogonal to the axis of symmetry, even though its position may vary during the entire loading process. In the condition of full plastic deformation (Figure 18.6(d)), we have

$$\sigma_P A_1 = \sigma_P A_2 \qquad (18.17)$$

577

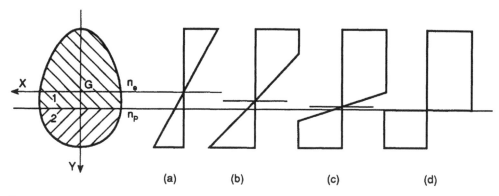

(a)   (b)   (c)   (d)

Figure 18.6

where $A_1 = A_2 = A/2$ are the areas of the portions of cross section which remain, respectively, above and below the plastic neutral axis n_P. Consequently, the plastic moment is

$$M_P = \sigma_P \frac{A}{2}(d_1 + d_2) \qquad (18.18)$$

where d_1 and d_2 are the distances of the plastic neutral axis n_P from the centroids of the two half-sections. When $M_e < M < M_P$, the neutral axis is between n_e and n_P, as is shown in Figures 18.6(b), (c). Whereas, therefore, the elastic neutral axis renders the static moments of the two portions into which it divides the section equal in absolute value, the plastic neutral axis makes the areas equal.

As regards the other static characteristics, the twisting moment applied to a circular section presents a behaviour altogether similar to that described previously for bending moment. The twisting moment of first plastic deformation, on the basis of equation (9.85), is

$$M_{ze} = \frac{I_p}{R}\tau_P = \frac{\pi}{2}R^3\tau_P \qquad (18.19a)$$

where, according to Tresca's criterion, $\tau_P = \frac{1}{2}\sigma_P$. The plastic twisting moment is, on the other hand, equal to the product of the yielding stress σ_P by the polar static moment of the cross section

$$M_{zP} = \frac{2\pi}{3}R^3\tau_P \qquad (18.19b)$$

The ratio M_{zP}/M_{ze} is hence 4/3.

The case of centred axial force is trivial and has already been dealt with in advance in Section 18.1. Clearly the plastic axial force is

$$N_P = \sigma_P A \qquad (18.20)$$

whilst the case of shearing force should be considered together with that of bending, although, generally speaking, in the framework of plastic calculation this characteristic has a negligible influence.

As regards the combined loading conditions, the case of eccentric axial force is noteworthy. For a rectangular cross section loaded by an axial force N,

578

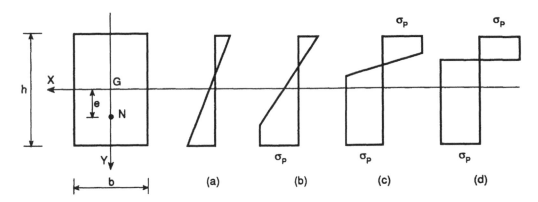

Figure 18.7

applied on the Y axis with eccentricity e (Figure 18.7), four different phases succeed one another as N increases. These phases are relative to the conditions represented in Figure 18.7:

(a) elastic;
(b) elastic–plastic, with yielding only at one edge;
(c) elastic–plastic, with yielding at both edges;
(d) full plasticity.

The diagram of Figure 18.7(d) can be split into two as shown in Figure 18.8, part (a) representing the resultant force N,

$$N = \sigma_P\, b(h - 2h') \qquad (18.21a)$$

and part (b) representing the moment $M = Ne$,

$$M = \sigma_P\, bh'(h - h') \qquad (18.21b)$$

On the basis of the plastic loadings

$$N_P = \sigma_P\, bh \qquad (18.22a)$$

$$M_P = \sigma_P\, b\frac{h^2}{4} \qquad (18.22b)$$

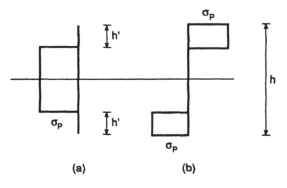

Figure 18.8

579

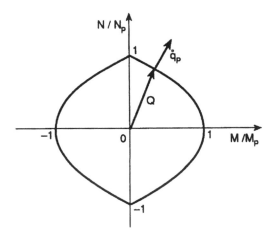

Figure 18.9

it is possible to define the following non-dimensional ratios:

$$\frac{N}{N_P} = 1 - 2\left(\frac{h'}{h}\right) \tag{18.23a}$$

$$\frac{M}{M_P} = 4\left(\frac{h'}{h}\right)\left(1 - \frac{h'}{h}\right) \tag{18.23b}$$

such that

$$\frac{M}{M_P} = 1 - \left(\frac{N}{N_P}\right)^2 \tag{18.24}$$

The plastic limit in the M–N plane is given by the closed curve of Figure 18.9, which is also called the **curve of interaction**. The couples M–N, which are internal to the domain, represent elastic–plastic conditions, whilst the couples that are on the boundary represent ultimate conditions of full plastic deformation (plastic flow of the cross section). These, as shall be seen in the sequel, occur with a localized rotation of the beam to which a localized axial dilation is added.

18.3 Incremental plastic analysis of beam systems

Let us consider a cantilever beam of length l, loaded by an orthogonal force F at the free end (Figure 18.10(a)). As the force increases, the plastic collapse of the cantilever beam is reached as soon as the fixed-end moment equals the plastic moment,

$$F_P l = M_P \tag{18.25}$$

and hence for $F_P = M_P / l$. At that point a localized rotation is produced in the fixed-end cross section, whilst the fixed-end moment cannot continue to grow and remains stationary at its limit value M_P. This situation is usually represented by inserting a hinge instead of the built-in constraint and by applying a moment M_P in the neighbourhood of the hinge (Figure 18.10(b)). The hinge

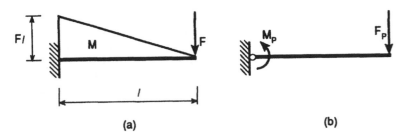

(a)   (b)

Figure 18.10

allows localized rotations, while the moment M_P represents the rotational reaction exerted by the fixed-end cross section. The system has thus become hypostatic but is in equilibrium on account of the particular loading condition. It should be noted that we have

$$F_P = \frac{3}{2} F_e \qquad (18.26)$$

where F_e denotes the maximum force applicable in the framework of the criterion of admissible stresses. The ratio 3/2 thus represents a sort of safety factor in the framework of the criterion of admissible stresses in regard to the ultimate plastic condition.

As a second elementary case, let us take that of a simply supported beam with a force applied in the centre (Figure 18.11(a)). As the force increases, plastic collapse is reached as soon as the moment in the centre equals the plastic moment,

$$\frac{1}{4} F_P l = M_P \qquad (18.27)$$

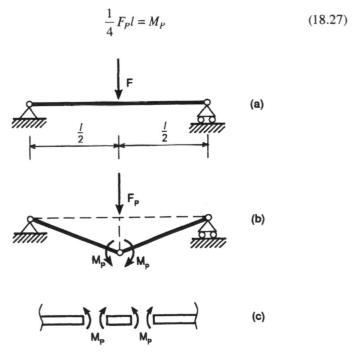

Figure 18.11

from which we obtain the collapse load $F_P = 4M_P / l$. At that point a **plastic hinge** is created in the centre, i.e. a hinge having a constant rotational reaction equal to M_P. Note that the moments M_P, acting in the scheme of Figure 18.11(b), tend to oppose the action of the external load and to cause the two arms to rotate in the direction opposite to that of the collapse mechanism. Also in this case, the mechanism is in equilibrium by virtue of the particular load condition. This equilibrium condition is neutral in the case of small displacements. If, on the other hand, the plastically deformed beam element is isolated, as shown in Figure 18.11(c), the moments M_P acting on the element and on the two arms of the beam, in all cases, stretch the lower longitudinal fibres. Here again relation (18.26) applies along with the observations deriving therefrom.

The safety factor, defined in accordance with equation (18.26), is 3/2 for all the statically determinate systems of deflected beams of rectangular cross section, once the contributions of the axial force and shearing force are neglected. The formation of a single plastic hinge in fact directly leads to the collapse of the system (Figure 18.12). In statically determinate trusses, or, at any rate, in systems made up of bars and hence subjected to axial force alone, this factor is obviously equal to unity. In the case of statically indeterminate systems of deflected beams, the safety factor is generally greater than 3/2. The formation of the first plastic hinge in fact does not bring about the collapse of the structure. In general, it may be stated that in a frame with n degrees of redundancy the number of plastic hinges that are activated for collapse to occur is less than or equal to $(n + 1)$.

Let us consider, for instance, the beam built in at both ends of Figure 18.13(a). As has already been established in Section 13.3, the fixed-end moments and the moment in the centre are equal to $Fl/8$, and stretch the upper and lower fibres, respectively. The fixed-end cross sections and the cross section at the centre hence reach full plasticity simultaneously

$$\frac{1}{8} F_P l = M_P \qquad (18.28)$$

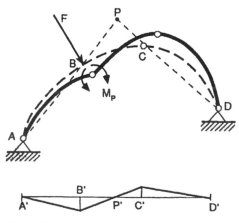

Figure 18.12

582

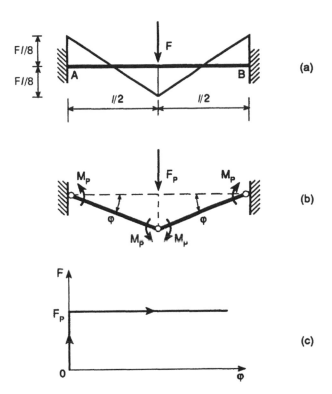

Figure 18.13

fiom which we obtain the collapse load $F_P = 8\, M_P / l$. Applying the Principle of Virtual Work to the collapse mechanism of Figure 18.13(b), it is possible to reobtain the previous value,

$$F_P\, \varphi \frac{l}{2} - 4M_P\, \varphi = 0 \qquad (18.29)$$

The absence of a strain-hardening phase in the loading process (Figure 18.13(c)) is due to the substantial statical determinacy of the structure (Section 13.3).

In the case of the closed framework of Figure 15.8(a), the maximum bending moment in the elastic phase is that found in the loaded cross sections. This equals $\frac{3}{16} Fl$, whereby the load producing the first two plastic hinges (Figure 18.14(a)) is

$$F_1 = \frac{16}{3} \frac{M_P}{l} \qquad (18.30)$$

The scheme of Figure 18.14(b) describes this situation, taking into account the double symmetry of the framework, whilst Figure 18.14(c) gives the corresponding moment diagram. The subsequent four hinges are formed simultaneously at A, B, C, D, when the moment in the nodes also attains the value M_P, as the external forces F increase,

$$M_A = \frac{Fl}{4} - M_P = M_P \qquad (18.31)$$

583

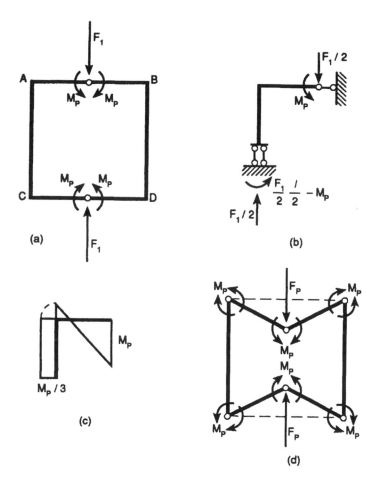

Figure 18.14

whence we find the collapse load

$$F_2 = F_P = 8M_P/l \tag{18.32}$$

It should be noted that this load is the same as that obtained in the case of the beam built in at both ends. This is due to the substantial identity of the corresponding mechanisms of collapse (Figure 18.13(b) and Figure 18.14(d)). The safety factor, in the framework of the criterion of admissible stresses and in regard to plastic collapse, is in this case

$$\frac{F_P}{F_e} = \frac{F_2}{\frac{2}{3}F_1} = \frac{3}{2}\frac{(8M_P/l)}{(16M_P/3l)} = \frac{9}{4} \tag{18.33}$$

In the case of the asymmetrical portal frame of Figure 15.12(a), the force that causes the formation of the first plastic hinge is (Figure 18.15(a))

$$F_1 = \frac{23}{18}\frac{M_P}{l} \tag{18.34}$$

584

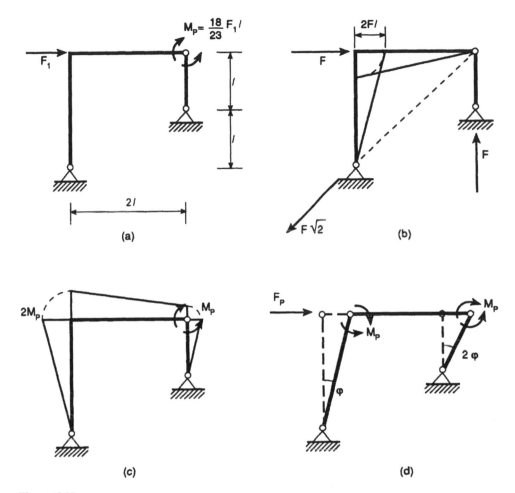

Figure 18.15

For $F > F_1$ the portal frame is transformed into a three-hinged arch, loaded by the external force F and by the two plastic moments M_P. The partial moment diagrams for the two loads are given in Figures 18.15(b), (c). The diagram corresponding to the plastic moments M_P is virtual since it shows on the cross member values $M > M_P$. A second plastic hinge is formed when the global moment in the left-hand node becomes equal to M_P (stretching the internal fibres)

$$2F_2 l - 2M_P = M_P \qquad (18.35)$$

from which there follows

$$F_2 = F_P = \frac{3}{2} \frac{M_P}{l} \qquad (18.36)$$

The value of F_P may alternatively be found by applying the Principle of Virtual Work to the collapse mechanism of Figure 18.15(d):

$$F_P(2l\varphi) - M_P\varphi - M_P(2\varphi) = 0 \qquad (18.37)$$

585

In this case the safety factor is

$$\frac{F_P}{F_e} = \frac{F_2}{\frac{2}{3}F_1} = \frac{3}{2}\frac{(3M_P/2l)}{(23M_P/18l)} = \frac{81}{46} \tag{18.38}$$

So far only concentrated loads have been considered. A first simple example of distributed load is furnished by the scheme of the beam built in at both ends shown in Figure 13.10(a). It is known from Section 13.3 that the maximum moment in the elastic regime is that of the built-in constraint, which is equal to $ql^2/12$ (Figure 13.10(c)). Therefore when the external load reaches the value

$$q_1 = 12\frac{M_P}{l^2} \tag{18.39}$$

two plastic hinges are formed at the built-in constraints. In line with the scheme of Figure 18.16(a), the third plastic hinge in the centre forms when

$$q_2\frac{l^2}{8} - M_P = M_P \tag{18.40}$$

whence we obtain the collapse load

$$q_2 = q_P = 16\frac{M_P}{l^2} \tag{18.41}$$

It is possible to arrive at this collapse load also by a simple application of the Principle of Virtual Work. The safety factor in this case is

$$\frac{q_P}{q_e} = \frac{q_2}{\frac{2}{3}q_1} = \frac{3}{2}\frac{(16M_P/l^2)}{(12M_P/l^2)} = 2 \tag{18.42}$$

The deflection in the centre for $q = q_1$ (Figure 18.16(a)) is given by the contributions of the external load q_1 and of the plastic moments M_P, respectively

$$\delta_1 = \frac{5}{384}\frac{q_1 l^4}{EI} - \frac{M_P l^2}{8EI} \tag{18.43}$$

which, via equation (18.39), becomes

$$\delta_1 = \frac{M_P l^2}{32EI} \tag{18.44}$$

On the other hand, the deflection in the centre for $q = q_2$ likewise is

$$\delta_2 = \frac{5}{384}\frac{q_2 l^4}{EI} - \frac{M_P l^2}{8EI} \tag{18.45}$$

and hence, inserting equation (18.41)

$$\delta_2 = \frac{M_P l^2}{12EI} \tag{18.46}$$

Plotting the points (δ_1, q_1) and (δ_2, q_2) on the non-dimensionalized plane of Figure 18.16(b), we immediately find the curve $\delta(q)$, i.e. the structural

586

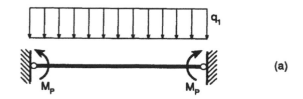

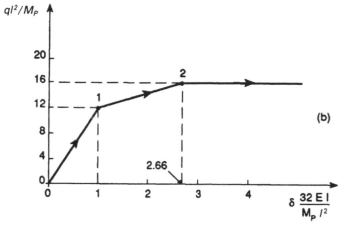

Figure 18.16

response as the external load increases. It may be noted that this response is elastic between points 0 and 1, strain-hardening between points 1 and 2, and finally perfectly plastic for $\delta > \delta_2$.

Let us now examine the case of the beam built in at one end and supported at the other, illustrated in Figure 13.4(a). In the elastic regime, the maximum bending moment occurs at the built-in constraint and is equal to $ql^2/8$. Hence a first plastic hinge is formed at the built-in constraint for $q_1 = 8 \, M_P/l^2$. At this point the structure becomes statically determinate and presents a globally strain-hardening behaviour, until the second and last plastic hinge is formed (Figure 18.17(a)). Whereas in the case of concentrated loadings it is straight-forward to identify the location of the subsequent plastic hinges, in the case of distributed loads such an identification is not usually immediate. In the case in point, for instance, which does not even present particular symmetries, it is necessary to calculate the maximum of the moment function in the hardening phase and to determine the value q_2 which makes this maximum equal to the plastic moment M_P. Expressed in formulas, this is

$$M(z) = -\frac{M_P}{l} z + \left(\frac{1}{2} qlz - \frac{1}{2} qz^2 \right) \qquad (18.47a)$$

$$T(z) = \frac{\mathrm{d}M}{\mathrm{d}z} = -\frac{M_P}{l} + \frac{1}{2} ql - qz \qquad (18.47b)$$

The shear vanishes for

$$z = \frac{l}{2} - \frac{M_P}{ql} \qquad (18.48)$$

587

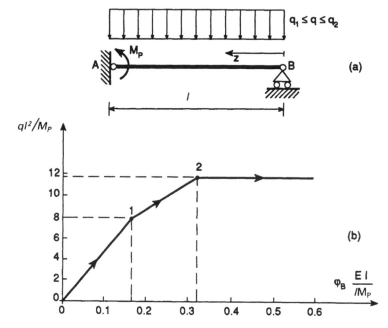

Figure 18.17

whence we obtain

$$M_{\text{max}} = \frac{1}{8} q l^2 \left(1 - \frac{2M_P}{q l^2}\right)^2 \tag{18.49}$$

Setting $M_{\text{max}} = M_P$, we obtain a second degree algebraic equation in the unknown q_2

$$\left(1 - \frac{2M_P}{q_2 l^2}\right)^2 = 4 \frac{2M_P}{q_2 l^2} \tag{18.50}$$

which, resolved, yields the two roots

$$\frac{2M_P}{q_2 l^2} = 3 \pm 2\sqrt{2} \tag{18.51}$$

Whereas the first root must be rejected, since it would imply $q_2 < q_1$, the second yields the collapse load

$$q_2 = q_P = \frac{M_P}{l^2} \frac{2}{3 - 2\sqrt{2}} \tag{18.52}$$

or

$$q_2 = q_P \simeq 11.6568 \frac{M_P}{l^2} \tag{18.53}$$

588

The safety factor is therefore

$$\frac{q_P}{q_e} = \frac{q_2}{\frac{2}{3}q_1} = \frac{3}{8(3-2\sqrt{2})} \simeq 2.19 \qquad (18.54)$$

To represent the structural response with the increase in the load q, it is necessary to choose a suitable kinematic parameter, for example the rotation of the end cross section B. In the elastic phase $(0 \leqslant q \leqslant q_1)$ we have

$$\varphi_B = \frac{ql^3}{24EI} - \left(q\frac{l^2}{8}\right)\frac{l}{6EI} = \frac{ql^3}{48EI} \qquad (18.55a)$$

whilst in the hardening phase $(q_1 \leqslant q \leqslant q_2)$ we obtain

$$\varphi_B = \frac{ql^3}{24EI} - M_P\frac{l}{6EI} = \frac{l^3}{24EI}\left(q - 4\frac{M_P}{l^2}\right) \qquad (18.55b)$$

Therefore the rotations for the notable loads q_1 and q_2 are respectively

$$\varphi_{B1} = \frac{1}{6}\frac{M_Pl}{EI} \qquad (18.56a)$$

$$\varphi_{B2} = \frac{11.6568 - 4}{24}\frac{M_Pl}{EI} \simeq 0.319\frac{M_Pl}{EI} \qquad (18.56b)$$

The diagram $\varphi_B(q)$ is given in non-dimensional form in Figure 18.17(b). Between the points 0 and 1 the global behaviour of the structure is elastic, between 1 and 2 it is hardening, while beyond this point it is perfectly plastic.

In the case of the continuous beam of Figure 13.33(a), the maximum bending moment is reached in the right-hand span, and hence it will be at this cross section that the first plastic hinge will be formed. Isolating the right-hand span (Figure 18.18(a)), we obtain

$$M(z) = \frac{3}{7}qlz - \frac{1}{2}qz^2 \qquad (18.57a)$$

$$T(z) = \frac{dM}{dz} = \frac{3}{7}ql - qz \qquad (18.57b)$$

The shear vanishes for $z = \frac{3}{7}l$, where we have

$$M_{max} = \frac{9}{98}ql^2 \qquad (18.58)$$

Setting $M_{max} = M_P$, we obtain the load that produces the first plastic hinge

$$q_1 = \frac{98}{9}\frac{M_P}{l^2} \simeq 10.89\frac{M_P}{l^2} \qquad (18.59)$$

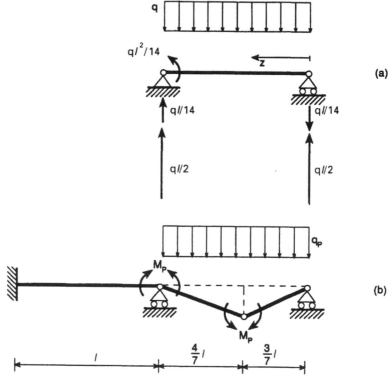

Figure 18.18

At this point it is easy to understand that the second and last plastic hinge will be formed at the central support (Figure 18.18(b)). Applying the Principle of Virtual Work to the collapse mechanism yields the equation

$$-M_P\left(\frac{3}{7}\varphi\right) - M_P\varphi + \left(\frac{4}{7}ql\right)\left(\frac{3}{7}\varphi\right)\left(\frac{2}{7}l\right) + \tag{18.60}$$

$$\left(\frac{3}{7}ql\right)\left(\frac{4}{7}\varphi\right)\left(\frac{3}{14}l\right) = 0$$

whence, performing the computation, we find

$$q_2 = q_P = \frac{70}{6}\frac{M_P}{l^2} \approx 11.6666\frac{M_P}{l^2} \tag{18.61}$$

Note that the collapse load of equation (18.61) is greater than the collapse load of equation (18.53) corresponding to the previously considered scheme by what, from an engineering standpoint, is a negligible amount (~1‰). The safety factor is, however, less, even though the structure in this case has two degrees of static indeterminacy:

$$\frac{q_P}{q_e} = \frac{q_2}{\frac{2}{3}q_1} = \frac{3}{2}\frac{\left(\frac{70}{6}\right)}{\left(\frac{98}{9}\right)} \approx 1.607 \tag{18.62}$$

This is due to the fact that, in order to bring about the collapse mechanism, the formation of two plastic hinges is sufficient, instead of three, as the global static determinacy would require in the proximity of collapse. Cases of this sort are referred to as instances of **partial collapse**, as opposed to complete collapse, whereby the entire structure becomes hypostatic.

As a final example, let us consider the portal frame with inclined stanchion of Figure 15.21(a). As has already been shown in Section 15.5, the maximum bending moment is reached, in the elastic phase, in the left-hand fixed joint-node (Figure 15.22(a)), whereby the load that produces the first plastic hinge in the same node is obtained from equation (15.45(a)).

$$q_1 = \frac{112}{16+3\sqrt{2}}\frac{M_P}{l^2} \approx 5.53\frac{M_P}{l^2} \tag{18.63}$$

Application of the Principle of Virtual Work yields the moment X_2 in the right-hand fixed joint-node, for $q > q_1$ (Figure 18.19 (a))

$$2M_P\varphi + 2X_2\varphi - ql\left(\frac{l}{2}\varphi\right) = 0 \tag{18.64}$$

whence we obtain

$$X_2 = \frac{1}{4}ql^2 - M_P \tag{18.65}$$

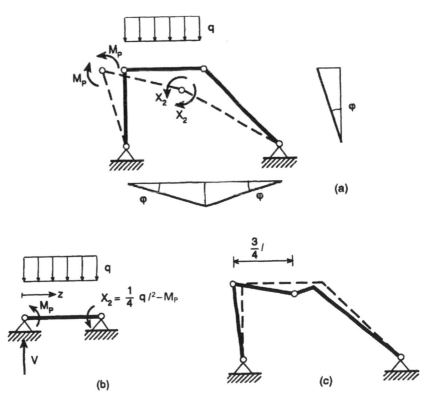

Figure 18.19

Equilibrium with regard to rotation of the cross member about the right-hand node yields the shear V, transmitted to the cross member by the left-hand stanchion (Figure 18.19(b))

$$Vl - \frac{1}{4}ql^2 - \frac{1}{2}ql^2 = 0 \tag{18.66}$$

whence we obtain

$$V = \frac{3}{4}ql \tag{18.67}$$

The shear on the cross member vanishes, and, therefore the moment is maximum, for $V = qz$, and hence for $z = \frac{3}{4}l$

$$M_{\max} = M\left(\frac{3}{4}l\right) = \frac{3}{4}Vl - M_P - \frac{1}{2}q\left(\frac{3}{4}l\right)^2 \tag{18.68}$$

or

$$M_{\max} = \frac{9}{32}ql^2 - M_P \tag{18.69}$$

The condition of second and ultimate plastic hinge formation is $M_{\max} = M_P$, from which we find the collapse load,

$$q_2 = q_P = \frac{64}{9}\frac{M_P}{l^2} \simeq 7.11\frac{M_P}{l^2} \tag{18.70}$$

whilst the mechanism of collapse consists of the articulated parallelogram of Figure 18.19(c). The safety factor is

$$\frac{q_P}{q_e} = \frac{q_2}{\frac{2}{3}q_1} \simeq 1.93 \tag{18.71}$$

18.4 Law of normality of incremental plastic deformation

As will be illustrated in this and the ensuing sections, it is possible to avoid the unwieldy incremental plastic calculation and focus attention on the ultimate condition of collapse, when the entire structure, or part thereof, undergoes large increments of displacement resulting from small increments of load. This can be achieved by means of the Theorems of Plastic Limit Analysis, which shall be demonstrated in the next section.

The present section will provide a preliminary demonstration of the two fundamental properties possessed, respectively, by the surface of plastic deformation in the space of principal stresses,

$$F(\sigma_1, \sigma_2, \sigma_3) = 0 \tag{18.72}$$

and the incremental plastic deformation.

As in the uniaxial condition, where the element of material is in the elastic state for $|\sigma| < \sigma_P$, likewise in the biaxial (plane stress) condition, the element

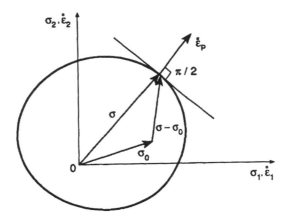

Figure 18.20

of material is in the elastic state for $F(\sigma_1, \sigma_2) < 0$. The function F was obtained in Section 8.12 according to Von Mises' criterion

$$F_M(\sigma_1, \sigma_2) = (\sigma_1^2 + \sigma_2^2 - \sigma_1\sigma_2) - \sigma_P^2 \qquad (18.73)$$

and Tresca's criterion

$$F_T(\sigma_1, \sigma_2) = \max\left\{|\sigma_1|, |\sigma_2|, |\sigma_1 - \sigma_2|\right\} - \sigma_P \qquad (18.74)$$

Whereas in the uniaxial condition the characteristics of plastic flow are evident (i.e. there is a dilation collinear to stress), in multiaxial conditions it is difficult to make out the mechanics of the deformation.

Let us consider an element of a two-dimensional solid subjected to the stress condition (Figure 18.20)

$$\{\sigma_0\} = [\sigma_1, \sigma_2]^T \qquad (18.75)$$

Suppose that an increment $\{\sigma\} - \{\sigma_0\}$ is applied to the same element, and that subsequently this increment is removed in a quasi-static manner. **Drucker's Postulate** states that the material may be defined as stable when the work performed in the loading cycle is non-negative. For a stress condition $\{\sigma\}$ lying on the surface of plastic deformation $F(\{\sigma\}) = 0$, and for each stress condition $\{\sigma_0\}$ which is admissible and thus contained within the elastic domain or lying on the boundary, we must have

$$\left(\{\sigma\} - \{\sigma_0\}\right)^T \{\dot{\varepsilon}_P\} \geq 0 \qquad (18.76)$$

where $\{\dot{\varepsilon}_P\}$ is the incremental plastic strain which occurs when the stress reaches $\{\sigma\}$. It is possible to give a highly significant geometrical interpretation, superposing the spaces $\{\sigma\}$ and $\{\dot{\varepsilon}_P\}$ (Figure 18.20): the scalar product (18.76) is always positive or at least zero. It thus follows that:

1. in each regular point of the surface of plasticity (single tangent plane), the incremental plastic strain $\{\dot{\varepsilon}_P\}$ is **normal** to the surface itself;
2. the surface of plasticity is **convex**.

593

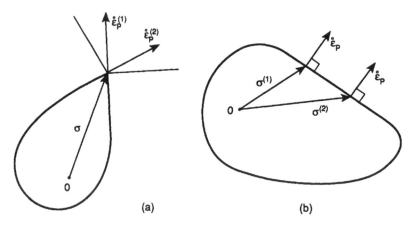

Figure 18.21

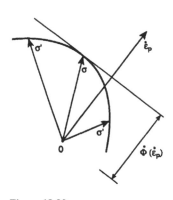

Figure 18.22

In the cusps of the surface of plasticity (Figure 18.21 (a)), $\{\dot{\varepsilon}_P\}$ cannot be external to the cone defined by the normals to the infinite tangent planes. In this case, more than one vector $\{\dot{\varepsilon}_P\}$ may correspond to a single vector $\{\sigma\}$. On the other hand, in the portions where the surface of plasticity is linear (i.e. not strictly convex), more than one vector $\{\sigma\}$ corresponds to a single vector $\{\dot{\varepsilon}_P\}$ (Figure 18.21(b)). These two conditions are both present in Tresca's hexagon.

The elastic domain includes the origin, and hence the inequality (18.76), when $\{\sigma_0\} = \{0\}$, becomes

$$\{\sigma\}^T\{\dot{\varepsilon}_P\} = \dot{\Phi}\left(\{\dot{\varepsilon}_P\}\right) > 0 \tag{18.77}$$

where $\dot{\Phi}$ represents the energy dissipated in the unit of volume and is a function solely of the incremental plastic strain. This consideration remains valid also when the surface of plastic deformation presents cusps and linear portions. Consequently, the following statement is equivalent to Drucker's Postulate: the energy dissipated in the unit of volume is a function only of the incremental plastic strain. Also from this it is possible to deduce the **law of normality** and the **convexity of the surface of plastic deformation**. Equation (18.77) shows in fact that each stress condition $\{\sigma\}$ capable of producing the incremental plastic strain $\{\dot{\varepsilon}_P\}$ must be on the plane normal to $\{\dot{\varepsilon}_P\}$ and distant $\dot{\Phi}(\{\dot{\varepsilon}_P\})$ from the origin (Figure 18.22). As $\{\dot{\varepsilon}_P\}$ is made to turn about the origin, all these planes envelop the surface of plastic deformation, which is thus convex.

If $\{\dot{\sigma}\}$ is the incremental stress vector corresponding to the incremental plastic strain vector $\{\dot{\varepsilon}_P\}$, we have

$$\{\dot{\sigma}\}^T\{\dot{\varepsilon}_P\} \geq 0 \tag{18.78}$$

assuming $\{\sigma\}$ to be the initial stress condition and applying equation (18.76). For an elastic–perfectly plastic material we have in particular

$$\{\dot{\sigma}\}^T\{\dot{\varepsilon}_P\} = 0 \tag{18.79}$$

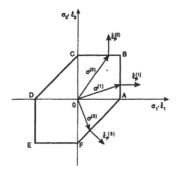

Figure 18.23

whereas for a strain-softening material we have

$$\{\dot{\sigma}\}^{\mathrm{T}}\{\dot{\varepsilon}_P\} < 0 \tag{18.80}$$

and Drucker's Postulate is violated.

Figure 18.23 represents Tresca's criterion in two dimensions and the corresponding mechanisms of plastic flow (incremental plastic strain). Along the sides *AB, BC, DE, EF* only one of the principal dilations $\dot{\varepsilon}_1$, $\dot{\varepsilon}_2$ is activated, whilst along the sides *CD* and *FA* one dilation is positive and the other is negative; these are activated simultaneously and with equal intensity.

18.5 Theorems of plastic limit analysis

Let us consider a rigid–perfectly plastic solid, subjected to a condition of proportional load, measured by the parameter λ (Figure 18.24). A stress field is said to be **statically admissible** when it is in equilibrium with the external load λ and at each point of the solid we have $F \le 0$. On the other hand, a collapse mechanism is said to be **kinematically admissible** when the external constraints are respected and the corresponding dissipated energy is positive.

Theorem of maximum dissipated energy
Given an incremental plastic strain $\{\dot{\varepsilon}_P\}$, the energy dissipated by the stress $\{\sigma\}$ corresponding to this strain (Figure 18.22) is greater than or equal to the energy dissipated by any other possible stress $\{\sigma'\}$:

$$\{\sigma\}^{\mathrm{T}}\{\dot{\varepsilon}_P\} \ge \{\sigma'\}^{\mathrm{T}}\{\dot{\varepsilon}_P\} \tag{18.81}$$

The foregoing inequality is valid at each point of the solid and therefore, on the basis of equation (18.77), it is possible to write

$$\int_V \dot{\Phi} \left(\{\dot{\varepsilon}_P\}\right)\mathrm{d}V \ge \int_V \{\sigma'\}^{\mathrm{T}}\{\dot{\varepsilon}_P\}\mathrm{d}V \tag{18.82}$$

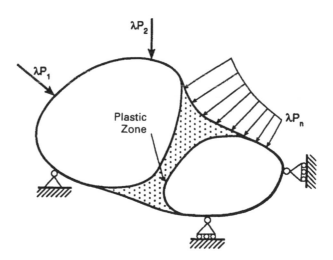

Figure 18.24

595

Static theorem (upper bound theorem)
The multiplier of the loads λ^- corresponding to any statically admissible stress field is less than or equal to the multiplier of collapse λ_P.

Let $\{\sigma^-\}$ be a statically admissible stress field and λ^- the corresponding multiplier of the external loads. Then let $\{\sigma\}$ be the stress field of collapse and $\{\dot{\eta}\}$, $\{\dot{\varepsilon}_P\}$ the incremental fields of displacement and of plastic strain, respectively, at the moment of collapse. Application of the Principle of Virtual Work yields the following relations:

$$\int_V \{\sigma^-\}^T \{\dot{\varepsilon}_P\} dV = \sum_i \lambda^- P_i \dot{\eta}_i \tag{18.83a}$$

$$\int_V \{\sigma\}^T \{\dot{\varepsilon}_P\} dV = \sum_i \lambda_P P_i \dot{\eta}_i \tag{18.83b}$$

where with P_i, $i = 1, 2, ..., n$ are indicated the external loads applied to the solid. Recalling the inequality (18.76), we obtain

$$\int_V \left(\{\sigma\} - \{\sigma^-\} \right)^T \{\dot{\varepsilon}_P\} dV \geq 0 \tag{18.84}$$

and hence

$$\lambda_P \geq \lambda^- \tag{18.85}$$

Kinematic theorem (lower bound theorem)
The multiplier of the loads λ^+ corresponding to any kinematically admissible collapse mechanism is greater than or equal to the multiplier of actual collapse λ_P.

Let $\{\dot{\eta}^+\}$, $\{\dot{\varepsilon}^+\}$ be the incremental fields, respectively, of displacement and of plastic strain, corresponding to a kinematically admissible mechanism of collapse. Then let $\{\sigma\}$ be the stress field of actual collapse. The multiplier of the external loads λ^+ corresponding to the kinematically admissible collapse mechanism is given by the following energy balance:

$$\int_V \dot{\Phi} \left(\{\dot{\varepsilon}^+\} \right) dV = \sum_i \lambda^+ P_i \dot{\eta}_i^+ \tag{18.86}$$

Application of the Principle of Virtual Work to the stress field of actual collapse $\{\sigma\}$ and to the kinematically admissible collapse mechanism $\{\dot{\varepsilon}^+\}$ yields

$$\int_V \{\sigma\}^T \{\dot{\varepsilon}^+\} dV = \sum_i \lambda_P P_i \dot{\eta}_i^+ \tag{18.87}$$

On the other hand, for the inequality (18.82) we have

$$\int_V \dot{\Phi} \left(\{\dot{\varepsilon}^+\} \right) dV \geq \int_V \{\sigma\}^T \{\dot{\varepsilon}^+\} dV \tag{18.88}$$

From relations (18.86), (18.87) and (18.88), there follows

$$\lambda^+ \geq \lambda_P \tag{18.89}$$

596

Mixed theorem

If the multiplier of the external loads λ corresponds to a statically admissible stress field and, at the same time, to a kinematically admissible collapse mechanism, then we have

$$\lambda = \lambda_P \qquad (18.90)$$

This statement follows immediately from the two theorems demonstrated previously, since the **actual collapse mechanism** represents a kinematically admissible mechanism and, at the same time, presupposes a statically admissible stress field.

Theorem of addition of material

A dimensional increment of a perfectly plastic solid cannot produce a decrement in the collapse load.

In fact, the sum of the stress field of collapse in the original solid and of an identically zero stress field in the portion of additional material constitutes a statically admissible stress field. This means that the collapse load of the new solid is greater than or equal to that of the original solid, and certainly not less.

The properties of **convexity** of the elastic domain and of **normality** of the plastic incremental deformation, as well as the **theorems of limit analysis**, just demonstrated for the three-dimensional solids, can readily be extended to two-dimensional solids (plates) and one-dimensional solids (beams), replacing the stress vector $\{\sigma\}$ with the static characteristics vector $\{Q\}$, and the vector of incremental plastic strains $\{\dot{\varepsilon}_P\}$ with the incremental vector of the deformation characteristics $\{\dot{q}_P\}$. As an example of convexity of the elastic domain and of normality of the plastic flow, consider the plastic limit of bending moment *vs.* axial force interaction in Figure 18.9. In the case where, as often occurs, only the bending moment M is considered as the active characteristic rather than the incremental vector $\{\dot{q}_P\}$, it is sufficient to consider the plastic increment of the curvature, $\dot{\chi}_P$ or, more simply, the relative rotation φ.

18.6 Beam systems loaded proportionally by concentrated forces

In the case of statically indeterminate systems of beams loaded proportionally by concentrated forces, application of the static theorem reduces the solution to that of a problem of **linear programming**. As an illustration of this, let us consider the continuous beam of Figure 18.25(a), constrained by three supports and a built-in constraint and loaded proportionally by two forces concentrated in the first and third spans. This structure has three degrees of redundancy, as is noted from an examination of the equivalent statically determinate system of Figure 18.25(b), and presents five critical cross sections for the formation of the plastic hinges; viz. the two central supports, the built-in constraint and the two sections in which the external forces are applied.

The total bending moment is expressible as the sum of four contributions, due to the external forces and to the redundant moments (Figure 18.25(b))

$$M(z) = \lambda M^{(0)} + \sum_{j=1}^{n} X_j M^{(j)} \qquad (18.91)$$

597

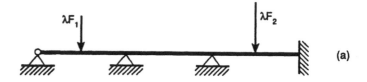

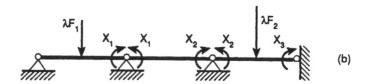

Figure 18.25

with $n = 3$. In the $m = 5$ critical cross sections we thus have

$$M_i = \lambda M_i^{(0)} + \sum_{j=1}^{n} X_j M_i^{(j)}, \quad \text{for } i = 1, 2, \ldots, m \qquad (18.92)$$

The static theorem states that the plastic collapse load is represented by the maximum value that the target function λ can assume in respect of the following $2m$ constraints:

$$- M_P \leqslant \lambda M_i^{(0)} + \sum_{j=1}^{n} X_j M_i^{(j)} \leqslant M_P, \quad \text{for } i = 1, 2, \ldots, m \qquad (18.93)$$

In the case of structures with many degrees of redundancy, this problem is resolvable with procedures of automatic calculation. It is in fact a problem of **linear programming** in the variables λ; $X_1, X_2, \ldots, X_n$.

As an example, let us reconsider the elementary case of a beam built in at either end and subjected to the vertical load λF in the centre (Figure 18.26(a)),

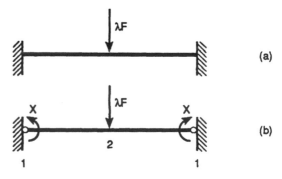

Figure 18.26

whereby $n = 1$, $m = 2$. Designating the built-in constraint moment as X (Figure 18.26 (b)), we have

$$M_1 = -X \tag{18.94a}$$

$$M_2 = -X + \frac{1}{4}\lambda Fl \tag{18.94b}$$

and hence the inequalities (18.93) in this case take on the following form:

$$-M_P \leqslant -X \leqslant M_P \tag{18.95a}$$

$$-M_P \leqslant \left(-X + \frac{1}{4}\lambda Fl\right) \leqslant M_P \tag{18.95b}$$

From relations (18.95) we deduce the following four inequalities:

$$X \geqslant -M_P \tag{18.96a}$$

$$X \leqslant M_P \tag{18.96b}$$

$$X \leqslant M_P + \frac{1}{4}\lambda Fl \tag{18.96c}$$

$$X \geqslant -M_P + \frac{1}{4}\lambda Fl \tag{18.96d}$$

which, on the plane X–λ, define the parallelogram represented in Figure 18.27. The maximum value of λ on this domain is given by the ordinate of point A

$$\lambda_{\mathrm{max}} = 8\frac{M_P}{Fl} \tag{18.97}$$

from which we again find the collapse load

$$F_P = \lambda_{\mathrm{max}} F = 8\frac{M_P}{l} \tag{18.98}$$

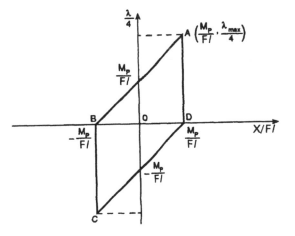

Figure 18.27

599

An alternative method for solving beam systems loaded proportionally by concentrated forces is that proposed by Neal and Symonds, which is also called the **method of combining mechanisms**. According to this method, each mechanism of collapse can be considered as the combination of a certain number of independent mechanisms. To each mechanism of collapse it is possible to apply the Principle of Virtual Work, so as to determine the corresponding multiplier of the loads λ. The actual mechanism of collapse is distinguished from amongst all the virtual mechanisms by the fact that, because of the kinematic theorem, it presents the minimum value of the multiplier λ. It is then a matter of examining the independent mechanisms with low values of the multiplier λ, and seeking to combine them to form mechanisms with values of λ still lower. To verify the validity of the result, it is then necessary to check its static admissibility.

The method proposed by Neal and Symonds will now be illustrated with reference to a simple portal frame, subjected to two equal forces, one horizontal and the other vertical (Figure 18.28(a)). Since the degrees of redundancy are $n = 3$, and the number of critical cross sections is $m = 5$, the number of supplementary equations of equilibrium and hence the number of independent mechanisms must be $m - n = 2$. As independent mechanisms, let the two represented in Figures 18.28(b) and 18.28(d) be chosen. These cause the horizontal force and the vertical force alternately to perform work. On the other hand, it may be demonstrated that both involve moment diagrams that are statically admissible (Figures 18.28(c), (e)). Applying the Principle of Virtual Work to the beam mechanism (Figure 18.28(b)) produces the equation

$$Fl\varphi - 4M_P\varphi = 0 \qquad (18.99)$$

from which we obtain

$$F = 4\frac{M_P}{l} \qquad (18.100)$$

Applying the Principle of Virtual Work to the sidesway mechanism (Figure 18.28(d)) produces, on the other hand, the same result.

If we now proceed to sum up algebraically (or combine) the two abovementioned mechanisms, we shall obtain the combined mechanism of Figure 18.28(f), with four plastic hinges in the cross sections 1, 3, 4, 5. The corresponding moment diagram (Figure 18.28(g)) proves to be statically admissible, and hence we may conclude that the collapse mechanism of Figure 18.28(f) is the actual one. On the other hand, application of the Principle of Virtual Work yields a collapse load smaller than the one corresponding to each of the two elementary mechanisms

$$2Fl\varphi - 6M_P\varphi = 0 \qquad (18.101)$$

whence we obtain

$$F_P = 3\frac{M_P}{l} \qquad (18.102)$$

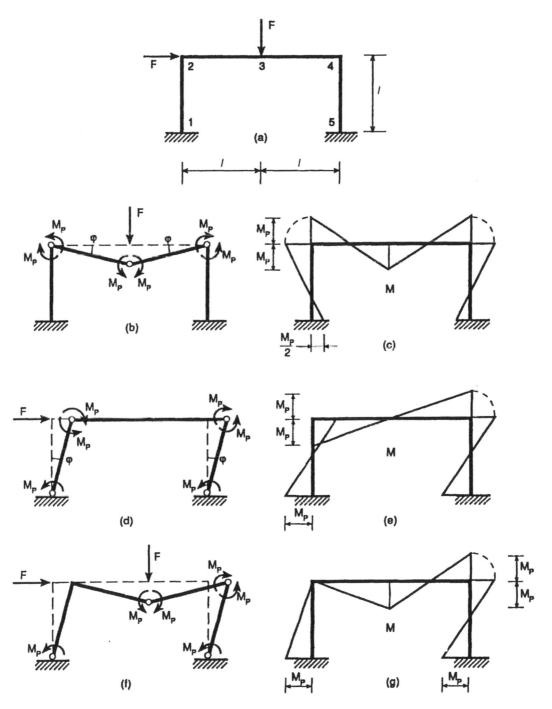

Figure 18.28

601

18.7 Beam systems loaded proportionally by distributed forces

The solution of statically indeterminate beam systems loaded proportionally also by distributed forces presents greater difficulties than does the solution of systems involving concentrated forces. This is due to the impossibility of identifying from the outset a finite number of critical cross sections. Since therefore there does not exist any systematic method, the procedure is one of trial and error, applying the kinematic and static theorems alternately.

Let us consider, for instance, the portal frame of Figure 18.29(a), subjected to a load uniformly distributed over the cross member and to a horizontal concentrated force of equal intensity. Let us take as collapse mechanism the actual one of the concentrated force scheme (Figure 18.28(a)) and apply the Principle of Virtual Work (Figure 18.29(b))

$$2ql\left(\frac{l}{2}\varphi\right) + 2ql(l\varphi) - 6M_P\varphi = 0 \tag{18.103}$$

from which we obtain the load

$$q = 2\frac{M_P}{l^2} \tag{18.104}$$

The constraint reactions at cross section 5 are obtained by assuming that, also in sections 3 and 4, the bending moment is equal to its plastic value M_P (Figure 18.29(c))

$$M_4 = -M_P + Hl = M_P \tag{18.105a}$$

$$M_3 = M_P - Hl + Vl - \frac{1}{2}ql^2 = M_P \tag{18.105b}$$

whence we obtain

$$H = 2\frac{M_P}{l} \tag{18.106a}$$

$$V = 3\frac{M_P}{l} \tag{18.106b}$$

The moment function on the beam is given by the sum of four contributions,

$$M(z) = M_P + 3\frac{M_P}{l}z - 2\frac{M_P}{l}l - \frac{1}{2}\left(2\frac{M_P}{l^2}\right)z^2 \tag{18.107}$$

$$= -M_P + 3\frac{M_P}{l}z - \frac{M_P}{l^2}z^2$$

whilst the shear is given by two contributions,

$$T(z) = \frac{\mathrm{d}M}{\mathrm{d}z} = 3\frac{M_P}{l} - 2\frac{M_P}{l^2}z \tag{18.108}$$

and vanishes for $z = \frac{3}{2}l$. The maximum moment is thus

$$M_{\max} = M\left(\frac{3}{2}l\right) = \frac{5}{4}M_P \tag{18.109}$$

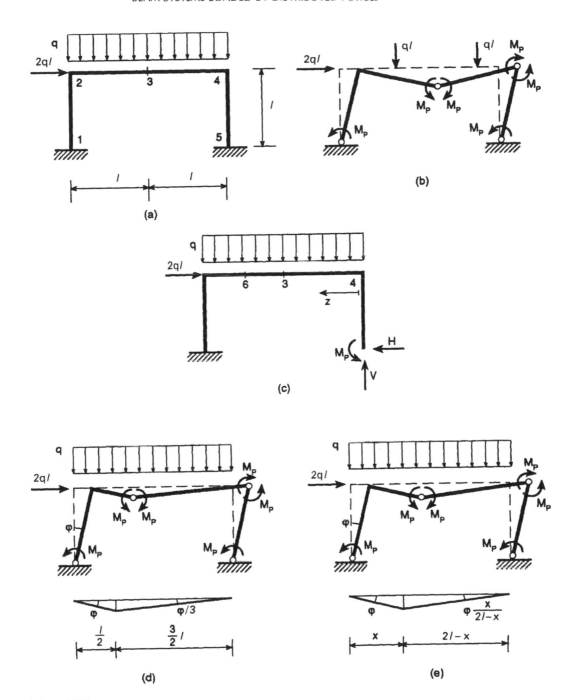

Figure 18.29

603

and, since it is greater than M_P, it reveals the static inadmissibility of the mechanism of Figure 18.29(b). On the other hand, dividing the load resulting from equation (18.104) by 5/4, we obtain a statically admissible scheme, and hence an application of the static and kinematic theorems leads to the conclusion that the actual collapse load must fall within the following interval:

$$1.6\frac{M_P}{l^2} < q_P < 2\frac{M_P}{l^2} \tag{18.110}$$

Since the interval given by relation (18.110) is still not sufficiently narrow, we assume as collapse mechanism of second approximation that presenting three plastic hinges, again in the sections 1, 4, 5, and the fourth hinge in the section which, in the previous scheme, was subjected to the maximum moment M_{max} (Figure 18.29(d)). Applying the Principle of Virtual Work yields the equation

$$q\frac{l}{2}\left(\frac{l}{4}\varphi\right) + \frac{3}{2}ql\left(\frac{3}{4}l\right)\left(\frac{1}{3}\varphi\right) + 2ql(l\varphi) - \tag{18.111}$$

$$4M_P\varphi - 2M_P\left(\frac{1}{3}\varphi\right) = 0$$

which presents the solution

$$q = \frac{28}{15}\frac{M_P}{l^2} \tag{18.112}$$

The constraint reactions at section 5 are obtained by assuming that, also in cross sections 4 and 6 (Figure 18.29(c)), the bending moment is equal to its plastic value M_P

$$M_4 = -M_P + Hl = M_P \tag{18.113a}$$

$$M_6 = M_P - Hl + V\frac{3}{2}l - \frac{1}{2}q\left(\frac{3}{2}l\right)^2 = M_P \tag{18.113b}$$

whence we obtain

$$H = 2\frac{M_P}{l} \tag{18.114a}$$

$$V = \frac{41}{15}\frac{M_P}{l} \tag{18.114b}$$

The bending moment and shearing force on the horizontal beam are therefore represented by the following functions:

$$M(z) = M_P + \left(\frac{41}{15}\frac{M_P}{l}\right)z - 2\frac{M_P}{l}l - \frac{1}{2}\left(\frac{28}{15}\frac{M_P}{l^2}\right)z^2 \tag{18.115a}$$

$$T(z) = \frac{dM}{dz} = \frac{41}{15}\frac{M_P}{l} - \frac{28}{15}\frac{M_P}{l^2}z \tag{18.115b}$$

The shear vanishes for $z = \frac{41}{28}l$, and the maximum bending moment is thus

$$M_{max} = M\left(\frac{41}{28}l\right) = \frac{841}{840}M_P \tag{18.116}$$

On the other hand, dividing the load obtained from equation (18.112) by 841/840, we obtain a statically admissible scheme, and thus an application of the static and kinematic theorems yields the following interval for the actual collapse load:

$$\frac{840}{841} \times \frac{28}{15} \frac{M_P}{l^2} < q_P < \frac{28}{15} \frac{M_P}{l^2} \qquad (18.117)$$

This interval is extremely narrow and, for engineering purposes, yields the actual collapse load with sufficient approximation ($\sim 1\ \text{‰}$).

To improve this approximation still further, it would suffice to consider a third mechanism with the hinge in $z = \frac{41}{28}l$, but this is not necessary, since it is possible to identify the actual collapse mechanism by minimizing the load q as the position of the plastic hinge on the horizontal beam varies (kinematic theorem).

Consider the mechanism of Figure 18.29(e), with the plastic hinge in an intermediate position of the cross member, at a distance x from the left-hand fixed joint-node. As the diagram of vertical displacements shows, the left-hand portion turns clockwise by the angle φ, whilst the right-hand portion turns counterclockwise by the angle

$$\vartheta = \varphi \frac{x}{2l - x} \qquad (18.118)$$

Application of the Principle of Virtual Work provides the following equation:

$$qx\left(\frac{x}{2}\varphi\right) + \frac{1}{2}q(2l - x)^2 \varphi \frac{x}{2l - x} + 2ql(l\varphi) - \qquad (18.119)$$

$$4M_P\varphi - 2M_P\varphi \frac{x}{2l - x} = 0$$

from which we obtain the load

$$q(x) = 2\frac{M_P}{l} \frac{4l - x}{4l^2 - x^2} \qquad (18.120)$$

The derivative of this load with respect to the coordinate x

$$\frac{dq}{dx} = \frac{-x^2 + 8lx - 4l^2}{(4l^2 - x^2)^2} \qquad (18.121)$$

vanishes for

$$x = 2l(2 \pm \sqrt{3}) \qquad (18.122)$$

The larger root is to be rejected, whereas if we substitute the value $x = 2l\,(2 - \sqrt{3})$ in expression (18.120), we obtain the actual collapse load

$$q_P = \frac{M_P}{l^2} \frac{3}{12 - 6\sqrt{3}} \qquad (18.123)$$

Rationalizing the ratio (18.123) up to the seventh decimal place, we have

$$q_P = 1.8660254 \frac{M_P}{l^2} \tag{18.124}$$

so that the inequalities (18.117) are verified:

$$1.8644471 < 1.8660254 < 1.8666667$$

Finally it should be noted that, if we put $2ql = F$ to make a comparison with the case where also the vertical load on the beam is concentrated (Figure 18.28(a)), the collapse load given by equation (18.124) can be expressed in the form

$$F_P = 2q_P l \simeq 3.73 \frac{M_P}{l} \tag{18.125}$$

and is thus greater than the collapse load given by equation (18.102) for the concentrated force in the centre.

The case of the portal frame with inclined stanchion of Figure 15.21(a) has thus been solved in Section 18.3 by means of an incremental analysis. If, instead, we had chosen to proceed by trial and error, applying the upper bound and lower bound theorems, we could have assumed the first-approximation mechanism illustrated in Figure 18.19(a), with two plastic hinges located in the fixed joint-nodes. Application of the Principle of Virtual Work yields the load

$$q = 8 \frac{M_P}{l^2} \tag{18.126}$$

which turns out to be greater than the collapse load given by equation (18.70) by virtue of the kinematic theorem. The fact of having imposed at the ends of the cross member the two plastic moments M_P means that the bending moment in the right-hand part of the cross member exceeds the value M_P (Figure 18.30 (a)). This is, on the other hand, a statically inadmissible situation.

To determine the maximum value of the moment, let us isolate the cross member (Figure 18.30 (b)) and identify the cross section in which the shear vanishes

$$T(z) = 2 \frac{M_P}{l} + \frac{1}{2} ql - qz = 0 \tag{18.127a}$$

$$\text{for } z = \frac{l}{2} + 2 \frac{M_P}{ql} = \frac{3}{4} l \tag{18.127b}$$

The maximum bending moment is thus

$$M_{\max} = \left(6 \frac{M_P}{l} \right) \left(\frac{3}{4} l \right) - M_P - \frac{1}{2} \left(8 \frac{M_P}{l^2} \right) \left(\frac{3}{4} l \right)^2 = \frac{5}{4} M_P \tag{18.128}$$

Then dividing the load given by equation (18.126) by 5/4, the system is reduced to a statically admissible one, whereby the following inequalities result:

$$6.4 \frac{M_P}{l^2} < q_P < 8 \frac{M_P}{l^2} \tag{18.129}$$

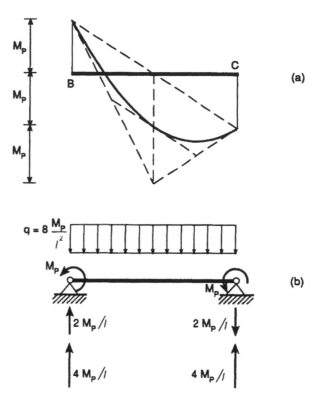

Figure 18.30

These inequalities are confirmed by solution (18.70). Finally, it should be noted that the second-approximation mechanism, with the hinge at a distance $z = \frac{3}{4}l$ from the left-hand node of the beam, is identified with the actual collapse mechanism (Figure 18.19 (c)).

As our last example of application of the limit analysis theorems, let us examine the case of the portal frame with a strut, illustrated in Figure 15.9(a). The moment and the shear on the portion *BF* are described by the following functions (Figure 18.31(a)):

$$M(z) = \frac{29}{124} qlz - \frac{3}{496} ql^2 - \frac{1}{2} qz^2 \qquad (18.130a)$$

$$T(z) = \frac{\mathrm{d}M}{\mathrm{d}z} = \frac{29}{124} ql - qz \qquad (18.130b)$$

The point of zero shear is given by

$$z = \frac{29}{124} l \qquad (18.131)$$

and hence the maximum bending moment is

$$M_{\max} = M\left(\frac{29}{124} l\right) = \frac{655}{30752} ql^2 \qquad (18.132)$$

607

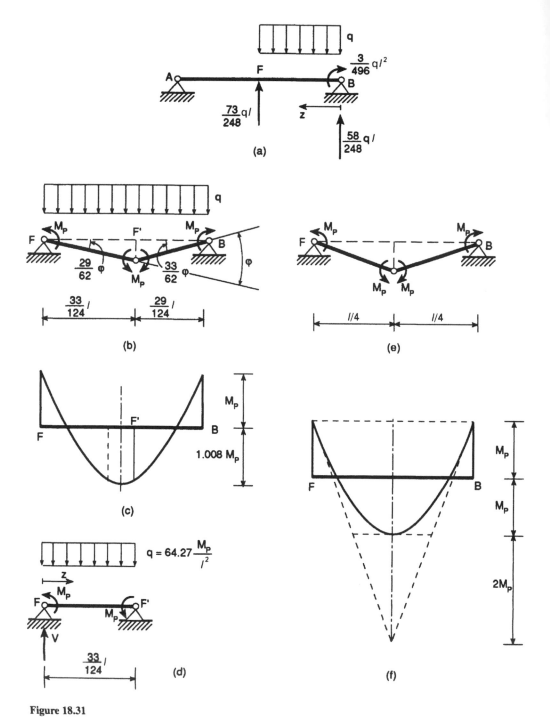

Figure 18.31

The load which produces the first plastic hinge is therefore

$$q_1 = \frac{30752}{655} \frac{M_P}{l^2} \simeq 46.95 \frac{M_P}{l^2} \qquad (18.133)$$

If we consider the mechanism of Figure 18.31(b), the Principle of Virtual Work yields the equation

$$\frac{1}{2} q \left(\frac{33}{124} l \right)^2 \left(\frac{29}{62} \varphi \right) + \frac{1}{2} q \left(\frac{29}{124} l \right)^2 \left(\frac{33}{62} \varphi \right) - 2 M_P \varphi = 0 \qquad (18.134)$$

from which we obtain the load

$$q \left(\frac{33^2 \times 29 + 33 \times 29^2}{124^2 \times 62} \right) = 4 \frac{M_P}{l^2} \qquad (18.135)$$

or

$$q \simeq 64.27 \frac{M_P}{l^2} \qquad (18.136)$$

On the other hand, the scheme of Figure 18.31(b) proves not to be statically admissible, since in an intermediate portion of the beam BF the moment assumes values greater than M_P (Figure 18.31(c)). In fact, isolating the portion of length $\frac{33}{124}l$, contained between the plastic hinges F and F' (Figure 18.31(d)), we find the shear at F:

$$V \simeq \frac{2 M_P}{(33l/124)} + 64.27 \frac{M_P}{l^2} \left(\frac{33}{248} l \right) \simeq 16.067 \frac{M_P}{l} \qquad (18.137)$$

The shear between F and F',

$$T(z) = V - qz \qquad (18.138)$$

vanishes for

$$z = \frac{V}{q} \simeq \frac{16.067}{64.27} l \approx 0.25 \, l \qquad (18.139)$$

from which we obtain the maximum bending moment

$$M_{max} = M \left(\frac{l}{4} \right) \qquad (18.140)$$

$$\simeq 16.067 \frac{M_P}{l} \left(\frac{l}{4} \right) - M_P - \frac{1}{2} \left(64.27 \frac{M_P}{l^2} \right) \left(\frac{l}{4} \right)^2$$

$$\simeq 1.008 M_P$$

which is not statically admissible.

Dividing the load given by equation (18.136) by 1.008, we obtain, on the other hand, a statically admissible system. The actual collapse load q_P must then be greater than

$$q = \frac{64.27}{1.008} \frac{M_P}{l^2} \qquad (18.141)$$

and consequently contained within the following interval:

$$63.76 \frac{M_P}{l^2} < q_P < 64.27 \frac{M_P}{l^2} \tag{18.142}$$

Let us assume as our second-approximation mechanism the one presenting the plastic hinges at B, F and in the centre of the portion BF, which proved to be the location of the maximum bending moment in the previous step (Figure 18.31(e)). The Principle of Virtual Work yields the load

$$q_P = 64 \frac{M_P}{l^2} \tag{18.143}$$

As the scheme of Figure 18.31(f) shows, the corresponding bending moment diagram is statically admissible, whence, by virtue of the mixed theorem, a mechanism that is both kinematically and statically admissible coincides with the actual collapse mechanism. The inequalities (18.142) are in fact verified by the collapse load given by equation (18.143).

18.8 Non-proportionally loaded beam systems

In the case of statically indeterminate beam systems, loaded non-proportionally by two or more concentrated forces, applying the kinematic theorem makes it possible to define the limit of collapse in the space of these forces.

In the case, for example, of the continuous beam of Figure 18.32(a), loaded by two non-proportional forces F_1 and F_2, the introduction of five plastic hinges in the critical sections makes it possible to define two different collapse mechanisms (Figure 18.32 (b)). Applying the Principle of Virtual Work to each of them (and to its converse) yields the following equations:

$$-3M_P \varphi \pm F_1 \frac{l}{2} \varphi = 0 \tag{18.144a}$$

$$-4M_P \varphi \pm F_2 \frac{l}{2} \varphi = 0 \tag{18.144b}$$

whence we obtain

$$F_1 = \pm 6 \frac{M_P}{l} \tag{18.145a}$$

$$F_2 = \pm 8 \frac{M_P}{l} \tag{18.145b}$$

The limit of collapse in the plane F_1–F_2 is thus represented by the rectangle of Figure 18.32 (c). In the case where

$$-\frac{4}{3} < \frac{F_2}{F_1} < \frac{4}{3} \tag{18.146}$$

we have the activation of the first mechanism; otherwise, we have the activation of the second mechanism. The properties of convexity of the surface of plastic deformation and of normality of the incremental plastic deformation in fact also apply in the case of beam systems loaded by two or more concentrated forces.

A second example may be represented by the portal frame, already examined in Sections 18.6 and 18.7, loaded in this case by two independent

610

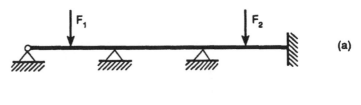

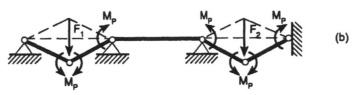

(a)

(b)

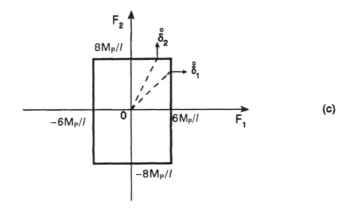

(c)

Figure 18.32

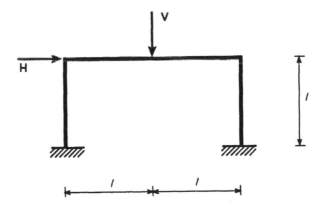

Figure 18.33

611

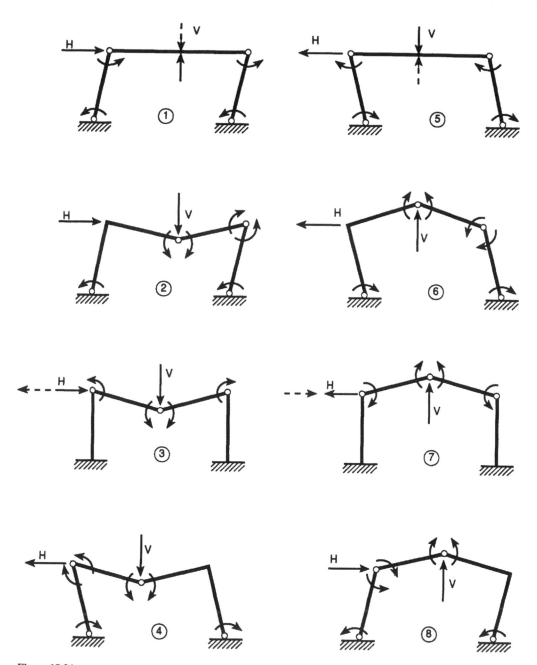

Figure 18.34

concentrated forces, *H* and *V* (Figure 18.33). The introduction of five plastic hinges in the critical sections enables the four different collapse mechanisms 1, 2, 3, 4 to be defined (Figure 18.34). Applying the Principle of Virtual Work to these four collapse mechanisms and to their respective opposites, 5, 6, 7, 8, yields the following equations:

612

$$- 4M_P\varphi \pm Hl\varphi = 0 \qquad\qquad (18.147)$$
$$- 6M_P\varphi \pm (Vl\varphi + Hl\varphi) = 0$$
$$- 4M_P\varphi \pm Vl\varphi = 0$$
$$- 6M_P\varphi \mp (Vl\varphi - Hl\varphi) = 0$$

Once the factor φ, which does not affect the problem, has been cancelled out, we then have the equations of the eight straight lines in the plane H–V, to which the respective sides of the boundary of collapse belong (Figure 18.35). The activations for each of the four pairs of mechanisms are given by

$$\left|\frac{V}{H}\right| < \frac{1}{2} \qquad\qquad (18.148\text{a})$$

$$\frac{1}{2} < \frac{V}{H} < 2 \qquad\qquad (18.148\text{b})$$

$$\left|\frac{V}{H}\right| > 2 \qquad\qquad (18.148\text{c})$$

$$-2 < \frac{V}{H} < -\frac{1}{2} \qquad\qquad (18.148\text{d})$$

respectively. Also in this case the properties of convexity of the plastic limit and of normality of the incremental deformation are both verified. In fact, when the ratio V/H is sufficiently small, the sidesway mechanism is activated

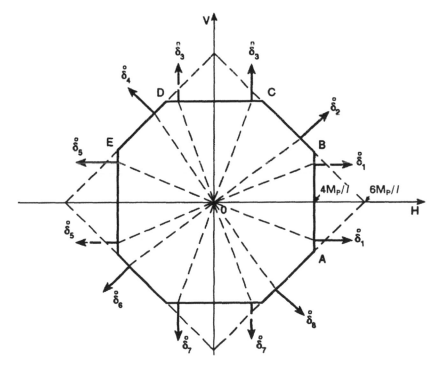

Figure 18.35

613

(1, 5). For intermediate values of V/H the combined mechanisms are activated (2, 4, 6, 8). For sufficiently high values of V/H the beam mechanism is activated (3, 7).

18.9 Cyclic loads and shake-down

Let us consider again the system of parallel connecting rods of Figure 18.1, and let us suppose that the rigid cross member is loaded repeatedly with a pulsating force (Figure 18.36). As was seen in Section 18.1, for

$$M_{max} = \frac{l}{8}ql^2\left(1 - \frac{2M_P}{ql^2}\right)^2 \qquad (18.149)$$

the behaviour of the system is elastic, and hence both loading and unloading occur along the segment 01 of Figure 18.37.

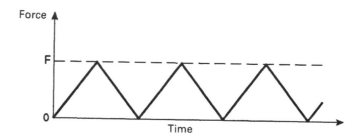

Figure 18.36

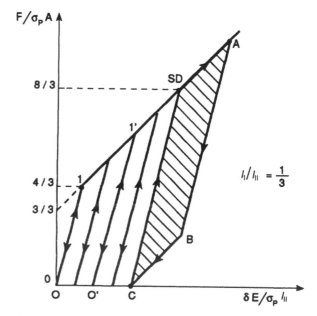

Figure 18.37

For greater values of the maximum load F, the central rod yields, so that, on unloading, this rod is found to be compressed by the lateral ones, which are assumed to obey an elastic constitutive law, devoid of yielding. On these hypotheses, a value F_{SD} of the load is shown to exist, such that the central rod also yields in compression.

When the external force is maximum, we have

$$X_I(\max) + 2X_{II}(\max) = F \tag{18.150a}$$

whereas, when the external force is zero, we have

$$X_I(\min) + 2X_{II}(\min) = 0 \tag{18.150b}$$

When the external force is maximum, the dilation of the central rod is

$$\varepsilon_I(\max) = \frac{l_{II}}{l_I}\varepsilon_{II}(\max) \tag{18.151}$$

and hence

$$\varepsilon_I(\max) = \frac{l_{II}}{l_I}\frac{X_{II}(\max)}{(EA/2)} \tag{18.152}$$

since the lateral rods are in elastic conditions.

On the other hand, when the external force is zero and in the case where inverse plastic deformation occurs, the dilation of the central rod is

$$\varepsilon_I(\min) = \varepsilon_I(\max) - \frac{2\sigma_P}{E} \tag{18.153}$$

Since it is also found that

$$\varepsilon_I(\min) = \frac{l_{II}}{l_I}\varepsilon_{II}(\min) \tag{18.154}$$

by virtue of the transitive law, we obtain

$$\frac{l_{II}}{l_I}\frac{X_{II}(\min)}{(EA/2)} = \frac{l_{II}}{l_I}\frac{X_{II}(\max)}{(EA/2)} - \frac{2\sigma_P}{E} \tag{18.155}$$

From equation (18.155) we deduce the reaction of the lateral rods at the maximum load

$$X_{II}(\max) = X_{II}(\min) + \sigma_P A \frac{l_I}{l_{II}} \tag{18.156}$$

On account of equilibrium of the rigid cross member, equation (18.150b) yields

$$X_{II}(\min) = -\frac{1}{2}X_I(\min) \tag{18.157}$$

615

The hypothesis of inverse plastic deformation of the central rod gives

$$X_I(\min) = -\sigma_P A \tag{18.158}$$

so that

$$X_{II}(\min) = \sigma_P \frac{A}{2} \tag{18.159}$$

From equation (18.156) we thus obtain

$$X_{II}(\max) = \sigma_P \frac{A}{2} + \sigma_P A \frac{l_I}{l_{II}} \tag{18.160}$$

Since at maximum load the central rod has yielded, we have

$$X_I(\max) = \sigma_P A \tag{18.161}$$

Finally, the substitution of equation (18.160) and (18.161) into equation (18.150a) yields the force of inverse plastic deformation

$$F_{SD} = \sigma_P A + 2\left(\sigma_P \frac{A}{2} + \sigma_P A \frac{l_I}{l_{II}}\right) \tag{18.162}$$

or

$$F_{SD} = 2\sigma_P A\left(1 + \frac{l_I}{l_{II}}\right) \tag{18.163}$$

Only for

$$F > F_{SD} \tag{18.164}$$

inverse plastic deformation of the central rod occurs. Note that this threshold load F_{SD} is exactly twice the load of first plastic deformation F_1 given by equation (18.4a).

Summarizing, it is possible to state:

1. For $0 \leq F \leq F_1$, the system behaves elastically and its representative point in the plane F–δ (Figure 18.37) oscillates on the segment 01.
2. For $F_1 < F \leq F_{SD}$, the central rod yields only in tension and the representative point of the system oscillates on the corresponding segment 0'1' (**shake-down**).
3. For $F > F_{SD}$, the central rod yields both in tension (loading) and in compression (unloading) and the representative point of the system traverses cyclically the parallelogram SD–A–B–C. More precisely, at the first loading the point moves to A; on subsequent elastic unloading, the point moves to B, where the plastic flow in compression of the central rod starts. This flow ceases at C. From here the second loading cycle begins, which develops first elastically along the segment C–SD, and then plastically along the segment SD–A, and so forth. The energy which is dissipated in each **hysteresis cycle** is equal to the area of the parallelogram SD–A–B–C. The

phenomenon of plastic dissipation just described is called **alternating plastic deformation**.

In the case where the cyclic load, instead of being pulsating (Figure 18.36), is alternating and symmetrical (Figure 18.38), it is possible to demonstrate how the hysteresis cycle assumes the appearance represented in Figures 18.39(a) and (b), for $F < F_{SD}$ and $F > F_{SD}$ respectively. In the cases of cyclic loads pulsating not from zero (Figure 18.40(a)), or of alternating non-symmetrical cyclic loads (Figure 18.40(b)), the hysteresis cycle will again take on the appearance of a parallelogram, with the sides parallel to those of the alternating and symmetrical cycles. On the other hand, it will not be symmetrical with respect to the origin (Figures 18.41(a), (b)).

Finally, it should be noted that in the cases where there is more than one hardening portion (Figure 18.42(a)), or in the cases where the plastic collapse of the system is reached (Figure 18.42(b)), the alternating plastic deformation

Figure 18.38

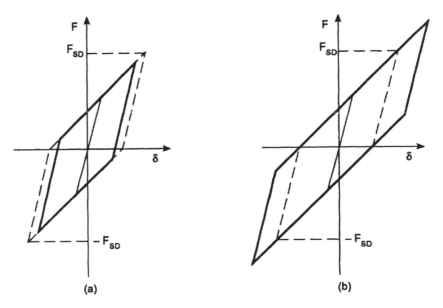

(a) (b)

Figure 18.39

617

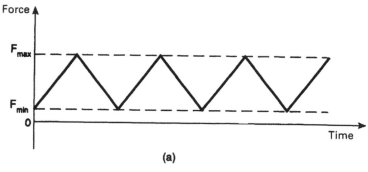

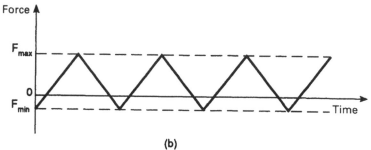

Figure 18.40

develops through polygonal cycles which are polar-symmetrical with respect to specific points of the plane F–δ.

18.10 Deflected plates

Let us consider the case of a deflected circular plate, consisting of elastic–perfectly plastic material. The indefinite equations of equilibrium (12.33b) can be written in explicit form as follows:

$$\frac{d}{dr}(rT_r) = qr \qquad (18.165a)$$

$$\frac{d}{dr}(rM_r) - M_\vartheta - rT_r = 0 \qquad (18.165b)$$

Differentiating equation (18.165b) and using equation (18.165a), we obtain the differential equation

$$\frac{d^2}{dr^2}(rM_r) - \frac{dM_\vartheta}{dr} - qr = 0 \qquad (18.166a)$$

which is valid for $0 \leqslant r \leqslant R$, if R is the radius of the plate, with the boundary condition

$$M_r(R) = 0 \qquad (18.166b)$$

if the plate is supported at the boundary. On account of the isotropy of the stress condition in the centre of the plate, we also have

$$M_r(0) = M_\vartheta(0) \tag{18.166c}$$

Let us assume, as a statically admissible régime, a condition of complete plastic deformation of the plate. According to equation (18.166c), the representative point of the characteristics that develop in the centre of the plate must fall in the vertex B of Tresca's hexagon illustrated in Figure 18.43. Moreover, since the condition (18.166b) holds good, it is legitimate to assume that the static regime is always represented by points belonging to the side BC of the same hexagon. Since on this side we have $M_\vartheta = M_P$, the static equation (18.166a) is transformed as follows:

$$\frac{d^2}{dr^2}(rM_r) - qr = 0 \tag{18.167}$$

and, on integration, yields

$$M_r = q\frac{r^2}{6} + C_1 + \frac{C_2}{r} \tag{18.168}$$

Applying the boundary conditions

$$M_r(0) = M_P \tag{18.169a}$$
$$M_r(R) = 0 \tag{18.169b}$$

the two constants of integration are determined

$$C_1 = M_P \tag{18.170a}$$
$$C_2 = 0 \tag{18.170b}$$

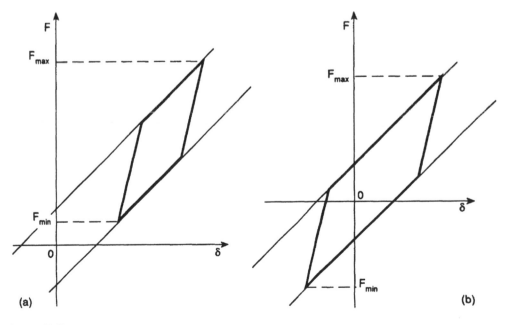

(a)

(b)

Figure 18.41

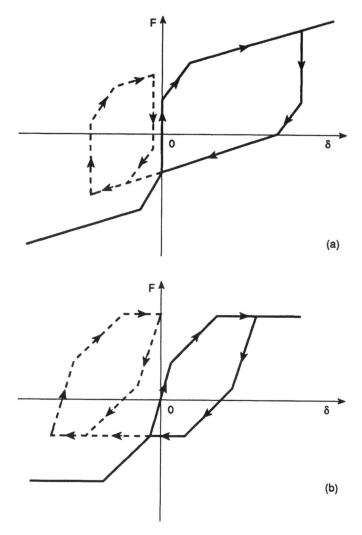

(a)

(b)

Figure 18.42

as well as the load

$$q = -6\frac{M_P}{R^2} \tag{18.171}$$

The radial bending moment thus becomes

$$M_r(r) = -M_P\left[\left(\frac{r}{R}\right)^2 - 1\right] \tag{18.172a}$$

whilst the circumferential bending moment has been assumed to be constant

620

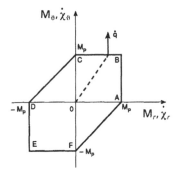

Figure 18.43

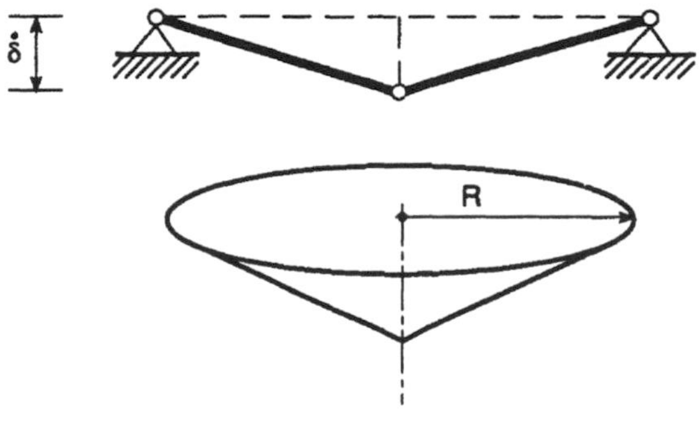

Figure 18.44

and equal to the plastic moment

$$M_\vartheta = M_P \qquad (18.172b)$$

By virtue of the static theorem, the load given by equation (18.171) represents a lower limit for the actual collapse load. On the other hand, on account of the property of normality of the incremental plastic deformation (Figure 18.43), the variations of the radial and circumferential curvatures must satisfy the conditions

$$\dot{\chi}_r = 0, \qquad \dot{\chi}_\vartheta \geqslant 0 \qquad (18.173)$$

Via the kinematic equations (12.33a), the corresponding conditions on the deflection are obtained:

$$\frac{d^2 \dot{w}}{dr^2} = 0, \qquad \frac{1}{r}\frac{d\dot{w}}{dr} \not> 0 \qquad (18.174)$$

It is possible to verify immediately that the mechanism of Figure 18.44

$$\dot{w} = \dot{\delta}\left(1 - \frac{r}{R}\right) \qquad (18.175)$$

where $\dot{\delta}$ is the incremental plastic deflection of the central point, satisfies equations (18.174) and may thus be associated with the static regime expressed by equation (18.172).

Since the load given by equation (18.171) corresponds both to a statically admissible condition and to a kinematically admissible mechanism, by virtue of the mixed theorem it is possible to state that this load represents the actual collapse load of the plate.

621

19 Plane stress and plane strain conditions

19.1 Introduction

This chapter will deal with two-dimensional elastic problems in the plane stress condition or the plane strain condition. The plane stress condition has been defined in Section 7.10, in reference to a single point. The stress condition at a point is said to be plane if the stress vector belongs to the same plane independently of the cross section chosen. Likewise, the strain condition is said to be plane if the displacement vector belongs to the same plane regardless of the direction chosen. As has already been seen, a necessary and sufficient condition for a state of stress or strain to be defined as plane at a point is that one of the three principal values (of stress on the one hand and of strain on the other) should be equal to zero.

Whilst it is possible for stress conditions that are plane at a point, but not globally, to exist, such as in the case of the Saint Venant solid, where the individual stress planes (Figure 9.40) for the different points of the solid are not necessarily parallel, in what follows we shall deal only with cases that are globally plane, i.e. ones having planes of stresses or strains which are all parallel.

Having first introduced a mathematical method which is generally suited to the situation of plane problems and which is based on the **Airy stress function**, we shall consider a number of notable cases, such as the deep beam, the thick-walled cylinder, the circular hole in a plate in tension and the concentrated force acting on an elastic half-plane. The subsequent introduction of an additional method, based on the theory of complex functions of a complex variable (Muskhelishvili's Method) will make possible the treatment of the elliptical hole in a plate in tension. This latter topic will, in turn, serve as an introduction to the problems of stress concentration and of fracture mechanics which will form the subject matter of the closing chapter. In the appendices will be illustrated the cases of a circular disk subjected to inertial forces (Appendix M) and to thermal stress (Appendix N).

19.2 Plane stress condition

The plane stress condition tends to occur in thin plates, loaded by forces contained in their own middle plane (Figure 19.1). There are originally five unknowns in plane stress problems: the three components of stress σ_x, σ_y, τ_{xy}, and the two components of displacement u and v. There are likewise five resolving equations, which are the two indefinite equations of equilibrium

$$\frac{\partial \sigma_x}{\partial x} + \frac{\partial \tau_{xy}}{\partial y} + \mathcal{F}_x = 0 \tag{19.1a}$$

$$\frac{\partial \tau_{xy}}{\partial x} + \frac{\partial \sigma_y}{\partial y} + \mathcal{F}_y = 0 \tag{19.1b}$$

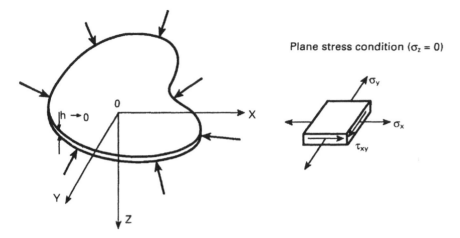

Figure 19.1

and the three elastic constitutive equations

$$\varepsilon_x = \frac{1}{E}\left(\sigma_x - v\sigma_y\right) \qquad (19.2a)$$

$$\varepsilon_y = \frac{1}{E}\left(\sigma_y - v\sigma_x\right) \qquad (19.2b)$$

$$\gamma_{xy} = \frac{2(1+v)}{E}\tau_{xy} \qquad (19.2c)$$

The equation of compatibility (7.45a)

$$\frac{\partial^2 \varepsilon_x}{\partial y^2} + \frac{\partial^2 \varepsilon_y}{\partial x^2} = \frac{\partial^2 \gamma_{xy}}{\partial x \partial y} \qquad (19.3)$$

on the basis of relations (19.2), may be expressed in stress terms

$$\frac{\partial^2}{\partial y^2}\left(\sigma_x - v\sigma_y\right) + \frac{\partial^2}{\partial x^2}\left(\sigma_y - v\sigma_x\right) = 2(1+v)\frac{\partial^2 \tau_{xy}}{\partial x \partial y} \qquad (19.4)$$

On the other hand, differentiating the indefinite equations of equilibrium (19.1), in order with respect to x and y, and adding together the resulting equations, we obtain

$$-\frac{\partial^2 \sigma_x}{\partial x^2} - \frac{\partial^2 \sigma_y}{\partial y^2} - \frac{\partial \mathcal{F}_x}{\partial x} - \frac{\partial \mathcal{F}_y}{\partial y} = 2\frac{\partial^2 \tau_{xy}}{\partial x \partial y} \qquad (19.5)$$

If we multiply both sides of equation (19.5) by $(1 + v)$, the right-hand side becomes the same as that of equation (19.4). By the transitive law, we have

623

$$\frac{\partial^2}{\partial y^2}\left(\sigma_x - v\sigma_y\right) + \frac{\partial^2}{\partial x^2}\left(\sigma_y - v\sigma_x\right) \tag{19.6}$$

$$= -(1+v)\left[\frac{\partial^2 \sigma_x}{\partial x^2} + \frac{\partial^2 \sigma_y}{\partial y^2} + \frac{\partial \mathscr{F}_x}{\partial x} + \frac{\partial \mathscr{F}_y}{\partial y}\right]$$

Finally, collecting terms, we obtain

$$\nabla^2\left(\sigma_x + \sigma_y\right) = -(1+v)\left[\frac{\partial \mathscr{F}_x}{\partial x} + \frac{\partial \mathscr{F}_y}{\partial y}\right] \tag{19.7}$$

If the body forces are zero, from equations (19.1) and (19.7) we obtain the following system of three differential equations in the three unknown functions $\sigma_x, \sigma_y, \tau_{xy}$:

$$\frac{\partial \sigma_x}{\partial x} + \frac{\partial \tau_{xy}}{\partial y} = 0 \tag{19.8a}$$

$$\frac{\partial \tau_{xy}}{\partial x} + \frac{\partial \sigma_y}{\partial y} = 0 \tag{19.8b}$$

$$\nabla^2\left(\sigma_x + \sigma_y\right) = 0 \tag{19.8c}$$

The elastic characteristics E, v of the material do not appear in the resolving equations (19.8). It follows therefore that the plane stress field does not depend in any way on the material, but only on the boundary conditions. Of course it is not possible to say the same of the strain field expressed in equations (19.2), and hence of the elastic displacements induced by it.

Let us assume that the components of the stress field are obtainable by derivation of an unknown function Φ, called the **Airy stress function**

$$\sigma_x = \frac{\partial^2 \Phi}{\partial y^2} \tag{19.9a}$$

$$\sigma_y = \frac{\partial^2 \Phi}{\partial x^2} \tag{19.9b}$$

$$\tau_{xy} = -\frac{\partial^2 \Phi}{\partial x \partial y} \tag{19.9c}$$

In this case the indefinite equations of equilibrium (19.8a,b) are identically satisfied, while the equation of compatibility (19.8c) is satisfied if and only if

$$\nabla^2 \nabla^2 \Phi = 0 \tag{19.10}$$

or

$$\nabla^4 \Phi = 0 \tag{19.11}$$

A function Φ which satisfies equation (19.11) is said to be **biharmonic**.

19.3 Plane strain condition

The plane strain condition tends to occur in cylindrical or prismatic solids of large thickness, loaded by orthogonal forces on the generatrices, the forces

624

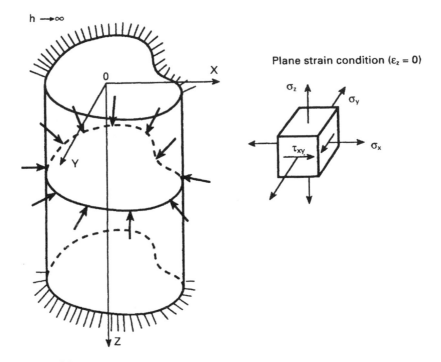

Figure 19.2

having constant distribution along these (Figure 19.2). If the end planes are considered as being constrained axially, it is legitimate to assume a zero dilation in the axial direction, $\varepsilon_z = 0$, and a strain condition which is repeated section by section.

The condition of annihilation of the axial dilation

$$\varepsilon_z = \frac{1}{E}\left(\sigma_z - v\sigma_x - v\sigma_y\right) = 0 \tag{19.12}$$

yields the axial stress as a function of the other two normal stresses

$$\sigma_z = v\left(\sigma_x + \sigma_y\right) \tag{19.13}$$

Substituting equation (19.13) in the elastic constitutive equations (8.73), we find

$$\varepsilon_x = \frac{1}{E}\left[\left(1 - v^2\right)\sigma_x - v(1+v)\,\sigma_y\right] \tag{19.14a}$$

$$\varepsilon_y = \frac{1}{E}\left[\left(1 - v^2\right)\sigma_y - v(1+v)\,\sigma_x\right] \tag{19.14b}$$

$$\gamma_{xy} = \frac{2(1+v)}{E}\,\tau_{xy} \tag{19.14c}$$

625

If we consider the so-called **stiffened elastic characteristics,**

$$E' = \frac{E}{1 - v^2} \tag{19.15a}$$

$$v' = \frac{v}{1 - v} \tag{19.15b}$$

equations (19.14) may assume the following form:

$$\varepsilon_x = \frac{1}{E'}\left(\sigma_x - v'\sigma_y\right) \tag{19.16a}$$

$$\varepsilon_y = \frac{1}{E'}\left(\sigma_y - v'\sigma_x\right) \tag{19.16b}$$

$$\gamma_{xy} = \frac{2(1+v')}{E'}\tau_{xy} \tag{19.16c}$$

which corresponds to that of relations (19.2).

The equation of compatibility, analogously to equation (19.7), can therefore be expressed as follows:

$$\nabla^2\left(\sigma_x + \sigma_y\right) = -(1 + v')\left[\frac{\partial \mathcal{F}_x}{\partial x} + \frac{\partial \mathcal{F}_y}{\partial y}\right] \tag{19.17}$$

and hence, on the basis of equation (19.15b)

$$\nabla^2\left(\sigma_x + \sigma_y\right) = -\frac{1}{1 - v}\left[\frac{\partial \mathcal{F}_x}{\partial x} + \frac{\partial \mathcal{F}_y}{\partial y}\right] \tag{19.18}$$

If the body forces are zero, we again obtain the three equations (19.8) that resolve the plane stress problem, plus a fourth equation

$$\frac{\partial \sigma_z}{\partial z} = 0 \tag{19.19}$$

which implies the constancy of axial stress along the thickness.

19.4 Deep beam

Let us consider a deep beam of rectangular cross section and unit base, in which the ratio of length l to depth h is not so high as to enable application of the elementary theory presented in Chapters 9 and 10 (Figure 19.3). Let this beam be supported at the ends and loaded by a constant distribution of vertical forces q. The boundary conditions on the upper and lower edges are

$$\tau_{xy}\left(y = \pm\frac{h}{2}\right) = 0 \tag{19.20a}$$

$$\sigma_y\left(y = \frac{h}{2}\right) = 0 \tag{19.20b}$$

$$\sigma_y\left(y = -\frac{h}{2}\right) = -q \tag{19.20c}$$

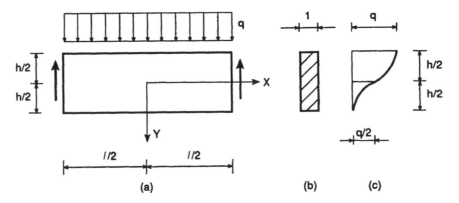

Figure 19.3

The conditions at the ends $x = \pm l/2$ are

$$\int_{-h/2}^{h/2} \tau_{xy} dy = \mp q \frac{l}{2} \qquad (19.21a)$$

$$\int_{-h/2}^{h/2} \sigma_x \, dy = 0 \qquad (19.21b)$$

$$\int_{-h/2}^{h/2} \sigma_x y \, dy = 0 \qquad (19.21c)$$

Equations (19.21b, c) impose the annihilation of the axial force and the bending moment on the end sections.

Let us assume for the stress components polynomial expressions with unknown coefficients

$$\sigma_x = a\left(x^2 y - \frac{2}{3} y^3\right) \qquad (19.22a)$$

$$\sigma_y = \frac{1}{3} a y^3 + by + c \qquad (19.22b)$$

$$\tau_{xy} = -axy^2 - bx \qquad (19.22c)$$

that satisfy equations (19.8a, b).

From the conditions expressed by equations (19.20) we obtain the system

$$-a\frac{h^2}{4} - b = 0 \qquad (19.23a)$$

$$\frac{1}{3} a \frac{h^3}{8} + b \frac{h}{2} + c = 0 \qquad (19.23b)$$

$$-\frac{1}{3} a \frac{h^3}{8} - b \frac{h}{2} + c = -q \qquad (19.23c)$$

627

which, on resolution, yields

$$a = -6\frac{q}{h^3} \tag{19.24a}$$

$$b = \frac{3}{2}\frac{q}{h} \tag{19.24b}$$

$$c = -\frac{q}{2} \tag{19.24c}$$

Noting that the moment of inertia of the rectangular cross section of unit base is $I = h^3/12$, we obtain $a = -q/2I$, and therefore equations (19.22) offer the following expressions:

$$\sigma_x = -\frac{q}{2I}\left(x^2 y - \frac{2}{3}y^3\right) \tag{19.25a}$$

$$\sigma_y = -\frac{q}{2I}\left(\frac{1}{3}y^3 - \frac{h^2}{4}y + \frac{h^3}{12}\right) \tag{19.25b}$$

$$\tau_{xy} = -\frac{q}{2I}\left(\frac{h^2}{4} - y^2\right)x \tag{19.25c}$$

These three components of stress also satisfy the conditions set by equations (19.21a, b). In order for the moments on the end cross sections also to become zero, it is necessary to superpose on the solution represented by expressions (19.25) a stress field of pure bending, $\sigma_x = yd$, $\sigma_y = \tau_{xy} = 0$, so as to determine the constant d via the condition (19.21c) for $x = \pm l/2$

$$\int_{-h/2}^{h/2} \sigma_x y \, dy = \int_{-h/2}^{h/2}\left[-6\frac{q}{h^3}\left(\frac{l^2}{4}y - \frac{2}{3}y^3\right) + yd\right]y \, dy = 0 \tag{19.26}$$

from which we find

$$d = \frac{3}{2}\frac{q}{h}\left(\frac{l^2}{h^2} - \frac{2}{5}\right) \tag{19.27}$$

whereby finally we have

$$\sigma_x = \frac{q}{2I}\left(\frac{l^2}{4} - x^2\right)y + \frac{q}{2I}\left(\frac{2}{3}y^3 - \frac{1}{10}h^2 y\right) \tag{19.28}$$

The first term on the right of equation (19.28) represents the stress given by the usual elementary theory of bending, whereas the second term represents its correction. This term does not depend on the abscissa x and is negligible only in the cases where the span of the beam is large compared with its depth. Note that expression (19.28) represents an exact solution, only if the axial stresses at the ends are distributed according to the following law:

$$\sigma_x\left(x = \pm\frac{l}{2}\right) = 6\frac{q}{h^3}\left(\frac{2}{3}y^3 - \frac{1}{10}h^2 y\right) \tag{19.29}$$

These stresses present both the resultant force and the resultant moment zero. Consequently, by virtue of Saint Venant's Principle, it is possible to deduce

DEEP BEAM

that their effect, at distances from the ends greater than the depth h of the beam, diminishes sensibly until it vanishes altogether.

From solution (19.25b) we detect the existence of compressive stresses σ_y, which instead are absent in the elementary theory. The distribution of these compressive stresses over the depth of the beam is shown in Figure 19.3(c). The distribution of the shearing stresses τ_{xy}, given by equation (19.25c), coincides instead with that furnished by the usual elementary theory.

Using relations (19.2), (19.25b, c) and (19.28), we obtain the components of strain

$$\varepsilon_x = \frac{q}{2EI}\left\{\left[\left(\frac{l^2}{4} - x^2\right)y + \left(\frac{2}{3}y^3 - \frac{1}{10}h^2 y\right)\right] + v\left[\frac{1}{3}y^3 - \frac{h^2}{4}y + \frac{h^3}{12}\right]\right\} \quad (19.30a)$$

$$\varepsilon_y = \frac{q}{2EI}\left\{-\left[\frac{1}{3}y^3 - \frac{h^2}{4}y + \frac{h^3}{12}\right] - v\left[\left(\frac{l^2}{4} - x^2\right)y + \left(\frac{2}{3}y^3 - \frac{1}{10}h^2 y\right)\right]\right\} \quad (19.30b)$$

$$\gamma_{xy} = -\frac{q}{EI}(1+v)\left(\frac{h^2}{4} - y^2\right)x \quad (19.30c)$$

Integrating the first two of these equations, we have

$$u = \frac{q}{2EI}\left\{\left[\left(\frac{l^2}{4}x - \frac{x^3}{3}\right)y + \left(\frac{2}{3}y^3 - \frac{1}{10}h^2 y\right)x\right] + \right. \quad (19.31a)$$

$$\left. vx\left[\frac{1}{3}y^3 - \frac{h^2}{4}y + \frac{h^3}{12}\right]\right\} + f(y)$$

$$v = \frac{q}{2EI}\left\{-\left[\frac{y^4}{12} - \frac{h^2}{8}y^2 + \frac{h^3}{12}y\right] - \right. \quad (19.31b)$$

$$\left. v\left[\left(\frac{l^2}{4} - x^2\right)\frac{y^2}{2} + \left(\frac{1}{6}y^4 - \frac{h^2}{20}y^2\right)\right]\right\} + g(x)$$

where f and g represent unknown functions of the coordinates y and x, respectively.

Equation (19.30c) then becomes

$$\frac{\partial u}{\partial y} + \frac{\partial v}{\partial x} = \frac{q}{2EI}\left\{\left[\left(\frac{l^2}{4}x - \frac{x^3}{3}\right) + \left(2y^2 - \frac{h^2}{10}\right)x\right] + \right. \quad (19.32)$$

$$\left. vx\left[y^2 - \frac{h^2}{4}\right]\right\} + \frac{\partial f}{\partial y} +$$

$$\frac{q}{2EI}\left\{-v\left[(-2x)\frac{y^2}{2}\right]\right\} + \frac{\partial g}{\partial x}$$

$$= -\frac{q}{EI}(1+v)\left(\frac{h^2}{4} - y^2\right)x$$

629

Collecting terms, we obtain

$$\frac{\partial g}{\partial x} + \frac{q}{2EI}\left\{x\left[\frac{l^2}{4} + h^2\left(\frac{2}{5} + \frac{v}{4}\right)\right] - \frac{x^3}{3}\right\} = -\frac{\partial f}{\partial y} \qquad (19.33)$$

Since the left-hand side is a function only of the variable x, whilst the right-hand side is a function only of the variable y, both must represent a constant C_1:

$$\frac{\partial g}{\partial x} = \frac{q}{2EI}\left\{\frac{x^3}{3} - x\left[\frac{l^2}{4} + h^2\left(\frac{2}{5} + \frac{v}{4}\right)\right]\right\} + C_1 \qquad (19.34)$$

Integrating equation (19.34), we find the function g:

$$g(x) = \frac{q}{2EI}\left\{\frac{x^4}{12} - \frac{x^2}{2}\left[\frac{l^2}{4} + h^2\left(\frac{2}{5} + \frac{v}{4}\right)\right]\right\} + C_1 x + C_2 \qquad (19.35)$$

The constants C_1 and C_2 can be inferred from the vertical displacement in the centre and from the conditions at the ends:

$$v(0,0) = \delta \qquad (19.36a)$$

$$v\left(\pm\frac{l}{2},0\right) = 0 \qquad (19.36b)$$

Applying equations (19.36) to equations (19.31b) and (19.35), we have

$$C_1 = 0 \qquad (19.37a)$$
$$C_2 = \delta \qquad (19.37b)$$

and hence the geometrical axis of the beam bends according to the following curve:

$$v(x,0) = \frac{q}{2EI}\left\{\frac{x^4}{12} - \frac{x^2}{2}\left[\frac{l^2}{4} + h^2\left(\frac{2}{5} + \frac{v}{4}\right)\right]\right\} + \delta \qquad (19.38)$$

Since the condition (19.36b) must hold, the vertical displacement in the centre is

$$\delta = \frac{5}{384}\frac{ql^4}{EI}\left[1 + \frac{48}{25}\frac{h^2}{l^2}\left(1 + \frac{5}{8}v\right)\right] \qquad (19.39)$$

It may be noted that the first term which contributes to equation (19.39) coincides with the vertical displacement (10.70a) deriving from the elementary theory. The second term, which on the other hand represents its correction, diminishes with the increase in the ratio l/h. For low values of this ratio, it may be seen how the cross sections of the beam do not remain plane, much less orthogonal to the centroidal axis. The effect of shear, according to the Saint Venant theory, can, on the other hand, be estimated by applying equation (9.168):

$$\delta^T = \int_0^{l/2} \gamma_y dx = \int_0^{l/2} \frac{6}{5}\frac{T_y}{Gh}dx \qquad (19.40)$$

Substituting the linear function of shear into equation (19.40), we obtain

$$\delta^T = \frac{6}{5Gh}\int_0^{l/2} q\left(\frac{l}{2} - x\right)dx = \frac{1+v}{40}\left(\frac{h}{l}\right)^2\frac{ql^4}{EI} \qquad (19.41)$$

and hence

$$\delta = \delta^M + \delta^T = \frac{5}{384} \frac{ql^4}{EI} \left[1 + \frac{48}{25} \frac{h^2}{l^2} (1 + v) \right] \qquad (19.42)$$

It is interesting to note that, for $v = 0$, equation (19.39) and equation (19.42) coincide, while, for $v > 0$, equation (19.42) overestimates the exact deflection.

19.5 Thick-walled cylinder

The indefinite equations of equilibrium in cylindrical coordinates were presented for the solid of revolution in Section 12.12. In the case of a plane stress condition, the components σ_z, τ_{rz}, $\tau_{\vartheta z}$, disappear, and the equations of equilibrium with regard to translation in the radial and circumferential directions are obtained directly from equations (12.95b)

$$\frac{\partial \sigma_r}{\partial r} + \frac{1}{r} \frac{\partial \tau_{r\vartheta}}{\partial \vartheta} + \frac{\sigma_r - \sigma_\vartheta}{r} + \mathcal{F}_r = 0 \qquad (19.43a)$$

$$\frac{1}{r} \frac{\partial \sigma_\vartheta}{\partial \vartheta} + \frac{\partial \tau_{r\vartheta}}{\partial r} + \frac{2}{r} \tau_{r\vartheta} + \widetilde{\mathcal{F}_\vartheta} = 0 \qquad (19.43b)$$

In the case where the body forces vanish, equations (19.43) are identically satisfied by

$$\sigma_r = \frac{1}{r} \frac{\partial \Phi}{\partial r} + \frac{1}{r^2} \frac{\partial^2 \Phi}{\partial \vartheta^2} \qquad (19.44a)$$

$$\sigma_\vartheta = \frac{\partial^2 \Phi}{\partial r^2} \qquad (19.44b)$$

$$\tau_{r\vartheta} = \frac{1}{r^2} \frac{\partial \Phi}{\partial \vartheta} - \frac{1}{r} \frac{\partial^2 \Phi}{\partial r \partial \vartheta} = -\frac{\partial}{\partial r} \left(\frac{1}{r} \frac{\partial \Phi}{\partial \vartheta} \right) \qquad (19.44c)$$

where Φ denotes the Airy stress function.

To obtain the biharmonic equation (19.11) in polar coordinates, consider the coordinate transformation

$$r^2 = x^2 + y^2 \qquad (19.45a)$$

$$\vartheta = \arctan \frac{y}{x} \qquad (19.45b)$$

from which there follow

$$\frac{\partial r}{\partial x} = \frac{x}{r} = \cos \vartheta \qquad (19.46a)$$

$$\frac{\partial r}{\partial y} = \frac{y}{r} = \sin \vartheta \qquad (19.46b)$$

$$\frac{\partial \vartheta}{\partial x} = -\frac{y}{r^2} = -\frac{\sin \vartheta}{r} \qquad (19.46c)$$

$$\frac{\partial \vartheta}{\partial y} = \frac{x}{r^2} = \frac{\cos \vartheta}{r} \qquad (19.46d)$$

631

The partial derivative of Φ with respect to x may be expressed, as is known, in the following way:

$$\frac{\partial \Phi}{\partial x} = \frac{\partial \Phi}{\partial r}\frac{\partial r}{\partial x} + \frac{\partial \Phi}{\partial \vartheta}\frac{\partial \vartheta}{\partial x} \qquad (19.47)$$

$$= \cos\vartheta \frac{\partial \Phi}{\partial r} - \frac{\sin\vartheta}{r}\frac{\partial \Phi}{\partial \vartheta}$$

The second partial derivative is thus equal to

$$\frac{\partial^2 \Phi}{\partial x^2} = \left(\cos\vartheta\frac{\partial}{\partial r} - \frac{\sin\vartheta}{r}\frac{\partial}{\partial \vartheta}\right)\left(\cos\vartheta\frac{\partial \Phi}{\partial r} - \frac{\sin\vartheta}{r}\frac{\partial \Phi}{\partial \vartheta}\right) \qquad (19.48)$$

$$= \cos^2\vartheta\frac{\partial^2 \Phi}{\partial r^2} - \cos\vartheta\,\sin\vartheta\frac{\partial}{\partial r}\left(\frac{1}{r}\frac{\partial \Phi}{\partial \vartheta}\right) -$$

$$\frac{\sin\vartheta}{r}\frac{\partial}{\partial \vartheta}\left(\cos\vartheta\frac{\partial \Phi}{\partial r}\right) + \frac{\sin\vartheta}{r^2}\frac{\partial}{\partial \vartheta}\left(\sin\vartheta\frac{\partial \Phi}{\partial \vartheta}\right)$$

Differentiating, we obtain

$$\frac{\partial^2 \Phi}{\partial x^2} = \cos^2\vartheta\frac{\partial^2 \Phi}{\partial r^2} + \sin^2\vartheta\left(\frac{1}{r}\frac{\partial \Phi}{\partial r} + \frac{1}{r^2}\frac{\partial^2 \Phi}{\partial \vartheta^2}\right) - \qquad (19.49a)$$

$$2\,\sin\vartheta\,\cos\vartheta\frac{\partial}{\partial r}\left(\frac{1}{r}\frac{\partial \Phi}{\partial \vartheta}\right)$$

In like manner we have

$$\frac{\partial^2 \Phi}{\partial y^2} = \sin^2\vartheta\frac{\partial^2 \Phi}{\partial r^2} + \cos^2\vartheta\left(\frac{1}{r}\frac{\partial \Phi}{\partial r} + \frac{1}{r^2}\frac{\partial^2 \Phi}{\partial \vartheta^2}\right) + \qquad (19.49b)$$

$$2\,\sin\vartheta\,\cos\vartheta\frac{\partial}{\partial r}\left(\frac{1}{r}\frac{\partial \Phi}{\partial \vartheta}\right)$$

$$-\frac{\partial^2 \Phi}{\partial x \partial y} = \sin\vartheta\,\cos\vartheta\left(\frac{1}{r}\frac{\partial \Phi}{\partial r} + \frac{1}{r^2}\frac{\partial^2 \Phi}{\partial \vartheta^2} - \frac{\partial^2 \Phi}{\partial r^2}\right) - \qquad (19.49c)$$

$$(\cos^2\vartheta - \sin^2\vartheta)\frac{\partial}{\partial r}\left(\frac{1}{r}\frac{\partial \Phi}{\partial \vartheta}\right)$$

Taking into account equations (19.49a, b) and equations (19.44a, b), equation (19.8c) may be cast in the following form:

$$\left(\frac{\partial^2}{\partial x^2} + \frac{\partial^2}{\partial y^2}\right)(\sigma_x + \sigma_y) \qquad (19.50)$$

$$= \left(\frac{\partial^2}{\partial r^2} + \frac{1}{r}\frac{\partial}{\partial r} + \frac{1}{r^2}\frac{\partial^2}{\partial \vartheta^2}\right)(\sigma_r + \sigma_\vartheta)$$

$$= \left(\frac{\partial^2}{\partial r^2} + \frac{1}{r}\frac{\partial}{\partial r} + \frac{1}{r^2}\frac{\partial^2}{\partial \vartheta^2}\right)^2 \Phi = 0$$

the sum of the normal stresses, $\sigma_x + \sigma_y = \sigma_r + \sigma_\vartheta$, being constant and equal to the first invariant J_I.

When there is polar symmetry, the Airy stress function depends only on the radial coordinate r, and the equation of compatibility (19.50) becomes

$$\left(\frac{d^2}{dr^2} + \frac{1}{r}\frac{d}{dr}\right)\left(\frac{d^2\Phi}{dr^2} + \frac{1}{r}\frac{d\Phi}{dr}\right) \qquad (19.51)$$

$$= \frac{d^4\Phi}{dr^4} + \frac{2}{r}\frac{d^3\Phi}{dr^3} - \frac{1}{r^2}\frac{d^2\Phi}{dr^2} + \frac{1}{r^3}\frac{d\Phi}{dr} = 0$$

It is possible to verify that the complete integral of equation (19.51) is

$$\Phi(r) = A\ \log r + Br^2\ \log r + Cr^2 + D \qquad (19.52)$$

The stress components are thus obtained from equations (19.44)

$$\sigma_r = \frac{1}{r}\frac{d\Phi}{dr} = \frac{A}{r^2} + B(1 + 2\log r) + 2C \qquad (19.53a)$$

$$\sigma_\vartheta = \frac{d^2\Phi}{dr^2} = -\frac{A}{r^2} + B(3 + 2\ \log r) + 2C \qquad (19.53b)$$

$$\tau_{r\vartheta} = 0 \qquad (19.53c)$$

In the case where there is not a hole at the origin, the only possible solution is that of uniform stress: $\sigma_r = \sigma_\vartheta = 2C$.

In the case of a thick-walled cylinder, subjected to uniform pressure both internally and externally (Figure 19.4), we have $B = 0$ with the boundary conditions

$$\sigma_r(r = R_1) = -p_i \qquad (19.54a)$$

$$\sigma_r(r = R_2) = -p_e \qquad (19.54b)$$

Imposing the conditions (19.54) on relation (19.53a), we obtain two equations in the two unknowns A and C:

$$\frac{A}{R_1^2} + 2C = -p_i \qquad (19.55a)$$

$$\frac{A}{R_2^2} + 2C = -p_e \qquad (19.55b)$$

from which we find

$$A = \frac{R_1^2 R_2^2 (p_e - p_i)}{R_2^2 - R_1^2} \qquad (19.56a)$$

$$2C = \frac{p_i R_1^2 - p_e R_2^2}{R_2^2 - R_1^2} \qquad (19.56b)$$

Substituting these two constants into equations (19.53), we have:

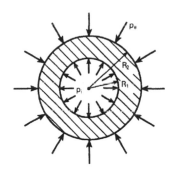

Figure 19.4

$$\sigma_r = \frac{R_1^2 R_2^2 (p_e - p_i)}{R_2^2 - R_1^2} \frac{1}{r^2} + \frac{p_i R_1^2 - p_e R_2^2}{R_2^2 - R_1^2} \qquad (19.57a)$$

$$\sigma_\vartheta = -\frac{R_1^2 R_2^2 (p_e - p_i)}{R_2^2 - R_1^2} \frac{1}{r^2} + \frac{p_i R_1^2 - p_e R_2^2}{R_2^2 - R_1^2} \qquad (19.57b)$$

It is interesting to note that the sum $(\sigma_r + \sigma_\vartheta)$ is constant throughout the thickness of the cylinder. Consequently the stresses σ_r and σ_ϑ produce a uniform dilation or contraction in the axial direction, so that the cross sections remain plane.

When the external pressure is zero, $p_e = 0$, and the cylinder is subjected to the internal pressure alone, equations (19.57) become

$$\sigma_r = \frac{p_i R_1^2}{R_2^2 - R_1^2}\left(1 - \frac{R_2^2}{r^2}\right) \qquad (19.58a)$$

$$\sigma_\vartheta = \frac{p_i R_1^2}{R_2^2 - R_1^2}\left(1 + \frac{R_2^2}{r^2}\right) \qquad (19.58b)$$

These equations show that σ_r is always compressive and σ_ϑ is always tensile. The latter is maximum on the internal surface of the cylinder, where

$$\sigma_\vartheta(\text{max}) = \frac{p_i\left(R_1^2 + R_2^2\right)}{R_2^2 - R_1^2} \qquad (19.59)$$

This tensile stress is greater than the pressure p_i for any value of the ratio R_2/R_1. In the extreme cases we have

$$\lim_{R_2/R_1 \to \infty} \sigma_\vartheta(\text{max}) = p_i \qquad (19.60a)$$

$$\lim_{R_2/R_1 \to 1} \sigma_\vartheta(\text{max}) = \frac{p_i\left(2R^2\right)}{t(2R)} = \frac{p_i R}{t} \qquad (19.60b)$$

where R designates the mean radius and t the thickness, in the case of a thin cylinder. The formula (19.60b) has already been obtained in Section 12.8, following another path.

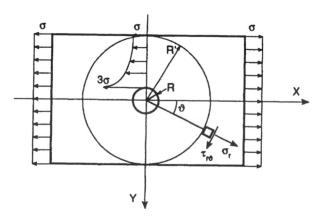

Figure 19.5

634

19.6 Circular hole in a plate subjected to tension

Let us consider an infinite plate, subjected to a uniaxial tensile stress σ in the X direction (Figure 19.5). In the case where a circular hole of radius R is present in the plate, the distribution of the stresses is perturbed in the neighbourhood of the hole itself. In the sequel we intend to calculate the effect of **stress concentration**, i.e. the amplification of the stresses on the edge of the hole.

Consider the portion of the plate which remains within the circumference of radius R', where $R' \gg R$. The stresses acting upon this circumference are approximately the same that would result in the absence of the hole and can therefore be deduced from Mohr's circle

$$\sigma_r(r=R') = \frac{1}{2}\sigma(1+\cos 2\vartheta) \tag{19.61a}$$

$$\tau_{,\vartheta}(r=R') = -\frac{1}{2}\sigma \sin 2\vartheta \tag{19.61b}$$

The radial stress is made up of two parts: the first is constant and produces a stress field, inside the ring, given by equations (19.57):

$$\sigma_r = -\frac{\sigma}{2}\frac{R^2 R'^2}{R'^2-R^2}\frac{1}{r^2} + \frac{\sigma}{2}\frac{R'^2}{R'^2-R^2} \tag{19.62a}$$

$$\sigma_\vartheta = +\frac{\sigma}{2}\frac{R^2 R'^2}{R'^2-R^2}\frac{1}{r^2} + \frac{\sigma}{2}\frac{R'^2}{R'^2-R^2} \tag{19.62b}$$

In the limit whereby $R' \to \infty$, equations (19.62) become

$$\sigma_r = \frac{\sigma}{2}\left(1-\frac{R^2}{r^2}\right) \tag{19.63a}$$

$$\sigma_\vartheta = \frac{\sigma}{2}\left(1+\frac{R^2}{r^2}\right) \tag{19.63b}$$

The second part of σ_r, $\frac{1}{2}\sigma \cos 2\vartheta$, together with the shearing stress, $-\frac{1}{2}\sigma \times \sin 2\vartheta$, produces a stress field that can be derived from an Airy stress function of the form

$$\Phi = f(r)\cos 2\vartheta \tag{19.64}$$

Substituting this function into the equation of compatibility (19.50), we obtain the following total derivative differential equation:

$$\left(\frac{d^2}{dr^2}+\frac{1}{r}\frac{d}{dr}-\frac{4}{r^2}\right)^2 f = 0 \tag{19.65}$$

the complete integral of which is

$$f(r) = Ar^2 + Br^4 + \frac{C}{r_2} + D \tag{19.66}$$

The components of stress are thus found from equations (19.44), (19.64) and (19.66)

$$\sigma_r = -\left(2A+\frac{6C}{r^4}+\frac{4D}{r^2}\right)\cos 2\vartheta \tag{19.67a}$$

$$\sigma_\vartheta = \left(2A + 12Br^2 + \frac{6C}{r^4}\right)\cos 2\vartheta \qquad (19.67\text{b})$$

$$\tau_{r\vartheta} = \left(2A + 6Br^2 - \frac{6C}{r^4} - \frac{2D}{r^2}\right)\sin 2\vartheta \qquad (19.67\text{c})$$

The four constants of integration can be determined from the boundary conditions on the external circumference, i.e. the two equations (19.61), and from the boundary conditions on the internal circumference, respectively

$$\sigma_r(R') = \frac{\sigma}{2}\cos 2\vartheta \qquad (19.68\text{a})$$

$$\tau_{r\vartheta}(R') = -\frac{\sigma}{2}\sin 2\vartheta \qquad (19.68\text{b})$$

$$\sigma_r(R) = 0 \qquad (19.68\text{c})$$

$$\tau_{r\vartheta}(R) = 0 \qquad (19.68\text{d})$$

From equations (19.67) and (19.68) we obtain the system of equations

$$2A + \frac{6C}{R'^4} + \frac{4D}{R'^2} = -\frac{\sigma}{2} \qquad (19.69\text{a})$$

$$2A + 6BR'^2 - \frac{6C}{R'^4} - \frac{2D}{R'^2} = -\frac{\sigma}{2} \qquad (19.69\text{b})$$

$$2A + \frac{6C}{R^4} + \frac{4D}{R^2} = 0 \qquad (19.69\text{c})$$

$$2A + 6BR^2 - \frac{6C}{R^4} - \frac{2D}{R^2} = 0 \qquad (19.69\text{d})$$

For $R' \to \infty$, we have

$$A = -\frac{\sigma}{4}, \quad B = 0, \quad C = -\frac{\sigma}{4}R^4, \quad D = \frac{\sigma}{2}R^2 \qquad (19.70)$$

Substituting these values into equations (19.67) and adding the contribution given by equations (19.63), produced by the uniform stress $\sigma/2$, we finally arrive at

$$\sigma_r = \frac{\sigma}{2}\left(1 - \frac{R^2}{r^2}\right) + \frac{\sigma}{2}\left(1 + 3\frac{R^4}{r^4} - 4\frac{R^2}{r^2}\right)\cos 2\vartheta \qquad (19.71\text{a})$$

$$\sigma_\vartheta = \frac{\sigma}{2}\left(1 + \frac{R^2}{r^2}\right) - \frac{\sigma}{2}\left(1 + 3\frac{R^4}{r^4}\right)\cos 2\vartheta \qquad (19.71\text{b})$$

$$\tau_{r\vartheta} = -\frac{\sigma}{2}\left(1 - 3\frac{R^4}{r^4} + 2\frac{R^2}{r^2}\right)\sin 2\vartheta \qquad (19.71\text{c})$$

This solution was obtained by Kirsch in 1898.

It may be verified how, for $r \to \infty$, the stress field, expressed by equations (19.71), reproduces the conditions at infinity (19.61), while on the edge of the hole, $r = R$, we have

$$\sigma_\vartheta = \sigma(1 - 2\cos 2\vartheta) \qquad (19.72)$$

with $\sigma_r = \tau_{r\vartheta} = 0$.

The circumferential stress σ_ϑ is maximum for $\vartheta = \pi/2$ and $\vartheta = \frac{3}{2}\pi$, i.e. at the extremities of the diameter orthogonal to the direction of tension (Figure 19.5):

$$\sigma_\vartheta(\max) = 3\sigma \qquad (19.73)$$

The so-called **stress concentration factor**, in the case of a circular hole, is therefore equal to 3, and is regardless of the radius of the hole.

On the other hand, the circumferential stress σ_ϑ is minimum for $\vartheta = 0$ and $\vartheta = \pi$, i.e. at the extremities of the diameter collinear to the direction of tension (Figure 19.5):

$$\sigma_\vartheta(\min) = -\sigma \qquad (19.74)$$

At these points a compression is thus expected, as in all the other points for which $-\frac{1}{6}\pi < \vartheta < \frac{1}{6}\pi$, and $\frac{5}{6}\pi < \vartheta < \frac{7}{6}\pi$.

On the section of the plate perpendicular to the axis X and passing through the origin of the axes, we have

$$\sigma_\vartheta = \frac{\sigma}{2}\left(2 + \frac{R^2}{r^2} + 3\frac{R^4}{r^4}\right) \qquad (19.75a)$$

$$\tau_{r\vartheta} = 0 \qquad (19.75b)$$

From an examination of expression (19.75a) the local character of stress concentration around the hole is evident. As r increases, the stress σ_ϑ tends rapidly to the value σ, as is shown by the diagram of Figure 19.5. At a distance from the edge of the hole equal to its diameter $2R$, the stress σ_ϑ is higher than the asymptotic value σ by 22%, whereas at a distance equal to twice the diameter it is higher only by 4%. The solution (19.71) is applicable also to a plate of finite width, provided that this width is not less than four times the diameter of the hole. In these cases the error does not exceed 6%.

In the cases of biaxial tensile and/or compressive loading, the stress field is obtained, by superposition, from solution (19.71). When, for instance, there are two orthogonal tensile stresses of intensity σ, i.e. a **uniform tensile condition** σ, the circumferential stress on the edge of the hole is also uniform and equal to 2σ. When, instead, there is a tensile stress σ in the X direction and a compressive stress $-\sigma$ in the Y direction, i.e. in the case of **pure shear**, $\tau = \sigma$, the stress is maximum and equal to 4σ (tensile) for $\vartheta = \pi/2$ and $\vartheta = \frac{3}{2}\pi$, whereas it is minimum and equal to -4σ (compressive) for $\vartheta = 0$ and $\vartheta = \pi$.

19.7 Concentrated force acting on the edge of an elastic half-plane

Consider a vertical force acting on the horizontal boundary of an elastic half-plane (Figure 19.6(a)). Let the distribution of the force along the thickness of the plate be uniform (Figure 19.6(b)), and let P denote the load per unit thickness.

The solution of this elastic problem may be found from the Airy stress function,

$$\Phi = -\frac{P}{\pi}r\vartheta\ \sin\vartheta \qquad (19.76)$$

Using equations (19.44) we obtain the stresses

$$\sigma_r = -\frac{2P}{\pi}\frac{\cos\vartheta}{r} \qquad (19.77a)$$

$$\sigma_\vartheta = \tau_{r\vartheta} = 0 \qquad (19.77b)$$

which produce a field of radial compression. The boundary conditions are satisfied by equation (19.77b), while the compression (19.77a) presents a singularity at the point of application of the force. The resultant of the forces which act on a cylindrical surface of radius r (Figure 19.6 (b)) is in equilibrium with the external force P. Summing up the vertical components, $\sigma_r r\mathrm{d}\vartheta \cos\vartheta$, of the elementary forces that act on each elementary portion $r\mathrm{d}\vartheta$ of the surface, we have

$$2\int_0^{\pi/2}\sigma_r r \,\cos\vartheta \,\mathrm{d}\vartheta = -\frac{4P}{\pi}\int_0^{\pi/2}\cos^2\vartheta \,\mathrm{d}\vartheta = -P \qquad (19.78)$$

If we consider a circumference of diameter d, with centre on the X axis and tangent to the Y axis at the point of application of the force (Figure 19.6 (a)),

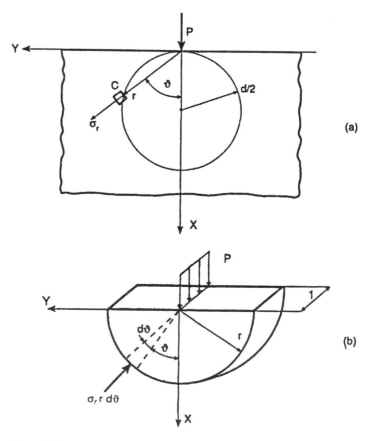

Figure 19.6

638

for each point C of this circumference we have $r = d\cos\vartheta$, and hence relation (19.77a) becomes

$$\sigma_r = -\frac{2P}{\pi d} \qquad (19.79)$$

This means that the radial stress is the same at each point of the circumference, with the exception of the point of application of the force.

Rendering explicit the strain components (12.95a) corresponding to the cylindrical geometry, we have

$$\varepsilon_r = \frac{\partial u}{\partial r} = -\frac{2P}{\pi E}\frac{\cos\vartheta}{r} \qquad (19.80a)$$

$$\varepsilon_\vartheta = \frac{u}{r} + \frac{1}{r}\frac{\partial v}{\partial \vartheta} = v\frac{2P}{\pi E}\frac{\cos\vartheta}{r} \qquad (19.80b)$$

$$\gamma_{r\vartheta} = \frac{1}{r}\frac{\partial u}{\partial \vartheta} + \frac{\partial v}{\partial r} - \frac{v}{r} = 0 \qquad (19.80c)$$

Integrating the first equation, we have

$$u = -\frac{2P}{\pi E}\cos\vartheta \ \log r + f(\vartheta) \qquad (19.81)$$

where f is a function only of ϑ. Substituting into equation (19.80b) and integrating, we find

$$v = \frac{2vP}{\pi E}\sin\vartheta + \frac{2P}{\pi E}\log r \ \sin\vartheta - \int f(\vartheta)\,d\vartheta + g(r) \qquad (19.82)$$

where g is a function only of r. Substituting equations (19.81) and (19.82) into equation (19.80c), we have finally

$$f(\vartheta) = -\frac{(1-v)P}{\pi E}\vartheta \ \sin\vartheta + A \ \sin\vartheta + B \ \cos\vartheta \qquad (19.83a)$$

$$g(r) = Cr \qquad (19.83b)$$

where A, B, C are constants of integration.

From equations (19.81) and (19.82), we obtain the field of displacements

$$u = -\frac{2P}{\pi E}\cos\vartheta \ \log r - \frac{(1-v)P}{\pi E}\vartheta \ \sin\vartheta + A \ \sin\vartheta + B \ \cos\vartheta \qquad (19.84a)$$

$$v = \frac{2vP}{\pi E}\sin\vartheta + \frac{2P}{\pi E}\log r \ \sin\vartheta - \frac{(1-v)P}{\pi E}\vartheta \ \cos\vartheta + \qquad (19.84b)$$

$$\frac{(1-v)P}{\pi E}\sin\vartheta + A \ \cos\vartheta - B \ \sin\vartheta + Cr$$

From symmetry we have $v = 0$ for $\vartheta = 0$, and hence we must find $A = C = 0$. With these values of the integration constants, the vertical displacements of the points of the X axis are

$$u(\vartheta = 0) = -\frac{2P}{\pi E}\log r + B \qquad (19.85)$$

639

To determine the constant B, it may be assumed that a point of the X axis at a distance d from the origin does not undergo a vertical displacement. From equation (19.85) we obtain

$$B = \frac{2P}{\pi E}\log d \qquad (19.86)$$

and hence

$$u(\vartheta = 0) = \frac{2P}{\pi E}\log\frac{d}{r} \qquad (19.87)$$

As may be noted, the solution thus determined predicts an infinite displacement of the point of application of the force, and likewise a non-integrable strain energy in the area around the same point

$$\int_{-\pi/2}^{\pi/2}\int_{\varepsilon}^{R}\frac{\sigma_r^2}{2E}\,r\,dr\,d\vartheta = \frac{2P^2}{\pi^2 E}\int_{-\pi/2}^{\pi/2}\cos^2\vartheta\,d\vartheta\int_{\varepsilon}^{R}\frac{dr}{r} \qquad (19.88)$$

$$= \frac{P^2}{\pi E}\log\frac{R}{\varepsilon}$$

The limit for $\varepsilon \to 0$ of expression (19.88) is infinite, but the same expression can be considered finite for $\varepsilon \neq 0$. This means that, if ideally we take away a portion of material around the point of application of the force, the incongruences pointed out above can in a sense be removed. This portion of material is the one which actually undergoes plastic deformation and flows under the action of the concentrated force.

19.8 Analytical functions

A function Z is said to be **complex** when it is made up of two parts, one real U and the other imaginary V:

$$Z = U + iV \qquad (19.89)$$

Z is also referred to as the **dependent variable**, and, in general, each of its two parts, U and V, is a function of the **independent variable**

$$z = x + iy \qquad (19.90)$$

where x is the real part and y is the imaginary part of that complex variable. It may be stated that $Z(z) = U(z) + iV(z)$ is a **complex function of a complex variable**.

The complex function of a complex variable $Z(z)$ is said to be **analytical** at a point in the complex plane when its derivative is unique and does not depend on the direction of the increment.

Conditions are shown to exist on the components U and V which imply that the function Z is analytical. As is usually done in the case of real functions, let the derivative of the function Z be defined as the limit of the difference quotient:

$$Z' = \lim_{\Delta z \to 0}\frac{\Delta Z}{\Delta z} = \frac{dZ}{dz} = \frac{\frac{\partial U}{\partial x}\,dx + \frac{\partial U}{\partial y}\,dy + i\left(\frac{\partial V}{\partial x}\,dx + \frac{\partial V}{\partial y}\,dy\right)}{dx + i\,dy} \qquad (19.91)$$

If we multiply both numerator and denominator by (dx–idy),

$$Z' = \frac{\left(\dfrac{\partial U}{\partial x}dx + \dfrac{\partial U}{\partial y}dy + i\dfrac{\partial V}{\partial x}dx + i\dfrac{\partial V}{\partial y}dy\right)(dx - idy)}{dx^2 + dy^2} \tag{19.92}$$

and designate as m the slope dy/dx, we
obtain

$$Z' = \frac{\dfrac{\partial U}{\partial x} + m\left(\dfrac{\partial U}{\partial y} + \dfrac{\partial V}{\partial x}\right) + m^2\dfrac{\partial V}{\partial y} + i\left[\dfrac{\partial V}{\partial x} + m\left(\dfrac{\partial V}{\partial y} - \dfrac{\partial U}{\partial x}\right) - m^2\dfrac{\partial U}{\partial y}\right]}{1 + m^2} \tag{19.93}$$

The function Z is analytical, and consequently its derivative is unique in z, if
and only if the following differential relations, referred to as **Cauchy–
Riemann conditions**, hold:

$$\frac{\partial U}{\partial x} = \frac{\partial V}{\partial y} \tag{19.94a}$$

$$\frac{\partial U}{\partial y} = -\frac{\partial V}{\partial x} \tag{19.94b}$$

If these conditions do hold, we have

$$Z' = \frac{\partial U}{\partial x} + i\frac{\partial V}{\partial x} \tag{19.95}$$

**If a function Z is analytical, its derivative is also analytical and *vice
versa*.** To demonstrate this, it is sufficient to differentiate relations (19.94)
with respect to x

$$\frac{\partial^2 U}{\partial x^2} = \frac{\partial^2 V}{\partial y \partial x} \tag{19.96a}$$

$$\frac{\partial^2 U}{\partial x \partial y} = -\frac{\partial^2 V}{\partial x^2} \tag{19.96b}$$

Taking into account equation (19.95), equations (19.96) can be transformed as
follows:

$$\frac{\partial}{\partial x}\text{Re } Z' = \frac{\partial}{\partial y}\text{Im } Z' \tag{19.97a}$$

$$\frac{\partial}{\partial y}\text{Re } Z' = -\frac{\partial}{\partial x}\text{Im } Z' \tag{19.97b}$$

which represent the Cauchy–Riemann conditions for the derivative function.
**If a function Z is analytical, its real and imaginary parts are harmonic
functions.** To demonstrate this, it is sufficient to derive relations (19.94a, b)
with respect to x and y

$$\frac{\partial^2 U}{\partial x^2} = \frac{\partial^2 V}{\partial x \partial y} \tag{19.98a}$$

641

$$\frac{\partial^2 U}{\partial y^2} = -\frac{\partial^2 V}{\partial x \partial y} \tag{19.98b}$$

and to apply the transitive law

$$\frac{\partial^2 U}{\partial x^2} + \frac{\partial^2 U}{\partial y^2} = 0 \tag{19.99}$$

Differentiating equation (19.94a) with respect to y, and equation (19.94b) with respect to x, we obtain, on the other hand

$$\frac{\partial^2 U}{\partial x \partial y} = \frac{\partial^2 V}{\partial y^2} \tag{19.100a}$$

$$\frac{\partial^2 U}{\partial x \partial y} = -\frac{\partial^2 V}{\partial x^2} \tag{19.100b}$$

and hence

$$\frac{\partial^2 V}{\partial x^2} + \frac{\partial^2 V}{\partial y^2} = 0 \tag{19.101}$$

Consider, for example, the function

$$Z(z) = z = x + iy$$

Since

$$\frac{\partial U}{\partial x} = \frac{\partial V}{\partial y} = 1$$

and

$$\frac{\partial U}{\partial y} = -\frac{\partial V}{\partial x} = 0$$

the Cauchy–Riemann conditions are satisfied at each point in the complex plane, and hence the function z is always analytical.

On the other hand, the function

$$Z(z) = \bar{z} = x - iy,$$

which associates with each point of the complex plane its conjugate, is not analytical at any point in the complex plane. We have in fact

$$\frac{\partial U}{\partial x} = 1, \quad \frac{\partial V}{\partial y} = -1$$

that is

$$\frac{\partial U}{\partial x} \neq \frac{\partial V}{\partial y}$$

The function

$$Z(z) = z^2 = (x + iy)(x + iy) = (x^2 - y^2) + 2ixy$$

is analytical because

$$\frac{\partial U}{\partial x} = \frac{\partial V}{\partial y} = 2x,$$

$$\frac{\partial U}{\partial y} = -\frac{\partial V}{\partial x} = -2y$$

More generally, all the polynomial functions are analytical.

The function

$$Z(z) = \frac{1}{z} = \frac{x}{x^2 + y^2} - i\frac{y}{x^2 + y^2},$$

since it is verified that

$$\frac{\partial U}{\partial x} = \frac{\partial V}{\partial y} = \frac{y^2 - x^2}{\left(x^2 + y^2\right)^2}$$

$$\frac{\partial U}{\partial y} = -\frac{\partial V}{\partial x} = -\frac{2xy}{\left(x^2 + y^2\right)^2}$$

is analytical at all points in the complex plane, excluding the origin, where $1/z$ is not definite.

Finally as regards the **modulus** function

$$Z(z) = |z|^2 = x^2 + y^2$$

since

$$\frac{\partial U}{\partial x} = 2x, \quad \frac{\partial V}{\partial y} = 0$$

$$\frac{\partial U}{\partial y} = 2y, \quad -\frac{\partial V}{\partial x} = 0$$

it is analytical only at the origin, while at the other points in the complex plane, it does not satisfy the Cauchy–Riemann conditions.

19.9   Kolosoff–Muskhelishvili method

Using the properties of the analytical functions, presented in the foregoing section, Kolosoff (1909), and subsequently Muskhelishvili (1933), developed a method for resolving plane elastic problems. According to this method, the Airy stress function becomes expressible as follows:

$$\Phi = \mathrm{Re}\left(\bar{z}\psi + \chi\right) \tag{19.102}$$

where ψ and χ are two analytical functions, called **complex potentials**. It may be demonstrated that the biharmonic equation (19.11) is satisfied by equation (19.102).

643

Equation (19.102) may be transformed as follows:

$$\Phi = \mathrm{Re}\left(z\bar{z}\frac{\psi}{z} + \chi\right)$$ (19.103)

whence we have

$$\Phi = \left(x^2 + y^2\right)\mathrm{Re}\frac{\psi}{z} + \mathrm{Re}\,\chi$$ (19.104)

Since $1/z$, ψ and χ are analytical functions, and the product of two analytical functions is itself an analytical function, as can very easily be demonstrated, Re (ψ/z) and Reχ are two harmonic functions. The procedure therefore is to demonstrate that the product $(x^2 + y^2)f$, with f as a harmonic function, represents a biharmonic function.

Applying the Laplace operator, we obtain

$$\nabla^2\left[(x^2 + y^2)f\right] = \frac{\partial}{\partial x}\left[\frac{\partial f}{\partial x}(x^2 + y^2) + 2xf\right] + \frac{\partial}{\partial y}\left[\frac{\partial f}{\partial y}(x^2 + y^2) + 2yf\right]$$ (19.105)

$$= (x^2 + y^2)\nabla^2 f + 4x\frac{\partial f}{\partial x} + 4y\frac{\partial f}{\partial y} + 4f$$

and hence, if f is harmonic

$$\nabla^2\left[(x^2 + y^2)f\right] = 4\left(x\frac{\partial f}{\partial x} + y\frac{\partial f}{\partial y} + f\right)$$ (19.106)

To demonstrate that the function is biharmonic, a second application of the Laplacian is necessary:

$$\nabla^2\nabla^2\left[(x^2 + y^2)f\right] = 4\left[\nabla^2\left(x\frac{\partial f}{\partial x}\right) + \nabla^2\left(y\frac{\partial f}{\partial y}\right)\right]$$ (19.107)

Performing the calculations, we have

$$\nabla^2\left(x\frac{\partial f}{\partial x}\right) = \frac{\partial}{\partial x}\left(\frac{\partial f}{\partial x} + x\frac{\partial^2 f}{\partial x^2}\right) + \frac{\partial}{\partial y}\left(x\frac{\partial^2 f}{\partial x\partial y}\right)$$ (19.108)

$$= 2\frac{\partial^2 f}{\partial x^2} + x\frac{\partial}{\partial x}\left(\frac{\partial^2 f}{\partial x^2} + \frac{\partial^2 f}{\partial y^2}\right)$$

Since f is a harmonic function, it follows that

$$\nabla^2\left(x\frac{\partial f}{\partial x}\right) = 2\frac{\partial^2 f}{\partial x^2}$$ (19.109a)

and likewise

$$\nabla^2\left(y\frac{\partial f}{\partial y}\right) = 2\frac{\partial^2 f}{\partial y^2}$$ (19.109b)

Finally, substituting equations (19.109) into equation (19.107), we obtain

$$\nabla^4\left[(x^2 + y^2)f\right] = 8\nabla^2 f = 0$$ (19.110)

644

Since the sum of a complex function and its conjugate is equal to twice its real part

$$f(z) + \overline{f}(z) = 2\operatorname{Re} f(z) \tag{19.111}$$

from equation (19.102) we find

$$2\Phi = \bar{z}\psi(z) + \chi(z) + z\overline{\psi}(z) + \overline{\chi}(z) \tag{19.112}$$

Applying to equation (19.112) the following rules of differentiation of composite functions:

$$\frac{\partial f}{\partial x} = \frac{df}{dz}\frac{\partial z}{\partial x} = \frac{df}{dz} = f' \tag{19.113a}$$

$$\frac{\partial f}{\partial y} = \frac{df}{dz}\frac{\partial z}{\partial y} = i\frac{df}{dz} = if' \tag{19.113b}$$

$$\frac{\partial \overline{f}}{\partial x} = \frac{d\overline{f}}{d\bar{z}}\frac{\partial \bar{z}}{\partial x} = \frac{d\overline{f}}{d\bar{z}} = \overline{f'} \tag{19.113c}$$

$$\frac{\partial \overline{f}}{\partial y} = \frac{d\overline{f}}{d\bar{z}}\frac{\partial \bar{z}}{\partial y} = -i\frac{d\overline{f}}{d\bar{z}} = -i\overline{f'} \tag{19.113d}$$

we obtain

$$2\frac{\partial \Phi}{\partial x} = \psi(z) + \bar{z}\psi'(z) + \chi'(z) + \overline{\psi}(z) + z\overline{\psi'}(z) + \overline{\chi'}(z) \tag{19.114a}$$

$$2\frac{\partial \Phi}{\partial y} = -i\psi(z) + i\bar{z}\psi'(z) + i\chi'(z) + i\overline{\psi}(z) - iz\overline{\psi'}(z) - i\overline{\chi'}(z) \tag{19.114b}$$

Multiplying equation (19.114b) by the imaginary unit, we have

$$2i\frac{\partial \Phi}{\partial y} = \psi(z) - \bar{z}\psi'(z) - \chi'(z) - \overline{\psi}(z) + z\overline{\psi'}(z) + \overline{\chi'}(z) \tag{19.115}$$

The sum of equations (19.114a) and (19.115) yields

$$\frac{\partial \Phi}{\partial x} + i\frac{\partial \Phi}{\partial y} = \psi(z) + z\overline{\psi'}(z) + \overline{\chi'}(z) \tag{19.116}$$

The partial derivation with respect to x and to y of expression (19.116) leads to the following equations:

$$\frac{\partial^2 \Phi}{\partial x^2} + i\frac{\partial^2 \Phi}{\partial x\partial y} = \psi'(z) + \overline{\psi'}(z) + z\overline{\psi''}(z) + \overline{\chi''}(z) \tag{19.117a}$$

$$\frac{\partial^2 \Phi}{\partial x\partial y} + i\frac{\partial^2 \Phi}{\partial y^2} = i\psi'(z) + i\overline{\psi'}(z) - iz\overline{\psi''}(z) - i\overline{\chi''}(z) \tag{19.117b}$$

Multiplying equation (19.117b) by the imaginary unit, we have

$$-\frac{\partial^2 \Phi}{\partial y^2} + i\frac{\partial^2 \Phi}{\partial x\partial y} = -\psi'(z) - \overline{\psi'}(z) + z\overline{\psi''}(z) + \overline{\chi''}(z) \tag{19.118}$$

The difference and the sum of equations (19.117a) and (19.118) give, respectively

$$\frac{\partial^2 \Phi}{\partial x^2} + \frac{\partial^2 \Phi}{\partial y^2} = 2\psi'(z) + 2\overline{\psi}'(z) = 4 \ \mathrm{Re}\,\psi'(z) \qquad (19.119a)$$

$$\frac{\partial^2 \Phi}{\partial x^2} - \frac{\partial^2 \Phi}{\partial y^2} + 2\mathrm{i}\frac{\partial^2 \Phi}{\partial x \partial y} = 2\left[z\overline{\psi}''(z) + \overline{\chi}''(z)\right] \qquad (19.119b)$$

Since the second partial derivatives of the Airy stress function represent, in agreement with equations (19.9), the stress components, equations (19.119) can be written as follows:

$$\sigma_x + \sigma_y = 4 \ \mathrm{Re}\,\psi'(z) \qquad (19.120a)$$

$$\sigma_y - \sigma_x - 2\mathrm{i}\tau_{xy} = 2\left[z\overline{\psi}''(z) + \overline{\chi}''(z)\right] \qquad (19.120b)$$

or

$$\sigma_x + \sigma_y = 4 \ \mathrm{Re}\,\psi'(z) \qquad (19.121a)$$

$$\sigma_y - \sigma_x + 2\mathrm{i}\tau_{xy} = 2\left[\bar{z}\psi''(z) + \chi''(z)\right] \qquad (19.121b)$$

Equations (19.121) may be cast in an alternative form, which allows the individual stress components to be separated. Subtracting and adding them, we obtain, respectively

$$2\left(\sigma_x - \mathrm{i}\tau_{xy}\right) = 4 \ \mathrm{Re}\,\psi'(z) - 2\left[\bar{z}\psi''(z) + \chi''(z)\right] \qquad (19.122a)$$

$$2\left(\sigma_y + \mathrm{i}\tau_{xy}\right) = 4 \ \mathrm{Re}\,\psi'(z) + 2\left[\bar{z}\psi''(z) + \chi''(z)\right] \qquad (19.122b)$$

whilst from equation (19.121b) we find

$$\tau_{xy} = \mathrm{Im}\left[\bar{z}\psi''(z) + \chi''(z)\right] \qquad (19.122c)$$

Isolating the real parts of equations (19.122a) and (19.122b), we obtain

$$\sigma_x = \mathrm{Re}\left[2\psi'(z)\right] - \mathrm{Re}\left[\chi''(z)\right] - x\,\mathrm{Re}\left[\psi''(z)\right] - y\,\mathrm{Im}\left[\psi''(z)\right] \quad (19.123a)$$

$$\sigma_y = \mathrm{Re}\left[2\psi'(z)\right] + \mathrm{Re}\left[\chi''(z)\right] + x\,\mathrm{Re}\left[\psi''(z)\right] + y\,\mathrm{Im}\left[\psi''(z)\right] \quad (19.123b)$$

whilst from equation (19.122c) we have

$$\tau_{xy} = \mathrm{Im}\left[\chi''(z)\right] - y\,\mathrm{Re}\left[\psi''(z)\right] + x\,\mathrm{Im}\left[\psi''(z)\right] \qquad (19.123c)$$

In the case of problems that are symmetrical with respect to the X axis, it must follow that

$$\tau_{xy}(x,0) = 0 \qquad (19.124)$$

and hence, from equations (19.122a, b)

$$\mathrm{Im}\left[\bar{z}\psi''(z) + \chi''(z)\right] = 0, \qquad \text{for } y = 0 \qquad (19.125)$$

646

On the real axis, since $z = \bar{z}$, the following relation also holds good:

$$\text{Im}\big[z\psi''(z) + \chi''(z)\big] = 0, \qquad \text{for} \quad y = 0 \tag{19.126}$$

If we extrapolate the foregoing condition to the entire complex plane, we can write

$$z\psi''(z) + \chi''(z) + B = 0, \qquad \forall z \in \mathcal{C} \tag{19.127}$$

where B is a real constant and C is the set of all the points of the complex plane. Note that B cannot be a real function of the complex variable z, since if it were it would not obey the Cauchy–Riemann conditions (19.94) at each point of the complex plane.

If the expression (19.127) is identically zero over the entire complex plane, so must its real and imaginary parts likewise be zero:

$$x\,\text{Re}\big[\psi''(z)\big] - y\,\text{Im}\big[\psi''(z)\big] + \text{Re}\big[\chi''(z)\big] + B = 0 \tag{19.128a}$$

$$x\,\text{Im}\big[\psi''(z)\big] + y\,\text{Re}\big[\psi''(z)\big] + \text{Im}\big[\chi''(z)\big] = 0 \tag{19.128b}$$

Substituting equations (19.128) in equations (19.123), we obtain finally

$$\sigma_x = \text{Re}\big[2\psi'(z)\big] - y\,\text{Im}\big[2\psi''(z)\big] + B \tag{19.129a}$$

$$\sigma_y = \text{Re}\big[2\psi'(z)\big] + y\,\text{Im}\big[2\psi''(z)\big] - B \tag{19.129b}$$

$$\tau_{xy} = -y\,\text{Re}\big[2\psi''(z)\big] \tag{19.129c}$$

The extrapolation of the condition (19.127) to the entire complex plane thus allows one of the two Muskhelishvili complex potentials to be eliminated. As shall be seen in the next chapter, the application of equations (19.129) leads to a relatively simple solution of the problem of a plate in tension with a rectilinear crack. This solution was obtained by Westergaard in 1939 on the basis of Muskhelishvili's treatment, which had previously been published in Russian in 1933 and was subsequently republished in English in 1953. However, in his original publication, Westergaard used a different notation, $Z_f(z)$ instead of $2\psi'(z)$, and disregarded the real constant B. The latter implicit and erroneous assumption was pointed out by Sih in 1966.

19.10 Elliptical hole in a plate subjected to tension

Let us consider an infinite plate in a condition of uniaxial tension σ in a direction that forms an angle β with the X axis (Figure 19.7). Let this uniaxial condition be disturbed by an elliptical hole having its major axis along the X axis and its minor axis along the Y axis.

Let X^*Y^* be the cartesian axes obtained from rotation of the XY axes by the angle β, such as to bring the X axis parallel to the tension σ (Figure 19.7). Evaluating equation (7.90), we obtain the following equations of transformation:

$$\sigma_x^* = \sigma_x \cos^2\beta + \sigma_y \sin^2\beta + 2\tau_{xy}\sin\beta\,\cos\beta \tag{19.130a}$$

$$\sigma_y^* = \sigma_x \sin^2\beta + \sigma_y \cos^2\beta - 2\tau_{xy}\sin\beta\,\cos\beta \tag{19.130b}$$

$$\tau_{xy}^* = (\sigma_y - \sigma_x)\sin\beta\cos\beta + \tau_{xy}(\cos^2\beta - \sin^2\beta) \tag{19.130c}$$

647

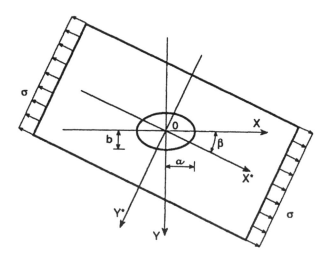

Figure 19.7

which, via the well-known trigonometric formulas, become

$$\sigma_x^* = \frac{1}{2}\left(\sigma_x + \sigma_y\right) + \frac{1}{2}\left(\sigma_x - \sigma_y\right)\cos 2\beta + \tau_{xy}\sin 2\beta \qquad (19.131a)$$

$$\sigma_y^* = \frac{1}{2}\left(\sigma_x + \sigma_y\right) - \frac{1}{2}\left(\sigma_x - \sigma_y\right)\cos 2\beta - \tau_{xy}\sin 2\beta \qquad (19.131b)$$

$$\tau_{xy}^* = -\frac{1}{2}\left(\sigma_x - \sigma_y\right)\sin 2\beta + \tau_{xy}\cos 2\beta \qquad (19.131c)$$

From equations (19.131) the following relations are easily obtained:

$$\sigma_x^* + \sigma_y^* = \sigma_x + \sigma_y \qquad (19.132a)$$

$$\sigma_y^* - \sigma_x^* + 2i\tau_{xy}^* = e^{2i\beta}\left(\sigma_y - \sigma_x + 2i\tau_{xy}\right) \qquad (19.132b)$$

Since at infinity we have

$$\sigma_x^* = \sigma, \qquad \sigma_y^* = \tau_{xy}^* = 0 \qquad (19.133)$$

equations (19.132) yield

$$\sigma_x + \sigma_y = \sigma \qquad (19.134a)$$

$$\sigma_y - \sigma_x + 2i\tau_{xy} = -\sigma e^{-2i\beta} \qquad (19.134b)$$

and hence, via equations (19.121), at infinity we have

$$4\,\mathrm{Re}\,\psi'(z) = \sigma \qquad (19.135a)$$

$$2\left[\bar{z}\,\psi''(z) + \chi''(z)\right] = -\sigma e^{-2i\beta} \qquad (19.135b)$$

648

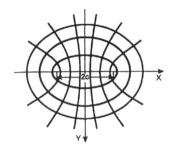

Figure 19.8

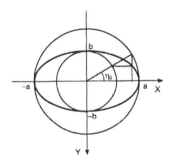

Figure 19.9

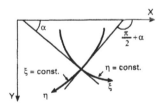

Figure 19.10

The elliptical coordinates ξ, η, shown in Figure 19.8, are defined by the following relation:

$$z = c \cosh \zeta \qquad (19.136a)$$

with

$$\zeta = \xi + i\eta \qquad (19.136b)$$

From equations (19.136) we have

$$x = c \ \cosh \xi \ \cos \eta \qquad (19.137a)$$

$$y = c \ \sinh \xi \ \sin \eta \qquad (19.137b)$$

The coordinate ξ is constant and equal to ξ_0, on an ellipse of semiaxes $c \cosh \xi_0$ and $c \sinh \xi_0$, just as the coordinate η is constant and equal to η_0 on a hyperbola which has the same focuses $(\pm c, 0)$ as the ellipse. It is sufficient in fact to take into account the relations

$$\cos^2 \eta + \sin^2 \eta = 1 \qquad (19.138a)$$

$$\cosh^2 \xi - \sinh^2 \xi = 1 \qquad (19.138b)$$

to obtain

$$\frac{x^2}{c^2 \cosh^2 \xi} + \frac{y^2}{c^2 \sinh^2 \xi} = 1 \qquad (19.139a)$$

$$\frac{x^2}{c^2 \cos^2 \eta} - \frac{y^2}{c^2 \sin^2 \eta} = 1 \qquad (19.139b)$$

Whereas then the semiaxes of the ellipse to which the point of **elliptical coordinates** (ξ_0, η_0) belongs are

$$\sigma_y^* = \sigma_x \sin^2 \beta + \sigma_y \cos^2 \beta - 2\tau_{xy} \sin \beta \ \cos \beta \qquad (19.130b)$$

η_0 represents the angular coordinate of the same point, on the basis of the scheme of Figure 19.9.

Since the ellipses $\xi = \xi_0$ and the hyperbolas $\eta = \eta_0$ are mutually orthogonal, it is possible to write a transformation analogous to equations (19.132),

$$\sigma_\xi + \sigma_\eta = \sigma_x + \sigma_y \qquad (19.141a)$$

$$\sigma_\eta - \sigma_\xi + 2i\tau_{\xi\eta} = e^{2i\alpha}\left(\sigma_y - \sigma_x + 2i\tau_{xy}\right) \qquad (19.141b)$$

where σ_ξ and σ_η are the normal stresses on the curves ξ = constant and η = constant, respectively, $\tau_{\xi\eta}$ the shearing stress along the same curves, and α the angle that the tangent to the curve η = constant forms with the X axis (Figure 19.10).

On the edge of the elliptical hole, of equation $\xi = \xi_0$, we have

$$\sigma_\xi = \tau_{\xi\eta} = 0 \qquad (19.142)$$

whereby, subtracting equation (19.141b) from equation (19.141a), we obtain

$$2\,\mathrm{Re}\,\psi'(z) - \left[\bar{z}\psi''(z) + \chi''(z)\right]e^{2i\alpha} = 0 \qquad (19.143)$$

649

for $\xi = \xi_0$.

The boundary conditions at infinity, given by equations (19.135), and on the edge of the hole, given by equation (19.143), can be satisfied by the following complex potentials:

$$4\psi(z) = Ac\cosh\zeta + Bc\sinh\zeta \tag{19.144a}$$

$$4\chi(z) = Cc^2\zeta + Dc^2\cosh2\zeta + Ec^2\sinh2\zeta \tag{19.144b}$$

where A, B, C, D, E are constants to be determined.

Substituting the foregoing forms given by equations (19.144) in the conditions (19.135), we obtain

$$\mathrm{Re}\,A + \mathrm{Re}\,B = \sigma \tag{19.145a}$$

$$2(D + E) = -\sigma e^{-2i\beta} \tag{19.145b}$$

just as, substituting the same equations in equation (19.143), we have

$$\mathrm{cosech}\,\overline{\zeta}\big[(2A + B\ \mathrm{cotanh}\,\zeta)\ \sinh\overline{\zeta} + \tag{19.146}$$

$$\left(\overline{B} + B\mathrm{cosech}^2\zeta\right)\cosh\overline{\zeta} +$$

$$(C + 2E)\mathrm{cosech}\,\zeta\,\mathrm{cotanh}\,\zeta -$$

$$4D\sinh\zeta - 4E\cosh\zeta\big] = 0$$

once having taken into account that

$$e^{2i\alpha} = \frac{\sinh\,\zeta}{\sinh\,\overline{\zeta}} \tag{19.147}$$

On the edge of the hole we have $\xi = \xi_0$ and $\overline{\zeta} = 2\xi_0 - \zeta$. If this expression for $\overline{\zeta}$ is substituted into equation (19.146), and the functions $\sinh(2\xi_0 - \zeta)$, $\cosh(2\xi_0 - \zeta)$ are suitably expanded, the same equation becomes

$$\left(2A\sinh2\xi_0 - 2i\ \mathrm{Im}\,B\cosh2\xi_0 - 4E\right)\cosh\zeta - \tag{19.148}$$

$$\left(2A\cosh2\xi_0 - 2i\ \mathrm{Im}\,B\sinh2\xi_0 + 4D\right)\sinh\zeta +$$

$$(C + 2E + B\ \cosh2\xi_0)\mathrm{cotanh}\,\zeta\ \mathrm{cosech}\,\zeta = 0$$

This equation is satisfied if the coefficients of $\cosh\zeta$, $\sinh\zeta$ and $\mathrm{cotanh}\,\zeta$ $\mathrm{cosech}\,\zeta$ become zero. We therefore find three equations, which, together with the two equations (19.145a, b), yield a system of five equations in the five unknowns A, B, C, D, E. The solution is given by the following expressions:

$$A = \sigma e^{2\xi_0}\cos2\beta \tag{19.149a}$$

$$B = \sigma\left(1 - e^{2\xi_0 + 2i\beta}\right) \tag{19.149b}$$

$$C = -\sigma\left(\cosh2\xi_0 - \cos2\beta\right) \tag{19.149c}$$

$$D = -\frac{1}{2}\sigma e^{2\xi_0}\cosh2\left(\xi_0 + i\beta\right) \tag{19.149d}$$

$$E = \frac{1}{2}\sigma e^{2\xi_0}\sinh2\left(\xi_0 + i\beta\right) \tag{19.149e}$$

The complex potentials (19.144) are given congruently by

$$4\psi(z) = \sigma c\left[e^{2\xi_0}\cos2\beta\ \cosh\zeta + \left(1 - e^{2\xi_0 + 2i\beta}\right)\sinh\zeta\right] \quad (19.150a)$$

$$4\chi(z) = -\sigma c^2\left[\left(\cosh2\xi_0 - \cos2\beta\right)\zeta + \right. \quad (19.150b)$$

$$\left.\frac{1}{2}e^{2\xi_0}\cosh2\left(\zeta - \xi_0 - i\beta\right)\right]$$

The normal stress σ_η along the contour of the hole may be obtained from the relation (19.141a), because there σ_ξ becomes zero,

$$\sigma_\eta = 4\ \mathrm{Re}\ \psi'(z) \quad (19.151)$$

whence, via equation (19.150a), we find

$$\sigma_\eta(\xi = \xi_0) = \sigma\frac{\sinh2\xi_0 + \cos2\beta - e^{2\xi_0}\cos2(\beta - \eta)}{\cosh2\xi_0 - \cos2\eta} \quad (19.152)$$

When the stress σ is orthogonal to the major axis, i.e. for $\beta = \pi/2$, equation (19.152) becomes

$$\sigma_\eta(\xi = \xi_0) = \sigma e^{2\xi_0}\left[\frac{\sinh2\xi_0\left(1 + e^{-2\xi_0}\right)}{\cosh2\xi_0 - \cos2\eta} - 1\right] \quad (19.153)$$

and the maximum value of σ_η is reached at the ends of the major axis (cos 2η = 1):

$$\sigma_\eta(\mathrm{max}) - \sigma\left(1 + 2\frac{a}{b}\right) \quad (19.154)$$

This value tends to infinity for $a/b \to \infty$, i.e. in the case where the ellipse becomes particularly eccentric, whilst it is equal to 3σ when $a = b$, i.e. in the case of a circular hole. This latter result has already been obtained in Section 19.6. The minimum value of σ_η is instead $-\sigma$ and, as in the case of the circular hole, is reached at the ends of the axis collinear to the external load (cos $2\eta = -1$).

When the stress σ is orthogonal to the minor axis, i.e. for $\beta = 0$, the maximum value of σ_η is reached at the ends of the minor axis, and is equal to $\sigma[1 + 2(b/a)]$. This value tends to σ in the case of very eccentric ellipses. At the ends of the major axis, on the other hand, the stress is $-\sigma$ for any value of the ratio a/b.

The case of **uniform tension** σ can be considered as the combination of the two cases previously considered. The maximum stress $2\sigma\,(a/b)$ is therefore reached at the ends of the major axis, while the minimum stress $2\sigma\,(b/a)$ is reached at the ends of the minor axis.

The perturbing effect of the elliptical hole on a condition of **pure shear**, $\tau = \sigma$, parallel to the axes XY, is obtainable by superposition of the two cases, that of tension σ, with $\beta = \frac{1}{4}\pi$, and that of compression $-\sigma$, with $\beta = \frac{3}{4}\pi$:

$$\sigma_\eta(\xi = \xi_0) = -2\sigma\frac{e^{2\xi_0}\sin2\eta}{\cosh2\xi_0 - \cos2\eta} \quad (19.155)$$

This stress vanishes at the ends of both the axes and presents the extreme values

$$\sigma_\eta \begin{pmatrix} \max \\ \min \end{pmatrix} = \pm\sigma \frac{(a+b)^2}{ab} \qquad (19.156)$$

at the points for which $\tan \eta = \mp b/a$.

When the ellipse is very eccentric, the stresses, given by equation (19.156), are very high, and the points where these are developed are very close to the ends of the major axis. When instead $a = b$, we find again the result for the circular hole, with a concentration factor equal to 4.

To summarize the conclusions both of Section 19.6 and the present section, Table 19.1 gives a complete presentation of the **stress concentration factors** for circular and elliptical holes.

Table 19.1

Scheme	Hole shape	Loading condition	Stress concentration factor
	Circular	Uniaxial	3
	Circular	Uniform	2
	Circular	Pure shear	4
	Elliptical	Uniaxial along major axis	$1 + 2b/a$
	Elliptical	Uniaxial along minor axis	$1 + 2a/b$
	Elliptical	Uniform	$2a/b$
	Elliptical	Pure shear	$(a+b)^2/ab$

652

20 Mechanics of fracture

20.1 Introduction

With the scientific advances of the last few decades in the field of **Material Mechanics** it has been realized that the classical concept of **strength**, understood as force per unit surface causing fracture, is in need of revision, especially in the cases where particularly large or particularly small structures are involved. The **strength** of the material must, that is, be compared against another characteristic, the **toughness** of the material, in order to define, via the **dimension** of the structure, the **ductility** or the **brittleness** of the structure itself. Two intrinsic characteristics of the material, plus a geometrical characteristic of the structure, are in fact the minimum basis for being able to predict the type of structural response. A foretaste of what will be dealt with in the present chapter has been provided in Section 8.11. In that section we defined the **fracture energy** $\mathscr{G}_{IC}$, one of the parameters capable of measuring the toughness of the material. We also described how the structural response to uniaxial tension varies as $\mathscr{G}_{IC}$ and/or the length of the bar longitudinally subjected to tension varies. In that case a tendency emerged towards a ductile behaviour in the case of short lengths of the bar and, on the other hand, a tendency towards a brittle behaviour (**snap-back**) in the case of greater lengths of the bar. This tendency will be encountered again, in the present chapter, also in the case of two- and three-dimensional solids, in such a way as to associate ductile behaviour with relatively small solids, and brittle behaviour with relatively large solids. Just as in structures acted upon prevalently by a compressive force (Figure 17.11), there is a transition from **plastic collapse** to **instability of elastic equilibrium** as **slenderness** increases, so in structures acted upon prevalently by a tensile force, there is a transition from **plastic collapse** to **brittle fracture** as the **size scale** increases.

Two extreme cases of the above properties are shown in Figure 20.1. The first (Figure 20.1(a)) depicts one of the hundreds of **Liberty ships** which, in the years of the Second World War, split into two parts, with extremely clean and brittle fractures and without the slightest evidence to forewarn of such an eventuality. What caused profound astonishment in the technicians who first looked into those accidents was, on the one hand, the extremely low stresses present in the hull at the moment of failure and, on the other, the contrast between the extreme brittleness of the failure and the considerable ductility shown in the laboratory by specimens of the same steel.

The second case (Figure 20.1(b)) depicts a microscopic filament of glass, used for fibre reinforcement of polymer materials. The filament is elastically bent with a large curvature, so that it undergoes a regime of large strains and stresses as much as two orders of magnitude greater than the tensile strength of glass, as measured in the laboratory with specimens of normal dimensions.

The two cases just examined bring out starkly and unequivocally how both strength and ductility are functions of the size scale, such as to lead to brittleness and low strength in enormous steel structures as well as ductility and high

653

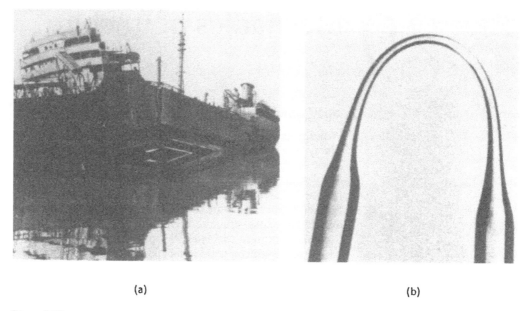

(a)   (b)

Figure 20.1

strength in microscopic structures made of glass. On the other hand, it is well-known that, in the size scale of the laboratory, steel is a particularly ductile material and glass a particularly brittle one.

It is not necessary, however, to consider extreme cases to realize how ductility is not a characteristic of the material, but rather a characteristic of the structure. Even at laboratory scale, the **ductile–brittle transition** with increase in size of the specimen has been brought to light (Figure 20.2). If the material and the geometrical shape are kept unvaried, increasing the size scale leads to a distinct transition towards a brittle type of behaviour accompanied by a sudden drop in the loading capacity and a rapid crack propagation which is in fact found for all materials, whether they be metal, polymer, ceramic or cement. On the other hand, with specimens of relatively modest dimensions, a ductile behaviour with slow crack propagation is encountered. In the case, for instance, of three-point bending, it is possible to witness the formation of a plastic hinge in the centre and the impossibility of separating the specimen into two distinct parts by applying a simple monotonic loading (Figure 20.2).

In this chapter, after a brief reference to the by now classic **Griffith's energy criterion** (1920), the major physico-mathematical theories which, between 1920 and 1950, paved the way to modern-day fracture mechanics will be presented; these are:

1. **Westergaard's method** (1939), or the **method of complex potentials**, which in fact is a simplification of Muskhelishvili's method (1933), already referred to in Section 19.9;
2. **Williams' method** (1952), or the **series expansion method**, which proves to be a new and more general approach.

Both these fundamental methods lead to the determination of the power of the stress singularity which is produced at the tip of the crack, and thus to the definition of the **stress-intensity factor**.

654

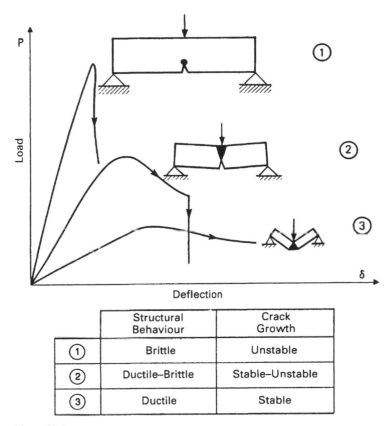

	Structural Behaviour	Crack Growth
(1)	Brittle	Unstable
(2)	Ductile–Brittle	Stable–Unstable
(3)	Ductile	Stable

Figure 20.2

In addition to the problem of the opening of a symmetrically loaded crack (Mode I), the problem of the same crack loaded skew-symmetrically by in-plane shear forces (Mode II) will also be addressed. In this more general context the more widely known branching criterion will be proposed, that of the maximum circumferential stress.

Subsequently the concept of fracture energy will be taken up again and directly correlated to the critical value of the stress-intensity factor. Finally, the plastic zone (or process zone), which always develops at the tip of each real crack, will be considered; the amplitude of the zone will be estimated, and a mathematical and numerical model will be proposed which is able to provide a continuous description of the ductile–brittle size-scale transition referred to above.

20.2 Griffith's energy criterion

Flaws in materials are often considered as the major causes of onset of brittle fractures. The effects of **stress concentration** in the vicinity of imperfections or irregularities have been well-known for a long time. Already in 1898 Kirsch provided a solution to the problem of an infinite plate with a circular hole, subjected to tension. As was shown in Section 19.6, the maximum stress on the edge of the hole is three times as great as that applied externally (Figure 20.3(a)). This means that the strength of a plate of dimensions much greater than the hole present in it is reduced to one third of that of the intact plate, regard-

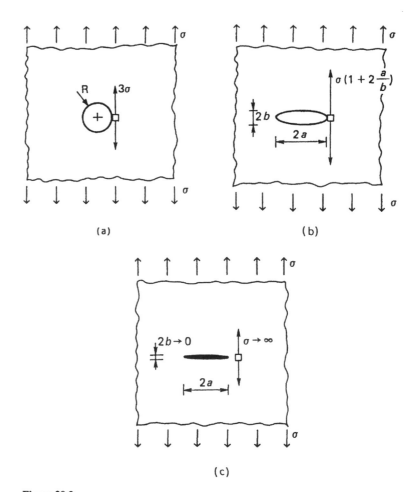

Figure 20.3

less of the size of the hole. Thus there is a compromise situation between the amount of material removed and the curvature of the hole. At the limit, even an infinitesimal hole causes a concentration factor equal to 3, even though the amount of material removed is practically nil. The radius of curvature of the hole is in fact in this case very small and creates conditions of particular severity.

Inglis (1913) extended investigations into stress concentration to the more general case of the elliptical hole (Figure 20.3(b)). As has been shown in Section 19.10, the maximum stress on the edge of the hole with major axis orthogonal to the external force is, in this case, multiplied by the factor $[1 + 2\,(a/b)]$. The strength of the plate with a hole thus comes to depend solely on the ratio between the semiaxes of the ellipse bounding the hole. The stress concentration factor increases with the increase in eccentricity of the ellipse. For $a/b \rightarrow \infty$, that is when the ellipse is very eccentric, the concentration factor tends to infinity. This model does not therefore prove useful for describing the critical condition of a crack of length $2a$ and initial width $2b$ tending to zero (Fig. 20.3(c)). In fact, very small external stresses suffice to exceed the tensile strength σ_u at the tip of the crack. In actual practice, instead, cracked solids can even stand up to considerable stress.

656

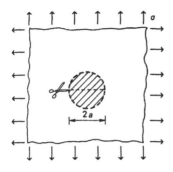

Figure 20.4

In 1920 Griffith, an aeronautics engineer engaged in the study of glass materials, thus felt the need to introduce energy considerations, and not only stress considerations, into the analysis of the fracture phenomenon. He showed that the elastic strain energy W_e, released by a uniformly extended plate of unit thickness, when a crack of length $2a$ is formed, and the displacements at infinity are maintained constant, is proportional to the energy contained in the circle of radius a before the crack originates (Figure 20.4),

$$W_e = \pi a^2 \frac{\sigma^2}{E} \tag{20.1}$$

where E is the elastic modulus of the material.

On the other hand, to create a crack of length $2a$ requires a **surface energy** equal to

$$W_s = 4a\gamma \tag{20.2}$$

where γ is the energy per unit surface.

Griffith supposed that, for a pre-existing crack of length $2a$ to extend, the elastic energy released in a virtual extension must be greater than or equal to that required by the new portion of free surface that is created:

$$\frac{dW_e}{da} \geq \frac{dW_s}{da} \tag{20.3}$$

The condition of instability is therefore the following:

$$2\pi a \frac{\sigma^2}{E} \geq 4\gamma \tag{20.4}$$

The foregoing inequality is valid both for the stress applied σ and for the half-length a of the crack. The pairs of values σ and a which fall beneath the curve of Figure 20.5 constitute stable cases, whereas the pairs that fall above it concern unstable cases. Rendering the condition (20.4) explicit with respect to the applied stress, we obtain in fact

$$\sigma \geq \sqrt{\frac{2\gamma E}{\pi a}} \tag{20.5}$$

Twice the value of the unit surface energy is usually termed **fracture energy**, $\mathcal{G}_{IC}$, so that equation (20.5) takes the form

$$\sigma \geq \sqrt{\frac{\mathcal{G}_{IC}E}{\pi a}} \tag{20.6}$$

The curve of Figure 20.5, which indicates the instability stress as a function of the half-length of the crack, presents two asymptotes. The horizontal asymptotic branch represents the decrease of plate strength with the increase in the crack length. When this length tends to infinity, the strength of the plate consistently tends to zero. The vertical asymptotic branch represents the increase of plate strength with the decrease in the crack length. For $a \to 0$, the strength tends to infinity. This result is not, on the other hand, consistent with the assumed existence of an intrinsic strength σ_P of the material of which the plate is made. A similar problem has already been met with in our discussion of the

657

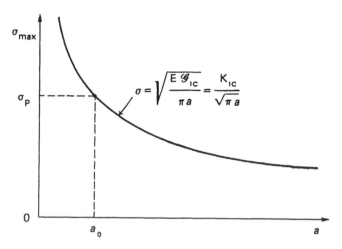

Figure 20.5

limits of validity of Euler's formula, in Section 17.4 (Figure 17.11). In that case a limit slenderness was defined, beneath which the compressive yielding of the bar precedes the instability of the elastic equilibrium. In the present case, analogously, it is possible to define a length $2a_0$ of the crack, beneath which tensile yielding of the entire plate precedes the unstable propagation of the crack:

$$a_0 = \frac{1}{\pi} \frac{\mathcal{G}_{IC} E}{\sigma_P^2} \tag{20.7}$$

The length $2a_0$ represents the equivalent length of the microcracks and of the flaws, pre-existing in the material of which the plate is made.

Equation (20.7) offers an explanation of the fact that glass filaments show a strength as much as two orders of magnitude greater than that found using macroscopic test specimens. The fibre cross section has in fact a diameter which is considerably less than the size of the flaws that are found in macroscopic test specimens. An increase of two orders of magnitude in the apparent strength σ_P derives in fact from the reduction in the characteristic length $2a_0$ by a factor of 10^4. In this perspective, the tensile strength ceases to represent a characteristic of the material and becomes a function of the length $2a_0$ of the preexisting microcracks.

20.3   Westergaard's method

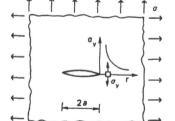

Figure 20.6

In this section we shall show how Westergaard (1939), taking up the treatment of the subject by Muskhelishvili (1933), identified the power of the singularity that the stresses present at the tips of the crack. It is known that if discontinuities or slits are considered instead of elliptical holes (Figure 20.3(c)), the stress that develops in the area around the tip tends to infinity as the distance from the tip itself tends to zero (Figure 20.6). At this point the stress field is not defined and, in actual practice, around it there forms a zone of plastic deformation, albeit small, even in the most brittle materials.

It is possible to demonstrate that each function of the form

$$\Phi = U_1 + x U_2 + y U_3 \tag{20.8}$$

658

where U_1, U_2, U_3 are harmonic real functions, satisfies the Airy equation (19.11). Obviously

$$\nabla^4 U_1 = \nabla^2(\nabla^2 U_1) = \nabla^2(0) = 0 \qquad (20.9)$$

On the other hand, we have

$$\nabla^2(xU_2) = \left(\frac{\partial^2}{\partial x^2} + \frac{\partial^2}{\partial y^2}\right)(xU_2) \qquad (20.10)$$

with

$$\frac{\partial^2}{\partial x^2}(xU_2) = \frac{\partial}{\partial x}\left[\frac{\partial}{\partial x}(xU_2)\right] = \frac{\partial}{\partial x}\left[U_2 + x\frac{\partial U_2}{\partial x}\right] = 2\frac{\partial U_2}{\partial x} + x\frac{\partial^2 U_2}{\partial x^2} \qquad (20.11a)$$

and

$$\frac{\partial^2}{\partial y^2}(xU_2) = x\frac{\partial^2 U_2}{\partial y^2} \qquad (20.11b)$$

Equation (20.10), on the basis of equations (20.11), becomes

$$\nabla^2(xU_2) = 2\frac{\partial U_2}{\partial x} + x\nabla^2 U_2 \qquad (20.12)$$

and, since the function U_2 is harmonic by hypothesis,

$$\nabla^2(xU_2) = 2\frac{\partial U_2}{\partial x} \qquad (20.13)$$

Further applying the Laplace operator to equation (20.13) and reversing the order of this with the partial derivative, we obtain finally

$$\nabla^4(xU_2) = \nabla^2\left(2\frac{\partial U_2}{\partial x}\right) = 2\frac{\partial}{\partial x}(\nabla^2 U_2) = 0 \qquad (20.14a)$$

Likewise, it is also possible to demonstrate that

$$\nabla^4(yU_3) = 0 \qquad (20.14b)$$

Consider an analytical function $\overline{\overline{Z}}(z)$, and its subsequent derivatives, themselves also analytical,

$$\frac{d\overline{\overline{Z}}}{dz} = \overline{Z}, \quad \frac{d\overline{Z}}{dz} = Z, \quad \frac{dZ}{dz} = Z' \qquad (20.15)$$

where the overbars represent Westergaard's original notation and have nothing to do with the symbol indicating the conjugate of a complex number. The rules of differentiation (19.113) in this case become

$$\frac{\partial\overline{\overline{Z}}}{\partial x} = \frac{d\overline{\overline{Z}}}{dz}\frac{\partial z}{\partial x} = \overline{Z} \qquad (20.16a)$$

$$\frac{\partial\overline{\overline{Z}}}{\partial y} = \frac{d\overline{\overline{Z}}}{dz}\frac{\partial z}{\partial y} = i\overline{Z} \qquad (20.16b)$$

and likewise for the subsequent derivatives.

659

From equations (20.16) there follow the rules of differentiation of the real part and the imaginary part of an analytical function:

$$\frac{\partial}{\partial x} \operatorname{Re} \overline{\overline{Z}} = \operatorname{Re} \frac{\partial \overline{\overline{Z}}}{\partial x} = \operatorname{Re} \overline{Z} \tag{20.17a}$$

$$\frac{\partial}{\partial y} \operatorname{Re} \overline{\overline{Z}} = \operatorname{Re} \frac{\partial \overline{\overline{Z}}}{\partial y} = -\operatorname{Im} \overline{Z} \tag{20.17b}$$

$$\frac{\partial}{\partial x} \operatorname{Im} \overline{\overline{Z}} = \operatorname{Im} \frac{\partial \overline{\overline{Z}}}{\partial x} = \operatorname{Im} \overline{Z} \tag{20.17c}$$

$$\frac{\partial}{\partial y} \operatorname{Im} \overline{\overline{Z}} = \operatorname{Im} \frac{\partial \overline{\overline{Z}}}{\partial y} = \operatorname{Re} \overline{Z} \tag{20.17d}$$

Notice that, by the transitive law, from the two pairs of relations (20.17a, d) and (20.17b, c) we find again the Cauchy–Riemann conditions (19.94).

The first of Westergaard's hypotheses concerns the Airy stress function, which is written in the form

$$\Phi_I = \operatorname{Re} \overline{\overline{Z}}_I + y \operatorname{Im} \overline{Z}_I + \frac{1}{2} B \left(y^2 - x^2 \right) \tag{20.18}$$

where the subscript I indicates a symmetrical situation with respect to the X axis. The real and imaginary parts of an analytical function are in fact harmonic functions, just as it is easy to verify that the function $\frac{1}{2} B(y^2 - x^2)$ with B as a real constant, is also harmonic.

The stress components are obtained by double derivation of the Airy stress function. Applying the rules of derivation (20.17) we have

$$\frac{\partial \Phi_I}{\partial x} = \operatorname{Re} \overline{Z}_I + y \operatorname{Im} Z_I - Bx \tag{20.19a}$$

$$\frac{\partial \Phi_I}{\partial y} = y \operatorname{Re} Z_I + By \tag{20.19b}$$

and hence, via equations (19.9)

$$\sigma_x = \frac{\partial^2 \Phi_I}{\partial y^2} = \operatorname{Re} Z_I - y \operatorname{Im} Z_I' + B \tag{20.20a}$$

$$\sigma_y = \frac{\partial^2 \Phi_I}{\partial x^2} = \operatorname{Re} Z_I + y \operatorname{Im} Z_I' - B \tag{20.20b}$$

$$\tau_{xy} = -\frac{\partial^2 \Phi_I}{\partial x \partial y} = -y \operatorname{Re} Z_I' \tag{20.20c}$$

Note that, as had already been anticipated in the foregoing chapter, equations (20.20) coincide with equations (19.129), once the following substitution of complex potential is made:

$$Z_I(z) = 2\psi'(z) \tag{20.21}$$

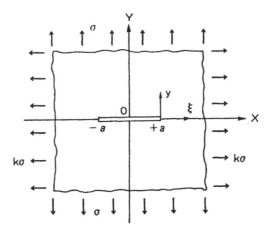

Figure 20.7

Now consider a rectilinear crack of length $2a$, along the X axis between $-a$ and $+a$ (Figure 20.7). The boundary conditions expressing the absence of stresses on the faces of the crack are

$$\sigma_y(x,0) = \tau_{xy}(x,0) = 0, \quad \text{for} -a < x < a \tag{20.22}$$

The second of Westergaard's hypotheses concerns the complex potential, which is written in the form

$$Z_I = \frac{g(z)}{\left[(z+a)(z-a)\right]^{1/2}} + B, \quad \forall z \in \mathscr{C} \tag{20.23}$$

where $g(z)$ is a real function of a complex variable and B is the real constant previously introduced in equation (20.18). The function (20.23) satisfies, via equations (20.20b, c), the boundary conditions (20.22). On the basis of equation (20.20a) we then find on the faces of the crack

$$\sigma_x(x,0) = 2B, \quad \text{for} - a < x < a \tag{20.24}$$

Carrying out in equation (20.23) the substitution of the variable

$$z = \zeta + a \tag{20.25}$$

and, therefore, considering a reference system centred in the right-hand tip of the crack (Figure 20.7), we obtain

$$Z_I = \frac{g(\zeta+a)/(\zeta+2a)^{1/2}}{\zeta^{1/2}} + B \tag{20.26}$$

In the area surrounding the right-hand tip of the crack, the function (20.26) can be approximated with

$$Z_I = \frac{g(a)/\sqrt{2a}}{\zeta^{1/2}} + B \tag{20.27}$$

661

If we put

$$\frac{g(a)}{\sqrt{a}} = \frac{K_I}{\sqrt{\pi}} \tag{20.28}$$

finally we obtain

$$Z_I = \frac{K_I}{\sqrt{2\pi\zeta}} + B \tag{20.29}$$

The real constant K_I represents the so-called **stress-intensity factor**.

To be able to understand the physical meaning of the factor K_I, it is necessary to introduce the polar coordinates into the study of the stresses given by equations (20.20). From complex analysis it is known that

$$\zeta = re^{i\vartheta} = r(\cos\vartheta + i\sin\vartheta) \tag{20.30a}$$

$$\zeta^{-1/2} = r^{-1/2}e^{-\frac{1}{2}i\vartheta} = r^{-1/2}\left(\cos\frac{\vartheta}{2} - i\sin\frac{\vartheta}{2}\right) \tag{20.30b}$$

$$\zeta^{-3/2} = r^{-3/2}e^{-\frac{3}{2}i\vartheta} = r^{-3/2}\left(\cos\frac{3}{2}\vartheta - i\sin\frac{3}{2}\vartheta\right) \tag{20.30c}$$

$$y = r\sin\vartheta = 2r\sin\frac{\vartheta}{2}\cos\frac{\vartheta}{2} \tag{20.30d}$$

The complex potential (20.29) can thus be expressed as follows:

$$Z_I = \frac{K_I}{\sqrt{2\pi r}}\left(\cos\frac{\vartheta}{2} - i\sin\frac{\vartheta}{2}\right) + B \tag{20.31}$$

having taken care to introduce also the constant B, which, at non-infinitesimal distances from the origin, cannot be considered negligible.

The derivative of the complex potential (20.29) may itself also be expressed in polar coordinates:

$$Z_I' = \frac{K_I}{\sqrt{2\pi}}\left(-\frac{1}{2}\right)\zeta^{-3/2} = -\frac{K_I}{2\sqrt{2\pi}(r^{3/2})}\left(\cos\frac{3}{2}\vartheta - i\sin\frac{3}{2}\vartheta\right) \tag{20.32}$$

Substituting equations (20.30d), (20.31) and (20.32) into equations (20.20), we obtain the stress field that is valid in the crack tip vicinity

$$\sigma_x = \frac{K_I}{\sqrt{2\pi r}}\cos\frac{\vartheta}{2} - 2r\sin\frac{\vartheta}{2}\cos\frac{\vartheta}{2}\frac{K_I}{2\sqrt{2\pi}(r^{3/2})}\sin\frac{3}{2}\vartheta + 2B \tag{20.33a}$$

$$\sigma_y = \frac{K_I}{\sqrt{2\pi r}}\cos\frac{\vartheta}{2} + 2r\sin\frac{\vartheta}{2}\cos\frac{\vartheta}{2}\frac{K_I}{2\sqrt{2\pi}(r^{3/2})}\sin\frac{3}{2}\vartheta \tag{20.33b}$$

$$\tau_{xy} = -2r\sin\frac{\vartheta}{2}\cos\frac{\vartheta}{2}\left(-\frac{K_I}{2\sqrt{2\pi}(r^{3/2})}\cos\frac{3}{2}\vartheta\right) \tag{20.33c}$$

Gathering common factors, we have finally

$$\sigma_x = \frac{K_I}{\sqrt{2\pi r}}\cos\frac{\vartheta}{2}\left(1 - \sin\frac{\vartheta}{2}\sin\frac{3}{2}\vartheta\right) + 2B \tag{20.34a}$$

$$\sigma_y = \frac{K_I}{\sqrt{2\pi r}} \cos\frac{\vartheta}{2}\left(1 + \sin\frac{\vartheta}{2}\sin\frac{3}{2}\vartheta\right) \qquad (20.34b)$$

$$\tau_{xy} = \frac{K_I}{\sqrt{2\pi r}} \sin\frac{\vartheta}{2}\cos\frac{\vartheta}{2}\cos\frac{3}{2}\vartheta \qquad (20.34c)$$

In regard to the stress components (20.34), the following observations may be made:

1. All three stress components (20.34) present a singularity $r^{-1/2}$ at the tip of the crack. The power $-1/2$ of this singularity depends only on the boundary conditions on the faces of the crack, and not on the conditions at infinity.
2. The angular profile of the stress field depends itself also on the boundary conditions on the faces of the crack, and not on the conditions at infinity.
3. The stress field in the crack tip vicinity is uniquely defined by the factor K_I, which is, on the other hand, a function of the conditions at infinity, or, in the case of plates of finite dimensions, a function of the conditions imposed on the external contour.
4. The physical dimensions of K_I are somewhat unusual: $[F][L]^{-3/2}$. It is precisely these dimensions that are the substantial cause of the size effects, both in fracture mechanics and, indirectly, in strength of materials.

The third of Westergaard's hypotheses regards the function $g(z)$, present in expression (20.23) of the complex potential, and is related to the conditions at infinity, which have so far been disregarded. Let it be assumed that the stress condition at infinity presents the principal directions parallel to the XY coordinate axes, with principal stresses equal, respectively, to $k\sigma$ and σ, k being a real constant (Figure 20.7). Setting

$$g(z) = \sigma z \qquad (20.35)$$

and hence

$$Z_I = \frac{\sigma z}{[(z+a)(z-a)]^{1/2}} + B, \quad \forall z \in \mathbb{C} \qquad (20.36)$$

the aforementioned conditions at infinity remain satisfied. From equations (20.20) we have in fact

$$\lim_{z \to \infty} \sigma_x = \sigma + 2B \qquad (20.37a)$$

$$\lim_{z \to \infty} \sigma_y = \sigma \qquad (20.37b)$$

$$\lim_{z \to \infty} \tau_{xy} = 0 \qquad (20.37c)$$

and the limit (20.37a) yields the value $k\sigma$ for

$$B = \frac{1}{2}\sigma(k-1) \qquad (20.38)$$

From positions (20.28) and (20.35) we obtain the expression of the **stress-intensity factor**

$$K_I = \sigma\sqrt{\pi a} \qquad (20.39)$$

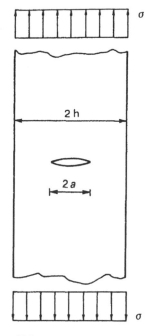

Figure 20.8

which turns out to depend on the stress at infinity orthogonal to the crack and on the half-length of the crack. The normal stress at infinity parallel to the crack does not enter into expression (20.39), since the latter is not a function of the factor k.

Moreover, from equations (20.24) and (20.38) we derive the value of the stress σ_x on the faces of the crack:

$$\sigma_x(x,0) = \sigma(k-1), \quad \text{for} -a < x < a \tag{20.40}$$

Whereas the radial and angular variation of the stress field around the crack tip is independent of the specific geometry under examination and is described by relations (20.34), the information on the geometry and on the external boundary conditions (loads and constraints) is summed up in the factor K_I. In the case, for instance, of a plate of finite width $2h$ with a centred crack of length $2a$, loaded at infinity with a stress σ orthogonal to the crack (Figure 20.8), we have

$$K_I = \sigma\sqrt{\pi a}\left(\sec\frac{\pi a}{2h}\right)^{1/2} \tag{20.41}$$

For $h/a \to \infty$, the foregoing expression tends to equation (20.39).

In the case of a three-point bending specimen, with a crack in the centre of length a, we have (Figure 20.9)

$$K_I = \frac{Pl}{th^{3/2}} f\left(\frac{a}{h}\right) \tag{20.42a}$$

with

$$f\left(\frac{a}{h}\right) = 2.9\left(\frac{a}{h}\right)^{1/2} - 4.6\left(\frac{a}{h}\right)^{3/2} + \tag{20.42b}$$

$$21.8\left(\frac{a}{h}\right)^{5/2} - 37.6\left(\frac{a}{h}\right)^{7/2} + 38.7\left(\frac{a}{h}\right)^{9/2}$$

where h is the depth, t the thickness and l the length of the plate, while P is the external force.

As regards the elastic **crack opening displacement** (COD), this may be found from the stress field, via the dilation

$$\varepsilon_y = \frac{\partial v}{\partial y} = \frac{1}{E}\left(\sigma_y - v\sigma_x\right) \tag{20.43}$$

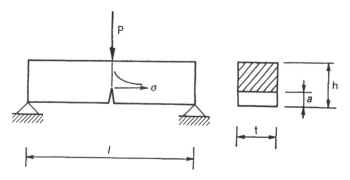

Figure 20.9

in the case of a plane stress condition. From equations (20.20a, b) we have

$$v = \int \varepsilon_y dy \qquad (20.44)$$

$$= \frac{1}{E} \int \left(\operatorname{Re} Z_I + y \operatorname{Im} Z_I' - B \right) dy - \frac{v}{E} \int \left(\operatorname{Re} Z_I - y \operatorname{Im} Z_I' + B \right) dy$$

It is easy to verify that the derivative of the following expression coincides with the integrand of equation (20.44):

$$v = \frac{2}{E} \operatorname{Im} \overline{Z}_I - \frac{1+v}{E} y \operatorname{Re} Z_I - \frac{1+v}{E} By \qquad (20.45)$$

From expression (20.29) of the complex potential we obtain by integration

$$\overline{Z}_I - \frac{K_I}{\sqrt{2\pi}} 2\zeta^{1/2} + B\zeta + C \qquad (20.46)$$

and, in polar coordinates

$$\overline{Z}_I = \frac{2K_I}{\sqrt{2\pi}} r^{1/2} \left(\cos \frac{\vartheta}{2} + i \sin \frac{\vartheta}{2} \right) + Br \left(\cos \vartheta + i \sin \vartheta \right) + C \qquad (20.47)$$

The displacements in the Y direction of the points belonging to the upper face of the crack are therefore

$$v(\vartheta = \pi) = 2 \left(\frac{2}{\pi} \right)^{1/2} \frac{K_I}{E} r^{1/2} \qquad (20.48a)$$

whereas, of course, the points belonging to the lower face present opposite values

$$v(\vartheta = -\pi) = -2 \left(\frac{2}{\pi} \right)^{1/2} \frac{K_I}{E} r^{1/2} \qquad (20.48b)$$

The relative displacement of crack opening in the vicinity of the tip is thus

$$\mathrm{COD} = v(\pi) - v(-\pi) = 4 \left(\frac{2}{\pi} \right)^{1/2} \frac{K_I}{E} r^{1/2} \qquad (20.49)$$

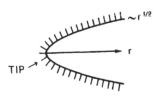

Figure 20.10

The crack opening displacement is directly proportional to the factor K_I (which in turn is always directly proportional to the external load) and inversely proportional to the elastic modulus E. It varies according to a parabolic law along the crack itself, presenting, of course, a null value in the tip (Figure 20.10). It is interesting to observe how the deformed configuration of the crack reveals a blunting with vertical tangent at the tip.

20.4 Mode II and mixed modes

Westergaard's treatment also concerns Mode II, i.e. those cases in which the crack undergoes skew-symmetrical loadings with respect to the X axis (Figure 20.11). It will be shown that, as with Mode I (symmetrical loadings), also with Mode II the stress field around the tip of the crack has a radial variation $r^{-1/2}$, with a singularity at the tip of an equal power $-1/2$, and that the angular variation does not depend on the geometry or on the boundary conditions.

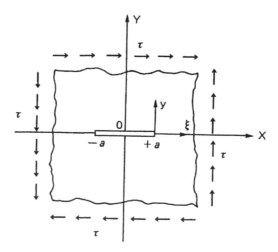

Figure 20.11

For the in-plane skew-symmetrical cases (Mode II) Westergaard chose an Airy stress function of the following form:

$$\Phi_{II} = -y \operatorname{Re} \overline{Z}_{II} \tag{20.50}$$

The stresses are obtained with a double derivation of equation (20.50)

$$\sigma_x = 2 \operatorname{Im} Z_{II} + y \operatorname{Re} Z'_{II} \tag{20.51a}$$

$$\sigma_y = -y \operatorname{Re} Z'_{II} \tag{20.51b}$$

$$\tau_{xy} = \operatorname{Re} Z_{II} - y \operatorname{Im} Z'_{II} \tag{20.51c}$$

The boundary conditions on the faces of the crack are still represented by equations (20.22), and are satisfied by a potential of the form

$$Z_{II} = \frac{f(z)}{\left[(z+a)(z-a)\right]^{1/2}}, \quad \forall z \in \mathscr{C} \tag{20.52}$$

with f as a real function.

The substitution of variable (20.25), in the vicinity of the right-hand tip of the crack, yields

$$Z_{II} = \frac{K_{II}}{\sqrt{2\pi\zeta}} \tag{20.53}$$

where

$$K_{II} = f(a)\sqrt{\frac{\pi}{a}} \tag{20.54}$$

is the **second stress-intensity factor**.

Differentiating the function (20.53) and expressing equations (20.51) in polar coordinates, we obtain

$$\sigma_x = -\frac{K_{II}}{\sqrt{2\pi r}} \sin\frac{\vartheta}{2}\left(2 + \cos\frac{\vartheta}{2}\cos\frac{3}{2}\vartheta\right) \tag{20.55a}$$

$$\sigma_y = \frac{K_{II}}{\sqrt{2\pi r}}\cos\frac{\vartheta}{2}\sin\frac{\vartheta}{2}\cos\frac{3}{2}\vartheta \qquad (20.55b)$$

$$\tau_{xy} = \frac{K_{II}}{\sqrt{2\pi r}}\cos\frac{\vartheta}{2}\left(1 - \sin\frac{\vartheta}{2}\sin\frac{3}{2}\vartheta\right) \qquad (20.55c)$$

The stress field given by equations (20.55) holds in the crack tip vicinity and is independent of the skew-symmetrical conditions at infinity, except for the factor K_{II}, which is instead a function thereof. In the particular condition of pure shear at infinity, parallel to the XY axes, with tension $\sigma = \tau$ at 45° and compression $\sigma = -\tau$ at $-45°$ (Figure 20.11), let us assume the function

$$f(z) = \tau z \qquad (20.56)$$

so that equation (20.52) becomes

$$Z_{II} = \frac{\tau z}{[(z+a)(z-a)]^{1/2}}, \quad \forall z \in \mathscr{C} \qquad (20.57)$$

This equation satisfies the conditions at infinity. In fact, via equations (20.51), we have

$$\lim_{z\to\infty} \sigma_x = 0 \qquad (20.58a)$$

$$\lim_{z\to\infty} \sigma_y = 0 \qquad (20.58b)$$

$$\lim_{z\to\infty} \tau_{xy} = \tau \qquad (20.58c)$$

Finally, from positions (20.54) and (20.56) we find the expression for the stress-intensity factor

$$K_{II} = \tau\sqrt{\pi a} \qquad (20.59)$$

which turns out to be analogous to equation (20.39).

Having resolved separately the symmetrical problem, with equations (20.34), and the skew-symmetrical problem, with equations (20.55), it is possible to demonstrate that each generic problem presents a solution, asymptotically valid in the crack tip vicinity, which is the sum of the two elementary modes, i.e. Modes I and II:

$$\{\sigma\} = \frac{K_I}{\sqrt{2\pi r}}\{\Theta_I(\vartheta)\} + \frac{K_{II}}{\sqrt{2\pi r}}\{\Theta_{II}(\vartheta)\} \qquad (20.60)$$

It is therefore sufficient to know the expressions of the factors K_I and K_{II} in order to define univocally the stress field around the tip of the crack. In the particular case of a stress condition set at infinity (Figure 20.12), with the principal directions inclined with respect to the crack, the Principle of Superposition furnishes the expression (20.60), with K_I and K_{II} given by equation (20.39) and equation (20.59), respectively.

So far we have considered the two elementary modes of loading the crack corresponding to plane stress or plane strain conditions: Mode I, or the **opening mode**, which is symmetrical with respect to the crack (Figure 20.13(a)); Mode II, or the **sliding mode**, which is skew-symmetrical with respect to the X axis (Figure 20.13(b)). There also exists a third elementary mode, corresponding to three-dimensional conditions: Mode III, or the **tearing mode**,

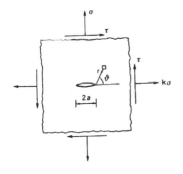

Figure 20.12

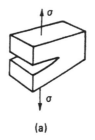
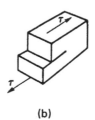

(a) (b) (c)

Figure 20.13

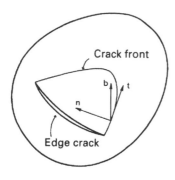

Figure 20.14

which is skew-symmetrical with respect to the *XZ* plane. This mode is characteristic of the tearing of a sheet of paper (Figure 20.13(c)). The above three modes represent all the modes that exist for subjecting a crack to stress, in the sense that, around a point belonging to the front of a skewed crack, in turn contained in a three-dimensional solid (Figure 20.14), the stress field, consisting of the five stress components, σ_n, σ_b, τ_{nb}, τ_{nt}, τ_{bt}, can be expressed as follows:

$$\{\sigma\} = (2\pi r)^{-1/2} \underset{5\times3}{[F(\vartheta,\varphi)]}\underset{3\times1}{\{K\}} \tag{20.61}$$

$$\underset{5\times1}{}$$

where *r* is the radial distance from the point of the crack front,

$$\{K\} = \begin{bmatrix} K_I \\ K_{II} \\ K_{III} \end{bmatrix} \tag{20.62}$$

is the vector of the stress-intensity factors for that same point, and $[F(\vartheta, \varphi)]$ is a (5×3) matrix which represents the angular profile of the asymptotic field, as a function of the latitude ϑ and the longitude φ, in the local reference system *tnb*, consisting of the tangent, the normal and the binormal to the crack front.

20.5 Williams' method

The problem of the determination of the stress field around the vertex of a **re-entrant corner** was tackled and solved by Williams in 1952. Five years later the same author extrapolated it to the limit case of the crack at the edge and so confirmed the stress singularity $r^{-1/2}$, already identified by Muskhelishvili and Westergaard. Williams' method is also known as the **series expansion method**, because the Airy stress function is expanded, as we shall see, in a series of functions.

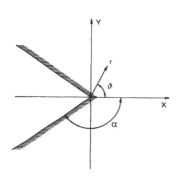

Figure 20.15

Consider a plane sector of elastic material with angular amplitude 2α, and a polar reference system centred at the vertex of this sector (Figure 20.15). Let *r* and ϑ be the radial and the angular coordinates, respectively, the angle ϑ being considered positive if it is counterclockwise and zero if it is in the direction of the bisector inside the elastic sector. Let it be assumed that a function series expansion may be carried out on the Airy stress function, and that each term of the series may be separated into a parabolic radial function with an *a priori* unknown exponent (**eigenvalue**) and an *a priori* unknown angular function (**eigenvector**)

$$\Phi(r,\vartheta) = \sum_n r^{\lambda_n+1} f_n(\vartheta) \tag{20.63}$$

668

Whereas the exponents $(\lambda_n + 1)$, and hence the power of the stress singularity, will be identified on the basis of the boundary conditions on the free edges of the sector, the functions f_n can be fully defined only on the basis of the loading conditions at infinity

Applying relations (19.44) to the series expansion (20.63), we obtain the stresses

$$\sigma_r = \sum_n r^{\lambda_n - 1}\left[f_n''(\vartheta) + (\lambda_n + 1)f_n(\vartheta)\right] \qquad (20.64a)$$

$$\sigma_\vartheta = \sum_n r^{\lambda_n - 1}\left[\lambda_n(\lambda_n + 1)f_n(\vartheta)\right] \qquad (20.64b)$$

$$\tau_{r\vartheta} = -\sum_n r^{\lambda_n - 1}\lambda_n f_n'(\vartheta) \qquad (20.64c)$$

where the prime indicates derivation with respect to ϑ. The equation of congruence (19.50) can therefore be evaluated by summing up the following terms:

$$\frac{1}{r^2}\frac{\partial^2}{\partial\vartheta^2}(\sigma_r + \sigma_\vartheta) = \sum_n r^{\lambda_n - 3}\left[f_n^{IV} + (\lambda_n + 1)^2 f_n''\right] \qquad (20.65a)$$

$$\frac{1}{r}\frac{\partial}{\partial r}(\sigma_r + \sigma_\vartheta) = \sum_n r^{\lambda_n - 3}(\lambda_n - 1)\left[f_n'' + (\lambda_n + 1)^2 f_n\right] \qquad (20.65b)$$

$$\frac{\partial^2}{\partial r^2}(\sigma_r + \sigma_\vartheta) = \sum_n r^{\lambda_n - 3}(\lambda_n - 1)(\lambda_n - 2)\left[f_n'' + (\lambda_n + 1)^2 f_n\right] \quad (20.65c)$$

Gathering common factors, finally we obtain

$$\nabla^2(\sigma_r + \sigma_\vartheta) = \sum_n r^{\lambda_n - 3}\left\{(\lambda_n - 1)^2\left[f_n'' + (\lambda_n + 1)^2 f_n\right] + \right. \qquad (20.66)$$

$$\left.\left[f_n^{IV} + (\lambda_n + 1)^2 f_n''\right]\right\} = 0$$

Equation (20.66) is identically satisfied by equating to zero the expression in the braces, which contains only angular functions:

$$f_n^{IV} + \left[(\lambda_n - 1)^2 + (\lambda_n + 1)^2\right]f_n'' + \left[(\lambda_n - 1)^2(\lambda_n + 1)^2\right]f_n = 0 \quad (20.67)$$

The fourth-order differential equation (20.67), and hence the congruence, are identically satisfied by the following trigonometric form:

$$f_n(\vartheta) = A_n\cos(\lambda_n + 1)\vartheta + B_n\cos(\lambda_n - 1)\vartheta + \qquad (20.68)$$

$$C_n\sin(\lambda_n + 1)\vartheta + D_n\sin(\lambda_n - 1)\vartheta$$

Whilst the first two terms of equation (20.68) represent the symmetrical solution (Mode I), the two remaining terms represent the skew-symmetrical solution (Mode II).

The boundary conditions on the edges of the elastic sector express the fact that the circumferential stress (and hence the one normal to the edge) and the shearing stress become zero,

669

$$\sigma_\vartheta(\pm\alpha)=0 \qquad (20.69a)$$

$$\tau_{r\vartheta}(\pm\alpha)=0 \qquad (20.69b)$$

for any radius $r>0$.

From equations (20.64b, c) we obtain

$$f_n(\pm\alpha)=0 \qquad (20.70a)$$

$$f_n'(\pm\alpha)=0 \qquad (20.70b)$$

Using solution (20.68), equations (20.70) become

$$A_n\cos(\lambda_n+1)\alpha+B_n\cos(\lambda_n-1)\alpha\pm C_n\sin(\lambda_n+1)\alpha\pm \qquad (20.71a)$$
$$D_n\sin(\lambda_n-1)\alpha=0$$

$$\pm A_n(\lambda_n+1)\sin(\lambda_n+1)\alpha\pm B_n(\lambda_n-1)\sin(\lambda_n-1)\alpha+ \qquad (20.71b)$$
$$C_n(\lambda_n+1)\cos(\lambda_n+1)\alpha+D_n(\lambda_n-1)\cos(\lambda_n-1)\alpha=0$$

These two equations can be separated, so as to obtain two systems of homogeneous linear algebraic equations in the unknowns A_n, B_n and C_n, D_n, respectively:

$$A_n\cos(\lambda_n+1)\alpha+B_n\cos(\lambda_n-1)\alpha=0 \qquad (20.72a)$$

$$A_n(\lambda_n+1)\sin(\lambda_n+1)\alpha+B_n(\lambda_n-1)\sin(\lambda_n-1)\alpha=0 \qquad (20.72b)$$

$$C_n\sin(\lambda_n+1)\alpha+D_n\sin(\lambda_n-1)\alpha=0 \qquad (20.72c)$$

$$C_n(\lambda_n+1)\cos(\lambda_n+1)\alpha+D_n(\lambda_n-1)\cos(\lambda_n-1)\alpha=0 \qquad (20.72d)$$

The first two equations correspond to the symmetrical problems (Mode I), whereas the last two correspond to the skew-symmetrical problems (Mode II). To obtain solutions different from the trivial one, the determinants of the coefficients of the two systems must become zero. The unknowns A_n and B_n will therefore be defined save for one factor, in the case of

$$(\lambda_n-1)\sin(\lambda_n-1)\alpha\cos(\lambda_n+1)\alpha- \qquad (20.73a)$$
$$(\lambda_n+1)\cos(\lambda_n-1)\alpha\sin(\lambda_n+1)\alpha=0$$

just as the unknowns C_n and D_n will be defined but for one factor, in the case of

$$(\lambda_n+1)\sin(\lambda_n-1)\alpha\cos(\lambda_n+1)\alpha- \qquad (20.73b)$$
$$(\lambda_n-1)\cos(\lambda_n-1)\alpha\sin(\lambda_n+1)\alpha=0$$

Note that in the case of A_1 and B_1 the aforementioned proportionality factor coincides with the stress-intensity factor K_I, just as in the case of C_1 and D_1 it coincides with K_{II}.

From equation (20.73a) and taking into account the well-known trigonometric relations

$$\sin x\cos y-\cos x\sin y=\sin(x-y) \qquad (20.74a)$$

$$\sin x\cos y+\cos x\sin y=\sin(x+y) \qquad (20.74b)$$

670

we find the condition

$$-\lambda_n \sin 2\alpha = \sin 2\lambda_n \alpha \qquad (20.75a)$$

Likewise, from equation (20.73b) we have

$$+\lambda_n \sin 2\alpha = \sin 2\lambda_n \alpha \qquad (20.75b)$$

Equations (20.75) are the eigenvalue equations for the elastic sector problem, from which the exponents $(\lambda_n + 1)$ of the series expansion (20.63) are obtainable. More precisely, from equation (20.75a) we obtain the eigenvalues of the symmetrical problem, whilst from equation (20.75b) we obtain the eigenvalues of the skew-symmetrical problem.

The terms of the series expansions (20.64) are finite or infinitesimal for $r \rightarrow 0^+$, should the corresponding eigenvalue satisfy the inequality

$$\lambda_n \geqslant 1 \qquad (20.76)$$

On the other hand, the strain energy contained in an infinitesimal circular area of radius R around the vertex of the sector is infinite if

$$\lambda_n \leqslant 0 \qquad (20.77)$$

Thus, we have

$$W(R) \propto \int_0^R r^{2(\lambda_n - 1)} r\, dr \qquad (20.78)$$

and the integral is divergent for $(2\lambda_n - 1) \leqslant -1$. Hence, for the analysis of the dominant singularity of the stress field at the vertex of the sector, the only eigenvalues of interest are those contained in the interval

$$0 < \lambda_n < 1 \qquad (20.79)$$

The eigenvalue equations (20.75) can be written in the following form:

$$\frac{\sin 2\lambda_n \alpha}{2\lambda_n \alpha} = \mp \frac{\sin 2\alpha}{2\alpha} \qquad (20.80)$$

with $0 \leqslant 2\alpha \leqslant 2\pi$.

From the graphical viewpoint, equations (20.80) may be resolved rather neatly by intersecting the oscillating function $y = \sin 2\lambda\alpha / (2\lambda\alpha)$ with the horizontal straight lines $y = \mp \sin 2\alpha/(2\alpha)$. In this way four principal cases can be distinguished (Figure 20.16).

1. $0 \leqslant 2\alpha \leqslant \pi$ (convex angle or wedge). The first eigenvalue of the symmetrical problem, λ_I, does not exist or else $\lambda_I \geqslant 1$. The first eigenvalue of the skew-symmetrical problem is $\lambda_{II} = 1$. Consequently there is no stress-singularity in the case where the elastic sector is convex.
2. $\pi < 2\alpha \leqslant 1.43\pi$ (obtuse concave angle). The first eigenvalue of the symmetrical problem is $\lambda_I < 1$, whilst we again have $\lambda_{II} = 1$. There is therefore only one symmetrical stress-singularity.
3. $1.43\pi < 2\alpha < 2\pi$ (acute concave angle). The first eigenvalues, whether of the symmetrical problem or of the skew-symmetrical one, are both less than one: $\lambda_I < 1$, $\lambda_{II} < 1$. We thus have both the symmetrical and the

671

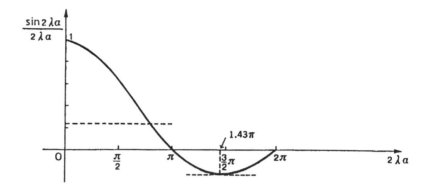

Figure 20.16

skew-symmetrical stress singularities, albeit the symmetrical one of a higher order.

4. $2\alpha = 2\pi$ (null concave angle or crack). In this case we have $\lambda_I = \lambda_{II} = 1/2$. Only in the case of a null concave angle (as well as that of a flat angle) is the first eigenvalue followed by a numerable infinity of other eigenvalues

$$\lambda_{In} = \lambda_{IIn} = \frac{1}{2}, 1, \frac{3}{2}, 2,... \qquad (20.81)$$

or equivalently

$$\lambda_n = \frac{n}{2}, \qquad n = \text{natural number} \qquad (20.82)$$

The power of the symmetrical stress singularity is represented in Figure 20.17 as a function of the notch angle γ. For $\gamma = 0$, the notch becomes a crack, and in fact we find again the classical singularity $r^{-1/2}$. As γ increases there is a transition, which up to $\gamma \simeq \pi/2$ is very slow, but subsequently, between $\pi/2$ and π, undergoes a rapid acceleration. Obviously, when $\gamma = \pi$, the notch disappears, just as the singularity of the stress field vanishes. When instead the re-entrant corner angle is a right angle, $\gamma = \pi/2$, we have the power $(1-\lambda_I) \simeq 0.45$.

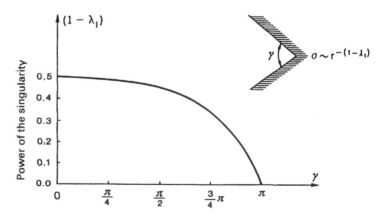

Figure 20.17

20.6 Relation between fracture energy $\mathcal{G}_{IC}$ and critical value K_{IC} of the stress-intensity factor

Griffith's criterion, discussed in Section 20.2, represents the first energy criterion of fracture mechanics. In the years that followed, between 1920 and 1950, the efforts of the research workers were all directed, as has been seen in the foregoing sections, towards defining the singular stress field around the tip of the crack. It was only in 1957 that Irwin made a direct correlation of the two different treatments: the energy approach of Griffith and the stress approach of Muskhelishvili, Westergaard and Williams.

As regards a more general energy criterion than that of Griffith, which had reference to a particular geometry (infinite plate with rectilinear crack, subjected to a loading condition uniform at infinity) and to a deformation-controlled loading process, we shall see how the concept of **total potential energy** allows a criterion to be defined which is independent of the control exercised over the loading process.

Let us consider an **imposed-force** loading process on a plate with an initial crack of length $2a$ (Figure 20.18(a)). For a certain critical value of the force F,

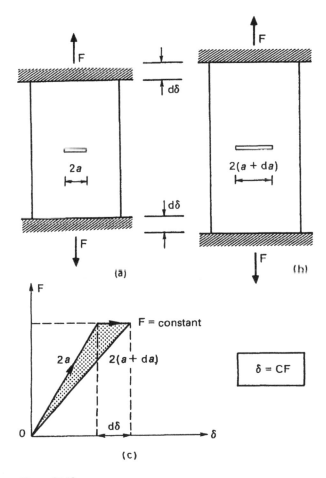

Figure 20.18

673

let the crack be assumed to extend for the length $2da$ (Figure 20.18(b)), so as to produce an increment of compliance dC and hence an incremental displacement of each end of the plate equal to (Figure 20.18(c))

$$d\delta = FdC \tag{20.83}$$

The variation in the total potential energy due to the infinitesimal propagation of the crack is

$$dW = dL - 2Fd\delta \tag{20.84}$$

where dL denotes the variation in elastic strain energy and the second term represents the variation in the potential energy of the external loads. By virtue of Clapeyron's Theorem and evaluating graphically the shaded area of the triangle of Figure 20.18(c), we have

$$dL = 2\left(\frac{1}{2}Fd\delta\right) \tag{20.85}$$

and hence, applying equation (20.83)

$$dL = F^2 dC \tag{20.86}$$

In conclusion, we therefore obtain a decrease in total potential energy

$$dW = -F^2 dC \tag{20.87}$$

Let us now consider an **imposed-displacement** loading process on the previously considered plate (Figure 20.19(a)). For a certain critical value of the displacement δ let us assume that the crack is extended by the length $2da$ (Figure 20.19(b)), so as to produce a decrement of stiffness dK and hence a decrement of the external force equal to (Figure 20.19(c))

$$dF = \delta \, dK \tag{20.88}$$

The variation in the total potential energy due to the infinitesimal propagation of the crack, in this second case, is equal to the variation in elastic strain energy, since the external loads by hypothesis do not perform incremental work:

$$dW = dL \tag{20.89}$$

By virtue of Clapeyron's Theorem and evaluating graphically the shaded area of the triangle of Figure 20.19(c), we have

$$dW = 2\left(\frac{1}{2}\delta \, dF\right) \tag{20.90}$$

and hence, applying equation (20.88),

$$dW = \delta^2 dK \tag{20.91}$$

Since stiffness is the inverse of compliance, we have

$$dK = d\left(\frac{1}{C}\right) = -\frac{1}{C^2}dC \tag{20.92}$$

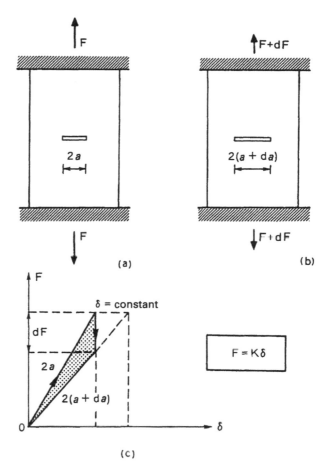

Figure 20.19

Substituting equation (20.92) into equation (20.91), we obtain

$$dW = -\frac{\delta^2}{C^2}dC \qquad (20.93)$$

and finally, since $\delta/C = F$, we once again obtain expression (20.87).

Differential calculus shows basically how the difference between the areas of the two shaded triangles of Figures 20.18(c) and 20.19(c) constitutes an infinitesimal of an order higher than the areas of the triangles themselves. We have therefore demonstrated how the total potential energy diminishes always by the same amount $F^2 dC$, following an infinitesimal extension of the crack, regardless of the control exercised over the loading process.

By virtue of the Principle of Conservation of Energy, the following balance between the variation in the total potential energy and the fracture energy must hold:

$$dW + 4\gamma\,da = 0 \qquad (20.94)$$

675

where $\gamma = \mathscr{G}_{IC}/2$ is the specific surface energy, i.e. the energy necessary for breaking the chemical and atomic bonds connecting two unit and contiguous surfaces of matter. Equation (20.94) represents the more general formulation of Griffith's criterion, expressed by equation (20.3).

Considering also **virtual**, and not only real, propagations of the crack, the concept of **strain energy release rate** is defined as

$$dW + \mathscr{G}_I dA = 0 \qquad (20.95)$$

where dA represents the incremental fracture area. The parameter $\mathscr{G}_I$ is thus defined as the total potential energy released per unit increment in fracture area,

$$\mathscr{G}_I = -\frac{dW}{dA} \qquad (20.96)$$

and is a positive quantity, dW always representing a decrement.

Brittle crack propagation occurs **really** when $\mathscr{G}_I$ reaches its critical value

$$\mathscr{G}_I = \mathscr{G}_{IC} \qquad (20.97a)$$

Since the stress field in the crack tip vicinity is univocally defined by the factor K_I, it is on the other hand legitimate to assume that the unstable propagation of the crack occurs when it attains its critical value

$$K_I = K_{IC} \qquad (20.97b)$$

It is thus evident how the two fracture criteria, the energy one (20.97a) and the stress one (20.97b), have completely different origins. The two critical values, $\mathscr{G}_{IC}$ of the variation in total potential energy (fracture energy), and K_{IC} of the stress-intensity factor, are not, however, independent, but linked by a fundamental relation, which will be described in the sequel.

A first simple way of arriving at the relation that links $\mathscr{G}_{IC}$ and K_{IC} is that of considering the case of the infinite cracked plate, loaded at infinity by a uniform stress condition. According to Griffith the condition of instability is given by equation (20.6), whilst according to Irwin and taking into account equation (20.39), it is

$$\sigma \geq \frac{K_{IC}}{\sqrt{\pi a}} \qquad (20.98)$$

Since the conditions expressed by equation (20.6) and equation (20.98) concern the same physical problem and both of them present the half-length a of the crack raised to the power $-1/2$, we immediately obtain

$$K_{IC} = \sqrt{\mathscr{G}_{IC} E} \qquad (20.99)$$

It may be noted that K_{IC} and $\mathscr{G}_{IC}$ are related via the elastic modulus E of the material. In Chapter 19 it was demonstrated that the plane stress fields do not depend upon E. There thus follows the independence of the factor K_I from the elastic modulus E, as well as from Poisson's ratio v. If, instead, one reasons in energy terms, and hence in terms of fracture energy, the influence of E emerges clearly.

Relation (20.99) concerns the critical values of the stress-intensity factor and the strain energy release rate. It can, however, also be extended to the generic values of these two parameters, using a demonstration due to Irwin.

Consider a cracked plate, subjected to a plane stress condition and to displacements imposed on its external border (**fixed grip condition**). Let a be the length of

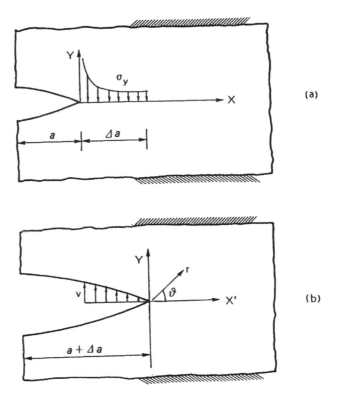

Figure 20.20

the crack and Δa the extension of the segment of the X axis on which the stresses σ_y are assumed to be known (Figure 20.20(a)). Consider then a virtual extension of the crack, so that it presents the incremented length $a + \Delta a$ (Figure 20.20(b)). Let v be the vertical displacements of the faces of the crack in this new configuration, which we take as being known on the same segment of extension Δa.

If we assume that the extension Δa is so small that on it Westergaard's asymptotic stress and displacement fields hold, and if we apply Clapeyron's Theorem to the phenomenon of crack reclosure (from scheme (b) to scheme (a) of Figure 20.20), we have the following variation in total potential energy:

$$\Delta W = 2 \int_0^{\Delta a} \frac{1}{2} \sigma_y v \, dr \qquad (20.100)$$

with

$$\sigma_y = \sigma_y(\vartheta = 0) = \frac{K_I}{\left[2\pi(\Delta a - r) \right]^{1/2}} \qquad (20.101)$$

from equation (20.34b), and

$$v = v(\vartheta = \pi) = 2\left(\frac{2}{\pi} \right)^{1/2} \frac{K_I}{E} r^{1/2} \qquad (20.102)$$

from equation (20.48a).

677

Substituting equations (20.101) and (20.102) into the integral (20.100), we have

$$\Delta W = \frac{2}{\pi}\frac{K_I^2}{E}\int_0^{\Delta a}\left(\frac{r}{\Delta a - r}\right)^{1/2}dr \qquad (20.103)$$

and evaluating the integral gives

$$\Delta W = \frac{K_I^2}{E}\Delta a \qquad (20.104)$$

On the other hand, from equation (20.95) we have

$$\Delta W = \mathcal{G}_I\Delta a \qquad (20.105)$$

omitting the negative algebraic sign, since the process of crack reclosure is exactly the inverse of the one so far considered.

The comparison between relations (20.104) and (20.105) gives finally the generalization of equation (20.99)

$$\mathcal{G}_I = \frac{K_I^2}{E} \qquad (20.106a)$$

which applies in cases where a **plane stress condition** obtains.

For **plane strain condition**s, it is not difficult to demonstrate, via a revision of the position (20.43), that the following relation instead holds:

$$\mathcal{G}_I = \frac{K_I^2}{E}\left(1 - v^2\right) \qquad (20.106b)$$

A check on the coherence of formulas (20.106) is afforded by dimensional analysis:

$$[\mathcal{G}_I] = \frac{[F]^2[L]^{-3}}{[F][L]^{-2}} = [F][L]^{-1} \qquad (20.107)$$

The physical dimension of fracture energy corresponds in fact to that of work per unit area, or force per unit length.

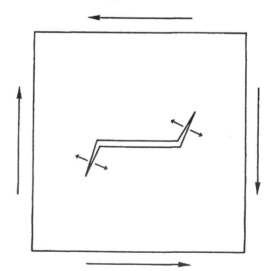

Figure 20.21

In the case of the **mixed mode** (Mode I + Mode II) condition, it is possible to extrapolate the foregoing reasoning:

$$\Delta W = 2 \int_0^{\Delta a} \frac{1}{2} \sigma_y v \, dr + 2 \int_0^{\Delta a} \frac{1}{2} \tau_{xy} u \, dr \qquad (20.108a)$$

$$\Delta W = \mathscr{G} \Delta a \qquad (20.108b)$$

Performing the calculations, we obtain

$$\mathscr{G} = \frac{K_I^2}{E} + \frac{K_{II}^2}{E} \qquad (20.109)$$

In this case $\mathscr{G}$ represents the variation in total potential energy per **virtual extension** of the crack. In fact, when subjected to a mixed mode loading, a crack does not extend collinearly to itself (**self-similar propagation**). In reality, as we shall see in the next section, it branches (Figure 20.21).

20.7 Crack branching criterion in mixed mode condition

As already mentioned in the previous section, Griffith's energy criterion applies consistently only to the case of collinear crack propagation, i.e. in the case of Mode I. It cannot be conveniently applied to situations where the crack branches out and changes direction, once subjected to biaxial load conditions. These conditions produce a superposition of Mode I and Mode II, which is conventionally termed **mixed mode**. The procedure then will be to determine all the pairs of values K_I and K_{II} which cause the critical condition around the crack tip and hence crack propagation.

The first branching criterion, chronologically speaking, is that of **maximum circumferential stress,** proposed by Erdogan and Sih in 1963. It is based on the hypothesis that the crack extends starting from its tip, in the direction normal to that of maximum circumferential stress σ_ϑ. Since the stresses around the crack tip are expressible as products of a radial function by an angular function, the above direction does not depend on the radius r of the circumference on which the maximum of the stress σ_ϑ is evaluated.

Translating expressions (20.34) and (20.55) into polar coordinates, and summing up the corresponding results, we obtain

$$\sigma_r = \frac{1}{(2\pi r)^{1/2}} \cos\frac{\vartheta}{2} \left[K_I \left(1 + \sin^2\frac{\vartheta}{2} \right) + K_{II} \left(\frac{3}{2} \sin\vartheta - 2\tan\frac{\vartheta}{2} \right) \right] \quad (20.110a)$$

$$\sigma_\vartheta = \frac{1}{(2\pi r)^{1/2}} \cos\frac{\vartheta}{2} \left[K_I \cos^2\frac{\vartheta}{2} - \frac{3}{2} K_{II} \sin\vartheta \right] \qquad (20.110b)$$

$$\tau_{r\vartheta} = \frac{1}{2(2\pi r)^{1/2}} \cos\frac{\vartheta}{2} \left[K_I \sin\vartheta + K_{II}(3\cos\vartheta - 1) \right] \qquad (20.110c)$$

The branching angle ϑ is obtained from the condition of stationarity

$$\frac{\partial \sigma_\vartheta}{\partial \vartheta} = -\frac{3}{4(2\pi r)^{1/2}} \left[K_I \sin\vartheta + K_{II}(3\cos\vartheta - 1) \right] \cos\frac{\vartheta}{2} \quad (20.111)$$

$$= -\frac{3}{2} \tau_{r\vartheta} = 0$$

which can be satisfied by setting $\cos\vartheta/2 = 0$, which corresponds to the zero shearing stress surface condition ($\tau_{r\vartheta} = 0$) for $\vartheta = \pm\pi$, or

$$K_I \sin\vartheta + K_{II}(3\cos\vartheta - 1) = 0 \qquad (20.112)$$

which yields the branching angle of the crack.

For a crack of length $2a$, subjected to a generic biaxial stress condition at infinity (Figure 20.22), the stress-intensity factors are

$$K_I = \sigma_\beta\sqrt{\pi a} \qquad (20.113a)$$

$$K_{II} = \tau_\beta\sqrt{\pi a} \qquad (20.113b)$$

where σ_β and τ_β are, respectively, the normal stress and the shearing stress with respect to the crack line, acting at infinity. The usual Mohr relations lead to the following expressions:

$$K_I = \left(\frac{\sigma_1 + \sigma_2}{2} + \frac{\sigma_1 - \sigma_2}{2}\cos 2\beta\right)\sqrt{\pi a} \qquad (20.114a)$$

$$K_{II} = \left(\frac{\sigma_2 - \sigma_1}{2}\sin 2\beta\right)\sqrt{\pi a} \qquad (20.114b)$$

where σ_1, σ_2 are the principal stresses at infinity and β is the angle of inclination of the crack (Figure 20.22). If we denote by m the ratio σ_1/σ_2, equations (20.114) can be recast in the following form:

$$K_I = \sigma_2\sqrt{\pi a}\left[m + (1 - m)\sin^2\beta\right] \qquad (20.115a)$$

$$K_{II} = \sigma_2\sqrt{\pi a}(1 - m)\sin\beta\cos\beta \qquad (20.115b)$$

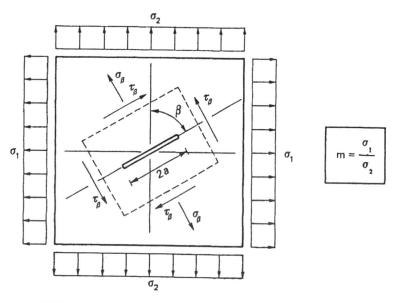

Figure 20.22

680

Equations (20.112) and (20.115) lead to a condition which relates the branching angle ϑ to the angle of inclination β

$$\left[m+(1-m)\sin^2\beta\right]\sin\vartheta+\left[\frac{1}{2}(1-m)\sin 2\beta\right](3\cos\vartheta-1)=0 \quad (20.116)$$

Equation (20.116) is equivalent to the following:

$$2(1-m)\sin 2\beta\left(\tan\frac{\vartheta}{2}\right)^2- \quad (20.117)$$

$$2\left[m+(1-m)\sin^2\beta\right]\left(\tan\frac{\vartheta}{2}\right)-(1-m)\sin 2\beta=0$$

The solution is represented in Figure 20.23 for various ratios m.

If $m=1$ (uniform stress at infinity), we always have $\vartheta=0$, and the extension of the crack is collinear by symmetry. On the other hand, if $m=0$ (uniaxial stress at infinity), a discontinuity occurs for $\beta=0$. In fact

$$\vartheta(m=0,\beta=0)=0 \quad (20.118a)$$

by symmetry, whereas instead

$$\lim_{\beta\to 0^+}\vartheta(m=0,\beta)\simeq 70° \quad (20.118b)$$

Then if m is small but other than zero, the discontinuity disappears and is replaced by a rapid variation, represented by a very steep branch in Figure 20.23. From a mathematical standpoint, this is a case of non-uniform convergence of the function $\vartheta(m,\beta)$ in $\beta=0$ for $m\to 0^+$.

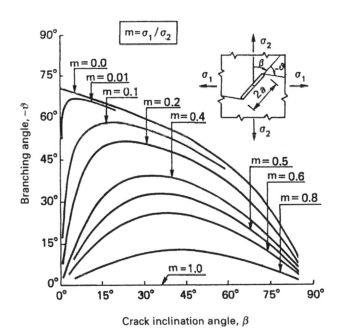

Figure 20.23

681

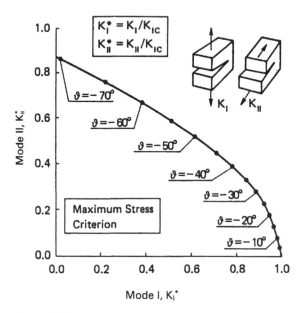

Figure 20.24

Whereas equation (20.111) defines the direction of maximum circumferential stress, the biaxial critical condition may be obtained from the comparison with the simple Mode I case:

$$\sqrt{2\pi r}\,\sigma_\vartheta = K_{IC} \qquad (20.119)$$

Introducing the nondimensional factors

$$K_I^* = K_I/K_{IC}, \quad K_{II}^* = K_{II}/K_{IC} \qquad (20.120)$$

the branching conditions given by equation (20.111) and equation (20.119) may be expressed as follows:

$$K_I^* \sin\vartheta + K_{II}^*(3\cos\vartheta - 1) = 0 \qquad (20.121a)$$

$$K_I^* \cos^2\frac{\vartheta}{2} - \frac{3}{2}K_{II}^* \sin\vartheta = \frac{1}{\cos(\vartheta/2)} \qquad (20.121b)$$

As the angle ϑ varies, all the points of the critical domain are thus defined in parametric form. These points are symmetrical with respect to the axis K_I^* and valid only in the half-plane $K_I^* \geqslant 0$ (Figure 20.24).

20.8 Plastic zone at the crack tip

The stress components around the tip of a real crack present a radial variation $r^{-1/2}$ only beyond a certain distance from that singular point. For smaller distances plastic phenomena occur which mean that the stresses are in fact smaller than those theoretically expected. In this way a plastic zone is created around the tip of the crack, which is the more extensive, the greater the ductility of the material. To a first approximation, since in front of the crack tip (Figure 20.6)

$$\sigma_y = \frac{K_I}{\sqrt{2\pi r}} \qquad (20.122)$$

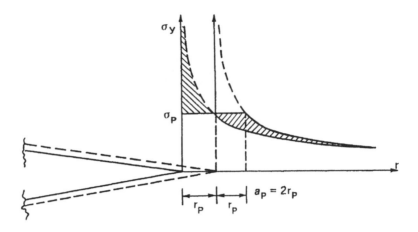

Figure 20.25

it follows that the radius r_P of the plastic zone can be estimated from (Figure 20.25)

$$\sigma_P = \frac{K_I}{\sqrt{2\pi r_P}} \qquad (20.123)$$

where σ_P is the yield stress of the material. At the moment of crack propagation we thus have the following estimation:

$$r_{PC} = \frac{1}{2\pi}\frac{K_{IC}^2}{\sigma_P^2} \qquad (20.124)$$

As will be understood better in the sequel, the ratio K_{IC}/σ_P therefore represents a measure of the material's **ductility**.

In actual fact, as Irwin observed in 1960, relation (20.124) provides only the order of magnitude of the plastic radius. A more accurate evaluation can be achieved by considering the redistribution of stresses, both elastic and plastic, that develop ahead of the crack. In other words, the singular stress distribution of Figure 20.25 is to be translated along the axis r, so that the integral of elastic and plastic stresses is equal to the integral of the aforesaid distribution. From the graphical viewpoint, therefore, the hatched areas of Figure 20.25 must be equal.

The integral of the singular stress distribution, between the crack tip and the plastic radius r_P, is

$$\int_0^{r_P} \frac{K_I}{\sqrt{2\pi r}}\, dr = \left(\frac{2}{\pi}\right)^{\frac{1}{2}} K_I r_P^{1/2} \qquad (20.125)$$

Finding K_I from equation (20.123) and inserting it into equation (20.125), we obtain

$$\int_0^{r_P} \frac{K_I}{\sqrt{2\pi r}}\, dr = 2\sigma_P r_P \qquad (20.126)$$

683

From relation (20.126) we deduce that the left-hand hatched area (Figure 20.25) is equal to that of the rectangle of sides σ_P, r_P. Also the right-hand hatched area, obtained with a translation r_P, is equal to that of the rectangle, since both of them are complementary of the same area. Finally, we obtain the following extension of the plastic zone at the moment of crack propagation, according to Irwin's evaluation:

$$a_{PC} = 2r_{PC} \qquad (20.127)$$

or, considering equation (20.124),

$$a_{PC} = \frac{1}{\pi}\frac{K_{IC}^2}{\sigma_P^2} \qquad (20.128)$$

A fracture may be defined as **brittle** when the plastic zone is much smaller than the initial crack and the solid containing it,

$$a_{PC} \ll a \qquad (20.129a)$$

$$a_{PC} \ll h \qquad (20.129b)$$

where h denotes a characteristic dimension of the cracked solid under examination.

From equations (20.129) and taking into account equation (20.128), we obtain the following limitations in non-dimensional form:

$$\frac{K_{IC}}{\sigma_P\sqrt{a}} \ll \sqrt{\pi} \qquad (20.130a)$$

$$\frac{K_{IC}}{\sigma_P\sqrt{h}} \ll \sqrt{\pi} \qquad (20.130b)$$

While equation (20.130a) may be obtained trivially from the condition

$$\sigma \ll \sigma_P \qquad (20.131)$$

once account is taken of relation (20.39) and the graph of Figure 20.5, equation (20.130b) is highly significant for structural purposes, and will be taken up again in the next section.

A different evaluation of the extension of the plastic zone is due to Dugdale (1960) and is based on the simulation of the plastic stresses via a constant distribution of forces directly applied to the faces of a **fictitious crack**, longer than the real one (Figure 20.26). The condition to be applied is that of making the total stress-intensity factor zero,

$$K_I(\sigma) + K_I(\sigma_P) = 0 \qquad (20.132)$$

where the first term is that of the stresses applied at infinity, whilst the second term represents the stress intensity due to the restraining stresses σ_P, applied orthogonally to the faces of the crack at distances from the tips less than or equal to a_P (Figure 20.26).

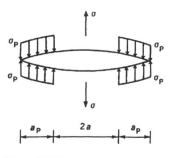

Figure 20.26

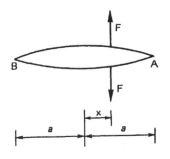

Figure 20.27

Since two concentrated forces F, applied orthogonally to the faces of the crack at a distance x from the centre, cause at the two tips, respectively near and far, the following stress-intensity factors (Figure 20.27):

$$K_I(A) = \frac{F}{\sqrt{\pi a}}\left(\frac{a+x}{a-x}\right)^{\frac{1}{2}} \tag{20.133a}$$

$$K_I(B) = \frac{F}{\sqrt{\pi a}}\left(\frac{a-x}{a+x}\right)^{\frac{1}{2}} \tag{20.133b}$$

where $2a$ is the length of the crack, integrating the effects of the plastic stresses σ_P (Figure 20.26), we have

$$-K_I(\sigma_P) = \frac{\sigma_P}{[\pi(a+a_P)]^{\frac{1}{2}}} \int_a^{a+a_P}\left(\frac{(a+a_P)+x}{(a+a_P)-x}\right)^{\frac{1}{2}} + \left(\frac{(a+a_P)-x}{(a+a_P)+x}\right)^{\frac{1}{2}} dx \tag{20.134}$$

Evaluating the integral, we obtain

$$-K_I(\sigma_P) = 2\sigma_P\left(\frac{a+a_P}{\pi}\right)^{\frac{1}{2}} \arccos\left(\frac{a}{a+a_P}\right) \tag{20.135}$$

while the factor corresponding to the external load σ equals

$$K_I(\sigma) = \sigma\sqrt{\pi(a+a_P)} \tag{20.136}$$

Substituting equations (20.135) and (20.136) into the condition (20.132), we have

$$\frac{a}{a+a_P} = \cos\frac{\pi\sigma}{2\sigma_P} \tag{20.137}$$

The limit cases of zero external stress, or external stress equal to the yield strength σ_P, consistently produce, according to Dugdale's model, a null plastic zone ($a_P = 0$), or a general yielding ($a_P \to \infty$), respectively.

Neglecting the terms of a higher order in the series expansion of the cosine, relation (20.137) is transformed as follows:

$$\frac{a}{a+a_P} = 1 - \frac{1}{2}\left(\frac{\pi\sigma}{2\sigma_P}\right)^2 \tag{20.138}$$

from which we obtain

$$\frac{a_P}{a+a_P} = \frac{\pi^2\sigma^2}{8\sigma_P^2} \tag{20.139}$$

Substituting equation (20.136) into the foregoing equation, we find finally

$$a_P = \frac{\pi}{8}\frac{K_I^2}{\sigma_P^2} \tag{20.140}$$

685

The extension of the plastic zone at the moment of crack propagation according to Dugdale's evaluation is therefore

$$a_{PC} = \frac{\pi}{8} \frac{K_{IC}^2}{\sigma_P^2} \qquad (20.141)$$

Also in this case, in estimating a_{PC}, the material's ductility ratio K_{IC}/σ_P is present.

The comparison between the plastic extensions according to Irwin (equation (20.128)) and Dugdale (equation (20.141)) shows how the two models, albeit notably different, lead to estimations that closely resemble one another. The plastic extension, as given by Dugdale, is greater than that given by Irwin by about 20%:

$$\frac{a_{PC}(\text{Dugdale})}{a_{PC}(\text{Irwin})} = \frac{\pi^2}{8} \simeq 1.23 \qquad (20.142)$$

20.9 Size effects and ductile-brittle transition

A first size effect has already been considered in Section 20.2, and is that corresponding to the length of the crack. As emerges clearly from Figure 20.5, for crack half-lengths greater than a_0, where a_0 is a characteristic length given by equation (20.7), the collapse due to brittle crack propagation precedes the plastic collapse of the plate. For $a < a_0$, the plastic collapse of the plate instead precedes the collapse due to brittle crack propagation. Recalling the fundamental relation (20.99) which links $\mathscr{G}_{IC}$ and K_{IC}, the characteristic length, given by equation (20.7), can also be expressed as a function of K_{IC}:

$$a_0 = \frac{1}{\pi} \frac{K_{IC}^2}{\sigma_P^2} \qquad (20.143)$$

The limitations defined by the inequalities (20.130) thus take the form

$$a \gg a_0 \qquad (20.144a)$$

$$h \gg a_0 \qquad (20.144b)$$

A second dimensional effect, which derives directly from the one just considered, is that corresponding to the dimensions of the cracked solid, once constant ratios are assumed between crack length and characteristic dimensions of the solid. In the case of the geometrically similar plates of Figure 20.28, the collapse stress will be only a function of the half-length of the crack, when the plates are sufficiently large to allow the effects of the free edge to be neglected:

$$\sigma = \frac{K_{IC}}{\sqrt{\pi a}}, \quad \text{for} \quad a \geqslant a_0 \qquad (20.145a)$$

$$\sigma = \sigma_P, \quad \text{for} \quad a < a_0 \qquad (20.145b)$$

Since, on the other hand, by virtue of the supposed geometrical similitude, the half-length a is proportional to the characteristic dimension h of the plate,

$$a = \xi h \qquad (20.146)$$

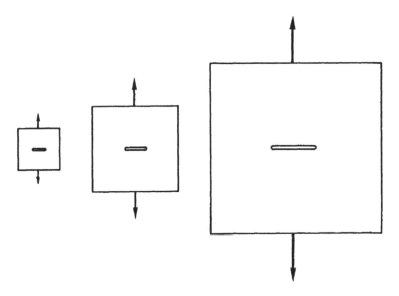

Figure 20.28

where ξ is the relative crack length, equations (20.145) can be recast in the form

$$\sigma = \frac{K_{IC}}{\sqrt{\pi\xi h}}, \quad \text{for} \quad h \geqslant \frac{a_0}{\xi} \tag{20.147a}$$

$$\sigma = \sigma_P, \quad \text{for} \quad h < \frac{a_0}{\xi} \tag{20.147b}$$

There thus exists a dimension of the plate, $h_0 = a_0/\xi$, below which the plastic collapse of the plate precedes the brittle propagation of the crack. This dimension depends not only on the geometrical shape of the plate and of the crack, but also on the ductility K_{IC}/σ_P of the material of which the plate is made.

From the simple example just dealt with, the absence of **physical similitude** in tensile collapse of geometrically similar solids is immediately inferred, once the existence of a crack of a length proportional to the dimension of the solid is assumed (Figure 20.28). As has already been observed, it is not possible to state the same thing in the case of an elliptical hole, for which the stress-concentration factor depends on the ratio between the semi-axes and not on their absolute dimensions.

The hypothesis of negligibility of edge effects can, on the other hand, be removed without vitiating the important conclusions displayed above; indeed, they are enriched with fresh insights. Let us consider a plate of finite width $2h$, with a crack of length $2a$, $0 < a/h < 1$, loaded at infinity by a stress σ orthogonal to the crack (Figure 20.8). Since the stress-intensity factor is given by relation (20.41), the brittle propagation of the crack occurs for

$$\sigma = \frac{K_{IC}}{\sqrt{\pi a}}\left(\cos\frac{\pi a}{2h}\right)^{1/2} \tag{20.148}$$

or, in non-dimensional form,

$$\frac{\sigma}{\sigma_P} = \frac{K_{IC}}{\sigma_P\sqrt{2h}}\left(\frac{\cos(\pi a/2h)}{\pi a/2h}\right)^{1/2} \qquad (20.149)$$

Denoting by

$$s = \frac{K_{IC}}{\sigma_P\sqrt{2h}} \qquad (20.150)$$

the so-called **brittleness number**, we obtain

$$\frac{\sigma}{\sigma_P} = s\left(\frac{\cos(\pi a/2h)}{\pi a/2h}\right)^{1/2} \qquad (20.151)$$

On the other hand, the plastic limit analysis carried out on the net section complementary to the crack, which is referred to as a **ligament**, provides a second collapse condition, different from equation (20.151):

$$\frac{\sigma}{\sigma_P} = 1 - \frac{a}{h} \qquad (20.152)$$

The diagrams of equations (20.151) and (20.152) are presented in Figure 20.29 as functions of the relative crack length a/h. Whilst the first of these equations gives a family of curves related to the non-dimensional number s,

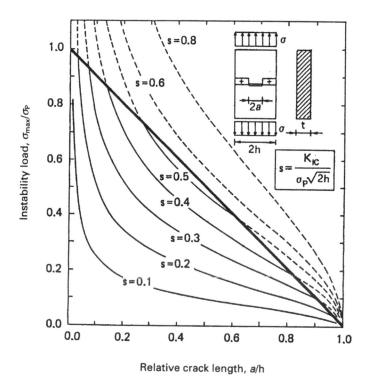

Figure 20.29

the second is represented by a single curve (thick line). When $s \lesssim 0.54$, it may be noted that the plastic collapse precedes the brittle crack propagation, both for sufficiently short cracks and for sufficiently long ones. Whilst the first tendency is by now familiar, starting from Section 20.2 onwards, the second represents a new, non-intuitive development. It is basically due to the unlikelihood of a singular stress distribution developing in the cases where there is an excessively reduced ligament. As the number s increases, the interval of a/h for which brittle propagation of the crack precedes plastic collapse contracts until it vanishes for $s = s_0 \simeq 0.54$, the value for which the corresponding fracture curve is tangential to the curve of plastic collapse. For $s \gtrsim 0.54$, plastic collapse precedes brittle crack propagation for any relative crack length, there existing no point of intersection between the fracture curve and the plastic collapse curve. Consequently the condition expressed by equation (20.130b) is reconfirmed following another path. This condition means that brittle types of collapse tend to occur with low material toughness, high yielding stress and/or large structural sizes. It is not the individual values of K_{IC}, σ_P and h that are responsible for the nature of the collapse mechanism, but rather only their function s (cf. equation (20.150)).

Also in the case of three-point bending of a plate (Figure 20.30), it is possible to arrive at the same conclusions. Recalling expression (20.42) of the

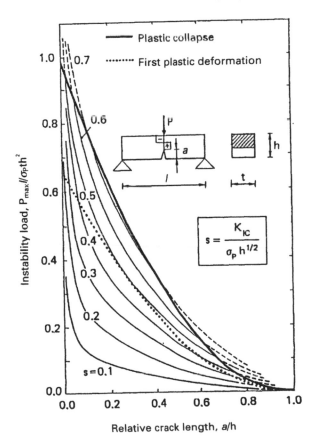

Figure 20.30

factor K_I at the moment of potential collapse due to brittle crack propagation, we have

$$K_{IC} = \frac{P_{max}l}{th^{3/2}} f\left(\frac{a}{h}\right)$$ (20.153)

from which, in non-dimensional form

$$\frac{P_{max}l}{\sigma_p th^2} = \frac{s}{f\left(\frac{a}{h}\right)}$$ (20.154)

where

$$s = \frac{K_{IC}}{\sigma_p h^{1/2}}$$ (20.155)

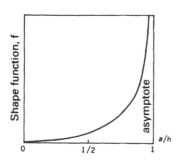

Figure 20.31

denotes the brittleness number which considers the plate depth h as the characteristic dimension, and where f is the shape function (20.42b), which like the function $((\pi a/2h) \sec(\pi a/2h))^{1/2}$ of equation (20.41) vanishes for $a/h = 0$ and tends to infinity for $a/h \rightarrow 1^-$ (Figure 20.31).

On the other hand, the force P which potentially produces plastic collapse can be held to be the one that generates a plastic hinge at the ligament

$$\frac{1}{4} P_{max}l = \sigma_p t \frac{(h-a)^2}{4}$$ (20.156)

from which follows, in non-dimensional form,

$$\frac{P_{max}l}{\sigma_p th^2} = \left(1 - \frac{a}{h}\right)^2$$ (20.157)

The diagrams of equations (20.154) and (20.157) are presented in Figure 20.30. For this structural geometry, the brittleness number that marks the transition from ductile collapse to brittle collapse is $s_0 \simeq 0.75$. For this value the fracture curve is tangential to the plastic collapse curve.

If the ductility of a material is therefore measurable via the ratio K_{IC}/σ_P, in order for the ductility of a structure to be defined, it is necessary that also a dimension of that structure be entered into the equation. The brittleness number s provided by equation (20.155) is certainly the most synthetic way of describing the degree of ductility of a structure. Plastic limit analysis therefore represents a reliable method of calculating only in the cases where the brittleness number of the structure being examined is not excessively low.

Table 20.1 gives indicative values of tensile strength σ_P and fracture toughness K_{IC} for some materials. The ratio σ_P/K_{IC} then provides a measurement

Table 20.1

	Strength $\sigma_P(\text{MN/m}^2)$	Toughness $K_{IC}(\text{MN/m}^{3/2})$	Brittleness $\sigma_P/K_{IC}(\text{m}^{-1/2})$
Concrete	3.57	1.96	1.8
Aluminium	500	100	5
Plexiglass	33	5.5	6
Glass	170	0.25	680

(in m$^{-1/2}$) of the brittleness of the material. Glass proves to be by far the most brittle, while concrete proves to be unexpectedly the most ductile. This ductility in the case of concrete cannot be put down to a hardening behaviour of the material but to its softening behaviour. On the other hand, in view of the dimension h in the brittleness number s, it is at this point easy to understand how glass can prove ductile for small structural dimensions and steel brittle for large structural dimensions (Figure 20.1). Basically, then, the size effects are caused by the different physical dimensions of **strength** and **toughness**.

Notice finally how, via definition (20.143) of the characteristic length of the microflaws, it is possible to give the brittleness number the following alternative form:

$$s = \left(\pi \frac{a_0}{h} \right)^{1/2} \tag{20.158}$$

20.10 Cohesive crack model and snap-back instability

One way of describing the behaviour of materials in a consistent manner is that of using a pair of constitutive laws:

1. a **stress–strain** relation that describes the elastic and hardening behaviour of the uncracked material up to the maximum stress σ_u, unloadings included (Figure 20.32(a));
2. a **stress–crack opening displacement** relation that describes the softening behaviour of the cracked material up to the critical opening w_c, beyond which the interaction between the crack faces becomes zero (Figure 20.32(b)).

The double constitutive law represented in Figure 20.32 has already been proposed in Section 8.11 for brittle materials having an elastic–softening behaviour, where the energy is dissipated exclusively on the crack surface. In the more general case of material presenting an elastic–hardening–softening behaviour (Figure 20.32), the energy is dissipated both in the volume of the uncracked material and on the surface of the crack. The energy J_V dissipated in unit volume is equal to the hatched area in Figure 20.32(a), while the energy J_S dissipated over unit surface is equal to the hatched area in Figure 20.32(b). If a bar is subjected to tension, cracks and eventually breaks into two

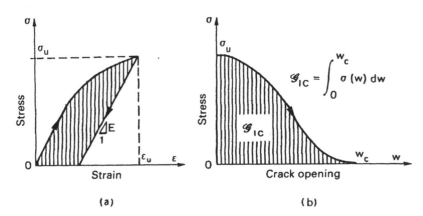

(a)   (b)

Figure 20.32

691

parts, the total dissipated energy is given by the sum of the energy dissipated in the volume of the bar plus that dissipated over the surface of the crack,

$$\text{Energy dissipated} = J_V \times \text{Area} \times l + J_S \times \text{Area} \qquad (20.159)$$

from which we find that the energy dissipated per unit surface of the crack, J_C, depends on the dimensions of the bar:

$$J_C = J_V l + J_S \qquad (20.160)$$

Only in the case of elastic–softening material, for which $J_V = 0$, does the so-called integral J_C not depend on the dimension of the bar:

$$J_C = J_S = \mathcal{G}_{IC} \qquad (20.161)$$

In the case of steels, the constitutive law is generally of the more complex type, i.e. elastic–hardening–softening, and consequently it is particularly difficult to find models to describe their behaviour, as any model must account for two different mechanisms of dissipation, on the surface and in the volume. However, in the case of concrete, rocks and ceramic materials, the simpler elastic–softening law is able to describe the actual behaviour consistently.

The so-called **cohesive crack** model is analogous to **Dugdale's model** of Figure 20.26, but with the difference that in the latter the distribution of the cohesive forces is not constant, but decreases as the crack opening increases, following a softening law like that of Figure 20.32(b). The zone ahead of the real crack tip appears damaged and presents microcracks. It represents a portion of the developing macrocrack, still, however, partially sutured by inclusions, aggregates or fibres (Figure 20.33(a)). This zone, in which nonlinear and dissipative phenomena of a microscopic nature occur, is termed the **process zone** (or **plastic zone**). If it is sufficiently small compared to the real crack, then the concepts of Linear Elastic Fracture Mechanics (LEFM) are fully applicable. On the other hand, if the extension of the process zone is

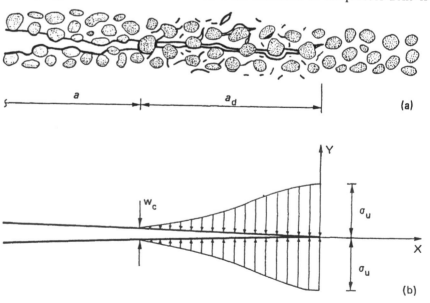

Figure 20.33

692

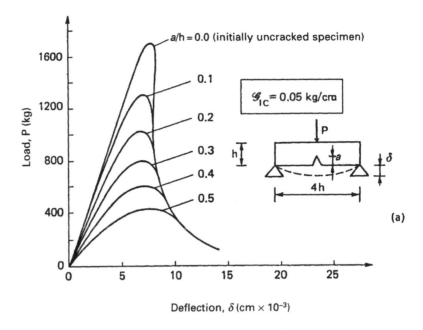

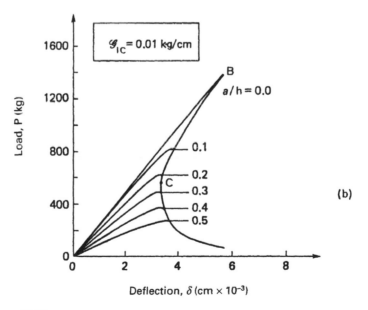

Figure 20.34

693

comparable with that of the real crack, then it must be appropriately modelled (Figure 20.33(b)). The tip of the **cohesive** (or **fictitious**) **crack** coincides with the tip of the process zone, in which the opening w goes to zero and the restraining stress is equal to the tensile strength σ_u. The tip of the **real crack** is found instead at the critical crack opening w_c, for which the interaction vanishes. In the intermediate points of the process zone, the pairs $\sigma-w$ are given by the diagram of Figure 20.32(b).

Figure 20.34(a) represents the structural response, in terms of load *vs.* deflection curve, of a three-point-bending concrete slab, as the relative depth a/h of the initial crack varies, for $\mathscr{G}_{IC} = 0.05$ kg/cm and $h = 15$ cm. The response predicted by the cohesive crack model is always of a softening type, and, as can be noted, with the increase in a/h, a decrease in stiffness and in loading capacity is found, together with an increase in ductility. The tail of the $P-\delta$ response proves insensitive to the length of the initial crack.

Figure 20.34(b) represents the structural response of the same slab for a lower toughness: $\mathscr{G}_{IC} = 0.01$ kg/cm. The trends are the same as in the previous case, but the responses, however, all appear more brittle, and especially the one corresponding to the initially uncracked slab ($a/h = 0.0$), which shows a marked phenomenon of snap-back. Snap-back disappears for $a/h \gtrsim 0.25$. This type of instability has already been encountered in Section 8.11, where brittle materials were dealt with, and again in Section 17.9, where mention was made of the instability of the elastic equilibrium of axially loaded cylindrical shells.

Figure 20.35 gives the diagrams of the load P as a function of the **crack mouth opening displacement** (CMOD). Whereas with the tougher material, the crack starts to open before the maximum load is reached, with the more brittle material, onset of crack opening corresponds exactly to the point of maximum load, and the crack continues to open in a monotonic way as the load diminishes in the softening stage. From this diagram it is possible to understand how, in order to detect the snap-back branch BC experimentally

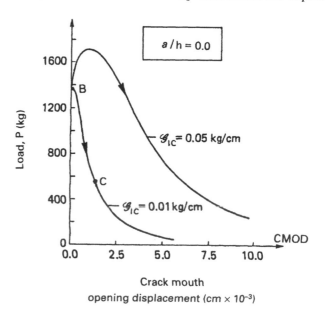

Crack mouth
opening displacement (cm × 10⁻³)

Figure 20.35

(Figure 20.34(b)), control is necessary via crack opening and not via deflection, which proves not to be a monotonic function of time and crack length.

The notable embrittlement of the structural response, which in Figure 20.34 is produced by a drop in toughness $\mathcal{G}_{IC}$ of the material, can equivalently be generated by a dilation of the size scale h. Figure 20.36 gives the structural responses of the previously considered slab, for $\mathcal{G}_{IC} = 0.05$ kg/cm and four different sizes: (a) $h = 10$ cm; (b) $h = 20$ cm; (c) $h = 40$ cm; (d) $h = 80$ cm. Whereas with $h = 10$ cm the response is of a softening type for each depth of the initial crack, with $h = 20$ cm there is a practically vertical drop in the loading capacity for the initially uncracked slab, and with $h = 40$ cm a clear instance of snap-back is recorded, which, with $h = 80$ cm, turns into a very sharply pointed cusp.

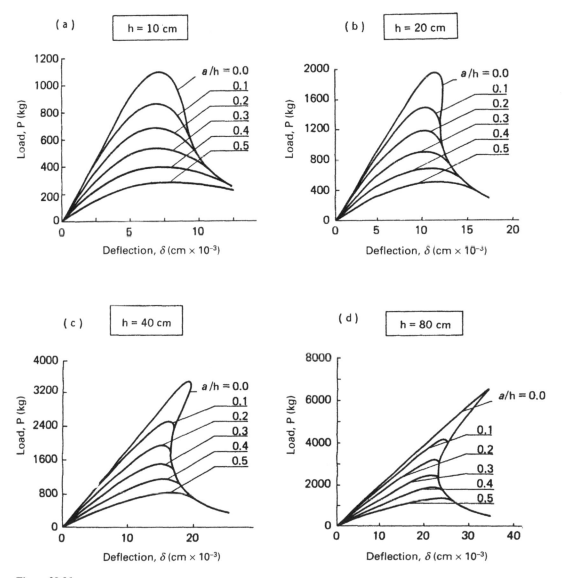

Figure 20.36

695

The embrittlement of the structural response, produced both by the decrease in toughness $\mathcal{G}_{IC}$ and by the increase in strength σ_u and/or in the size h, can be described in a unitary and synthetic manner via the variation in the following dimensionless number:

$$s_E = \frac{\mathcal{G}_{IC}}{\sigma_u h} \qquad (20.162)$$

Whereas the brittleness number (20.155) is of a stress type, the brittleness number (20.162) is of an energy type. Taking into account the fundamental relation (20.99) which links K_{IC} and $\mathcal{G}_{IC}$, it is possible to demonstrate that between the two brittleness numbers referred to above there is the following relation:

$$s_E = \varepsilon_u s^2 \qquad (20.163)$$

where $\varepsilon_u = \sigma_u/E$ represents the ultimate tensile dilation. It may be shown that there is a perfect physical similarity in the failure behaviour, when two of the three pure numbers s, s_E, ε_u are equal.

Figure 20.37(a) gives the load *vs.* deflection response in non-dimensional form, for $a/h = 0.1$, $\varepsilon_u = 0.87 \times 10^{-4}$, $v = 0.1$, $l = 4h$, as the brittleness number s_E varies. It is clearly evident that, as s_E varies through four orders of magnitude, the shape of the non-dimensional curve changes totally, from ductile to brittle. For $s_E \lesssim 10.45 \times 10^{-5}$, the softening branch acquires, at least for a portion, a positive slope, and hence the phenomenon of snap-back is seen to occur.

The area bounded by each individual curve of Figure 20.37(a) and the horizontal axis represents the product between fracture energy $\mathcal{G}_{IC}$ and initial area

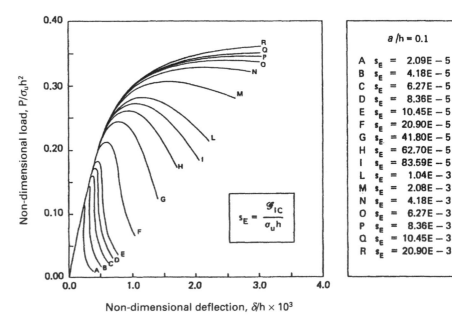

Figure 20.37a

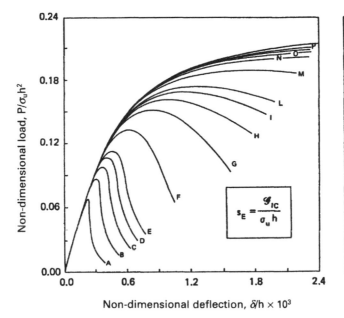

Figure 20.37b

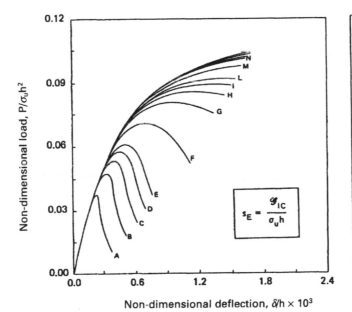

Figure 20.37c

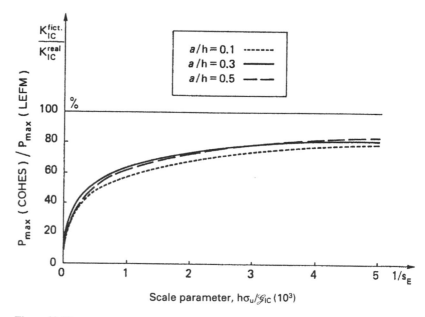

Figure 20.38

of the ligament $(h - a)t$. The areas under the non-dimensional P–δ curves are therefore proportional to the respective brittleness numbers s_E. This simple result is made possible by the hypothesis that the energy dissipation occurs exclusively on the fracture surface and not in the volume of the slab. Figures 20.37(b), (c) illustrate the cases $a/h = 0.3$ and 0.5, respectively, which show a greater ductility.

The maximum load deriving from the cohesive crack model can be compared with the load of brittle crack propagation, expressed by equation (20.154). The values of their ratio are presented in the graphs of Figure 20.38 as functions of the inverse of the brittleness number s_E. This ratio can be regarded as the ratio between **fictitious toughness** and **actual toughness**. Fictitious toughness is always lower than actual toughness, because, for high values of s_E, plastic collapse tends to precede potential collapse due to brittle crack propagation. It is evident that, for $s_E \to 0$, the results of the cohesive crack model tend to converge with those deriving from Linear Elastic Fracture Mechanics, and that consequently the phenomenon of snap-back tends to represent the classical Griffith's instability. The cohesive crack model thus manifests its ability to describe the ductile-brittle transition, the existence of which has already been revealed in the foregoing sections.

Another demonstration of the fact that the cohesive crack model tends to approach asymptotically Linear Elastic Fracture Mechanics is provided by the diagrams of Figure 20.39, which give the relative depth of the fictitious crack at maximum load as a function of the inverse of the number s_E. For $s_E \to 0$, this depth tends to that of the initial crack and hence, at the point of snap-back instability, there is the absence of the process zone, or a completely brittle-type fracture. On the other hand, for $s_E \to \infty$, the process zone at maximum load invades the entire ligament.

698

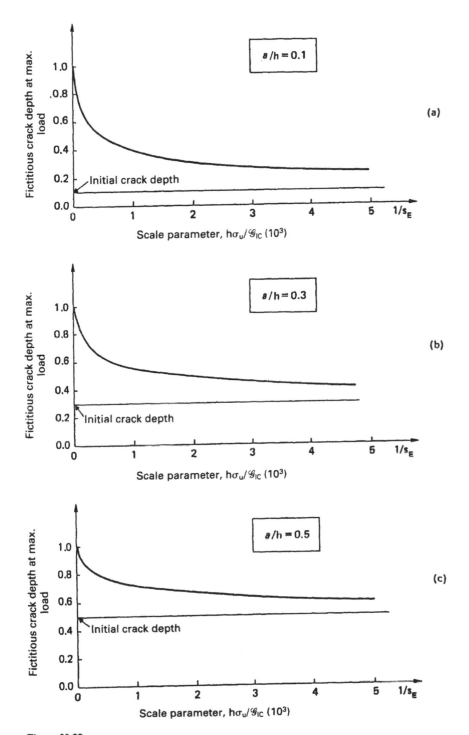

Figure 20.39

699

Appendix A Calculation of the internal reactions in a circular arch subjected to a radial hydrostatic load

A.1 Analytical method

The differential equation (5.9) in the case of Figure 5.4, whereby

$$p = m = 0 \tag{A.1a}$$

$$q(\vartheta) = -\gamma R(1 - \cos \vartheta) \tag{A.1b}$$

reduces to the following form:

$$\frac{d^3 M}{d\vartheta^3} + \frac{dM}{d\vartheta} = \gamma R^3 \sin \vartheta \tag{A.2}$$

For the calculation of the complete integral, i.e. of the integral of the associated homogeneous equation, consider the characteristic equation

$$\lambda^3 + \lambda = 0 \tag{A.3}$$

which presents the following roots:

$$\lambda_1 = 0, \ \lambda_2 = +i, \ \lambda_3 = -i$$

where i is the imaginary unit. As is known from mathematical analysis, the complete integral therefore takes the form

$$M_g(\vartheta) = C_1 + C_2 \cos \vartheta + C_3 \sin \vartheta \tag{A.4}$$

As regards the particular solution, since we are considering a case in which the known term is of the sort

$$b(\vartheta) = P_m(\vartheta) e^{\alpha \vartheta} \tag{A.5}$$

where $P_m(\vartheta)$ indicates a polynomial of the mth order, with $m = 0$, and $\alpha = i$, this may be sought in the form

$$M_0(\vartheta) = \vartheta^r R_m(\vartheta) e^{\alpha \vartheta} \tag{A.6}$$

where $r = 1$ is the multiplicity with which $\alpha = i$ is the solution of the characteristic equation, and $R_m(\vartheta)$ is a polynomial in ϑ of the order $m = 0$, and thus in this case is a constant. We have therefore

$$M_0(\vartheta) = a\vartheta \sin \vartheta \tag{A.7}$$

Substituting equation (A.7) in equation (A.2), we obtain the coefficient a. Differentiating equation (A.7) sequentially, we have in fact

$$\frac{dM_0}{d\vartheta} = a(\vartheta \cos \vartheta + \sin \vartheta) \tag{A.8a}$$

700

$$\frac{d^2 M_0}{d\vartheta^2} = a(-\vartheta \sin \vartheta + 2 \cos \vartheta) \qquad \text{(A.8b)}$$

$$\frac{d^3 M_0}{d\vartheta^3} = a(-\vartheta \cos \vartheta - 3 \sin \vartheta) \qquad \text{(A.8c)}$$

and thus substituting equations (A.8a, c) in equation (A.2)

$$-2a \sin \vartheta = \gamma R^3 \sin \vartheta \qquad \text{(A.9)}$$

or

$$a = -\frac{\gamma R^3}{2} \qquad \text{(A.10)}$$

Summing the complete integral (A.4) and the particular solution (A.7), with the coefficient a given by equation (A.10), we can write finally

$$M(\vartheta) = C_1 + C_2 \cos \vartheta + C_3 \sin \vartheta - \frac{\gamma R^3}{2} \vartheta \sin \vartheta \qquad \text{(A.11)}$$

The arbitrary constants C_1, C_2, C_3 are calculated using the following boundary conditions:

$$M_A = 0 \qquad \text{(A.12a)}$$

$$T_A = 0 \qquad \text{(A.12b)}$$

$$M_B = 0 \qquad \text{(A.12c)}$$

since in A there is a roller support and in B a hinge (Figure 5.4). Recalling the relation (5.6), which links shear and bending moment, we have

$$M(0) = C_1 + C_2 = 0 \qquad \text{(A.13a)}$$

$$T(0) = \frac{C_3}{R} = 0 \qquad \text{(A.13b)}$$

$$M\left(-\frac{\pi}{2}\right) = C_1 - C_3 - \frac{\pi \gamma R^3}{4} = 0 \qquad \text{(A.13c)}$$

The linear algebraic system (A.13) admits of the following roots:

$$C_1 = \frac{\gamma R^3}{4} \pi \qquad \text{(A.14a)}$$

$$C_2 = -\frac{\gamma R^3}{4} \pi \qquad \text{(A.14b)}$$

$$C_3 = 0 \qquad \text{(A.14c)}$$

Finally we obtain from equation (A.11)

$$M(\vartheta) = \frac{\gamma R^3}{4} (\pi - \pi \cos \vartheta - 2\vartheta \sin \vartheta) \qquad \text{(A.15)}$$

701

Applying equations (5.6) and (5.8), we obtain also the shearing force and the axial force

$$T(\vartheta) = -\frac{\gamma R^2}{4}\left[(2-\pi)\sin\vartheta + 2\vartheta\cos\vartheta\right] \tag{A.16}$$

$$N(\vartheta) = \frac{\gamma R^2}{4}\left[-4 + \pi\cos\vartheta + 2\vartheta\sin\vartheta\right] \tag{A.17}$$

A.2 Direct method

The equations of equilibrium to horizontal translation, vertical translation and rotation about point B, respectively, appear as follows (Figure A.1):

$$H_A + H_B = \int_0^{\pi/2} |q(\vartheta)|\sin\vartheta\ R d\vartheta \tag{A.18a}$$

$$V_B = \int_0^{\pi/2} |q(\vartheta)|\cos\vartheta\ R d\vartheta \tag{A.18b}$$

$$H_A R = \int_0^{\pi/2} |q(\vartheta)|\cos\vartheta\ R^2 d\vartheta \tag{A.18c}$$

Computing the integrals, we obtain a linear algebraic system in the three unknowns H_A, H_B, V_B,

$$H_A + H_B = \frac{\gamma R^2}{2} \tag{A.19a}$$

$$V_B = \gamma R^2\left(1-\frac{\pi}{4}\right) \tag{A.19b}$$

$$H_A = \gamma R^2\left(1-\frac{\pi}{4}\right) \tag{A.19c}$$

which gives the solution

$$H_A = \gamma R^2\left(1-\frac{\pi}{4}\right) \tag{A.20a}$$

$$H_B = \gamma R^2\left(\frac{\pi}{4}-\frac{1}{2}\right) \tag{A.20b}$$

$$V_B = \gamma R^2\left(1-\frac{\pi}{4}\right) \tag{A.20c}$$

The bending moment acting in a generic cross section of angular coordinate ϑ is obtained by summing the contributions which precede the cross section itself (Figure A.1)

$$M(\vartheta) = -H_A R(1-\cos\vartheta) + \int_0^\vartheta |q(\omega)|R^2\sin(\vartheta-\omega)d\omega \tag{A.21}$$

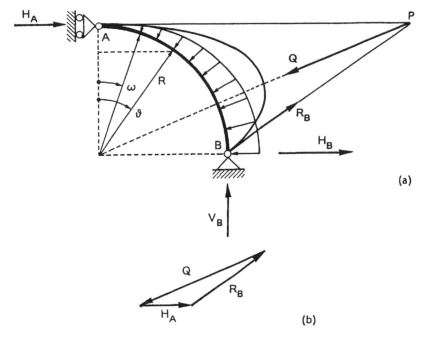

(a)

(b)

Figure A.1

Substituting the distribution (A.1b) into the integral, we have

$$\int_0^\vartheta |q(\omega)| R^2 \sin(\vartheta - \omega)\mathrm{d}\omega \tag{A.22}$$

$$= \gamma R^3 \int_0^\vartheta \sin(\vartheta - \omega)\mathrm{d}\omega - \gamma R^3 \int_0^\vartheta \cos\omega \sin(\vartheta - \omega)\mathrm{d}\omega$$

Applying the well-known trigonometric formulas, we obtain

$$\int_0^\vartheta |q(\omega)| R^2 \sin(\vartheta - \omega)\mathrm{d}\omega \tag{A.23}$$

$$= \gamma R^3 \int_0^\vartheta (\sin\vartheta \cos\omega - \cos\vartheta \sin\omega)\mathrm{d}\omega -$$

$$\gamma R^3 \int_0^\vartheta \cos\omega (\sin\vartheta \cos\omega - \cos\vartheta \sin\omega)\mathrm{d}\omega$$

$$= \gamma R^3 \left\{ \sin\vartheta [\sin\omega]_0^\vartheta + \cos\vartheta [\cos\omega]_0^\vartheta - \right.$$

$$\left. \frac{1}{2}\sin\vartheta \left[\omega + \frac{1}{2}\sin 2\omega\right]_0^\vartheta - \frac{1}{4}\cos\vartheta [\cos 2\omega]_0^\vartheta \right\}$$

$$= \gamma R^3 \left\{ \sin^2\vartheta + \cos\vartheta(\cos\vartheta - 1) - \right.$$

$$\left. \frac{1}{2}\sin\vartheta\left(\vartheta + \frac{1}{2}\sin 2\vartheta\right) - \frac{1}{4}\cos\vartheta(\cos 2\vartheta - 1) \right\}$$

703

Finally

$$M(\vartheta) = -\gamma R^3\left(1 - \frac{\pi}{4}\right)(1 - \cos\vartheta) + \tag{A.24}$$

$$\gamma R^3\left(1 - \cos\vartheta - \frac{1}{2}\vartheta\sin\vartheta - \frac{1}{4}\sin\vartheta\sin 2\vartheta - \right.$$

$$\left.\frac{1}{4}\cos\vartheta\cos 2\vartheta + \frac{1}{4}\cos\vartheta\right)$$

The expression (A.24) reduces to equation (A.15), just as equations (A.16) and (A.17) for shearing force and axial force can be found again using the direct method.

Figures A.2(a), (b), (c) represent the variations of $M(\vartheta)$, $T(\vartheta)$, $N(\vartheta)$, respectively. It may be noted how the bending moment in each case stretches the fibres at the intrados, the configuration of the pressure line being that of Figure A.1, and how it presents a maximum absolute value for $\vartheta \simeq 62°$, where the shear vanishes. It is further to be noted how the axial force is always

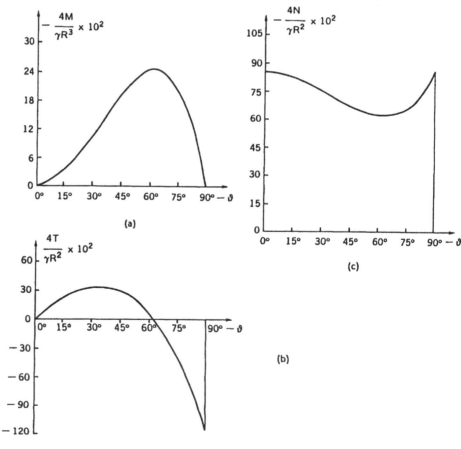

Figure A.2

compressive, with a minimum absolute value for $\vartheta \simeq 60°$ and equal maximum absolute values at the two extremes.

Appendix B Calculation of the internal reactions in a circular arch subjected to a uniformly distributed vertical load

B.1 Analytical method

The differential equation (5.9) in the case of Figure 5.8, whereby

$$m = 0 \qquad \text{(B.1a)}$$

$$p = q_0 \cos \vartheta \sin \vartheta \qquad \text{(B.1b)}$$

$$q = -q_0 \cos^2 \vartheta \qquad \text{(B.1c)}$$

reduces to the following form:

$$\frac{\mathrm{d}^3 M}{\mathrm{d}\vartheta^3} + \frac{\mathrm{d}M}{\mathrm{d}\vartheta} = -3R^2 q_0 \sin \vartheta \cos \vartheta \qquad \text{(B.2)}$$

The complete integral is the same as in the case studied in Appendix A, equation (A.4), while the particular solution is to be sought in the form

$$M_0(\vartheta) = a \cos 2\vartheta \qquad \text{(B.3)}$$

the known term being expressible as

$$b(\vartheta) = -\frac{3}{2} R^2 q_0 \sin 2\vartheta \qquad \text{(B.4)}$$

Differentiating equation (B.3) sequentially we obtain in fact

$$\frac{\mathrm{d}M_0}{\mathrm{d}\vartheta} = -2a \sin 2\vartheta \qquad \text{(B.5a)}$$

$$\frac{\mathrm{d}^2 M_0}{\mathrm{d}\vartheta^2} = -4a \cos 2\vartheta \qquad \text{(B.5b)}$$

$$\frac{\mathrm{d}^3 M_0}{\mathrm{d}\vartheta^3} = 8a \sin 2\vartheta \qquad \text{(B.5c)}$$

and thus substituting equations (B.5a, c) in equation (B.2)

$$6a \sin 2\vartheta = -\frac{3}{2} R^2 q_0 \sin 2\vartheta \qquad \text{(B.6)}$$

705

or

$$a = -\frac{q_0 R^2}{4} \tag{B.7}$$

The solution thus appears as follows:

$$M(\vartheta) = C_1 + C_2 \cos \vartheta + C_3 \sin \vartheta - \frac{q_0 R^2}{4} \cos 2\vartheta \tag{B.8}$$

The arbitrary constants C_1, C_2, C_3 are calculated via the boundary conditions (A.12):

$$M(0) = C_1 + C_2 - \frac{q_0 R^2}{4} = 0 \tag{B.9a}$$

$$T(0) = \frac{C_3}{R} = 0 \tag{B.9b}$$

$$M\left(-\frac{\pi}{2}\right) = C_1 - C_3 + \frac{q_0 R^2}{4} = 0 \tag{B.9c}$$

The linear algebraic system (B.9) possesses the following roots:

$$C_1 = -\frac{q_0 R^2}{4} \tag{B.10a}$$

$$C_2 = \frac{q_0 R^2}{4} \tag{B.10b}$$

$$C_3 = 0 \tag{B.10c}$$

so that the solution becomes

$$M(\vartheta) = -\frac{q_0 R^2}{4} + \frac{q_0 R^2}{2} \cos \vartheta - \frac{q_0 R^2}{4} \cos 2\vartheta \tag{B.11}$$

or

$$M(\vartheta) = -\frac{q_0 R^2}{2} \left(1 - \cos \vartheta - \sin^2 \vartheta\right) \tag{B.12}$$

Applying equations (5.6) and (5.8), we obtain also the shearing force and the axial force

$$T(\vartheta) = -\frac{q_0 R}{2} \left(\sin \vartheta - 2 \sin \vartheta \cos \vartheta\right) \tag{B.13}$$

$$N(\vartheta) = -\frac{q_0 R}{2} \left(\cos \vartheta + 2 \sin^2 \vartheta\right) \tag{B.14}$$

B.2 Direct method

The equations of equilibrium with regard to horizontal translation, vertical translation and rotation about point B, respectively, appear as follows:

$$H_A = H_B \qquad (B.15a)$$

$$V_B = q_0 R \qquad (B.15b)$$

$$H_A R = q_0 \frac{R^2}{2} \qquad (B.15c)$$

from which the constraint reactions are obtained:

$$H_A = \frac{1}{2} q_0 R \qquad (B.16a)$$

$$H_B = \frac{1}{2} q_0 R \qquad (B.16b)$$

$$V_B = q_0 R \qquad (B.16c)$$

The bending moment acting in a generic section of angular coordinate ϑ is obtained by summing the contributions which precede the section itself (Figure 5.8):

$$M(\vartheta) = -H_A R(1 - \cos \vartheta) + \frac{1}{2} q_0 R^2 \sin^2 \vartheta \qquad (B.17)$$

Substituting equation (B.16a) in equation (B.17) we obtain again equation (B.12).

In the same way the shearing force is

$$T(\vartheta) = -H_A \sin \vartheta + \left(q_0 R \sin \vartheta \right) \cos \vartheta \qquad (B.18)$$

just as the axial force is

$$N(\vartheta) = -H_A \cos \vartheta - \left(q_0 R \sin \vartheta \right) \sin \vartheta \qquad (B.19)$$

Equations (B.18) and (B.19) coincide with equations (B.13) and (B.14), respectively.

Figures B.2(a), (b), (c) represent, in order, the variations of $M(\vartheta)$, $T(\vartheta)$, $N(\vartheta)$. It may be noted that the bending moment in each case stretches the fibres at the extrados, the configuration of the pressure line being that of Figure B.1, and that it presents a maximum for $\vartheta \simeq 60°$, where the shear vanishes. It should further be noted how the axial force is always compressive, with a minimum absolute value at the end A and a maximum for $\vartheta \simeq 75°$.

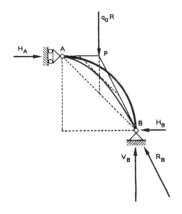

Figure B.1

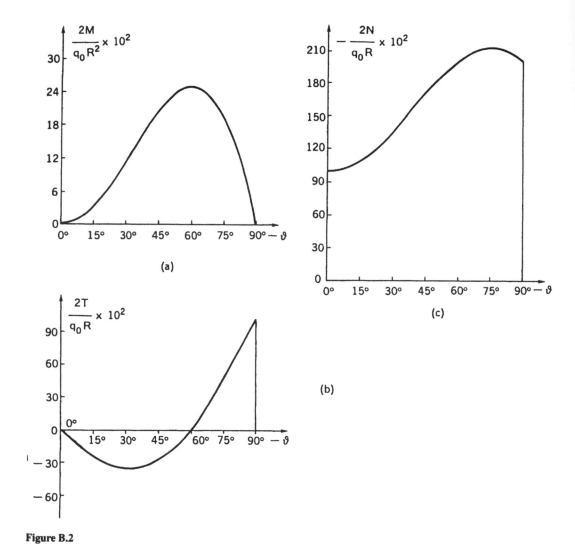

Figure B.2

Appendix C Anisotropic material

C.1   Anisotropic elastic constitutive law

The matrix expression of elastic potential is given by equation (8.44a), which represents a quadratic form in the components of strain, just as the elastic constitutive law is given by equation (8.44b), which links components of stress and components of strain. In both formulas there appears the Hessian matrix [H], which, in the case of isotropic material, has been rendered explicit in the expression (8.74). The expression of complementary elastic potential, on the other hand, is given by equation (8.49), just as the inverse constitutive law is formally represented by equation (8.46) and rendered explicit for isotropic material in equation (8.73).

Whereas in the case of isotropic material the inverse matrix $[H]^{-1}$ is a function only of the two parameters E and ν, in the case of anisotropic material the independent parameters can even amount to 21, as has been seen in Chapter 8. There are, however, a number of intermediate cases, related to the properties of symmetry of the material. It is possible to show how, if there is a plane of symmetry $z = 0$, the material in the point under consideration is called **monoclinic**, and the relation (8.46) reduces to the following:

$$\begin{bmatrix} \varepsilon_x \\ \varepsilon_y \\ \varepsilon_z \\ \gamma_{xy} \\ \gamma_{xz} \\ \gamma_{yz} \end{bmatrix} = \begin{bmatrix} C_{11} & C_{12} & C_{13} & C_{14} & 0 & 0 \\ C_{12} & C_{22} & C_{23} & C_{24} & 0 & 0 \\ C_{13} & C_{23} & C_{33} & C_{34} & 0 & 0 \\ C_{14} & C_{24} & C_{34} & C_{44} & 0 & 0 \\ 0 & 0 & 0 & 0 & C_{55} & C_{56} \\ 0 & 0 & 0 & 0 & C_{56} & C_{66} \end{bmatrix} \begin{bmatrix} \sigma_x \\ \sigma_y \\ \sigma_z \\ \tau_{xy} \\ \tau_{xz} \\ \tau_{yz} \end{bmatrix} \tag{C.1}$$

with 13 independent parameters.

If there are two perpendicular planes of symmetry, it is possible to demonstrate that there exists then a third plane of symmetry which is perpendicular to both. The relation (8.46), in the reference system oriented according to the principal directions of the material, takes the following form:

$$\begin{bmatrix} \varepsilon_x \\ \varepsilon_y \\ \varepsilon_z \\ \gamma_{xy} \\ \gamma_{xz} \\ \gamma_{yz} \end{bmatrix} = \begin{bmatrix} C_{11} & C_{12} & C_{13} & 0 & 0 & 0 \\ C_{12} & C_{22} & C_{23} & 0 & 0 & 0 \\ C_{13} & C_{23} & C_{33} & 0 & 0 & 0 \\ 0 & 0 & 0 & C_{44} & 0 & 0 \\ 0 & 0 & 0 & 0 & C_{55} & 0 \\ 0 & 0 & 0 & 0 & 0 & C_{66} \end{bmatrix} \begin{bmatrix} \sigma_x \\ \sigma_y \\ \sigma_z \\ \tau_{xy} \\ \tau_{xz} \\ \tau_{yz} \end{bmatrix} \tag{C.2}$$

with nine independent parameters, and the material at the point under consideration is said to be **orthotropic**. It should be noted that, in this case, there is no interaction between dilations and shearing stresses, just as there is none between normal stresses and shearing strains.

If the material is **transversely isotropic**, i.e. if the properties are the same in all the directions that define one of the three principal planes, for example the plane $z = 0$, the relation (8.46) then presents only five independent parameters:

$$\begin{bmatrix} \varepsilon_x \\ \varepsilon_y \\ \varepsilon_z \\ \gamma_{xy} \\ \gamma_{xz} \\ \gamma_{yz} \end{bmatrix} = \begin{bmatrix} C_{11} & C_{12} & C_{13} & 0 & 0 & 0 \\ C_{12} & C_{11} & C_{13} & 0 & 0 & 0 \\ C_{13} & C_{13} & C_{33} & 0 & 0 & 0 \\ 0 & 0 & 0 & 2(C_{11}-C_{12}) & 0 & 0 \\ 0 & 0 & 0 & 0 & C_{55} & 0 \\ 0 & 0 & 0 & 0 & 0 & C_{55} \end{bmatrix} \begin{bmatrix} \sigma_x \\ \sigma_y \\ \sigma_z \\ \tau_{xy} \\ \tau_{xz} \\ \tau_{yz} \end{bmatrix} \tag{C.3}$$

If, finally, the planes of symmetry are infinite, i.e. the material is transversely isotropic on any plane, we have

$$
\begin{bmatrix} \varepsilon_x \\ \varepsilon_y \\ \varepsilon_z \\ \gamma_{xy} \\ \gamma_{xz} \\ \gamma_{yz} \end{bmatrix} = \begin{bmatrix} C_{11} & C_{12} & C_{12} & 0 & 0 & 0 \\ C_{12} & C_{11} & C_{12} & 0 & 0 & 0 \\ C_{12} & C_{12} & C_{11} & 0 & 0 & 0 \\ 0 & 0 & 0 & 2(C_{11}-C_{12}) & 0 & 0 \\ 0 & 0 & 0 & 0 & 2(C_{11}-C_{12}) & 0 \\ 0 & 0 & 0 & 0 & 0 & 2(C_{11}-C_{12}) \end{bmatrix} \begin{bmatrix} \sigma_x \\ \sigma_y \\ \sigma_z \\ \tau_{xy} \\ \tau_{xz} \\ \tau_{yz} \end{bmatrix}
$$

$$(C.4)$$

the material being completely **isotropic** at the point under consideration, with $C_{11} = 1/E$ and $C_{12} = -v/E$, equation (8.73).

C.2 Orthotropic material

Fibre-reinforced materials, which are by now extensively used in all manner of engineering sectors, are generally orthotropic, or, at least, transversely isotropic. The principal planes of the material are naturally defined by the directions of the reinforcing fibres. The technical constants of these materials are the normal elastic modulus and the shear modulus, as well as the Poisson ratios, according to the following explicit compliance matrix:

$$
[H]^{-1} = \begin{bmatrix} \dfrac{1}{E_1} & -\dfrac{v_{21}}{E_2} & -\dfrac{v_{31}}{E_3} & 0 & 0 & 0 \\[2mm] -\dfrac{v_{12}}{E_1} & \dfrac{1}{E_2} & -\dfrac{v_{32}}{E_3} & 0 & 0 & 0 \\[2mm] -\dfrac{v_{13}}{E_1} & -\dfrac{v_{23}}{E_2} & \dfrac{1}{E_3} & 0 & 0 & 0 \\[2mm] 0 & 0 & 0 & \dfrac{1}{G_{12}} & 0 & 0 \\[2mm] 0 & 0 & 0 & 0 & \dfrac{1}{G_{13}} & 0 \\[2mm] 0 & 0 & 0 & 0 & 0 & \dfrac{1}{G_{23}} \end{bmatrix}
$$

$$(C.5)$$

The Poisson ratio v_{ij} represents the transverse dilation in the j direction, when the material is stressed in the i direction

$$ v_{ij} = -\frac{\varepsilon_j}{\varepsilon_i} \tag{C.6}$$

By virtue of the symmetry of the compliance matrix we have

$$ \frac{v_{ij}}{E_i} = \frac{v_{ji}}{E_j}, \quad \text{for} \quad i,j = 1,2,3 \tag{C.7}$$

so that there are only nine independent parameters, as is known. The symmetry of the matrix $[H]^{-1}$ ensures, on the other hand, that Betti's Reciprocal Theorem is satisfied.

The elements of the stiffness matrix $[H]$ are obtained by inverting the compliance matrix, and are as follows:

$$H_{11} = \frac{1 - \nu_{23}\nu_{32}}{E_2 E_3 \Delta} \tag{C.8a}$$

$$H_{12} = \frac{\nu_{21} + \nu_{31}\nu_{23}}{E_2 E_3 \Delta} = \frac{\nu_{12} + \nu_{32}\nu_{13}}{E_1 E_3 \Delta} \tag{C.8b}$$

$$H_{13} = \frac{\nu_{31} + \nu_{21}\nu_{32}}{E_2 E_3 \Delta} = \frac{\nu_{13} + \nu_{12}\nu_{23}}{E_1 E_2 \Delta} \tag{C.8c}$$

$$H_{22} = \frac{1 - \nu_{13}\nu_{31}}{E_1 E_3 \Delta} \tag{C.8d}$$

$$H_{23} = \frac{\nu_{32} + \nu_{12}\nu_{31}}{E_1 E_3 \Delta} = \frac{\nu_{23} + \nu_{21}\nu_{13}}{E_1 E_2 \Delta} \tag{C.8e}$$

$$H_{33} = \frac{1 - \nu_{12}\nu_{21}}{E_1 E_2 \Delta} \tag{C.8f}$$

$$H_{44} = G_{12} \tag{C.8g}$$

$$H_{55} = G_{13} \tag{C.8h}$$

$$H_{66} = G_{23} \tag{C.8i}$$

with

$$\Delta = \frac{1 - \nu_{12}\nu_{21} - \nu_{23}\nu_{32} - \nu_{31}\nu_{13} - 2\nu_{21}\nu_{32}\nu_{13}}{E_1 E_2 E_3} \tag{C.9}$$

As in the case of isotropic material, also in the more general case of orthotropic material the elastic constants must respect certain conditions, so as to render both the stiffness matrix and the compliance matrix positive definite. From equations (C.8) there follows

$$(1 - \nu_{23}\nu_{32}), (1 - \nu_{13}\nu_{31}), (1 - \nu_{12}\nu_{21}) > 0 \tag{C.10a}$$

$$\overline{\Delta} = 1 - \nu_{12}\nu_{21} - \nu_{23}\nu_{32} - \nu_{31}\nu_{13} - 2\nu_{21}\nu_{32}\nu_{13} > 0 \tag{C.10b}$$

Via the relations of symmetry (C.7), the inequalities (C.10a) may be reproposed in the form

$$|\nu_{21}| < \left(\frac{E_2}{E_1}\right)^{1/2}, \quad |\nu_{12}| < \left(\frac{E_1}{E_2}\right)^{1/2} \tag{C.11a}$$

$$|\nu_{32}| < \left(\frac{E_3}{E_2}\right)^{1/2}, \quad |\nu_{23}| < \left(\frac{E_2}{E_3}\right)^{1/2} \tag{C.11b}$$

$$|\nu_{13}| < \left(\frac{E_1}{E_3}\right)^{1/2}, \quad |\nu_{31}| < \left(\frac{E_3}{E_1}\right)^{1/2} \tag{C.11c}$$

711

just as equation (C.10b) can alternatively be expressed thus:

$$v_{21}v_{32}v_{13} < \frac{1 - v_{21}^2\left(\dfrac{E_1}{E_2}\right) - v_{32}^2\left(\dfrac{E_2}{E_3}\right) - v_{13}^2\left(\dfrac{E_3}{E_1}\right)}{2} < \frac{1}{2} \tag{C.12}$$

As an example, it is possible to mention that of a composite material made up of boron fibres embedded in an **epoxy** polymer matrix, for which $v_{12} \simeq 2$, $E_1/E_2 \simeq 10$, so that the second of inequalities (C.11a) is satisfied. The coefficient of transverse contraction, though appearing surprisingly high, is consistent with the conditions obtained previously. On the other hand, the mutual coefficient $v_{21} \simeq 0.22$ satisfies the first of inequalities (C.11a).

C.3 Stress–strain relations for plane stress conditions

In the case of a plane stress condition (in the plane $z = 0$) the relation (C.2) reduces to the following:

$$\begin{bmatrix} \varepsilon_1 \\ \varepsilon_2 \\ \gamma_{12} \end{bmatrix} = \begin{bmatrix} \dfrac{1}{E_1} & -\dfrac{v_{21}}{E_2} & 0 \\ -\dfrac{v_{12}}{E_1} & \dfrac{1}{E_2} & 0 \\ 0 & 0 & \dfrac{1}{G_{12}} \end{bmatrix} \begin{bmatrix} \sigma_1 \\ \sigma_2 \\ \tau_{12} \end{bmatrix} \tag{C.13a}$$

whilst its inverse is

$$\begin{bmatrix} \sigma_1 \\ \sigma_2 \\ \tau_{12} \end{bmatrix} = \begin{bmatrix} H_{11} & H_{12} & 0 \\ H_{12} & H_{22} & 0 \\ 0 & 0 & H_{44} \end{bmatrix} \begin{bmatrix} \varepsilon_1 \\ \varepsilon_2 \\ \gamma_{12} \end{bmatrix} \tag{C.13b}$$

with

$$H_{11} = \frac{E_1}{1 - v_{12}v_{21}} \tag{C.14a}$$

$$H_{12} = \frac{v_{12}E_2}{1 - v_{12}v_{21}} = \frac{v_{21}E_1}{1 - v_{12}v_{21}} \tag{C.14b}$$

$$H_{22} = \frac{E_2}{1 - v_{12}v_{21}} \tag{C.14c}$$

$$H_{44} = G_{12} \tag{C.14d}$$

Frequently the principal directions of orthotropy 12 do not coincide with the directions of the *XY* coordinate axes, which are the geometrically natural directions for solving the problem (Figure C.1). For this reason it is necessary to determine a relation that can connect stresses and strains in the principal system of the material with stresses and strains in the coordinate system of the body.

712

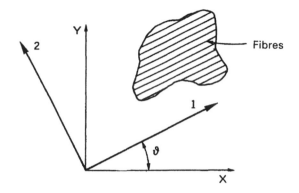

Figure C.1

Recalling relations (2.33) introduced in the framework of the geometry of areas, the transformation equation which expresses the stresses in the system 12 as functions of the stresses in the system XY is as follows:

$$
\begin{bmatrix} \sigma_1 \\ \sigma_2 \\ \tau_{12} \end{bmatrix} = \begin{bmatrix} \cos^2 \vartheta & \sin^2 \vartheta & -2\sin\vartheta\cos\vartheta \\ \sin^2 \vartheta & \cos^2 \vartheta & 2\sin\vartheta\cos\vartheta \\ \sin\vartheta\cos\vartheta & -\sin\vartheta\cos\vartheta & \cos 2\vartheta \end{bmatrix} \begin{bmatrix} \sigma_x \\ \sigma_y \\ \tau_{xy} \end{bmatrix} \quad \text{(C.15)}
$$

where ϑ is the angle between the axes X and 1 (Figure C.1). The transformation corresponding to strain is analogous

$$
\begin{bmatrix} \varepsilon_1 \\ \varepsilon_2 \\ \frac{1}{2}\gamma_{12} \end{bmatrix} = \begin{bmatrix} \cos^2 \vartheta & \sin^2 \vartheta & -2\sin\vartheta\cos\vartheta \\ \sin^2 \vartheta & \cos^2 \vartheta & 2\sin\vartheta\cos\vartheta \\ \sin\vartheta\cos\vartheta & -\sin\vartheta\cos\vartheta & \cos 2\vartheta \end{bmatrix} \begin{bmatrix} \varepsilon_x \\ \varepsilon_y \\ \frac{1}{2}\gamma_{xy} \end{bmatrix} \quad \text{(C.16)}
$$

The two foregoing transformations can be put in a compact form

$$
\{\sigma\}_{12} = [T]\{\sigma\}_{XY} \quad \text{(C.17a)}
$$

$$
\{\varepsilon\}_{12} = [R][T][R]^{-1}\{\varepsilon\}_{XY} \quad \text{(C.17b)}
$$

where

$$
\{\varepsilon\}_{XY}^{T} = \left[\varepsilon_x, \varepsilon_y, \gamma_{xy}\right] \quad \text{(C.18a)}
$$

$$
\{\varepsilon\}_{12}^{T} = \left[\varepsilon_1, \varepsilon_2, \gamma_{12}\right] \quad \text{(C.18b)}
$$

and $[R]$ is the so-called **Reuter matrix**

$$
[R] = \begin{bmatrix} 1 & 0 & 0 \\ 0 & 1 & 0 \\ 0 & 0 & 2 \end{bmatrix} \quad \text{(C.19)}
$$

713

which serves to consider the shearing strain and not its half, which appears as an off-diagonal term in the strain tensor.

Casting equation (C.13a) in the form

$$\{\varepsilon\}_{12} = [C]\{\sigma\}_{12} \tag{C.20}$$

and substituting equations (C.17) in equation (C.20), we obtain

$$[R][T][R]^{-1}\{\varepsilon\}_{XY} = [C][T]\{\sigma\}_{XY} \tag{C.21}$$

Premultiplying both sides of equation (C.21) by $[R][T]^{-1}[R]^{-1}$, we have

$$\{\varepsilon\}_{XY} = [R][T]^{-1}[R]^{-1}[C][T]\{\sigma\}_{XY} \tag{C.22}$$

Since it is possible to show that

$$[R][T]^{-1}[R]^{-1} = [T]^{T} \tag{C.23}$$

finally we can write

$$\{\varepsilon\}_{XY} = [T]^{T}[C][T]\{\sigma\}_{XY} \tag{C.24}$$

or

$$\{\varepsilon\}_{XY} = [C^*]\{\sigma\}_{XY} \tag{C.25}$$

The compliance matrix, rotated, may be cast in the following form:

$$[C^*] = \begin{bmatrix} \dfrac{1}{E_x} & -\dfrac{v_{yx}}{E_y} & \dfrac{\eta_{x,xy}}{G_{xy}} \\ -\dfrac{v_{xy}}{E_x} & \dfrac{1}{E_y} & \dfrac{\eta_{y,xy}}{G_{xy}} \\ \dfrac{\eta_{xy,x}}{E_x} & \dfrac{\eta_{xy,y}}{E_y} & \dfrac{1}{G_{xy}} \end{bmatrix} \tag{C.26}$$

which is symmetrical by virtue of Betti's Reciprocal Theorem. Herein there appear **Lekhnitski's coefficients**, with the following physical meaning: $\eta_{ij,i}$, dilation in the i direction caused by the shearing stress τ_{ij},

$$\eta_{ij,i} = \frac{\varepsilon_i}{\gamma_{ij}} \tag{C.27}$$

$\eta_{i,ij}$, shearing strain of the axes ij caused by the normal stress in the i direction,

$$\eta_{i,ij} = \frac{\gamma_{ij}}{\varepsilon_i} \tag{C.28}$$

714

Note that the rotated compliance matrix $[C*]$ is a fully-populated matrix, unlike the principal compliance matrix $[C]$. Notwithstanding this, the independent parameters remain four $(E_1, E_2, v_{12}, G_{12})$

$$\frac{1}{E_x} = \frac{1}{E_1}\cos^4\vartheta + \left(\frac{1}{G_{12}} - \frac{2v_{12}}{E_1}\right)\sin^2\vartheta\cos^2\vartheta + \frac{1}{E_2}\sin^4\vartheta \qquad \text{(C.29a)}$$

$$v_{xy} = E_x\left[\frac{v_{12}}{E_1}\left(\sin^4\vartheta + \cos^4\vartheta\right) - \left(\frac{1}{E_1} + \frac{1}{E_2} - \frac{1}{G_{12}}\right)\sin^2\vartheta\cos^2\vartheta\right] \text{(C.29b)}$$

$$\frac{1}{E_y} = \frac{1}{E_1}\sin^4\vartheta + \left(\frac{1}{G_{12}} - \frac{2v_{12}}{E_1}\right)\sin^2\vartheta\cos^2\vartheta + \frac{1}{E_2}\cos^4\vartheta \qquad \text{(C.29c)}$$

$$\frac{1}{G_{xy}} = 2\left(\frac{2}{E_1} + \frac{2}{E_2} + \frac{4v_{12}}{E_1} - \frac{1}{G_{12}}\right)\sin^2\vartheta\cos^2\vartheta + \qquad \text{(C.29d)}$$

$$\frac{1}{G_{12}}\left(\sin^4\vartheta + \cos^4\vartheta\right)$$

$$\eta_{xy,x} = E_x\left[\left(\frac{2}{E_1} + \frac{2v_{12}}{E_1} - \frac{1}{G_{12}}\right)\sin\vartheta\cos^3\vartheta - \qquad \text{(C.29c)}\right.$$

$$\left.\left(\frac{2}{E_2} + \frac{2v_{12}}{E_1} - \frac{1}{G_{12}}\right)\sin^3\vartheta\cos\vartheta\right]$$

$$\eta_{xy,y} = E_y\left[\left(\frac{2}{E_1} + \frac{2v_{12}}{E_1} - \frac{1}{G_{12}}\right)\sin^3\vartheta\cos\vartheta - \qquad \text{(C.29f)}\right.$$

$$\left.\left(\frac{2}{E_2} + \frac{2v_{12}}{E_1} - \frac{1}{G_{12}}\right)\sin\vartheta\cos^3\vartheta\right]$$

The apparent parameters (C.29) are plotted as functions of the angle ϑ, in Figure C.2, in the case of the **epoxy–boron** composite material, already considered. It may be noted that:

1. the shear modulus G_{xy} is maximum for $\vartheta = 45°$;
2. the coefficient $\eta_{xy,x}$ vanishes, obviously, for $\vartheta = 0°$ and $\vartheta = 90°$;
3. the normal elastic modulus E_y varies identically as E_x, exchanging ϑ with $90°-\vartheta$ (the same applies for v_{yx} and $\eta_{xy,y}$);
4. E_x may be less than both E_1 and E_2, just as it may be greater than both (cf. what occurs in the case where $\vartheta \simeq 60°$). In other words, the extreme values of the parameters are not found necessarily in the principal directions of the material.

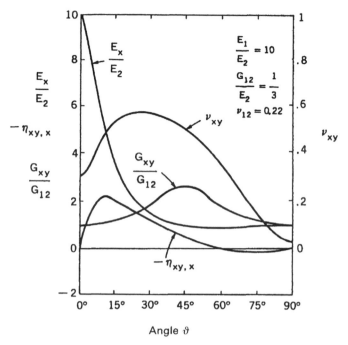

Figure C.2

C.4 Strength criteria for orthotropic materials

Since in orthotropic materials strength varies with the variation of direction, the direction of maximum stress may not be the most dangerous one.

Let different properties be assumed in tension and in compression (Figure C.3):

X_t = tensile strength in the direction 1;
X_c = compressive strength in the direction 1;
Y_t = tensile strength in the direction 2;
Y_c = compressive strength in the direction 2;
S = shear strength.

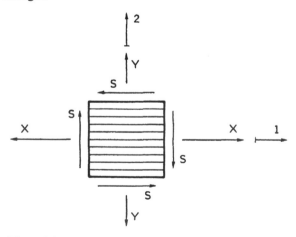

Figure C.3

716

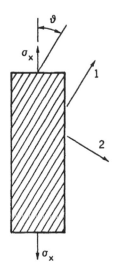

Figure C.4

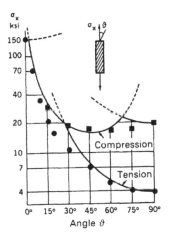

Figure C.5

Note that the foregoing strengths have been defined in the principal directions of the material, and that of course they vary as the coordinate axes vary.

The **criterion of maximum stress** requires that all the following inequalities should be satisfied at the same time:

$$-X_c < \sigma_1 < X_t \qquad \text{(C.30a)}$$

$$-Y_c < \sigma_2 < Y_t \qquad \text{(C.30b)}$$

$$|\tau_{12}| < S \qquad \text{(C.30c)}$$

If only one of equations (C.30) is not satisfied, the critical condition of the material is assumed according to the mechanism of rupture associated with X_c, X_t, Y_c, Y_t or S. Hence interaction is not assumed to exist between the various modes of rupture.

Let us consider a material that is fibre-reinforced in one direction, submitted to a condition of uniaxial stress inclined at an angle ϑ with respect to the fibres (Figure C.4). The stresses in the principal reference system of the material are obtained from equations (C.15):

$$\sigma_1 = \sigma_x \cos^2 \vartheta \qquad \text{(C.31a)}$$

$$\sigma_2 = \sigma_x \sin^2 \vartheta \qquad \text{(C.31b)}$$

$$\tau_{12} = \sigma_x \sin \vartheta \cos \vartheta \qquad \text{(C.31c)}$$

Substituting equations (C.31) in inequalities (C.30), we obtain three mutually competing criteria:

$$-\frac{X_c}{\cos^2 \vartheta} < \sigma_x < \frac{X_t}{\cos^2 \vartheta} \qquad \text{(C.32a)}$$

$$-\frac{Y_c}{\sin^2 \vartheta} < \sigma_x < \frac{Y_t}{\sin^2 \vartheta} \qquad \text{(C.32b)}$$

$$-\frac{S}{\sin \vartheta \cos \vartheta} < \sigma_x < \frac{S}{\sin \vartheta \cos \vartheta} \qquad \text{(C.32c)}$$

These criteria are plotted in Figure C.5 for a **glass–epoxy** composite having the following properties:

$$X_c = 150 \text{ ksi}$$
$$X_t = 150 \text{ ksi}$$
$$Y_c = 20 \text{ ksi}$$
$$Y_t = 4 \text{ ksi}$$
$$S = 6 \text{ ksi}$$

Hence the uniaxial strength is represented as a function of the angle ϑ in Figure C.5 (some experimental results are also given). The criterion of maximum stress, both tensile and compressive, consists in fact of three curves, the bottom one of which in each case governs the rupture phenomenon.

The **Tsai–Hill criterion** instead consists of a single curve devoid of cusps which presents the following general form:

$$(G+H)\sigma_1^2 + (F+H)\sigma_2^2 + (F+G)\sigma_3^2 - \qquad \text{(C.33)}$$
$$2H\sigma_1\sigma_2 - 2G\sigma_1\sigma_3 - 2F\sigma_2\sigma_3 +$$
$$2L\tau_{12}^2 + 2M\tau_{13}^2 + 2N\tau_{23}^2 = 1$$

717

The parameters F, G, H, L, M, N are correlated with the strengths X, Y, S introduced previously. If only τ_{12} acts, we have in fact

$$2L = \frac{1}{S^2} \tag{C.34}$$

just as, if only σ_1 acts, or, respectively σ_2 or σ_3

$$G + H = \frac{1}{X^2} \tag{C.35a}$$

$$F + H = \frac{1}{Y^2} \tag{C.35b}$$

$$F + G = \frac{1}{Z^2} \tag{C.35c}$$

where Z indicates the strength in the direction 3, normal to the stress plane. From equations (C.35) it follows that

$$2H = \frac{1}{X^2} + \frac{1}{Y^2} - \frac{1}{Z^2} \tag{C.36a}$$

$$2G = \frac{1}{X^2} + \frac{1}{Z^2} - \frac{1}{Y^2} \tag{C.36b}$$

$$2F = \frac{1}{Y^2} + \frac{1}{Z^2} - \frac{1}{X^2} \tag{C.36c}$$

If the body is assumed to be transversely isotropic in the plane 23, then we have $Y = Z$, and thus equation (C.33) reduces to the following form:

$$\frac{\sigma_1^2}{X^2} + \frac{\sigma_2^2}{Y^2} - \frac{\sigma_1\sigma_2}{X^2} + \frac{\tau_{12}^2}{S^2} = 1 \tag{C.37}$$

Finally, substituting equations (C.31) in equation (C.37), we obtain

$$\frac{\cos^4\vartheta}{X^2} + \frac{\sin^4\vartheta}{Y^2} + \left(\frac{1}{S^2} - \frac{1}{X^2}\right)\sin^2\vartheta\cos^2\vartheta = \frac{1}{\sigma_x^2} \tag{C.38}$$

The criterion (C.38) is represented in Figure C.6 for the **glass–epoxy** composite, both in the case of tension and in that of compression. It is possible to note an excellent agreement between theory and experimentation. In particular, a notable improvement has been obtained in comparison with the criterion of maximum stress for $\vartheta \simeq 30°$, where the latter fails by approximately 100% (Figure C.5).

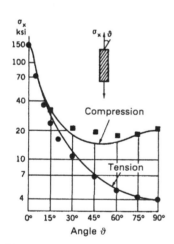

Figure C.6

Appendix D Heterogeneous beam

D.1 Multilayer beam in flexure

Let us consider a beam having rectangular cross section, consisting of n layers of different materials (Figure D.1(a)). If, as in the case of homogeneous mate-

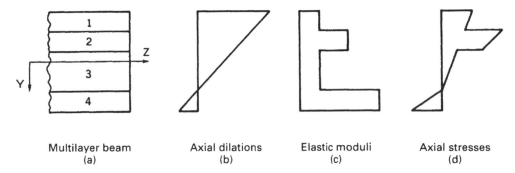

Multilayer beam   Axial dilations   Elastic moduli   Axial stresses
    (a)   (b)   (c)   (d)

Figure D.1

rial, we assume the conservation of the plane sections and thus the linear variation of axial dilation

$$\varepsilon_z = \frac{1}{E_1}(by + c) \tag{D.1}$$

and consider the conditions of equivalence (9.16a, d), we have

$$\sum_{i=1}^{n} \int_{A_i} (by+c)\frac{E_i}{E_1}\,dA = b\sum_{i=1}^{n} \frac{E_i}{E_1}\int_{A_i} y\,dA + c\sum_{i=1}^{n} \frac{E_i}{E_1}\int_{A_i} dA = 0 \tag{D.2a}$$

$$\sum_{i=1}^{n} \int_{A_i} (by+c)y\frac{E_i}{E_1}\,dA = b\sum_{i=1}^{n} \frac{E_i}{E_1}\int_{A_i} y^2\,dA + c\sum_{i=1}^{n} \frac{E_i}{E_1}\int_{A_i} y\,dA = M_x \tag{D.2b}$$

Equations (D.2) can take the form

$$bS_x + cA = 0 \tag{D.3a}$$
$$bI_x + cS_x = M_x \tag{D.3b}$$

where the usual symbols introduced in Chapter 2, corresponding to the geometry of areas, must be translated into the ones corresponding to the geometry of masses, each elementary area dA being weighed via the ratio of the moduli of elasticity E_i/E_1. If we define the **centroid of the cross section** as the point of the Y axis for which $S_x = 0$, results are obtained that are formally analogous to those obtained in Section 9.4:

$$\varepsilon_z^i = \frac{1}{E_1}\frac{M_x}{I_x}y \tag{D.4a}$$

$$\sigma_z^i = \frac{E_i}{E_1}\frac{M_x}{I_x}y \tag{D.4b}$$

Whilst equation (D.4a) represents the supposed linear function (Figure D.1(b)), equation (D.4b) represents a step function, the stresses being greater in more rigid materials (Figure D.1(c)), as well as in the points farther from the neutral axis (Figure D.1(d)).

719

As regards the strain condition of the beam, it may be stated, in an analogous way to the case of homogeneous material, that the curvature of the centroidal axis is

$$\chi_x = \frac{M_x}{E_1 I_x} \tag{D.5}$$

where of course the moment of inertia is the one corresponding to the elementary areas weighed via the moduli of elasticity:

$$I_x = \sum_{i=1}^{n} \frac{E_i}{E_1} \int_{A_i} y^2 dA \tag{D.6}$$

D.2 Reinforced concrete

Reinforced-concrete beams may be considered as multilayer beams. Steel bars have the function of withstanding tensile forces and are thus usually embedded in the concrete on the side of the fibres in tension. Concrete and steel present excellent adherence and the same coefficient of thermal expansion, so that their dilation tends to occur without differential displacements or discontinuities.

The basic hypotheses for the statics of reinforced concrete are the following:

1. Concrete behaves like a linear elastic material in compression, whilst it is **non-traction-bearing**. In other words, concrete presents a zero elastic modulus in tension.
2. Steel behaves as a linear elastic material both in compression and in tension.
3. The steel bars cannot slip inside the concrete.
4. The cross section of the beam remains plane.

In bending, the neutral axis divides the section into two parts: one part is in compression with elastic modulus E_c, while the other is in tension with zero elastic modulus (Figure D.2). The steel bars are below the neutral axis, in the part in tension, and present an elastic modulus E_s. On the other hand, the position of the neutral axis and thus the area of material with elastic modulus E_c is *a priori* unknown. Hence this area has to be identified first, and then the formulas seen in the previous section are applied.

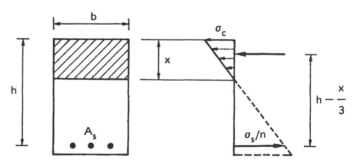

Figure D.2

The condition of axial equivalence for the rectangular cross section in Figure D.2 is

$$\sigma_s A_s = \sigma_c bx/2 \qquad (D.7)$$

where σ_s is the tensile stress in the steel, σ_c is the maximum compressive stress in the upper edge of the concrete, A_s is the area of steel, b is the width of the beam, and x is the unknown distance of the neutral axis from the upper edge of the beam.

The condition of equivalence corresponding to the bending moment is written

$$M = \sigma_s A_s \left(h - \frac{x}{3} \right) \qquad (D.8)$$

where M is the moment applied to the section and $h - x/3$ the arm contained between the two resultants of the tensile and compressive forces, respectively. Note that the thickness of the concrete cover does not enter into this calculation.

The condition of linear variation of the axial dilations gives

$$\frac{\varepsilon_s}{\varepsilon_c} = \frac{h - x}{x} \qquad (D.9)$$

where ε_s is the dilation of the bars, whilst ε_c is the dilation of the concrete at the upper edge of the section (Figure D.2).

Introducing the stresses we have

$$\sigma_s = n\sigma_c \frac{h - x}{x} \qquad (D.10)$$

where n is the ratio between the elastic modulus of the steel and the elastic modulus of the compressed concrete,

$$n = \frac{E_s}{E_c} \simeq 10 \qquad (D.11)$$

Combining equations (D.7) and (D.10), we obtain

$$n\sigma_c A_s \frac{h - x}{x} = \frac{\sigma_c bx}{2} \qquad (D.12)$$

or

$$\frac{1}{2} bx^2 - nA_s(h - x) = 0 \qquad (D.13)$$

The positive root x of the quadratic equation (D.13) gives the position of the neutral axis.

From equations (D.7) and (D.8) it is possible to obtain the stresses in the concrete and in the steel as functions of the distance x

$$\sigma_s = \frac{M}{A_s \left(h - \frac{x}{3} \right)} \qquad (D.14a)$$

$$\sigma_c = \frac{2M}{bx\left(h - \dfrac{x}{3}\right)} \tag{D.14b}$$

Equations (D.14) resolve the problem of verifying the strength of the reinforced section. If, instead, we are faced with the design problem, the unknowns to be determined are h, x, A_s. Equation (D.10) then transforms as follows:

$$x = \frac{n\sigma_c h}{\sigma_s + n\sigma_c} \tag{D.15}$$

Equations (D.7) and (D.15) yield on the other hand

$$A_s = \frac{n\sigma_c^2}{2\sigma_s(\sigma_s + n\sigma_c)}bh \tag{D.16}$$

Equation (D.8), via equations (D.15) and (D.16), becomes

$$M = \left(1 - \frac{\alpha}{3}\right)\beta\sigma_s bh^2 \tag{D.17}$$

with

$$\alpha = \frac{n\sigma_c}{\sigma_s + n\sigma_c} \tag{D.18a}$$

$$\beta = \frac{n\sigma_c^2}{2\sigma_s(\sigma_s + n\sigma_c)} \tag{D.18b}$$

From equation (D.17) we obtain finally

$$h = \left(\frac{M}{\left(1 - \dfrac{\alpha}{3}\right)\beta\sigma_s b}\right)^{\frac{1}{2}} \tag{D.19}$$

The admissible values of σ_c and σ_s having been assigned, equation (D.19) gives the depth of the beam, whilst equations (D.15) and (D.16) give the position of the neutral axis and the area of steel, respectively.

Appendix E Heterogeneous plate

Let us consider a **laminate**, i.e. a multilayer plate, in which each layer (or lamina) is orthotropic in a particular principal orientation (Figure E.1). The stress–strain relation in the principal coordinates of each layer is of the type represented by equations (C.13b) and (C.14). In an external reference system the inverse relation of equation (C.25) is presented as follows:

$$\{\sigma\}_{XY} = [H^*]\{\varepsilon\}_{XY} \tag{E.1}$$

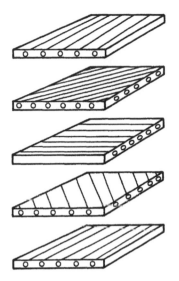

Figure E.1

with

$$H_{11}^* = H_{11}\cos^4\vartheta + 2(H_{12} + 2H_{44})\sin^2\vartheta\cos^2\vartheta + H_{22}\sin^4\vartheta \qquad \text{(E.2a)}$$

$$H_{12}^* = (H_{11} + H_{22} - 4H_{44})\sin^2\vartheta\cos^2\vartheta + H_{12}(\sin^4\vartheta + \cos^4\vartheta) \qquad \text{(E.2b)}$$

$$H_{22}^* = H_{11}\sin^4\vartheta + 2(H_{12} + 2H_{44})\sin^2\vartheta\cos^2\vartheta + H_{22}\cos^4\vartheta \qquad \text{(E.2c)}$$

$$H_{14}^* = (H_{11} - H_{12} - 2H_{44})\sin\vartheta\cos^3\vartheta + (H_{12} - H_{22} + 2H_{44})\sin^3\vartheta\cos\vartheta$$
$$\text{(E.2d)}$$

$$H_{24}^* = (H_{11} - H_{12} - 2H_{44})\sin^3\vartheta\cos\vartheta + (H_{12} - H_{22} + 2H_{44})\sin\vartheta\cos^3\vartheta$$
$$\text{(E.2e)}$$

$$H_{44}^* = (H_{11} + H_{22} - 2H_{12} - 2H_{44})\sin^2\vartheta\cos^2\vartheta + H_{44}(\sin^4\vartheta + \cos^4\vartheta) \quad \text{(E.2f)}$$

Also in the case of the multilayer plate it is possible to formulate Kirchhoff's hypothesis, already described in Section 10.10. In the case where the plate presents a membrane regime, in addition to a flexural regime, equations (10.158) are completed as follows:

$$\begin{bmatrix} \varepsilon_x \\ \varepsilon_y \\ \gamma_{xy} \end{bmatrix} = \begin{bmatrix} \varepsilon_x^0 \\ \varepsilon_y^0 \\ \gamma_{xy}^0 \end{bmatrix} + z \begin{bmatrix} \chi_x \\ \chi_y \\ \chi_{xy} \end{bmatrix} \qquad \text{(E.3)}$$

where ε_x^0, ε_y^0, γ_{xy}^0 are the strains of the middle plane. Substituting equation (E.3) in equation (E.1), the stresses in the kth layer can then be expressed as functions of the strains and curvatures of the middle plane:

$$\{\sigma\}_{XY}^k = [H^*]^k\left(\{\varepsilon^0\}_{XY} + z\{\chi\}_{XY}\right) \qquad \text{(E.4)}$$

Since the stiffness matrix $[H^*]^k$ can vary from layer to layer, the variation of the stresses through the thickness of the laminate is not necessarily linear, although the variation of the strains is (Figure D.1).

The internal forces and moments acting in the laminate are obtained by integration of the stresses that develop in each lamina

$$\begin{bmatrix} N_x \\ N_y \\ N_{xy} \end{bmatrix} = \int_{-h/2}^{+h/2} \begin{bmatrix} \sigma_x \\ \sigma_y \\ \tau_{xy} \end{bmatrix} dz = \sum_{k=1}^{n} \int_{z_{k-1}}^{z_k} \begin{bmatrix} \sigma_x \\ \sigma_y \\ \tau_{xy} \end{bmatrix}^k dz \qquad \text{(E.5a)}$$

$$\begin{bmatrix} M_x \\ M_y \\ M_{xy} \end{bmatrix} = \int_{-h/2}^{+h/2} \begin{bmatrix} \sigma_x \\ \sigma_y \\ \tau_{xy} \end{bmatrix} z\,dz = \sum_{k=1}^{n} \int_{z_{k-1}}^{z_k} \begin{bmatrix} \sigma_x \\ \sigma_y \\ \tau_{xy} \end{bmatrix}^k z\,dz \qquad \text{(E.5b)}$$

where z_k and z_{k-1} are defined in Figure E.2, with $z_0 = -h/2$. Substituting equation (E.4) in equations (E.5), we obtain

$$\{N\} = \sum_{k=1}^{n} [H^*]^k\left(\{\varepsilon^0\}_{XY}\int_{z_{k-1}}^{z_k} dz + \{\chi\}_{XY}\int_{z_{k-1}}^{z_k} z\,dz\right) \qquad \text{(E.6a)}$$

723

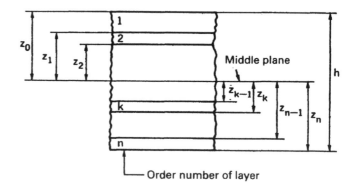

Figure E.2

$$\{M\} = \sum_{k=1}^{n} [H^*]^k \left(\{\varepsilon^0\}_{XY} \int_{z_{k-1}}^{z_k} z \, dz + \{\chi\}_{XY} \int_{z_{k-1}}^{z_k} z^2 \, dz \right) \tag{E.6b}$$

Finally we can write

$$
\begin{bmatrix} N_x \\ N_y \\ N_{xy} \\ \hline M_x \\ M_y \\ M_{xy} \end{bmatrix}
=
\left[
\begin{array}{ccc:ccc}
A_{11} & A_{12} & A_{14} & B_{11} & B_{12} & B_{14} \\
A_{12} & A_{22} & A_{24} & B_{12} & B_{22} & B_{24} \\
A_{14} & A_{24} & A_{44} & B_{14} & B_{24} & B_{44} \\ \hdashline
B_{11} & B_{12} & B_{14} & D_{11} & D_{12} & D_{14} \\
B_{12} & B_{22} & B_{24} & D_{12} & D_{22} & D_{24} \\
B_{14} & B_{24} & B_{44} & D_{14} & D_{24} & D_{44}
\end{array}
\right]
\begin{bmatrix} \varepsilon_x^0 \\ \varepsilon_y^0 \\ \gamma_{xy}^0 \\ \hline \chi_x \\ \chi_y \\ \chi_{xy} \end{bmatrix}
\tag{E.7}
$$

with

$$A_{ij} = \sum_{k=1}^{n} H_{ij}^{*k} \left(z_k - z_{k-1} \right) \tag{E.8a}$$

$$B_{ij} = \frac{1}{2} \sum_{k=1}^{n} H_{ij}^{*k} \left(z_k^2 - z_{k-1}^2 \right) \tag{E.8b}$$

$$D_{ij} = \frac{1}{3} \sum_{k=1}^{n} H_{ij}^{*k} \left(z_k^3 - z_{k-1}^3 \right) \tag{E.8c}$$

Expressions (E.8a) represent membrane stiffness, expressions (E.8c) represent flexural stiffness, whilst expressions (E.8b) represent the mutual stiffness or coupling stiffness, which produces flexure and/or torsion of the laminate subjected to tension, as well as stretching of the middle plane when the laminate is bent.

Appendix F Finite difference method

F.1 Torsion of beams of generic cross section ($\nabla^2 \omega = 0$)

As we have seen in Section 9.7, the warping function, introduced in treating the problem of torsion of beams of generic cross section, is **harmonic** and thus satisfies Laplace's equation (9.98) with the boundary condition (9.106). Here we shall mention the solution based on the **Finite Difference Method**, which is a numerical method of discretization that is useful for dealing with problems for which closed-form solutions are not possible.

If a regular function $y(x)$ presents, in a series of equidistant points, the values $y_0, y_1, y_2, \ldots,$ for $x = 0, x = \delta, x = 2\delta, \ldots,$ the first differential at these points can be approximated as follows:

$$\left(\Delta y\right)_{x=0} = y_1 - y_0 \tag{F.1a}$$

$$\left(\Delta y\right)_{x=\delta} = y_2 - y_1 \tag{F.1b}$$

$$\left(\Delta y\right)_{x=2\delta} = y_3 - y_2 \tag{F.1c}$$

$$\vdots$$

Dividing the differentials (F.1) by the length δ of the intervals, we obtain an approximate value of the first derivative at the corresponding points, in the form of an incremental ratio:

$$\left(\frac{dy}{dx}\right)_{x=0} \simeq \frac{y_1 - y_0}{\delta} \tag{F.2a}$$

$$\left(\frac{dy}{dx}\right)_{x=\delta} \simeq \frac{y_2 - y_1}{\delta} \tag{F.2b}$$

$$\left(\frac{dy}{dx}\right)_{x=2\delta} \simeq \frac{y_3 - y_2}{\delta} \tag{F.2c}$$

$$\vdots$$

It is then possible to approximate the second differentials using the first differentials

$$\left(\Delta^2 y\right)_{x=\delta} = \left(\Delta y\right)_{x=\delta} - \left(\Delta y\right)_{x=0} = y_2 - 2y_1 + y_0 \tag{F.3a}$$

$$\left(\Delta^2 y\right)_{x=2\delta} = y_3 - 2y_2 + y_1 \tag{F.3b}$$

$$\left(\Delta^2 y\right)_{x=3\delta} = y_4 - 2y_3 + y_2 \tag{F.3c}$$

$$\vdots$$

so that the second derivatives, calculated again as incremental ratios, appear thus:

$$\left(\frac{d^2 y}{dx^2}\right)_{x=\delta} \simeq \frac{\left(\Delta^2 y\right)_{x=\delta}}{\delta^2} = \frac{y_2 - 2y_1 + y_0}{\delta^2} \tag{F.4a}$$

$$\left(\frac{d^2 y}{dx^2}\right)_{x=2\delta} \simeq \frac{y_3 - 2y_2 + y_1}{\delta^2} \tag{F.4b}$$

$$\left(\frac{d^2 y}{dx^2}\right)_{x=3\delta} \simeq \frac{y_4 - 2y_3 + y_2}{\delta^2} \qquad (F.4c)$$

$$\vdots$$

If we have a function of two variables $\omega(x, y)$, it is possible to calculate the partial derivatives using expressions similar to equations (F.2) and (F.4). Let us consider, for instance, a generic compact cross section (Figure F.1) and the superposition of a regular square-mesh grid of nodal points. We can then approximate the values of the partial derivatives of the function ω in the generic point 0:

$$\frac{\partial \omega}{\partial x} \simeq \frac{\omega_1 - \omega_0}{\delta}, \quad \frac{\partial \omega}{\partial y} \simeq \frac{\omega_2 - \omega_0}{\delta} \qquad (F.5a)$$

$$\frac{\partial^2 \omega}{\partial x^2} \simeq \frac{\omega_1 - 2\omega_0 + \omega_3}{\delta^2}, \quad \frac{\partial^2 \omega}{\partial y^2} \simeq \frac{\omega_2 - 2\omega_0 + \omega_4}{\delta^2} \qquad (F.5b)$$

Using the foregoing expressions and similar ones, the differential equation (9.98) is transformed into a system of algebraic finite difference equations of the type

$$\frac{1}{\delta^2}\left(\omega_1 + \omega_2 + \omega_3 + \omega_4 - 4\omega_0\right) = 0 \qquad (F.6)$$

As many equations as there are unknowns ω_i will thus be obtained. For each node of the grid, in fact, we have one unknown and one equation. This equation is the field one (F.6) if the node is internal, while it represents the discretized form of the integrodifferential boundary condition (9.106) if the node belongs to the boundary.

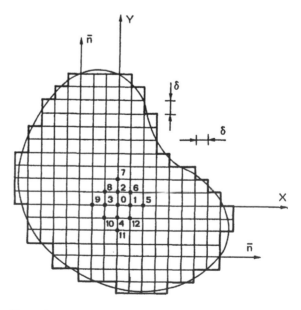

Figure F.1

726

F.2 Plates in flexure ($\nabla^4 w = q/D$)

The Finite Difference Method is applied to advantage also for the approximate numerical solution of the plate equation (10.184), which is a fourth-order differential equation, like the equation governing the Airy stress function Φ for plane problems, which is introduced in Chapter 19.

Consider once more the grid of Figure F.1. The second partial derivatives in the points 0, 1, 3 may be approximated, respectively, as follows:

$$\left(\frac{\partial^2 w}{\partial x^2}\right)_0 \simeq \frac{1}{\delta^2}\left(w_1 - 2w_0 + w_3\right) \tag{F.7a}$$

$$\left(\frac{\partial^2 w}{\partial x^2}\right)_1 \simeq \frac{1}{\delta^2}\left(w_5 - 2w_1 + w_0\right) \tag{F.7b}$$

$$\left(\frac{\partial^2 w}{\partial x^2}\right)_3 \simeq \frac{1}{\delta^2}\left(w_0 - 2w_3 + w_9\right) \tag{F.7c}$$

so that the fourth partial derivative becomes

$$\left(\frac{\partial^4 w}{\partial x^4}\right)_0 = \frac{\partial^2}{\partial x^2}\left(\frac{\partial^2 w}{\partial x^2}\right)_0 \tag{F.8a}$$

$$\simeq \frac{1}{\delta^2}\left[\left(\frac{\partial^2 w}{\partial x^2}\right)_1 - 2\left(\frac{\partial^2 w}{\partial x^2}\right)_0 + \left(\frac{\partial^2 w}{\partial x^2}\right)_3\right]$$

$$\simeq \frac{1}{\delta^4}\left(6w_0 - 4w_1 - 4w_3 + w_5 + w_9\right)$$

Likewise we have

$$\left(\frac{\partial^4 w}{\partial y^4}\right)_0 \simeq \frac{1}{\delta^4}\left(6w_0 - 4w_2 - 4w_4 + w_7 + w_{11}\right) \tag{F.8b}$$

$$\left(\frac{\partial^4 w}{\partial x^2 \partial y^2}\right)_0 \simeq \frac{1}{\delta^4}\left[4w_0 - 2\left(w_1 + w_2 + w_3 + w_4\right) + w_6 + w_8 + w_{10} + w_{12}\right] \tag{F.8c}$$

Substituting equations (F.8) in equation (10.181), we obtain the finite difference equation corresponding to the node 0:

$$20w_0 - 8\left(w_1 + w_2 + w_3 + w_4\right) + 2\left(w_6 + w_8 + w_{10} + w_{12}\right) + \tag{F.9}$$
$$w_5 + w_7 + w_9 + w_{11} = q_0/D$$

Also in this case, as many algebraic equations will be obtained as there are unknowns. For each node within the grid we have in fact a field equation (F.9), whilst for the boundary nodes it is possible to write two kinematic conditions:

$$w_i = 0, \quad \frac{\partial w_i}{\partial n} = 0 \tag{F.10}$$

which represent a built-in edge of the plate, or three static conditions corresponding to the free-edge loadings, which are the shearing force, given by equations (10.167), as well as the bending and twisting moments, given by

727

equations (10.180). These loadings can be expressed as linear combinations of the second and third partial derivatives of the function w, and thus involve a further two fictitious nodes, which are outside the domain of interest. Likewise, the second of the two kinematic conditions (F.10) involves a fictitious supplementary node, outside the domain.

Appendix G Torsion of multiply-connected thin-walled cross sections

The problem of torsion of **doubly-connected** closed thin-walled sections has been dealt with in Section 9.9 and solved by applying equations (9.136) and (9.140).

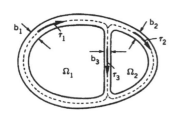

Figure G.1

Consider, instead, the case of a **triply-connected** cross section consisting of a tubular element with a diaphragm (Figure G.1). Let τ_i and b_i be the shearing stress and the thickness in each of the portions that make up the cross section. From the **hydrodynamic analogy**, the products $\tau_1 b_1$, $\tau_2 b_2$, $\tau_3 b_3$, are constant in each point of each portion, and

$$\tau_1 b_1 = \tau_2 b_2 + \tau_3 b_3 \tag{G.1}$$

If $h(s)$ denotes the distance of the centroid from the generic tangent to the mid-line, by equivalence we have

$$\tau_1 b_1 \int_1 h(s)\,\mathrm{d}s + \tau_2 b_2 \int_2 h(s)\,\mathrm{d}s + \tau_3 b_3 \int_3 h(s)\,\mathrm{d}s = M_z \tag{G.2}$$

Using equation (G.1) and indicating with Ω_1 and Ω_2 the areas enclosed in the circuits 1–3 and 2–3, we obtain

$$2\left(\tau_1 b_1 \Omega_1 + \tau_2 b_2 \Omega_2\right) = M_z \tag{G.3}$$

The unit angle of torsion is in general expressed by equation (9.109). Introducing the factor of torsional rigidity (9.140) and expressing the thickness $b(s)$ on the basis of equation (9.136), we have

$$\Theta = \frac{M_z}{4G\Omega^2}\oint_c \frac{\mathrm{d}s}{\left(M_z/2\Omega\tau_{zs}\right)} = \frac{1}{2G\Omega}\oint_c \tau_{zs}\,\mathrm{d}s \tag{G.4}$$

or

$$\oint_c \tau_{zs}\,\mathrm{d}s = 2G\Theta\Omega \tag{G.5}$$

This relation applies to any closed line, also in the case of multiply-connected cross sections. If it is applied to circuits 1–3 and 2–3, in the case where the thicknesses b_1, b_2, b_3, and therefore also the stresses τ_1, τ_2, τ_3, are constant in each of the three portions 1, 2, 3, we have

$$\tau_1 s_1 + \tau_3 s_3 = 2G\Theta\Omega_1 \tag{G.6}$$

$$\tau_2 s_2 - \tau_3 s_3 = 2G\Theta\Omega_2 \tag{G.7}$$

The linear algebraic system consisting of the four equations (G.1), (G.3), (G.6), (G.7), in the four unknowns τ_1, τ_2, τ_3, Θ, provides the solution to the problem.

Appendix H Shape functions

H.1 Rectangular finite elements: Lagrange family

A recursive and relatively simple method for generating shape functions of any order is that of multiplying appropriate polynomials which go to zero at the mesh nodes. Let us take, for instance, the element of Figure H.1, where a series of nodes, both internal and boundary ones, is located on a regular grid. Suppose we have to define the shape function of the node indicated by the double circle. Of course, the shape function sought will be given by the product of a fifth-order polynomial in ξ, having a value of unity in the second column of nodes and zero in all the others, and a fourth-order polynomial in η, having a value of unity in the first row of nodes and zero in all the others.

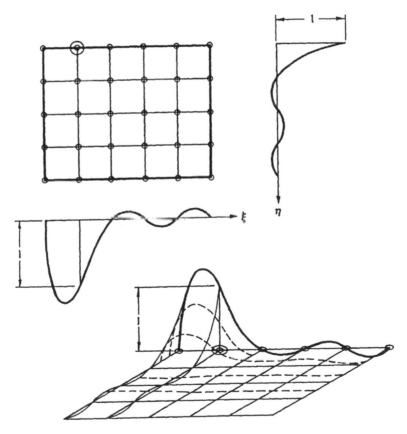

Figure H.1

729

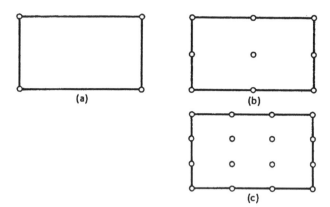

Figure H.2

Polynomials in one variable which present the above properties, are termed **Lagrange polynomials.** For the nodes of abscissa ξ_i, we have the following polynomial:

$$L_i^{(n)} = \frac{(\xi - \xi_1)(\xi - \xi_2)\ldots(\xi - \xi_{i-1})(\xi - \xi_{i+1})\ldots(\xi - \xi_n)}{(\xi_i - \xi_1)(\xi_i - \xi_2)\ldots(\xi_i - \xi_{i-1})(\xi_i - \xi_{i+1})\ldots(\xi_i - \xi_n)} \qquad (H.1)$$

The shape function of the node of coordinates (ξ_i, η_j) is thus given by the product

$$N_{ij}(\xi, \eta) = L_i^{(n)}(\xi) L_j^{(m)}(\eta) \qquad (H.2)$$

where n and m represent the number of subdivisions in each direction.

Figure H.2 shows some elements of this unlimited family. The shape functions of the element (b), which presents an internal node, are shown in Figure 11.5. Notwithstanding the ease with which such shape functions may be generated, their use is limited on account of the high number of internal nodes and the poor ability of higher-order polynomials to approximate the solutions. The next section will outline a method of obviating such drawbacks.

H.2 Rectangular finite elements: Serendipity family

Consider the elements of Figure H.3, which present the nodal points, spaced at equal intervals, only on the boundary sides. In the case of element (a) equation (H.1) yields

$$N_{11}(\xi, \eta) = \frac{1}{4}(1 - \xi)(1 + \eta) \qquad (H.3a)$$

$$N_{12}(\xi, \eta) = \frac{1}{4}(1 + \xi)(1 + \eta) \qquad (H.3b)$$

$$N_{21}(\xi, \eta) = \frac{1}{4}(1 - \xi)(1 - \eta) \qquad (H.3c)$$

$$N_{22}(\xi, \eta) = \frac{1}{4}(1 + \xi)(1 - \eta) \qquad (H.3d)$$

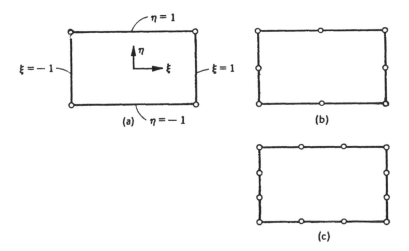

Figure H.3

In the case of element (b), which presents intermediate nodes on the sides, we have

$$N_{ij}(\xi,\eta) = \frac{1}{4}(1+\xi_0)(1+\eta_0)(\xi_0+\eta_0-1) \qquad \text{(H.4a)}$$

at the corner nodes, and

$$N_{ij}(\xi,\eta) = \frac{1}{2}(1-\xi^2)(1+\eta_0), \quad \text{for} \quad \xi_i = 0 \qquad \text{(H.4b)}$$

$$N_{ij}(\xi,\eta) = \frac{1}{2}(1+\xi_0)(1-\eta^2), \quad \text{for} \quad \eta_j = 0 \qquad \text{(H.4c)}$$

at the mid-side nodes, having introduced the new coordinates

$$\xi_0 = \xi\xi_i, \quad \eta_0 = \eta\eta_j \qquad \text{(H.5)}$$

It is possible to verify that in the case of the element of Figure H.3(c) we have

$$N_{ij}(\xi,\eta) = \frac{1}{32}(1+\xi_0)(1+\eta_0)\left[-10+9(\xi^2+\eta^2)\right] \qquad \text{(H.6a)}$$

at the corner nodes, and

$$N_{ij}(\xi,\eta) = \frac{9}{32}(1+\xi_0)(1-\eta^2)(1+9\eta_0), \quad \text{for} \quad \xi_i = \pm1, \eta_j = \pm\frac{1}{3} \;\text{(H.6b)}$$

$$N_{ij}(\xi,\eta) = \frac{9}{32}(1-\xi^2)(1+\eta_0)(1+9\xi_0), \quad \text{for} \quad \xi_i = \pm\frac{1}{3}, \eta_j = \pm1 \;\text{(H.6c)}$$

at the mid-side nodes.

It is interesting to note how the Lagrangian and Serendipity finite elements are identical only in their linear form (Figures H.2(a) and H.3(a)), whereas they differ, by the existence or otherwise of the central node, in their quadratic forms (Figures H.2(b) and H.3(b)). The corresponding shape functions are

731

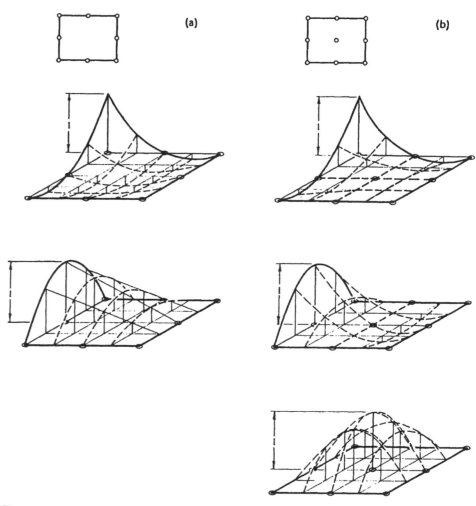

Figure H.4

shown in Figure H.4(a) (Serendipity element) and Figure H.4(b) (Lagrangian element).

H.3 Triangular finite elements

It is well-known how the triangular shape is the simplest and most appropriate for constructing meshes on complex-shape plane structural elements. Whereas cartesian coordinates constitute the most natural choice for the rectangular element, the most appropriate choice for the triangular element is represented by **area coordinates** (Figure H.5)

$$L_1 = \frac{\text{Area } P23}{\text{Area } 123} \qquad (H.7a)$$

$$L_2 = \frac{\text{Area } 1P3}{\text{Area } 123} \qquad (H.7b)$$

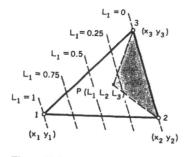

Figure H.5

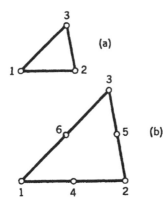

Figure H.6

$$L_3 = \frac{\text{Area } 12P}{\text{Area } 123} \tag{H.7c}$$

For the triangular element of Figure H.6(a), the shape functions are given simply by the area coordinates

$$N_1 = L_1, \quad N_2 = L_2, \quad N_3 = L_3 \tag{H.8}$$

For the triangular element of Figure H.6(b), which also presents mid-side nodes, we have

$$N_1 = (2L_1 - 1)L_1 \tag{H.9a}$$

$$N_2 = (2L_2 - 1)L_2 \tag{H.9b}$$

$$N_3 = (2L_3 - 1)L_3 \tag{H.9c}$$

at the corner nodes, and

$$N_4 = 4L_1L_2 \tag{H.9d}$$

$$N_5 = 4L_2L_3 \tag{H.9e}$$

$$N_6 = 4L_1L_3 \tag{H.9f}$$

at the mid-side nodes.

The area coordinates are, on the other hand, linked to the cartesian coordinates by the following relations:

$$x = L_1x_1 + L_2x_2 + L_3x_3 \tag{H.10a}$$

$$y = L_1y_1 + L_2y_2 + L_3y_3 \tag{H.10b}$$

$$1 = L_1 + L_2 + L_3 \tag{H.10c}$$

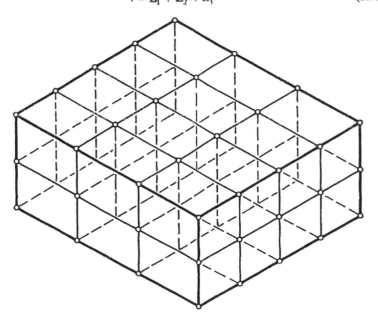

Figure H.7

733

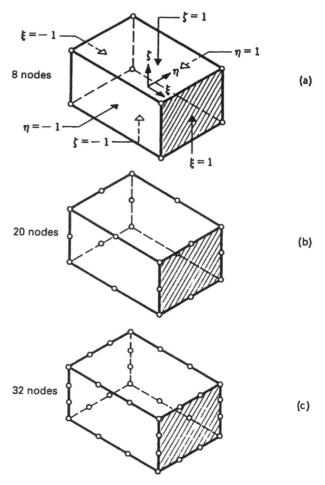

8 nodes (a)

20 nodes (b)

32 nodes (c)

Figure H.8

H.4 Three-dimensional finite elements

In the case of three-dimensional finite elements, shape functions altogether similar to the foregoing ones may be generated by simply adding a dimension.

In the case of **rectangular prisms**, the Lagrange family of functions (Figure H.7) is generated from the product of three polynomials. Extending the notation of equation (H.2), we have

$$N_{ijk}(\xi,\eta,\zeta) = L_i^{(n)}(\xi)L_j^{(m)}(\eta)L_k^{(l)}(\zeta) \qquad (H.11)$$

On the other hand, the family of elements shown in Figure H.8 is altogether analogous to that of Figure H.3 (Serendipity). For eight-node linear elements we have (Figure H.8(a))

$$N_{ijk}(\xi,\eta,\zeta) = \frac{1}{8}(1+\xi_0)(1+\eta_0)(1+\zeta_0) \qquad (H.12)$$

734

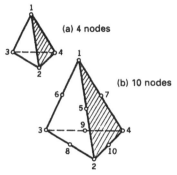

(a) 4 nodes

(b) 10 nodes

Figure H.9

whereas for 20-node quadratic elements we have (Figure H.8(b))

$$N_{ijk}(\xi,\eta,\zeta) = \frac{1}{8}(1+\xi_0)(1+\eta_0)(1+\zeta_0)(\xi_0+\eta_0+\zeta_0-2) \quad \text{(H.13a)}$$

at the corner nodes, and

$$N_{ijk}(\xi,\eta,\zeta) = \frac{1}{4}(1-\xi^2)(1+\eta_0)(1+\zeta_0), \text{for } \xi_i=0, \eta_j=\pm 1, \zeta_k=\pm 1 \quad \text{(H.13b)}$$

at four typical mid-side nodes.

Finally for **tetrahedral elements** (Figure H.9) the properties are similar to those of triangular plane elements. Introducing volume coordinates analogous to those of equations (H.7)

$$L_1 = \frac{\text{Volume } P234}{\text{Volume } 1234}, \text{ etc.} \quad \text{(H.14)}$$

we obtain linear shape functions coinciding with the corresponding volume coordinates (Figure H.9(a)), while the quadratic shape functions (Figure H.9(b)) for the 10-node tetrahedron are

$$N_i = (2L_i - 1)L_i, \quad i=1,2,3,4 \quad \text{(H.15)}$$

for the corner nodes, and

$$N_5 = 4L_1L_2 \quad \text{(H.16a)}$$
$$N_6 = 4L_1L_3 \quad \text{(H.16b)}$$
$$N_7 = 4L_1L_4 \quad \text{(H.16c)}$$
$$N_8 = 4L_2L_3 \quad \text{(H.16d)}$$
$$N_9 = 4L_3L_4 \quad \text{(H.16e)}$$
$$N_{10} = 4L_2L_4 \quad \text{(H.16f)}$$

for the mid-side nodes.

Appendix I   Application of the Finite Element Method to diffusion problems

The phenomenon of **heat conduction** in solid bodies can be described by a set of quantities and of equations which correspond exactly to those introduced in Chapter 8 for studying elastic solids. These equations can, moreover, be discretized in the form described in Chapter 11.

Let us take as principal unknown the **temperature** T. This is a scalar quantity and corresponds to the displacement vector in the elastic case. On the other hand, the quantity that corresponds to the deformation characteristics vector is the **temperature gradient**

$$\{\text{grad } T\} = [\partial] \ T \quad \text{(I.1)}$$
$$\underset{(3\times1)}{} \quad \underset{(3\times1)(1\times1)}{}$$

735

which is a three-component vector.

The 'kinematic' operator $[\partial]$ in this case is the following differential vector:

$$[\partial] = \begin{bmatrix} \dfrac{\partial}{\partial x} \\[2mm] \dfrac{\partial}{\partial y} \\[2mm] \dfrac{\partial}{\partial z} \end{bmatrix} \tag{I.2}$$

The quantity corresponding to the static characteristics vector is the **heat flux**, which is a vector with three components that, according to the corresponding **coefficients of thermal conductivity**, are proportional to the respective components of the temperature gradient. Expressed in formulas, we have the following 'constitutive' equation:

$$\begin{bmatrix} q_x \\ q_y \\ q_z \end{bmatrix} = - \begin{bmatrix} k_x & 0 & 0 \\ 0 & k_y & 0 \\ 0 & 0 & k_z \end{bmatrix} \begin{bmatrix} \dfrac{\partial T}{\partial x} \\[2mm] \dfrac{\partial T}{\partial y} \\[2mm] \dfrac{\partial T}{\partial z} \end{bmatrix} \tag{I.3}$$

or, in compact form

$$\underset{(3\times1)}{\{q\}} = - \underset{(3\times3)}{[k]} \underset{(3\times1)}{\{\text{grad } T\}} \tag{I.4}$$

The energy balance, for the infinitesimal element in the **steady state regime** yields, on the other hand, the following scalar equation:

$$\text{div}\{q\} = \dot{Q} \tag{I.5}$$

where $\dot{Q}$ denotes the **power generation per unit volume**, and the divergence operator can be represented in matrix form thus:

$$\underset{(1\times3)}{\text{div} = [\partial]^{\mathrm{T}}} \tag{I.6}$$

where $[\partial]$ is the differential operator (I.2).

Finally, combining the 'static' equation (I.5) with the 'constitutive' equation (I.4) and the 'kinematic' equation (I.1), we obtain the operator equation

$$\left(\underset{(1\times3)\,(3\times3)\,(3\times1)}{[\partial]^{\mathrm{T}} [k] [\partial]} \right) T + \dot{Q} = 0 \tag{I.7}$$

736

which has the same form as the Lamé's equation for elastic problems. At this point it is clear how, whereas the gradient and the divergence correspond respectively to the kinematic and static operators, the coefficients of thermal conductivity correspond to the stiffness, the power generation to the external force, and the temperature to the displacement.

Since the energy balance in the **transient regime** yields, instead of equation (I.5), the following equation:

$$\operatorname{div}\{q\} + c\frac{\partial T}{\partial t} = \dot{Q} \qquad (\text{I.8})$$

where c is the **thermal capacity** of the material, and t represents time, equation (I.7) can be generalized in this regime as follows:

$$\left([\partial]^{\mathrm{T}}[k][\partial]\right)T + \dot{Q} = c\frac{\partial T}{\partial t} \qquad (\text{I.9})$$

As regards the boundary conditions, these may be of two kinds, as in the elastic problems. In fact on the boundary it is possible to assign the **temperature** or the **normal heat flux**

$$T = T_0, \quad \forall P \in S_T \qquad (\text{I.10a})$$

$$\left(\underset{(1\times3)}{[\mathcal{N}]^{\mathrm{T}}}\ \underset{(3\times3)}{[k]}\ \underset{(3\times1)}{[\partial]}\right)T = -q_n, \quad \forall P \in S_q \qquad (\text{I.10b})$$

The matrix $[\mathcal{N}]$ represents the normal unit vector on the external surface

$$[\mathcal{N}] = \begin{bmatrix} n_x \\ n_y \\ n_z \end{bmatrix} \qquad (\text{I.11})$$

and corresponds to equation (I.2) in the spirit of Green's Theorem.

The Finite Element Method can therefore be applied to discretize heat conduction problems and, more generally, all **diffusion problems** which are governed by equations altogether analogous to those introduced hitherto. Referring to Table 11.1, for such problems we have $g = 1$, $d = 3$. This means that the principal unknown is scalar, while the characteristic vector represents a flux that is, in isotropic cases ($k_x = k_y = k_z = k$), proportional to the gradient of the scalar unknown.

Examples of diffusion problems of applicational importance are: **infiltration of fluids in porous media**, for which the scalar is the **fluid pressure**, whilst the constant k represents the **permeability** of the medium; **electrical conduction**, for which the scalar is the **potential**, whilst the constant k represents the **electrical conductivity**.

Finally, it should be noted that, in the case of a thermally isotropic material, equation (I.9) simplifies to the well-known form

$$k\nabla^2 T + \dot{Q} = c\frac{\partial T}{\partial t} \qquad (\text{I.12})$$

737

Appendix J   Initial strains and residual stresses

Initial strains $\{\varepsilon_0\}$ may be due to non-mechanical internal causes, such as temperature variations, shrinkage and phase transformations. The stresses will result from the difference between the actual and the initial strains. On the other hand, at the outset of analysis, the body could be stressed by some known field of initial **residual stresses** $\{\sigma_0\}$, due, for instance, to imposed displacements, constraint settlements, assemblage defects and welding effects. These stresses must simply be added on to the general definition.

If a general elastic behaviour is assumed, the relationship between stresses and strains will be linear and of the form

$$\{\sigma\} = [H]\big(\{\varepsilon\} - \{\varepsilon_0\}\big) + \{\sigma_0\} \tag{J.1}$$

where $[H]$ is the Hessian of the elastic potential energy. Substituting the constitutive equation (J.1) and the kinematic equation (8.9) into the static equation (8.11), we get

$$\big([\partial]^{\mathrm{T}}[H][\partial]\big)\{\eta\} - [\partial]^{\mathrm{T}}[H]\{\varepsilon_0\} + [\partial]^{\mathrm{T}}\{\sigma_0\} + \{\mathscr{F}\} = \{0\} \tag{J.2}$$

or, in compact form

$$[\mathscr{L}]\{\eta\} = -\{\mathscr{F}\} - \{\mathscr{F}_{\sigma_0}\} + \{\mathscr{F}_{\varepsilon_0}\} \tag{J.3}$$

where $\{\mathscr{F}_{\sigma_0}\}$ and $-\{\mathscr{F}_{\varepsilon_0}\}$ represent equivalent body forces, due to residual stresses and initial strains, respectively.

If the effects of residual stresses and initial strains are merged with the body forces, the Principle of Minimum Total Potential Energy may be demonstrated as in Section 11.3, and likewise the Finite Element Method may be defined in the same manner as in Sections 11.3 and 11.4. In particular, the vectors of the equivalent nodal forces, according to equation (11.44), are

$$\{F_e\} = \int_{V_e} [\eta_e]^{\mathrm{T}}\{\mathscr{F}\}\,\mathrm{d}V \tag{J.4a}$$

$$\left\{F_e^{\sigma_0}\right\} = \int_{V_e} [B_e]^{\mathrm{T}}\{\sigma_0\}\,\mathrm{d}V \tag{J.4b}$$

$$\left\{F_e^{\varepsilon_0}\right\} = -\int_{V_e} [B_e]^{\mathrm{T}}[H]\{\varepsilon_0\}\,\mathrm{d}V \tag{J.4c}$$

where the matrix $[B_e]$ is given in equation (11.39). After the expansion and assemblage procedures, the finite element equation (11.55) may be generalized as follows:

$$[K]\{\delta\} = \{F\} + \left\{F^{\sigma_0}\right\} + \left\{F^{\varepsilon_0}\right\} \tag{J.5}$$

738

Appendix K Dynamic behaviour of elastic solids with linear damping

When the dynamic behaviour of elastic solids with linear **damping** is considered, two sets of additional body forces are called into play. The first is the inertial force, which for an acceleration $\partial^2\{\eta\}/\partial t^2$, can be represented by its static equivalent

$$-[\rho]\frac{\partial^2}{\partial t^2}\{\eta\} \qquad (K.1)$$

using the well-known d'Alembert Principle (Section 11.7).

The second force is that due to frictional resistances opposing the motion. Usually a linear, viscous-type force is taken into account with its static equivalent

$$-[\mu]\frac{\partial}{\partial t}\{\eta\} \qquad (K.2)$$

where $[\mu]$ is the matrix of the damping coefficients.

The equivalent static problem, at any instant of time, may be discretized in the manner of Chapter 11 by replacing the static Lamé's operator with its equivalent dynamic and damping operator:

$$[\mathscr{L}]-[\mu]\frac{\partial}{\partial t}-[\rho]\frac{\partial^2}{\partial t^2} \qquad (K.3)$$

The difference between the problem of initial strains and residual stresses and the problem of dynamic damping is manifest. In the former case (Appendix J) we have to consider additional equivalent body forces, whereas in the latter case we have to assume a different operator where additional equivalent terms appear.

In the presence of inertial forces and viscous forces, relation (11.58) becomes

$$[K_{edd}]=-\int_{V_e}[\eta_e]^T[\mathscr{L}][\eta_e]dV+\int_{V_e}[\eta_e]^T[\mu][\eta_e]dV\cdot\frac{\partial}{\partial t}+ \qquad (K.4)$$

$$\int_{V_e}[\eta_e]^T[\rho][\eta_e]dV\cdot\frac{\partial^2}{\partial t^2}+\int_{S_e}[\eta_e]^T[\mathscr{L}_0][\eta_e]dS$$

or, in compact form

$$[K_{edd}]=[K_e]+[C_e]\frac{\partial}{\partial t}+[M_e]\frac{\partial^2}{\partial t^2} \qquad (K.5)$$

with

$$[C_e]=\int_{V_e}[\eta_e]^T[\mu][\eta_e]dV \qquad (K.6a)$$

which represents the **local damping matrix**, and

$$[M_e]=\int_{V_e}[\eta_e]^T[\rho][\eta_e]dV \qquad (K.6b)$$

739

which represents the **local mass matrix**.

At this point, expanding and assembling the local matrices, we obtain

$$\sum_e \left[K^{edd} \right] = \sum_e \left[A_e \right]^{\mathrm{T}} \left[K_{edd} \right] \left[A_e \right] \qquad \text{(K.7)}$$

a relation which is altogether analogous to equations (11.51) and (11.52), and provides the **global matrices**

$$[C] = \sum_e \left[A_e \right]^{\mathrm{T}} \left[C_e \right] \left[A_e \right] \qquad \text{(K.8a)}$$

$$[M] = \sum_e \left[A_e \right]^{\mathrm{T}} \left[M_e \right] \left[A_e \right] \qquad \text{(K.8b)}$$

Finally we obtain then the equation

$$[K]\{\delta\} + [C]\{\dot{\delta}\} + [M]\{\ddot{\delta}\} = \{F\} \qquad \text{(K.9)}$$

which is formally analogous to the equation of an oscillator with one degree of freedom, subjected to viscous forces and forcing loads. It should be noted that $[C]$ and $[M]$ are generally not diagonal matrices.

Appendix L Plane elasticity with couple stresses

In the classical theory of elasticity due to Cauchy it is assumed that the action of the material on one side of an elementary surface upon the material on the other side of the same surface is equivalent to a force (Figure 7.13). In the **couple-stress theory**, introduced by Cosserat, the interaction is assumed to be equivalent to a force and a couple. The **couple stresses** are taken to be moments per unit area, just as the body couples are moments per unit volume.

For the plane problem, the indefinite equations of equilibrium in the case of a medium that can support couple stresses are (Figure L.1)

$$
\begin{bmatrix}
\dfrac{\partial}{\partial x} & 0 & 0 & \dfrac{\partial}{\partial y} & 0 & 0 \\[2ex]
0 & \dfrac{\partial}{\partial y} & \dfrac{\partial}{\partial x} & 0 & 0 & 0 \\[2ex]
0 & 0 & +1 & -1 & \dfrac{\partial}{\partial x} & \dfrac{\partial}{\partial y}
\end{bmatrix}
\begin{bmatrix}
\sigma_x \\[1ex] \sigma_y \\[1ex] \tau_{xy} \\[1ex] \tau_{yx} \\[1ex] m_{xz} \\[1ex] m_{yz}
\end{bmatrix}
+
\begin{bmatrix}
\mathscr{F}_x \\[1ex] \mathscr{F}_y \\[1ex] \mathscr{M}_z
\end{bmatrix}
=
\begin{bmatrix}
0 \\[1ex] 0 \\[1ex] 0
\end{bmatrix}
\qquad \text{(L.1)}
$$

Accordingly, for non-constant couple stresses ($\partial m_{xz}/\partial x \neq 0$, $\partial m_{yz}/\partial y \neq 0$), the shearing stresses are not necessarily equal (i.e. $\tau_{xy} \neq \tau_{yx}$).

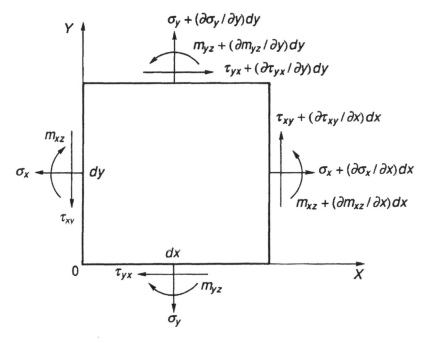

Figure L.1

In the definition of the deformation characteristics, it is possible to follow a heuristic procedure and consider the adjoint of the static operator

$$
\begin{bmatrix}
\varepsilon_x \\[4pt]
\varepsilon_y \\[4pt]
\gamma_{xy} \\[4pt]
\gamma_{yx} \\[4pt]
\chi_{xz} \\[4pt]
\chi_{yz}
\end{bmatrix}
=
\begin{bmatrix}
\dfrac{\partial}{\partial x} & 0 & 0 \\[8pt]
0 & \dfrac{\partial}{\partial y} & 0 \\[8pt]
0 & \dfrac{\partial}{\partial x} & -1 \\[8pt]
\dfrac{\partial}{\partial y} & 0 & +1 \\[8pt]
0 & 0 & \dfrac{\partial}{\partial x} \\[8pt]
0 & 0 & \dfrac{\partial}{\partial y}
\end{bmatrix}
\begin{bmatrix}
u \\[4pt]
\upsilon \\[4pt]
\varphi_z
\end{bmatrix}
\tag{L.2}
$$

Observe that, just as in the static equations (L.1) also rotational equilibrium is contemplated, so in the kinematic equations (L.2) also the dual generalized displacement appears, namely the rotation φ_z. Consistently with this, in addition to the dilations, two shearing strains and two curvatures form the vector of the deformation characteristics. It is evident how the structure of the equations (L.1) and (L.2) is a combination of the equations (8.8) and (8.10) for the elastic solid, and the equations (10.19) and (10.20) for the elastic beam.

The constitutive equations may be written as follows:

$$
\begin{bmatrix} \sigma_x \\ \sigma_y \\ \tau_{xy} \\ \tau_{yx} \\ m_{xz} \\ m_{yz} \end{bmatrix} = \begin{bmatrix} A & B & 0 & 0 & 0 & 0 \\ B & A & 0 & 0 & 0 & 0 \\ 0 & 0 & G & 0 & 0 & 0 \\ 0 & 0 & 0 & G & 0 & 0 \\ 0 & 0 & 0 & 0 & C & 0 \\ 0 & 0 & 0 & 0 & 0 & C \end{bmatrix} \begin{bmatrix} \varepsilon_x \\ \varepsilon_y \\ \gamma_{xy} \\ \gamma_{yx} \\ \chi_{xz} \\ \chi_{yz} \end{bmatrix}
\tag{L.3}
$$

with

$$
A = \frac{E}{1-v^2}, \quad B = \frac{Ev}{1-v^2}, \quad G = \frac{E}{2(1+v)}
$$

where E is Young's modulus and v is Poisson's ratio.

The three elements of the elasticity matrix, A, B, G, have the dimensions of stress, whereas C, i.e. the bending modulus, has the dimensions of a force. By considering the ratio $C/G \sim l^2$, we can define a length scale l, which represents the scale at which local rotations φ_z take place. Above this scale, rotations may be neglected as in the classical theory of elasticity.

Appendix M Rotating circular disk

Equation (12.97) can be written in the form

$$
\frac{\mathrm{d}}{\mathrm{d}r}(r\sigma_r) - \sigma_\vartheta + \rho\omega^2 r^2 = 0
\tag{M.1}
$$

where the body force $\mathscr{F}$ is set equal to the inertial force $\rho\omega^2 r$, where ρ is the material density and ω the angular velocity of the rotating circular disk.

The strain components in the case of symmetry are, from equations (12.95a)

$$
\varepsilon_r = \frac{\mathrm{d}u}{\mathrm{d}r}
\tag{M.2a}
$$

$$
\varepsilon_\vartheta = \frac{u}{r}
\tag{M.2b}
$$

The constitutive elastic equations then are

$$
\sigma_r = \frac{E}{1-v^2}\left(\frac{\mathrm{d}u}{\mathrm{d}r} + v\frac{u}{r}\right)
\tag{M.3a}
$$

$$
\sigma_\vartheta = \frac{E}{1-v^2}\left(\frac{u}{r} + v\frac{\mathrm{d}u}{\mathrm{d}r}\right)
\tag{M.3b}
$$

When the stresses (M.3) are substituted in equation (M.1), we find that u must satisfy

$$r^2 \frac{d^2u}{dr^2} + r\frac{du}{dr} - u = -\frac{1-v^2}{E}\rho\omega^2 r^3 \tag{M.4}$$

The general solution of this equation is

$$u = \frac{1}{E}\left[(1-v)Cr - (1+v)C_1\frac{1}{r} - \frac{1-v^2}{8}\rho\omega^2 r^3\right] \tag{M.5}$$

where C and C_1 are arbitrary constants. The corresponding stress components are now found from equations (M.3):

$$\sigma_r = C + C_1\frac{1}{r^2} - \frac{3+v}{8}\rho\omega^2 r^2 \tag{M.6a}$$

$$\sigma_\vartheta = C - C_1\frac{1}{r^2} - \frac{1+3v}{8}\rho\omega^2 r^2 \tag{M.6b}$$

The integration constants C and C_1 are determined from the boundary conditions.

For a **solid disk** we must take $C_1 = 0$ to have $u = 0$ at the centre. The constant C is determined from the condition at the periphery ($r = b$) of the disk:

$$\sigma_r(r = b) = C - \frac{3+v}{8}\rho\omega^2 b^2 = 0 \tag{M.7}$$

from which

$$C = \frac{3+v}{8}\rho\omega^2 b^2 \tag{M.8}$$

The stress components (M.6) then take the following form:

$$\sigma_r = \frac{3+v}{8}\rho\omega^2\left(b^2 - r^2\right) \tag{M.9a}$$

$$\sigma_\vartheta = \frac{3+v}{8}\rho\omega^2 b^2 - \frac{1+3v}{8}\rho\omega^2 r^2 \tag{M.9b}$$

These stresses are greatest at the centre of the disk:

$$\sigma_r(\max) = \sigma_\vartheta(\max) = \frac{3+v}{8}\rho\omega^2 b^2 \tag{M.10}$$

In the case of a **disk with a circular hole** of radius a at the centre, the constants of integration in equations (M.6) are obtained from the conditions at the inner and outer boundaries:

$$\sigma_r(r = a) = \sigma_r(r = b) = 0 \tag{M.11}$$

The calculation gives

$$C = \frac{3+v}{8}\rho\omega^2\left(a^2 + b^2\right) \tag{M.12a}$$

$$C_1 = -\frac{3+v}{8}\rho\omega^2 a^2 b^2 \tag{M.12b}$$

Substituting in equations (M.6) we have

$$\sigma_r = \frac{3+v}{8}\rho\omega^2\left(a^2 + b^2 - \frac{a^2 b^2}{r^2} - r^2\right) \tag{M.13a}$$

$$\sigma_\vartheta = \frac{3+v}{8}\rho\omega^2\left(a^2 + b^2 + \frac{a^2 b^2}{r^2} - \frac{1+3v}{3+v}r^2\right) \tag{M.13b}$$

We find the maximum radial stress at $r = \sqrt{ab}$,

$$\sigma_r(\max) = \frac{3+v}{8}\rho\omega^2(b-a)^2 \tag{M.14a}$$

and the maximum circumferential stress at the inner boundary:

$$\sigma_\vartheta(\max) = \frac{3+v}{4}\rho\omega^2\left(b^2 + \frac{1-v}{3+v}a^2\right) \tag{M.14b}$$

The latter is larger than σ_r (max).

When the radius a of the hole approaches zero, the maximum circumferential stress approaches a value twice as great as that for a solid disk (M.10); i.e. by making a small circular hole at the centre of a rotating disk we double the maximum stress. This is a phenomenon of stress concentration similar to that discussed in Section 19.6.

Appendix N Thermal stress in a circular disk

The thermal stresses σ_r and σ_ϑ satisfy equation (12.97) with $\mathcal{F}_r = 0$, in the case of a circular disk with a temperature distribution symmetrical about the centre. The strain is due partly to stress and partly to thermal expansion:

$$\varepsilon_r = \frac{1}{E}(\sigma_r - v\sigma_\vartheta) + \alpha T \tag{N.1a}$$

$$\varepsilon_\vartheta = \frac{1}{E}(\sigma_\vartheta - v\sigma_r) + \alpha T \tag{N.1b}$$

Solving equations (N.1) for σ_r, σ_ϑ, we find

$$\sigma_r = \frac{E}{1-v^2}\left[\varepsilon_r + v\varepsilon_\vartheta - (1+v)\alpha T\right] \tag{N.2a}$$

$$\sigma_\vartheta = \frac{E}{1-v^2}\left[\varepsilon_\vartheta + v\varepsilon_r - (1+v)\alpha T\right] \tag{N.2b}$$

Equation (12.97) then becomes

$$r\frac{d}{dr}(\varepsilon_r + v\varepsilon_\vartheta) + (1-v)(\varepsilon_r - \varepsilon_\vartheta) = (1+v)\alpha r\frac{dT}{dr} \tag{N.3}$$

Substituting equations (M.2) in equation (N.3), we obtain

$$\frac{d^2 u}{dr^2} + \frac{1}{r}\frac{du}{dr} - \frac{u}{r^2} = (1+v)\alpha\frac{dT}{dr} \tag{N.4}$$

which may be written

$$\frac{d}{dr}\left[\frac{1}{r}\frac{d(ur)}{dr}\right] = (1+v)\alpha\frac{dT}{dr} \tag{N.5}$$

Integration of this equation yields

$$u = (1+v)\alpha\frac{1}{r}\int_a^r Tr\,dr + C_1 r + \frac{C_2}{r} \tag{N.6}$$

where a is the inner radius for a disk with a hole, or zero for a solid disk.

The stress components are found by using the solution (N.6) in equations (M.2) and substituting the results in equations (N.2)

$$\sigma_r = -\alpha E\frac{1}{r^2}\int_a^r Tr\,dr + \frac{E}{1-v^2}\left[C_1(1+v) - C_2(1-v)\frac{1}{r^2}\right] \tag{N.7a}$$

$$\sigma_\vartheta = \alpha E\frac{1}{r^2}\int_a^r Tr\,dr - \alpha ET + \frac{E}{1-v^2}\left[C_1(1+v) + C_2(1-v)\frac{1}{r^2}\right] \tag{N.7b}$$

The constants C_1 and C_2 are determined by the boundary conditions.

For a **solid disk** ($a = 0$), the constant C_2 must be equal to zero in order that u may be zero at the centre, since

$$\lim_{r\to 0}\frac{1}{r}\int_0^r Tr\,dr = \lim_{r\to 0}\left(\frac{1}{2}T_0 r\right) = 0 \tag{N.8}$$

T_0 being the temperature at the centre.

The boundary condition $\sigma_r(r = b) = 0$ gives

$$C_1 = (1-v)\frac{\alpha}{b^2}\int_0^b Tr\,dr \tag{N.9}$$

so that the final expressions for the stresses are consequently

$$\sigma_r = \alpha E\left(\frac{1}{b^2}\int_0^b Tr\,dr - \frac{1}{r^2}\int_0^r Tr\,dr\right) \tag{N.10a}$$

$$\sigma_\vartheta = \alpha E\left(-T + \frac{1}{b^2}\int_0^b Tr\,dr + \frac{1}{r^2}\int_0^r Tr\,dr\right) \tag{N.10b}$$

These give finite values at the centre, since

$$\lim_{r\to 0}\frac{1}{r^2}\int_0^r Tr\,dr = \frac{1}{2}T_0 \tag{N.11}$$

Further reading

Baldacci, R., Ceradini, G. and Giangreco, E. (1974) *Plasticità*, Cisia, Tamburini, Milan.

Bathe, K.J. and Wilson, E.L. (1976) *Numerical Methods in Finite Element Analysis*, Prentice-Hall, Englewood Cliffs, New Jersey.

Belluzzi, O. (1941–48) *Scienza delle Costruzioni*, Zanichelli, Bologna.

Benvenuto, E. (1991) *An Introduction to the History of Structural Mechanics*, Springer-Verlag, New York.

Boresi, A.P. and Chong, K.P. (1987) *Elasticity in Engineering Mechanics*, Elsevier, New York.

Boscotrecase, L. and Di Tommaso, A. (1976) *Statica Applicata alle Costruzioni*, Patron, Bologna.

Bushnell, D. (1985) *Computerized Buckling Analysis of Shells*, Martinus Nijhoff Publishers, Dordrecht.

Capurso, M. (1971) *Lezioni di Scienza delle Costruzioni*, Pitagora, Bologna.

Carpinteri, A. (1986) *Mechanical Damage and Crack Growth in Concrete: Plastic Collapse to Brittle Fracture*, Martinus Nijhoff, Dordrecht.

Carpinteri, A. (1992) *Scienza delle Costruzioni*, Vol 1 and 2, Pitagora, Bologna.

Colonnetti, G. (1941) *Scienza delle Costruzioni*, Einaudi, Turin.

Corradi, L. (1978) *Instabilità delle Strutture*, Clup, Milan.

Di Tommaso, A. (1981) *Fondamenti di Scienza delle Costruzioni*, Patron, Bologna.

Fung, Y.C. (1965) *Foundations of Solid Mechanics*, Prentice-Hall, Englewood Cliffs, New Jersey.

Gavarini, C. (1978) *Dinamica delle Strutture*, Esa, Rome.

Gurtin, M.E. (1981) *An Introduction to Continuum Mechanics*, Academic Press, New York.

Jones, R.M. (1975) *Mechanics of Composite Materials*, McGraw-Hill, New York.

Levi-Civita, T. and Amaldi, U. (1965) *Compendio di Meccanica Razionale*, Zanichelli, Bologna.

Malvern, M. (1969) *Introduction to the Mechanics of a Continuous Medium*, Prentice-Hall, Englewood Cliffs, New Jersey.

Massonnet, C. and Save, M. (1977) *Calcolo a Rottura delle Strutture*, Zanichelli, Bologna.

Neal, B.G. (1977) *The Plastic Methods of Structural Analysis*, Chapman & Hall, London.

Novozhilov, V.V. (1961) *Theory of Elasticity*, Pergamon Press, Oxford.

Novozhilov, V.V. (1970) *Thin Shell Theory*, Noordhoff, Groningen.

Pignataro, M., Rizzi, N. and Luongo, A. (1991) *Stability, Bifurcation and Postcritical Behaviour of Elastic Structures*, Elsevier, Amsterdam.

Timoshenko, S.P. (1940) *Strength of Materials*, Van Nostrand, New York.

Timoshenko, S.P. and Goodier, J.N. (1970) *Theory of Elasticity*, McGraw-Hill, Tokyo.

Viola, E. (1985) *Esercitazioni di Scienza delle Costruzioni*, Vol 1 and 2, Pitagora, Bologna.

Zienkiewicz, O.C. (1971) *The Finite Element Method in Engineering Science*, McGraw-Hill, London.

Index

Page numbers appearing in **bold** refer to figures, those in *italic* to tables.

747